Algae Based Polymers, Blends, and Composites

Algae Based Polymers, Blends, and Composites

Chemistry, Biotechnology and Material Sciences

Edited by

Khalid Mahmood Zia

Mohammad Zuber

Muhammad Ali

ELSEVIER

Elsevier
Radarweg 29, PO Box 211, 1000 AE Amsterdam, Netherlands
The Boulevard, Langford Lane, Kidlington, Oxford OX5 1GB, United Kingdom
50 Hampshire Street, 5th Floor, Cambridge, MA 02139, United States

Notices

Knowledge and best practice in this field are constantly changing. As new research and experience broaden our understanding, changes in research methods, professional practices, or medical treatment may become necessary.

Practitioners and researchers must always rely on their own experience and knowledge in evaluating and using any information, methods, compounds, or experiments described herein. In using such information or methods they should be mindful of their own safety and the safety of others, including parties for whom they have a professional responsibility.

To the fullest extent of the law, neither the Publisher nor the authors, contributors, or editors, assume any liability for any injury and/or damage to persons or property as a matter of products liability, negligence or otherwise, or from any use or operation of any methods, products, instructions, or ideas contained in the material herein.

Library of Congress Cataloging-in-Publication Data

A catalog record for this book is available from the Library of Congress

British Library Cataloguing-in-Publication Data

A catalogue record for this book is available from the British Library

ISBN: 978-0-12-812360-7

Publisher: Matthew Deans
Acquisition Editor: David Jackson
Editorial Project Manager: Jennifer Pierce
Production Project Manager: Susan Li
Designer: Greg Harris

Typeset by TNQ Books and Journals

Dedicated to the one who gave me life and brought me up!
This particular copy of the book belongs to you. Every copy, however, belongs to the most holy person, MOTHER, who learnt to keep believing in dreams which seemed impossible to me!
Mrs. Hanifan Bibi (1939–2015)
Being the first educator and the mentor, a mother truly determines the happiness, future greatness, courteous ways, learning and judgment, understanding, patience, and faith of their little ones to let people understand that dreams can come true.
Love You Mom. Thanks for your unbiased support!

Contents

List of Contributors

Ifrah Afzal
Government College University Faisalabad, Faisalabad, Pakistan

Muhammad Afzal
National Institute of Biotechnology and Genetic Engineering, Faisalabad, Pakistan

Muhammad Afzal
Government College University Faisalabad, Faisalabad, Pakistan

Sajid Hamid Akash
Government College University Faisalabad, Faisalabad, Pakistan

Nadia Akram
Government College University Faisalabad, Faisalabad, Pakistan

Muhammad Ali
Government College University Faisalabad, Faisalabad, Pakistan

Muhammad A. Ali
University of Agriculture Faisalabad, Faisalabad, Pakistan

Shahzad Ali Shahid Chatha
Government College University Faisalabad, Faisalabad, Pakistan

Muhammad N. Anjum
Government College University Faisalabad, Faisalabad, Pakistan

Anbreen Anjum
Government College University Faisalabad, Faisalabad, Pakistan

Shazia Anjum
Islamia University Bahawalpur, Bahawalpur, Pakistan

Muhammad R. Asi
Nuclear Institute of Agriculture and Biology, Faisalabad, Pakistan

Nosheen Aslam
Government College University Faisalabad, Faisalabad, Pakistan

Makshoof Athar
University of the Punjab, Lahore, Pakistan

Muhammad Azeem
Government College University Faisalabad, Faisalabad, Pakistan

Farrukh Azeem
Government College University Faisalabad, Faisalabad, Pakistan

Mehdi Barikani
Iran Polymer & Petrochemical Institute, Tehran, Iran

Fatima Batool
Government College University Faisalabad, Faisalabad, Pakistan

Ismat BiBi
Islamia University Bahawalpur, Bahawalpur, Pakistan

Shazia A. Bukhari
Government College University Faisalabad, Faisalabad, Pakistan

Tuncer Caykara
Gazi University, Beşevler, Turkey

Ameer Fawad Zahoor
Government College University Faisalabad, Faisalabad, Pakistan

Fazal-ur-Rehman
Government College University Faisalabad, Faisalabad, Pakistan

Yolanda Freile-Pelegrín
CINVESTAV-IPN, Unidad Mérida, Mérida, Mexico

Takahiro Hirotsu
National Institute of Advanced Industrial Science and Technology (AIST), Takamatsu, Japan

Zakir Hussain
National University of Sciences & Technology (NUST), Islamabad, Pakistan

Muhammad Ibrahim
Government College University Faisalabad, Faisalabad, Pakistan

Abdullah Ijaz Hussain
Government College University Faisalabad, Faisalabad, Pakistan

Ikram-ul-Haq
Government College University Lahore, Lahore, Pakistan

Saiqa Ikram
Jamia Millia Islamia, New Delhi, India

Shahzad Z. Iqbal
Government College University Faisalabad, Faisalabad, Pakistan

Naeem Iqbal
Government College University Faisalabad, Faisalabad, Pakistan

Rehana Iqbal
Bahauddin Zakariya University, Multan, Pakistan

Farukh Jabeen
North Dakota State University, Fargo, ND, United States

Mudassir Jabeen
Government College University Faisalabad, Faisalabad, Pakistan

Tahir Jamil
University of the Punjab, Lahore, Pakistan

Shagufta Kamal
Government College University Faisalabad, Faisalabad, Pakistan

Sundas Khalid
National University of Sciences & Technology (NUST), Islamabad, Pakistan

Awais Khatri
Mehran University of Engineering and Technology, Jamshoro, Pakistan

Muhammad K. Khosa
Government College University Faisalabad, Faisalabad, Pakistan

Tianzhong Liu
Microalgae Biotechnology Centre, Qingdao Institute of Bio-energy & Bioprocess Technology, Chinese Academy of Sciences, Qingdao, People's Republic of China

Tomás J. Madera-Santana
Centro de Investigación en Alimentación y Desarrollo A.C. Hermosillo, México

Noeen Malik
New York University, New York, NY, United States

Sana Malik
Government College University Faisalabad, Faisalabad, Pakistan

Kausar Malik
University of the Punjab, Lahore, Pakistan

Muhammad Asim Mansha
Government College University Faisalabad, Faisalabad, Pakistan

Muhammad A. Mehmood
Government College University Faisalabad, Faisalabad, Pakistan

Ho S. Min
INTI International University, Nilai, Malaysia

Mohsen Mohammadi
Qom University of Technology (QUT), Old Tehran, Iran

Anam Munawar
National Institute for Biotechnology and Genetic Engineering, Faisalabad, Pakistan

Neelma Munir
Lahore College for Women University, Lahore, Pakistan

Saima Muzammil
Government College University Faisalabad, Faisalabad, Pakistan

Habibullah Nadeem
Government College University Faisalabad, Faisalabad, Pakistan

Mirza Nadeem Ahmad
Government College University Faisalabad, Faisalabad, Pakistan

Rehana Naseer
Government College Women University, Faisalabad, Pakistan

Muhammad Nawaz
Government College University Faisalabad, Faisalabad, Pakistan

Shagufta Naz
Lahore College for Women University, Lahore, Pakistan

Zill-i-Huma Nazli
Government College Women University Faisalabad, Faisalabad, Pakistan

Aqdas Noreen
Government College University Faisalabad, Faisalabad, Pakistan

Razia Noreen
Government College University Faisalabad, Faisalabad, Pakistan

Toshi Otsuki
IHI Corporation, Yokohama, Japan

Ijaz Rasul
Government College University Faisalabad, Faisalabad, Pakistan

Azhar Rasul
Government College University Faisalabad, Faisalabad, Pakistan;
RIKEN Center for Sustainable Resource Science (CSRS), Saitama, Japan

Waqar Rauf
National Institute of Biology & Genetics Engineering (NIBGE), Faisalabad, Pakistan

Aneeza Razzaq
Government College University Faisalabad, Faisalabad, Pakistan

Kanwal Rehman
University of Agriculture, Faisalabad, Pakistan

Maryam Rehman
Government College University Faisalabad, Faisalabad, Pakistan

Saima Rehman
Government College University Faisalabad, Faisalabad, Pakistan

Muhammad J. Saif
Government College University Faisalabad, Faisalabad, Pakistan

Yasar Saleem
Biotechnology Research Center, PCSIR, Laboratories Complex, Lahore, Pakistan

Mahwish Salman
Government College University Faisalabad, Faisalabad, Pakistan

Ayesha Shahid
Government College University Faisalabad, Faisalabad, Pakistan

Nadia Sharif
Lahore College for Women University, Lahore, Pakistan

Yury Shchipunov
Russian Academy of Sciences, Vladivostok, Russia

Bo Shi
Kimberly-Clark Corporation, Neenah, WI, United States

Muhammad H. Siddique
Government College University Faisalabad, Faisalabad, Pakistan

Hadi Sobhani
Hamadan University of Technology, Hamadan, Iran

Neelam Sultan
Government College University Faisalabad, Faisalabad, Pakistan

Salma Sultana
Government College University Faisalabad, Faisalabad, Pakistan

Tayyaba Sultana
Government College University Faisalabad, Faisalabad, Pakistan

Shazia Tabasum
Government College University Faisalabad, Faisalabad, Pakistan

Adil Usman
National University of Sciences & Technology (NUST), Islamabad, Pakistan

Atif Usman
SEGi University, Petaling Jaya, Malaysia

Yanmei Wang
University of Science and Technology of China (USTC), Hefei, China

Qun Wang
Iowa State University, Ames, IA, United States

Nazia Yaqoob
Government College Women University Faisalabad, Faisalabad, Pakistan

Aqib Zafar Khan
Government College University Faisalabad, Faisalabad, Pakistan

Fatima Zia
Government College University Faisalabad, Faisalabad, Pakistan

Khalid M. Zia
Government College University Faisalabad, Faisalabad, Pakistan

Muhammad A. Zia
University of Agriculture Faisalabad, Faisalabad, Pakistan

Mohammad Zuber
Government College University Faisalabad, Faisalabad, Pakistan

About the Editors

Khalid Mahmood Zia, PhD (Chemistry)

Khalid M Zia is the Director, Office of Research Innovation and Commercialization (ORIC), at Government College University, Faisalabad. He worked as a research scholar at the University of Agriculture, Faisalabad, Pakistan, and obtained his doctoral degree in 2008. He has 6½ years of experience of working in industrial sectors ranging from powerhouse to textiles industries. He has over 10 years of teaching, research, and administrative experience. He understands the interests of the industrial sector as well as academic institutions. His current interest involves synthetic polymer functionalized by biomolecules. He was awarded with the Best Researcher Award in the field of science and technology by the Higher Education Commission, Pakistan, in 2014. He received the highest ranking Research Productivity Award consecutively for last 5 years. He has authored 82 publications and 4 books.

Mohammad Zuber, PhD

Prof Mohammad Zuber is a professor of Polymer Sciences and the Vice Chancellor at MNS University of Engineering & Technology, Multan, Pakistan. He did his MSc and MPhil in Organic Chemistry from Quaid-i-Azam University, Islamabad, Pakistan, in 1981 and 1984, respectively. He obtained his doctorate degree in Polymer Science from Case Western Reserve University, the United States, in 1991. He has over 25 years of teaching, research, and administrative experience. He has published 108 papers in national and international journals with an Impact Factor over 190. He has supervised 114 BSc, MSc, MPhil, and PhD students. His current research interests involve biomaterials, synthesis of biomolecules-based water dispersible polymers used as thickeners, sizing agents, and adhesives, synthesis and characterization of urethane–acrylate copolymers for textile applications, synthesis of shape memory polyurethanes for nonbiomedical and biomedical applications, liquid crystalline polyurethanes, their synthesis and characterization, and synthesis, application, and study of dyeing behavior of different synthetic dyes. He is the organizer of four national and international conferences.

Muhammad Ali, PhD

Prof Muhammad Ali is the Vice Chancellor at Government College University, Faisalabad. He obtained his MSc in Zoology from University of the Punjab, Lahore, Pakistan, in 1987 with distinction. He obtained his doctorate degree in Biological Sciences (Fisheries, Molecular Biology) from University of Wales, UK, in 1999. He did his Postdoc from UM, Columbia, the United States, in 2009. He has 25 years of teaching, research, and administrative experience. He was the recipient of Best University Teacher Award in 2002, 2005, and 2011. He is the recipient of Research Productivity Award for the last 5 years. He has received Prof. Dr A.R. Shakoori Gold Medal, Agha Hasan Ali Abidi Gold Medal, and Prof. Dr Mirza Azhar Beg Gold Medal. He was awarded the National Civil Award "Tamgha-e-Imtiaz" on August 14, 2015, by the President of Islamic Republic of Pakistan. He has 167 publications and 2 patents in national and international journals with an impact factor of 171.534. He has participated in 73 conferences, symposia, seminars, workshops, and trainings.

Foreword

There is a distinct lack of consensus among the broader scientific community with respect to acceptance of a precise definition for the term *algae*. It is thus considered an informal term for a large, diverse group of simple, nonflowering, typically aquatic, polyphyletic, photosynthetic organisms. Although containing chlorophyll, algae lack many of the distinct cells and tissue types found in land plants, such as stomata, xylem, and phloem, and thus do not possess true stems, roots, and leaves. The simplest algae are unicellular genera, such as *Chlorella* (a common green alga of both terrestrial and aquatic habitats, responsible for turning stagnant water into an opaque green) and the diatoms (the most common types of phytoplankton). Although diatoms are unicellular, they can form colonies with a variety of shapes, such as filaments or ribbons (e.g., *Fragilaria*), fans (e.g., *Meridion*), zigzags (e.g., *Tabellaria*), or stars (e.g., *Asterionella*). The largest and most complex multicellular marine algae are the seaweeds, while the most complex freshwater algae are the *Charophyta* (a division of green algae which includes *Spirogyra* and stoneworts). The lack of consensus is due to some definitions excluding all prokaryotes, which therefore do not consider cyanobacteria (blue-green algae) as algae.

Many seaweeds are edible and have been used in foodstuffs for thousands of years, such as the red algae (*Rhodophyta*), which include carrageen moss (*Mastocarpus stellatus*), dulse (*Palmaria palmata*), Irish moss (*Chondrus crispus*), laver (*Porphyra laciniate*/*Porphyra umbilicalis*), nori (*Pyropia yezoensis*/*Pyropia tenera*); green algae, such as sea grapes (*Caulerpa lentillifera*) and sea lettuce (various species of the genus *Ulva*); brown algae (*Phaeophyceae*), such as the kelps (*Laminariaceae*), which include arame (*Eisenia bicyclis*) and wakame (*Undaria pinnatifida*), and the fucales, such as bladderwrack (*Fucus vesiculosus*). Such seaweeds are often excellent dietary sources of carbohydrates (especially fiber), proteins, and minerals. Specific historical applications of seaweeds include their use as animal feed, fertilizers, as a source of soda for glass production (in the early 18th to mid-19th centuries), and as a source of iodine (mid- to late-19th century).

With the continued development of scientific techniques in the late 19th and early 20th centuries came considerable advancement in the understanding of the composition and structure of seaweed and its components. The British chemist E. C. C. Stanford first described alginate (the preparation of algic acid from brown algae) with a patent dated January 12, 1881. Such advancements ultimately led to the birth of seaweed-derived carbohydrate polymer industries, such as the carrageenan (late-19th century) and alginate (early-20th century) industries. The so-called seaweed farming developed in many parts of Europe, providing alternative incomes to coastal communities that had traditionally relied on fishing (particularly in parts of Ireland and Scotland in the United Kingdom). Such European industries ultimately went into decline in the latter part of the 20th century, partly due to markets being dominated by cheaper imports from the Far East, where seaweed farming still thrives today. Traditionally, such products were used predominantly by the food industry (for thickening, gelling, stabilizing, etc.) and in the paper, printing, and textile industries.

A large, and structurally diverse, number of carbohydrate polymers can be found in algae/seaweeds, some being relatively simple in structure, such as agar/agarose (galactan), carrageenans (sulfated galactans), porphyran (sulfated galactan), alginate (glycuronan), laminaran (glucan), paramylon (glucan), and chitin/chitosan (aminoglucan). Others are composed of more complex, varying structures containing a range of different residues, such as fucans (glycuronogalactofucans, fucoidans, xylofucoglycuronans), ulvans (sulfated rhamnoxyloglycuronans), spirulan (rhamnoxyloglycuronan), and naviculan (complex sulfated heteropolysaccharide containing galactose, xylose, rhamnose, fucose, and mannose residues). In recent years attention has focused upon the isolation, structural characterization, and bio-/physico-/chemical-property testing of such algal biopolymers. Components with antioxidant, antibacterial, antifungal, antihyperlipidemic, antiviral, anticancer, antithrombotic, anticoagulant, anti-inflammatory, antiadhesive, anti-obesity, antiangiogenic, immunostimulatory, and immunomodulating activities have been identified. This should lead to the development of new products in many biomedical application areas, such as medical textiles, hydrogels, complexing agents/biopurification, encapsulation/drug delivery/controlled release, etc. Another important application area is the use of algal feedstock as biorefineries for the production of bioenergy (bioalcohol, biodiesel, biogas, biohydrogen) and bioproducts (pigments, proteins, fatty acids, amino acids, animal feeds, and biomedical products).

This informative volume aims to bring together information on a wide variety of relatively diverse topics associated with the entire spectrum/lifecycle of algal polymers, from cultivation and processing, through extraction, characterization, and property assessment, to end use (individual carbohydrate polymers, blends, composite materials, and bioplastics). It is hoped that this volume will prove to be an invaluable resource to all individuals, in both academia and industry, with research interests covering any aspects of algae-derived polymeric materials and their application. The importance of algae as a sustainable source of biomaterials will continue to gain momentum as more emphasis and resources are focused on the need to identify/develop new biologically active products (particularly as antibiotic resistance continues to grow), and also to develop renewable alternatives to synthetic fossil fuel–derived materials (which have a decreasing finite global feedstock). Thus, the future for algae is bright, or colorful (red, brown, green, yellow-green, blue-green, and golden to be precise)!

John F. Kennedy
Editor-in-Chief
Carbohydrate Polymers
International Journal of Biological Macromolecules

Preface

Progress in the field of "Polymer Science" has established a good reputation in scientific community, especially after the introduction of bio-based polymers. Algal biomass has potential to produce useful polymers and polymer feedstocks. Use of algae as sustainable biofuels and energy source and their attractive features have frequently been described in many publications and books. However, it is worth mentioning that no book is available to cover algae-combined polymeric materials and their biocompatibility and biofunctionality in particular.

The book presents a critical review on present, past, and future trends, including several characteristic aspects of algae-based polymer blends, and composites expected to provide a resource to assist significant research activities in this domain. The various chapters in this book actually summarize the effective work of prominent researchers of both industry and academia and thus provide contemporary record of the prominent results in this subject area. This book will be able to fulfill the requirement of university and college professionals/students, researchers, post-doctoral and senior graduate fellows, and for persons in R&D laboratories, interested in working on algae-based polymer blends and composites.

The first part (Chapters 1–9) of the book emphasizes the new challenges and opportunities in the extraction, physicochemical and biological properties, and functional applications of algae polymers and their derived compounds using biorefining and biotechnology.

The second part (Chapters 10–17) of the book overviews the blends/hybrids, composites, and nanocomposites of algae-based various natural polymers, synthetic polymers (polyurethanes, polyesters, polyolefin, polyvinyl alcohol and poly(*N*-vinyl-2-pyrrolidone), polyether and poly (ethylene oxide)), and inorganic materials (graphene, titanium dioxide) for food, water treatment, electronics and sensing devices, tissue engineering, drug-delivery systems and encapsulation, and biomedical uses.

Chapters 18–20 focus on the processing and characterization methods for algae-derived materials with their future prospects, providing exciting information for both the experienced and newer researchers to explore further.

The editors acknowledge the sincere efforts of all the contributors to this book, whose genuine collaboration ensured the thriving accomplishment of the project. We are thankful for their devotion and faithful attitude, and the sincerity they have put forward toward the attainment of this goal. The compliance of the book seemed to be impossible if their enthusiastic support was not there. We would like to say thanks to all the reviewers who dedicated their precious time to highlight the critical points in each chapter. Lastly, we show gratitude to the publisher, Elsevier, for considering our idea for such a book by understanding the significance of the area of algae–polymer blends and composites. Any comments/suggestions toward improving the text would be most welcome.

Khalid Mahmood Zia
Mohammad Zuber
Muhammad Ali

CHAPTER

Algal-Based Biopolymers

1

Muhammad Azeem[1], Fatima Batool[1], Naeem Iqbal[1], Ikram-ul-Haq[2]

Government College University Faisalabad, Faisalabad, Pakistan[1]; Government College University Lahore, Lahore, Pakistan[2]

1.1 INTRODUCTION

Polymers are solid macromolecules consisting of discrete building blocks linked together to form long chains of molecules. Simple building blocks, which are called monomers, arrange themselves in an intricate fashion to yield more complicated ones known as repeating units [1]. Polymers are used in every field of life [2,3]. During previous decades, increased population across the globe, improved living conditions, and modified life style have led to a dramatic increase in polymer consumption predominantly of plastic-based polymers [4]. These are intertwined with our consumer society in such a way that it is hard to imagine a modern society without polymers [5]. Each year approximately 140 million tons of synthetic polymers are produced [6]. However, all of them cannot be utilized for industrial purpose.

The synthetic polymers are resistant to chemical and physical degradation [7]. Their stability and nondegradable behavior results in deterioration of the environmental quality and posing a challenge to solid wastes treatment plants [8]. Thus major problem is disposal of synthetic wastes. Landfill is the most common method for disposing of such stuff along with municipal solid waste. In advanced countries such as the United States, Australia, and Germany, synthetic polymers are estimated to be approximately 20%–25% of the volume of municipal solid wastes [9,10]. Since 1990 the plastic industry has made huge investments to support increased recycling and to educate communities. Environmental challenges and overexploitation of nonrenewable resources, such as coal, fossil fuel, and oil, [11] have drawn the attention of substitute synthetic polymers with biodegradable types [12].

Biopolymers are obtained from different renewable natural sources such as plants, algae, animals, and microorganisms, or chemically synthesized from natural products, such as sugars, starch, natural fats, and oils [13]. A biopolymer usually does not evoke an inflammatory/toxic response disproportionate to its beneficial effect. Biopolymers are eco-friendly with acceptable shelf-life, easy to be sterilized, and biodegradable [14]. Most of the plant-based polymers include pectin, starch, chitosan, xylan, galactoglucomannan, and lignin [15]. These are usually derived from plant oil, cellulose, cornstarch, potato starch, sugarcane, weeds, and hemp,

Algae Based Polymers, Blends, and Composites. http://dx.doi.org/10.1016/B978-0-12-812360-7.00001-X

etc. [16,17]. One of the good examples of plant-based biopolymer is bioplastics. Approximately, 200,000 tons bioplastic are produced each year worldwide. The utilization of biopolymers helps to conserve fossil resources and reduce greenhouse gas emissions [17]. Considering the eco-friendly nature, use of biopolymers is gaining importance day by day [18].

1.2 APPLICATION AND PRODUCTION OF BIO-BASED POLYMERS

Bio-based materials are mainly obtained either by the direct production of polymers or by producing bio-based monomers followed by their biochemical polymerization. Major sources of the direct production of biopolymers include a vast diversity of biological objects ranging from microorganisms, algae, and higher plants. For instance, polyhydroxyalkanoates (PHAs) are produced from microorganism [19], alginate from algae, and pectin from higher plants [20]. There are certain biopolymers such as cellulose, which are not only produced by higher plants but also by bacteria [21]. Another example is chitin and chitosan, which are synthesized by crustaceans and fungi [22]. Biopolymers obtained from any source possess thermo-mechanical properties, which results in superior quality compared to synthetic polymers. Such properties are directly linked with the nature of its constituent monomers. Thus quality of bio-based polymer production directly depends upon the selection of monomer units. Size of the monomer used is another factor, which determines the quality of biopolymer produced. The regulation of these particular factors is really complicated as several microorganisms produce the same monomer units; however, this can be modulated by modifying their feeding substrates to make microorganisms, or other bioproducers, biosynthesize the desired product in a consistent fashion. Application and production of bio-based polysaccharides, PHAs, and others are discussed below.

1.2.1 APPLICATION AND PRODUCTION OF BIO-BASED POLYSACCHARIDES

Algae, plants, and animals are the major producers of polysaccharides and oligosaccharides. In addition, a wide variety of microorganisms produces a diverse range of these substances.

1.2.1.1 Polysaccharides Derived From Algae

Algae are the fast growing and dominant photosynthetic organisms. Most of the algal phototrophs contain long chains of polysaccharides such as cellulose, glucan, and galactan. Due to presence of diverse natural compounds and rapid growth, algae can be an excellent source of many natural products including biopolymers [23]. Algal phototrophs provide an excellent feedstock for biopolymer production owing to many advantages, such as high yield and ability to grow in diverse environments [24]. The yield of cultivated algae is much greater (22 kg m^2/year) than land plants (0.5−4.4 kg m^2/year), a most favorable feature making them an ideal candidate for utilization in polymer production [25,26]. Algae-based plastics have recently gained more importance [27] compared to traditional methods of utilizing feedstock. Algae-derived natural fibers are now commercially available [19].

Photosynthetic organisms, especially algae, are being explored as an excellent source of different polysaccharides and other natural bioactive products such as fatty acids, proteins, vitamins, and mineral elements [28]. These compounds have potential applications in medicine, industry, food, and agriculture.

Red algae are rich source of sulfated galactans and carrageenans that possess broad spectrum of therapeutic characteristics such as anticoagulant [29,30], antitumor [31], antiviral [32], and antioxidant properties [33]. Brown seaweed-derived polysaccharides such as fucoidan, laminaran, alginates, and mannitol are studied due to their effectiveness as anticoagulant, antitumor, antithrombotic, antiinflammatory, contraceptive, and antiviral agents [33,34]. Brown algae such as *Hizikia fusiforme* and *Laminaria japonica* are tremendously used for various medical aspects [35]. Fucoidans, or sulfated fucans, constituting up to 25%–30% of the seaweed dry weight [36], are used to treat various disorders such as hepatitis, herpes, and human immunodeficiency syndrome (AIDS) viruses, anticoagulant heparin inflammation, cell proliferation, adhesion, and fertilization functions [37,38]. Alginates are used to treat various disorders, such as gastric ulcers, lowering of cholesterol levels, and allergic reactions [39]. Alginates are also used as pill disintegrators, as bandages, and important material of dental impression [40,41]. Akamatsu and colleagues [42] reported effective anti-influenza viral compound known as MC26, a fucose polysaccharides isolated from marine brown algae species called *Sargassum piluliferum*. Fucoidan is a sulfated L-fucose polysaccharide present in the extracellular matrix of many brown algae such as mozuku, komby, limu moui, bladder wrack, wakame, hijiki, and sea cucumber [43]. Red algae contain small amounts of porphyran and xylans. *Porphyra* species contains a sulfated polysaccharide called porphyran, working as antitumor agent against Meth-A fibrosarcoma [44]. A Japan-based company has developed a medicinal composition from different polysaccharides isolated from red algae, especially *Gelidium amansii*, *Geum japonicum*, *Gastroclonium pacificum*, *Gelidium subcostatum*, *Pteocladia tennis*, and *Hypneaceae* species, to cure cancer, diabetes, and rheumatism [45,46]. Different mixtures of alginic acid and alginates are used to cure epigastric burning [47,48].

Latest research showed that ulvan from *Ulva lactuca*, rhamnan from rhamnose, and oligomers from *Monostroma* ulvans has anti-influenza activities and can be used for the treatment of gastric ulcers [49]. Carrageenan has good biological activities including anticoagulant, antiviral, and antitumor properties [50] used to treat various disorders such as respiratory problems lowering cholesterol, influenza, hepatitis A, herpes, and dengue [51]. Alginate immobilized cells are mostly used in cell transplantation, where alginate acts as a barrier between the transplant and the immune system of host. Many different cells are used for gel immobilization, e.g., parathyroid cells are used to treat hypocalcaemia and dopamine-producing cells are used to treat Parkinson disease [52].

Algal-based polymers are widely used in industry. Due to their high nutritional value, seaweeds are used in various industrial applications [53]. Marine algae are the only sources for industrially important phycocolloids such as agar, carrageenan, and alginate, which are used as stabilizer, viscosifier, gelling, and emulsifying agents [35]. Alginates are widely used in industries for consistency (viscosity) or to form gels. For example, they are used in baked foods, textile prints, welding rods, pill disintegrators, bandages, and dental impression material, among many others. Humans have been using red algae for at least 2800 years. *Gelidium*, *Pterocladiella*, and *Acanthopeltis*, members of red algae, are used in the pulp and paper-making industries [54]. Seaweeds are the excellent source of different biopolymers such as phycocolloids, alginate, agar, and carrageenan, which have different industrial applications in food, pharmaceutical, and biotechnology industries [55]. Agar and carrageenans both produce thermoreversible gels.

Many algal-based polymers such as alginates are widely used in different industrial applications, such as in dairy products and in canned foods, as they help retain moisture and improve the texture of baked foods. In textile, alginates are used as ink-paste thickeners for different textile prints, in the paper industry for a smooth paper surface, and as a material for welding rods [56,57]. Currently, the main producers of such algal-based polymers are China, followed by Norway, France, Japan, and Chile.

Some algal-based biopolymers are used to treat wastewater and as plant growth promoters [58–62]. These can also be used as metal interchangers [63,64], yeast immobilization [65], and microbeads [66]. Residual materials obtained from these biopolymers by different extraction methods are good source of fertilizers [67] and bioethanol production [68,69]. Seaweeds are essential food for East Asians for centuries. Especially in China, these are used as a staple food [70]. Sulfated galactans such as carrageenans and agarans are used as food thickeners, stabilizers, and gels, e.g., yogurt [71]. About 90% of the agar produced is extensively used in food industry and the remaining 10% in various bacteriological and biotechnological applications [72,73]. Agar has high fiber content and is used as a vegetarian gelatin. It has a high nutritional value, and is widely used in cheeses and sweets. Sugar-free products have gained much attention recently, due to increasing health problems such as obesity and diabetes. Sugars can be replaced by low calories inulin fructans of agar [74]. Carrageenan is mostly used in preserving meat. Agar holds on to the gelatin replacement market in canned meats and spices with a market share of 150 tons during 2009. Until the late 1970s carrageenan has been used in food products, mainly milk products, due to its unique ability to interact with casein in milk. Its low concentration causes solidification of milk, e.g., especially used in ice cream, gummy candies, fruit gels, fruit juices, marmalades, cheese products, cocoa suspensions and chocolate milk products, tooth paste, bakery products, and slow-release capsules. Potassium alginate is an important by-product of brown seaweed used in cosmetics and pharmaceutical industry as thickener. Following are the biopolymers being obtained from algae.

1.2.1.1.1 Agar

Agar, the first food ingredient approved by the FDA [75], is a polymer consisting of agarose and agropectin polysaccharide, having similar physical properties as carrageenans [76]. Algal phototrophs, such as *Gelidium*, *Gracilaria*, *Gracilariopsis*, *Gelidiella*, and *Pterocladia* species, are rich sources of agar [77–79]. A major part of world's total agar production (53%) is obtained from two main species, *Gracilaria* and *Gracilariopsis* [80,81]. *Gracilaria* species has 31% agar of their total biomass [82]. *Gracilaria*-based agar polysaccharides are characteristically more sulfated compared to those obtained from *Gelidium*. Agar causes lowering of blood glucose level, has an anti-aggregation effect on red blood cells [83] and antitumor activity [81]. Agar is the most expensive colloid (US$18/kg), followed by alginates and carrageenan (US$12/kg and US$10.5/kg, respectively) [72].

Agar (Fig. 1.1) consists of disaccharide-repeating units of 3-linked D-galactose and 4-linked 3,6-anhydro-L-galactose (3,6-AG) residues, with possible occurrence of sulfate, methoxyl, and/or pyruvate substituents at various positions in the polysaccharide chain [84]. The sulfation pattern is dominated by the esterification of C-6 of the linked galactose L-unit. This L-galactose 6-sulfated residue is initially synthesized as a precursor of the 3,6-anhydro-L-galactose. However, afterward sulphohydrolases converts it to an anhydrous form [71]. The enzymatic activity of sulphohydrolases is found to be lower than that in *Gelidium* and *Pterocladia*. That is why *Gracilaria* agar exhibits non-gelling or weak gel–forming polysaccharides. The gel strength, gelling, melting temperatures, and certain other physical properties of agar are directly associated to their chemical structures. Therefore by increasing the proportion of 3,6-anhydrogalactose, gelling ability of agars extracted from most of the *Gracilaria* species can be considerably enhanced by an alkali treatment before extraction. In a strong alkaline media, if high temperature is given to polysaccharides having 6-sulfated-L-galactose, the 3-OH group will be ionized, resulting in an intramolecular nucleophilic displacement of the sulfate group at position 6. This reaction is highly specific and does not affect any other sulfate group [85]. Helical conformation of agar polysaccharides usually mediates the agar gelation. These helices can be grouped by hydrogen bonds. Presence of any charged group will interfere with intermolecular hydrogen bonding. As a result, typical hysteresis of the thermoreversible order–disorder transition of agar will be disturbed [86]. It is observed that methoxyl groups affect the gelling temperature, depending on the algal species [87]. In particular, the methoxylation rate of agar polymers of

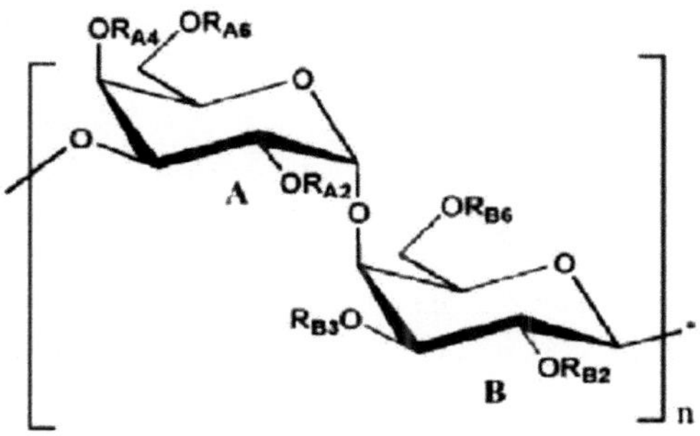

FIGURE 1.1

Structure of agar.

Gracilaria-based agars is significantly higher compared to their *Gelidium* and *Pterocladia*-based counterparts. That is why gelling temperature ranges from 40 to 42°C for *Gracilaria*-based agar and from 34 to 36°C for *Gelidiaceae*-based agar. To get maximum agar production, technology modification is devised to include multiple steps such as pretreatment, extraction, filtration, concentration, and dehydration depending on the algal species used. Agar is usually obtained from two genera of seaweeds, *Gelidium* and *Gracilaria*. For *Gracilaria*, the alkali treatment is performed before extraction to increase gel strength. In case of *Gelidiu*, it involves initial treatment with a mild alkaline solution that eliminates phycoerythrine and prepares the seaweed for a more efficient extraction.

1.2.1.1.2 Ulvan

Green seaweeds are excellent source of ulvan (Fig. 1.2), the sulfated polysaccharides. The sea lettuce (*Ulva* species) is an important source of food rich in carbohydrates, vitamins, essential amino acids, minerals, and dietary fibers. Many different species of *Ulva* has ulvan, the sulfated heteropolysaccharide with average molecular weight from 189 to 8200 kDa [88]. Ulvan extracted from cell wall of *Ulva* species are good polymers. In addition to water-soluble ulvan, *Ulva* species has three main polysaccharides, insoluble cellulose, an alkali-soluble linear xyloglucan, and glucuronan [85]. Ulvans are rich source of sugars, such as iduronic acid [89], that act as anticoagulant substance [90]. Rhamnose, a rare sugar and an important component of ulvans, is a precursor for the synthesis of aroma compounds. *Monostroma* (species of Ulvales found in Japan) is an excellent source of rhamnose [85]. Ulvan extraction is mostly performed with hot water solutions and can be further improved by the presence of calcium chelating agents in acidic or alkaline solutions. Purification of the polymer to eliminate pigments, lipids, amino acids, and peptides is reported by using different methods. A polysaccharide with improved purity is obtained by precipitation with organic solvents, frequently ethanol. Finally, ulvan aqueous extract can be concentrated in a rotary evaporator or dried by freeze drying or hot air drying [90].

1.2.1.1.3 Galactan

Sulfated galactan is an extracellular polysaccharide obtained from red algae. This particular biopolymer consists of linear chains of galactoses with a chain of alternating 3-β-D-galactopyranose G units and 4-α-D-galactopyranose residues,

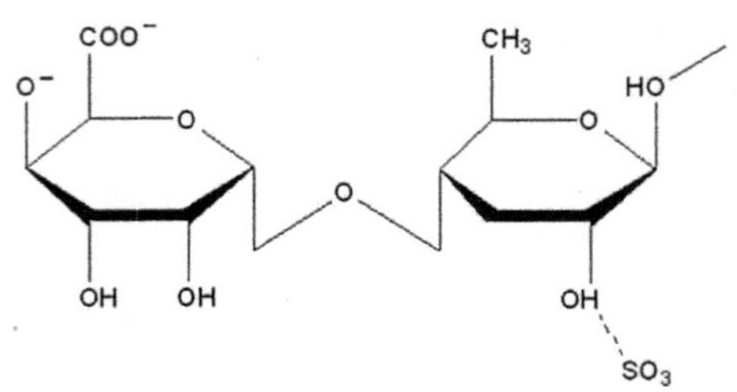

FIGURE 1.2

Structure of ulvan.

or 4-3,6-anhydrogalactopyranose residues as shown in Fig. 1.3 [91]. Galactan is obtained from red algae species and marine green algae species [92,93].

1.2.1.1.4 Fucan

Fucans, found in cell wall of brown algae, are sulfated polysaccharides with high molecular weight [94]. There are three major types of fucans including glycuronogalactofucans, fucoidans, and xylofucoglycuronans. Sulfated fucans are of medicinal significance containing antiviral and antimicrobial properties [95]. Some fucans extracted from seaweeds species, such as *Dictyota mertensii*, *Lobophora variegata*, *Fucus vesiculosus*, and *Spatoglossum schroederi*, are being used to prevent HIV infections [96]. MC26, fucose polysaccharide, is an effective anti-influenza viral compound and is isolated from *S. piluliferum*, a marine brown algae species [42].

Fucoidans, heterogenic mixtures of structurally related polysaccharides, are present in the extracellular matrix of all species of brown algae [97]. These polymers are present in the intercellular tissues or mucilaginous matrix of brown algae; their structure varies from species to species [98,99]. Minor amount of this polysaccharide was found in other algal groups such as green algae, red algae, and golden algae [100]. In 1913 Kylin reported fucoidan for the first time as a complex sulfated polysaccharide from marine brown algae. Initially it was called "fucoidin," but later on International Union of Pure and Applied Chemistry (IUPAC) named it as fucoidan. Molecular weight of fucoidan varies from 13 to 950 kDa. Structurally fucoidan possess two different types of backbone chains as shown in Fig. 1.4. This includes the following types:

1. Repeating (1-3)-linked α-L-fucopyranose residues.
2. Alternating (1-3)- and (1-4)-linked α-L-fucopyranose residues.

These chains have L-fucopyranose and α-D-glucuronic acid as carbohydrate substituents, and sulfate and acetyl groups as noncarbohydrate substituents. However, this chemistry varies with the environmental conditions and method of extraction. In addition to fucose and sulfate, fucoidan may also contain additional sugar constituents, including mannose, galactose, glucose, xylose, uronic acids, and even acetyl groups

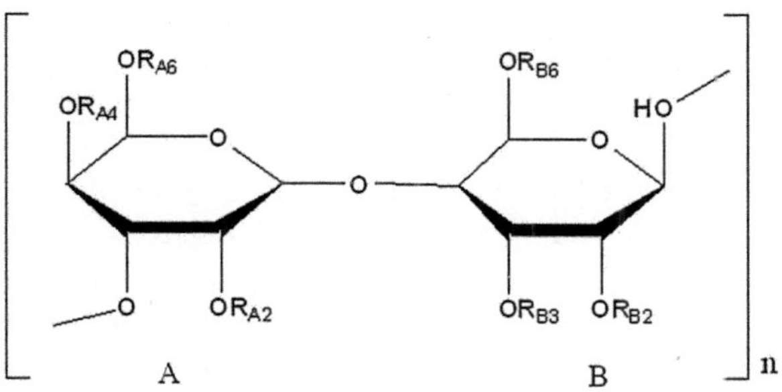

FIGURE 1.3

Structure of galactan (B units in its D configuration R_{A2}: H_2SO_3; R_{A4}: H_2SO_3; pyruvic acid (cyclic ketal with O_6) R_{A6}: H, CH_3, SO_3; pyruvic acid (cyclic ketal with O_4); R_{B2}: H, CH_3, SO_3; R_{A6}: H; RB: H, SO_3).

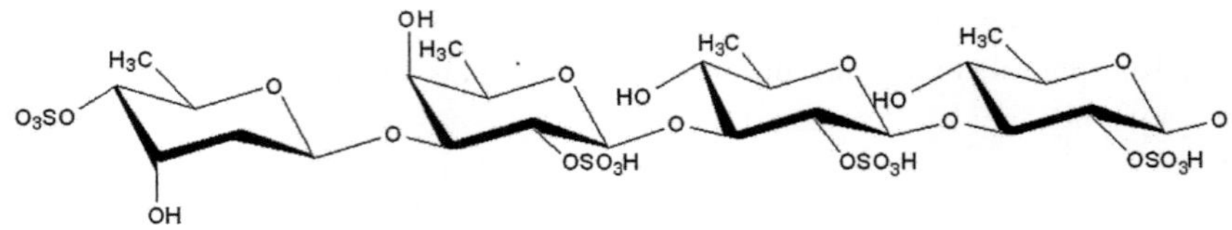

FIGURE 1.4

Structure of fucan.

[101]. The presence of these additional components, sometimes in appreciable amounts, is not yet established as a regular phenomenon. On the other hand, certain similarity in the backbone structure of different fucoidan molecules is observed regarding the positions of inter-glycosidic linkages. Many studies show that several representatives of the orders Chordariales and Laminariales contain fucoidan with a linear backbone composed of (1-3)-linked L-fucose residues. However, fucoidan isolated from algae belonging to the order Fucales mostly displays a backbone composed of alternating (1-3)- and (1-4)-linked L-fucose residues [102].

Fucoidan does not develop highly viscous solutions [103]. Therefore unlike many other polysaccharides, the polymer is not industrially used as thickening or gelling agent. It was observed that partially purified fucoidan from *Laminaria religiosa*, *Undaria pinnatifida*, *H. fusiforme*, and *Sargassum fulvellum* produced aqueous solutions of low viscosity with pseudoplastic flow behavior [104].

Fucoidan obtained from *F. vesiculosus* exhibited Newtonian behavior and had the highest viscosity, when compared to that obtained from *Saccharina longicruris* and *Ascophyllum nodosum* species [105]. Fucoidan has gelling and film-forming properties as compared to ulvan and carrageenan [106]. *F. vesiculosus* is a rich source of fucoidan polymer [107], an excellent inhibitor of leukocyte migration during inflammatory response, owing to its interaction with P- and L-selectins [108,109]. Recent research showed that sulfated fucans from the Fucales and Laminariales orders inhibit leukocyte recruitment in an inflammation model in rats [110,111]. The cell walls of pheophytes consist of an amorphous matrix of acid polysaccharides, linked to each other by proteins. These acid polysaccharides are mainly composed of fucoidan and alginic acid, which confer structural toughness and flexibility to seaweed [112]. Fucoidan usually constitutes about 5%−10% of dry algal biomass, depending on the species, the part of the thallus used, and the harvesting period [113].

There are several methods mainly used for the extraction of fucoidan from algae. For instance, one method reported in a study [114] involves the treatment of algal biomass with deionized water, boiled and centrifuged. The fucoidan is then extracted by series of steps. The methods of fucoidan extraction from brown macroalgae are described in many studies [114−117]. Algal powder is mixed with 85% ethanol and stirred for at least 12 h at room temperature to remove lipids and pigments. The solution is then centrifuged. The remaining sediment is then washed with acetone and left to dry at room temperature overnight.

1.2.1.1.5 Alginates

Alginates or phycocolloids are found in the cell walls of seaweeds and provide flexibility and mechanical strength. Structurally alginate is a linear polysaccharide copolymer of two different monomer units, (1-4)-linked β-D-mannuronic acid (M), and α-L-guluronic acid (G), arranged randomly [113,118]. Generally, three types of segments are present within the alginate:

1. Two homopolymeric, MM and GG
2. One heterogeneous or alternating, MG

Species of brown seaweed *Laminaria*, *Ascophyllum*, and *Lessonia* are major source of alginates. Alginates are isolated from red algae species belonging to the genus *Gelidium* [119,120]. The word alginate is derived from alginic acid that either occurs naturally in certain brown seaweeds (alginophyter) or is produced from the natural derivatives. It is reported that all species of the brown algae contain alginates [121]. Alginates are obtained by pretreatment of algal mass with HCl at pH 4.0. Extraction of alginate with Na_2CO_3 solution can be carried out by maintaining the extraction medium at pH 10 at 80°C for 2 h, followed by dilution and filtration in a vacuum rotary filter. Alginates are then recovered as insoluble calcium-alginate. These insoluble alginates are then converted to soluble by treating the fiber first with acid, and then with Na_2CO_3 resulting in sodium alginate. Quality control of alginates is focused on particle size distribution, viscosity, pH, ash, and calcium content.

Extracting the alginate from algal bodies can also be achieved by McHugh method [122]. The dry alga is soaked in formaldehyde (0.4%) for 30 min, rinsed with water, and then dipped into HCl solution. The samples are then rinsed again in distilled water before extraction in 2% sodium carbonate solution. The alginate extraction is usually carried at 60°C by soaking. The alga residues are then separated from sample by filtration and precipitation with ethanol [123].

1.2.1.1.6 Agarose

Agarose is a polysaccharide obtained mostly from seaweed. Chemically, agarose is a linear polymer (Fig. 1.5) made up of the agarobiose monomers. The monomer

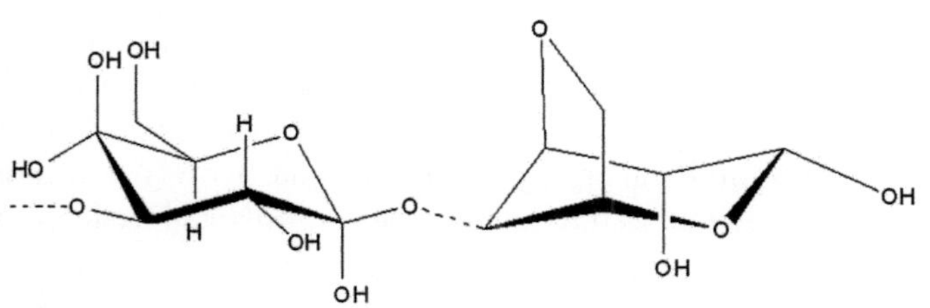

FIGURE 1.5

Structure of agarose.

is a disaccharide made up of D-galactose and 3,6-anhydro-L-galactopyranose. Agarose is purified from agar by removing other components, e.g., agaropectin. Extraction of agarose from red algae is a complicated process. Dried powder of algae is treated with preheated ionic liquids (ILs). The mixture containing algae + ILs is stirred at 80/100°C for 2 h. To reduce the viscosity of IL-algae solution, 10% boiled water is added. Hot water treated mixture is stirred to get free-flow homogenization, and is filtered using thin cotton under vacuum. Filtrates are treated with methanol to precipitate agarose dissolved in ILs. Several methanol washings are given to ensure the complete removal of ILs from extracted agarose, which is then dried under the vacuum [124].

1.2.1.1.7 Carrageenans

The word "carrageenan" derived from Irish word carrageen means "rock moss" was introduced in 1829 [125]. Carrageenan is the most abundant cell wall constituent of red algae. It is mainly composed of polysaccharide based on linear chains of repeating galactose units in D configuration (D-sugar) and 3,6-anhydro-galactose copolymer joined in alternating manner (Fig. 1.6). Average molecular weight of commercially available carrageenans ranges from 100 to 1000 kDa [50]. Different types of carrageenan have slight variation in the chemical structure [124]. The most industrially important carrageenans are κ, ι, and λ forms; *Kappaphycus alvarezii*, a red algae, is a potent source of these carrageenans [126]. Several species of red seaweeds, such as *Chondrus crispus*, *Gigartina*, *Eucheuma cottonii*, and *Spinosum*, are also excellent source of carrageenans [127,128]. Carrageenans consist of linear polysaccharide chains with sulfate half-esters attached to the sugar unit and are soluble in water [86]. Their biological activity [129], antimicrobial activity, and antitumor anticoagulant properties [130,131] can be attributed to a heterogeneous sulfate group. Kappa, mainly obtained from farmed *K. alvarezii*, and iota, from farmed *Eucheuma denticulatum* [132], carrageenans were also recorded recently. *C. crispus* and *Kappaphycus* species have 71% and 88% of carrageenan polymer [85,88]. These have diverse applications in canned foods, dessert mousses, salad dressings, bakery fillings, ice cream, instant desserts, canned pet-foods, and brewing industry [73]. *C. crispus*, known as "Irish moss," is the major source of carrageenan production [133]. Most carrageenan is now extracted from *K. alvarezii* and *E. denticulatum* but many species of *Gigartina*, *Iridae*, *Hypnea*, and *Furcellaria* genera were explored to obtain different types of carrageenan extracts [133,134]. Extraction of carrageenan involves different steps, such as purification, concentration, precipitation, and drying [135]. There are two main methods for producing carrageenan based on different principles. One method involves extraction of carrageenan in aqueous solution. Carrageenan is usually recovered from the solution by addition of an alcohol to induce precipitation. Finally, the precipitate is separated, dried, and milled, resulting in a refined carrageenan. In the second method, carrageenan is not actually extracted from the seaweed. Instead, the principle is to wash out the residual minerals, soluble protein, and fat from the seaweed, leaving behind carrageenan and other insoluble matter. This insoluble residue, consisting

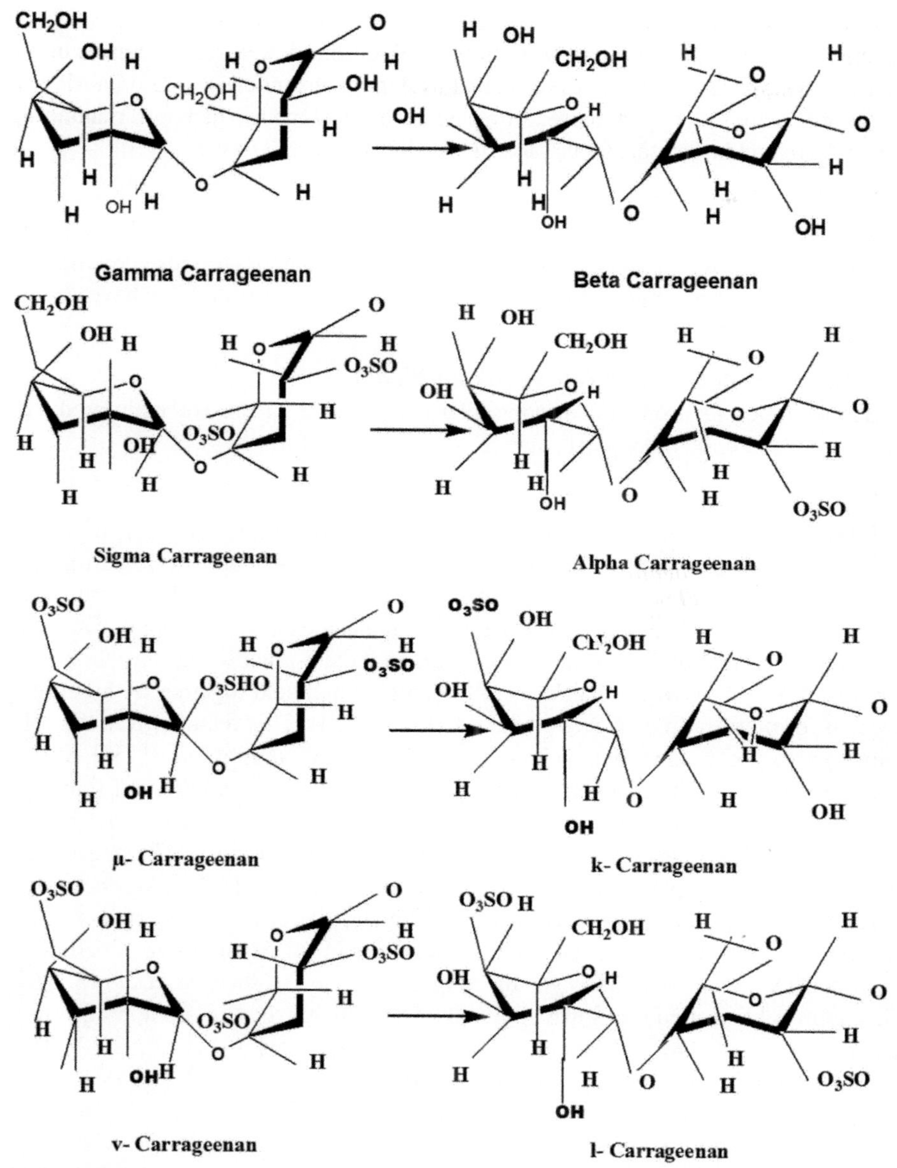

FIGURE 1.6

Structure of carrageenans.

largely of carrageenan and cellulose, is then dried and sold as semi-refined carrageenan. For nonfood gelling applications, this procedure is much shorter and cheaper as compared to first method [136]. Some other methods such as enzyme-treated or fungal-treated extractions are used for specific gelation properties [127].

1.2.1.1.8 Calcium Spirulan

Calcium spirulan (Ca-SP) is a novel sulfated polysaccharide reported in marine blue-green algae, *Spirulina platensis* now called as *Arthrospira platensis* [77]. Ca-SP has antiviral activities and is composed of uronic acids, 3-*O*-methylrhamnose rhamnose, 2,3-di-*O*-methylrhamnose, sulfate, and 3-*O*-methylxylose [137].

1.2.1.1.9 Naviculan

Diatom species such as *Navicula directa* have significant amount of naviculan sulfated polysaccharide, which mainly consists of galactose, xylose, rhamnose, fucose mannose, and sulfate [138]. Naviculan is generally considered to have strong antiviral activities.

1.2.1.2 Polysaccharides Derived From Plants

The most important producers of polysaccharides are the plants. The main products include cellulose, hemicellulose, starch, inulin, and pectin.

1.2.1.2.1 Cellulose

Cellulose is a homopolymer composed by only D-glucose units. It is a linear glucan chain of unbranched (1-4)-β-linked D-glucose in which every other glucose residue is rotated at 180 degrees with respect to its two neighbors and contrasts with other glucan polymers such as starch (1-4-α-glucan) and callose (1-3-β-glucan). Thus the basic repeating unit of the cellulose molecule is not the D-glucose but the cellobiose. Groups of 30—40 of these chains laterally combine and held together by hydrogen bond to form crystalline or para-crystalline microfibrils. This feature gives the cellulose a network structure, which holds and supports any external force [139]. Structurally cellulose exhibits remarkably high strength, stiffness, and thermal stability. The coefficient of thermal expansion of cellulose is very low (0.1 ppm/K) [140]. Young's modulus of its single fibril is very high (114 GPa) [141]. The degrees of crystallinity and polymerization are also high (89% and 14,400, respectively) [142,143]. The high specific surface area (37 m^2/g) of cellulose also makes it an attractive candidate for industrial commercialization [144]. Smaller size of cellulose in suspension can be obtained by subjecting the biomass to shear forces [145,146]. This approach is called top-down approach. Alternatively, bottom-up approach can also be used where biosynthesis of cellulose can be achieved by bacteria [147]. The *Acetobacter xylinum* is observed to be the most effective bacterial species for this purpose. The bacteria-based cellulose is considered to be superior over other sources owing to the purity of the product. This cellulose is not contaminated with wax, lignin, pectin, hemicelluloses, and other impurities unlike cellulose derived from plants where such stuff is present in more concentration. Another advantage of bacteria-based cellulose is the possibility of effective modulation of its molecular weight and its repeating units during the course of fermentation. Unfortunately, from commercial point of view, the cost-to-benefit ratio of bacteria-based cellulose is relatively high and hence possesses limitations to its commercial use. This factor of cost effectiveness makes the plant-based cellulose more promising candidate in commercial market.

1.2.1.2.2 Starch

Starch is homopolymer composed of only D-glucose units. Starch having linear glucan chains of unbranched (1-4)-α-linked D-glucose is called pectin. The same backbone structure but with (1-6)-α-linkage is called amylopectin. Starch has different proportions of amylose and amylopectin are present in different proportion ranging from 10% to 20% amylose and 80%–90% amylopectin [148]. However, this proportion is usually source dependent. A lot of hydroxyl groups are present on starch chains including two secondary hydroxyl groups at C-2 and C-3 of each glucose unit. In addition, due to non-linkage structure, one primary hydroxyl group is also present at C-6. The presence of these hydroxyl groups on the starch chains results in their potential of being oxidized or reduced, and their involvement in hydrogen bonds, ethers, and esters formation [149]. Amylose is water soluble and forms a helical structure [150]. As the short-branched amylopectin chains are able to form helical structures, which are crystallized, starch occurs in a discrete granular form with hydrophilic properties and strong intermolecular association via hydrogen bonding or ester linkage due to presence of the hydroxyl groups. As a result of these properties, water molecules readily changes the starch granules. The presence of water content determines its glass transition temperature (Tg) and mechanical properties. For instance, if the water weight fraction ranges from 0.12 to 0.14, the Tg of native starch becomes 60–80°C. This feature makes starch the most suitable candidate for thermoplastic polymers in the presence of water [151]. Another important aspect of starch is its biodegradability in diverse environment. It can be done by microorganism or even enzymes can hydrolyze it into glucose, and ultimately metabolize it into carbon dioxide and water [152]. The resultant carbon dioxide is recaptured by plants, and contributes as starting material in Kelvin cycle ultimately resulting in starch synthesis. Unlike cellulose, which needs both physical and chemical steps to be hydrolyzed, starch is a digestible product and hence is used in food and feed by humans. In addition, starch is also used in textile and paper industries. Another application of starch is its use in biodegradable packaging material [153,154]. However, limitations in the direct use of native starch include its poor processability, dimensional stability, and mechanical properties of its end products [155].

1.2.1.2.3 Inulin

Inulin is also a plant-based homopolymer composed of only fructose units. Inulin may also occasionally possess a glucose moiety initially. Chicory and artichokes are the main sources of inulin. It can also be synthesized bio-catalytically as fructo-oligosaccharides [156]. Owing to its nutritional value it is mainly used in food industry. However, some of the nonfood applications are also reported. Many strategies have been devised to obtain diverse functions including modified inulin synthesis. For instance, carboxymethyl inulin (CMI) is a modified inulin used as dispersing agent. Dicarboxy-inulin (DCI) is another modified inulin that has replaced the polyacrylates in detergent formulation [157].

1.2.1.2.4 Pectins

Pectins are also plant-based biopolymers and chemically polysaccharides chiefly composed of methyl-esterified galacturonic acid and rhamnose units [158]. Pectins are either rhamnogalacturonan-I (RG-I) or rhamnogalacturonan-II (RG-II). In case of RG-I, the basic unit structure is composed of a backbone of alternating (1-4)-α-linked D-GalA and (1-2)-α-linked L-rhamnosyl residues, which forms a contorted rod. The side chains of neutral arabinan, galactan, and arabinogalactan are linked to the C-4 and/or C-3 of the rhamnose. Although neutral sugars found in pectin includes D-galactose, L-arabinose, and D-xylose, the proportion usually varies with the pectin source. It may contain both methyl esters and acetyl ester groups. The ratio of esterified to nonesterified galacturonic acid modulates the pectin behavior in food applications. RG-II is an extremely conserved molecule and provides a good example of the complexity that can result from the ability of polysaccharides to form branched structures [159]. This pectin is composed of a backbone consisting of nine linear (1-4)-α-linked D-galacturonosyl residues to which four side chains are attached. It is rich in rhamnose, which is either (1-3)- and (1-2,3,4)-linked or as terminal units. RG-II contains 11 different sugars including fucose, arabinose, galactose, and apiose. This type of pectin possess molecular weight of 60–130,000 g/mol [160].

Pectin, as obvious from its complex structure, is a mixture of polysaccharides that constitute one-third of the cell wall of dry substance of higher plants. Not all plants possess this specific polysaccharide. Grasses either lack or have smaller proportions of these substances in their cell walls [161]. Rest of the angiosperms especially fruit have higher proportion. It is not homogenously distributed inside the cell. The highest concentration is present in the middle lamella of cell while the lowest in primary cell wall, and almost negligible in cell membrane. Among all plant sources, citrus peel and apple pomace are extensively used to get pectin under mildly acidic conditions [162–164]. Citrus peel and apple pomace contains 20%–30% and 10%–15% of dry biomass, respectively [165]. As per fresh weight the pectin level in apples is 1%–1.5% followed by apricot (1%), cherries (0.4%), oranges (0.5%–3.5%), carrots (1.4%), and citrus peel (30%) [166]. Thus, fruits having thicker skin such as apples, plums, gooseberries, quince, oranges, and other citrus fruits usually contain more pectin as compared to fruits with soft skin such as cherries, grapes, and strawberries [166]. Physical properties of pectins derived from both of these sources are also variable. Color of pectin obtained from citrus peel is light cream or light tan while apple pectins are comparatively darker [166]. Sources other than fruit include sugar beet waste, sunflower heads, and mango waste [167,168]. Non-plant sources of pectins are very limited. Direct boiling or microwave extraction is recommended to get maximum pectin from the source. Chemical modification or gene technology can also be used to obtain more homogeneous polysaccharides [169,170].

1.2.1.3 Polysaccharides Derived From Animals

Chitin, the second prevalent form of polymerized carbon in nature, is a linear polysaccharide made up of (1-4)-linked 2-acetamido-2-deoxy-β-D-glucopyranose units (Fig. 1.7). Although organisms producing cellulose lack chitin, chemically it

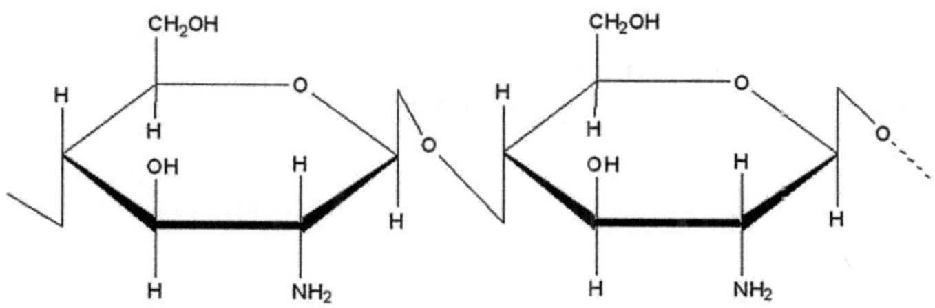

FIGURE 1.7

Structure of chitosan.

is considered as a cellulose derivative. Its structure is similar to cellulose except at the C-2 position where acetamide group (−$NHCOCH_3$) is present [171]. It is estimated that approximately 10 billion tons of chitin is produced every year by mollusks, crustaceans, insects, fungus, algae, and related organisms. Chitin is a potential candidate to be declared as biopolymer owing to its environment-friendly and bio-functional nature [172,173]. Moreover, it is bio-renewable and biodegradable [174]. It is highly hydrophobic and hence has very low solubility. Chitin is commercially used as a chelating agent, additive, wound-healing agent, drug carrier, and in water treatment and adhesive tape [175,176]. Chemical processing of the waste fraction of the shellfish industry is the current major source of chitin [22]. Collected waste fraction is subjected to decalcification in dilute aqueous HCL solution followed by deproteination in dilute aqueous NaOH solution. The desired product is obtained after decolorization. Chitosan derivative of chitin is prepared by subjecting the native chitin to deacetylation and depolymerization followed by partial deacetylation under alkaline conditions or enzymatic hydrolysis in the presence of an enzyme named as chitin deacetylase [177].

1.2.1.4 Polysaccharides Derived From Microorganisms

A wide range of microorganisms excrete polysaccharide substances that can be used as biopolymer. Thus microorganism-based polysaccharide is extracellular contrary to one derived from other organisms. Bacteria are the major source of such substances. Some of the well-known biopolymers derived from bacteria include cellulose, curdlan, dextran and derivatives, hyaluronan and hyaluronic acid, xanthan, glycogen, succinoglycan, alginate, glucuronan, sphingan, alternan, levan, murein, and teichoic acid [178−181]. To obtain maximum production of these biopolymers, a vast diversity of bacteria has been explored. *Acetobacter, Alcaligenes, Agrobacterium, Streptococcus, Leuconostoc, Streptococcus, Pasteurella, Xanthomonas, Escherichia coli, Clostridia, Bacillus,* and *Streptomyces* are reported to be the most productive microbial sources. These biopolymers have wide range of commercial applications especially in paper, textile, food, pharmaceutical, agriculture, cosmetics, and oil-drilling [182−184].

1.2.2 APPLICATION AND PRODUCTION OF BIO-BASED POLYHYDROXYALKANOATES

PHAs, although have not yet occupied a big market share, are becoming more promising candidates among biodegradable polymers in diverse environments [185]. PHAs can be used in many applications, including packaging, films and performance additives, molded goods, nonwoven fabrics, paper coatings, and adhesives [186].

PHAs are linear polyesters, which are naturally produced by microorganisms and cultivated on agronomical raw materials. They can be processed to make a variety of useful products, where their biodegradability is quite beneficial, particularly in single-use packaging and agriculture. PHAs were first reported as energy storage materials in 1926 by Maurice Lemoignez while working on *Bacillus megatherium*, *Bacillus mesentericus*, and *Bacillus vulgatis*. These bacteria accumulated PHAs in intracellular granules under conditions of a nutrient limitation other than the carbon source [187]. Their molecular weight ranges from about 50,000 Da to well over a million daltons, but depends on the bioproducer, growth conditions, and extraction method [188].

The PHA family is diverse family with vast differences in structure, which ultimately lead to their diverse thermo-mechanical properties. The whole family is divided into three subfamilies:

1. PHA_{SCL} (short-chain length PHA)
2. PHA_{MCL} (medium-chain length PHA)
3. Rarer PHAs

1.2.2.1 PHA_{SCL} (Short-Chain Length Polyhydroxyalkanoates)

PHA_{SCL} family includes polyhydroxybutanoates (PHB), polyhydroxy valerate (PHV), and poly(hydroxybutanoate-*co*-valerate) (PHBV). PHA_{SCL} are brittle and stiff semi crystalline isotactic stereo-regular polymers with a melting point of 160–180°C, and a glass transition temperature of −5 to 0°C. These show 100% R configuration that allows a high level of degradability [189].

PHA_{SCL} is produced when microorganisms under physiological stress use glucose or starch to assimilate carbon. When other energy sources are diminished, this energy storing molecule is metabolized. Microbial biosynthesis of PHA_{SCL} involves the condensation of two molecules of acetyl-CoA. As a result, acetoacetyl-CoA is produced, which is later reduced to hydroxybutyryl-CoA. Hydroxybutyryl-CoA is used as a monomer to polymerize PHA_{SCL} [17]. The PHA_{SCL} production was improved by exploring more and more microorganisms and was successfully achieved during last few decades. Extensive work on *Cupriavidus* genera, previously called as *Ralstonia*, made it predominant candidate in PHA_{SCL}-producing microorganisms [19]. In addition to this *Bacillus cereus*, *Bacillus megaterium, Haloferax mediterranei*, *Sphingopyxis macrogoltabia*, *Brevundimonas vesicularis*, *Nostoc muscorum*, *Synechocystis* sp., *Herbaspirillum seropedicae*, and certain other organisms were also explored [190–193]. The wild-type strains of *Cupriavidus* genera produced 121 g/L of PHB [194]. In addition to wild-type, engineering of certain *E. coli* strains also resulted in 161 g/L of PHB [194]. Modified feeding strategy has also proved to be 33 times more efficient [195]. The raw and agricultural products such as vinasse, oil palm frond juice, and soybean oil were also used as growth medium for different biopolymer producers; however, the data is not much promising. For example, use of soybean oil as substrate resulted in the production of only 67.2 g/L of PHB [196]. Certain bioprocess such as down-stream processing (DSP), solid-state fermentation (SSF), and continuous and two-stage culture systems were also used to obtain maximum production with improved quality.

1.2.2.2 PHA_{MCL} (Medium-Chain Length Polyhydroxyalkanoates)

Compared to HA_{SCL}, PHA_{MCL} possess more interesting thermo-mechanical properties. Synthesis of PHA_{MCL} involves copolymers of two, three, or even more monomers, which are usually obtained by β-oxidation of fatty acids [197]. *Pseudomonas* is the major source of PHA_{MCL} [198].

Many strategies were devised to get improved yield of PHA_{MCL} involving exploration of different *Pseudomonas* strains. The strains used for the study includes *Pseudomonas oleovorans*, *Pseudomonas putida*, *Comamonas testosteroni*, etc. [199]. It was observed that one strain, i.e., *Pseudomonas mendocina* was able to synthesize pure homopolymers of poly-3-hydroxyoctanoates (PHO) [200]. In addition to strain engineering, certain other strategies, for instance, batch and chemostat strategies, multiple nutrient limitations, were applied but results were not promising. For example, *C. testosteroni* showed only 53%–58% conversions of raw materials when grown on a medium rich in vegetable oil. PHA_{MCL} production from *P. oleovorans* when grown on a medium containing octanoic acid remained approximately 0.2 g/L h [201].

1.2.2.3 Rarer Polyhydroxyalkanoates

These PHAs were obtained by synthesizing PHA_{SCL}-*co*-PHA_{MCL}. The product usually contains the properties of both parent species. This hybrid is obtained from a bacterium named *Aeromonas caviae*. One of such example is copolymers of poly(3-hydroxybutanoate-*co*-3-hydroxyhexanoate) (P(HB-*co*-HHx)) [202]. Other PHAs of this subfamily include P3HA-*co*-P4HA and thiopolyesters polyhydroxybutanoate-*co*-polymercaptoprionate (PHB-*co*-PMP) [203].

Some strategies were devised to obtain improved yield of rarer PHA involving engineering of the strains. For example, *Cupriavidus necator* grown on kernel oil, when engineered with *Rhodococcus aetherivorans* PHA synthase, resulted in up to 70% of 3-HHx contents production.

1.2.3 APPLICATION AND PRODUCTION OF BIO-BASED PROTEINS AND POLY(AMINO ACID)S

Proteins are naturally occurring heteropolymers made up of amino acids, which are linked through peptide linkage. The protein-based biopolymers have been successfully applied in pharmaceutical [204] and nutraceutical [205]. ε-poly-L-lysine has specified antimicrobial activities and hence became the most suitable candidate biopolymer for the food industry [206]. Similarly, poly-α,β-aspartic acid is usually used in detergents as a polydispersant [207].

Some proteins were studied for a long time for their material applications: Some widely used protein-based biopolymers include collagen, wheat gluten, and soy protein [208–211].

Biosynthetic pathways for protein production are either ribosomal or nonribosomal–multienzyme paths. Although variety of amino acids contributes in protein synthesis, three poly(amino acid)s can be produced by using multienzyme pathway. These include poly-α,β-aspartic acid, aspartic acid–arginine dipeptide (cyanophycin), and ε-poly-L-lysine.

1.2.4 APPLICATION AND PRODUCTION OF BIO-BASED LIGNINS

Phenolic compound has an −OH group directly attached to a benzene ring. The primary cell walls of many, if not all, higher plants contain polymer-bound phenolics [212]. Generally, phenolics undergo oxidative-coupling reactions. Such coupling reactions, which are probably catalyzed within the cell wall itself by extracellular peroxidases, could play an important structural role by cross-linking the polymers to which the phenolics are bound [213]. This leads to lignification of the cell wall. Lignin has a complex, cross-linked, polyphenolic structure and is the third main constituent of plants after cellulose and hemicellulose. Initially considered as a waste, lignins found their place in commercial markets being the sole abundant source of bio-based aromatic compounds. However, its purification and de- and re-polymerization are not easy processes, which limit its widespread applicability [214].

1.2.5 APPLICATION AND PRODUCTION OF BIO-BASED MONOMERS

Controlled polymerization is an important tool to obtain the product with desired quality and characteristics. It seems more convenient to produce monomers from biological bodies instead of getting the biopolymers directly. The polymers can be produced from these bio-based monomers and are supposed to give more beneficial commercial products. One of such products is polyester. The major studied monomers include lactic acid and an α-hydroxy acid. These monomers can self-condense and hence produce some interesting products, for instance, polylactic acid (PLA) and 1,3-propanediol (PDO) when condensed with terephthalic acid and succinic acid, result in "Dupont's Sorona" [215]. Some other bio-based monomers, which have recently gained attention, are ethanol, caprolactam acid, muconic acid, and isobutylene [216,217]. These monomers can be used to produce polyethylene (PE), polyamides (PA), polyisobutylene, and several other bio-based polymers.

1.2.5.1 Lactic Acid

PLA, a bio-based bioplastic, is biodegradable and is a potential candidate to replace petroleum-based polymers in packaging materials and applications. This polymer is synthesized from bio-based monomer named lactic acid, which is produced by anaerobic fermentation in many bacteria. Lactic acid is traditionally used in food industry but its condensation reportedly leads to production of PLA [218].

Usually simple condensation of lactic acid leads to synthesis of many short oligomers instead of PLA. Hence several strategies were tried to obtain desired product including modification in processing techniques. Lactic acid on oligomerization results in its dimer, lactide, which upon ring-opening polymerization (ROP) results in PLA [219]. Recycling is another approach for PLA production instead of composting. Several approaches were attempted for this, including de- and re-polymerization techniques. However, most of the time, it resulted in racemization. Enzyme-based biocatalytic discrimination and desymmetrization approaches were also not suitable, as the former could not give any product while the latter causes stereoselective oxidation, which leads to significant loss of the product. Kinetic resolution using (R)-myrtenol, however, was reported to give some promising results but still intensive work is required [220].

1.2.5.2 Succinic Acid

Anaerobic fermentation of *Escherichia*, *Anaerobiospirillum*, and *Propionibacterium* results in succinic acid, which was reported to be the monomer of major biopolymer. It has been widely used in the production of certain organic acids such as malic acid, fumaric acid, itaconic acids, etc. [221]. Succinic acid was also found to be involved in production of certain polymers, such as polybutylene succinate, PBS, and others. Several strategies were developed to get maximum production of succinic acid including exploration of diversity of strains and also the biological engineering of certain strains, especially of *E. coli* or *S. cerevisiae* [222].

1.2.5.3 1,3-Propanediol

Another product of anaerobic fermentation is 1,3-propanediol (PDO). Large number of bacteria, for instance, *Klebsiella*, *Citrobacter*, *Clostridium*, *Enterobacter*, and several others synthesize PDO from glycerol [223–227]. Glycerol dehydolase converts glycerol to 3-hydroxypropionaldehyde, which in the presence of oxyoreductase results into 1,3-propanediol. Several strategies were tried to achieve good yield of PDO. Biological engineering was tried to get direct production of PDO, which however did not give desired results [228]. Alternatively, overexpression of glycerol in certain bacterial strains gave promising output and is used in industry. Later the polymerization of PDO with terephthalic acid resulted in production of Sorona [215].

1.3 FUTURE PERSPECTIVE

Recent advances in coping with the environmental pollution hazards being biodegradable, bio-based products attracted the researchers in academia as well as industry. These biomaterials were once criticized for their poor thermo-mechanical properties. The current research has extensively addressed such issues to achieve valuable commercial products. Recycling appears to be more sustainable than composting as an end-of-life solution. This twist is extremely important for future research and industry directions. Thus, low-added value applications, such as packaging, are hardly considered any longer. Exploration of biodiversity and genetic engineering are the basic tools, which can further introduce new organism and new versions of bio-based products.

REFERENCES

[1] Rao MG, Bharathi P, Akila RM. A comprehensive review on biopolymers. Sci Rev Chem Commun 2014;4(2):61–8.

[2] Shimao M. Biodegradation of plastics. Curr Opin Biotechnol 2001;12:242.

[3] Matsubara K, Matsuura Y, Bacic A, Liao ML, Hori K, Miyazawa K. Anticoagulant properties of a sulfated galactan preparation from a marine green alga, *Codium cylindricum*. Int J Bio Macromol 2001;28(5):395–9.

[4] Pawar PA, Purwar AH. Biodegradable polymers in food packaging. Am J Eng Res 2013;2(5):151–64.

[5] Hamad K, Kaseem M, Deri F. Recycling of waste from polymer materials: an overview of the recent works. Polym Degrad Stab 2013;98:2801–12.

[6] Leja K, Lewandowicz G. Polymer biodegradation and biodegradable polymers – a review. Pol J Environ Stud 2010;19(2):255–66.

[7] Bhadra J, Madi NK, Al-Thani NJ. Polyaniline/polyvinyl alcohol blends: effect of sulfonic acid dopants on microstructural, optical, thermal and electrical properties. Synth Met 2014;191:26–134.

[8] Kyrikou J, Briassoulis D. Biodegradation of agricultural plastic films: a critical review. J Polym Environ 2007;15:125.

[9] Fritz HG, Seidenstűcker T, Bolz U, Juza M, Schroeter J, Enders H-J. Study on production of thermoplastics and fibers based mainly on biological materials. Agro Ind Res Div 1994:392.

[10] Jayasekara R, Harding I, Bowater I, Lonergan G. Biodegradability of selected range of polymers and polymer blends and standard methods for assessment of biodegradation. J Polym Environ 2005;13:231–51.

[11] Premraj R, Doble M. Biodegradation of polymers. Indian J Biotechnol 2005;4: 194–200.

[12] Taboada C, Millan R, Miguez I. Composition, nutritional aspects and effect on serum parameters of marine algae *Ulva rigida*. J Sci Food Agric 2010;90(3). ISSN: 1097-0010:445–9.

[13] Davis L, U.S. Department of Agriculture, Cooperative State Research Service, Office of Agricultural Materials; 1993 [Personal communication].

[14] Nilani P, Raveesha P, Rahul B, Nandkumar Kasthuribai N, Duraisamy B, Dhamodaran P, Elango K. Formulation and evaluation of polysaccharide based biopolymer – an ecofriendly alternative for synthetic polymer. Pharm Sci Res 2010; 2(3):178–84.

[15] Bird CJ, Saunders GW, McLachlan J. Biology of *Furcellaria lumbricalis* (Hudson) Lamouroux (Rhodophyta: Gigartinales), a commercial carrageenophyte. J Appl Phycol 1991;3:61−82.

[16] Reddy RL, Reddy VS, Gupta GA. Study of bio-plastics as green & sustainable alternative. Int J Emerg Technol Adv Eng 2013;3(5):82−9.

[17] Bugnicourt E, Cinelli P, Lazzeri A, Alvarez V. Polyhydroxyalkanoate (PHA): review of synthesis, characteristics, processing and potential applications in packaging. Express Polym Lett 2014;8(11):791−808.

[18] Athanasiou KA, Agrawal CE, Barber FA, Burkhart SS. Orthopaedic applications for PLA-PGA biodegradable polymers. Arthrosc J Arthrosc Relat Surg 1998;14(7): 726−37.

[19] Laycock B, Halley P, Pratt S, Werker A, Lant P. The chemomechanical properties of microbial polyhydroxyalkanoates. Prog Polym Sci 2013;38:536−83.

[20] Griffiths M, Harrison STL, Smit M, Maharajh D. Major commercial products from micro- and macroalgae. Algae Biotechnol 2016:269−300.

[21] Scandola M, Ceccorulli G, Pizzoli M. Miscibility of bacterial poly(3-hydroxybutyrate) with cellulose esters. Macromolecules 1992;25:6441−6.

[22] Muzzarelli RA, Jeuniaux C, Gooday GW. Chitin in nature and technology. 1st ed. New York: Plenum Press; 1986.

[23] Velde F, Pereira L, Rollema HS. The revised NMR chemical shift data of carrageenans. Carbohydr Res 2004;339:2309−13.

[24] Wei N, Quarterman J, Jin Y. Marine macroalgae: an untapped resource for producing fuels and chemicals. Trends Biotechnol 2013;31:2.

[25] Masarin F, Roberto Paz Cedeno F, Chavez EGS, Ezequiel de Oliveira L, Gelli VC, Monti R. Chemical analysis and biorefinery of red algae *Kappaphycus alvarezii* for efficient production of glucose from residue of carrageenan extraction process. Biotechnol Biofuels 2016;9(122). http://dx.doi.org/10.1186/s13068-016-0535-9.

[26] Teixeira RE. Energy-efficient extraction of fuel and chemical feedstocks from algae. Green Chem 2012;14:419−27.

[27] Abe M, Fukaya Y, Ohno H. Extraction of polysaccharides from bran with phosphonate or phosphinate derived ionic liquids under short mixing time and low temperature. Green Chem 2010;12:1274−80.

[28] Pérez R. Ces algues qui nous entourent. In: Arbault S, Barbaroux O, Phliponeau P, Rouxel C, editors. Aquaculture. Plouzané (France): Editions IFREMER; 1997. p. 65−178.

[29] Melo FR, Pereira MS, Foguel D, Mourão PA. Antithrombin-mediated anticoagulant activity of sulfated polysaccharides. J Biol Chem 2004;279(20):20824−35.

[30] Farias W, Valente AP, Pereira M, Mourão P. Structure and anticoagulant activity of sulfated galactans. J Biol Chem 2000;275:29299−307.

[31] Yuan H, Song J, Li X, Dai J. Immunomodulation and antitumor activity of carrageenan oligosaccharides. Cancer Lett 2006;243:228−34.

[32] Yan XJ, Zheng L, Chen HM, Lin W, Zhang WW. Enriched accumulation and biotransformation of selenium in the edible seaweed *Laminaria japonica*. J Agric Food Chem 2004;52(21):6460−4.

[33] Rocha de Souza MC, Marques CT, Dore CMG, da Silva FRF, Rocha HAO, Leite EL. Antioxidant activities of sulfated polysaccharides from brown and red seaweeds. J Appl Phycol 2007;19:153−60.

[34] Nabors LOB. Alternative sweeteners. Agro Food Ind Hi-Tech 2004;15(4):39–41.

[35] Balakrishnan CP, Jenifer P, Esakkilingam M. Algal documentation and phytochemical studies of red algae *Gracilaria corticata* of Manapad Coast, Tamil Nadu. J Pharmacogn Phytochem 2013;2(4):193–7.

[36] Kusaykin MI, Chizhov AO, Grachev AA, Alekseeva SA, Bakunina IY, Nedashkovskaya O, Sova VV, Zvyagintseva TN. Comparative study of specificity of fucoidanases from marine microorganisms and invertebrates. J Appl Phycol 2006; 18:369–73.

[37] Ellouali M, Boissonvidal C, Durand P, Jozefonvicz J. Antitumor-activity of low molecular weight fucans extracted from brown seaweed *Ascophyllum nodosum*. Anticancer Res 1993;13(6A):2011–9.

[38] Wijesinghe WAJP, Jeon YJ. Biological activities and potential industrial applications of fucose rich sulfated polysaccharides and fucoidans isolated from brown seaweeds: a review. Carbohydr Polym 2012;88:13–20.

[39] Nagaoka M, Shibata H, Kimura-Takagi I, Hashimoto S, Kimura K, Makino T, Aiyama R, Ueyama S, Yokokura T. Structural study of fucoidan from *Cladosiphon okamuranus* TOKIDA. Glycoconj J 1999;16:19–26.

[40] Mcpeak RH, Glanz D. Harvesting California's kelp forest. Oceanus 2014;27(1): 19–26.

[41] Reyes-Tisnado R, López-Gutiérrez F, Hernández-Carmona G, Vernon-Carter J, Castro-Moroyoqui P. Sodium and potassium alginates extracted from *Macrocystis pyrifera* algae for use in dental impression materials. Cienc Mar 2004;30(3): 189–1991.

[42] Akamatsu E, Shimanaga M, Kamei Y. Isolation of an anti-influenza virus substance, MC26 from a marine brown alga, *Sargassum piluliferum* and its antiviral activity against influenza virus, vol. 1. Coastal Bioenvironment-Saga University; 2003. p. 29–34.

[43] Murano E, Gilli R, Navarini L, Toffanin R, D'Agnolo E, Paoletti S, Rizzo R. Ion-driven gelation of highly sulfated agar. J Mar Biotechnol 1995;3:143–5.

[44] Noda H. Health benefits and nutritional properties of nori. J Appl Phycol 1993;5(2): 255–8.

[45] Enoki T, Sagawa H, Tominaga T, Nishiyama E, Komyama N, Sakai T, et al. Drugs, foods or drinks with the use of algae-derived physiologically active substances. US Patent. 2003.

[46] Smit AJ. Medicinal and pharmaceutical uses of seaweed natural products: a review. J Appl Phycol 2004;16(4):245–62.

[47] Klinkenberg-Knol EC, Festen HP, Meuwissen SG. Pharmacological management of gastro-oesophageal reflux disease. Drugs 1995;49:495–710.

[48] Nishide E, Anzai H, Uchida N. Effects of alginates on the ingestion and excretion of cholesterol in the rat. J Appl Phycol 1993;5(2):207–11.

[49] Nagaoka M, Shibata H, Kimura I, Hashimoto S. Oligosaccharide derivatives and process for producing the same. United States Patent 6,645,940. 2003.

[50] Necas J, Bartosikova L. Carrageenan: a review. Vet Med 2013;58(4):187–205.

[51] Luescher-Mattli M. Algae, a possible source for new drugs in the treatment of HIV and other viral diseases. Curr Med Chem 2003;2:219–25.

[52] Aebisher P, Goddard M, Treseco PA. Cell encapsulation for the nervous system. In: Goosen MFA, editor. Fundamental of animal cell encapsulation and immobilization; 1993. p. 197–224. Boca Ratan.

[53] MacArtain P, Gill CIR, Brooks M, Campbell R, Rowland IR. Nutritional value of edible seaweeds. Nutr Rev 2007;65:535–43.

[54] Seo Y, Lee Y, Lee C, You H. Red algae and their use in papermaking. Bioresour Technol 2010;10:2549–53.

[55] Kinch J, Bagita J, Bate M. Exploring the potential for seaweed farming in Milne Bay Province, Papua New Guinea. SPC Fish Newsl 2003;104:25–31.

[56] Vartiainen J, Vähä-Nissi M, Harlin A. Biopolymer films and coatings in packaging applications—a review of recent developments. Mater Sci Appl 2014;5:708–18. Published Online August 2014 in Sci Res.

[57] Wang C-Y, Chen Y-C. Extraction and characterization of fucoidan from six brown macroalgae. J Mar Sci Technol 2016;24(2):319–28.

[58] Iwasaki K, Matsubara Y. Purification of alginate oligosaccharides with root growth-promoting activity toward lettuce. Biosci Biotechnol Biochem 2000;64:1067–70.

[59] Yabur R, Bashan Y, Hernández-Carmona G. Alginate from the macroalgae *Sargassum sinicola* as a novel source for microbial immobilization material in wastewater treatment and plant growth promotion. J Appl Phycol 2007;19(1):43–53.

[60] Abdel-Raouf N, Al-Homaidan AA, Ibraheemb IBM. Microalgae and wastewater treatment. Saudi J Biol Sci 2012;19(3):257–75.

[61] Coppi G, Iannuccelli V, Leo E, Bernabei MT, Cameroni R. Chitosan-alginate micro particles as a protein carrier. Drug Dev Ind Pharm 2001;27(5):393–400.

[62] Evonne PY. Tang polar cyanobacteria versus green algae for tertiary wastewater treatment in cool climates. J Appl Phycol 1997;9:371–81.

[63] Davis TA, Ramirez M, Mucci A, Larsen B. Extraction, isolation and cadmium binding of alginate from *Sargassum* spp. J Appl Phycol 2004;16(4):275–84.

[64] De Stefano C, Gianguzza A, Piazzese D, Sammartano S. Modelling of proton and metal exchange in the alginate biopolymer. Anal Bioanal Chem 2005;383(4):587–96.

[65] Pajic-Lijakovic I, Plavsic M, Nedovic V, Bugarski B. Investigation of Ca-alginate hydrogel rheological behavior in conjunction with immobilized yeast cell growth dynamics. Microencapsulation 2007;24(5):420–9.

[66] Schuldt U, Hunkeler D. Alginate-cellulose sulphate-oligocation microcapsules: optimization of mass transport and mechanical properties. J Microencapsul 2007;24(1): 1–10.

[67] Tang JC, Taniguchi H, Chu H, Zhou Q, Nagata S. Isolation and characterization of alginate degrading bacteria for disposal of seaweeds waste. Lett Appl Microbiol 2009;48:38–43.

[68] Klinke HB, Thomsen A, Ahring BK. Potential inhibitors from wet oxidation of wheat straw and their effect on growth and ethanol production by *Thermoanaerobacter mathranii*. Appl Microbiol Biotechnol 2001;57:631–8.

[69] Horn SJ, Aasen IM, Stgaard K. Ethanol production from seaweed extract. J Ind Microbiol Biotechnol 2000;25:249–54.

[70] Tseng CK. Algal biotechnology industries and research activities in China. J Appl Phycol 2001;13:375–80.

[71] Murano E. Chemical structure and quality of agars from *Gracilaria*. J Appl Phycol 1995;7(3):245–54.

[72] Armisen R, Galatas F. Agar. In: Phillips G, Williams P, editors. Handbook of hydrocolloid; 2000. Boca Raton (Florida).

[73] Bixler HJ, Porse H. A decade of change in the seaweed hydrocolloids industry. J Appl Phycol 2011;23:321–35.

[74] Kronberga M, Karklina D, Murniece I, Kruma Z. Changes of agar-agar gel properties after replacing sucrose by inulin syrup. In: Straumite E, editor. Proceedings of 6th Baltic conference on food science and technology FOODBALT-2011, Latvia; 2011.

[75] Carmona GH, Freile-Pelegrin Y, Garibay EH. Conventional and alternative technologies for the extraction of algal polysaccharides. In: Functional ingredients from algae for foods and nutraceuticals. Cambridge: Woodhead Published Limited; 2013. p. 475–516.

[76] Craigie JS. Cell walls. In: Biology of the red algae, vol. 517. Cambridge University Press; 1990. p. 221–58.

[77] Jeon YJ, Athukorala Y, Lee J. Characterization of agarose product from agar using DMSO. Algae 2005;20:61–7.

[78] Rasmussen RS, Morrissey MT. Marine biotechnology for production of food ingredients. Adv Food Nutr Res 2007;52:237–92.

[79] Kakita H, Kamishima H, Chirapart A, Ohno M. Marine biopolymers from the red algae, *Gracilaria* spp. Recent Adv Mar Biotechnol 2003;9:79–109.

[80] Villanueva RD, Sousa AMM, Gonçalves MP, Nilsson M, Hilliou L. Production and properties of agar from the invasive marinealga, *Gracilaria vermiculophylla* (Gracilariales, Rhodophyta). J Appl Phycol 2010;22:211–20.

[81] Chen HM, Zheng L, Yan XJ. The preparation and bioactivity research of agaro-oligosaccharides. Food Technol Biotechnol 2005;43(1):29–36.

[82] Murata M, Nakazoe J. Production and use of marine algae in Japan. JARQ – Japan Agric Res Q 2001;35(4):281–90.

[83] Koltai K, Feher G, Kesmarky G, Keszthelyi Z, Czopf L, Toth K. The effect of blood glucose levels on hemorheological parameters, platelet activation and aggregation in oral glucose tolerance tests. Clin Hemorheol Microcirc 2006;35(4):517–25.

[84] Rees DA. Structure, conformation, mechanism in the formation of polysaccharide gels and networks. Adv Carbohydr Chem Biochem 1969;24:267–332.

[85] Lahaye M, Robic A. Structure and functional properties of ulvan, a polysaccharide from green seaweeds. Biomacromolecules 2007;8(6):1765–74.

[86] Lahaye M, Rochas C. Chemical structure and physico-chemical properties of agar. Hydrobiologia 1991;221:137–48.

[87] Guiseley KB. The relationship between methoxyl content and gelling temperature of agarose. Carbohydr Res 1970;13:247–56.

[88] Lahaye M. NMR spectroscopic characterisation of oligosaccharides from two *Ulva rigida* ulvan samples (Ulvales, Chlorophyta) degraded by a lyase. Carbohydr Res 1998;314(1–2):1–12.

[89] Lahaye M, Ray B. Cell-wall polysaccharides from the marine green alga *Ulva rigida* (Ulvales, Chlorophyta) – NMR analysis of ulvan oligosaccharides. Carbohydr Res 1996;283:161–73.

[90] Paradossi G, Cavalieri F, Chiessi E. A conformational study on the algal polysaccharide ulvan. Macromolecules 2002;35:6404–11.

[91] Delattre C, Fenoradosoa T, Michaud A, Galactans P. An overview of their most important sourcing and applications as natural polysaccharides. Braz Arch Biol Technol 2011;54:1075–92.

[92] Jiao G, Yu G, Zhang J, Ewart HS. Chemical structures and bioactivities of sulfated polysaccharides from marine algae. Mar Drugs 2011;9(2):196–223.

[93] Silva TH, Alves A, Popa EG, Reys LL, Gomes ME, Sousa RA, Silva SS, Mano JF, Reis RL. Marine algae sulfated polysaccharides for tissue engineering and drug delivery approaches. Biomatter 2012;2(4):278–89.

[94] Mabeau S, Kloareg B. Isolation and analysis of the cell walls of brown algae: *Fucus spiralis, F. ceranoides, F. vesiculosus, F. serratus, Bifurcaria bifurcata and Laminaria digitata*. Oxf J Sci Math – J Exp Bot 1987;38(9):1573–80.

[95] Queiroz KCS, Medeiros VP, Queiroz LS, et al. Inhibition of reverse transcriptase activity of HIV by polysaccharides of brown algae. Biomed Pharmacother 2008;62(5): 303–7.

[96] Li B, Lu F, Wei X, Zhao R. Fucoidan: structure and bioactivity. Molecules 2008;13: 1671–95. http://dx.doi.org/10.3390/molecules13081671.

[97] Pomin VH, Mourao PAS. Structure, biology, evolution, and medical importance of sulfated ucans and galactans. Glycobiology 2008;18:1016–27.

[98] Dietrich CP, Farias GGM, de Abreu LRD, Leite EL, da Silva LF, Nader HB. A new approach for the characterization of polysaccharides from algae: presence of four main acidic polysaccharides in three species of the class Phaeophycea. Plant Sci 1995;108(2):143–53.

[99] Cumashi A, Ushakova NA, Preobrazhenskaya ME, et al. A comparative study of the anti-inflammatory, anticoagulant, antiangiogenic, and antiadhesive activities of nine different fucoidans from brown seaweeds. Glycobiology 2007;17:541–52.

[100] Lamote M, Johnson LE, Lemoine Y. Photosynthetic abilities of fucoid algae during low tide: differing responses of adults and young stages, vol. 37; 2001. p. 4–57.

[101] Li B, Xin JW, Sun JL, Xu SY. Structural investigation of a fucoidan containing a fucose-free core from the brown seaweed *Hizikia fusiforme*. Carbohydr Res 2006; 341:1135–46.

[102] Chevolot L, Foucault A, Chaubet F, Kervarec N, Sinquin C, Fisher AM, Boisson VC. Further data on the structure of brown seaweed fucans: relationships with anticoagulant activity. Carbohydr Res 1999;319:154–65.

[103] Ruparez P, Ahrazem O, Leal JA. Potential antioxidant capacity of sulfated polysaccharides from the edible marine brown seaweed *Fucus vesiculosus*. J Agric Food Chem 2002;50:840–5.

[104] Rioux LE, Turgeon SL, Beaulieu M. Characterization of polysaccharides extracted from brown seaweeds. Carbohydr Polym 2007;69:530–7.

[105] Bellion C, Brigand G, Prome JC, Welti D, Bociek S. Identification et caractérisation des précurseurs biologiques des carraghénanes par spectroscopie de RMN ^{13}C. Carbohydr Res 1983;119:31–48.

[106] Udani J, Hesslink R. The potential use of fucoidans from brown seaweed as a dietary supplement. J Nutr Food Sci 2012;2:10.

[107] Morel F, Dounassiere J, Vignais PV. The functional expression of p47-*phox* and p67-*phox* may contribute to the generation of superoxide by an NADPH oxidase like system in human fibroblasts. Eur J Biochem 1991;201:523–46.

[108] Silva ABL, Dias KS, Marques MS, Menezes IAC, Santos TC, Mello ICM, Lisboa ACCD, Cavalcanti SCH, Marçal RM, Antoniolli AR. Avaliação do efeito antinociceptivo e da toxicidade aguda do extrato aquoso da *Hyptis fruticosa* Salmz. ex Benth. Rev Bras Farmacogn 2006;16:475–9.

[109] Cardoso ML, Bezerra MEB, Paiva AAO, Carvalho MGF, Benevides NMB, Rocha FA, Leite EL. Assessment of arthritis in a rat model using sulfated polysaccharides. Planta Med 2010;76:113–9.

[110] Zang XW, Liu Q, Wang Y, Thorlacius. CXC chemokines, MIP-2 and KC, induce P-selectin-dependent neutrophil rolling and extravascular migration in vivo. Br J Pharm 2001;133:413–21.

[111] Holtkamp A, Kelly S, Ulber R, Lang S. Fucoidans and fuc-oidanases—focus on techniques for molecular structure elucidation and modification of marine polysaccharides. Appl Microbiol Biotechnol 2009;82:1–11.

[112] Skriptsova AV, Shevchenko NM, Zvyagintseva TN, Imbs TI. Monthly changes in the content and monosaccharide composition of fucoidan from *Undaria pinnatifida* (Laminariales, Phaeophyta). J Appl Phycol 2009;22:79–86.

[113] Haug A. Composition and properties of alginate [thesis]. Trondheim: Norwegian Institute of Technology; 1964.

[114] Lee JB, Koizumi S, Hayashi K, Hayashi T. Structure of rhamnan sulfate from the green alga *Monostroma nitidum* and its anti-herpetic effect. Carbohydr Polym 2010;81(3): 572–7.

[115] Yang CD, Chung IS, Shin H, Lee J, Kim Y, Lee, You S. Effects of molecular weight and hydrolysis conditions on anticancer activity of fucoidans from sporophyll of *Undaria pinnatifida*. J Biol Macromol 2008;43:433–7.

[116] Rodriguez-Jasso RM, Mussatto SI, Pastrana L, Aguilar CN, Teixeira JA. Microwave-assisted extraction of sulfated polysaccha-rides (fucoidan) from brown seaweed. Carbohydr Polym 2011;86:1137–44.

[117] Foley SA, Mulloy B, Tuohy MG. An unfractionated fucoidan from *Ascophyllum nodosum*: extraction, characterization, and apoptotic effects in vitro. J Nat Prod 2011;74: 1851–61.

[118] Haug A, Larsen B. A study on the constitution of alginic acid by partial acid hydrolysis. Proc Int Seaweed Symp 1966;5:271–7.

[119] Jenson A. Industrial utilization of seaweeds in the past, present and future. Proc Int Seaweed Symp 1979;9:17–34.

[120] Chapman ARO, Chapman DJ. Seaweeds and their uses. Can J Bot 1980;61:1654–9.

[121] Viswanathan S, Nallamuthu T. Extraction of sodium alginate from selected seaweeds and their physiochemical and biochemical properties. Int J Innovative Res Sci Eng Tech 2014;3(4):10998–1003.

[122] Mchugh DJ. Production and utilization of products from commercial seaweeds. FAO fisheries technical paper, vol. 288; 1987.

[123] Kleinübing SJ, Gaia F, Bertagnolli C, da Silva MGC. Extraction of alginate biopolymer present in marine alga *Sargassum filipendula* and bioadsorption of metallic ions. Mater Res 2013;16(2):481–8.

[124] Tushar J, Trivedi T, Kumar A. Efficient extraction of agarose from red algae using ionic liquids. Green Sustainable Chem 2014;4:190–201.

[125] Mitchell ME, Guiry MD. Carrageen: a local habitation or a name? J Ethnopharmacol 1983;9:347–51.

[126] Pereira L, Amado AM, Critchley AT, van de Velde F, Ribeiro-Claro PJA. Identification of selected seaweed polysaccharides (phycocolloids) by vibrational spectroscopy (FTIR-ATR and FT-Raman). Food Hydrocoll 2009;23:1903–9.

[127] Tuvikene R, Truus K, Vaher M, Kailas T, Martin G, Kersen P. Extraction and quantification of hybrid carrageenans from the biomass of the red algae *Furcellaria lumbricalis* and *Coccotylus truncatus*. Proc Est Acad Sci Chem 2006;55:40–53.

[128] Hoefler AC. Hydrocolloids. St. Paul: Eagan Press; 2004. p. 7–25.

[129] van de Velde F, De Ruiter GA. Carrageenan. In: Vandamme EJ, Baets SD, Steinbüchel A, editors. Biopolym 6 polysaccharides II. Polysaccharides from eukaryotes. Weinheim: Wiley; 2002. p. 245–74.

[130] Shanmugam M, Mody KH. Heparinoid-active sulphated polysaccharides from marine algae as potential blood anticoagulant agents. Curr Sci 2000;79:1672–83.

[131] Vera J, Castro J, Gonzalez A, Moenne A. Seaweed polysaccharides and derived oligosaccharides stimulate defense responses and protection against pathogens in plants. Mar Drugs 2011;9:2514–25.

[132] Painter TJ. Algal polysaccharides. In: Aspinall GO, editor. The polysaccharides. New York: Academic Press; 1983. p. 195–285.

[133] McHugh DJ. A guide to the seaweed industry. FAO fisheries technical paper, vol. 441; 2003. Rome.

[134] De Ruiter GA, Rudolph B. Carrageenan biotechnology. Trends Food Sci Technol 1997;8:389–95.

[135] Schweiger RJ, Guardado-Puentes J, Hernandez-Garibay E, Ones CN, Bautista-Alcantar J. Planta Piloto de Carragenanos. Reporte interno. Centro Regional de Investigación Pesquera de Ensenada; 1992. INP.

[136] Dewi EN, Java C. Characterization and quality of semi refined carrageenan (SCR) products from different coastal waters based on fourier transform infrared technique. J Coast Dev 2012;16:25–31.

[137] Nakashima Y, Kido N, Kobayashi Y, Motoki M, Neushul Yamamoto N. Purification and characterization of an avian myeloblastosis and human immunodeficiency virus reverse transcriptase inhibitor, sulfated polysaccharides extracted from sea algae. Antimicrob Agents Chemother 1987;31(10):1524–8.

[138] Lee JB, Hayashi K, Hirata M, et al. Antiviral sulfated polysaccharide from *Navicula directa*, a diatom collected from deep-sea water in Toyama Bay. Biol Pharm Bull 2006;29(10):2135–9.

[139] Novak BM. Hybrid nano composite materials-between in organic glasses and organic polymers. Adv Mater 1993;5:422–33.

[140] Nakagaito AN, Yano H. Novel high-strength biocomposites based on micro-fibrillated cellulose having nano-order-unit web-like network structure. Appl Phys A Mater Sci Process 2005;80:155–9.

[141] Hsieh YC, Yano H, Nogi M, Eichhorn SJ. An estimation of the Young's modulus for bacterial cellulose filaments. Cellulose 2008;15:572–82.

[142] Czaja W, Romanovicz D, Brown RM. Structural investigations of microbial cellulose produced in stationary and agitated culture. Cellulose 2004;11:403–11.

[143] Watanabe K, Tabuchi M, Morinaga Y, Yoshinaga F. Structural features and properties of bacterial cellulose produced in agitated culture. Cellulose 1998;5:187–200.

[144] Kim DY, Nishiyama Y, Kuga S. Surface acetylation of bacterial cellulose. Cellulose 2002;9:361–7.

[145] Wu C. Production and characterization of optically transparent nanocomposite film. Faculty of Forestry. Toronto: University of Toronto; 2010.

[146] Sain MM, Bhatnagar A. Manufacturing process of cellulose nanofibers from renewable feed stocks. US Patent 20080146701. 2008.

[147] Juntaro J. Environmentally friendly hierarchical composites. London: Department of Chemical Engineering and Chemical Technology, University of London; 2009.

[148] Tomasik P, Schilling CH. Chemical modification of starch. Adv Carbohydr Chem Biochem 2004;59:175–403.

[149] Ramesh M, Mitchell JR, Harding SE. Amylose content of rice starch. Starch 1999;51:311–3.

[150] Wallace RA, King JL, Sanders GP. Biology - the science of life. California: Goodyear Publishing Company; 1981.

[151] Stepto RFT. The processing of starch as a thermoplastic. Macromol Symp 2003;201: 203–12.

[152] Primarini D, Ohta Y. Some enzyme properties of raw starch digesting amylases from *Streptomyces* sp. Starch 2000;52(4):28–32.

[153] Fanta GF, Burr RC, Doane WM, Russell CR. Graft polymerization of vinyl acetate onto starch. Saponification to starch-g-poly(vinyl alcohol). J Appl Polym Sci 1979; 23:229–40.

[154] Simi CK, Abraham TE. Hydrophobic grafted and cross linked starch nanoparticles for drug delivery. Bioprocess Biosyst Eng 2007;30:173–80.

[155] Samaha SH, Nasr HE, Hebeish A. Synthesis and characterization of starch-poly(vinyl acetate) graft copolymer and their saponified form. J Polym Res 2005;12:343–53.

[156] Franck A, De Leenheer L. Inulin. In: De Baets S, Vandamme EJ, Steinbüchel A, editors. Biopolymers: polysaccharides II. Weinheim: Wiley-VCH; 2002. p. 439–53.

[157] Kleesem B, Sykura B, Zunft HJ. Effects of inulin and lactose on fecal microflora, microbial activity, and bowel habit in elderly constipated persons. Am J Clin Nutr 1997; 65:1397–402.

[158] Rees DA, Wright AW. Polysaccharide conformation, part VII, model building computations for α-1,4-galacturonan and the kinking function of L-rhamnose residues in pectic substances. J Chem Soc 1971;2(Sect B):1366–72.

[159] Sriamornsak P. Chemistry of pectin and its pharmaceutical uses: a review. J Pharm Pharm Sci 1998;44:207–28.

[160] Mukhiddinov ZK. Isolation and structural characterization of a pectin homo and ramnogalacturonan. Talanta 2000;53:171–6.

[161] Manish S, Bhatia MS, Deshmukh R, Choudhari P, Bhatia NM. Chemical modification of pectins, characterization and evaluation for drug delivery. Sci Pharm 2008;76: 775–84.

[162] Carpita NC, Gibeaut DM. Structural models of primary cell walls in flowering plants: consistency of molecular structure with the physical properties of the walls during growth. Plant J 1993;3:1–30.

[163] Rouse HA, Crandall PG. Nitric acid extraction of pectin from citrus peel. Proc Fla State Hortic Soc 1976;89:166–8.

[164] Boonrod D, Kamonrad R, Hataichanoke N. Extraction and physicochemical characteristics of acid-soluble pectin from raw papaya (*Carica papaya*) peel. Chiang Mai J Sci 2006;33(1):129–35.

[165] Helene M, Heloísa CS, Fertonani CR, Waszczynskyj N, Wosiacki G. Extraction of pectin from apple pomace. Braz Arch Biol Technol Curitiba March 2005;48:2–3.

[166] Thakur BR. Chemistry and uses of pectin – a review. Crit Rev Food Sci Nutr 1997;37: 47–73.

[167] Rolin C. Pectin. In: Whistler RL, BeMiller JN, editors. Industrial gums. 3rd ed. New York: Academic Press; 1993.

[168] Ralet MC, Bonnin E, Thibault JF. In: De Baets S, Vandamme EJ, Steinbüchel A, editors. Biopolymers: polysaccharides II. Weinheim: Wiley-VCH; 2002. p. 345.

[169] Portenoy RK, Burton AW, Gabrail N, Taylor D. A multicenter, placebo-controlled, double-blind, multiple-crossover study of Fentanyl Pectin Nasal Spray (FPNS) in the treatment of breakthrough cancer pain. Pain 2010;151:617–24.

[170] Kumar MNVR. A review of chitin and chitosan applications. React Funct Polym 2000; 46(1):1–27.

[171] Zargar V, Asghari M, Amir Dashti A. A review on chitin and chitosan polymers: structure, chemistry, solubility, derivatives, and applications. Chem Biol Eng 2015:1–24.

[172] Hirano S, Noishiki Y, Kinugawa J, Higashijima H, Hayashi T. Advances in biomedical polymers. Berlin: Springer; 1987.

[173] Gallardo A, Aguilar MR, Elvira C, Peniche C, San Román J. Biodegradable systems in tissue engineering and regenerative medicine. 1st ed. Boca Raton (FL): CRC Press; 2005.

[174] Lejardi A, Hernández R, Criado M, Santos JI, Etxeberria A, Sarasua JR, Mijangos C. Carbohydr Polym 2014;103:267–73.

[175] Mitsumata T, Suemitsu Y, Fujii K, Fujii T, Taniguchi T, Koyama K. pH-response of chitosan. Polymer 2003;44(23):7103–11.

[176] Delattre C, Michaud P, Mati-Baouche N, Elchinger PH, de Baynast H, Pierre G. Eur Polym J 2014;60:198–212.

[177] Hudson S, Smith C. Biopolymers from renewable resources. (Berlin): Springer; 1998.

[178] Heidrich C, Vollmer W. Murein. In: De Baets S, Vandamme EJ, Steinbüchel A, editors. Biopolymers: polysaccharides II. Weinheim: Wiley-VCH; 2002 p. 431.

[179] Lazarevic V, Pooley HM, Mauel C, Karamata D. In: De Baets S, Vandamme EJ, Steinbüchel A, editors. Biopolymers: polysaccharides II. Weinheim: Wiley-VCH; 2002. p. 465.

[180] El-Boshy ME, El-Ashram AM, AbdelHamid FM, Gadalla HA. Fish Shellfish Immun 2010;28:802–8.

[181] Schizophyllan RU. In: De Baets S, Vandamme EJ, Steinbüchel A, editors. Biopolymers: polysaccharides II. Weinheim: Wiley-VCH; 2002. p. 61.

[182] Bielecki S, krystynowicz A, Turkiewicz M, Kalinowska H. In: De Baets S, Vandamme EJ, Steinbüchel A, editors. Biopolymers: polysaccharides II. Weinheim: Wiley-VCH; 2002. p. 37.

[183] Leathers TD. Pullulan. In: De Baets S, Vandamme EJ, Steinbüchel A, editors. Biopolymers: polysaccharides II. Weinheim: Wiley-VCH; 2002. p. 2–23.

[184] Born K, Langendorff V, Boulenguer P. In: De Baets S, Vandamme EJ, Steinbüchel A, editors. Biopolymers: polysaccharides II. Weinheim: Wiley-VCH; 2002. p. 259.

[185] Schuster E, Schlegel HG. Chemolithotrophic growth of *Hydrogenomonas* H-16 using electrolytic production of hydrogen and oxygen in a chemostat. Arch Microbiol 1967; 58(4):380–409.

[186] Schlegel HG, Lafferty R. Growth of knallgas bacteria (*Hydrogenomonas*) using direct electrolysis of the culture medium. Nature 1965;205:308–9.

[187] Tian PY, Shang L, Ren H, Mi Y, Fan DD, Jiang M. Biosynthesis of polyhydroxyalkanoates: current research and development. Afr J Biotechnol 2009;8:709–14.

[188] Suriyamongkol P, Weselake R, Narine S, Moloney M, Shah S. Biotechnological approaches for the production of polyhydroxyalkanoates in microorganisms and plants – a review. Biotechnol Adv 2007;25:148–75.

[189] Akaraonye E, Keshavarz T, Roy I. Production of polyhydroxyalkanoates: the future green materials of choice. J Chem Technol Biotechnol 2010;85:732–43.

[190] Cavalheiro IMBT, Raposo RS, Almeida MCMD, Cesario MT, Sevrin C, Grandfils C, Fonseca MMR. Effect of cultivation parameters on the production of poly(3-hydroxybutyrate-co-4-hydroxybutyrate) and poly(3-hydroxybutyrate-4-hydroxybutyrate-3-hydroxyvalerate) by *Cupriavidus necator* using waste glycerol. Bioresour Technol 2012;111:391–7.

[191] Garcia IL, Lopez JA, Dorado MP, Kopsahelis N, Alexandri M, Papalikolaou S, Villar MA, Koutinas AA. Evaluation of by-products from the biodiesel industry as

fermentation feedstock for poly(3-hydroxybutyrate-co-3-hydroxyvalerate) production by *Cupriavidus necator*. Bioresour Technol 2013;130:16–22.

[192] Spoljaric IV, Lopar M, Koller M, Muhr A, Salerno A, Reiterer A, Malli K, Angerer H, Strohmeier K, Schober S, Mittelbach M, Horvat P. Mathematical modeling of poly [(R)-3-hydroxyalkanoate] synthesis by *Cupriavidus necator* DSM 545 on substrates stemming from biodiesel production. Bioresour Technol 2013;133:482–94.

[193] Valappil SP, Misra SK, Boccaccini AR, Keshavarz T, Bucke C, Roy I. Large-scale production and efficient recovery of PHB with desirable material properties, from the newly characterized *Bacillus cereus* SPV. J Biotechnol 2007;132:251–8.

[194] Tokiwa Y, Ugwu CU. Biotechnological production of (R)-3-hydroxybutyric acid monomer. J Biotechnol 2007;132:264–72.

[195] Borman EJ, Leißner M, Roth M, Beer B, Metzner K. Appl Microbiol Biotechnol 1998; 50:604–7.

[196] Cruz Pradella JG, Ienczak JL, Delgado CR, Taciro MK. Carbon source pulsed feeding to attain high yield and high productivity in poly(3-hydroxybutyrate) (PHB) production from soybean oil using *Cupriavidus necator*. Biotechnol Lett 2012;34:1003–7.

[197] Rai R, Yunos DM, Boccaccini AR, Knowles JC, Barker IA, Howdle SM, Tredwell GD, Keshavarz T, Roy I. Poly-3-hydroxyoctanoate P(3HO), a medium chain length polyhydroxyalkanoate homopolymer from *Pseudomonas mendocina*. Biomacromolecules 2011;12:2126–36.

[198] Hoffmann N, Steinbüchel A, Rehm BH. The *Pseudomonas aeruginosa* phaG gene product is involved in the synthesis of polyhydroxyalkanoic acid consisting of medium-chain-length constituents from non-related carbon sources. FEMS Microbiol Lett 2000;184:253–9.

[199] Kabilan S, Ayyasamy M, Jayavel S, Paramasamy G. *Pseudomonas* sp. as a source of medium chain length polyhydroxyalkanoates for controlled drug delivery: perspective. Int J Micro 2012:1–10.

[200] Sudesh K, Abe H, Doi Y. Synthesis, structure and properties of polyhydroxyalkanoates: biological polyesters. Prog Polym Sci 2000;25:1503–55.

[201] Foster LJR, Russel RA, Sanguanchaipaiwong V, Stone DJM, Hook JM, Holden PJ. Biosynthesis and characterization of deuterated polyhydroxyoctanoate. Biomacromolecules 2006;7:1344–9.

[202] Cavalheiro JMBT, Almeida MCMD, Fonseca MMR, Carvalho CCCR. Effect of cultivation parameters on the production of poly(3-hydroxybutyrate-co-4-hydroxybutyrate) and poly(3-hydroxybutyrate-4-hydroxybutyrate-3-hydroxyvalerate) by *Cupriavidus necator* using waste glycerol. J Biotechnol 2012;164:309–17.

[203] Lutke-Eversloh T, Steinbuchel A. Novel precursor substrates for polythioesters (PTE) and limits of PTE biosynthesis in *Ralstonia eutropha*. FEMS Microbiol Lett 2003;221: 191–6.

[204] Carrillo-Navas H, Cruz-Olivares J, Varela-Guerrero V, Alamilla-Beltŕan L, Vernon-Carter EJ, Pérez-Alonso C. Rheological properties of a double emulsion nutraceutical system incorporating chia essential oil and ascorbic acid stabilized by carbohydrate polymer–protein blends. Carbohydr Polym 2012;87:1231–5.

[205] Hoffman AS, Stayton PS. Conjugates of stimuli-responsive polymers and proteins. Prog Polym Sci 2007;32:922.

[206] Yoshida T, Hiraki J, Nagasawa T. In: Fahnestock SR, Steinbüchel A, editors. Biopolymers: polyamides and complex proteinaceous materials I. Weinheim: Wiley-VCH; 2003. p. 107.

[207] Joengten W, Müller N, Mitschker A, Scmidt A. In: Fahnestock SR, Steinbüchel A, editors. Biopolymers: polyamides and complex proteinaceous materials I. Weinheim: Wiley-VCH; 2003. p. 175.
[208] Sionkowa A. Prog Polym Sci 2011;36:1254.
[209] Tchmutin I, Ryvkina N, Saha P. Study on biodegradability of protein filled polymer composites using dielectric measurements. Polym Degrad Stab 2004;86:411−7.
[210] Mohanty AK, Liu W, Tummula P, Drza LT, Manjusri M, Narayan R. In: Mohanty AK, Misra M, Drzal LT, editors. Natural fibers, biopolymers, and biocomposites. Boca Raton: Taylor & Francis; 2005. p. 1−36.
[211] Jönsson AS, Nordin AK, Wallberg O. Concentration and purification of lignin in hardwood kraft pulping liquor by ultrafiltration and nanofiltration. Chem Eng Res Des 2008;86:1271−80.
[212] Monties B, Fukushima Y. Occurrence, function and biosynthesis of lignins. In: Hofrichter M, Steinbüchel A, editors. Biopolymers1. Weinheim (Germany): Wiley-VCH; 2001. p. 1−64.
[213] Brunow G. In: Hofrichter M, Steinbüchel A, editors. Biopolymers: lignin, humic substances and coal. Weinheim: Wiley-VCH; 2001. p. 89−116.
[214] Mohamad Ibrahim MN, Azian H, Mohd Yusop MR. Separation and characterization of the vanillin compound from soda lignin. J Technol 2006;44:83−94.
[215] Kelsey DR. US5705575. 1998.
[216] Curran KA, Leavitt JM, Karim AS, Alper S. Metabolic engineering of muconic acid production in *Saccharomyces cerevisiae*. Metab Eng 2013;15:55−66.
[217] Guzman DICIS. Green chemicals. 2012.
[218] Lunt J. Large-scale production, properties and commercial applications of polylactic acid polymers. Polym Degrad Stab 1998;59:142−52.
[219] Carlotta D. A literature review of poly(lactic acid). J Polym Environ 2002;9:63−84.
[220] Berezina N, Landercy N, Mariage PA, Moreau B. Separation of the enantiomers of lactide, lactic acid dimer. J Org Chem 2013;1:20−3.
[221] Lee JW, Kim HU, Choi S, Yi J, Lee SY. Microbial production of building block chemicals and polymers. Curr Opin Biotechnol 2011;22:758−67.
[222] Delwiche EA. A biotin function in succinic acid decarboxylation by *Propionibacterium pentosaceum*. J Bacteriol 1950;59(3):439−42.
[223] Homan T, Tag C, Biebl H, Deckwer WD, Schink B. Fermentation of glycerol to 1,3-propanediol by *Klebsiella* and *Citrobacter* strains. Appl Microbiol Biotechnol 1990; 33:121−6.
[224] Drozdzynska A, Leja K, Czaczyk K. Biotechnological production of 1,3-propanediol from crude glycerol. J Biotechnol Comput Biol Bionanotechnol 2011;92:92−100.
[225] Stols L, Donnelly MI. Production of succinic acid through overexpression of NAD^+-dependent malic enzyme in an *Escherichia coli* mutant. Appl Environ Microbiol 1997; 63:2695−701.
[226] Ahmed H, Naouel O, Djilali B. 1,3-Propanediol production by *Clostridium butyricum* in various fed-batch feeding strategies. J Biotechnol Biomater 2012;2:134.
[227] Rizzarelli P, Carroccio S. Thermo-oxidative processes in biodegradable poly(butylene succinate). Polym Degrad Stab 2009;94:1825−38.
[228] Nakamura CE, Gatenby AA, Hsu AK-H, La Reau RD, Haynie SL, Diaz-Torres M, Trimbur DE, Whited GM, Nagarajan V, Payne MS, et al. Method for the production of 1,3-propanediol by recombinant microorganisms. Patent US 6013494. 2000.

CHAPTER

Synthetic Materials and the Problems They Pose

2

Mirza Nadeem Ahmad, Ameer Fawad Zahoor, Shahzad Ali Shahid Chatha, Abdullah Ijaz Hussain, Muhammad Asim Mansha

Government College University Faisalabad, Faisalabad, Pakistan

2.1 INTRODUCTION

The type of waste varies depending on the site and country, as the humans have been discarding the waste on the land and into oceans. The term "harmful waste" creeps in our minds because the waste is hazardous to human health and life (Fig. 2.1).

The usage of synthetic and plastic materials has expanded at an alarming pace since past 30 years (Table 2.1). The quantity of plastic debris going into the environment has significantly increased. However, the degradation of most of this stuff is very slow. These materials float on the ocean surface and remain suspended for a long time. Such debris does not decompose and stays in the bottom even for decades. The accumulated debris poses a significant threat for seabirds, turtles, mammals, crustaceans, and fish. The threats are clear-cut and mechanical in principal. These animals may ingest plastic materials or get tangled in opening of perched or sunken waste or loops. Ingested plastics might lessen the food digestion, injure stomach linings, or block digestive tract. The environmental waste containing such materials is a risk for the aquatic life, and oceans have become the reservoir of potentially injurious plastics waste. Mounting information suggested that the motorized effect of the materials distresses most of the aquatic species in many oceanic areas. It also affects the rationalized detection of relentless plastics.

The complexity in identifying this relentless plastic garbage as a considerable marine contamination is probably due to these following reasons:

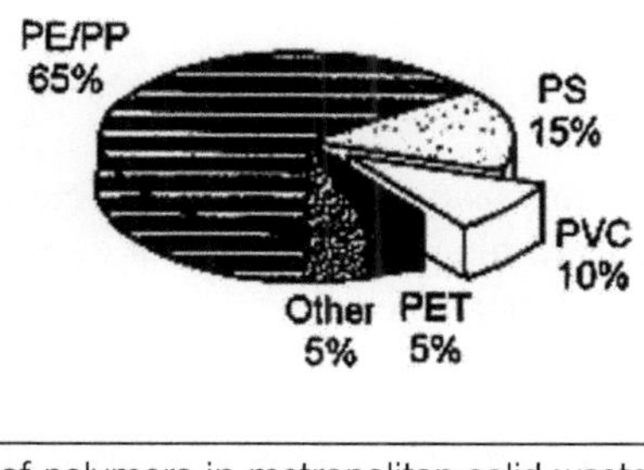

FIGURE 2.1

Percentages of few types of polymers in metropolitan solid waste [1].

Algae Based Polymers, Blends, and Composites. http://dx.doi.org/10.1016/B978-0-12-812360-7.00002-1

Table 2.1 Applications of Synthetic Plastics [1]

Plastics	Uses
Polyethylene	Plastic bags, milk and water bottles, food packaging films, toys, irrigation and drainage pipes, and motor oil bottles
Polystyrene	Disposable cups, packaging materials, laboratory ware, and electronics
Polyurethane	Tires, gaskets, bumpers, refrigerator insulation, sponges, furniture cushioning, and life jackets
Polyvinyl chloride	Automobile seat covers, shower curtains, raincoats, bottles, visors, shoe soles, garden hoses, and electricity pipes
Polypropylene	Bottle caps, drinking straws, medicine bottles, car seats, car batteries, bumpers, disposable syringes, carpets, backings
Polyethylene terephthalate	Beverages bottles, processed meat packages, peanut butter jars, pillows, sleeping bag filling, and textile fibers
Nylon	Small bearings, speedometer gears, windshield wipers, water hose nozzles, football helmets, racehorse shoes, inks, clothing parachute fabrics, rainwear, and cellophane
Polycarbonate	Nozzles for paper making machinery, street lighting, safety visors, rear lights of cars, baby bottles, sky lights, roofs of greenhouses, sun rooms, and verandas
Polytetraflouroethylene	Specialized chemical plant, electronics and bearings, nonstick coating on kitchen utensils, such as sauce pans, and frying pans

1. The ways how waste is affecting oceanic life are apparent and this is a reality.
2. Oceans cover nearly three-fourth of earth, and the chance of an aquatic creature encountering a lost plastic waste would appear to be occasional or not likely to be happening.
3. Countless aquatic animals are strangling, drown, starved, and suffocated due to man-made materials, and it seems to become more severe in future.
   The development of information technology reveals the origin of problems associated with plastic debris, which cause deaths for several marine animals, like seals, sea birds, turtles, fish, and crustaceans.

2.2 PLASTIC DEBRIS

2.2.1 AESTHETICS

According to WHO, dirt-free seaside is the main characteristics demanded by tourists. The harmful effects of waste involve the lack of tourists, which directly affects the tourism business and also spoils the fishery activities. These things were faced in New Jersey (1987) and Long Island (1988), USA. The medical waste, such as syringes, plastic catheters, and vials, was found disposed everywhere, resulting in a loss between US$121 and 327 million/day by the seaside including a loss of

US$5.4 million for tourism-related outlays (Table 2.2). To keep the seaside clean, debris was removed free of charge. On International Cleanup Day, six million volunteers participated from 100 countries and 100 million pounds of garbage was removed from 170,000 miles of beach and inland waterways. It is reported that International Cleanup Day efforts are expanding every year to remove larger quantity of debris. During 2006 to 1996, a total of 310 items were removed at Escondido

Table 2.2 US Trade of Plastics in Million Pound in 1988 [2]

Resin	Sales	%Age of Total Sale
Thermoplastic Resins		
Low-density polyethylene (LDPE)	9,865	17.3
Polyvinyl chloride and copolymers (PVC)	8,323	14.6
High-density polyethylene (HDPE)	8,244	14.5
Polypropylene (PP)	7,304	12.0
Polystyrene (PS)	5,131	9.0
Thermoplastic polyester (including polyethylene terephthalate (PET) and polybutylene terephthalate (PBT)	2,007	3.5
Acrylonitrile:butadiene:styrene (ABS)	1,238	2.2
Other styrenics	1,220	2.1
Other vinyls	958	1.7
Nylon	558	1.0
Acrylics	697	1.2
Thermoplastic elastomers	495	0.9
Polycarbonate	430	0.8
Polyphenylene-based alloys	180	0.3
Styrene:acrylonitrile (SAN)	137	0.2
Polyacetal	128	0.2
Cellulosics	90	0.2
Total thermoplastic	47,005	82.6
Thermosetting Resins		
Phenolic	3,032	5.3
Polyurethane	2,905	5.1
Urea and melamine	1,515	2.7
Polyester, unsaturated	1,373	2.4
Epoxy	470	0.8
Alkyd	320	0.6
Total thermosetting	9,615	16.9
Others	288	0.5
Total	56,908	100.0

seaside, California, while 182 of these were removed in 2005, which is 59% of total recovered in last 10 years. In the fourth quarter of 2005, 136 items were detached, and in second quarter of the 2006 alone, 189 items were removed at Torrey Pines seaside, California [3].

2.2.2 ENTANGLEMENT

In 1980 the scientists predicted that there were about one million deaths of aquatic mammals annually in North Pacific due to plastic netting and fish line [4]. The use of digitally traumatized marine mammals has enhanced manifold. Their photos showed that they were entangled in the nets and lines. Lost and dumped fishing nets are called "ghost nets" to continue fishing and destruction of resources. It is reported by the Canadian Food and Agriculture Organization that 10% of all static gear loss can lead to the failure of 10% of the targeted fish stocks. Sincere efforts are needed to eliminate this mechanism. The cost to remove the waste is enormous. The cost can be a threat to the fiscal feasibility of fishing business. Seabirds and other marine species get entangled in these plastic nets and bottles leading to their death and disintegration of the environment. The composition of polymers is chemically altered by absorbing UV radiations from the sunlight, which generates toxins. Plastic debris also decreases the penetration of UV radiation that is beneficial for marine species. Moreover, this problem becomes long lasting due to nondegradable nature of the waste [5].

2.2.3 INGESTION OF PLASTICS

Floating on marine and aquatic environment, plastic debris seems to look like marine food. Jellyfish and sea turtles, particularly leatherbacks, regularly consume the skinny plastic shopping bags and water balloons [6,7]. Almost all food sources, which exploded tons of multicolored plastic waste in the aquatic environment, produce imitation in unlimited ways. It is revealed that crustacean zooplankton trunk coregent intake the polystyrene beads (2−6 mm) plus algae as combination of nonselective food [8].

Plastic filter intake appears to be an optimal size class, smaller than 1 mm in diameter, though bigger particles were also found in a few individuals. Many particles (27,448) of plastics float up in the North Pacific trawling circulation. AMRF 1999 study shows that nearly 1 mm particle size (9470), 0.5 mm particles (4646), and 0.3 mm particles (2626) were eliminated by this system [9]. There are small particles in water with size 0.2 mm in diameter called microscopic plastic particles [10]. Intake of used microscopic plastic tubes is another problem. Adverse outcomes of these types of particles were also reported [8]. The survey established that manure containing plastics particle from the sea near the active night time was consumed by myctophids (lantern fish) and similar species.

2.2.4 THE THREATS FROM PLASTIC POLLUTION TO MARINE BIOTA

Synthetic polymers have been entering into aquatic environment parallel to the production quantity levels for the past half century [11]. Moreover, for the last two decades of 20th century, the speed of deposition accelerated along with the pace of production of plastics. They are the most ordinary and unrelenting pollutants of the marine water as well as to the beaches around the world. For last 30 years the main argument of the plastic industries was that "plastic" garbage is an extremely small part of the whole litter. So, it does not damage the environment but is just an eyesore [12].

From 1960 to 2000 world's manufacturing of plastic resins increased 25 times, although the recycling of the materials remains less than 5%. From 1970 to 2003 plastic has become the fastest rising section of the flow of municipal waste in the United States, an increase by nine times. While aquatic plastic debris is currently 60%–80%, it is reaching to 90%–95% in a few areas. Aquatic plastic waste can be divided into two categories: macro (45 mm) and micro (0.5 mm). Macro particles can be traced by identifying the source. Micro litters are the particles consisting of two basic varieties, the broken-down fragments of larger objects, and polymer granules and powders from main industries of thermoplastic materials. Plastic debris ingestion by small filter-feeders at the bottom of the food web has started, but is not computed yet.

Uptake and degradation fragments of plastic particles cause toxicity problems. Adsorption of plastics is a well-known hydrophobic contaminant. Bioplastics desorbed in the ocean pose many physiological effects, which are investigated. Twenty years ago, it was found that the intake of large plastic by seagulls was positively related to PCBs and egg fat. The growth of such waste may inhibit gas exchange between overlying water and the sediment pore water, and disrupt or suffocate resident bentos. The degree of trouble and its implications are now getting investigated.

2.3 IMPACTS OF PLASTICS ON AGRICULTURE

Plastics have major and adverse impact on agriculture sector globally. Since people are not alert about plastic recycling and throw plastics all over the places. Inevitably, these useless plastics will eventually go into the field and mix with other material to decompose. But, plastic does not decompose in the soil, instead remains in farming soils and blocks the system. It also slows down the growth and progress of the agricultural crops. Plastics, although very thin, have this intrinsic property to block plant food. Actually, the roots of the plants are not able to penetrate in array and move out of the territory of the natural nutrients. Thus plastics cause enormous damage to agricultural production.

Due to some indifferent factors, about 96% of all plastics are discarded in landfills [13]. It is acknowledged that plastics are nonbiodegradable. However, plastics do decompose gradually. Actually, plastic bags can take up to 1000 years in ground, preventing degradation of other biodegradable materials [14]. Plastic shopping bags cause the extra damage to environment as they are easily moved by air and get stuck in trees and plants brushwood, which makes the environment unattractive. Plastic bags have filled roadside ditches, public places, and float on the oceans and rivers. An example is Cape Town, South Africa, where every kilometer of road was covered by more than 3000 plastic grocery bags [15].

The impact of plastics on agriculture is damaging in more aspects, and therefore these plastics are becoming even more dangerous to our lives. Most of the plastic bags are dispersed on land as landfills, the strategic surface of the planet containing a huge volume of plastic bags, and as a result it is the only trash above and below the surface of earth [16]. It will stay in soil by the order of extremely extended period of time to cause unbelievable damage to the agricultural sector. Agricultural crops could not be grown wherein the plastic bags are dumped, as the roots are unable to tear down the plastic bags to reach the ground below. This is truly wonderful that the skinny plastic bags are so sturdy that the tree roots are not able to penetrate to find its ways into top soil to get the nutrients. On other hand, we find plenty of plants and trees in the mountains due to the obvious lack of plastic bags.

2.4 SUBSTITUTE OF PLASTIC BAGS

2.4.1 NATURAL FIBER BAGS

Jute bags made up of natural fiber can be used for packaging. A wide variety of industrial and agricultural products such as sugar, cereals, salt, oilseeds, and others can be packed in jute bags according to their appropriate size and pattern. These products are in great demand in the packaging industry, in agricultural sectors, and in the manufacturing and packaging of agricultural products. Jute bags are also appropriate for the purpose of shopping as well as groceries. Jute bag is biodegradable containing vegetable fibers [16].

Although jute bag is replaceable with man-made material, natural environmental processes can take some advantages of the biodegradable nature of jute, but synthetics are not suitable for this. Jute bag for agricultural products are free from harmful effects of hydrocarbons which are present in plastic bags. They are also known as free hydrocarbon bags. Due to their biodegradability, jut bags have become an eco-friendly alternative [16].

2.4.2 BENEFITS AND ADVANTAGES OF JUTE BAGS

- Lush jute, the raw material for jute bags, is obtained from plants and provides the much needed oxygen in the atmosphere that helps to maintain the balance of the ecosystem.
- Jute bags are not only biodegradable and reusable but also convert manure into fertilizer for plants.
- The manufacturing method for jute bags is simple and unlike synthetic bags, as it does not use toxic chemicals which are associated with unwanted and dangerous by-products.

Jute bags are used to store cereals and grains such as rice and wheat. The bag has the intrinsic property of ventilation and secures storage space purposes. These are stable and do not slip downward as they are stacked. Jute bag is easy to hold by hand and mechanically strong [15]. Jute bag has always been offered as the environment-friendly alternative to plastic bags. Before the manufacturing of synthetic bags, jute bags were usually used for shopping. Reusable sources and natural fibers can be used for positive impact. For example, paper bags are now recycled to a great extent which were earlier discarded to landfill [16]. It is scientifically demonstrated that paper bags are not as damaging to the environment as synthetic bags [17,18]. A Scottish study has shown that a heavy tax on plastic bags can lead to enlargement of the paper bag utilization [18]. The tax on plastic bags is a strategy for improving the environment.

2.4.3 BIODEGRADABLE PLASTIC BAGS

Biodegradable plastic bags are also made in plastic industry. These are less injurious to the soil and environment. Production of nonbiodegradable plastic bags involves toxic components and is dangerous to the environment [19,20]. Therefore substitute biodegradable bags are mostly welcomed instead of plastic bags. Sufficient research is needed to replace plastic bags. The Irish government encourages people to use reusable shopping bags. Such bags will decrease the use of plastic bags and will contribute to keep the environment safe from toxins [21]. Biodegradable plastics are supposed to be a substitute to habitual toxic plastic bags. This provides a positive image of the similar environment-friendly natural fibers and biodegradation with paper or jute bags. Therefore there is no need to prevent the use of biodegradable plastic bags. The public pressure on the oil reserves is growing and the plastic war has come a long way. It is inculcating the public and encourages the use of biodegradable plastics for sustainable development. Therefore reducing the extensive use of traditional plastic is mandatory. Moreover, efforts must be made to increase the potential market for the production of new biodegradable plastic products for global use [22].

The sale of biodegradable plastic bags is an environment-friendly choice. It can definitely control the environmental problems, such as soil litter, recovery flow, increase the social impact of litter pollution, and reduce environmental functions [18]. Biodegradable plastic is known to be safer by the same ratio as the nonbiodegradable plastics unsafe. However, there are many varieties of degradable and biodegradable plastics. It makes the problem more difficult since not all the biodegradable and alternative composite plastics are environment-friendly. The term biodegradable product marketing has become a popular word nowadays. It is not essentially meant that the raw materials are natural or hazardous. The term biodegradable means "can be decayed by bacteria or other natural devices" [22]. Biodegradable synthetic plastic materials are the synthetic or starch-based plastics, which could be decomposed by biological processes, degradation, or decay blend. This is not essentially meant that the degradation of all products or by-products can be simply riveted into the natural environmental cycle. Degradable plastics from biological actives can still be put down to be discarded by extra efforts. Degradable plastic products vary among diverse varieties of biodegradable plastics [18].

Table 2.3 Fuel Properties and ASTM Technique for Characterization [23–28]

Properties	Method
Moisture content	ASTM E 790-87
Ash content	ASTM E 830-87
High heating value	ASTM D 3286-7
Volatile matter	ASTM E 897-88
Boiling point distribution	ASTM 7169-05
Viscosity	ASTM D 445-95
Pour and cloud points	ASTM D-5949

Kumar et al. reported on the quality of the fuel, the waste heat from the catalytic cracking of HDPE production, and diesel hybrid [23]. Kaolin is used as a catalyst in oil production from HDPE. As the performance of diesel blend reduced, the NO_x emission is increased, while CO_2 emission rate is lower than in all other diesel blends.

Sharma et al. pyrolyzed HDPE (shopping bags) in a 2 L batch reactor. The properties of hydrocarbons produced via pyrolysis were analyzed by using different characterization techniques including simulated distillation [25]. Ultra-low sulfur diesel (ULSD) and biodiesel hydrocarbons blend performance from cracking pyrolysis production was studied. The paraffin oil, having 96.8% purity, 46.16 MJ/kg calorific value and free from any oxidation products, makes it appropriate for petroleum diesel blend. The imitated refining consequence showed the existence of just about 20% motor gasoline, 41% diesel #1, 23% diesel #2, and 16% vacuum gas oil. For ease the ASTM methods are used to evaluate the above fuel properties as shown in Table 2.3.

Conversion of low-density polyethylene plastic into fuel by hydrocracking using bifunctional catalyst was studied scientifically. The natural zeolite from Sukabumi Indonesia was dried in 6N HCl for 30 min in an oven. Then activated natural zeolite (Z) was prepared and refluxed. The natural zeolite obtained from Sukabumi contains crystalline mordenite. For bifunctional catalysts, Ni, Ni-loaded-Mo, Co and Co–Mo metal natural zeolite were used to increase the activity and selectivity of hydrocracking process. The catalyst sample was heated in a microwave and reduced by hydrogen gas flow, produced by Ni/Z, Ni–Mo/Z, cobalt/Z, and cobalt–molybdenum/Z. Hydrocracking process was carried out at 350°C with hydrogen gas flow at the rate of 20 mL/min for 1 h. Higher selectivity of low-density polyethylene plastic waste at 350°C by the Co–Mo/Z transformation of gasoline produced 71.49% conversion. GCMS data based on the liquid yield from the atomic chain C_6 to C_{19} hydrocarbon compounds indicated the fuel chemical structure, such as paraffins, olefins, and naphthenic. By using this technique, plastic waste could be converted into fuel ethanol in order to reduce the pollution of the environment (Fig. 2.2). It also supports the utilization of soil and enhances energy storage.

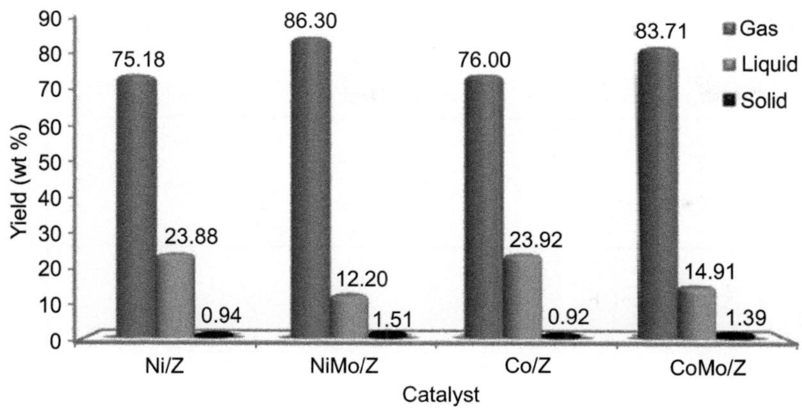

FIGURE 2.2

Conversion yield of plastic waste by different catalysts [29].

2.5 PYROLYSIS

In this study, models of polyethylenes (HDPE and LDPE) and polypropylenes (PP) are discussed because these polymers are present in various commercial waste products (packaging films, bags, tubes, food retailers, etc.) (Table 2.4). These polymers were pyrolyzed with a catalyst in FCC having following characteristics. The total surface area: 178.4 m^2/g; zeolite area: 58.5 m^2/g, Z/M: 0.49, UCS: 24.26 Å, Ni 150 ppm, and V 367 ppm (Figs. 2.3 and 2.4).

In such cases, three kinds of the municipal plastic waste feedstock, i.e., with a bag of polyethylene (PE bag, 2) and (PE bag, 1), were broken off and washed along with polyethylene (HDPE) waste. These were taken from the final discarding place. A small plastic recycling company is located in Indonesia. The demonstration of the feedstock is illustrated in Fig. 2.5. The diameter of the pellet, as binder, was 1.6 mm, containing 20% aluminum oxide. Natural zeolite Klaten collected from Indonesia was calcinated for 3 h at 500°C to remove some of the volatiles [30].

2.6 BIODEGRADATION OF THERMOPLASTIC POLYOLEFINS

2.6.1 POLYETHYLENE

Biodegradation of polyethylenes is reported to take place by following mechanisms: Hydro-biodegradable and oxo-biodegradable [32]. These mechanisms involve two of the additives used in the synthesis of biodegradable polyethylene by modification with starch and pro-oxidant. Starch-blend polyethylene material is hydrophilic, and therefore, you have a continuous starch phase to be catalyzed by the enzyme amylase. Microorganisms can easily attack and you can remove this part. Thus the matrix and hydrophilic polyethylene carries on to become hydro-biodegradable.

Table 2.4 Mass Balance of Crude Oil, Residue and Gas Yields on Pyrolysis of Polyethylene, Polypropylene and Polystyrene Using Various Temperatures [24]

Investigation	Plastics	Reactor	Pyrolysis Temperature (°C)	Catalyst	Crude Oil (wt%)	Residue	Gas (wt%)
William et al.	PE	Parr mini benchtop	500	None	93	0.0	7.0
	PP				95	0.0	5.0
	PS				71	27	2.0
	PET				15	5.0	5.0
Sarkar et al.	Mixed	Proprietary (natural state research Inc.)	370–420	None	90	5.0	5.0
Alston et al.	Mixed		800	None	73	23.5	30.0
Sharma et al.	HDPE	2 L batch reactor	440	None	74	17.0	9
Buekens et al.	PP		740	None	48.8	1.6	49.6
	PE		760		42.4	1.8	55.8
	PS		581		24.6	0.6	9.90
Beltrame et al.	PE		400	None	44.0	30.0	20.0
				H-Y-Zeolite	91.0	0.0	9.90
				Silica–alumina	93.0	7.0	0.0
			600	None	29.0	65.0	6.0
Scott et al.	PE	Activated carbon bed	515–795	None	88–96		
Miskolczi et al.	HDPE	Continues reactor	520	ZSM-5			
	PP			NONE			
Uemichi et al.	LDPE	Fixed-bed tubular flow reactor	425	HZSM-5SiO_2-Al_2O_3			

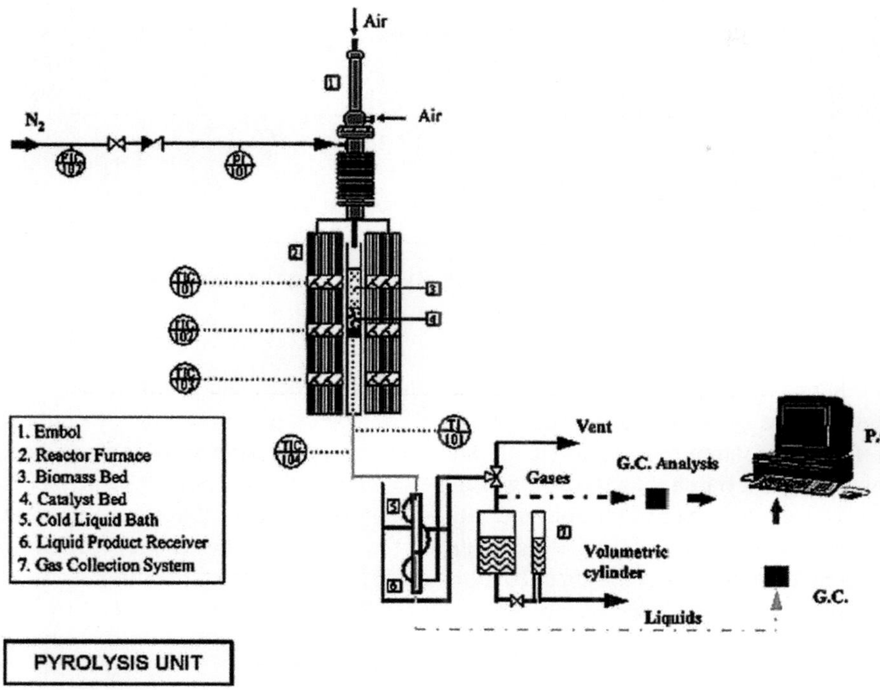

FIGURE 2.3

Fixed-bed reactor system [30].

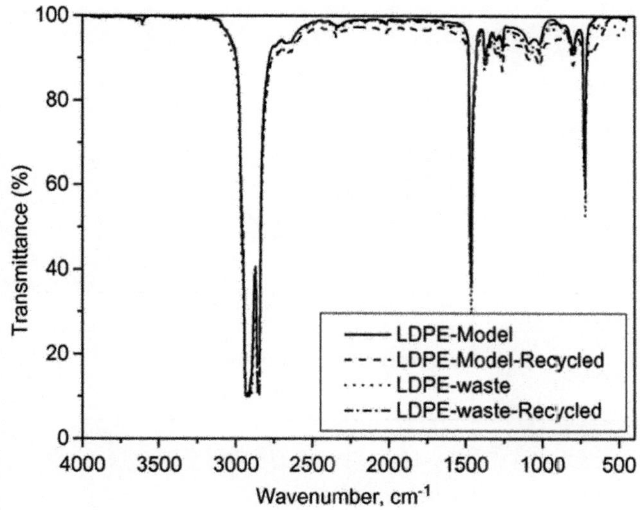

FIGURE 2.4

Fourier transform infrared spectra of polyethylene before and after recycling [30].

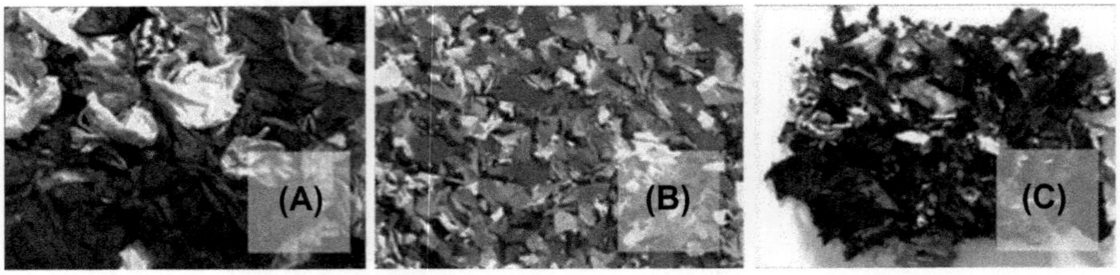

FIGURE 2.5

The feedstock used in the experiments: (A) PE bag 1; (B) High-density polyethylene waste; and (C) PE bag 2 [31].

In other cases, the oxidative degradation, light degradation, and chemical degradation can also assist the biodegradation (Fig. 2.6) [32–34]. We examined the attack of fungi and streptomycin strain. It attacked the degradable polyethylene bag, containing 6% starch. In a study, it was mixed with low-density polyethylene pieces sewage sludge, buried in the soil, indicating the possible use of plastic as a nutrient source. The adhesion of fungi was found on the surface of the plastic which was examined microscopically after 10 months [35]. The inaccessible fungal strain was recognized as *Fusarium* sp. AF4, *Aspergillus terreus* AF5, and *Penicillium*

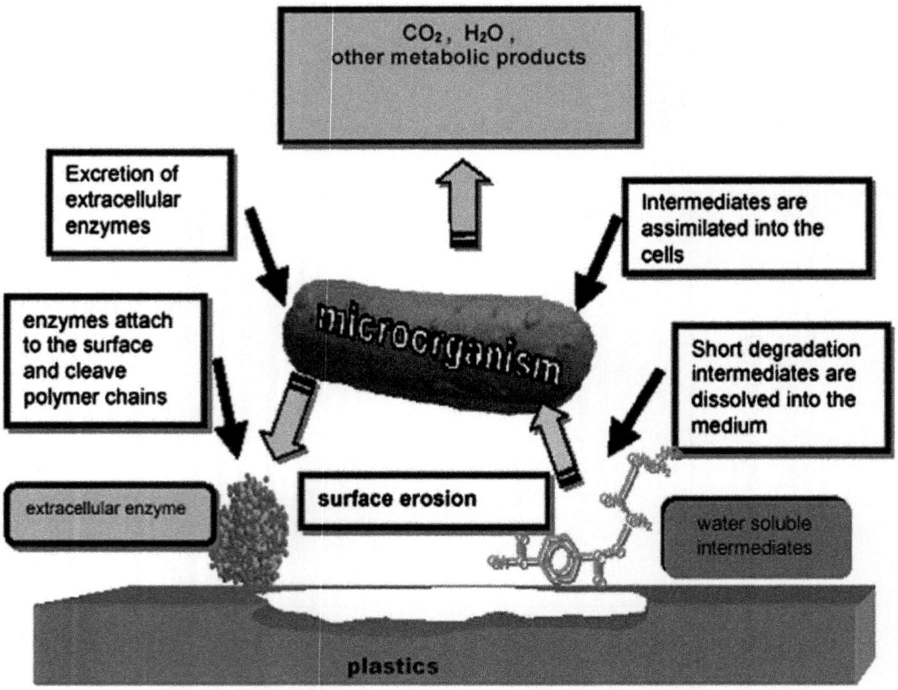

FIGURE 2.6

General mechanism of plastic biodegradation [37].

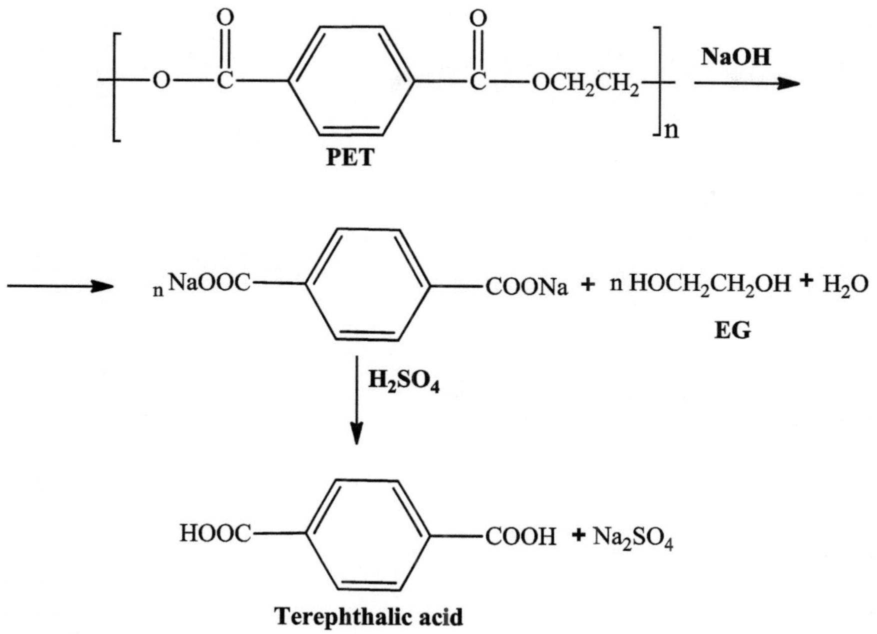

SCHEME 2.1

Alkaline hydrolysis of polyethylene terephthalate [39].

sp. AF6. The fungus strain and the ability of biofilm formation of polyethylene credited to the gradual decrease in the surface hydrophobicity [36].

2.7 RECYCLING OF POLYETHYLENE TEREPHTHALATE

Alkaline hydrolysis of PET (Scheme 2.1) is mostly performed using an alkaline solution of KOH or NaOH at a concentration of 4%–20% by weight (Fig. 2.7) [39]. A technique of alkaline hydrolysis of PET by 18% by weight solutions of NaOH [40] was developed, and the best results are obtained with a PET:NaOH weight ratio of 1:20 at about 100°C in 2 h. Lazarus et al. reported the process in Fig. 2.8 [42].

TPA and other monomeric components can be recovered from a polymer mixture of PET/nylon 6. The reactivity of the solutions of NaOH and sodium *tert*-butoxide solution in *tert*-butanol was compared [43]. They verified that sodium ethoxide in ethanol is more reactive than solution of NaOH. Oku et al. reported the addition of ethereal solvents (tetrahydrofuran or dioxine) as a mixture of solvent with an alcohol (ethanol/methanol), which accelerates the reactions [44].

2.7.1 SOLVENT-ASSISTED GLYCOLYSIS

It is related to PET degradation by ethylene glycol as a solvent present in the reaction medium. Güçlü et al. added xylene to the zinc acetate catalyzed PET glycolysis

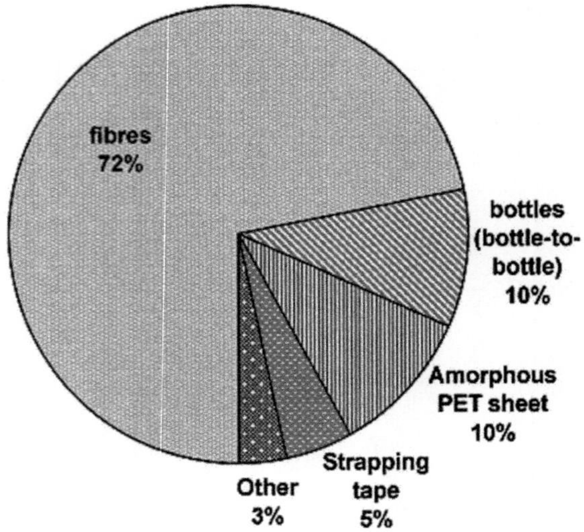

FIGURE 2.7

Applications of recycled polyethylene terephthalate flake [38].

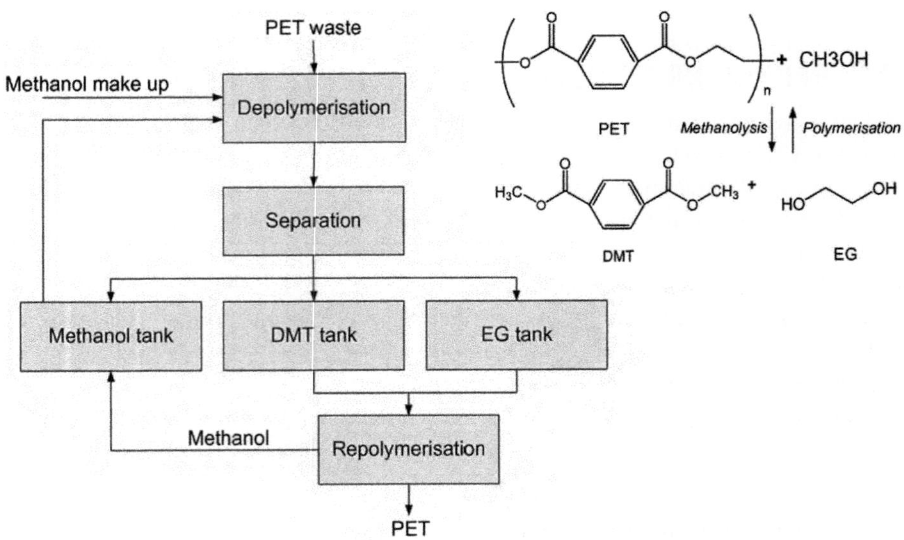

FIGURE 2.8

Chemical recycling of PET via the methanolysis process [41].

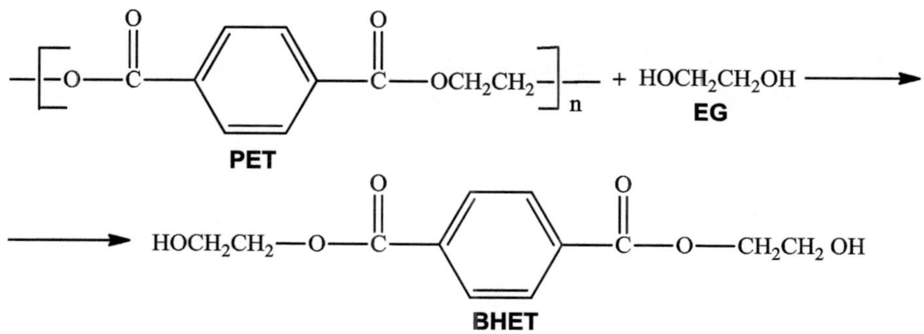

SCHEME 2.2

Glycolysis of PET [45].

reactions, and the yield of bis(2-hydroxyethyl) terephthalate (BHET) obtained has increased than that without xylene (Scheme 2.2) [45]. The main purpose of xylene is to provide a mixture of miscible PET-glycol. It is carried out at a temperature between 170 and 225°C to easily dissolve the PET and EG which is sparingly soluble in xylene. On the other hand, glycolysis product is soluble in xylene. Therefore as the reaction proceeds, glycolysis product moves the reaction in the direction of the depolymerization shift, i.e., the xylene phase from the PET-glycol phase. Large-scale utility of this method is prevented due to the reason that it is harmful for environment because of organic solvents, which is not an appealing idea [45].

2.7.2 CHEMICAL RECYCLING OF POLYETHYLENE TEREPHTHALATE

Methanolysis of PET under 2–4 MPa pressure and 180–280°C temperature will give the depolymerization with DMT and EG methanol in the presence of a catalyst. The reactions mixture was frozen and the DMT prepared well from the mixture by sedimentation, crystallization, centrifugation [41]. Fig. 2.9 showed the flow-sheet diagram for chemical recycling of PET. It is carried out by means of the methanolysis of recycled polymer followed by spinning and finishing processes for transforming into a fiber.

A current copyright illustrated the depolymerized PET with EG to produce BHET [47]. It is then divided more downward in the DMT with methanol [48]. The procedure is known to be cost-effectively reliable than direct methanolysis of PET to DMT (Table 2.5) [49].

Repolymerization is actually the polymerization procedure and is technically identical to make the virgin PET. It is reported in the earlier publications of AMPE [50]. We assume that recycled PET by DMT or EG repolymerization has the same energy needs.

2.8 RECYCLING OF POLYVINYL CHLORIDE

Table 2.6 gives a detail comparison of different methods to dispose PVC waste.

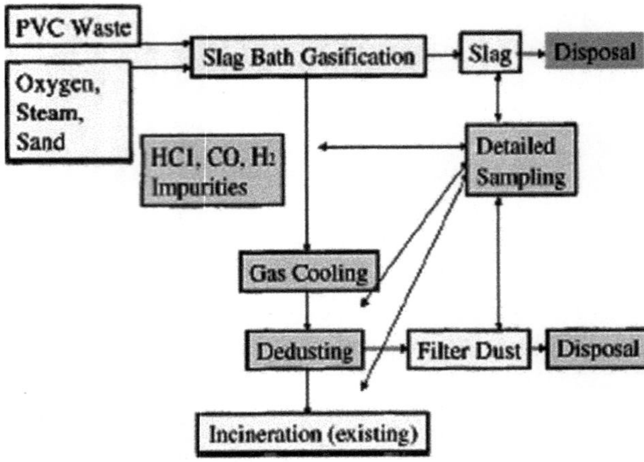

FIGURE 2.9

Pilot plant for chemical recycling of polyvinyl chloride [46].

Table 2.5 Data and Assumptions for Chemical Recycling Through Methanolysis

	Low Case	High Case	Average
NREU of depolymerization (GJ/t DMT)[a]		15.78	
GWP100a of depolymerization (tCO_2 equiv./tDMT)[a]		1.29	
Overall PET loss (by weight)[b]	0	10%	5%
Overall MeOH makeup (by weight)[c]	0	10%	5%
EG recovery (by weight)[c]	0	100%	50%
External EG required (by weight)[c]	100%	0	50%
Repolymerization[d]			
Fuel (GJ/t)		1.63	
Electricity (GJ/t)		0.70	
Steam (t/t)		0.94	

[a] *Patagonia. The allocation is based on mass [51].*
[b] *Assumption based on Marathe et al. [52].*
[c] *Own estimated assumption or estimate [38].*
[d] *Boustead [50].*

2.8.1 CHEMICAL RECYCLING

In addition to PVC material, several recovery attempts have been made by chemical or heat treatment to produce low molecular weight products from PVC. Mostly recommend use of PVC is alkaline dehydrochlorination in heat medium. The

Table 2.6 The Comparison of Different Approaches for Disposing of Polyvinyl Chloride Wastes [53]

Method of Disposing	Sensitivities of Impurity	Degree of Pollution Generation	Cost	Recycled Products	Properties of Recycled Material	Number of Plants in Operation in World	Accepting by Countries (During the Recent Decade, Especially for Developed Countries)
Land filling	Nonsensitive	Very high	Low-cost	No material	–	Large	Nonacceptable
Incineration	Usually none sensitive	Low	Usually low cost	Energy	Energetically not efficient	Large	None-acceptable
Mechanical recycling	Highly sensitive	Low	Middle-cost	PVC	Dependent on feed material and processing variable of recycling	Fair	Highly acceptable
Chemical recycling	Relatively sensitive	Usually low	High cost	Diverse raw materials	Dependent on feed material and processing variable of recycling	Small	Low acceptable

$$2[-CH_2-CHCl-] + O_2 + 2H_2O \longrightarrow 2HCl + 4H_2 + 4CO$$

SCHEME 2.3

Chemical illustration of pilot plant [46].

derivative extrusion is based on the PVC degradation. It is carried out in an extruder by heating and mechanical action in the presence of steam, oxygen, or catalysts [54]. The chief degradation product HCl, for example, can be reused [55,56]. The major yields are oxalic acid and carbon dioxide, produced at optimum reaction conditions and the concentration of alkalis. Highest percentage of oxalic acid was 45% while 42% chlorine content was recovered as HCl. Another way for recycling PVC feedstock for recovery and organic chlorine products is also reported [57,58]. Chlorine, reusable as energy form or as a feedstock, was found in the petrochemical refining of HCl, hydrocarbons, and salts. The results of these investigations are to examine the processing of PVC-rich stream [59], the potential of technology and technical recommendations, and the chemical recycling of PVC is currently installed as new pilot plant Tuvalu (France) (Scheme 2.3). The processing based upon slag bath gasification is industrialized by Lined AG. PVC waste was converted, in accordance, with the following formula with oxygen, steam, and sand. HCl and the resulting synthesis gas would be used for the manufacture of new PVC.

2.8.2 POLYVINYL CHLORIDE TO FUEL

Though there are a lot of studies that show the prospective of a single type of plastic pyrolysis to be applied, there is a requirement to understand the performance of other types of plastic mixture methods. This understanding is important for the recycling of waste plastic, where the waste is processed as a whole. According to the literature, PP and PVC products can be treated as polyethylene pyrolysis. The polymers have to be depolymerized into monomers and oligomers, which were then repolymerized into larger aliphatics [60]. A further revision of pyrolysis of polymer mixture comprising PS, PVC, and PP were reported with an elevated aromatic content (470%). Olefenic compounds (420%) in liquid yield, possibly caused by extended reaction time and elevated temperatures, were employed at the interface between plastics [61]. At higher temperatures (600°C), an increase of the aromatic content of the product due to formation of PAH was observed. After 15–30 min of reaction time, the liquid and gaseous products were suitable for use as a fuel. It was sufficient to achieve the desired conversion of PVC-waste pyrolysis. But polymer contains chlorine compounds as a potential problem with the production of HCl. It leads the pyrolysis technique to corrosion, reducing the feasibility of a liquid product, which is used as fuel [62,63]. As a result, it is important to avoid the probable chlorine contents in products and need dechlorination. There are many possible ways for dechlorination.

1. Catalytic pyrolysis, where the catalyst prevents the pattern of HCl pyrolysis. The use of ZSM-5 is enough for the goal, but the liquid pyrolysis product of this

method is heavier in carbon number. This happens because of the decrease in activity of the catalyst. Hydrochloric acid absorbent is known as carbonates, to avert its existence in pyrolysis products.

2. Absorption also exhibits pure catalytic outcome on the pyrolysis.
3. Batch pyrolysis, where the polymer scrap is chlorinated at 300°C before pyrolysis. The HCl gas produced during dechlorination can be captured and neutralize in the washing tower. The possible deactivation of the catalyst can be decreased as this is not concerned in dechlorination. At elevated temperatures and large reaction time, it is used for dechlorination. Thus the heavier liquid stuffs were obtained in terms of the number of carbon atoms.

2.9 CONCLUSION AND FUTURE PROSPECTS

The use of synthetic and plastic materials has been increased at alarming rate. After the use, it becomes waste which is a serious risk for environment. Therefore the role of degradable or decomposable material is significant as they are the substitute of conventional plastic materials. Various recycling methods have been developed to control the growing issues of synthetic materials to convert them into useful and environment-friendly products. In future more efforts are needed to promote the use of degradable materials and cost-efficient methods for recycling of plastic waste.

REFERENCES

[1] Vona IA, Costanza JR, Cantor HA, Robert WJ. Manufacture of plastics, vol. 1(66). New York: Wiley; 1965. p. 141–2.

[2] Brilly M. Monitoring bathing waters—a practical guide to the design and implementation of assessments and monitoring programmes, Jamie Bartram and Gareth Rees; E&FN Spon Press, London and New York, 2000, 337 pp, ISBN 0-419-24380-1 HBK, ISBN 0-419-24370-4 PB. Urban Water June 30, 2002;4(2):212.

[3] Ocean Conservancy. National marine debris monitoring program. September 2007. Final Report, Washington, DC.

[4] Wallace N. Debris entanglement in the marine environment: a review. In: Proc. of the ninth ann. work. on sea turtle cons. and bio., SA; 1985.

[5] Gregory MR, Andrady AL. Plastics in the marine environment. Plast Environ February 20, 2003:379–402.

[6] Lutcavage ME, Plotkin P, Witherington B, Lutz PL. Human impacts on sea turtle survival. In: The biology of sea turtles, vol. 1; 1997. p. 387–409.

[7] Bern L. Size-related discrimination of nutritive and inert particles by freshwater zooplankton. J Plankton Res January 1, 1990;12(5):1059–67.

[8] Eriksson C, Burton H. Origins and biological accumulation of small plastic particles in fur seals from Macquarie Island. AMBIO: J Hum Environ September 2003;32(6):380–4.

[9] Moore CJ, Moore SL, Leecaster MK, Weisberg SB. A comparison of plastic and plankton in the north Pacific central gyre. Mar Pollut Bull December 31, 2001; 42(12):1297–300.

[10] Thompson RC, Olsen Y, Mitchell RP, Davis A, Rowland SJ, John AW, McGonigle D, Russell AE. Lost at sea: where is all the plastic? Science May 7, 2004;304(5672):838.
[11] Moore CJ. Synthetic polymers in the marine environment: a rapidly increasing, long-term threat. Environ Res October 31, 2008;108(2):131–9.
[12] Derraik JGB. The pollution of the marine environment by plastic debris: a review. Mar Pollut Bull 2002;44(9):842–52.
[13] Williamson LJ. It's not my bag, baby! On earth: environmental politics people, vol. 25(2); June 2003. p. 32–4.
[14] Stevens ES. Green plastics: an introduction to the new science of biodegradable plastics. Princeton University Press; 2002.
[15] Ryan PG, Swanepoel D, Rice N, Preston GR. The 'free' shopping-bag debate: costs and attitudes. S Afr J Sci 1996;92(4):163–4.
[16] Jalil MA, Mian MN, Rahman MK. Using plastic bags and its damaging impact on environment and agriculture: an alternative proposal. Int J Learn Dev August 17, 2013;3(4): 1–4.
[17] ACG. Allen Consulting Group [Ireland]; 2006.
[18] Though R. Plastic shopping bags: environmental impacts and policy options. Victoria University of Wellington; 2007.
[19] Matlack A. Introduction to green chemistry. New York, NY: Marcel Dekker; 2001.
[20] Miller GT. Sustaining the earth: an integrated approach. Pacific Grove, CA: Thomson Brooks/Cole; 2005.
[21] Reusable Bags Facts and figures regarding the true cost of plastic bags; 2005. [Online] Available from: http://www.reusablebags.com/facts.php.
[22] Collins English Dictionary and Thesaurus also known as primary biodegradation, ultimate biodegradation is the mineralization or incorporation of materials in biomass. 2002.
[23] Kumar S, Prakash R, Murugan S, Singh RK. Performance and emission analysis of blends of waste plastic oil obtained by catalytic pyrolysis of waste HDPE with diesel in a CI engine. Energy Convers Manag 2013;74:323–31.
[24] Kunwar B, Cheng HN, Chandrashekaran SR, Sharma BK. Plastics to fuel: a review. Renew Sustain Energy Rev February 29, 2016;54:421–8.
[25] Bartolome L, Cho BG, Kim DH, Imran M, Al-Masry WA. Recent developments in the chemical recycling of PET. INTECH Open Access Publisher; 2012.
[26] Sharma BK, Rashid U, Anwar F, Erhan SZ. Lubricant properties of Moringa oil using thermal and tribological techniques. J Therm Anal Calorim June 1, 2009;96(3): 999–1008.
[27] Sørum L, Grønli MG, Hustad JE. Pyrolysis characteristics and kinetics of municipal solid wastes. Fuel July 31, 2001;80(9):1217–27.
[28] Vardon DR, Sharma BK, Scott J, Yu G, Wang Z, Schideman L, Zhang Y, Strathmann TJ. Chemical properties of biocrude oil from the hydrothermal liquefaction of *Spirulina* algae, swine manure, and digested anaerobic sludge. Bioresour Technol September 30, 2011;102(17):8295–303.
[29] Sriningsih W, Saerodji MG, Trisunaryanti W, Armunanto R, Falah II. Fuel production from LDPE plastic waste over natural zeolite supported Ni, Ni-Mo, Co and Co-Mo metals. Procedia Environ Sci December 31, 2014;20:215–24.
[30] Achilias DS, Roupakias C, Megalokonomos P, Lappas AA, Antonakou EV. Chemical recycling of plastic wastes made from polyethylene (LDPE and HDPE) and polypropylene (PP). J Hazard Mater November 19, 2007;149(3):536–42.

[31] Syamsiro M, Saptoadi H, Norsujianto T, Noviasri P, Cheng S, Alimuddin Z, Yoshikawa K. Fuel oil production from municipal plastic wastes in sequential pyrolysis and catalytic reforming reactors. Energy Procedia December 31, 2014;47:180–8.

[32] Bonhomme S, Cuer A, Delort AM, Lemaire J, Sancelme M, Scott G. Environmental biodegradation of polyethylene. Polym Degrad Stab December 31, 2003;81(3):441–52.

[33] El-Shafei HA, El-Nasser NH, Kansoh AL, Ali AM. Biodegradation of disposable polyethylene by fungi and *Streptomyces* species. Polym Degrad Stab November 30, 1998; 62(2):361–5.

[34] Yamada-Onodera K, Mukumoto H, Katsuyaya Y, Saiganji A, Tani Y. Degradation of polyethylene by a fungus, *Penicillium simplicissimum* YK. Polym Degrad Stab May 31, 2001;72(2):323–7.

[35] Shah, AA. Role of microorganisms in biodegradation of plastics [Doctoral dissertation, Quaid-i-Azam University, Islamabad].

[36] Orr IG, Hadar Y, Sivan A. Colonization, biofilm formation and biodegradation of polyethylene by a strain of *Rhodococcus ruber*. Appl Microbiol Biotechnol July 1, 2004; 65(1):97–104.

[37] Mueller RJ. Biodegradability of polymers: regulations and methods for testing. In: Steinbüchel A, editor. Biopolymers, vol. 10; 2003.

[38] Shen L, Worrell E, Patel MK. Open-loop recycling: a LCA case study of PET bottle-to-fibre recycling. Resour, Conserv Recycl November 30, 2010;55(1):34–52.

[39] Karayannidis GP, Chatziavgoustis AP, Achilias DS. Adv Polym Technol 2002;21: 250–9.

[40] Pitat J, Holcik V, Bacak M. A method of processing waste of polyethylene terephthalate by hydrolysis. GB Patent. 1959. 822:834.

[41] Paszun D, Spychaj T. Chemical recycling of poly (ethylene terephthalate). Ind Eng Chem Res April 2, 1997;36(4):1373–83.

[42] Lazarus SD, Twilley JC, Snider OE. Simultaneous depolymerization of polycaproamide and polyester with recovery of caprolactam. US Patent 3317519A. 1967.

[43] Namboori CG, Haith MS. Steric effects in the basic hydrolysis of poly (ethylene terephthalate). J Appl Polym Sci September 1, 1968;12(9):1999–2005.

[44] Oku A, Hu LC, Yamada E. Alkali decomposition of poly (ethylene terephthalate) with sodium hydroxide in nonaqueous ethylene glycol: a study on recycling of terephthalic acid and ethylene glycol. J Appl Polym Sci January 31, 1997;63(5):595–601.

[45] Güçlü G, Kasgöz A, Özbudak S, Özgümüs S, Orbay M. J Appl Polym Sci 1998;69: 2311–9.

[46] Information of Arbeitsgemeinschaft PVC und Umwelt e.V., D-53113 Bonn.

[47] Nakao T, Chikatsune T, Nakashima M, Suzuki M, Nagano H. Method for recycling PET bottle. Patent WO/2003/033581. 2003.

[48] Delattre J, Raynaud R, Thomas C, inventor; Rhone-Poulenc-Textile, assignee. Process for obtaining dimethyl terephthalate from polyester scrap. US Patent 4463860; 1976.

[49] Lorenzetti C, Manaresi P, Berti C, Barbiroli G. Chemical recovery of useful chemicals from polyester (PET) waste for resource conservation: a survey of state of the art. J Polym Environ January 1, 2006;14(1):89–101.

[50] Boustead I. Eco-profiles of the European plastics industry, polyethylene terephthalate. A report for the European Centre for Plastics in the Environment, Brussels. 2002.

[51] Patagonia. Patagonia's common threads garment recycling program: a detailed analysis. Available from: http://www.patagonia.com/pdf/en US/common threads whitepaper. 2005;pdf.

[52] Marathe MN, Dabholkar DA, Jain MK. Process for the recovery of dimethyl terephthalate from polyethylene terephthalate. GB Patent. 1980. 2(041):916.

[53] Sadat-Shojai M, Bakhshandeh GR. Recycling of PVC wastes. Polym Degrad Stab April 30, 2011;96(4):404−15.

[54] Michaeli W, Lackner V. Annu Tech Conf—Soc Plast Engng 1994:2901.

[55] Yoshioka T, Yasuda S, Kawamura K, Sato T, Okuwaki A. Oxidation of poly (vinyl chloride) powder by molecular oxygen in alkaline solutions at high temperatures. Nippon Kagaku Kaishi (Japan) May 1992;(5):534−41.

[56] Shin Shin SM, Yoshioka T, Okuwaki A. Dehydrochlorination behavior of rigid PVC pellet in NaOH solutions at elevated temperature. Polym Degrad Stab December 31, 1998;61(2):349−53.

[57] Yoshioka T, Furukawa K, Sato T, Okuwaki A. Chemical recycling of flexible PVC by oxygen oxidation in NaOH solutions at elevated temperatures. J Appl Polym Sci October 3, 1998;70(1):129−35.

[58] Widmer A. Kunstst Synth 1994;25(12):28.

[59] Bühl R. Options for poly (vinyl chloride) feedstock recycling. Results of research project on feedstock recycling processes. Plast, Rubber Compos March 1, 1999; 28(3):131−5.

[60] Sarker M, Rashid MM, Molla M. Waste polypropylene plastic conversion into liquid hydrocarbon fuel for producing electricity and energies. Environ Technol December 1, 2012;33(24):2709−21.

[61] López A, De Marco I, Caballero BM, Laresgoiti MF, Adrados A. Influence of time and temperature on pyrolysis of plastic wastes in a semi-batch reactor. Chem Eng J September 1, 2011;173(1):62−71.

[62] López A, De Marco I, Caballero BM, Laresgoiti MF, Adrados A. Dechlorination of fuels in pyrolysis of PVC containing plastic wastes. Fuel Process Technol February 28, 2011;92(2):253−60.

[63] Fukushima M, Wu B, Ibe H, Wakai K, Sugiyama E, Abe H, Kitagawa K, Tsuruga S, Shimura K, Ono E. Study on dechlorination technology for municipal waste plastics containing polyvinyl chloride and polyethylene terephthalate. J Mater Cycles Waste Manag June 1, 2010;12(2):108−22.

CHAPTER

3 Microalgae: A Promising Feedstock for Energy and High-Value Products

Ifrah Afzal[1], Ayesha Shahid[1], Muhammad Ibrahim[1], Tianzhong Liu[2], Muhammad Nawaz[1], Muhammad A. Mehmood[1]

Government College University Faisalabad, Faisalabad, Pakistan[1]; Microalgae Biotechnology Centre, Qingdao Institute of Bio-energy & Bioprocess Technology, Chinese Academy of Sciences, Qingdao, People's Republic of China[2]

3.1 MICROALGAE: POTENTIAL AND PROPERTIES

Algae are a diverse group of organisms found almost everywhere on the earth ranging from microscopic blue-green algae to complex kelps and seaweeds with 200,000 species worldwide. Algal storage products make algae industrially useful organisms. On the basis of their storage products and pigment compositions algae are classified into 12 major groups [1]. Microalgae have received huge scientific attention in recent years among other different groups. They are present at the bottom of the food chain and contain 50% essential amino acids, 65% vegetable proteins, up to 70% lipids, and up to 60% carbohydrates. Though small in size microalgae are among the most nutritious food sources which also help to maintain body health [2].

The 1950s surge of the world's population led to search for alternative food sources. Owing to their nutritional value, microalgae biomass appeared as a possible choice for this purpose at that time. The history reveals that the commercial production of microalgae started with a culture of *Chlorella* in early 1960s in Japan. In 1970s a large-scale culturing facility of *Spirulina* species was established in Texcoco Lake, Mexico. It is reported that in late 1980s there were 46 factories producing microalgae of more than 1000 kg of biomass. Western Biotechnology Company Ltd. established the third leading microalgal industry of *Dunaliella salina* species which is a rich source of β-carotene. Thus, in a short period of time microalgae-based biotechnology industry was grown and is still growing significantly [3].

Microalgae offer a diversified range of products [4]. Species such as *Nostoc*, *Spirulina*, and *Aphanizomenon* have been used as protein supplement and food additives for thousands of years. In cosmetic industry, extracts from *Arthrospira* and *Chlorella* are used in antiaging creams as they exert a skin tightening effect. Penta Pharm (a Swiss Company) has launched various skin care products based on microalgal extracts. Microalgae also contain pigments such as carotenoids (up to 14%), chlorophyll (0.5%−1%), and phycobilin proteins (20%−50% on the dry weight basis) and therefore can be used to produce biopigments on large scale. Moreover, they play

Algae Based Polymers, Blends, and Composites. http://dx.doi.org/10.1016/B978-0-12-812360-7.00003-3

remarkable role in pharmaceutical industry owing to their antioxidants, antihypertensive, anticancerous, anticoagulant, and immune-modulatory functions [5]. They offer remarkable sequestration of atmospheric CO_2 from industrial flue through biofixation due to their fast growth rates and biomass productivities [6]. Moreover, it is believed that microalgae have potential to mitigate the nitrogen, phosphorus, and heavy metals from municipal and industrial wastewater. Algal biomass may also be used in poultry feed, as it contains high quantity of digestible proteins and omega-3 fatty acids [7], as organic fertilizers, and in the making biochar [8].

3.2 MICROALGAE: POTENTIAL FEEDSTOCK FOR BIOENERGY

In addition to food, cosmetics, and health, microalgae are also recognized as potential feedstock for renewable energy sources. Microalgal biomass is viable for biofuel production such as biodiesel, bioalcohol, and biogas. Microalgae-based biofuel contains small amount of SO_x and their photosynthetic activity can help to reduce the greenhouse gas emission (GHGs) [6], and may yield 300 times more oil as compared to traditional energy crops productivity. Their oil contents may go up to 75%−80% for some special oleaginous species, of which more than 50% are triacylglycerides which are ideal to convert into biodiesel [fatty acids methyl esters (FAME)]. It was estimated that 30% oil in algal dry biomass may yield about 58,7000 L/ha year [9], and that oil extracted from microalgae cultivated on 30 million acres would have capacity to replace the USA Imported oil supply mainly from Canada, Saudi Arabia, Mexico, Colombia, and Venezuela [10]. Algal oil often contains four unsaturated fatty acids including palmitolleic acid, oleic acid, linoleic, and linolenic acid along with many saturated fatty acids in small quantities [11]. Special microalgae species such as *Ankistrodesmus*, *Isochrysis*, *Nannochloris*, and *Nitzschia* also contain polyunsaturated fatty acids [12]. Microalgal produced biodiesel, with a higher cetane number, is proved to offer better performance than fossil diesel in the internal combustion engines [13].

3.2.1 BIOALCOHOLS

Alcohols derived exclusively from the fermentation of biomass are referred to as bioalcohols. Aliphatic alcohols such as methanol, ethanol, propanol, and butanol are the best candidates to be used as fuels. Production of bioalcohols is desirable due to their environmental friendly nature, ease of synthesis by chemical or biological methods, and characteristics suitable to be used in current engines. Higher octane rating, increased fuel efficiency, and low energy density are some valuable advantages associated with alcohol fuels [14]. Longer carbon chain alcohols such as butanol (Biobutanol) have more suitable properties, are a direct replacement of gasoline as they have similar energy density, and are more suitable for current fuel distribution infrastructure [15]. Out of all alcoholic fuels bioethanol is the most common worldwide, especially in Brazil, the United States, and Europe [16] as it reduces atmospheric CO_2, sulfur, lead, and particulate emission [17].

Biomass for bioalcohol production includes organic material which is grown, collected, and harvested for sustainable energy purposes. Feedstock for bioalcohols mainly consists of sucrose containing biomass, starchy material, and lignocellulosic biomass [16]. First generation feedstock included corn or sugarcane starch that encountered problems of slow growth rate and fuel versus food dilemma. The second generation feedstock such as woody or agricultural wastes, lignocellulosic biomass, and bagasse [18] lack economic viability when applied at large scale due to the recalcitrant nature of lignocellulosic content [19,20]. Bioalcohol production from biomass includes several steps: pretreatment, saccharification, fermentation, recovery, and distillation. This intensive processing contributes to ~40% increase in the overall cost of production of bioalcohol [21].

Algae are considered as a source of high yield bioethanol due to higher photon conversion efficiency and higher starch and glycogen contents along with little amount of cellulose which can be fermented to ethanol [22]. Microalgal carbohydrates are highly fermentable sugars and have low lignin content and usually do not require pretreatment steps [23], and algal biomass can be directly subjected to simultaneous saccharification and fermentation. Owing to their higher photosynthetic efficiency, higher starch and glycogen contents, and easy saccharification, the microalgae look promising to replace food-based feedstock [22]. Microalgae species such as *Chlamydomonas reinhardtii* UTEX 90, *Chlorococcum infusionum*, *Chlorella vulgaris*, and many other are reported suitable for bioethanol production [18]. *C. vulgaris* possess 35%–50% carbohydrates (Table 3.1) of its total dry weight [27,28]. Some examples of bioalcohol production using microalgae are discussed below.

Sequential treatment of *C. reinhardtii* UTEX 90 with glucoamylase and α-amylase transforms the algae to ethanol by simultaneous saccharification and fermentation process [29]. In another study *Chlorococum* sp. was fermented using *Saccharomyces bayanus* to produce around 3.83 g/L ethanol with 38% (w/w) productivity [30]. *Spirogyra* has shown a yield potential of 14%–17%. Moreover, a conversion efficiency up to 65% can be achieved by using *C. vulgaris* [17]. Methanol production by gasification process was reported for *Spirulina*. In this process,

Table 3.1 Microalgae-Based Bioethanol Production

Microalgae Species	Starch g/Dry Weight (%)	References
Chlorella vulgaris	37	[24]
Chlorella sp. TISTR 8262	21.5	[25]
Chlorella sp. TISTR 8485	27.0	[25]
Chlorococcum sp. TISTR8583	26.0	[25]
Nostoc sp. TISTR 8872	30.7	[25]
Nostoc sp. TISTR 8873	32.9	[25]
Spirulina fusiformis	37–56	[26]

microalgae were concentrated and centrifuged at first and was later oxidized into gas. The composition of gas determines the methanol yield. Maximum 0.64 g methanol yield was obtained per gram of algae at 1008°C [17].

3.2.2 BIODIESEL

A fuel comprising alkyl esters of long chain fatty acids derived from renewable biological sources, such as vegetable oils, animal fats, or algae, is considered as biodiesel. Biodiesel is recognized as a clean and attractive alternative fuel, and its use as motor fuel has grown dramatically in recent years [31]. It is equivalent to petroleum-based diesel in terms of performance but is better than the latter in terms of environmental safety due to its very low sulfur and aromatic contents. Moreover, it has lower ignition point and higher flash point and is compatible with all conventional diesel engines [32]. There are four ways reported to produce biodiesel: direct use and blending, microemulsions, pyrolysis, and transesterification [33]. But the most common way to produce biodiesel is to transesterify the lipids into FAME or fatty acids ethyl esters along with glycerin as a by-product.

The United States and Europe are the current global leaders in biodiesel production [34], but the developing countries such as Brazil and Argentina are also producing biodiesel to meet their increasing energy requirements. It is expected that biodiesel production would increase up to 121 million tons in this year [12]. Commercialized biodiesel can be produced either from vegetable oils obtained from soybean, palm, canola, rapeseed, sunflower, or from waste cooking oil. An in situ esterification method was developed to decrease the amount of acid catalyst and alcohols for production of biodiesel using rice bran as feedstock, with reported yield and purity of biodiesel up to 18% and 67%, respectively [31]. In another study, a two-step esterification process was established to overcome the reaction between free fatty acids and alkaline catalyst which interferes with the separation of glycerol from methyl esters [35]. Biodiesel was also produced from rubber-seed oil having viscosity close to fossil diesel and was declared as a viable alternative [35]. To lower down the cost of biodiesel produced from virgin vegetable oils, scientists have investigated the production of biodiesel from soap stock, using saponification of soap followed by acidulation into free fatty acid oil subsequent production of FAME. Biodiesel extracted from soap stock has shown to be similar to biodiesel extracted from soybean oil [36]. Moreover, Jatropha seeds contain up to 37% of oil content and may produce good quality biodiesel. Besides, FAME obtained from transesterification of Jatropha oil has good stability, low viscosity, and high cetane number [37].

Microalgae had received scientific attention in previous decades as an alternative feedstock for biodiesel production due to several associated benefits of microalgae. Microalgae do not compete with traditional food culture and can be harvested daily. Estimated biodiesel production from microalgae is 20,000−80,000 L/acre year, which is 31 times higher than the oleaginous seeds [38,39]. Microalgae species accumulate up to 70% lipids in their cellular dry mass, which in some species reaches approximately 90%. Microalgae lipids contain saturated and unsaturated fatty acids, and produce valuable biodiesel [6]. Some algal species have high fatty acid profile such as steroids allowing to produce biodiesel with high oxidation stability [39]. Among 4000 identified microalgal species, diatoms and green microalgae are most abundant group, and contain 40%–75% (Table 3.2) lipids in their dry mass.

There are several factors affecting the storage of lipids in microalgal cell factories. For instance, nitrogen-limited nutrients conditions are believed to enhance the lipid content in algal biomass [12]. Moreover, the effect of storage temperature on lipid content of *Scenedesmus* spp. is shown to increase the free fatty acid content in microalgae up to 62%, and these fatty acids are then converted into biodiesel by using two-step catalytic conversion (preesterification and transesterification) [40]. A comparative study of three microalgal species *Cladophora* sp., *Oedogonium* sp., and *Spirogyra* sp. has shown that *Cladophora* sp. has greater potential for biodiesel production due to its higher biomass production and sediments (glycerin, water, and pigments) [41]. Growth rate is another parameter which affects the lipid storage. It is reported that fast-growing microalgae contain lower lipid content as compared to slow-growing. It is also found that higher iron concentrations are source of high lipid accumulation in microalgae [6]. Under stress or unfavorable conditions, such as nitrogen/phosphorus starvation, light irradiation, and pH and temperature stress the synthesis and accumulation of lipids were reported to be high [42]. Moreover, microalgae cultivated on wastewater have shown to produce more biomass and lipid oil along with the removal of pollutants from the water [43].

Table 3.2 Lipid Content of Microalgae Species

Microalgae Species	Lipid Content (Dry Weight Biomass, %)	References
Ankistrodesmus sp.	24–31	[9]
Botryococcus braunii	25–75	[12]
Chlamydomonas reinhardtii	21	[12]
Chlorella emersonii	25–63	[6]
Nannochloropsis spp.	29–65	[6]
Chlorella minutissima	25–63	[9]
Chlorella protothecoides	14–57	[9]
Chlorella vulgaris	5–58	[9]
Dunaliella tertiolecta	16–71	[9]
Euglena gracilis	14–20	[12]
Phaeodactylum tricornutum	18–57	[6]
Neochloris oleoabundans	29–65	[6]
Schizochytrium sp.	50–77	[6]
Nitzschia sp.	45–47	[12]

With the advent of modern genetic modification and synthetic biology approaches, genetic engineering can also be used to enhance the photosynthetic activity along with biomass and improved lipid biosynthesis [18]. For instance, targeted knockdown of acyl transferases/lipases has shown to increase the lipid content in microalgae and to produce high cetane number biodiesel [44]. Similarly, overexpression of endogenous thioesterase has shown to be effective for lipid biosynthesis in *C. rienhardtii* [45].

3.2.3 BIOGAS

Another clean, valuable, and renewable alternative energy source (especially for rural area) is biogas. It can directly be exploited in its raw form for syngas and hydrogen gas production. Biogas mainly contains carbon dioxide and methane; however, depending upon its source, small quantities of contaminants such as carbon dioxide, ammonia, hydrogen sulfide, siloxanes, dust, and water vapors are also present [46]. Although beneficial, it has various disadvantages associated to it, for example, low production rate during winter, higher hydraulic retention time, presence of contaminants that can cause corrosion or fouling, and emission of harmful environmental compounds. Hence, there is a need to enhance the production rate as well as addition of purification steps prior to its final use [47,48]. The oldest and established biogas production process is by using activated sludge. At industrial scale, manures and municipal solid wastes are used for biogas production [49]. Different biomasses containing fats, proteins, cellulose, and hemicellulose can also be used for biogas production with the only exception of strong lignified wood or woodlike biomass because of its slow anaerobic decomposition rate [50]. Common agricultural biogas plants use manure from cows, pigs, and chickens with additional cosubstrates to increase organic material content resulting in higher gas yield, where agriculture-related organic waste, energy crops, municipal biowaste, and food waste are common cosubstrates [50,51].

Recently, microalgae were proved to be a potential feedstock for the biogas production (Table 3.3). It is suggested that one can exploit the high energy content of microalgae via anaerobic fermentation along with the additional benefit of wastewater treatment using algal cultivation [58,59]. Biogas production efficiency depends on the selected algal species, cell degradation efficiency, and presence or absence of methanogenic archaea inhibiting molecules [59]. Limitations such as inefficient degradation, land area requirement, low carbon/nitrogen ratio, and infrastructure requirement are the hindrances in the use of algae for biogas production at large scale [54,59]. These challenges are required to be addressed in future research.

Recently the biogas productivity potential was evaluated sporadically for microalgae such as *Spirulina* sp., *C. vulgaris*, *C. reinhardtii*, *Scenedesmus* sp. [60]. A novel strategy of combining cultivation and fermentation with progressing nitrogen starvation conditions was used on *C. reinhardtii*, *Parachlorella kessleri*, and *Scenedesmus obliquus* to improve biomass composition. It was shown that nitrogen starvation could increase C/N ratio by 24- to 26-folds and algal degradation efficiencies

Table 3.3 Biogas Yield and Productivity of Different Microalgae

Microalgae Strain	Biogas Yield (mL/g VS)	Biogas Productivity (mL/L)	Methane (mL/g VS)	References
Chlamydomonas reinhardtii	587	–	387	[52]
Chlorella vulgaris Str. 211–1e	510	1944.8	–	[53]
Chlorella kessleri	335	–	218	[54]
Chlorella minutissima	340	–	166	[55]
Chlorella pyrenoidosa	464	–	265	[55]
Chloroidium saccharophilum Str. 2149	304	495.52	–	[53]
Dunaliella armatus Str. 276–4d	518	1319.17	–	[53]
Dunaliella salina	505		–	[52]
Haematococcus pluvialis Str. 192.80	413	1261.03	–	[53]
Nannochloropis salina	–	–	233–549	[56]
Parachlorella kessleri	706	–	449	[57]
Scenedesmus. obliquus Str. 276–1	433	548.47	–	[53]
S. obliquus	287	–	401	[57]
Stichococcus. sp. Str. 2118	576	1384.32	–	[53]

by 35%–100% [54]. Moreover, biogas yields were improved by 65%, 94%, and 106% for *C. reinhardtii*, *P. kessleri*, and *S. obliquus*, respectively, mainly due to nitrogen limitation treatment [54]. Therefore, this strategy may be used for algal biomass production, and to improve methane production through anaerobic fermentation without facing ammonia inhibition effects in fermenter [54]. In another study, biogas production was evaluated using *Hydrodictyon reticulatum* (a filamentous algae), where biomass was produced using ultrasound-disinfected wastewater. Algal cells were disintegrated by ultrasound prior to batch experiments for biogas production. It was shown that ultrasound pretreatment improved degradation rate and biogas production by 2.3 folds as compared to nontreated biomass [61]. Similarly, 587 mL of biogas was produced using 1-g volatile solid (VS) of *C. reinhardtii* as a substrate [52]. Interestingly, an 80% reduction in biogas production was observed in this case, if predried biomass is used. Moreover, when *C. reinhardtii* was used to hydrogen production followed by anaerobic fermentation for biogas, an overall 123% increase was observed in biogas yield [52].

3.2.4 BIOHYDROGEN

Biological hydrogen is considered as an excellent substitute of fossils fuels and an energy carrier [54] because of its nontoxic and odorless nature as it produces only water on combustion preventing environmental pollution [62]. Hydrogen fuel produced from nonrenewable sources is not feasible and requires high energy input. Moreover, these nonrenewable sources including natural gas, coal, naphtha, and heavy oil are already scarce [63]. Alternatively, biohydrogen becomes more promising as it is produced at normal temperature and pressure, and requires minimal energy input [62]. It can be produced via direct biophotolysis, indirect biophotolysis, photo-fermentation, and dark fermentation. Among these, dark fermentation process has great potential as it is easy to handle, requires low amount of biomass, and yields high amount of biohydrogen. Substrate used for its production contains lignocellulose or starch waste, palm oil, glycerol, food, and waste from paper milling industries [64].

Recently, bacterial cultures were used for production of biohydrogen. It is documented that employment of mesophile and thermophile cultures may increase the production of hydrogen as they degrade complex sugars easily [65]. It was also reported that the employment of mixed culture of hydrogen-producing strains *Clostridium butyricum*, *Clostridium saccharobutylicum*, *Clostridium tertium*, and *Clostridium perfringens* in batch fermentation enhances the biohydrogen yield using wastewater as substrate [66]. Cyanobacteria are considered as the most promising organisms for biohydrogen production, where biohydrogen synthesis reaction is catalyzed by nitrogenase and hydrogenase enzymes [67]. Cyanobacteria use organic compounds as electron donor, i.e., catabolism of carbohydrates like glycogen releases electrons that are supplied to bidirectional [Ni-Fe] hydrogenase for hydrogen synthesis [68]. An attempt was made to manipulate the metabolic pathway to increase the level of internal NADPH pool, thus hydrogen yields also increases up to threefold [69].

Microalgae are an interesting and alternative way for biohydrogen production because of their unique features as discussed already. Gaffron and Rubin discovered in 1942 that microalgae *Scenedesmus* have the potential to produce hydrogen. There are different approaches used for biohydrogen production from microalgae, such as production in anaerobic conditions or dark fermentations. Aerobic conditions can be chemically and physically induced into microalgal cultures to enhance the production of hydrogen. It was studied in the last decade that microalgae can produce hydrogen in the presence of sunlight conditions when cells deprived of sulfur enter hydrogen production stage [70]. In another study, carbohydrate-rich biomass of *C. vulgaris* was used for biohydrogen production via dark fermentation reaction [71]. *Nannochloropsis* spp. was also shown to be an attractive feedstock for biohydrogen production. *Nannochloropsis* biomass residues, after oil extraction and carotenoid production, were used for biohydrogen production via dark fermentation by the action of *Enterobacter aeruginosa,* a kind of biorefinery approach, and biohydrogen yield of 60.6 mL/g dry biomass was obtained [72]. Moreover, a comparative study of hydrothermal gasification of microalgae revealed that macroalgae may be a better choice for biohydrogen production, if hydrothermal gasification is to be employed as method of choice [73]. Most methods use dry microalgal biomass. However, it does not seem compulsory and wet algal biomass can also be used for the same purpose. Both wet and dry biomass of *S. obliquus* strain were subjected to dark fermentation reaction, and it was shown that both could give equal yields and purities of biohydrogen [74]. Thus, the energy intensive drying process may not be necessary to include. Moreover, it was also shown that instead of sulfur-deprivation step, which is costly too, mixed culture of microalgae and bacteria may be employed in the fermentation process to get higher yield of biohydrogen, and the residual biomass can be subjected to biogas production [74]. Conclusively, biohydrogen production using microalgae as feedstock is more promising and cheaper than using cyanobacteria and other microbes because microalgae does not require stringent conditions and selective media for its growth, which significantly lowers the production cost.

3.2.5 ADVANCED BIOFUELS

In recent years, synthetic biology and metabolic engineering were performed to develop the heterologous strains of microorganisms and photosynthetic organisms for production of advanced biofuels [75]. Drop-in biofuels are considered to be a promising option to address the current challenges in the commercialization of biofuels (bioalcohols and biodiesel). Generally considered, advanced biofuels have similar energy content and combustion properties like current transportation fuel, and hence can be used as gasoline and jet fuel replacements [76]. Advanced biofuels include isoprenoids, butanol, fatty acid ethyl esters, alkanes, and alkenes also termed as "petroleum replica" fuels [77]. Some microorganisms can naturally produce certain petroleum replica compounds. For instance, *Clostridium* species produce isopropanol, and *Pseudomonas putida* and *Bacillus subtilis* produce butanol naturally [78]. However, these compounds are secreted in trace amounts as microorganisms suffer low growth rates and intolerance to toxicity of these biofuels. To overcome this, industrially used microorganisms are genetically modified to get the increased amount of alkyl esters, alkanes, alkenes, and terpenes [79]. Most commonly manipulated microorganisms are *Escherichia coli*, *Zymomonas mobilis*, *Saccharomyces cerevisiae*, and cyanobacteria [78].

The microalgae had received great scientific attention in previous decade because of its ability to hyperaccumulate triterpene oil and other forms of hydrocarbons considered to be advanced biofuels. These hydrocarbons do not require transesterification reaction. Among the various microlagae, *Botryococcus braunii* accumulates up to 86% of hydrocarbon oils and is a major contributor of triterpenes

[80]. All three races (A, B, and L) of *B. braunii* have been extensively discussed in literature for advanced biofuel production. Race A produces alkanes and alkadienes, race B synthesizes two types of "terpenoids," and race L accumulates tetraterpenoid known as lycopadiene [81]. Among these races, B is more suitable for production of advanced biofuels as it synthesizes triterpenes "botryococcene and squalene" in its extracellular bodies. These terpene oils can be used to produce octane and kerosene upon hydrocracking. The major problems encountered in the commercialization of microalgae-based drop-in biofuels are slow growth and higher contamination risk [82,83]. Alternatively, algal-derived metabolic pathways were introduced into fast growing industrial use of microorganisms or photosynthetic organisms to synthesize the advanced biofuel molecules. Triterpene synthesis pathway isolated from *B. braunii* was introduced into wild-type and transgenic lines of tobacco plant for high level triterpene accumulation [84]. In another study, triterpene synthesis pathway from *B. braunii* was introduced into *Rhodobacter capsulatus* to explore the increased production of squalene and botryococcene [85]. Overall synthetic biology holds promising future for commercialization of microalgae-based advanced biofuels.

3.3 BIOLOGICAL PIGMENTS AND MEDICINES

Any colored material which is naturally produced by plants, animals, fungi, and microalgae is considered as biological pigment or natural biopigment. Main consumption of biopigments is believed to be in paints, clothing, crimson ink, healthcare, cosmetics, and food colors due to their nontoxic, noncarcinogenic, and biodegradable nature. For these reasons, biopigments are gaining increasing market over synthetic pigments. Traditional natural colorants, which include flavonoids, antharaquinones, and β-carotenoides, are being used in routine life in textile, food, and paper industry [86]. The demand for biological pigments is also increasing dramatically because of their unique features such as safe use, antioxidant properties, and stability under variable light, pH, and temperature conditions [87].

Plants are ancient source for biopigment production. Plants produce and accumulate chlorophyll, carotenoids, anthocyanins, and betalains in various parts to absorb different wavelengths to perform various ecological features. These pigments give color to fruits, flowers, and leaves [88], which are further important to carry on various biological functions or can be a way the plants communicate with other species, for instance the pollinating insects. Similarly, bacteria and fungi are also widely known to produce various important biological pigments [89,90].

Microalgae have gathered much attention as important source of bioactive molecules due to wide range of pigments and higher productivity when compared to plants and microbes [91]. Microalgae-based pigments are grouped into chlorophylls, carotenoids, and phycobiliproteins [91]. Phycocyanin, a phycobilin protein, extracted from *Spirulina platensis* is used as colorant in food industry such as chewing gums, popsicles, dairy products, jellies; and in cosmetic industry such as lipsticks and eyeliners [92]. Similarly, phycoerythrin derived from *Porphyridium aerugineum* (a microalga) is also used in food and cosmetic industries [93]. Chlorophyll pigments derived from microalgae help in wound healing and have anticancerous and antiinflammatory properties [94]. β-carotene extracted from *D. salina* was commercialized worldwide in 1986 [95]. Astaxanthin, another pigment produced by *Hematococcus* microalgae, has potential role in aquaculture, poultry feeds, and nutraceutical market [96].

Market demand for biological pigments is continuously increasing. To meet the global demand, mass cultivation systems are required to be optimized and established in cost-effective manner and on sustainable grounds [97]. For instance, *Spirulina* species are considered as best source for large-scale production of carotenoids because of their cheap growth conditions [98]. However, other growth conditions are still to be determined for optimum pigment yields. In this context, growth of *S. platensis* was studied using different media under different temperature ranges for carotenoid production, and it was found that high temperature retarded the growth and pigment productivity in *S. plantensis*. Moreover, it was revealed that optimum temperature for its growth is 35−40°C and pH is 10 [98]. C-phycocyanin (C-PC), a blue pigment mainly produced by cyanobacteria and rhodophytes with fluorescent and antioxidant properties, has applications in pharmaceutical, food, and cosmetic industry [92], and can be obtained with high productivity of C-PC from heterotrophic cultivation of *Galdieria sulphuraria* under controlled conditions [92]. In another study, it was shown that mixotrophic cultivation of *C. vulgaris* and *Chlorella sorokiniana*, immobilized on alginate beads in combination with *Azospirillum brasilense* (a microalgae growth promoting bacterium), could offer higher yields of chlorophyll a, chlorophyll b, lutein, and violaxanthin [99].

Microalgae have been widely investigated for health benefits, especially after the probiotic supplement introduction [6]. Various types of sterols are found in microalgae. *Spirulina* spp. are considered as a good source of clionasterol which is helpful to prevent cardiovascular diseases and increases plasminogen activating factors [100]. Moreover, *Spirullina* spp. are considered best for cancer and viral infection prevention due to their immune-boosting ability. Different antioxidant compounds including mycosporine-like amino acids and carotenoids may be isolated from *Muriellopsis* spp., and can be used to prevent oxidative stress and to treat degenerative diseases [6]. Extract of *Chlorella* spp. has several health benefits, including blood hemoglobin enhancer, hepatoprotective agent, hypocholesterolemic and blood sugar reducing agent, immune stimulator, and blood lipid reducer [4]. In 19th century, *Chlorella* cells were used to treat leprosy, as dietary aid to increase health, weight, and energy of the patients [100]. It is postulated that carotenoids from *Dunaliella* spp. and *Spirulina* spp. are more potent anticancer agents as compared to β-carotene [6]. Table 3.4 provides a list of various pigments along with their functions.

Table 3.4 List of Pigments and Their Functions in Daily Life

Pigments/ Molecules	Applications	Biological Source	References
α-Tocopherol	Antioxidant, food preservation, cosmetics, and sunscreens	*Euglena gracilis*, *Dunaliella tertiolecta*	[101]
β-Carotene	Food colorant, vitamin precursor	*Dunaliella* spp.	[4]
Canthaxanthin	Antioxidant	*Chlorella protothecoides*	[102]
Carotenoids	Food colorant	*Dunaliella salina*, *Hematococcus pluvialis*, *Chlorella zofingiensis*, and *Chlorella vulgaris*	[103]
Phycocyanin	Dietary supplements	*Arthrospira platensis*	[103]
Phycobilins	Natural colorants, fluorescent agent, and in pharmaceuticals	Cyanobacteria and green algae	[103]
Phycoerythrin	Fluorescent agent	*Porphyridium* spp.	[103]
Astaxanthin	Animal feed and in pharmaceutical industry	*H. pluvialis*	[4]
Ziaxanthin	In pharmaceutical industry	*Chlorella protothecoides*	[102]

3.4 BIOCHAR PRODUCTION

Biochar is an organic material which is produced by pyrolysis or charring process by heating biomass at higher temperature usually ranging from 250 to 500°C under anaerobic conditions [104]. It is a rich source of carbon, phosphorous, magnesium, calcium, and sometimes nitrogen. From centuries biochar is mixed with soil [105]. Globally, biochar has been evaluated for soil amendment, soil fertility improvement, and environmental mitigation [106]. It is also known that biochar improves the soil's nutrient level, water-holding capacity, microbial activity, agricultural productivity, and production from biomass waste [106,107]. Biochar is a highly porous, fine-grained charcoal substance and is differentiated from other charcoals for its use in soil amendment. This charcoal is produced under specific conditions that optimize certain characteristics deemed useful in agriculture. These characteristics include low amounts of residual resins, high surface area per unit volume, and longer stability in soil (~100–1000 years) [108]. The higher surface area per unit volume is a consequence of heat treatment of organic material and its ability to persist in soil with very little decomposition [109]. Because of its nature and composition, it takes care of nutrition of plant and soil microorganisms, serving as a catalytic agent that enhances uptake of water and nutrients to the plants. The characteristic properties of biochar, namely high surface area and higher porosity, make it a suitable adsorbent to retain nutrients and water molecules, thus making it a good habitat for microorganisms to flourish [109–111].

The use of biochar is not very recent and is reported to be used in Amazon Basin in unusually fertile soils [112]. The world soil is known to contain char substances. These char substances were deposited as a result of natural processes, e.g., grassland fires [113]. Biochar properties are affected by oxygen supply, type of organic material used, and temperature conditions. These ultimately determine the suitability of biochar as a soil modifier [114,115]. The production of biochar under specific conditions imparts specific properties to the char produced, and its quality is related to its adsorption capacity and cation exchange capacity along with low amount of tarr, resin, etc. [116]. The biochar production methods also release heat, oil, and gas, which can be recovered for subsequent uses (power generation).

Algal biochar is assumed to have comparatively higher nutrient and mineral levels compared to others [117,118]. Algal biochar had over 50% higher nitrogen content than lignocellulosic and carbon content [104]. Algae (macroalgae and microalgae) have higher rate of bioextraction of minerals from wastewater [119]; therefore, it is believed that biochar produced from algal biomass will be contained higher nutrients and minerals when compared to other biomass sources [108]. This makes algae a considerable feedstock for biochar along with attractive carbon sequestration and soil amelioration option, providing a high-value product. However, the proportion and composition of nutrients is expected to vary among different algal species and cultivation conditions ranging from fresh through brackish to saline environments [108].

3.5 AMINO ACIDS AND POULTRY FEED

Amino acids are at the basis of all life processes and play an important role in human and animal health maintenance. Biotechnological processes have been used for industrial production of amino acids for almost 20 years. Industrially produced feed amino acids L-lysine, DL-methionine, L-threonine, and L-tryptophan cover almost up to 60% total amino acid market [120].

Microalgae species are excellent producers of essential and nonessential amino acids. *Spirulina* and *Chlorella* species are also rich sources of essential amino acids. *Chlorella* species contain higher amounts of lysine which is found in limited quantity in cereals [121]. It is investigated that leucine is the most frequent amino acid present in the biomass of *Spirulina* species. Among the nonessential amino acids, aspartic and glutamic acids are most abundant in microalgae species and are used in industrial production of aspartame (artificial sweetener) and monosodium glutamate (a flavor enhancer) [122]. Microalgae having high level of essential amino acids can be used as food ingredient to improve the amino acid profile of food,

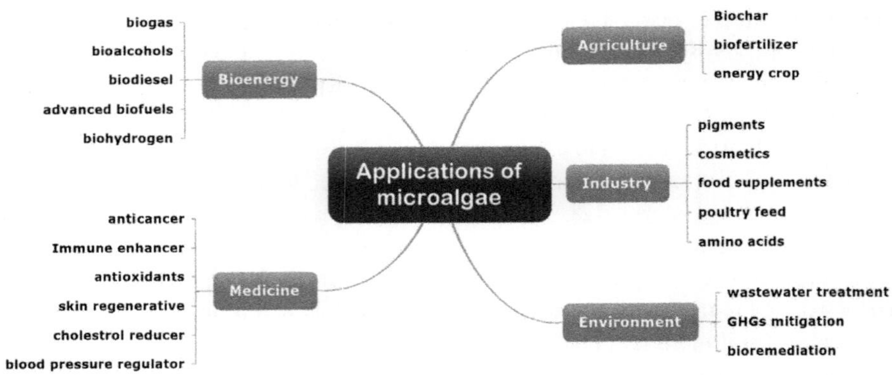

FIGURE 3.1

Applications of microalgae at a glance.

subject to the availability of fully characterized and identified potential strains/species of the microalgae.

Poultry feed is commonly produced as a mixture of cereal grains, fats, plant protein sources, animal by-products, vitamin and mineral supplements, and crystalline amino acids. In developing countries, increasing cost and limited supply of feed ingredients are expected to compel the future expansion of poultry production [123]. Microalgae seem to be a promising alternative source for poultry and animal feed. *Spirulina* has been extensively used as an animal feed supplement as it contains B-complex vitamins, minerals, proteins, γ-linolenic acid, antioxidants, and a number of bioactive compounds [123]. Microalgae serves a remarkable replacement for conventional proteins in poultry feed. Microalgae feed also have an important role in metabolism of chicken, as it reduces the cholesterol level of egg yolk up to 10% [124]. In aquaculture, fish powder is one of the important nutrients of aquaculture fodder which usually contains 20%–30% proteins. The shortage of fish powder (the global production amount is about 2 M tones) limited the sustainable development of aquaculture. In this area, microalgae, such as *Spirulina* and *Chlorella* may play an important role to supplement the nutrition value of feed for aquaculture. A graphical representation of applications of microalgae is presented in Fig. 3.1.

REFERENCES

[1] Radmer RJ. Algal diversity and commercial algal products. Bioscience 1996;46(4):263–70.

[2] Draaisma RB, Wijffels RH, Slegers PE, Brentner LB, Roy A, Barbosa MJ. Food commodities from microalgae. Curr Opin Biotechnol 2012;24(2):169–77.

[3] Borowitzka MA. Commercial production of microalgae: ponds, tanks, tubes and fermenters. J Biotechnol 1999;70(1):313–21.

[4] Spolaore P, Joannis-Cassan C, Duran E, Isambert A. Commercial applications of microalgae. J Biosci Bioeng 2006;101(2):87–96.

[5] Samarakoon K, Jeon Y-J. Bio-functionalities of proteins derived from marine algae — a review. Food Res Int 2012;48(2):948—60.

[6] Mata TM, Martins AA, Caetano NS. Microalgae for biodiesel production and other applications: a review. Renew Sustain Energy Rev 2010;14(1):217—32.

[7] Benemann J. Microalgae for biofuels and animal feeds. Energies 2013;6(11):5869—86.

[8] Wang K, Brown RC, Homsy S, Martinez L, Sidhu SS. Fast pyrolysis of microalgae remnants in a fluidized bed reactor for bio-oil and biochar production. Bioresour Technol 2013;127:494—9.

[9] Chisti Y. Biodiesel from microalgae. Biotechnol Adv 2007;25(3):294—306.

[10] Um B-H, Kim Y-S. Review: a chance for Korea to advance algal-biodiesel technology. J Ind Eng Chem 2009;15(1):1—7.

[11] Meng X, Yang J, Xu X, Zhang L, Nie Q, Xian M. Biodiesel production from oleaginous microorganisms. Renew Energy 2009;34(1):1—5.

[12] Deng X, Li Y, Fei X. Microalgae: a promising feedstock for biodiesel. Afr J Microbiol Res 2009;3(13):1008—14.

[13] Ahmad A, Yasin NM, Derek C, Lim J. Microalgae as a sustainable energy source for biodiesel production: a review. Renew Sustain Energy Rev 2011;15(1):584—93.

[14] Shah YR, Sen DJ. Bioalcohol as green energy — a review. Int J Curr Sci Res 2011;1(2):57—62.

[15] Weber C, Farwick A, Benisch F, Brat D, Dietz H, Subtil T, et al. Trends and challenges in the microbial production of lignocellulosic bioalcohol fuels. Appl Microbiol Biotechnol 2010;87(4):1303—15.

[16] Balat M, Balat H. Recent trends in global production and utilization of bio-ethanol fuel. Appl Energy 2009;86(11):2273—82.

[17] Bahadar A, Khan MB. Progress in energy from microalgae: a review. Renew Sustain Energy Rev 2013;27:128—48.

[18] Daroch M, Geng S, Wang G. Recent advances in liquid biofuel production from algal feedstocks. Appl Energy 2013;102:1371—81.

[19] Himmel ME, Ding S-Y, Johnson DK, Adney WS, Nimlos MR, Brady JW, et al. Biomass recalcitrance: engineering plants and enzymes for biofuels production. Science 2007;315(5813):804—7.

[20] Zaldivar J, Nielsen J, Olsson L. Fuel ethanol production from lignocellulose: a challenge for metabolic engineering and process integration. Appl Microbiol Biotechnol 2001;56(1—2):17—34.

[21] Tan KT, Lee KT, Mohamed AR. Role of energy policy in renewable energy accomplishment: the case of second-generation bioethanol. Energy Policy 2008;36(9):3360—5.

[22] John RP, Anisha G, Nampoothiri KM, Pandey A. Micro and macroalgal biomass: a renewable source for bioethanol. Bioresour Technol 2011;102(1):186—93.

[23] Chen C-Y, Zhao X-Q, Yen H-W, Ho S-H, Cheng C-L, Lee D-J, et al. Microalgae-based carbohydrates for biofuel production. Biochem Eng J 2013;78:1—10.

[24] Hirano A, Ueda R, Hirayama S, Ogushi Y. CO_2 fixation and ethanol production with microalgal photosynthesis and intracellular anaerobic fermentation. Energy 1997;22(2):137—42.

[25] Rodjaroen S, Juntawong N, Mahakhant A, Miyamoto K. High biomass production and starch accumulation in native green algal strains and cyanobacterial strains of Thailand. Kasetsart J — Nat Sci 2007;41:570—5.

[26] Rafiqul I, Hassan A, Sulebele G, Orosco C, Roustaian P, Jalal K. Salt stress culture of blue green algae *Spirulina fusiformis*. Pak J Biol Sci 2003;6(7):648–50.
[27] Ho S-H, Huang S-W, Chen C-Y, Hasunuma T, Kondo A, Chang J-S. Bioethanol production using carbohydrate-rich microalgae biomass as feedstock. Bioresour Technol 2013;135:191–8.
[28] Nigam PS, Singh A. Production of liquid biofuels from renewable resources. Prog Energy Combust Sci 2011;37(1):52–68.
[29] Choi SP, Nguyen MT, Sim SJ. Enzymatic pretreatment of *Chlamydomonas reinhardtii* biomass for ethanol production. Bioresour Technol 2010;101(14):5330–6.
[30] Harun R, Danquah MK, Forde GM. Microalgal biomass as a fermentation feedstock for bioethanol production. J Chem Technol Biotechnol 2010;85(2):199–203.
[31] Zullaikah S, Rakhadima YT, Rachimoellah M, Widjaja T, Sumarno S. An efficient method for the production of biodiesel from rice bran. IPTEK J Proc Ser 2015;1(1).
[32] Chang Y-C, Lee W-J, Wang L-C, Yang H-H, Cheng M-T, Lu J-H, et al. Effects of waste cooking oil-based biodiesel on the toxic organic pollutant emissions from a diesel engine. Appl Energy 2014;113:631–8.
[33] Ma F, Hanna MA. Biodiesel production: a review. Bioresour Technol 1999;70(1): 1–15.
[34] Zhang Y, Dube M, McLean D, Kates M. Biodiesel production from waste cooking oil: 2. Economic assessment and sensitivity analysis. Bioresour Technol 2003;90(3): 229–40.
[35] Ramadhas AS, Jayaraj S, Muraleedharan C. Biodiesel production from high FFA rubber seed oil. Fuel 2005;84(4):335–40.
[36] Haas MJ. Improving the economics of biodiesel production through the use of low value lipids as feedstocks: vegetable oil soapstock. Fuel Process Technol 2005; 86(10):1087–96.
[37] Koh MY, Ghazi TIM. A review of biodiesel production from *Jatropha curcas* L. oil. Renew Sustain Energy Rev 2011;15(5):2240–51.
[38] Demirbas A, Demirbas MF. Importance of algae oil as a source of biodiesel. Energy Convers Manage 2011;52(1):163–70.
[39] Gouveia L, Oliveira AC. Microalgae as a raw material for biofuels production. J Ind Microbiol Biotechnol 2009;36(2):269–74.
[40] Chen L, Liu T, Zhang W, Chen X, Wang J. Biodiesel production from algae oil high in free fatty acids by two-step catalytic conversion. Bioresour Technol 2012;111: 208–14.
[41] Khola G, Ghazala B. Biodiesel production from algae. Pak J Bot 2012;44(1):379–81.
[42] Sharma KK, Schuhmann H, Schenk PM. High lipid induction in microalgae for biodiesel production. Energies 2012;5(5):1532–53.
[43] Ji M-K, Abou-Shanab RA, Kim S-H, Salama E-S, Lee S-H, Kabra AN, et al. Cultivation of microalgae species in tertiary municipal wastewater supplemented with CO_2 for nutrient removal and biomass production. Ecol Eng 2013;58:142–8.
[44] Trentacoste EM, Shrestha RP, Smith SR, Gle C, Hartmann AC, Hildebrand M, et al. Metabolic engineering of lipid catabolism increases microalgal lipid accumulation without compromising growth. Proc Natl Acad Sci 2013;110(49):19748–53.
[45] Gimpel JA, Specht EA, Georgianna DR, Mayfield SP. Advances in microalgae engineering and synthetic biology applications for biofuel production. Curr Opin Chem Biol 2013;17(3):489–95.

[46] Dueblein D, Steinhauser A. Biogas. In: Biogas from waste and renewable resources. KGaA: Wiley-VCH Verlag GmbH and Co.; 2008.

[47] Abatzoglou N, Boivin S. A review of biogas purification processes. Biofuels Bioprod Biorefin 2009;3(1):42–71.

[48] Sreekrishnan T, Kohli S, Rana V. Enhancement of biogas production from solid substrates using different techniques——a review. Bioresour Technol 2004;95(1):1–10.

[49] Taherzadeh MJ, Karimi K. Pretreatment of lignocellulosic wastes to improve ethanol and biogas production: a review. Int J Mol Sci 2008;9(9):1621–51.

[50] Weiland P. Biogas production: current state and perspectives. Appl Microbiol Biotechnol 2010;85(4):849–60.

[51] Al Seadi T. Biogas handbook. Syddansk Universitet; 2008.

[52] Mussgnug JH, Klassen V, Schlüter A, Kruse O. Microalgae as substrates for fermentative biogas production in a combined biorefinery concept. J Biotechnol 2010;150(1): 51–6.

[53] Mudimu O, Rybalka N, Bauersachs T, Born J, Friedl T, Schulz R. Biotechnological screening of microalgal and cyanobacterial strains for biogas production and antibacterial and antifungal effects. Metabolites 2014;4(2):373–93.

[54] Yasin NHM, Mumtaz T, Hassan MA. Food waste and food processing waste for biohydrogen production: a review. J Environ Manage 2013;130:375–85.

[55] Prajapati SK, Malik A, Vijay VK. Comparative evaluation of biomass production and bioenergy generation potential of *Chlorella* spp. through anaerobic digestion. Appl Energy 2014;114:790–7.

[56] Montingelli M, Tedesco S, Olabi A. Biogas production from algal biomass: a review. Renew Sustain Energy Rev 2015;43:961–72.

[57] Klassen V, Blifernez-Klassen O, Hoekzema Y, Mussgnug JH, Kruse O. A novel one-stage cultivation/fermentation strategy for improved biogas production with microalgal biomass. J Biotechnol 2015;215:44–51.

[58] Collet P, Hélias A, Lardon L, Ras M, Goy R-A, Steyer J-P. Life-cycle assessment of microalgae culture coupled to biogas production. Bioresour Technol 2011;102(1): 207–14.

[59] Jones CS, Mayfield SP. Algae biofuels: versatility for the future of bioenergy. Curr Opin Biotechnol 2012;23(3):346–51.

[60] Debowski M, Zieliński M, Grala A, Dudek M. Algae biomass as an alternative substrate in biogas production technologies—review. Renew Sustain Energy Rev 2013; 27:596–604.

[61] Lee K, Chantrasakdakul P, Kim D, Kong M, Park KY. Ultrasound pretreatment of filamentous algal biomass for enhanced biogas production. Waste Manage 2014;34(6): 1035–40.

[62] Azwar M, Hussain M, Abdul-Wahab A. Development of biohydrogen production by photobiological, fermentation and electrochemical processes: a review. Renew Sustain Energy Rev 2014;31:158–73.

[63] Hay JXW, Wu TY, Juan JC. Biohydrogen production through photo fermentation or dark fermentation using waste as a substrate: overview, economics, and future prospects of hydrogen usage. Biofuels Bioprod Biorefin 2013;7(3):334–52.

[64] de Sa LRV, Cammarota MC, de Oliveira TC, Oliveira EMM, Matos A, Ferreira-Leitão VS. Pentoses, hexoses and glycerin as substrates for biohydrogen production: an approach for Brazilian biofuel integration. Int J Hydrogen Energy 2013;38(7): 2986–97.

[65] Elsharnouby O, Hafez H, Nakhla G, El Naggar MH. A critical literature review on biohydrogen production by pure cultures. Int J Hydrogen Energy 2013;38(12):4945–66.

[66] Sivagurunathan P, Sen B, Lin C-Y. Batch fermentative hydrogen production by enriched mixed culture: combination strategy and their microbial composition. J Biosci Bioeng 2014;117(2):222–8.

[67] Tiwari A, Pandey A. Cyanobacterial hydrogen production – a step towards clean environment. Int J Hydrogen Energy 2012;37(1):139–50.

[68] Srirangan K, Pyne ME, Chou CP. Biochemical and genetic engineering strategies to enhance hydrogen production in photosynthetic algae and cyanobacteria. Bioresour Technol 2011;102(18):8589–604.

[69] Kumaraswamy GK, Guerra T, Qian X, Zhang S, Bryant DA, Dismukes GC. Reprogramming the glycolytic pathway for increased hydrogen production in cyanobacteria: metabolic engineering of NAD^+-dependent GAPDH. Energy Environ Sci 2013;6(12):3722–31.

[70] Sakjanes K, Rebours C, Lindblad P. Potential for green microalgae to produce hydrogen, pharmaceuticals and other high value products in a combined process. Crit Rev Biotechnol 2013;33(2):172–215.

[71] Liu C-H, Chang C-Y, Cheng C-L, Lee D-J, Chang J-S. Fermentative hydrogen production by *Clostridium butyricum* CGS5 using carbohydrate-rich microalgal biomass as feedstock. Int J Hydrogen Energy 2012;37(20):15458–64.

[72] Nobre BP, Villalobos F, Barragan BE, Oliveira A, Batista AP, Marques P, et al. A biorefinery from *Nannochloropsis* sp. microalgae – extraction of oils and pigments. Production of biohydrogen from the leftover biomass. Bioresour Technol 2013;135:128–36.

[73] Onwudili JA, Lea-Langton AR, Ross AB, Williams PT. Catalytic hydrothermal gasification of algae for hydrogen production: composition of reaction products and potential for nutrient recycling. Bioresour Technol 2013;127:72–80.

[74] Wirth R, Lakatos G, Maroti G, Bagi Z, Minarovics J, Nagy K, et al. Exploitation of algal-bacterial associations in a two-stage biohydrogen and biogas generation process. Biotechnol Biofuels 2015;8(1):1.

[75] Lee SY, Kim HM, Cheon S. Metabolic engineering for the production of hydrocarbon fuels. Curr Opin Biotechnol 2015;33:15–22.

[76] Zhang F, Rodriguez S, Keasling JD. Metabolic engineering of microbial pathways for advanced biofuels production. Curr Opin Biotechnol 2011;22(6):775–83.

[77] Heeres AS, Picone CS, van der Wielen LA, Cunha RL, Cuellar MC. Microbial advanced biofuels production: overcoming emulsification challenges for large-scale operation. Trends Biotechnol 2014;32(4):221–9.

[78] Singh V, Mani I, Chaudhary DK, Dhar PK. Metabolic engineering of biosynthetic pathway for production of renewable biofuels. Appl Biochem Biotechnol 2014;172(3):1158–71.

[79] Rabinovitch-Deere CA, Oliver JW, Rodriguez GM, Atsumi S. Synthetic biology and metabolic engineering approaches to produce biofuels. Chem Rev 2013;113(7):4611–32.

[80] Niehaus TD, Kinison S, Okada S, Yeo Y-S, Bell SA, Cui P, et al. Functional identification of triterpene methyltransferases from *Botryococcus braunii* race B. J Biol Chem 2012;287(11):8163–73.

[81] Li Y, Moore R, Qin J, Scott A, Ball A. Extractable liquid, its energy and hydrocarbon content in the green alga *Botryococcus braunii*. Biomass Bioenerg 2013;52:103–12.

[82] Al-Hothaly KA, Adetutu EM, May BH, Taha M, Ball AS. Towards the commercialization of *Botryococcus braunii* for triterpenoid production. J Ind Microbiol Biotechnol 2015;42(10):1415–8.

[83] Yoshimura T, Okada S, Honda M. Culture of the hydrocarbon producing microalga *Botryococcus braunii* strain Showa: optimal CO_2, salinity, temperature, and irradiance conditions. Bioresour Technol 2013;133:232–9.

[84] Jiang Z, Kempinski C, Bush CJ, Nybo SE, Chappell J. Engineering triterpene and methylated triterpene production in plants provides biochemical and physiological insights into terpene metabolism. Plant Physiol 2015. http://dx.doi.org/10.1104/pp.15.01548.

[85] Khan NE, Nybo SE, Chappell J, Curtis WR. Triterpene hydrocarbon production engineered into a metabolically versatile host – *Rhodobacter capsulatus*. Biotechnol Bioeng 2015;112(8):1523–32.

[86] Venil CK, Zakaria ZA, Ahmad WA. Bacterial pigments and their applications. Process Biochem 2013;48(7):1065–79.

[87] Malik K, Tokkas J, Goyal S. Microbial pigments: a review. Int J Microb Resour Technol 2012;1:361–5.

[88] Mlodzinska E. Survey of plant pigments: molecular and environmental determinants of plant colors. Acta Biol Cracov Ser Bot 2009;51(1):7–16.

[89] Das A, Yoon S-H, Lee S-H, Kim J-Y, Oh D-K, Kim S-W. An update on microbial carotenoid production: application of recent metabolic engineering tools. Appl Microbiol Biotechnol 2007;77(3):505–12.

[90] Dufosse L, Fouillaud M, Caro Y, Mapari SA, Sutthiwong N. Filamentous fungi are large-scale producers of pigments and colorants for the food industry. Curr Opin Biotechnol 2014;26:56–61.

[91] Mulders KJ, Lamers PP, Martens DE, Wijffels RH. Phototrophic pigment production with microalgae: biological constraints and opportunities. J Phycol 2014;50(2): 229–42.

[92] Eriksen NT. Production of phycocyanin – a pigment with applications in biology, biotechnology, foods and medicine. Appl Microbiol Biotechnol 2008;80(1):1–14.

[93] Dufosse L, Galaup P, Yaron A, Arad SM, Blanc P, Murthy KNC, et al. Microorganisms and microalgae as sources of pigments for food use: a scientific oddity or an industrial reality? Trends Food Sci Tech 2005;16(9):389–406.

[94] Ferruzzi MG, Blakeslee J. Digestion, absorption, and cancer preventative activity of dietary chlorophyll derivatives. Nutr Res 2007;27(1):1–12.

[95] Begum H, Yusoff FM, Banerjee S, Khatoon H, Shariff M. Availability and utilization of pigments from microalgae. Crit Rev Food Sci Nutr 2015;56(13):2209–22.

[96] Lorenz RT, Cysewski GR. Commercial potential for *Haematococcus* microalgae as a natural source of astaxanthin. Trends Biotechnol 2000;18(4):160–7.

[97] Del Campo JA, Garcia-Gonzalez M, Guerrero MG. Outdoor cultivation of microalgae for carotenoid production: current state and perspectives. Appl Microbiol Biotechnol 2007;74(6):1163–74.

[98] Thirumala M. Optimization of growth of *Spirulina platensis* LN1 for production of carotenoids. Int J Life Sci Biotechnol Pharm Res 2012;1(2):152–7.

[99] de-Bashan LE, Bashan Y, Moreno M, Lebsky VK, Bustillos JJ. Increased pigment and lipid content, lipid variety, and cell and population size of the microalgae *Chlorella* spp. when co-immobilized in alginate beads with the microalgae-growth-promoting bacterium *Azospirillum brasilense*. Can J Microbiol 2002;48(6):514–21.

[100] Barrow C, Shahidi F. Marine nutraceuticals and functional foods. CRC Press; 2007.

[101] Survase SA, Bajaj IB, Singhal RS. Biotechnological production of vitamins. Food Technol Biotechnol 2006;44(3):381.

[102] de Jesus Raposo MF, de Morais RMSC, de Morais AMMB. Health applications of bioactive compounds from marine microalgae. Life Sci 2013;93(15):479–86.

[103] Markou G, Nerantzis E. Microalgae for high-value compounds and biofuels production: a review with focus on cultivation under stress conditions. Biotechnol Adv 2013;31(8):1532–42.

[104] Chaiwong K, Kiatsiriroat T, Vorayos N, Thararax C. Study of bio-oil and bio-char production from algae by slow pyrolysis. Biomass Bioenergy 2013;56:600–6.

[105] Lehmann J, Joseph S. Biochar for environmental management: science, technology and implementation. Routledge; 2015.

[106] Clough TJ, Condron LM, Kammann C, Muller C. A review of biochar and soil nitrogen dynamics. Agronomy 2013;3(2):275–93.

[107] Domene X, Mattana S, Hanley K, Enders A, Lehmann J. Medium-term effects of corn biochar addition on soil biota activities and functions in a temperate soil cropped to corn. Soil Biol Biochem 2014;72:152–62.

[108] Bird MI, Wurster CM, de Paula Silva PH, Bass AM, De Nys R. Algal biochar production and properties. Bioresour Technol 2011;102(2):1886–91.

[109] Lehmann J, Rondon M. Bio-char soil management on highly weathered soils in the humid tropics. In: Biological Approaches to Sustainable Soil Systems. Boca Raton (FL): CRC Press; 2006. p. 517–30.

[110] Glaser B, Lehmann J, Zech W. Ameliorating physical and chemical properties of highly weathered soils in the tropics with charcoal – a review. Biol Fertil Soils 2002;35(4): 219–30.

[111] Warnock DD, Lehmann J, Kuyper TW, Rillig MC. Mycorrhizal responses to biochar in soil – concepts and mechanisms. Plant Soil 2007;300(1–2):9–20.

[112] O'Neill B, Grossman J, Tsai M, Gomes J, Lehmann J, Peterson J, et al. Bacterial community composition in Brazilian anthrosols and adjacent soils characterized using culturing and molecular identification. Microb Ecol 2009;58(1):23–35.

[113] Skjemstad JO, Reicosky DC, Wilts AR, McGowan JA. Charcoal carbon in US agricultural soils. Soil Sci Soc Am J 2002;66(4):1249–55.

[114] Lehmann J. Bio-energy in the black. Front Ecol Environ 2007;5(7):381–7.

[115] McClellan AT, Deenik J, Uehara G, Antal M. Effects of flashed carbonized macadamia nutshell charcoal on plant growth and soil chemical properties. ASA-CSSA-SSA Int Annu Meet 2007;80(100):120.

[116] Liang B, Lehmann J, Solomon D, Sohi S, Thies JE, Skjemstad JO, et al. Stability of biomass-derived black carbon in soils. Geochim Cosmochim Acta 2008;72(24): 6069–78.

[117] Ahmad M, Rajapaksha AU, Lim JE, Zhang M, Bolan N, Mohan D, et al. Biochar as a sorbent for contaminant management in soil and water: a review. Chemosphere 2014; 99:19–33.

[118] Cayuela M, Van Zwieten L, Singh B, Jeffery S, Roig A, Sanchez-Monedero M. Biochar's role in mitigating soil nitrous oxide emissions: a review and meta-analysis. Agric Ecosyst Environ 2014;191:5–16.

[119] Gill SS, Mehmood MA, Rashid U, Ibrahim M, Saqib A, Tabassum MR. Waste-water treatment coupled with biodiesel production using microalgae: a bio-refinery approach. Pak J Life Soc Sci 2013;11(3):179–89.

[120] Xie N-Z, Liang H, Huang R-B, Xu P. Biotechnological production of muconic acid: current status and future prospects. Biotechnol Adv 2014;32(3):615–22.
[121] Misurcova L, Bunka F, Vavra Ambrozova J, Machu L, Samek D, Kracmar S. Amino acid composition of algal products and its contribution to RDI. Food Chem 2014;151: 120–5.
[122] Gutierrez-Salmean G, Fabila-Castillo L, Chamorro-Cevallos G. Nutritional and toxicological aspects of *Spirulina* (*Arthrospira*). Nutr Hosp 2015;32(1):34–40.
[123] Gerrard CL, Smith J, Nelder R, Bright A, Colley M, Clements R, et al. 100% organic poultry feed: can algae replace soybean expeller in organic broiler diets? Org Farming 2015;1(1):38–45.
[124] Yaakob Z, Ali E, Zainal A, Mohamad M, Takriff MS. An overview: biomolecules from microalgae for animal feed and aquaculture. J Biol Res 2014;21(6):1–10.

CHAPTER

Origin of Algae and Their Plastids

4

Nadia Sharif[1], Neelma Munir[1], Shagufta Naz[1], Rehana Iqbal[2], Waqar Rauf[3]

Lahore College for Women University, Lahore, Pakistan[1]; Bahauddin Zakariya University, Multan, Pakistan[2]; National Institute of Biology & Genetics Engineering (NIBGE), Faisalabad, Pakistan[3]

4.1 INTRODUCTION

Algae (alga, singular) can be defined as nonvascular plants that have chlorophyll *a* as their primary photosynthetic pigment and lack a sterile covering of cells around the reproductive cells. Algae occur in a wide variety of environments including brackish, marine, freshwaters, hot and cold springs, desert soils, and on bare rocks. Diverse unicellular and multicellular algae groups are chlorarachniophytes, haptophytes, heterokonts, cryptophytes, rhodophytes, and dinoflagellates [1,2]. The term plastid (greek word Plastikas i-e formed or moulded) was applied by Schimper (1985) [2a] to the organelles that are primarily concerned with food storage. Plastid is a major double-membrane cytoplasmic semiautonomous organelles, it synthesize or store various types of organic compounds. Plastids are derived from endosymbiotic event through a complex mix of movement, loss and replacement, non-photosynthetic plastids are being found in many non-photosynthetic lineages. In other cases, photosynthetic lineages have evolved from ancestors with a plastid of different origin, so an ancestral plastid has been replaced with a new one. Different compounds such as fatty acids, heme, amino acids, and aromatic isoprenoids are synthesized by plastids through photosynthetic and biochemical pathways [3].

Plastids are mostly similar in structure in different groups of algae, and are assumed to be closely related. Preliminary ultrastructural investigations complimented this view by showing that similarities between the plastids of two algal groups often differed with similarities between cytoplasmic features shared by nonalgal and eukaryotes algae. This might have been obvious in the euglenids, as their plastids share characteristics with green algae [4,5]. The molecular systematics reinforced these contradictions that correlate the trypanosomes and euglenids with the exclusion of green algae. This complexity is a result of endosymbiosis. Hence in the context of endosymbiosis and eukaryotes tree, the organisms, plastids, and their individual genes have somewhat diverse, distinctive, and superimposed phylogeny [4].

Algae Based Polymers, Blends, and Composites. http://dx.doi.org/10.1016/B978-0-12-812360-7.00004-5

4.2 EVOLUTION OF OXYGENIC PHOTOSYNTHESIS AND PRIMARY ENDOSYMBIOSIS

The ancestors of cyanobacteria are the first organisms that developed oxygenic photosynthesis. The cyanobacteria are considered as the only organisms initially responsible for producing an oxidizing environment from a reducing one. The cyanobacteria created the oxygenated atmosphere due to which aerobic organisms evolved and used oxygen as potent electron acceptor. As more energy could be garnered per electron transfer, aerobes soon succeeded to sustain considerable dynamic ecosystems. Subsequently, the oceanic primary production increased an order of magnitude that permitted the adaptation of novel biochemical pathways and evolution of complex life forms [6,7].

The rise in atmospheric oxygen is directly involved in eukaryogenesis and activated cell compartmentalization. By constraining the number and size of the outer domains of transmembrane proteins, the environmental oxygen is considered to have compelled the ancient transmembrane proteins topology. At the point when oxygen levels increased, the limitation was likely reduced, and further communication-related transmembrane proteins and the subsequent compartmentalization were permitted. The rising levels of oxygen conjectured to have advanced cellular compartmentalization is also involved in securing the reactive oxygen species in the cellular environment and the metabolic activities of the plasma membrane. The time-aligned phylogenies and fossil record propose that the major eukaryotic genealogies veered in the Paleoproterozoic Era as of now Ref. [8].

Primary endosymbiosis is the original uptake and retention of a cyanobacterium by a eukaryote. The internal and external membranes of the Gram-negative cyanobacterium form the two membranes of the plastids (Fig. 4.1). Thus the endosymbiosis supports the hypothesis that the invasions of prokaryote cell, loss of much of its

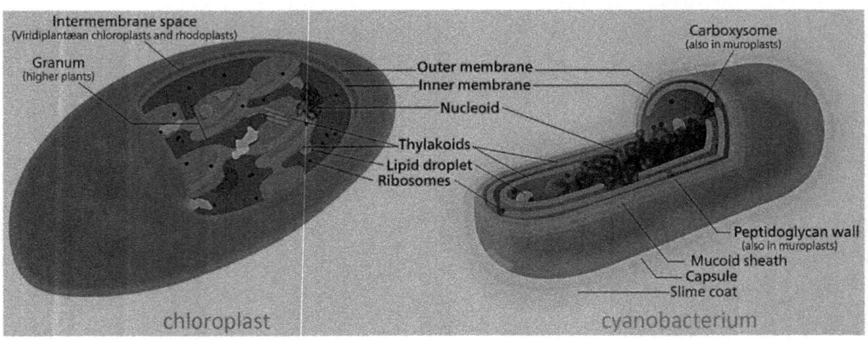

FIGURE 4.1

Comparison of a chloroplast and cyanobacterium: cyanobacteria and chloroplasts both have a double membrane, ribosomes, thylakoids, and DNA.

Courtesy of wikipedia commons. https://upload.wikimedia.org/wikipedia/commons/9/92/Chloroplast-cyanobacterium_comparison.svg, Kelvinsong (creative common).

genome, and the transfer of most of the remaining genes to the nucleus of its host produced the modern-day chloroplast as eukaryotic organelles [9].

The bulk of the evidence converged that all primary plastids can be traced back to a single endosymbiotic event. Like the presence of common light-harvesting complex proteins, the characteristics of plastid genome structure in red and green algae and the molecular phylogenetic data from plastid-encoded genes in cyanobacteria have supported a single origin [8]. It is difficult to say what sort of cyanobacterium may have offered ascend to plastids, on the grounds that the relationship of plastids to the different cyanobacterial groups has proved more difficult. It still needs to be explored whether plastids are derived from distinct but closely related cyanobacteria. Contrary to plastid gene analysis, the nucleus-encoded cytoplasmic proteins accompany certain doubts on the single origin of primary plastids. Generally, the monophyletic clade of red, green, and glaucophyte algae cannot be resolved from the nuclear gene trees [10].

The consideration and interpretation of multiple independent plastid origins or more complex events is acknowledged due to the lack of primary algal monophyly. The greater parts of these trees are ineffectively bolstered, in any case, and an inability to demonstrate a relationship between essential algal heredities is particularly not quite the same, as if really showing that they are not identified with each other. The gene for the largest subunit of RNA polymerase II is the initial strong support for isolating green and red alga. Increased sampling eroded support for their separation, though in recent phylogenetic trees these groups are joined with weak support [2,11].

The monophyly of green and red algae are now strengthened by the mitochondrial genes, and relationship between the host components of primary algae is also shown by some nuclear gene phylogenies. Though it is restricted to green and red algae, glaucophytes also fall within other primary algal lineages in concatenated gene trees as shown in Fig. 4.2 [4,8].

From all three genomes there is an evidence for a monophyletic origin of green, red, and glaucophyte algae, also no strong evidence against monophyletic origin of algae (that support single origin of primary plastids similar to super group of plants) is present. A common ancestor for primary algae and the single origin of primary plastids bring up the question, "Which lineage came first?" as the molecular analyses bolster each of the three lineages as the first primary algal group that have diverged. Nevertheless, the molecular trees support the prior branching of glaucophytes, than the green and red algae divergence from each other, which is also strengthened by the gene analyses of nucleus plastid [2,4].

Furthermore, the ancient divergence of the glaucophytes is supported by plastid-targeted fructose bisphosphate aldolase (FBA) gene. The green and red algae enzymes specify its replacement in their common ancestor. However, this is not true for FBA of the glaucophytes. In red and green algae lineages, the plastid-targeted FBA gene is as a result of endosymbiotic gene replacement event, which involves the duplication and replacement of nuclear gene by the glycolytic and cytosolic enzymes with plastid-targeted Calvin cycle enzyme. Acquisitively, it is seen that the glaucophytes have a prior divergence than the divergence of red and green algae from each other, but it needs further evidences as the evidence for the evolution of FBA needs the

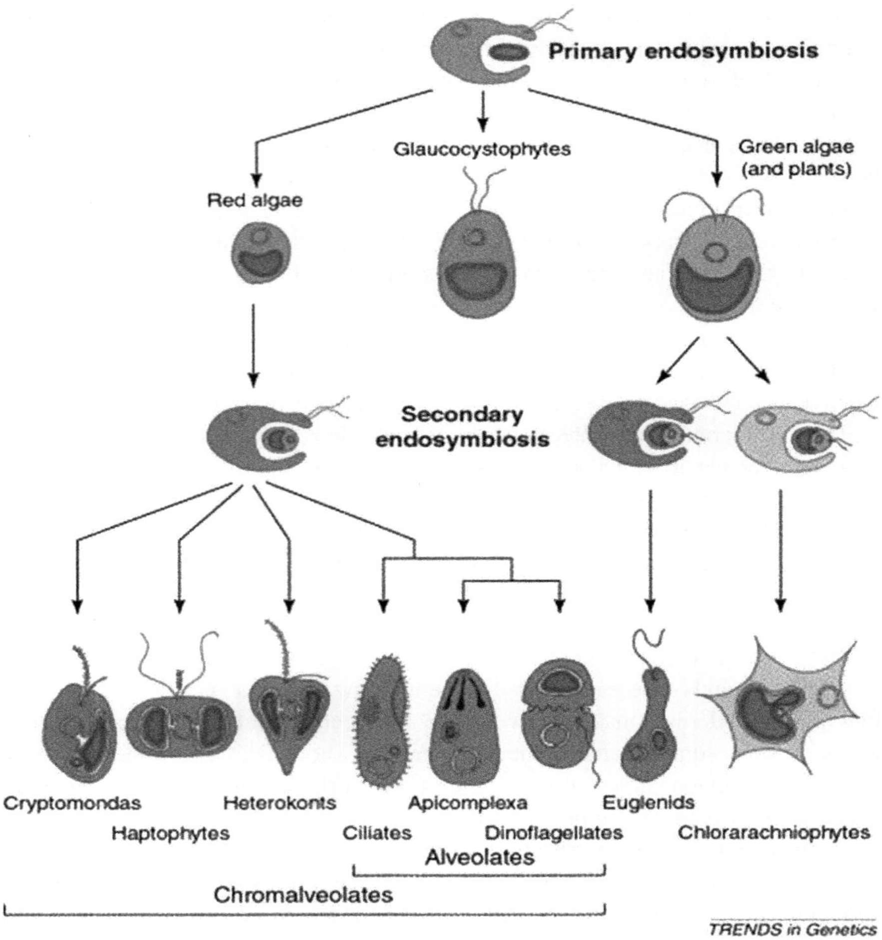

FIGURE 4.2

Algae and their plastids evolution by endosymbiosis [4].

http://vignette3.wikia.nocookie.net/mmg-233-2013-genetics-genomics/images/7/79/1-s2.0-S0168952502027774-gr3.jpg/revision/latest?cb=20131213012330, https://www.researchgate.net/profile/Jacopo_Paini/publication/257890285/figure/fig1/AS:297617331376128@1447968909096/Figure-6-Algal-evolution-and-the-origin-and-spread-of-plastids-by-endosymbiosis.png.

interpretation of complex events, and the previous concatenated analyses are based on limited sampling of taxon [4].

If it is considered that the primary plastids are evolved only once, then it requires certain concerns to be addressed. *Paulinella chromatophora* is euglyphid marine testate amoeba; it could be classified as an independently acquired plastid. It has two kidney-shaped cyanobacterial endosymbionts that are analogous to *Synechococcus*. It is closely related to the heterotrophic *Porcellio ovalis*, which lacks endosymbionts and auspiciously feeds on *Synechococcus* with filose pseudopods distinctive to this group of amoebae. However, *P. chromatophora* does not feed, instead its endosymbionts actively photosynthesize and transfer the photosynthate to the host. Additionally, the endosymbionts cannot be cultivated autonomously and appeared to divide synchronously with the host [2,4].

The accurate basis that recognizes endosymbiont from organelle, to a great extent, is a matter of feeling, yet the relationship between *P. chromatophora* and its endosymbionts is firmly demonstrative that the two are adequately integrated for the endosymbionts to be easily considered organelles and, by extension, freely procured plastids. Primary plastids are present in chlorophytes, rhodophytes, and glaucophytes lineages and together they make Archaeplastida. The glaucophytes hold a significant position in the evolution of photosynthesis in eukaryotes though not common in nature. In the transition from endosymbiont to plastid, the glaucophytes act as intermediates. They are one of a kind among plastids in holding the prokaryotic peptidoglycan layer between their two membranes. The plastids of the glaucophytes have phycobilins, chlorophyll *a*, phycobilisomes, photosynthetic pigments, and small particles that are organized on the outer face of thylakoid membranes as in cyanobacteria. Phycobilisomes, chlorophyll *a*, and phycobilins are distinctive to the plastids of red algae, while green algae plastids contain chlorophylls *a* and *b* [8].

4.3 SECONDARY ENDOSYMBIOSIS

Albeit glaucophyte, red, and green algae represent enormous diversity, they still record for a few differences of photosynthetic eukaryotes. Remaining plastid-containing eukaryotes developed the plastids through secondary endosymbiosis, which involves the uptake and retention of a primary plastid-containing alga by a second eukaryote. Secondary plastids are surrounded by the additional membranes, mostly by three or four membranes, instead of the two membranes bounding a primary plastid. The extra membranes are formed due to the phagocytosis of the primary alga. The two inner membranes of the secondary plastid are corresponding to the primary plastid membranes; third membrane is from the inside in plastid; fourth membrane corresponds to the cytoplasmic envelope of the primary algal endosymbiont, but the outermost membrane is a derivative of the phagocytotic vacuole in which the algal prey was taken up, and it is a part of the secondary host endomembrane system. The glaucophyte is relatively infrequent in nature and therefore less likely to be indulged in secondary plastid evolution; hence a secondary plastid derived from a glaucophyte was never observed, but both green and red algae were involved in secondary endosymbiotic events [2,4,12].

The primary algal nucleus is highly reduced or more commonly altogether lost in the secondary plastids of both red and green origins, and therefore the majority of the plastid-targeted proteins encoded in the primary algal nucleus were transferred to the secondary host nucleus. The secondary plastid is not physically located in the cytoplasm and thus the primary plastids target the plastid-targeted proteins to the secondary plastid in the lumen of the endomembrane system by an extra step. Consequently, secondary plastids containing algal plastid—targeted proteins encode bipartite leaders with a signal peptide, directing them to the endomembrane system, following a transit peptide as in primary algae [4,13].

The transport of proteins from the endomembrane system to the plastid can be divided into different groups, e.g., all secondary algae in general, and plastids containing three-membrane-bounded algae in particular, use this basic strategy to target proteins. It signifies a remarkable convergence to get about the tricky protein-trafficking problem that defines the independent cooption of the same existing cellular machine in several groups [2,4]. Since secondary plastids are believed to be a divergence of red and green algae, it clarifies the occurrence of this process more than once; however, the explanation of diversity of plastids and how many secondary endosymbiotic events take place need an extensive debate [2,4,13].

According to recent hypothesis about the evolution and history of plastids, a common endosymbiosis is involved in chlorarachniophyte and euglenid plastids; however, it is not supported by evidence. No evidence is found for the common ancestry of these two groups, mainly because euglenids are evidently members of the trypanosomes-related discicristates, while chlorarachniophytes are indeed members of the cercozoa. Likewise, a small number of genes analyzed from the plastid genome also do not support a common ancestry between these two groups. Euglenid analysis made on the basis of their feeding strategies, phylogeny, and cytoskeleton designates that their plastids ascended after the divergence of euglenids from other euglenozoa. It has also been proposed that a number of plastid-derived genes are present in trypanosomes that support more ancient origin of euglenid plastids; conversely, strong evidences of the presence of plant-like genes in *Trypanosoma* reject this hypothesis; hence this hypothesis could have had a lack of sampling in the analyses. Thus there is no real evidence for a plastid in the history of trypanosomes [2,4].

Though it is difficult to prove that they are closely related, it could be possibly done through the comprehensive phylogeny of green plastid genes, together with representation from numerous chlorarachniophytes, euglenids, and green algal lineages. Their divergence from a common endosymbiosis would be possibly clear if their plastids are shown to be related to the exclusion of all other green plastids, otherwise they must have been acquired independently [2,4,14].

The number of groups involved in the secondary endosymbiosis of red algal make their history more complicated. Diversity of the red algal secondary plastids is great, and numerous endosymbiotic events were hypothesized to explain this diversity. It has been acclaimed that the gene-rich nature of red algal plastid genomes can trace the diversity of red algal secondary plastids; since as compared to green algae, red algal plastid genomes encode many more genes. Red algal plastid genomes are hypothesized to be more adroit at integrating into a new host environment [2,4].

Regardless of great diversity of the red algal secondary plastids, it is clear from the evidence that in the ancestor of all chromalveolates red alga can be traced to a single endosymbiosis. It shows that as compared to green algae, red algae are involved less often in secondary endosymbiotic events. A variety of observations have supported this conclusion. Certain physical and chemical characteristics are shared by few or all plastids, such as the similarities in membrane organization in haptophytes, heterokont, and cryptomonad plastids, unique presence of chlorophyll *c* in all, and the presence of nonphotosynthetic plastids in apicomplexa [4].

In this respect, the molecular sequence analysis provided the most compelling evidence. Except dinoflagellates and alveolate apicomplexa, initial molecular trees were not able to show the close relationship concerning plastid or nuclear genes from these organisms. The current nuclear and plastid analysis supports the relationship between members of the chromalveolates. The SSU rRNA steadily demonstrates the relationship among heterokonts and alveolates, as have several individual and concatenated protein-coding genes, whereas nucleus-encoded genes for RNAs and cytosolic proteins never supported the chromalveolates [2,4].

An analysis of six concatenated genes supports the heterokonts and alveolates strongly and the cryptomonads and haptophytes weakly. These analyses still do not support a specific position for cryptomonads or haptophytes. The plastid-encoded genes indicate this type of relationship: a recent analysis of five concatenated genes showed support for the union of heterokonts, cryptomonads, and haptophytes. Either due to unavailability of data or lack of resolution, none of these analyses show all chromalveolate groups together, but all analyses together support the chromalveolates as a whole [4].

The nucleus-encoded, plastid-targeted gene, utmost conspicuously, glyceraldehyde-3-phosphate dehydrogenase (GAPDH) is able to address the entire group at once. The GAPDH (like plastid metabolic genes), in algae and plant nuclear genomes, is present in two copies, one targeted to the plastid and one expressed in the cytosol. Generally, the plastid-targeted copy would be cyanobacterial in origin, but the plastid GAPDH genes of dinoflagellates, haptophytes, apicomplexa, cryptomonads, and heterokonts all originated from the duplication of cytosolic gene. The endosymbiotic gene replacement is quite rare in other genes, while the common origin of all chromalveolates plastid GAPDH genes appeal the evidence for the divergence of their plastids from single common endosymbiotic event. Remarkably, similar result is observed for fructose bisphosphate aldolase (FBA) plastid enzyme. As compared to the red algae plastids, heterokont plastid-targeted FBA is an entirely different class of enzyme, and the plastid FBA genes of dinoflagellates, cryptomonads, and haptophytes are now categorized and correlated to the heterokont genes [4,15,16].

FBA and GAPDH augment the chromalveolates common ancestry, and the single origin of their plastids. Nevertheless, further persuasive analyses are required, principally, to find the relationship between the host lineages using concatenated nuclear gene analyses. Furthermore, the dinoflagellate and apicomplexan plastids remain to spark controversy [4]. Secondary plastids possessing seven major lineages are recognized presently as shown in Fig. 4.2.

4.3.1 EUGLENIDS

Euglenids are the diverse group of common freshwater and marine flagellates; half of them encompass three-membrane-bounded green algal plastids specialized for chlorophylls *a* and *b*. Remaining half are heterotroph or osmotroph groups, which feed on certain eukaryotic bacteria. Euglenids are diligently linked to the parasitic trypanosomes; with diplonemids, they make Euglenozoa, which is considered to be linked with heterolobosean amoeboflagellates, usually deliberated as the part of the excavates super group [2,4,5].

4.3.2 CHLORARACHNIOPHYTES

Chlorarachniophytes are rare group of marine flagellates and amoeba–flagellates. They have green algal plastid specialized for chlorophylls *a* and *b* and are bounded by four membranes. Chlorarachniophyte is one of the two groups, which keep a small relict nucleus known as nucleomorph instead of complete absence of primary algal nucleus. Chlorarachniophyte nucleomorphs have highly reduced genomes (380–455 kbp), which are divided into three chromosomes. These are compacted genomes with divergent sequences having various signatures of green algal origin. Till now the host component of chlorarachniophytes were an enigma; nevertheless, they are now part of the super-group rhizaria [2,4,5].

4.3.3 CRYPTOMONADS

Cryptomonads are a profuse group of freshwater and marine flagellates, they have red algal plastid specialized for chlorophylls *a* and *c*, and are bounded by four membranes. It is another group with a nucleomorph, and shares several superficial resemblances with chlorarachniophytes. Cryptomonad nucleomorphs have larger genomes (450–710 kbp), though they are divided into three small chromosomes and are still highly reduced. Nucleomorphs' presence and origin from red and green algae nuclei is helpful in studying the parallel events of genome reduction. One nucleomorph genome of cryptomonad is complete now, while that of chlorarachniophyte is in progress. The host component of cryptomonads also need to be mentioned; however, it is clear from the evidence that this and the remaining groups are all members of the chromalveolates super group [2,4,11].

4.3.4 HAPTOPHYTES

Haptophytes that include prymnesiophytes or coccolithophorids are copious, red algae originated from marine flagellate group; they are specialized for chlorophylls *a* and *c* and have four-membrane plastids. The outermost membrane is continuous with the nuclear envelope endoplasmic reticulum (ER). They are substantial primary bloom producers as well as economically important.

4.3.5 HETEROKONTS

Heterokonts were separately classified into fungi, protozoa, and algae. They are enormously diverse nonphotosynthetic and photosynthetic groups. The plastids of

Heterokont, wherever present, are structurally analogous to haptophytes plastids, and are specialized for the chlorophylls *a* and *c*. Heterokont algae contain ecologically significant forms of macroalgae (e.g., kelps) and microalgae (e.g., diatoms) [2,4,9].

4.3.6 DINOFLAGELLATES

One more large and diverse group is dinoflagellates, about half of which have red algae–derived renowned plastids that are specialized for chlorophylls *a* and *c*. Similar to euglenids, the dinoflagellates' main plastid is bounded by three membranes and is eminent for the peridinin pigment. Plastids having peridinin pigments contain unusual genomes, with only a handful of genes, and each one is encoded on a single-gene minicircle. The dinoflagellates are unusual because they lost and replaced their plastid on numerous occasions [2,4,15].

4.3.7 APICOMPLEXA

Apicomplexa are obligate intracellular parasites containing large groups, they have a relict plastid (or apicoplast) that is bounded by four membranes. They are involved in causing certain diseases like malaria. They are of major concern due to possible drug target and evolutionary novelty. The evidence available up to now related this group with red alga and their sister group dinoflagellates [2,4,13].

4.4 TERTIARY AND SERIAL SECONDARY ENDOSYMBIOSIS

The dinoflagellates are capable of doing things differently; therefore it is perchance not unexpected that infrequent diversity of plastids types is recognized as a result of their ability to lose, replace, or gain new plastids. Dinoflagellates have different types of plastids; most of them are peridinin-containing plastids bounded by three membranes, among the other types of plastids; some plastids are derived by uptaking a new primary plastid-containing endosymbiont through serial secondary endosymbiosis [2,4,11].

Some groups are derived by uptaking of a secondary plastid–containing endosymbiont through tertiary endosymbiosis. Tertiary plastids occur in the toxic *Dinophysis*, some are cryptomonad-derived two-membrane plastids; the toxic *Karenia*, which contains two- to four-membrane-bounded haptophyte-derived plastid; and *Kryptoperidinium* and its close relatives, which have diatom-derived, five-membrane plastid having a diatom nucleus of unknown intricacy. Apparently three-membrane-bounded original peridinin plastid of *Kryptoperidinium* is not completely lost, and converted it into an eyespot. Alternatively, in *Lepidodinium* and its close relatives a serial secondary plastid is found, in which the peridinin plastid is apparently replaced by a secondary plastid derived from a green alga, which is surrounded by two membranes. Each of these cases, should they prove to be fully integrated organelles, presents a fascinating problem for plastid targeting and could represent several novel solutions to protein trafficking [2,4,11].

The study done at the molecular level for these organisms is poor, so the impacts of these events on the host and plastid genomes are only starting to emerge. The first targeted gene of *Karenia* was studied recently. It is unexpectedly haptophyte in nature and suggests that plastid replacement perhaps involves replacement of many or most of the nucleus-encoded genes for plastid-targeted proteins. Furthermore, short-term plastids that are stolen from their food source (kleptochloroplasts) are confined in a number of dinoflagellates; therefore sometimes it becomes difficult to discriminate between permanent replaced plastids and these plastids. *Kryptoperidinium* and *Karenia* plastids are proved permanent, nevertheless a big debate is due whether the plastid of *Dinophysis* is kleptochloroplast or truly integrated. The plastid of the Lepidodinium group is virtually uncharacterized. On the whole, the dinoflagellates have an unusual ability to take up new plastids, though these events are rare [2,4,5].

4.5 APICOMPLEXA AND DINOFLAGELLATES PLASTIDS

Within the tree of eukaryotes, the apicomplexa (members of the alveolates) and dinoflagellates are closely related sister groups. Therefore, after the plastid discovery in apicomplexa, the discussion was opened for if the apicomplexa acquired the plastids through the endosymbiosis similar to dinoflagellates. The single origin of all chromalveolate plastids conferred stipulates their common origin, though, some interpreted data display the whole phenomenon to be more complex. On the whole, the dinoflagellate and apicomplexan are odd, and as they are odd in a different way, they are difficult to compare [2,4,11].

The most recently discovered plastid is the apicomplexan plastid, and was something of a surprise given the obligate intracellular nature of these parasites. In this plastid, the characterization of function and trafficking systems have quickly surpassed the understanding of numerous other secondary plastids, driven by the probable importance of the plastid in treating parasites, such as *Plasmodium*, and the novelty of the organelle. Due to the commercial and medical importance of apicomplexan parasites, many tools are present for the study of their plastids. Incongruously, the first completely sequenced nuclear genome of any alga was the malarial parasite *Plasmodium*. Numerous apicomplexan plastid genomes have also been fully sequenced. They have lost all photosynthesis-related genes, nevertheless, they resemble to reduced red algal plastid genomes in content and structure otherwise [2,4].

Phylogenetic studies based on plastid-targeted genes have also inadequately supported a red algal origin for apicomplexan plastids; however, three genes analyses have argued for a green algal origin. The gene trees of RNA polymerase and plastid *tufa* demonstrate a weak affinity between the green alga and apicomplexa; however, in both cases the apicomplexan homologue is greatly divergent and branches with other divergent homologues. The mitochondrial *cox2* was also proposed to indirectly show a green algal origin of the apicomplexan plastid, because the chlorophyte green algae and apicomplexan both have a nucleus-encoded *cox2*, which is split into two subunits, while a single mitochondrion-encoded gene is present in other eukaryotes. Therefore two assumptions can be made for the divergence of apicomplexan plastid from green algae: first that *cox2* split and moved to the nuclear genome and second that the donor was the same alga that gave rise to the plastid. However, further evaluations indicate that the split and nuclear location may not be independent events and that they may have occurred in parallel in apicomplexa and chlorophytes. Additionally, *cox2* is not a plastid protein; therefore even if there were explicit signal for a green algal origin of the apicomplexan *cox2*, it would not be necessarily despicable that the plastid was also green algal [2,4,13].

Dissimilarly, no question arises for the presence of a plastid in dinoflagellates, as about half are well studied as primary producers, producers of toxic blooms, and photosynthetic endosymbionts. On the other hand, because of the lack of related molecular data of dinoflagellates, their three-membrane, peridinin-containing plastid remained mysterious [2,4].

The first genes from the dinoflagellate plastid were characterized recently, and were found to be located on unusual single-gene minicircles relatively than the usual plastid chromosome. Phylogenetic analyses of several of these genes established the already widely believed red algal origin of the peridinin-containing plastids, based on the presence of chlorophyll *c*. Nevertheless, analyses of photosystem genes headed to the hypothesis that the peridinin-containing plastid was not ancestral to dinoflagellates, but was instead attained by tertiary endosymbiosis with a haptophyte [2,4].

The precedent for tertiary endosymbiosis in dinoflagellates certainly advances integrity to this probability; it is clear from the reanalysis of the data that instead of tertiary endosymbiosis, the tree is the result of a base-composition bias. Consequently, on one side we have the chromalveolate hypothesis, which states that a common endosymbiosis results in all red algal secondary plastids derivation, while on the other hand, it is suggested that the apicomplexan plastid is consequential from a green alga, and that the peridinin-containing plastid is derived from a haptophyte [2,4].

With the chromalveolate hypothesis, these alternate suggestions are not irreconcilable. Indeed, all have been plaited together by the proposal that the dinoflagellates and apicomplexa ancestrally enclosed chromalveolate plastids, but that the plastids of one or both were replaced. To see if they are closely related, the most direct test of these alternatives would be to analyze their plastid genomes. Inopportunely, it is impossible to compare the plastid genomes of these two groups. In the plastid genome of any peridinin-containing dinoflagellate, merely 16 genes have been characterized; remaining appears to have been transferred to the nuclear genome. This transfer is important since almost all of the few genes that do remain encode proteins related to photosynthesis that is lost in apicomplexan plastid [2,4].

Consequently, only LSU rRNA and SSU genes have been analyzed from both dinoflagellates and apicomplexa. These two groups cannot be easily compared as both rRNAs are highly divergent, though they do inadequately support a common origin. Although no definitive answers are currently available for the origin of dinoflagellate and apicomplexan plastids, the pretentious description is that the chromalveolates and their plastids share a common ancestor. A repercussion of this would be that the nucleus-encoded genes for both plastid-targeted and cytosolic proteins in these groups should be apt to be meticulously linked within the limits of phylogenetic reconstruction [2,4].

We know little about the fate of nucleus-encoded genes of plastid-targeted proteins when a plastid is replaced, but in the one instance it was studied, the protein comes from the new plastid. Hence, if either one apicomplexan or dinoflagellate plastids have diverged though replacement, then from the aforementioned prediction, a substantial fraction of their plastid-targeted proteins should differ: they should be phylogenetically related distantly. Furthermore, these genes should all come from a single source, as it was recently revealed that the genes for plastid-targeted proteins in a chlorarachniophyte can come from variable sources, with no indication to do anything with the replacement of plastid [2,4,11].

Origin of plastid by primary endosymbiosis between a eukaryotic host and a cyanobacterium was integrated, and reduced endosymbiosis involved the transfer of hundreds of genes cyanobacterium/plastid to the eukaryotic host nucleus (top). The descended groups from this endosymbiosis were Glaucophytes, red algae, and green algae.

Plastids spread to other eukaryotic groups by secondary endosymbiosis (middle), two groups, chlorarachniophytes and euglenids, diverged from green algae, while chromalveolates group evolved from red alga. Protein genes in the plastid genomes of various lineages with secondary red plastids are represented by approximate numbers. The dinoflagellates involving other chromalveolates ancestral plastids were substituted numerous times through tertiary endosymbiosis and in green alga through serial secondary endosymbiosis (bottom).

4.6 CHARACTERISTICS AND PROPERTIES OF ALGAL STRAINS

4.6.1 TYPES OF ALGAE

Algae being thallophytic and photosynthetic include both prokaryotic (Cyanophyta) and eukaryotic forms. Based on its size, algae can be divided into microalgae and macroalgae, while other divisions are on the basis of color (red algae, green algae, yellow green algae, blue green algae, etc.) and form (unicellular, multicellular, filamentous, and colonial). On the basis of origin of algae and their plastids, the four distinct groups of algae include prokaryotes, the cyanobacteria; chloroplast-containing eukaryotic algae with one membrane of chloroplast ER; chloroplast-containing eukaryotic algae surrounded by two membranes of chloroplast ER; and chloroplasts containing eukaryotic algae surrounded by the two membranes of the chloroplast envelope [17,18].

4.6.1.1 Microalgae

Microalgae are single-cell organisms that lack stems, leaves, and roots. In various forms, colors, and simple cell makeup, they are efficient energy producers. About 200,000−800,000 species of microalgae exist in many different genera, of which about 50,000 species are defined. Microalgae are able to grow on wastewater, range from 2 to 200 μm in cell size, and may have autotrophic (require light CO_2 and inorganic nutrients) or heterotrophic (require an organic carbon source) feeding strategies. Algae use photosynthetic pigments and liberate O_2 [19]. Microalgae harvesting period is very short with minimal land requirements, and they are easy to process for biodiesel fabrication as compared to macroalgae. Nutritional composition of microalgae contains lipids, protein, minerals, trace elements, and vitamins. It can be fed to humans and animals by replacing synthetic components. Microalgae serve as thickening operators in desserts, as cleanser, and can be utilized as medications to avoid maladies. People having microalgae as their staple food have low rates of goiter [20,21]. Microalgae are used in the treatments of a lot of illnesses of stomach, throat, and kidney, and hypertension-related issues especially in Asia. Several antioxidants of phenolic and carotenoid nature obtained from algae are currently applied in anticancer and antitumor treatments, most of which are of cyanobacteria origin. Another recent application of microalgae is forensic medicine [22]. Unlike other crops, microalgae do not require the fertilizers and insecticides thereby saving an additional cost. Nutrients requirements for microalgae are minimum and they can be grown in saline, industrial, pond, or wastewaters [23]. Microalgae produce TAGs and hydrocarbons by terpene biosynthesis pathway. Uses and chemical properties of these microalgal products are quite different. As compared to hydrocarbons, that are similar to crude oil and could be added in refining process, the lipids are converted to biodiesel by transesterification process [24,25]. Microalgae oil productivity is evaluated at 20,000−80,000 L/acre that is 7−31 times higher than traditional crops [26]. Furthermore microalgae provide the clean blazing biofuel [27]. Some microalgae examples are as follows (Fig. 4.3):

Botryococcus is a member of chlorophyceae (chlorophyta), and is a unicellular photosynthetic microalga. It is widespread in brackish and freshwaters of all continents. It is regarded as a source of renewable fuel due to its ability to produce large amount of hydrocarbons. Morphologically, colonial cells of *Botryococcus* usually have a gelatinous coat; four daughter cells that are derived from their maternal cell are connected with filaments wall.

Its cell body is spherical, elliptical, or heart-shaped, with a single plate-like or cup-shaped chloroplast containing a pyrenoid. It reproduces asexually by autospores and sexually by fertilization of egg and sperm. The use of *Botryococcus braunii* for the treatment of municipal sewage and conversion of the biomass to energy can reduce global warming and eutrophication [28].

Chlorella: *Chlorella* is a green, spherical, single-celled organism that can occur in both fresh and marine water. The genus *Chlorella* sp. is represented only by

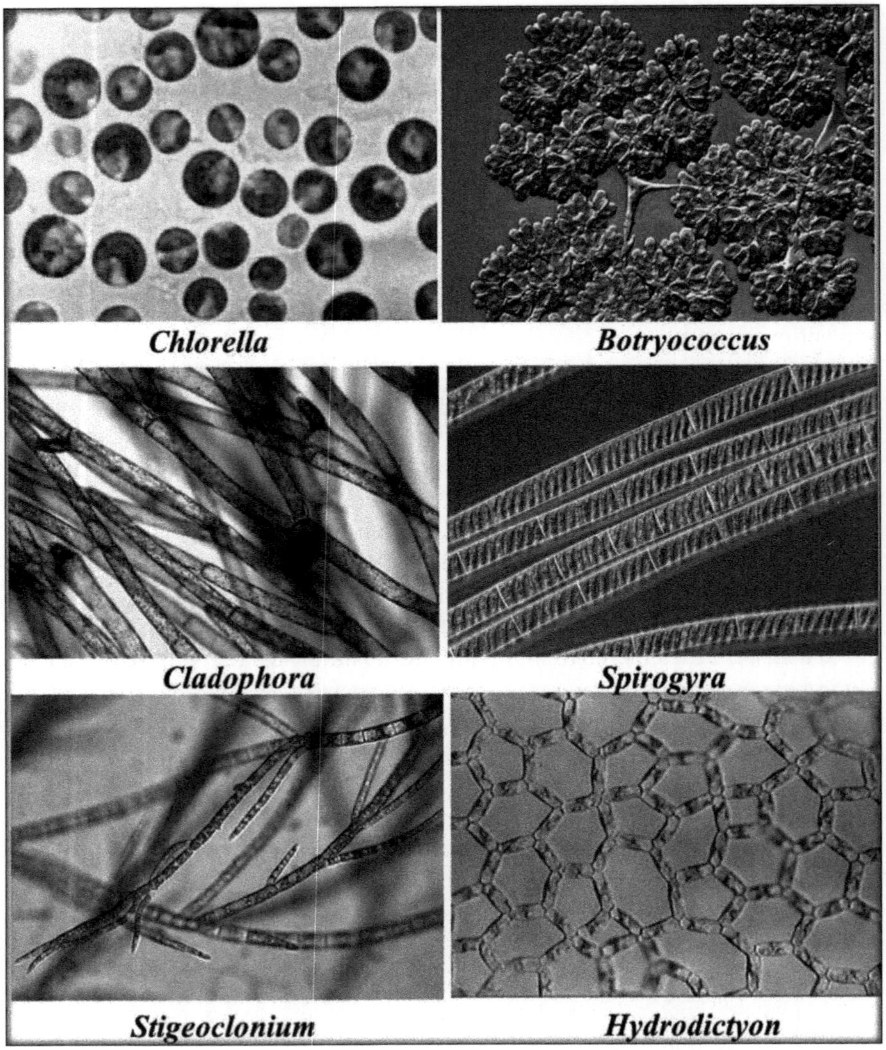

FIGURE 4.3

The morphology of some representative species of microalgae. (A) *Chlorella*, (B) *Botryococcus*, (C) *Cladophora*, (D) *Spirogyra*, (E) *Stigeoclonium*, and (F) *Hydrodictyon*.

Courtesy of wikipedia commons. (A) https://fr.wikipedia.org/wiki/Chlorella, (B) https://en.wikipedia.org/wiki/Botryococcus_braunii, (C) https://commons.wikimedia.org/wiki/File:British_fresh-water_algae,_exclusive_of_Desmidieae_and_Diatomaceae_(1882-1884)_(20230064960).jpg, (D) https://en.wikipedia.org/wiki/Spirogyra, (E) https://en.wikipedia.org/wiki/Stigeoclonium, and (F) https://en.wikipedia.org/wiki/Water_net.

eight species. Their shapes are spherical (globular or ellipsoidal) with microcellulosic wall membrane, ranging from 2.0 to 10.0 μm. Microalgae oil content can range from 10% to 80% in its dried biomass. *Chlorella* sp. may have a variable nutritional composition due to numerous factors. It is important to remember that species, methods of cultivation, and water quality of the production system will influence the microalgae composition. *Chlorella* sp. can be found spread out in many different locations such as freshwater, air, and soil [29]. *Cladophora* is typically fresh/turbulent water, branched, filamentous algae. Its cells have large parietal net chloroplast with distal ends branching that contains 50 nuclei. Under utmost suitable conditions *Cladophora* grow quite profusely. *Cladophora* blooms at phosphate-rich places. It deposits lime in the cellulose strands of wall and therefore strands mats are slimy, rough to touch, and are unable to produce mucilage. Not all but many *Cladophora* show quadriflagellate zoospores, biflagellate isogametes, and isomorphic alternation of generation [30,31].

Hydrodictyon of chlorophyceae are joined in hexagons in the form of colonial net bags, and when it is in colony form it is unable to form new cells for colony. In its place the cytoplasm is accreted by nuclei of the cell and then discriminated in biflagellate zoospore. These zoospores do not leave the parent cell, swim, and arrange hexagonally. They emerge from the parent cell by asexual cycle. In sexual cycle, zoospores resembling cells function as isogametes. A zygote is formed by the fusion of compatible mating-type gametes. The resting cyst zygospore evolves again when the environment becomes favorable. Sexual life history of *Hydrodictyon* is zygotic meiosis, haplontic, and isogamous [32].

Microspora are attached to substratum by holdfast and have unbranched filaments common in small freshwater bodies. Cells are bulbous or barrel-shaped, and H-shaped wall section is the chief diagnostic character. Pyrenoid is absent and dense net-like chloroplast fills the cell. The cell wall is thin, gelatinous, with overlapping H pieces and is difficult to distinguish from *Binuclearia*. They have brown, darkened bands between adjacent cells. Cells are 1–3 times as long as broad whereas diameter is 5–30 μm. Microspores are abundant in cooler times of the year and could be tangled in moss or other vegetation [33–35].

Numerous *Stigeoclonium* species are heterotrichous, and the plants are frequently surrounded in a broad, gelatinous, watery covering. The chloroplast is band-like and often does not fill the entire cell, particularly in the thallus older parts. The aerial part bears branches that terminate in a colorless hair, the degree and nature of the branching depending upon illumination, nutrition, and the rate of water flow [31]. Vegetative reproductions occur by fragmentation, while asexual and sexual reproduction is the same as in *Draparnaldia*, except that there is only one macrozoospore produced per cell. However, a third type of biflagellate swarmer is known in the two species, and hence reproduction in these species is comparable to that in *Ulothrix* [31,36].

Spirogyra is an unbranched filamentous algae present in a wide range of habitats. Its long, green filaments are formed by end-to-end connection of cells. The unbranched bright green filaments have a foamy look. They form slimy threads known as "Water-silk." They are present in the form of free-floating mats as well as occur in periphyton form. *Spirogyra* may occur in response to sources of nutrients in slow-flowing backwaters, pristine mountain rivers, or lowland streams [37]. They are most abundant in spring or autumn, the latter due to germination of a percentage of the spring zygospores. The cell wall is convex, generally less or more flat. According to various investigators, there are two cellulose layers, while others maintain that there is only an inner cellulose layer with an outer cuticle [38]. The whole filament is enclosed in mucilage. In the middle of the large central vacuole, a single nucleus is suspended by means of protoplasmic threads that radiate out to the parietal protoplasm. Each cell of *Spirogyra* contains overlapping spiral or spiral, long ribbon-like chloroplasts. Chloroplasts may be occasionally branched, are T- or U-shaped in cross section, and contain numerous pyrenoids, which project into the vacuole on the inner side, the majority of which arise de novo at cell division. In frozen algal biomass a dense mass is formed at the center of each cell due to chloroplasts shrinkage [39].

4.6.1.2 Macroalgae

Macroalgae are commonly called seaweeds because of their size, attachment to firm substrata, and multicellular construction. Different divisions include Rhodophyta, Phaeophyta, and Chlorophyta. The three major divisions of seaweeds contain taxa that have more fundamental (e.g., cytological, chemical, and life histories) differences between one another than with the vascular plants [40].

Dring [41] has listed 900 Chlorophycean, 997 Phaeophycean, and 2540 Rhodophycean marine species worldwide. Thus there are fewer marine macroalgal than microalgal species. Within the Chlorophyta alone, the majority of species are unicellular or microscopic filamentous forms, whereas in the brown and red algal divisions, almost all the species range from filamentous forms to large thalloid plants. The studies that deal with seaweed biology and taxonomy include Bold and Wynne [42], and van den Hoek et al. [43]. Seaweeds probably evolved in about 900–600 million ybp (years before present) in the late Precambrian. It is interesting that seagrasses, which evolved around 100 m ybp in the Cretaceous, have not “displaced” seaweeds as the dominant marine vegetation. Thus, although primitive in terms of plant body structure, algae have dominated the benthic marine communities since the end of the Precambrian period [44].

Foremost, macroalgal growth can occur by the division of one or more *apical cells* (cells at the tip of a branch, e.g., the brown alga *Dictyota*), which can form an *apical meristem* (a group of apical cells, e.g., most red algae *Eucheuma*). By contrast, seaweed growth may occur throughout the plant and not be restricted to branch tips, and this is called diffused growth (no defined area of cell division, e.g., the brown alga *Hummia*) or *intercalary growth* (defined areas of cell division, e.g., the brown alga *Laminaria*). *Trichothallic growth* is a form of intercalary cell division that produces a hair or filament in addition to the production of new branch tissue (e.g., the brown alga *Sporochnus*). Trichothallic growth results in the production of tufts of hairs at the tip of each branch that may be long lasting or only temporary [45].

Macroalgal construction can range from *filamentous* (one or more rows of cells, e.g., the green alga), to *foliose* (flattened or membranous blades, e.g., the red alga *Halymenia*), to *tubular* (cylindrical or terete, e.g., *Eucheuma*). Tubular construction may be *multiaxial* (central core or *medulla* being filamentous, e.g., *Eucheuma*) or *pseudoparenchymatous* (central core cells cuboidal or spherical in shape but still in filaments, e.g., *Hummi*). Finally, branches can exhibit *parenchymatous* construction, where the cells are isodiametric and not filamentous as found in *Dictyota* [45].

Each species has differences in the nucleotides that make up the DNA and differences in nucleotides that can be used to produce an evolutionary history of the cell. The DNA nucleotides that are most commonly sequenced to produce phylogenies are those of ribosomal DNAs (rDNAs). These rDNA nucleotides make up the genes that code for the rRNAs. The rDNA for the 18S rRNA of the small subunit of the eukaryotic ribosome is the form of rDNA usually sequenced to determine the phylogeny of eukaryotic organisms. The nucleotides coding for all of the ribosomal subunits are encompassed within a single operon and transcribed by a single RNA polymerase. The rDNA for the 5S rRNA is also used in phylogeny studies. Although less nucleotide are in the rDNA coding for 5S rRNA, making it easier to sequence, the data was suspected because of large deviations in the nucleotides. The DNA coding for other molecules, such as ribulose bisphosphate carboxylase/oxygenase was also used in determining phylogeny [44,45]. Some macroalgae are as follows:

Monostroma belongs to the family Monostromaceae of the order Ulvales and the class Chlorophyceae, which consists of 40 species including six species of *Monostroma nitidum*, *Monostroma latissimum*, *Monostroma crassifolia*, *Monostroma arcticum*, and *Monostroma undulatum* (Fig. 4.4). *Monostroma* is an important source of food or chemical derivatives since it contains abundant nutrient components and many active substances, such as sulfated polysaccharide [46].

Enteromorpha is considered to be part of the genus *Ulva* and considered as the most common fouling green algae. The fronds of the *Enteromorpha prolifera* are tubular, though often more or less flattened, from little to much-branched. The arrangement of the cells, longitudinal and transverse rows in the central part of the frond, is characteristic of this species, as are the cylindrical chloroplasts seeming to fill the cell and the usually single, central pyrenoids [47]. *E. prolifera* includes sexual, asexual, and vegetative reproduction [48].

Ulva lactuca is common green algae in the division *Chlorophyta* found in Africa, Australia, Europe, the Caribbean Islands, New Zealand, Indian Ocean Islands, South-West Asia, China, Pacific Islands, and North and South America. A large quantity of *U. lactuca* is washed up on beaches every year Decomposition of *U. lactuca* produces hydrogen sulfide, methane, and other gases that are not environment-friendly. Polysaccharides are the main structural chemicals of alga and *U. lactuca* is no exception as a potentially renewable source of useful polysaccharides [49].

Sargassum, also called gulfweed, is a brown algae found normally attached to rocks along coasts in temperate regions. The Sargasso Sea is characterized by a free-floating mass of seaweed, predominately *Sphaerotilus natans* and *Sargassum*

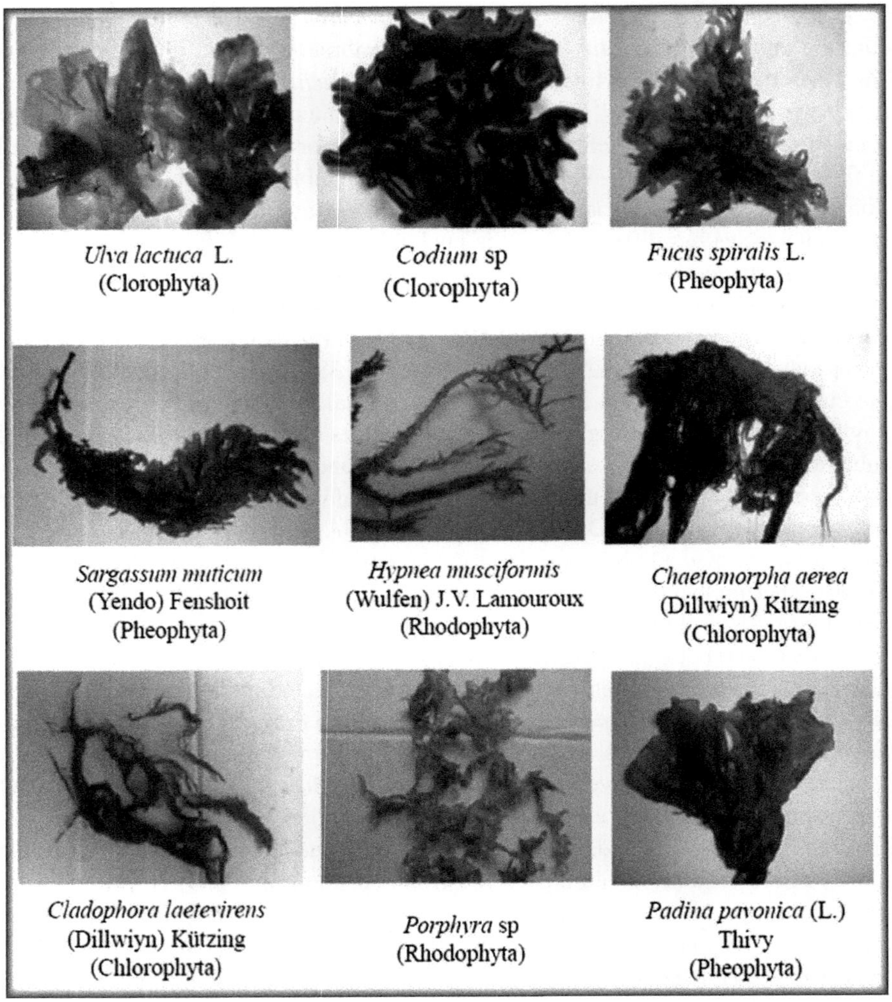

FIGURE 4.4

The morphology of some representative species of macroalgae.

http://www.academicjournals.org/journal/SRE/article-full-text/DBFDC3348776.

fluitans, in the western Atlantic Ocean. Structurally, it is highly branched thallus with hollow, berry-like floats (pneumatocysts), and many leaf-like sawtooth-edged blades. It is used as fertilizer in New Zealand. Most species reproduce sexually, but *S. natans* reproduces only by fragmentation [50].

Fucus spiralis is a species of seaweed, a brown alga (Heterokontophyta, Phaeophyceae), and is the uppermost species of *Fucus* that occurs on the shore. The frond lacks bladders (*Fucus vesiculosus*) and does not have a serrated edge (*Fucus serratus*); it is sometimes spiraled. They are monoecious with both sexes occurring on one plant; most are dioecious with each sex being found on different plants. Some monoecious species may have both sexes in one conceptacle while others may have them in separate conceptacles [51].

4.7 METABOLITES FROM ALGAE

4.7.1 POLYSACCHARIDES

Polysaccharides are utmost bounteous among the characteristic natural bioactives produced by algae, microorganisms, and plants. Polysaccharides are polymeric carbohydrate structures that are formed by monosaccharides linked by different glucosidic linkages. Distinctive functional properties of polysaccharides are because of their structure organized by their building blocks. Sulfated polysaccharides are widespread in nature, and seaweeds are their major natural source. The complex mixtures of molecules show wide variations in their activities and structures. In the cell-wall matrix of various brown seaweeds, fucoidan with sulfate ester and of L-fucose are commonly found. Its fundamental subunit is fucose, and it is extensively studied for its biological activities and gained more researchers' interest for new drugs [52].

Comparison of the *Ecklonia cava* polysaccharide compounds with commercial counterparts (BHA and BHT) found that due to the high amount of polyphenol contents it can suppress cancer cell proliferation. Calcium *Spirulina* (Ca-SP) is a polysaccharide of *Spirulina platensis*, which inhibits tumor invasion and metastasis. The acidic polysaccharides in *S. platensis* are involved in macrophage tumor necrosis factor (TNF)−dependent tumoricidal activity. The *Spirulina* polysaccharides suppress the glioma cell (murine RSV-M) growth via partial regulation of interleukin-17 production and downregulating angiogenesis [52].

Polysaccharides-fucoidans consist of sulfated L-fucose with short of what 10% of different monosaccharides. Fucoidans are present in the cell walls of brown seaweed. Fucoidan has antiproliferative activities, although its complete mechanism is not yet known. Tumor size and growth could be controlled by the polyanionic polysaccharides due to their antiproliferation and anti-angiogenesis actions toward tumor cells. These substances are also involved in anticancer activity through ERK and caspase pathways against human HS-Sultan cells. Fucoidans are also able to enhance the immunomodulatory activity to restrain the diffusion and development of tumor cells as they mediate the destruction of tumors by NK cell and type-1 T-helper (Th1) cell responses [52].

4.7.1.1 High-Molecular-Weight Compounds

The Rhodophyta and Chlorophyta use a common pathway in the synthesis of starch. Myxophycean starch present in Cyanophyta has a similar structure to glycogen. In the Chlorophyta, starch is composed of amylose and amylopectin. It occurs inside the chloroplast in the form of starch gains. Starch has an unusually high content of amylose in Cryptophyta and occurs as grains between the chloroplast envelope and the chloroplast ER. In the Dinophyta also, starch occurs outside the chloroplast in the cytoplasm. The Laminarin occurs as an oil-like liquid outside the chloroplasts, generally in a vesicle surrounding the pyrenoid in Phaeophyceae. Chrysolaminarin occurs in the vesicles outside the chloroplast in Bacillariophyceae, Chrysophyceae, and Prymnesiophyta, and has more glucose residues per molecule than laminarin. In Euglenophyta, Xanthophyceae, and Prymnesiophyta (*Pavlova mesolychnon*), paramylon occurs as water-soluble, single-membrane-bounded inclusion of various shapes and dimensions outside the chloroplast. Paramylon consists solely of 1,3 linked glucose residues, and the molecule is about as large as that of chrysolaminarin. *Acetabularia* (Chlorophyta) has an insulin-like storage product consisting of a series of 1,2-linked fructose units terminated by a glucose end group [53,54].

4.7.1.2 Low-Molecular-Weight Compounds

Chlorophyta and Euglenophyta form sucrose as a reserve product; trehalose is found in Cyanophyta and at low levels in the Rhodophyta. Glycerol glycosides, floridoside, and isofloridoside are widely distributed in the Rhodophyta. Mannitol occurs in Rhodophyta and Phaeophyceae, and is also present in lower green algae, where it replaces sucrose as a photosynthetic product. Free glycerol occurs widely in the algae and is an important photosynthetic product in several Zooxanthellae (endosymbiotic algae in animals) and in some marine Volvocales, especially *Dunaliella*. The ability of algal cells to adjust to changes in the salinity of the medium is an important aspect of the physiology of these cells. In the anterior end of a cell, most algal flagellates have two contractile vacuoles. These flagellates use a combination of contractile vacuoles and osmoregulation to control the water content of their cells [1].

4.7.2 PHENOLICS AND PHLOROTANNINS

The term "phenolic compounds" describe several hundred molecules with benzene ring having one hydroxyl group. Phenolic compounds are generally produced by plants including seaweeds and act as antioxidants by chelating metal ions, prevent radical formation, and improve the antioxidant endogenous system. Flavonoids, tannins, lignins, tocopherols, and phenolic acids represent a diverse class of polyphenol compounds [55].

A widespread group of phenolic compounds is flavonoids; they have a broad spectrum of biological and chemical activities including free-radical scavenging and antioxidant properties. This group contains isoflavonoids, anthocyanidins, flavones, proanthocyanidins, flavonols, and catechins. Phlorotannins, characteristic to brown algae, have up to eight consistent rings. The free-radical scavenging activity of algal phlorotannins is more potent than polyphenols of terrestrial plants [55].

4.7.3 PROTEIN, PEPTIDES, AND ESSENTIAL AMINO ACIDS

Algae are a rich source of proteins. There is an increasing interest in peptides and bio-functional proteins from both microalgae and macroalgae. The protein content of *Spirulina* is higher (65%) than any natural food (fish flesh, 15%–25%; soybeans, 35%; dried milk, 35%; and peanuts, 25%). Protein contents are generally reduced under the nitrogen and phosphorous limitations Klok et al. [56].

4.7.4 LIPIDS

Biofuel from algae is the answer for clean and abundant fuel. Algae have the more potential of amassing lipids than oil crops that they can even secrete in media. The oil productivity of algae is evaluated at 20,000–80,000 L/acre that is 7–31 times higher than traditional crops. Furthermore, microalgae provide the clean blazing biofuel. The traits that dramatically make algae as a solution for fuel are large cell size and fast growth for huge biomass production. Different groups of algal lipids include glycolipids, nonpolar glycerolipids (neutral lipids), betaine, and phospholipids. The degree of unsaturation and chain length of algal lipids is considerably higher than plants [26,27].

4.7.4.1 Phospholipids

Phospholipids have a phosphate group at *sn*-3 position that is linked to a hydrophilic head group to classify the individual phospholipid molecules. Phosphatidylglycerol (PG), phoshatidic acid (PA), phosphatidylserine (PS), phosphatidylcholine (PC), phosphatidylinositol (PI), and phosphatidylethanolamine (PE) are the major phospholipids found in algae (Fig. 4.5). With the exception of PG that occurs in thylakoid membranes, the phospholipids are located in extra-chloroplast membranes. The amphiphilic nature of phospholipids helps in maintaining the selective permeability and structural integrity of cell membranes, while the stability of the photosynthetic apparatus is maintained by PG along with glycolipids [57].

PC is the dominant phospholipid (>60%) in red algae, in green algae the PG accounts for 20%–47%, while in brown algae PC and PE are dominant (11.3%–29.3%). However, PC is often supplanted by DGTA in brown algae and by its homologue DGTS in green algae. DPG and PA are present as minor components while PS and PI are found in appreciable amounts [57].

4.7.4.2 Glycolipids

Glycolipids have 1,2-diacyl-*sn*-glycerol moiety that have mono- or oligosaccharide groups at the *sn*-3 position of the glycerol backbone. Monogalactosyldiacylglycerol (MGDG), digalactosyldiacylglycerol (DGDG), and sulfoquinovosyl diacylglycerol (SQDG) are the typical algal glycolipids. Their structures are shown in Fig. 4.6. MGDG have one while DGDG have two galactose molecules, and they are uncharged at physiological pH. Due to the sulfonic acid residue at sixth position of the monosaccharide moiety, the SQDG have a negative charge. Primarily with reference to the positions of glycolipids in photosynthetic membranes, DGDG are restricted to extraplastidial membranes, whereas MGDG and SQDG are located in thylakoid membranes of the chloroplast [57,58].

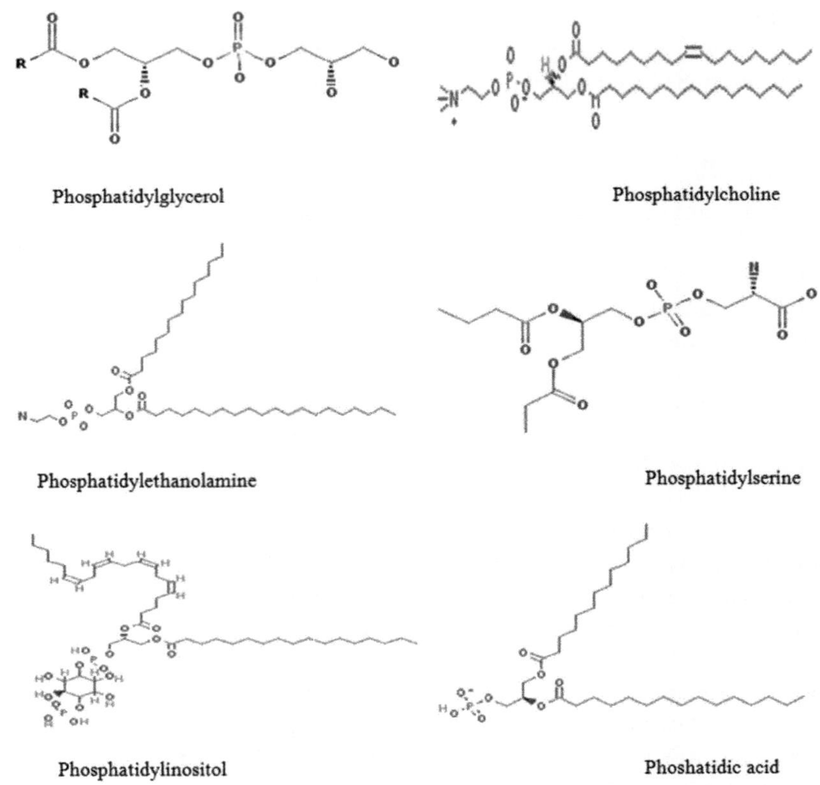

FIGURE 4.5

Chemical structures of phospholipids. (A) Phosphatidylglycerol, (B) phosphatidylcholine, (C) phosphatidylethanolamine, (D) phosphatidylserine, (E) phosphatidylinositol, and (F) phoshatidic acid.

Courtesy of wikipedia commons. *(A). https://en.wikipedia.org/wiki/Phosphatidylglycerol, (B).https://en.wikipedia.org/wiki/Phosphatidylcholine#/media/File:Phosphatidylcholine.jpg, and (F). https://en.wikipedia.org/wiki/Lysophosphatidic_acid.*

Glycolipids have high *n*-3 polyunsaturated fatty acids (PUFA) contents similar to higher plants. In brown algae the DGDG, and in green and red algae the MGDG is the most unsaturated glycolipid; however, the most saturated one is SQDG. The FA composition of these glycolipids revealed the presence of a mixture of eukaryotic and prokaryotic types of FAs [58].

4.7.4.3 Betaine Lipids

Instead of carbohydrate or phosphorus, betaine lipids have a betaine moiety as a polar group that is linked to *sn*-3 position of glycerol by an ether bond with fatty acids esterified at *sn*-1 and *sn*-2 positions. There are three types of betaine lipids present in algae, 1,2-diacylglyceryl-3-*O*-2-β-alanine (DGTA), 1,2-diacylglyceryl-3-*O*-carboxy-choline

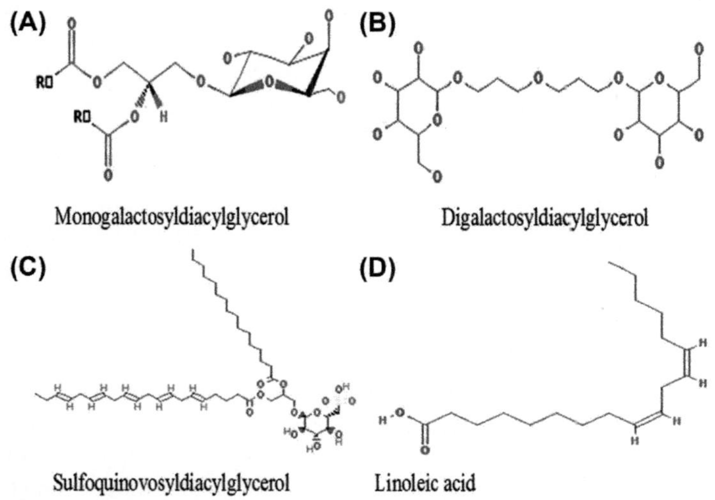

FIGURE 4.6

Chemical structures of glycerolipids. (A) Monogalactosyldiacylglycerol, (B) digalactosyldiacylglycerol, (C) sulfoquinovosyl diacylglycerol, and (D) linoleic acid.

Courtesy of wikipedia commons.

(DGCC), and 1,2-diacylglyceryl-3-*O*-homoserine (DGTS). DGTA abundantly occurs in brown algae whereas DGTS in green algae. Due to the presence of quaternary ammonium group, the betaine lipids resemble PC in structure. In freshwater algae, little DGTS is present beside PC while the DGTS are esterified to long-chain PUFAs at the *sn*-1 and *sn*-2 positions in marine algae. The major FAs of DGTA contain AA, EPA, LA, ALA, myristic, oleic, and palmitic acid. DGTA plays essential part in DGTS and galactolipids biosynthesis in lipid-linked desaturation of FAs and acyl chains redistribution. DGCC is distinctive to Haptophyceae, and it has oleic, palmitic, stearic, AA, DHA, DPA, and EPA as major FAs [57,58].

4.7.4.4 Nonpolar Glycerolipids (Neutral Lipids)

In neutral lipids, *sn*-1, *sn*-2, and *sn*-3 (one, two, or all the three positions) are esterified to the hydrophobic acyl groups that may be saturated or unsaturated, and form monoacylglycerol (MAG), diacylglycerol (DAG), and triacylglycerol (TAG), respectively (Fig. 4.7). Among these TAG is the most predominant neutral lipid accumulated in algae as energy reservoir storage and product. TAG contents in algae vary from 1% to 97%. TAGs are generally produced in light, deposited in cytosolic lipid bodies, and reutilized for polar lipid biosynthesis in the dark. Algal lipids are generally categorized by monounsaturated and saturated fatty acids nevertheless numerous oleaginous algae are likely to amass long-chain PUFAs. TAGs that are rich in PUFA are FAs reservoirs, and in adverse conditions, they donate acyl groups for the biosynthesis of polar lipids, particularly, when de novo syntheses of PUFAs is impaired [57,58].

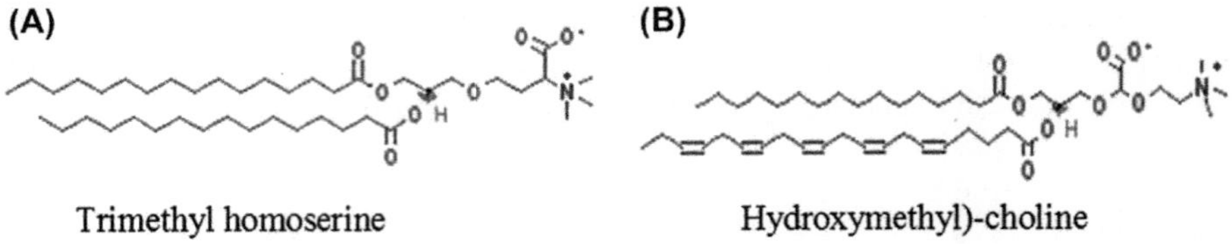

FIGURE 4.7

Chemical structures of glycerolipids. (A) Trimethyl homoserine and (B) (hydroxymethyl)-choline.

4.7.5 TERPENOIDS AND STEROIDS

Sterols are amphipathic compounds that originate in isoprenoid biosynthesis; they are vital auxiliary segments of cell membranes ubiquitously found in all eukaryotic organisms, and regulate the membrane permeability and fluidity. Sterols form a group of triterpenes with a tetracyclic cyclopenta (α) phenanthrene structure and a side chain at C-17. The basic sterol from which other sterol structures are defined is the 5-α-cholestan-3-β-ol (Fig. 4.8). The classification of sterols is based on their biosynthesis or structure, such as 4α-monomethyl sterols have 1 methyl group at C-4, 4,4-dimethyl sterols have 2 methyl groups at C-4, and 4-desmethyl sterols have no substituent at C-4. Their subdivision is based on the number of bonds in ring. In terms of isoprenoid biosynthesis, these sterols are extremely diverse in methyl-D-erythritol 4-phosphate (MEP) and mevalonate (MVA) pathways. The regulation and pathway choice depends on their evolutionary history. For the algal groups that arise from primary endosymbiosis except red algae, *Galdieria sulphuraria* and *Cyanidium caldarium* synthesize sterols through MEP pathway, though cyanobacteria use MVA. Algal groups of haptophytes, bacillariophytes, chlororachinophytes, euglenophytes, chrysophytes, and pelagophytes those arising from secondary endosymbiosis have MEP pathway active in plastids and MVA pathway in cytosol [57].

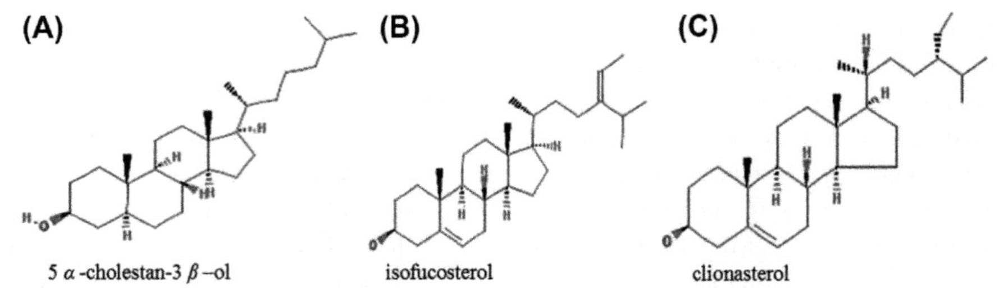

FIGURE 4.8

Chemical structures of steroids and terpenoids of algae. (A) 5α-cholestan-3β-ol, (B) isofucosterol, and (C) clionasterol.

Courtesy of wikipedia commons.

In macroalgae, fucosterol is dominant in Phaeophyta, while cholesterol is the dominant sterol in all the Rhodophyta. Within Chlorophyta orders sterol seems to vary, such as in Siphonocladales and Bryopsidales the clionasterol, and in Ulvales the isofucosterol is the dominant sterol [57].

Among microalgae the diatoms have 4-desmethyl-23,24-dimethyl steroids; the pelagophytic algae have 24-propylidenecholesterol, the stigmasterols; dinophytes have dinosterols; and chlororachinophytes have crinosterols dinostanols. Whereas the haptophytes are characterized by the presence of an unusual pavlovols sterol. The algal sterols have beneficial health impacts, for instance, antihypertensive, anticancer, antiinflammatory, antidiabetic, antioxidant, and hypercholesterolemic responses. 1,9-*O*-β-D-*N*-acetyl-2-aminoglucopyranosyl-19-hydroxy-cholest-4-en-3-one and 19-*O*-β-D-glucopyranosyl-19-hydroxy-cholest-4-en-3-one from *Peyssonnelia* sp. and Fijian red alga are characterized for displaying anticancer properties [57].

4.7.6 VITAMINS

Recent investigations have revealed that synthetic vitamins are less effective than natural vitamins; algae are the natural source of vitamins. The major vitamins of algae include vitamin A, B1, B2, B6, B12, C, E, pantothenic acid, biotin, folic acid, and nicotinate, although the biodispensibility of algal vitamin B12 in humans is contradictory. The overall vitamin contents of algae are 10%–20% higher than plants. For vegetarian individuals, the high levels of iron and vitamin B12 in some algae species make them more appropriate as nutritional supplements. Among different algal strains the *Spirulina* is recommended as a whole food alternative for vitamin supplements, also the nutrient profile of this alga is more potent than any other food. The genotype, environment, and culture conditions affect the vitamin contents of an alga. Vitamin content is amenable by the strain selection, changing the culture conditions, and through mutagenesis [59].

4.7.7 MINERALS

Algae including both microalgae and macroalgae could be used as a food supplement to meet the recommended daily intake of trace elements and essential minerals. Calcium, iodine, magnesium, and potassium are the main minerals of algae. Algae have potential as mineral additives in the animal feed industry. Microalgae comprise relatively higher levels of macrominerals (Mg, Na, Ca, K, and P) and trace elements (Mn, Zn, Al, Fe, and Cu) as compared to willow coppice, miscanthus, and oat straw. As compared to edible land plants, some edible macroalgae have higher amounts of macrominerals (Mg, K, Na, and Ca) and trace elements (Mn, Fe, Cu, and Zn) [60].

Land vegetables that are natural sources of minerals might lack trace elements all of which are present in edible marine seaweeds. Algae contain potassium, nitrogen, phosphorus, and other minerals that are used as soil conditioner and needed for healthy soil growth. Blue green algae are used as a source of calcium and magnesium, whereas *Spirulina* and blue green algae are rich in trace minerals that the modern diets lack. Algae also provide minerals and nutrients for the skin. Algae extracts in the form of rich minerals source are available in North America, Europe, Australia, and Asia, under the name of Aquamin and Algaecal [60,61].

4.7.8 COMMON ALGAL PIGMENTS

Chlorophylls, carotenoids (carotenes and xanthophylls), and phycobilins are three major classes of photosynthetic pigments occurring among the algae.

4.7.8.1 Chlorophylls

Chlorophyll is composed of a porphyrin ring system that is similar to hemoglobin but has a magnesium atom instead of an iron atom. The algae have four types of chlorophyll, *a*, *b*, *c* (*c*1 and *c*2), and *d* (Fig. 4.9). Chlorophyll *a* is the primary

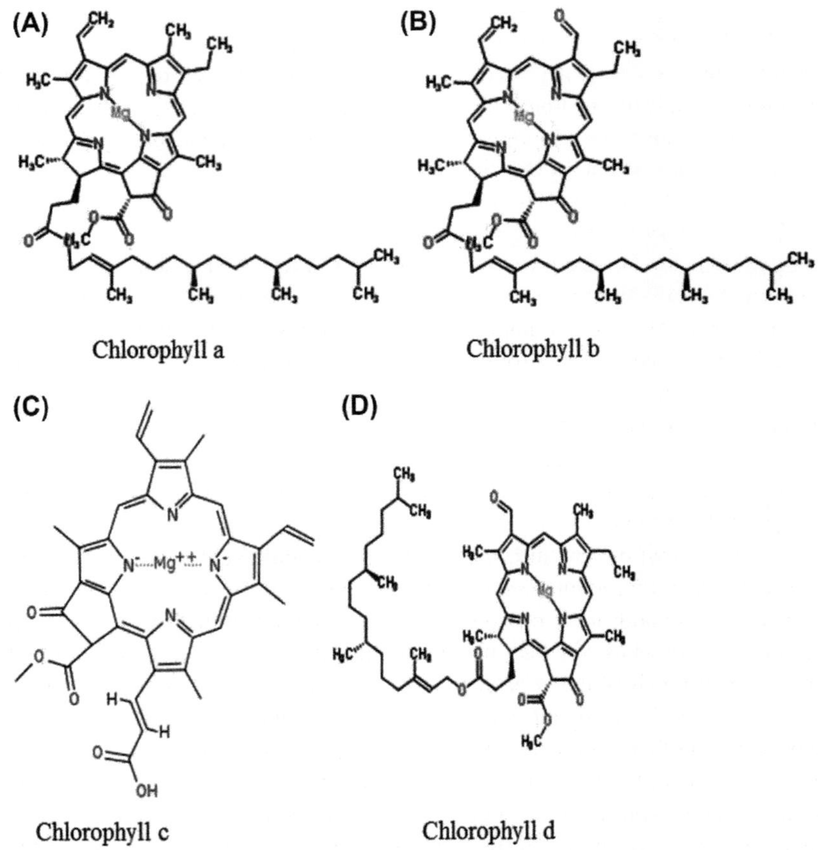

FIGURE 4.9

Chemical structures of algal chlorophylls. (A) Chlorophyll *a*, (B) chlorophyll *b*, (C) chlorophyll *c*, and (D) Chlorophyll *d*.

Courtesy of wikipedia commons.

photosynthetic pigment in all photosynthetic algae and ranges 0.3%–3.0% dry weight. It is soluble in benzene, alcohol, acetone, and diethyl ether but insoluble in water and petroleum ether. Among the two main in vitro absorption bands of pigment, one band near infrared region at 663 nm and the other at 430 nm. Chlorophyll *a* is found in all photosynthetic algae, while the other algal chlorophylls, as accessory photosynthetic pigments, have more limited function and distribution. Chlorophyll *b* is distributed in Chlorophyta and Euglenophyta; it functions photosynthetically to harvest light and transfer the absorbed light energy to chlorophyll *a* [62].

Chlorophyll *a* to chlorophyll *b* ratio varies from 2:1 to 3:1. In in vitro observation, chlorophyll *b* has two main absorption maxima in methanol or acetone, one at 435 nm and the other at 645 nm. The solubility characteristics of chlorophyll *a* are similar to chlorophyll *b*. Chlorophyll *c* is found in the Dinophyta, Cryptophyta, and most of the Heterokontophyta. Chlorophyll *c* has two spectrally different components: chlorophyll *c*1 and *c*2. Chlorophyll *c*2 is always present, but chlorophyll *c*1 is absent in the Dinophyta and Cryptophyta. Chlorophyll *c* to chlorophyll *a* ratio ranges from 5.5:1 to 1.2:2. To photosystem II (PSII), the Chlorophyll *c* ratio functions as an accessory pigment. The pigment is insoluble in petroleum ether and water but soluble in methanol, ethyl acetate, acetone, and ether. Chlorophyll *d* occurs in some cyanobacteria, and has three main absorption bands at 696, 456, and 400 nm [62].

4.7.8.2 Carotenoids

Carotenoids are the pigments responsible for orange, yellow, and bright red hues, normally they are present inside the plastids though may also occur outside in some instances. The energy absorbed by carotenoids can be transferred to chlorophyll *a* for photosynthesis. Carotenoids in different groups of algae are siphonaxanthin (Ulvophyceae and Chlorophyceae); fucoxanthin (Bacillariophyceae, Haptophyta, Chrysophyceae, Phaeophyceae, and Raphidophyceae); peridinin (Dinophyta); violaxanthin and vaucheriaxanthin (Eustigmatophyceae); diadinoxanthin and vaucheriaxanthin (Xanthophyceae); diadinoxanthin (Euglenophyta); and alloxanthin (Cryptophyta) [63,64].

Zeaxanthin, myxol pentosides, β-carotene, and echinenone are the main carotenoids of cyanobacteria. Some species contain oscillol dipentoside, canthaxanthin, and nostoxanthin-like additional carotenoids, while others lack part of these. In comparison to the chloroplasts of algae, the carotenoids might be substantially restructured during the symbiosis of cyanobacteria to eukaryotic cell. Consequently, the composition of carotenoids of cyanobacteria is very different. On the basis of carotenoid composition, the Rhodophyta could be divided into two groups. One is β-carotene and zeaxanthin containing unicellular-type; other is the macrophytic-type that contains additional α-carotene and lutein. Alongside the α-carotene and its acetylenic derivatives, the Cryptophyta also contains specific monadoxanthin and crocoxanthin. Acetylenic carotenoids of diatoxanthin and diadinoxanthin are produced in Dinophyta, Heterokontophyta, and Haptophyta groups, in addition to β-carotene and its derivatives [63].

Among different groups, fucoxanthin and its derivatives are found in Dinophyta, Heterokontophyta, and Haptophyta. Peridinin and its derivatives are generally characterized only in Dinophyta, while 9′-*cis* neoxanthin, violaxanthin, lutein, and β-carotene are also characterized in Chlorophyta, Euglenophyta, and Chlorarachniophyta. The derivatives of lutein such as prasinoxanthin, loroxanthin, and siphonaxanthin are also present in some classes as additional carotenoids. In general, there are two naturally occurring carotenoids: one is the oxygen-free hydrocarbons carotenes and the other is xanthophylls, an oxygenated derivative. β-carotene is the most widespread carotene in algae. Limited information regarding hydroxylation of α-carotene is also available. Among the different xanthophylls, Chlorophyta xanthophylls most closely resemble to those in higher plants. The characteristic color of the golden-brown algae is due to fucoxanthin [63,64].

Allenic carotenoids are very limited in algae and their enzymes and pathways are still not known. *Trans*-neoxanthin may be changed to dinoxanthin, diadinoxanthin, vaucheriaxanthin, peridinin, and fucoxanthin through their chemical structures. Cryptophyta, Chlorarachniophyta, Chlorophyta, Euglenophyta, and Rhodophyta are the algal groups having lutein and its derivatives. For photosynthesis chlorophylls and carotenoids are required to form the pigment–protein complexes in the thylakoid membrane. Whereas in cyanobacteria, the glycosides some carotenoids and myxol are located in cytoplasmic membrane for protection from high-light. The zeaxanthin is found in light-harvesting complexes (LHC) of PSI in some red algae, whereas β-carotene is found in the most divisions of LHC, the reaction–center complexes (RC) of PSI, and also in core LHC and RC of PSII. Alternatively, depending on the classes, in the peripheral LHC of PSII, the bound carotenoids are heterogeneous. In both RCs, the β-carotene may have protective functions, while the carotenoids in the peripheral LHC of PSII may have light-harvesting functions [63,65].

4.7.8.3 Phycobilins

The cyanobacteria and chloroplasts of the Rhodophyta and Cryptophyta have evolved membrane-peripheral antenna complexes containing phycobiliproteins that are responsible for the transfer of light energy to PSII reaction centers. Similar to chlorophyll *b*/*c*/*d*, the phycobiliproteins expand the range of light energy that can be utilized in photosynthesis. As the light courses down the water column, it tends to become blue green, and in comparison to chlorophyll *a* this light is better absorbed by the phycobiliproteins. The phycobiliproteins are the water-soluble red or blue pigments located *inside* (Cryptophyta) or *on* (Rhodophyta, Cyanophyta) thylakoids of algal chloroplasts. Phycobiliproteins are described as chromoproteins (colored proteins) in which the prosthetic group or chromophore is a tetrapyrole known as phycobilin. The prosthetic group is tightly bound to its apoprotein by covalent linkages. Because it is difficult to separate the pigment from the apoprotein, the term phycobiliprotein is used. The basic unit of the phycobiliproteins is formed by two different apoproteins, α and β. To either α or β are attached the colored chromophores. The major "blue" chromophore in allophycocyanin and phycocyanin is phycocyanobilin, and the key "red" chromophore in phycoerythrin is phycoerythrobilin. The classification is based on the absorption spectra of the phycobiliproteins. There are three types of phycoerythrin: C-phycoerythrin in the Cyanophyta; B-phycoerythrin; and R-phycoerythrin in the Rhodophyta. There are three types of phycocyanin: allophycocyanin, C-phycocyanin from the Cyanophyta, and R-phycocyanin from the Rhodophyta. In addition, the Cryptophyta have three spectral types of phycocyanin and three spectral types of phycoerythrin [64,66].

The apoproteins α and β as the basic subunits of a phycobilisome are attached to a chromophore. The α and β subunits are attached to allophycocyanins in the core of the phycobilisome, while in energy transfer pathway the allophycocyanins are closest to chlorophyll. Both α and β subunits are also reattached to phycocyanin in the outer rods. The α and β subunits make α_1 and β_1 hexamers. Hexamers making the core of the phycobilisome and the hexamers of the rods radiating from the core are amassed in pairs. The linker polypeptides are involved in joining these hexamers together. The acidic nature of hexamers and basic nature of linker polypeptides in the assembly of phycobiliproteins shows the importance of electrostatic interactions. The high-molecular-weight polypeptides are involved in the attachment of phycobilisome to the thylakoid membrane (containing the reaction center and associated chlorophylls). The pathway of energy transfer is in intact cells and the efficiency for the energy transfer to chlorophyll *a* in the thylakoids from the phycobilisome exceeds 90%. Chromatic adapters change their pigment components under different light wavelengths. For example, the cyanobacterium *Synechocystis* grown in green light produces phycoerythrin (red in color), phycocyanin (blue), and allophycocyanin (blue green) in a molar ratio of about 2:2:1; when it is grown in red light, the ratio is about 0.4:2:1. The phycobilisome structure changes appropriately, with the peripheral rods having more phycoerythrin hexamers under green light, and less phycocyanin hexamers. The allophycocyanin core hexamers stay the same. Ordered degradation of phycobilisomes occurs in nitrogen-depriving cells, along with progressive degradation of hexamer rod, linker polypeptides, and core peptides. Rapid synthesis of phycobilisomes occurs upon the nitrogen addition to the medium. Phycobilisomes are, thus, an important source of internal nitrogen and offer the nitrogen to algae that have phycobilisomes (cyanobacteria, cryptophytes, and rhodophytes) predominantly in nitrogen-limited ecological open ocean environment [64,66].

4.7.9 TOXINS

All eukaryotic toxic algae contain chloroplast ER around the chloroplast and have evolved through a secondary endosymbiosis. Toxins are classified on the basis of their nature and classification of algal strains, structures are given in Fig. 4.10 [1,3].

4.7.9.1 Classification of Toxins by Algal Strains

4.7.9.1.1 Cyanophyceae (Cyanobacteria)

Cyanobacteria produce neurotoxins and hepatotoxins. *Anatoxin* and *saxitoxin* neurotoxins block the transmission of signal from neuron to neuron. These alkaloids block influx of Na^+ by binding to the voltage-activated Na^+ channels and preventing the generation of an action potential. Hepatotoxins include *microcystin* and *nodularin* that inhibit the protein phosphatases 1 and 2A [67,68].

4.7.9.1.2 *Dinophyceae* (Dinoflagellates)

Yessotoxins, *macrolide toxins*, and *okadaic acid* cause diarrhetic shellfish poisoning. *Dinophysistoxins* and okadaic acid are the inhibitors of protein phosphatases.

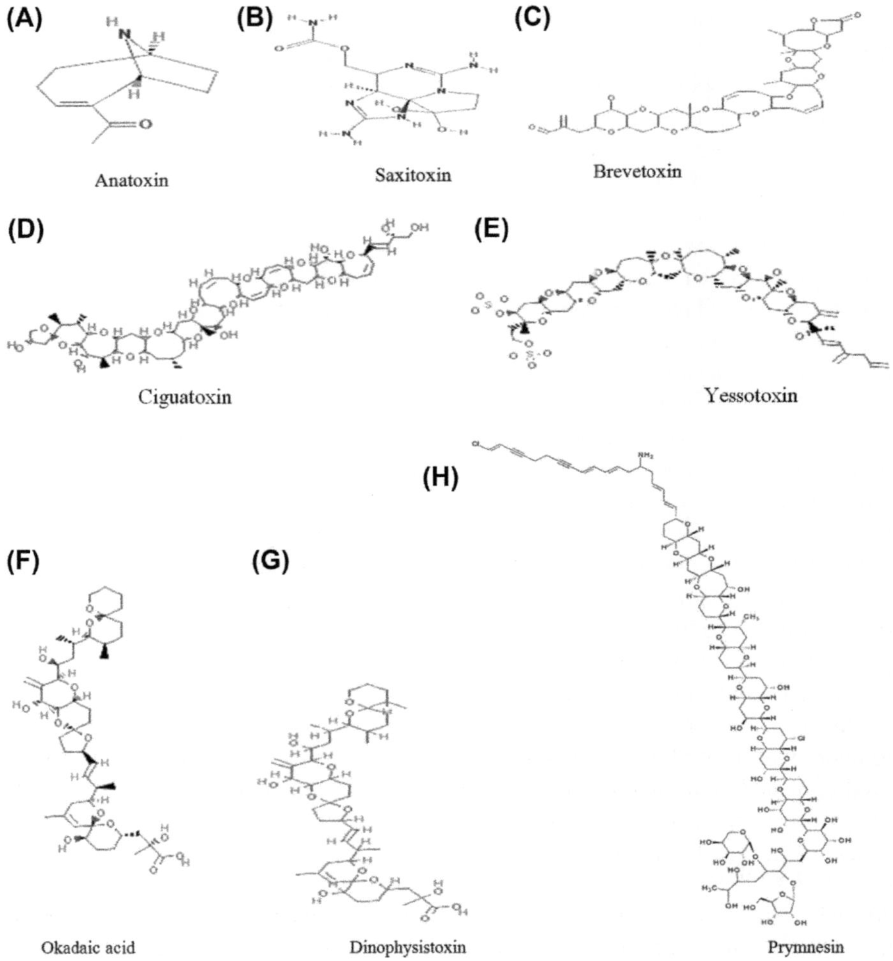

FIGURE 4.10

Chemical structures of algal toxins. (A) Anatoxin, (B) saxitoxin, (C) brevetoxin, (D) ciguatoxin, (E) yessotoxin, (F) okadaic acid, (G) dinophysistoxin, (H) prymnesin, (I) spirolode, (J) gymnodimine, (K) gonyautoxin, (L) pinnatoxins, (M) microcystin, (N) nodularin, (O) pteriatoxin, and (P) maitotoxin.

Courtesy of wikipedia commons. https://en.wikipedia.org/wiki/Anatoxin-a#/media/File:Anatoxin-a.png, https://upload.wikimedia.org/wikipedia/commons/thumb/b/b7/Saxitoxin.svg/2000px-Saxitoxin.svg.png.

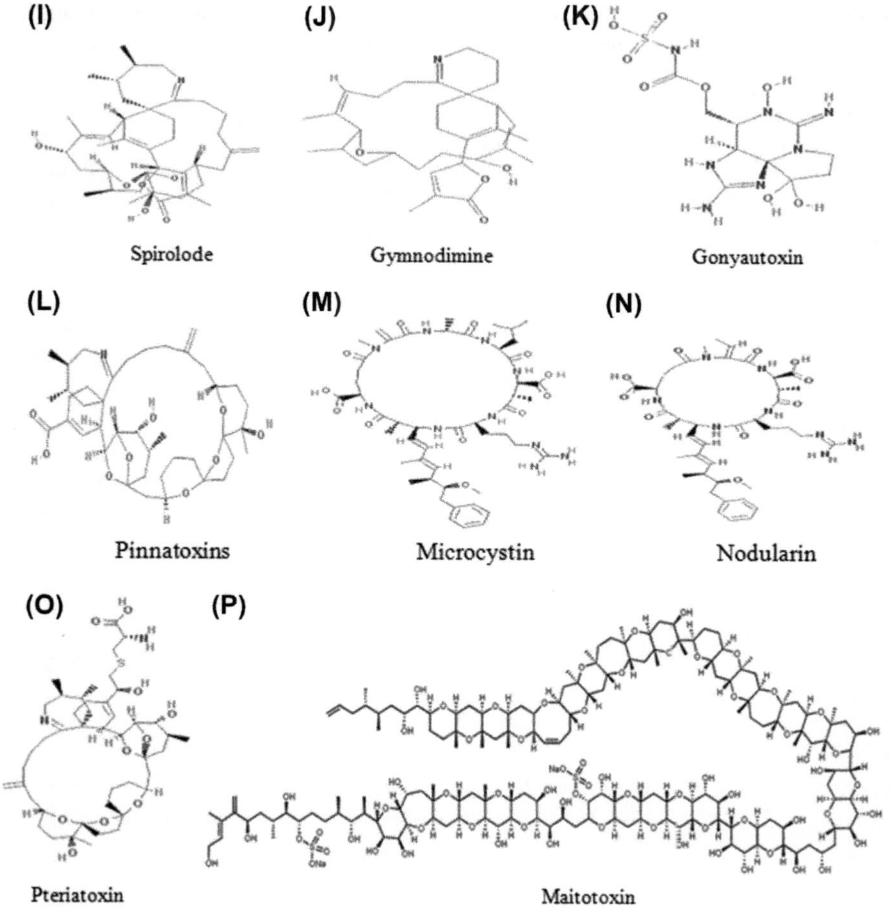

FIGURE 4.10 cont'd.

Yessotoxin may interfere with cyclic AMP in cells causing cytotoxicity. *Brevetoxins*, *ciguatoxin*, and *maitotoxins* cause the ciguatera fish poisoning. These compounds act as ion-channel disrupters expanding the penetrability of the cell layer to positive ions, and causing membrane depolarization [69]. Brevetoxins require a lactone ring for activity. Ciguatoxin and brevetoxin bind to the voltage-sensitive channels, increasing the permeability of the cell membrane, causing membrane depolarization. Maitotoxin initiates Ca^{2+}-channel activation, causing Ca^{2+}-influx and activation of calmodulin. This is followed by promotion of phospholipase A2 activity and ultimately cell membrane disruption. Paralytic shellfish poisoning is caused by saxitoxin and approximately two dozen naturally occurring analogues. The saxitonins bind to voltage-activated Na^{+} channels, block the influx of Na^{+}, and inhibit the generation of an action potential. With no discharge or very little spillage from cells, the saxitoxins are retained principally inside the cells. These highly water-soluble, but stable, compounds must be released during senescence or cell lysis, which coincides with the decline of a bloom. *Pinnatoxins*, *pteriatoxin*, *spirolides*, and *gymnodimine* are cyclic imines of spirolide poisoning [67,70].

4.7.9.1.3 *Bacillariophyceae* (Diatoms)

The neurotoxin domoic acid causes amnesic shellfish poisoning. It functions as a *glutamate agonist* and causes extensive depolarization in areas rich in glutamate receptors (e.g., the hippocampus) in humans. The biosynthesis of domoic acid is inducible and sustained only under stress conditions [68,71].

4.7.9.1.4 Raphidophyceae (Chloromonads)

Members of the Raphidophyceae produce *brevetoxins*, neurotoxins as red tide poisoning that cause the influx of Na^{+}-ions and membrane depolarization by binding to voltage-activated Na^{+} channels [68,72].

4.7.9.1.5 Prymnesiophyceae (Haptophytes)

The phycotoxin *prymnesin* causes hemolysis of red blood cells by disrupting the pores in the cell membrane, causing influx of positive ions, and membrane depolarization [72].

4.7.9.2 Classification of Toxins by Their Chemical Structure

4.7.9.2.1 Amino Acid Analogues

Domoic acid is a glutamate receptor agonist produced by the diatom *Pseudo-nitzschia* spp. [68].

4.7.9.2.2 Peptides

Microcystins and nodularin phycotoxins are produced by cyanobacteria. Microcystin is formed by species of *Microcystis*, *Oscillatoria, Nostoc*, *Nodularia*, and *Anabaena*, while *Nodularia* species produce nodularins [68].

4.7.9.2.3 Purine Derivatives

There are about two dozen analogues of saxitoxins, gonyautoxins, and anatoxin with their potency varying over several orders of magnitude. Saxitoxins are produced by some cyanobacteria and dinoflagellates *Alexandrium* spp., *Pyrodinium bahamense*, and *Gymnodinium catenatum*. Anatoxin is produced by the cyanobacteria *Anabaena*, *Aphanizomenon*, *Oscillatoria*, and *Trichodesmium* [68].

4.7.9.2.4 Cyclic Imines

The spirolides and gymnodimine are produced by the marine dinoflagellate *Alexandrium ostenfeldii*, while gymnodimine is produced by the dinoflagellate *Karenia selliformis* [68].

4.7.9.2.5 Non-Nitrogen Compounds Containing Linear and Macrocyclic Polyethers

Okadaic acid dinophysistoxins and prymnesium dinophysistoxins are produced by the benthic dinoflagellate *Prorocentrum* and the planktonic forms of *Dinophysis*. *Prymnesium* is formed by species of the haptophyte *Prymnesium* [71].

4.7.9.2.6 Non-Nitrogen Compounds Containing Ladder-Frame Polyethers

Brevetoxins are formed by species of the dinoflagellate *Karenia* (*Gymnodinium*) and by the raphidophytes *Chattonella* and *Heterosigma*. Ciguatoxins are produced by species of the tropical and subtropical benthic dinoflagellate *Gambierdiscus*. Yessotoxins are produced by the dinoflagellate *Protoceratium* [71].

4.7.10 DEFENSE CHEMICALS

Algae have evolved chemical defense mechanisms that can be defined as constitutive or inducible.

4.7.10.1 Constitutive Defense

The chemicals involved in defense are present all the time. An example is the toxic "red tide" dinoflagellates that have reduced grazing of copepods. Coccolithophorids is another example. *Emiliana huxleyi* has an increased amount of dimethylsulfoniopropionate (DMSP) when it is grazed by invertebrates. DMSP lyase breaks down DMSP into dimethylsulfide and acrylate. Water containing the latter compound is avoided by grazing protozoans [70].

4.7.10.2 Inducible Defense

The defense chemicals are only produced when the alga is under pressure from a consuming organism. These are costly for the alga to produce and it is more economical for the algal cells to produce them only when in danger of being consumed. Examples of inducible defense chemicals include the increased production of phlorotannins by brown algae, and the increased production of halogenated furones by red algae, upon grazing by invertebrates. Seriochemicals are defense chemicals, which act between individuals of the same species. Pheromones that function in sexual reproduction are examples of seriochemicals. Defense chemicals allelochemicals act between members of different species, e.g., kairomones and synomones. Kairomones are chemicals secreted by a predator that induce changes in the behavior, morphology or life history of the prey. Kairomones can be noxious compounds produced by the prey to act as feeding deterrents via olfactory or gustatory responses of the predator at the preingestive stage. Such contact signals could be sequestered on the surface of the prey cells or released into the medium. Often their main purpose would not be to intoxicate the predator, but to discourage "tasting" or to initiate rapid release of the prey following physical handling and capture. For example, grazing experiments exposing the tintinnid *Favella ehrenbergii* to cells of the toxic dinoflagellate *Alexandrium* resulted in the ciliate swimming in retrograde (avoidance) manner once a threshold of dinoflagellate cells was reached [71,73].

Kairomones are the toxic compounds released from ingested cells that result in physical incapacitation or mortality of the predator. The prymnesiophyte alga *Phaeocystis* contains large amounts of DMSP. Ingestion of the alga releases the intracellular DMSP, resulting in reduced grazing by the heterotrophic dinoflagellates

Amphidinium, *Gymnodinium*, *Oxyrrhis*, and the ciliate *Coxliella*. Post-digestive fecundity is reduced in adult predators by low acute toxicity of kairomones. Post-digestive deterrence would obviously be ineffective for the protection of the individual ingested cell, but group defense would be maintained on a community basis. For example, the ingestion of high concentrations of diatoms by copepods results in the low viability of eggs of the nauplii (larvae) of the copepods. Synomones are produced by prey species to attract predators of their predators at the next level of the food web [73].

REFERENCES

[1] Lee RE. Phycology. Cambridge University Press; 2008.
[2] Bhattacharya D, Medlin L. Algal phylogeny and the origin of land plants. Plant Physiol 1998;116:9−15.
[2a] Schimper AFW. Untersuchungen über die Chlorophyllkörner und die ihnen homologen Gebilde. Jahrb wiss Bot 1985;16:158−72.
[3] Nelson DL, Lehninger AL, Cox MM. Lehninger principles of biochemistry. Macmillan; 2008.
[4] Keeling PJ. Diversity and evolutionary history of plastids and their hosts. Am J Bot 2004;91:1481−93.
[5] Keeling PJ. Chromalveolates and the evolution of plastids by secondary endosymbiosis. J Eukaryot Microbiol 2009;56:1−8.
[6] Canfield DE. The early history of atmospheric oxygen: homage to Robert M. Garrels. Annu Rev Earth Planet Sci 2005;33:1−36.
[7] Maguire DJ, Goodchild MF, Rhind DW. Principles and applications. 1991.
[8] De Clerck O, Bogaert KA, Leliaert F. Diversity and evolution of algae: primary endosymbiosis. Adv Bot Res 2012;64:55−86.
[9] Lane N. Energetics and genetics across the prokaryote-eukaryote divide. Biol Direct 2011;6:35.
[10] Delwiche CF, Palmer JD. The origin of plastids and their spread via secondary symbiosis. Springer; 1997.
[11] Roger AJ, Hug LA. The origin and diversification of eukaryotes: problems with molecular phylogenetics and molecular clock estimation. Philos Trans R Soc Lond B Biol Sci 2006;361:1039−54.
[12] Coelho SM, Simon N, Ahmed S, Cock JM, Partensky F. Ecological and evolutionary genomics of marine photosynthetic organisms. Mol Ecol 2013;22:867−907.
[13] Keeling PJ. The endosymbiotic origin, diversification and fate of plastids. Philos Trans R Soc Lond B Biol Sci 2010;365:729−48.
[14] Keeling PJ. The number, speed, and impact of plastid endosymbioses in eukaryotic evolution. Annu Rev Plant Biol 2013;64:583−607.
[15] Yoon HS, Hackett JD, Pinto G, Bhattacharya D. The single, ancient origin of chromist plastids. J Phycol 2002;38:40.
[16] Cavalier-Smith T. The origin, losses and gains of chloroplasts, origins of plastids. Springer; 1992. p. 291−348.
[17] Barsanti L, Gualtieri P. Algae: anatomy, biochemistry, and biotechnology. CRC Press; 2014.

[18] Kim S-K. Handbook of marine microalgae: biotechnology advances. Academic Press; 2015.

[19] Sharma YC, Singh B, Korstad J. A critical review on recent methods used for economically viable and eco-friendly development of microalgae as a potential feedstock for synthesis of biodiesel. Green Chem 2011;13:2993−3006.

[20] Albert OA, Deborah VR, Gavin WM, Robert J, John JB. South African seaweed aquaculture: a sustainable development example for other African coastal countries. Afr J Agric Res 2013;8:5268−79.

[21] Misurcova L, Machu L, Orsavova J. Seaweed minerals as nutraceuticals. Adv Food Nutr Res 2011;64:371−90.

[22] Sawadogo WR, Schumacher M, Teiten M-H, Cerella C, Dicato M, Diederich M. A survey of marine natural compounds and their derivatives with anti-cancer activity reported in 2011. Molecules 2013;18:3641−73.

[23] Rawat I, Ranjith Kumar R, Mutanda T, Bux F. Biodiesel from microalgae: a critical evaluation from laboratory to large scale production. Appl Energy 2013; 103:444−67.

[24] Serrano-Ruiz JC, Dumesic JA. Catalytic routes for the conversion of biomass into liquid hydrocarbon transportation fuels. Energy Environ Sci 2011;4:83−99.

[25] Naik S, Goud VV, Rout PK, Dalai AK. Production of first and second generation biofuels: a comprehensive review. Renew Sustain Energy Rev 2010;14:578−97.

[26] Demirbas A, Demirbas MF. Importance of algae oil as a source of biodiesel. Energy Convers Manag 2011;52:163−70.

[27] Mutanda T, Ramesh D, Karthikeyan S, Kumari S, Anandraj A, Bux F. Bioprospecting for hyper-lipid producing microalgal strains for sustainable biofuel production. Bioresour Technol 2011;102:57−70.

[28] Banerjee A, Sharma R, Chisti Y, Banerjee U. *Botryococcus braunii*: a renewable source of hydrocarbons and other chemicals. Crit Rev Biotechnol 2002;22:245−79.

[29] Phukan MM, Chutia RS, Konwar B, Kataki R. Microalgae *Chlorella* as a potential bioenergy feedstock. Appl Energy 2011;88:3307−12.

[30] Zulkifly SB, Graham JM, Young EB, Mayer RJ, Piotrowski MJ, Smith I, Graham LE. The genus *Cladophora* Kützing (Ulvophyceae) as a globally distributed ecological engineer. J Phycol 2013;49:1−17.

[31] Chapman VJ. An introduction to the study of algae. Cambridge University Press; 2013.

[32] Van den Hoek C, Mann D, Jahns H. Algae: an introduction to phycology. Cambridge: Cambridge University Press; 1995.

[33] Hoffmann WE, Tilden JE. *Basicladia*, a new genus of cladophoraceae. Bot Gaz 1930: 374−84.

[34] Novis PM. A taxonomic survey of *Microspora* (Chlorophyceae, Chlorophyta) in New Zealand. N Z J Bot 2004;42:153−65.

[35] Junior ON, Spezamiglio DN, Branco CCZ. Taxonomy and ecological distribution of the genus *Microspora* (Microsporaceae, Chlorophyta) in lotic ecosystems of Sao Paulo state, southeastern Brazil. Archiv für Hydrobiologie 2002:39. Supplement volume.

[36] Krishnamurthy V. Reproductive biology of eukaryotic algae, reproductive biology of plants. Springer; 2001. p. 57−95.

[37] Pal R, Choudhury AK. A brief introduction to phytoplanktons, an introduction to phytoplanktons: diversity and ecology. Springer; 2014. p. 1−41.

[38] Johnston H. The biological and economic importance of algae. Part 4: the industrial culturing of algae. *Tuatara* J Biol Soc 1976;22:1−105.

[39] Miller SR, Castenholz RW. The evolution of thermotolerance in hot spring cyanobacteria of the genus *Synechococcus*. J Phycol 2000;36:48.

[40] Santelices B. Patterns of reproduction, dispersal and recruitment in seaweeds. Oceanogr Mar Biol Annu Rev 1990;28:177–276.

[41] Dring M. Photosynthesis of intertidal brown algae during and after periods of emersion: a renewed search for physiological causes of zonation. Mar Ecol Prog Ser 1982;8: 301–8. Oldendorf.

[42] Wynne BH, Bold H. Introduction to the algae. Structure and reproduction. Englewood Cliffs (New Jersey, USA): Prentice-Hall, Inc.; 1985.

[43] Den Hoek V, Mann D, Johns H. Algae: an introduction to phycology. Cambridge: Cambridge University Press; 1995.

[44] Dawes CJ. Marine botany. John Wiley & Sons; 1998.

[45] Lüning K, Yarish C, Kirkman H. Seaweeds: their environment, biogeography, and ecophysiology. John Wiley & Sons; 1990.

[46] Hua W-H, Xie E-Y, Ma J-H. Life history of *Monostroma latissimum*. Acta Bot Sin 2004; 46:457–62.

[47] Li D, Chen L, Zhao J, Zhang X, Wang Q, Wang H, Ye N. Evaluation of the pyrolytic and kinetic characteristics of *Enteromorpha prolifera* as a source of renewable bio-fuel from the Yellow Sea of China. Chem Eng Res Des 2010;88:647–52.

[48] Lin A, Shen S, Wang J, Yan B. Reproduction diversity of *Enteromorpha prolifera*. J Integr Plant Biol 2008;50:622–9.

[49] Tian H, Yin X, Zeng Q, Zhu L, Chen J. Isolation, structure, and surfactant properties of polysaccharides from *Ulva lactuca* L. from South China Sea. Int J Biol Macromol 2015; 79:577–82.

[50] Liu L, Heinrich M, Myers S, Dworjanyn SA. Towards a better understanding of medicinal uses of the brown seaweed *Sargassum* in traditional Chinese medicine: a phytochemical and pharmacological review. J Ethnopharmacol 2012;142:591–619.

[51] Mabeau S, Kloareg B. Isolation and analysis of the cell walls of brown algae: *Fucus spiralis*, *F. ceranoides*, *F. vesiculosus*, *F. serratus*, *Bifurcaria bifurcata* and *Laminaria digitata*. J Exp Bot 1987;38:1573–80.

[52] Wijesinghe W, Jeon Y-J. Biological activities and potential industrial applications of fucose rich sulfated polysaccharides and fucoidans isolated from brown seaweeds: a review. Carbohydr Polym 2012;88:13–20.

[53] Patron NJ, Keeling PJ. Common evolutionary origin of starch biosynthetic enzymes in green and red Algae. J Phycol 2005;41:1131–41.

[54] Stephen AM. Food polysaccharides and their applications. CRC Press; 1995.

[55] Munir N, Sharif N, Naz S, Manzoor F. Algae: a potent antioxidant source. Sky J Microbiol Res 2013;1:22–31.

[56] Klok AJ, Martens DE, Wijffels RH, Lamers PP. Simultaneous growth and neutral lipid accumulation in microalgae. Bioresour Technol 2013;134:233–43.

[57] Kumari P, Kumar M, Reddy C, Jha B. Algal lipids, fatty acids and sterols. Cambridge (UK): Woodhead Publishing Limited; 2013.

[58] Guschina IA, Harwood JL. Algal lipids and effect of the environment on their biochemistry, lipids in aquatic ecosystems. Springer; 2009. p. 1–24.

[59] Aaronson S, Dhawale S, Patni N, DeAngelis B, Frank O, Baker H. The cell content and secretion of water-soluble vitamins by several freshwater algae. Arch Microbiol 1977; 112:57–9.

[60] Sivasubramanian P. Marine non-renewable resources. In: Proceedings of National Seminar on Marine Resources NSMR'16; 2016. p. 52.

[61] Grobbelaar JU. Algal nutrition–mineral nutrition, handbook of microalgal culture: biotechnology and applied phycology. 2004. p. 95–115.

[62] Murakami A, Miyashita H, Iseki M, Adachi K, Mimuro M. Chlorophyll d in an epiphytic cyanobacterium of red algae. Science 2004;303:1633.

[63] Takaichi S. Carotenoids in algae: distributions, biosyntheses and functions. Mar Drugs 2011;9:1101–18.

[64] Pal R, Choudhury AK. An introduction to phytoplanktons: diversity and ecology. Springer; 2014.

[65] van Amerongen H, van Grondelle R. Understanding the energy transfer function of LHCII, the major light-harvesting complex of green plants. J Phys Chem B 2001; 105:604–17.

[66] Samsonoff WA, MacColl R. Biliproteins and phycobilisomes from cyanobacteria and red algae at the extremes of habitat. Arch Microbiol 2001;176:400–5.

[67] Wang D-Z. Neurotoxins from marine dinoflagellates: a brief review. Mar Drugs 2008;6: 349–71.

[68] Molgó J, Girard E, Benoit E. 18 cyclic imines: an insight into this emerging group of bioactive marine toxins, phycotoxins: chemistry and biochemistry. Wiley-Blackwell; 2007. p. 319–35.

[69] Tubaro A, Dell'Ovo V, Sosa S, Florio C. Yessotoxins: A toxicological overview. Toxicon 2010;56:163–72.

[70] Tillmann U, John U. Toxic effects of *Alexandrium* spp. on heterotrophic dinoflagellates: an allelochemical defence mechanism independent of PSP-toxin content. Mar Ecol Prog Ser 2002;230:47–58.

[71] Cembella AD. Chemical ecology of eukaryotic microalgae in marine ecosystems. Phycologia 2003;42:420–47.

[72] Rasmussen SA, Andersen AJC, Andersen NG, Nielsen KF, Hansen PJ, Larsen TO. Chemical diversity, origin, and analysis of phycotoxins. J Nat Prod 2016;79:662–73.

[73] Selander E. Chemical ecology of paralytic shellfish toxin producing dinoflagellates. Kristineberg: Department of Marine Ecology, Göteborg University; 2007.

CHAPTER

5 Algal Polysaccharides, Novel Application, and Outlook

Adil Usman[1], Sundas Khalid[1], Atif Usman[2], Zakir Hussain[1], Yanmei Wang[3]

National University of Sciences & Technology (NUST), Islamabad, Pakistan[1]; SEGi University, Petaling Jaya, Malaysia[2]; University of Science and Technology of China (USTC), Hefei, China[3]

5.1 INTRODUCTION

Polysaccharides are polymeric molecules of carbohydrates, which are composed of long chain monosaccharides units bound together by glycosidic bonds. Upon hydrolysis, they breakdown into their constituent oligosaccharides or monosaccharides. Their structure ranges from highly branched to linear. Algal polysaccharides are obtained from algae. Marine algae contain high amounts of polysaccharides including mucopolysaccharides, storage, and cell wall—structured polysaccharides [1,2]. Polysaccharides have numerous commercial applications in industrial products such as thickeners, emulsifiers, feed, stabilizers, food, and beverages [3]. Polysaccharides content in some seaweed species ranges from 4% to 76% of their total dry weight (Table 5.1). Highest polysaccharide contents are found in species such as *Ascophyllum*, *Palmaria*, and *Porphyra*. *Ulva*, which are green seaweed species have 65% of dry weight polysaccharide content.

From nutrition point of view, seaweeds are low in calories and lipid content. Although they are rich in carbohydrates, this carbohydrate content majorly consists of dietary fibers which human body cannot take up directly. These dietary fibers are considered good for human health as they make an excellent digestive environment in the intestine [4].

Polysaccharides of cell walled species majorly consist of cellulose, hemicellulose, and neutral polysaccharides. They play an important role in physically supporting the thallus in water. These building blocks that support thalli of the seaweed in water are low in strength in comparison to the terrestrial trees and plants. Seaweed species contain cellulose and hemicellulose content up to 2%—10% and 9% of their dry weight, respectively. Lignin is found only in green seaweed species such as *Ulva* up to 3% of their dry weight (Table 5.2).

Storage and cell wall polysaccharides are species-specific. Some examples of their concentrations are given in Table 5.3. Brown algae contain laminarin (β-1,3 glucan), alginic acid, sargassan, and fucodin (sulfated fucose). Red algae contain carrageenans, agars, xylans, water-soluble sulfated galactan, floridean starch (amylopectin-like

Algae Based Polymers, Blends, and Composites. http://dx.doi.org/10.1016/B978-0-12-812360-7.00005-7

Table 5.1 Total Polysaccharide Contents (% of Dry Weight) and Structural and Dietary Fibers (% of Dry Weight) in Seaweed Species of North-West Europe

		Brown Algae					Green Algae				
		Laminaria and Saccharina	**Fucus (%)**	**Ascophyllum (%)**	**Undaria (%)**	**Sargassum (%)**	**Ulva (%)**	**Chondrus (%)**	**Porphyra (%)**	**Gracilaria (%)**	**Palimaria (%)**
Polysaccharides	Total	38%	62	42–64	35–45	4	15–65	55–66	40	36	38–74
		48%	66	44		68	18		41	62	50
		58%		70			42–46		50–76	63	66
		61%							54		
Structural and dietary fibers	Total	36%			35–46	49–62	38		35–49		
	Soluble			38	30	33	21		18		
	Lignin						3				
	Cellulose	10% in stipe 4.5%–9%	2–4.5	2 3.5–4.6			9				
	Hemicellulose						9				

Adapted from Kraan, S. Algal polysaccharides, novel applications and outlook. In: Chang CF., editor. Carbohydrates – comprehensive studies on glycobiology and glycotechnology, InTech; 2012.

Table 5.2 Most Commonly Produced Genera of Seaweeds in Aquaculture

Class	Genus	Uses
Chlorophyta	*Monostroma*	Edible, human food
	Enteromorpha	Edible, human food
Phaeophyta	*Laminaria*	Alginates, edible human food
	Undaria	Edible, human food
	Cladosiphon	Edible, human food
Rhodophyta	*Asparagopsis*	Medical applications
	Gelidiella	Agar, food and medical
	Gelidiopsis	Agar, food and medical
	Gracilaria	Agar, food and medical
	Pterocladia	Agar, food and medical
	Chondrus	Agar, food and medical
	Eucheuma	Carrageenan, human food
	Kappaphycus	Carrageenan, human food
	Gigartina	Carrageenan, human food
	Hypnea	Carrageenan, human food
	Iridaea	Carrageenan, human food
	Palmaria	Human and animal feed
	Porphyra	Human food

Adapted from Kraan S, Algal polysaccharides, novel applications and outlook. In: Chang CF., editor. Carbohydrates – comprehensive studies on glycobiology and glycotechnology, InTech; 2012.

glucan), and porphyrin as a mucopolysaccharide which is located in intercellular spaces. Green algae contain sulfated galactans, xylans, and sulfuric acid polysaccharides (Table 5.3) [1,2].

Polysaccharide contents, both total and species-specific, show extreme seasonal variations. For example, laminarin shows variation throughout the year with no production or very less amount from February to July and maximum production from September to November [5,6]. Fronds of *Laminaria* and *Saccharina* produce

Table 5.3 Global Production of Phycocolloids, Retail Price, Approximate Gross Value, and Usage Percentage in Pharmacy and Food

Product	Global Production (ton/year)	Retail Price US$/kg	Approximate Gross Market Value US$ million/year	Amount Used Food (%)	Pharmacy (%)
Agar	9,600	18	173	80	c.15
Alginates	26,500	12	318	30	5
Carrageenans	50,000	10.5	525	80	10

Adapted from Kraan S, Algal polysaccharides, novel applications and outlook. In: Chang CF., editor. Carbohydrates – comprehensive studies on glycobiology and glycotechnology, InTech; 2012.

maximum amount of mannitol during summer and also in autumn from June to November. Maximum amount of alginic acid in the fronds of *Laminaria* and *Saccharina* species is found from March to June and minimum in the months of September to October [5]. Another study on *Laminaria* species shows that highest amount of alginic acid is produced during winter [6].

Upon further investigation, hydrolysates obtained from some brown algae were found to contain complex mixture of monosaccharides including glucose, galactose, fructose, xylose, mannose, arabinose, and fucose. Glucose content amounted to 65%, 30%, and 20% in *Saccharina*, *Fucus*, and *Ascophyllum*, respectively, of total sugars found in them during autumn [6].

5.2 ALGINATES

E. C. C. Stanford, a British Pharmacist, discovered alginates in 1880s. Their industrial production begun in 1929 in California. They are extracted from brown seaweed. Alginates are available in both salt and acid forms. Salt form is an important component of the cell wall of all brown seaweed species, consisting up to 40%–47% of their total dry algal biomass weight, whereas acid form consists of linear polyuronic acid known as alginic acid [7,8]. Chemically, alginates are chemically anionic in nature containing linear blocks of covalently linked (β 1-4) C-5 epimer α-L-guluronate with D-mannuronate. These blocks could contain one or both of these monomers. The ratio and number of these monomers vary from species to species.

Alginates are used in manufacturing of processed food, cosmetic creams, cardboard and paper, and pharmaceutical industry [9]. It is the most important seaweed colloid in terms of volume. Various grades of alginates are available for different applications having different prices. For example, food grade sodium alginate is valued at US$6.5–11.0/kg while pharmaceutical grade is valued at US$13–15.5/kg. In Korea and Japan, *Laminaria japonica* is in very high demand which has resulted in its high price in addition to importing alginates [10]. Alginates annual growth rate is approximately 2%–3%. Applications such as textile printing account for nearly half of the global market. Medical and pharmaceutical applications account for about 20% value of global market with annual growth rates of 2%–4% based on use of alginates in wound healing applications and continuous developments in controlled release technologies. Paper industry amounts to 5% of the global market [3] (Table 5.4).

5.2.1 CHEMICAL STRUCTURE

Chemically alginates belong to a family of binary linear copolymers. They consist of α-L-guluronic acid (G) and (1 → 4)-linked β-D-mannuronic acid residues (Fig. 5.1). Chemical sequence and composition vary extremely based on algae species, time of the year they are harvested, and various parts of algae [11]. Alginates have extremely high amount of guluronic acid, which is very vital for the mechanical properties of

Table 5.4 Tonnage, Value and Sector Growth of Alginate Markets Over Last Decade

Sector	Demand (%)	Sector Growth (%)	Tons	Value US$ million
Food and pharmaceutical	49	60	13,000	150
Textile printing and technical grades	41.5	Flat	11,000	140
PGA	5.7	−25	1,500	10
Animal feed	3.8	−75	1,000	18
Total	100		26,500	318

Adapted from Kraan S, Algal polysaccharides, novel applications and outlook. In: Chang CF., editor. Carbohydrates – comprehensive studies on glycobiology and glycotechnology, InTech; 2012.

alginate gel, and these can obtained from special algal tissue such as the outer cortex of old stipule of the species *Laminaria hyperborea* [11]. On the other hand, alginates containing up to 100% mannuronic acid can be extracted from bacteria [12]. Alginates can be sequentially post-polymerized through enzymatic modification by

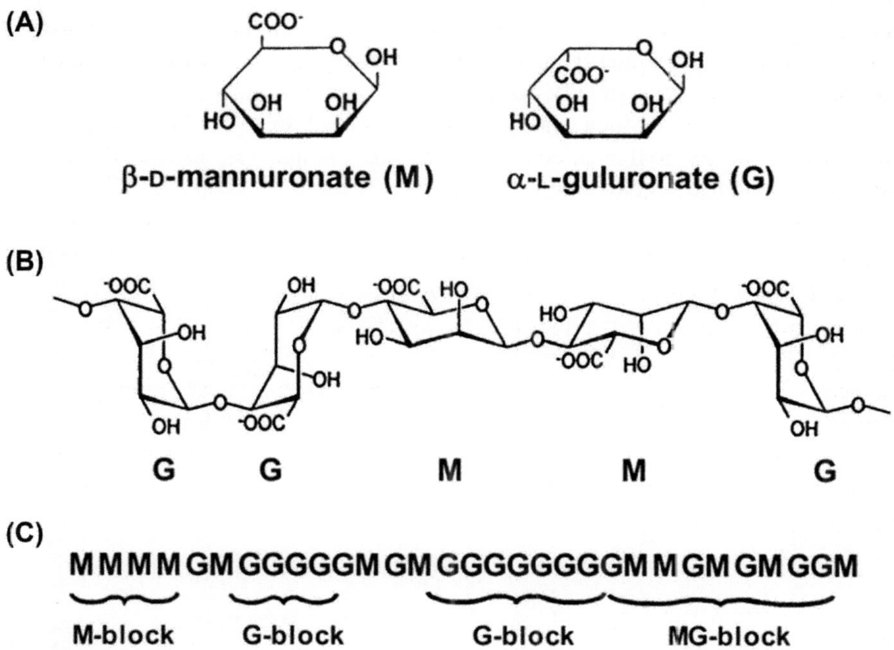

FIGURE 5.1

Structural chemical characteristics of alginates: (A) alginate monomers, (B) chain conformation, and (C) block distribution.

Adopted from Draget KI, Taylor C. Chemical, physical and biological properties of alginates and their biomedical implications. Food Hydrocoll 2011; 25(2):251–6.

applying mannuronan C-5 epimerase isolated from *Azobacter vinelandii* [13]. This enzyme converts M residues to G residues within the polymer chain. *A. vinelandii* encodes a family of seven exocellular isoenzymes, which have the capability to epimerize alginates into a wild range of polymer sequences which range from polyalternating alginates, to very long G-blocks, to short G-blocks [14]. Hence alginates having tailored chemical and physical properties can be obtained.

5.2.2 EXTRACTION METHODS

Global alginate production is approximately 25,500 tons, valuing US$318 million annually. Alginates are mostly extracted from wild brown seaweed. Brown seaweeds grow in cold water and those growing at a temperature up to 20°C are more useful. *Microcystis*, *Laminaria*, and *Ascophyllum* are major commercial sources of brown algae for alginate production. *Lessonia*, *Ecklonia*, *Turbinaria*, *Durvillea*, and *Sargassum* are some other minor sources of alginates from brown algae (Fig. 5.2) [15].

Brown seaweeds found in warm waters are not very suitable for alginate production. They are not commonly used as food because of presence of undesirable chemical compounds such as terpenes in them. A wide variety of species harvested from southern and northern hemispheres are used for alginate production. Argentina, Austria, Chile, Canada, China, Mexico, Japan, Norway, Scotland, the United States, South Africa, and the United Kingdom are the major alginate producing countries. All the species used are generally harvested from natural resources. China, however, cultivates these species which is an expensive way of alginate production. In China *Laminaria* is mostly used as a food, and if surplus amounts are cultivated then it is also used for alginate production.

Hydrocolloids of seaweeds such as carrageenans, agar, and alginate compete with cheaper plant gums (locust bean and guar) and cellulose derivatives [methyl cellulose, carboxy methyl cellulose (CMC)] [3].

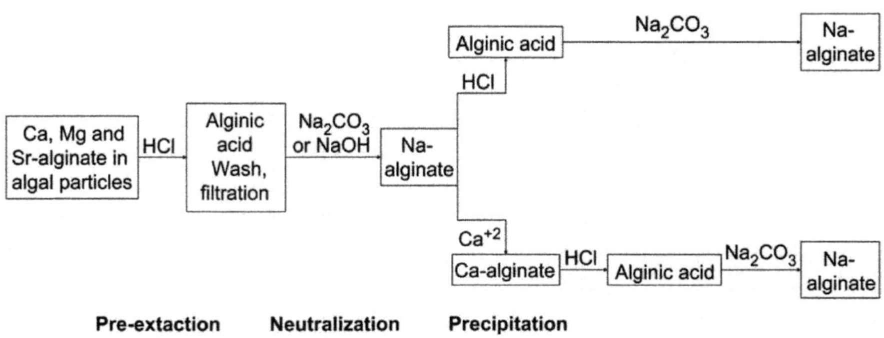

FIGURE 5.2

Schematic diagram showing alginate extracting procedures from algae.

Adapted from Pawar SN, Edgar KJ. Alginate derivatization: a review of chemistry, properties and applications. Biomaterials 2012;33(11):3279–305.

5.2.3 BIOMEDICAL APPLICATIONS OF ALGINATES

Alginic acid decreases the cholesterol concentration in the blood and has an antihypertensive effect. It also prevents the absorption of toxic chemical compounds. One of its major uses is as dietary fiber which helps in maintaining both animal and human health [2,17]. Terrestrial plants lack these dietary polysaccharides which help in protecting against potential carcinogens found in the body. They also clear the digestive system and protect surface membranes of the intestine and stomach. They also help in absorbing the substances such as cholesterol which are then excreted out of the body [18,19], and result in hypolipidemic and hypocholesterolemic responses in the body [20–22]. These effects are accompanied with hypoglycemic response and increase in the cholesterol content in the feces [23,24].

Alginates were also tested against nine bacterial strains including *Staphylococcus*, *Listeria*, *Salmonella*, and *Escherichia coli* and found to be effective against *Staphylococcus* and *E. coli*. Among alginate compounds, sodium alginate showed the most dominant antibacterial properties. It has bacteriostatic as well as bacteriocidic characteristics. Some alginate extracts also possess antiinflammatory effect and up till now no toxic effects have been found in humans. Moreover, alginates with higher molecular weights 50 kDa or more can also prevent hypocholesterolemia, obesity, and diabetes [25]. A clinical study showed that when a drug containing alginic acid was administered to 25%–30% overweight volunteers, their bodyweights decreased considerably [26]. Treatment of diabetes type-II combined with 5 g sodium alginate daily helped preventing a postprandial increase in insulin, glucose, and C-peptide levels with reduced rate of gastric transit [27]. Another study conducted on pigs showed that when their meal was supplemented with 5% alginates obtained from kelp decreased their glucose absorption for about 8 h. Identical studies have been carried out on rats and humans [28]. The binding property of alginic acid has another health effect. Degree of precipitation or gelation is correlated to the binding property of alginic acid in the following order $Ba < Pb < Cu < Sr < Cd < Sa < Zn < Ni < Co < Mn < Fe < Mg$. Alginic acid cannot be digested by intestinal enzymes, which means that if heavy metals get inside the human body, they are either rendered insoluble or get gelated by alginic acid in the intestine and hence the body tissue cannot absorb them [7]. Many patents have been filed in countries such as the United States, Japan, Canada, Belgium, and Germany for the usage of alginic acid along with its derivatives for treating gastritis and gastroduodenal ulcers along with its usage as antiulcer agent [19]. Several drugs containing alginate have shown to effectively suppress postprandial binding of bile acids, acidic refluxes, and duodenal ulcers in humans. Some examples of alginate drugs used by humans are "Gastralgin" (alginic acid, sodium alginate, aluminum hydroxide, magnesium hydroxide, and calcium carbonate), "Algitec" (sodium alginate and cimetidine) and "Gaviscon" (sodium bicarbonate, sodium alginate, and calcium carbonate) [29,30]. Sodium alginate promotes regeneration of stomach mucous membrane, suppressed inflammation, eliminated *Helicobacter pylori* colonies in the mucous membrane, and normalized nonspecific resistance in children aged between 4 and 15 in some clinical trials. Sodium alginate also promotes the restoration of the intestinal biocoenosis [31]. Some other studies also show its positive dietary effect on microbial fauna in feces, changes in concentration of acids and compounds, and prebiotic properties which improves health [32,33].

Sodium alginate is used on septic wounds as a pure powder or mixed with other drugs. Its polysaccharide base stimulates the reparative process by preparing the wound for scarring. It also shows protective coating effects by shielding the mucous membrane and damaged skin irritations from unfavorable conditions. Calcium alginate stimulates the proliferation of fibroblasts and stops the proliferation of keratinocytes and microvascular endothelial cells [34,35]. Alginate–gelatin sponge infused with antibiotics and antiseptics is also reported to have profound wound healing effects [36].

Absorbing hemostatic effect which is exploited in surgery is another use of alginates. Cotton swabs, gauze dressings, and swabs impregnated with sodium alginate solution are externally used for applying onto the bleeding points during surgical abdominal procedures of parenchymatous organs [29]. The effect of alginate on prothrombotic coagulation of blood and activation of platelets is dependent on the ratio between guluronic and mannuronic chains present in the molecule and calcium concentration. Alginate containing a zinc ion has more profound hemostatic effect [37]. Severe gingivostomatitis that is complicated by ulcers and hemorrhagic erosions can also be treated by polaprezinc–sodium alginate suspension [38]. Alginate fibers swell to form a gel-like substance when applied to the tooth surface forming a matrix for coagulation. Tooth cavities, fistulas, and sinuses are also packed by alginate dressings [39].

Moreover, alginates also have anticancer properties [2]. Detoxal, containing calcium alginate, is a bioactive food additive with antitoxic effects against hepatitis. It also decreases lipid peroxidation products and brings the concentration of glycogen and lipid to normal in the liver [29].

5.3 ULVAN

The original terms for ulvan are ulvin and ulvacin which were introduced by Kylin. These terms were introduced in context of different fractions of sulfated and water-soluble polysaccharides of *Ulva lactuca*. Ulvan is now used for polysaccharides obtained from members of Ulvales, especially *Ulva* sp.

5.3.1 CHEMICAL STRUCTURE OF ULVAN

Ulvans are highly charged polyelectrolytes in nature. They are majorly composed of xylose, uronic acid, and rhamnose as their main monomer sugars. They also contain a constituting disaccharide known as aldobiuronic acid (→4)-D-glucuronic acid-(1→4)-L-rhamnose-3-sulfate-(1→) which is common among them. Another constituting sugar is iduronic acid. Ulvans have average molecular weight ranging from 189 to 8200 kDa [40]. About 38%–54% of the dry algal weight of ulvans is the cell wall polysaccharide [40]. Insoluble cellulose-like and water soluble are the two common types of ulvan.

Ulvans are a key component of green algae cell walls consisting up to 8%–29% of their dry weight [41]. Normally ulvans are sulfated polysaccharides in nature. They consist of central disaccharide backbone units in which L-rhamnose 3-sulfate in binded with (1) a D-glucuronic acid (ulvabiouronic acid A) residue, (2) an L-iduronic acid (ulvabiouronic acid B), (3) a D-xylcse 4-sulfate (ulvabios unit A), or (4) D-xylose (ulvabios unit B) residue [42–44] as shown in Fig. 5.3. The most recurring disaccharide units found in ulvans are ulvabiouronic acid 3-sulfate consisting of either iduronic $((\rightarrow 4)\text{-}\alpha\text{-L-Idop-}(1\rightarrow 4)\text{-}\alpha\text{-L-Rhap-3S-}(1\rightarrow))_n$ or glucuronic acid $((\rightarrow 4)\text{-}\beta\text{-D-Glcp-}(1\rightarrow 4)\text{-}\alpha\text{-L-Rhap-3S-}(1\rightarrow))_n$ [41,45].

5.3.2 BIOMEDICAL APPLICATIONS OF ULVANS

The gel-forming mechanism of ulvans is unique among all other hydrogels composed of polysaccharides. This mechanism is quiet complex and is not yet completely understood. Ulvan solutions are very viscous and their viscosity can be compared with Arabic gum. It is also not yet established whether ulvans also contain other functional groups as those of Arabic gum. pH, divalent cations, and boric acid affect the gelling properties of ulvans. Ulvans are also thermoreversible without the presence of thermal hysteresis. The gelling property of ulvan is of great interest in precisely controlled gel formation (by temperature or pH) applications, e.g., releasing of the trapped particles or molecules under specific conditions

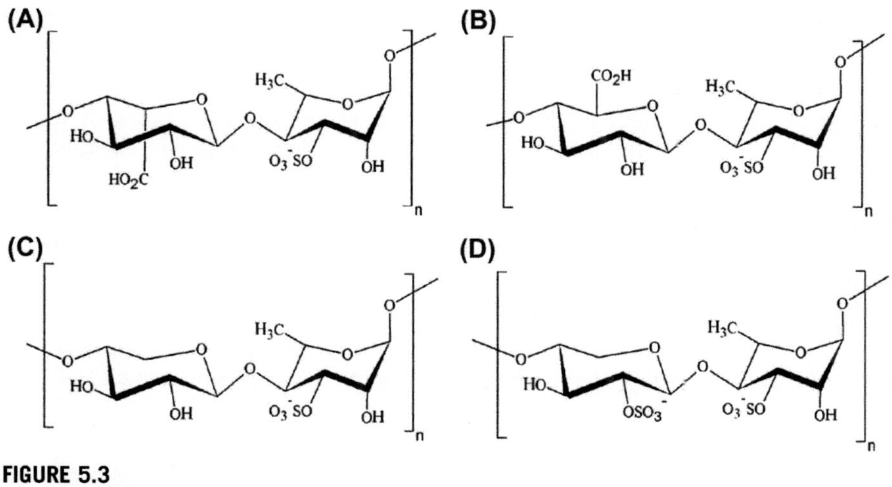

FIGURE 5.3

Disaccharide units in ulvans. (A) Ulvabiouronic acid A, (B) ulvabiouronic acid B, (C) ulvabiose A, and (D) ulvabiose B.

Adopted from Cunha L, Grenha A. Sulfated seaweed polysaccharides as multifunctional materials in drug delivery applications. Mar Drugs 2016;14(3).

[40,47]. Biodegradable hydrocolloid wound dressings which are highly absorbent limit the secretions from wounds and minimize the microbial contamination. Polysaccharide fibers are readily biodegradable when they come in contact with the wound. In prion contamination diseases, such as mad cow disease, macromolecular materials such as ulvan from algal biomass can present a low-cost and effective substitute to meat-derived products such as gelatin. This is because of their gelling and rheological properties which make them a suitable alternative for gelatin and other related compounds [36].

Ulvans are used as a source of sugar for synthesizing fine chemicals. In particular, they are the source of iduronic acid which is a rare uronic acid. Iduronic acid is also used for the synthesizing heparin fragment analogues to have antithrombotic activity. Heparin fragments are obtained by a lengthy synthetic process, which can be replaced by a more cost-effective and natural process, in the form of iduronic acid fragments [40]. Oligosaccharides obtained from *Ulva* can also be used as reference compounds for the analysis of biologically active domains of heparin, such as glycosaminoglycans. The oligosaccharides of ulvans are expected to be used as anticoagulanting agents because other sulfated polysaccharides, e.g., fucan or dermatan sulfate of brown algae, have demonstrated this antithrombotic activity. Regular oligomers of ulvans are believed to provide better tolerance to antithrombinic drugs [49].

Rhamnose which is a rare sugar and a major component of ulvans is used for synthesizing aroma compounds as a precursor. Another example for the use of L-rhamnose is in pharmaceutical industry by forming combinatorial libraries in glycopeptide mimetics. The production of rhamnose from a Japanese species of Codiales, *Monostroma* is patented. Rare sulfated sugars such as xylose-2-sulfate and rhamnose-3-sulfate are also of great interest [41].

The biological properties of ulvan polymers and oligomers also have some potential applications. Some recent studies have shown that ulvans and oligosaccharides of ulvans were able to modify the proliferation and adhesion process of normal and tumoral colonic cells of humans, expression of surface glycosyl markers, and transforming growth factors (TGF) which are related to differentiation of cells [41]. Antiinfluenza activity of ulvan against strain-specific influenza virus is obtained from *U. lactuca* and its oligomers, rhamnan, and rhamnose for treatment of gastric ulcers from desulfated *Monostroma* ulvans have been patented [50,51].

Oligomers obtained from seaweed species such as *Fucus serratus* or *Laminaria* sp. are also being studied as plant elicitors. These natural plant compounds stimulate the natural defense mechanism of plants. Several brown algae–derived products are already being marketed worldwide. The success of these products is due to their availability and size compared to their chemical composition. Ulvans are major contributor with production of 2.8–3.77 meq/g polysaccharide while *Ulva* cell walls bind the heavy metals. Ulvans are also used as bioindicators for detecting heavy metal pollution in coastal areas due to their ion-exchange property [52] (Table 5.5).

Table 5.5 Sugar Composition and Bioactivities of Ulvans

Chlorophytes		
Species	**Sugar Composition**	**Bioactivities**
Ulva pertusa	Rhamnose, xylose, glucose, glucuronic acid, iduronic acid, and sulfate. Traces of mannose, arabinose, and galactose were also present	Antioxidant activities
Ulva spp.	Uronic acid, rhamnose, xylose, galactose, and sulfate groups	Defense mechanism in plants. Activating the jasmonic acid signaling pathway
Ulva lactuca	Rhamnose, xylose, glucose, uronic acid, and sulfate	Inhibit Japanese encephalitis virus (JEV)
		Antihyperlipidemic activities
Ulva clathrata	Rhamnose, glucuronic acid, glucose, xylose, iduronic acid, and galactose (presence of sulfate)	Inhibit viral attachment/entry and cell–cell fusion, in NDV infection
U. pertusa	Sulfated ulvan derivative	Antihyperlipidemia abilities

Adopted from Nazia Auckloo B, Wu B. Structure, biological properties and applications of marine-derived polysaccharides. Curr Org Chem 2016;20(19):2002–12.

5.4 CARRAGEENANS

A group of biomolecules which consist of linear polysaccharide chains having sulfate half-esters bonded to sugar units are known as carrageenans. Their chemical nature allows carrageenans to dissolve in water completely forming a highly viscous solution which remains stable over a wide range of pH. There are three forms of carrageenans kappa, lambda, and iota (κ, λ and ι) in general with each having its own gelling property. ι is mostly obtained from farmed *Eucheuma denticulatum* and κ is mostly obtained from farmed *Kappaphycus alvarezii* [8].

Global carrageenan market is worth US$527 million. Semi-refined grade carrageenan is mostly used as a human food which constitutes about 90% of global market value while the rest is used as pet food in the same form. *Kappaphycus* sp. and *Chondrus crispus* species contain up to 88% and 71% of carrageenan, respectively [53,54]. Carrageenans (E-407) have many food applications including desert mousses, canned food, ice cream, bakery fillings, instant desserts, salad dressings, and canned pet foods. A highly energy-efficient process was developed in Philippines during 1970s to make strong-gelling, low-cost κ and weak-gelling ι carrageenan. These semi-refined carrageenans soon replaced the refined carrageenans as gelling agent in canned meat in pet food. This process requires lower capital investment as compared to standard carrageenan refineries. The semi-refined extracts could be profitably sold for about two-third of the price of conventionally refined carrageenan [15]. Carrageenans also have diversified industrial applications in the form of refined extracts such as in brewing industry for clarifying wine, beers, and honey [3].

5.4.1 CHEMICAL STRUCTURE

Carrageenans are the major component of red algae cell walls constituting 30%–75% of their dry weight [55]. Carrageenans are hydrophilic, high-molecular-weight sulfated polysaccharides consisting of D-galactopyranose bound with alternate α-1,3 and β-1,4 linkages or 3,6-anhydro-α-galactopyranose units. Among 15 different structures of carrageenans [56], 3 are mainly used for commercial purposes, namely κ, ι, and λ. κ-carrageenans are derived from *K. alvarezii* and have the D-galactose-4-sulfate bound to anhydrogalactose structural configuration [57,58]. *Eucheuma spinosum* is the leading source of ι-carrageenan [59] and this polysaccharide is composed of linking D-galactose 4-sulfate with anhydrogalactose 2-sulfate [58,60]. λ-carrageenans are mainly extracted from *Chondrus* and *Gigartina* generas [59,61] and contain 3-sulfate groups in which D-galactose 2-sulfate is bound with D-galactose 2,6, sulfate [58] (Fig. 5.4).

5.4.2 BIOMEDICAL APPLICATIONS OF CARRAGEENANS

It is being reported that carrageenans possess antiviral and antitumor properties [62–65]. Moreover, carrageenan or Irish moss (*C. crispus* and *Mastocarpus stellatus*) has numerous medical applications, dating back to 1830s. In Ireland, carrageenans are still used in making traditional medicinal teas and cough medicines which are used in treating bronchitis, cold, and chronic coughs. Carrageenans are

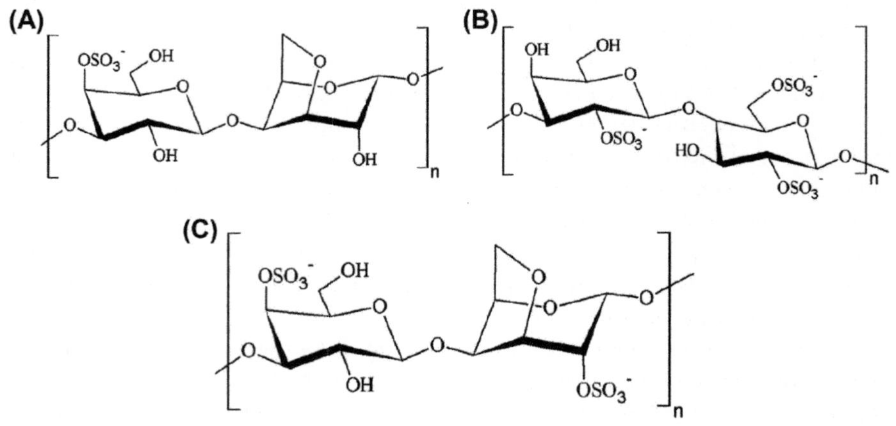

FIGURE 5.4

Structure of (A) κ-carrageenan, (B) λ-carrageenan, and (C) ι-carrageenan.

Adopted from Cunha L, Grenha A. Sulfated seaweed polysaccharides as multifunctional materials in drug delivery applications. Mar Drugs 2016;14(3).

particularly useful in dislodging mucous and having antiviral properties. They are also used as stabilizers and suspension agents in some other drugs, medicinal creams, and lotions. Some other medical applications include treating of bowel problems such as dysentery, diarrhea, and constipation. They are also used in controlling stomach ulcers by making internal poultices [66].

Recent researches showed that application of carrageenan gels obtained from *C. crispus* has certain biocide properties which may prove helpful in blocking the transmission of HIV and other STD viruses including genital warts, gonorrhea, and herpes simplex virus (HSV) as well [67–71]. Because carrageenans do not show significant anticoagulant activity, or levels of cytotoxicity therefore, they are an excellent candidate to be used as vaginal microbicide [72,73]. The results obtained from carrageenans used as a sexual lubricating gel, or condoms lubricated with such type of gel, showed that they are helpful in preventing the spread of human papilloma virus (HPV) which causes genital warts and cervical cancer. In comparison to IC50s, in vitro studies have demonstrated that these carrageenan gels are about 1000 times more effective against HIV-1 and HSV [69,71]. Carraguard, a carrageenan containing vaginal microbicide, was demonstrated to prevent HIV and other STDs in in vitro studies. Population Control Centre began massive clinical trials in two severely affected African countries, South Africa and Botswana, in 2002. Carraguard has also entered into phase-III clinical trials which involve 600 HIV negative and nonpregnant women in Botswana and South Africa in 2003 [74].

Numerous reports exist regarding inhibited platelet aggregation and anticoagulant activities of carrageenans [75–77]. In all the carrageenan types, λ-carrageenan obtained from *C. crispus* shows twice the activity as compared to unfractioned carrageenans, and almost four times the activity of κ-carrageenans obtained from *E. spinosum* and *Eucheuma cottoni*. The most active carrageenan has nearly one-fifteenth of the activity of heparin [75], whereas the sulfated galactan obtained from *Grateloupia indica*, collected from Indian waters, has as potent anticoagulant activity as heparin [78]. Antithrombotic property is the principal basis of the anticoagulant activity of carrageenans. λ-carrageenans depicted increased antithrombotic activity compared to κ-carrageenans because of higher sulfate content. However, the activity of unfractioned material remained between the two of them. The highest anticoagulant activity results in blood were obtained from λ-carrageenans of *Phyllophora brodiaei* [78]. The toxicity of carrageenans depends upon their molecular weight rather than their sulfate content. Moreover, the hypoglycemic effects of carrageenan are also useful in the management and prevention of metabolic conditions such as diabetes [23].

Carrageenans are being used for food as long as 600 years. Due to this long history of use, carrageenans are normally considered as safe by the US Food and Drug Administration (FDA) experts (21 CFR 182.7255) and are also approved as food additive (21 CFR 172.260). The Joint Export Committee of Food Additives of World Health Organization (WHO) has also determined that there is no need for specifying the daily intake limit of carrageenans [79]. Although carrageenans are being used commonly as a food ingredient, they are also being used in experimental animal

Table 5.6 Bioactivities of Three Main Types of Carrageenans

Rhodophytes	
Types of Carrageenans	**Bioactivities**
κ-Carrageenans, λ-carrageenans, ι-carrageenans	Excellent inhibitor of a wide range of sexually transmitted human papillomavirus
	Lubricant gels for vagina and condom coatings having antimicrobial effects
κ-Carrageenans, ι-carrageenans	Act as a vaginal microbicide
κ/λ-Carrageenans	Lowers blood cholesterol
ι-Carrageenans	Potent against dengue virus (DENV) replication
λ-Carrageenan	Antiviral activities against HSV-1

Adopted from Nazia Auckloo B, Wu B. Structure, biological properties and applications of marine-derived polysaccharides. Curr Org Chem 2016;20(19):2002–12.

research to induce ulceration or pleurisy of the colon [80]. Moreover, carrageenans also cause reproducible, acute inflammatory reaction and remain as a standard irritant to examine acute inflammation to study the effect of antiinflammatory drugs. There are two test systems widely used for evaluating nonsteroidal cyclooxygenase activity and antiinflammatory drugs: (1) carrageenan-induced rat paw edema assay, and (2) carrageenan air pouch model [81]. The role of carrageenans in proliferation of colorectal ulcer formation is quiet controversial and apparently seems to be dependent upon the molecular weight of carrageenan. Carrageenan classification is degraded by the international agency for research on cancer as a possible human carcinogen. On the other hand, native carrageenan remains unclassified in comparison to a human colon cancer causative agent, and it has been mentioned that it contains status as GRAS [82,83] (Table 5.6).

5.5 FUCOIDAN

Fucoidans belong to a group of polysaccharides fucans, which are chiefly composed of sulfated L-fucose. Other monosaccharides compose less than 10% of them. They are only found in brown seaweed cell walls, and no other algae or higher plants contain them [84].

Fucoidans are major cell wall reinforcing molecule. They are also associated with protecting the seaweed from desiccation when it is exposed to air at low tide. Fucoidans have interesting bioactivities and were first isolated about a century ago by Kylin [85]. *Fucus vesiculosus* species contains the highest contents (up to 20% of their dry weight) of fucoidans. Fucanoids, on the other hand, constitute over 40% of the dry weight of isolated algal cell walls. Fucanoids can also be separated easily by either using an acid solution or hot water [84]. They are viscous in nature and the viscosity increases at lower concentrations. They are also susceptible to breakdown by diluted bases and acids.

5.5.1 CHEMICAL STRUCTURE OF FUCOIDANS

Algae produce fucoidans as complex and heterogeneous polysaccharides that contribute in forming intercellular mucilage. The structural complexity of fucoidans varies in the degree of type of linkage, substituents, branching, and sulfation. Finer structures depend upon the source of polysaccharides. Although the composition of fucoidans varies with geographical origin and species, they always contain fucose, galactose, mannose, arabinose, xylose, and sulfate with small proportions of uronic acids. There are two types of homofucose backbones, of which algal fucoidans contain one. One of them is repeated (1 → 3)-linked α-L-fucopyranosyl residues or alternating (1 → 3) and (1 → 4)-linked α-L-fucopyranosyl residues [86]. Sulfonato and acetyl groups are also present on α-D-glucuronopyranosyl, α-L-fucopyranosyl groups along with some other sugar residues which may be present at O-2 or O-4 of L-fucose units of the constituting backbone. First type of backbone is found in fucoidans extracted from seaweeds *Chorda filum*, *Laminaria digitata*, *Cladosiphon okamuranus*, and *Saccharina latissima*. Fucoidans containing second type of backbone are found in *Ascophyllum nodosum*, *Fucus distichus*, and *Fucus evanescence* (Fig. 5.5).

5.5.2 BIOMEDICAL APPLICATIONS OF FUCOIDANS

Marinova Pty Ltd., a Tasmanian company, supplies fucoidan extract and their derivatives in commercial volumes with up to 95% purity. All fucoidans obtained from *Lessonia* sp., *Undaria* sp., *Laminaria* sp., *Macroystis* sp., *Fucus* sp., *Ascophyllum* sp., *Durvillea* sp., *Cladosiphon* sp., *Alaria* sp., and *Ecklonia* sp. are Kosher and Halal certified. Marinova extracts fucoidans from a vast variety of species and also

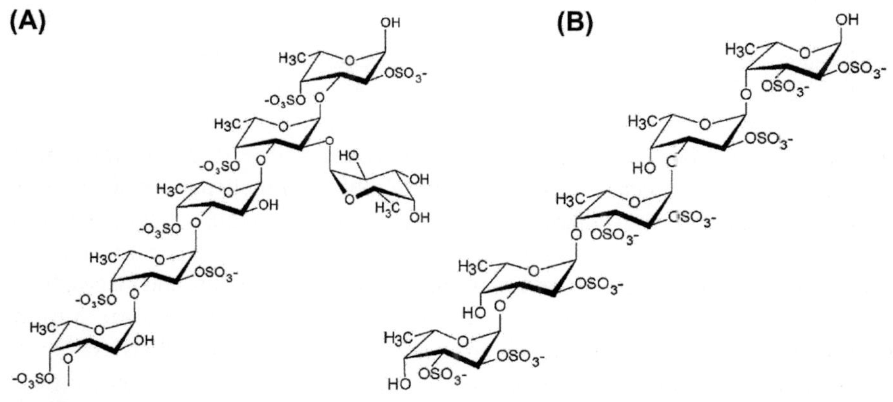

FIGURE 5.5

(A) Fucoidan structure from *Chorda filum* and (B) oligofucoidan from *Cladosiphon okamuranus* (R = sulfated fucose).

Adopted from Ale MT, Meyer AS. Fucoidans from brown seaweeds: an update on structures, extraction techniques and use of enzymes as tools for structural elucidation. RSC Adv 2013;3(22):8131–41.

provides characterized fractions for nutraceutical applications, investigational research, or as a cosmetic. Different therapeutic profiles of fucoidan exist primarily because of its molecular structure. Martitech coldwater extraction process has also been developed by the company in which integrity of fucoidans is maintained and nature-equivalent, high-molecular-weight fucoidans with optimal bioactivity are produced. Solvent-based extraction is commonly used which causes degradation of fucoidans and limits their activity in biological assays.

The physiological role of fucans in algae is yet to be understood, but they possess various biological properties having potential to be explored in human health applications [84]. There is a long list of fucoidans' bioactivity related to human health. Fucoidans extracted from seaweed species such as *Laminaria* and *Undaria* show anticancer, antiviral, and anticoagulant properties [88,89].

Fucoidans are originated from plant matter and are less likely to have infectious agents such as viruses; hence fucoidan preparations are also proposed as an alternative to injectable anticoagulant heparin [84]. When rats were administered with 300 mg/kg of their body weight with fucoidans, no toxicological effects were seen, but when the dose was increased threefold, significantly long blood-clotting times were observed [90]. The biological activity of sulfated polysaccharides (e.g., anticoagulant or antioxidant) is not only related to sulfated ester content (role in charging of the molecule) and molecular weight but also to fucose and glucuronic acid contents, along with the positioning of sulfate group on sugar residues [84,90,91]. Fucoidan requires a long sugar chain to be able to bind with thrombin which is a natural coagulation protein in blood; hence a large molecular weight of fucoidans is needed to achieve this anticoagulant activity. Some studies, reported the required fucoidan' molecular weight for anticoagulant activity as 100 kDa, whereas in others report it to be 1600 kDa [39]. Fucoidan extracted from *Lessonia vadosa* shows good anticoagulant activity at 320 kDa in comparison to that showed by a small depolymerized fraction having molecular weight of 32 kDa [92]. Hence it is concluded that for having certain specific activities, fucoidans require some structural features.

Fucoidans stimulate the immune system in different ways. Several important biological effects are analogues to fucoidan's ability to alter the surface properties of cells [93]. Fucoidans present in dietary brown seaweed if taken orally might have the protective effects by directly inhibiting the viral replication and stimulates the immune system both adaptive and innate [94]. In immunosuppressed mice, fucoidans are found to restore their immune functions, by acting as an immunomodulator directly on B cells, T lymphocytes, natural killer (NK) cells, and macrophages [95]. This mechanism promotes the recovery of immunological function in irradiated cells [96]. Moreover, in vitro production of interferon-γ (INF-γ) and interleukin-1 (IL-1) is induced by fucoidans, and also the primary antibody response is promoted in red blood cells of sheep in vivo [97].

The antiviral activity mechanism of fucoidans involves inhibiting the viral-induced syncytium formation by inhibiting the viral sorption [98]. Sulfate is needed to show the antiviral activity and, in this regard, the sulfate located at C-4 of (1-3)-linked fucopyranosyl unit is believed to be playing a very important role in antiherpetic activity shown by fucoidans [98]. Antiviral properties exhibited by fucans were reported previously, e.g., inhibition of HSV, infection of HIV [71,98,99], adenovirus III, coxsackie A16, coxsackie B3, poliovirus III and ECH06 virus [100], bovine viral diarrhea virus, and cytomegalovirus [99].

Fucoidans are also known to have antitumor effects but their mode of action is not yet fully understood. A study conducted by Alekseyenko et al. showed that when 10 mg/kg fucoidan when administered in mice having Lewis lung adenocarcinoma transplanted on them, moderate antimetastatic and antitumor effects were seen [101,102]. The polyanionic polysaccharides of fucoidans have antiproliferation and antiangiogenesis effects against tumor cells. They reduce the tumor size and inhibit tumor growth [92,103,104], inhibit the adhesion of tumor cells on various substrates [105], and show direct anticancer activity against human HS-Sultan cells by ERK and caspase pathways [106].

Fucoidans restrain the diffusion and development of tumor cells in addition to directly inhibiting the growth of tumor cells. This is achieved by enhancing the immunomodulatory activities of the body. This is due to the fact that fucoidans mediate tumor destruction via NK cells or type-I helper T cell (Th1) responses [107]. Fucoidans at a dose of 25 mg/kg show the potential to neutralize the toxic effects of cyclophosphamide which is used to treat autoimmune disorders and different types of cancer in humans [101].

It has been reported in many studies that fucoidans have potential to be used as antiinflammatory agents. In a study conducted on mice, it was shown that less severe symptoms of *Staphylococcus aureus*—induced arthritis were developed in them, while on the other hand delayed phagocyte recruitment and lower level of bacterial clearance were also observed [108]. Moreover, a fucoidan injection administered to sensitized mice reduced contact hypersensitivity reactions in them before the heparin challenge [109]. In a meningitis model, recruitment of leukocytes into cerebrospinal fluid is reduced by fucoidan [110] as IL-1 production in similar modals [111].

In L- or P-selectins, which can interact with sulfated oligosaccharides having therapeutically beneficial physiological consequences, fucoidans can act as ligands [112]. Selectins are a group of sugar-binding proteins called lectins that interact with oligosaccharides attached to the cell surface during the rolling and margination of leukocytes before their migration, extravasation, and adhesion at the site of infection [113].

In some in vitro experiments, fucoidan presented excellent antioxidant activity showing it has significant natural antioxidant potential. Recent studies demonstrated fucoidans in the form of sulfated polysaccharides, extracted from marine algae *Ulva pertusa* [114,115], *L. japonica* [116], *Porphyra haitanensis* [117], *Ecklonia kurome* [118], and *F. vesiculosus* [119], to possess antioxidant activity. However, there are only a few studies that give the details about the relationship between antioxidant activity and structure of sulfated polysaccharides extracted from marine algae. Fucans show lesser antioxidant activity in comparison to fucoidans [120]. As mentioned above, the molecular weight and the ratio of fucose/sulfated contents are the effective indicators of antioxidant activity shown by fucoidans [121]. Fucoidans have the potential for preventing aging process and free-radical mediated diseases such as Alzheimer. Three fractions of sulfated polysaccharide were isolated by anion-exchange chromatography from a commercially important algae *L. japonica* and their antioxidant potential were examined in an in vitro study along with their chelating ability, hydroxyl radical scavenging ability, and reducing power [121]. All these extracted fractions were found to be more effective than unprocessed fucoidans. Two galactose-rich fractions had the most prominent scavenging activity against superoxide generated in the PMS-NADH cycle and hydroxyl radicals having EC_{50} values of 1.7 μg/mL and 1.42 mg/mL, respectively. One of the fractions also demonstrated the highest ferrous ion chelating ability at the concentration of 0.76 mg/mL [121]. Furthermore, heterofucnas obtained from *Padina gymnospora* and homofucans from *F. vesiculosus* showed inhibitory effects in the formation of superoxide and hydroxyl radicals [120]. A highly growing area of research in pharmaceutical biotechnology is the healing of dermal wounds by macromolecular agents such as natural polymers. Fucoidans have demonstrated modulating the effects of growth factors by mechanisms which are thought to be quite similar to heparin action. Fucoidan—chitosan blend films promote the contraction and reepithelization of wounds. Fucoidan—chitosan blend films are also suitable for using in hydrogel formulations for treating dermal burns [122] (Table 5.7).

Table 5.7 Bioactivities of Some Fucans

Phaeophytes		
Species	**Sugar Composition**	**Bioactivities**
Ascophyllum nodosum	Fucose, sulfate activity	Antithrombotic activities
Fucus vesiculosus	L-fucose (presence of sulfate groups on C-4), traces of galactose and xylose, one unidentified sugar	Anticoagulant activities
	A homopolymer of L-fucose	Antiinflammatory activities
	Fucose (presence of sulfate groups on C-4), glucose, galactose, xylose, uronic acids	Antioxidant activity
Turbinaria conoides	Fucose, sulfate groups	Anticancer activity
	D-galacton group	
Undaria pinnatifida	α-fucose, β-galactose, xylose, mannose	Antitumor activity
Alaria angusta	Fucose, galactose residues	Anticancer activity
Cocophora langsdorfii	α-L-fucopyranose (presence of sulfate at C-2 and C-4)	Anticancer activity
Sargassum cristaefolium	Fucose, galactose	Immunomodulating activity

Adopted from Nazia Auckloo B, Wu B. Structure, biological properties and applications of marine-derived polysaccharides. Curr Org Chem 2016;20(19):2002–12.

5.6 LAMINARIN

5.6.1 ISOLATION, SOURCE, AND OCCURRENCE

Laminarin (also known as laminaran or leucosin) was first isolated by Schmiedeberg in 1885 from the family Laminariaceae of the phylum Phaeophyta (brown algae) [123]. It is basically a storage polysaccharide obtained as a product of photosynthesis and is ubiquitously found in vacuoles of cells. Laminarin is mostly acquired from *Laminaria* and *Alaria* sp. (22%–24% of dry weight of these plants with up to 49% dry weight in *Laminaria* species [124]). The percentage dry weight of laminarin extracted from a species depends on the source and harvesting season. Table 5.8 shows the main algal sources for laminarin.

5.6.2 CHEMICAL STRUCTURE

The chemical structure of laminarin consists of branched glucans made up of β(1 → 3)- and β(1 → 6)-glucose residues with a typical β(1 → 3):β(1 → 6) ratio of 3:1. Most laminarins are low molecular weight compounds with 20–25 glucose moieties, but the weight of the molecule depends on the algal source, degree of polymerization [125,126], and extraction conditions [127]. There are two types of laminarin chains depending upon the type of sugar at the reducing end. M chains have a terminal 1-*O*-substituted D-mannitol, whereas G chains have a terminal glucose at the reducing end (Fig. 5.6). In laminarins obtained from *Laminaria* and *Fucus* species, 40%–75% of the chains are M type [128]. In addition to molecular weight and ratio of types of laminarin chains, environmental factors such as frond age and nutritive salts, water temperature, salinity, oceanic currents, waves, and depth of immersion also alter the structure and composition of laminarin [129]. Moreover, factors such as choice of extraction method also influence the biofunctional activity of laminarin [128].

Table 5.8 Main Sources of Laminarian

Seaweed	Content
Saccharina latissima	0%–33% of dry weight
Laminaria hyperborea	0%–32% of dry weigh
Laminaria digitata	14% of dry weight
L. digitata	0%–35% of dry weight, based on season
Fucus vesiculosus	84% of total sugars
Undaria pinnatifida	3% of dry weight
Ascophyllum nodosum	4.5% of dry weight

Adopted from Kadam SU, Tiwari BK, O'Donnell CP. Extraction, structure and biofunctional activities of laminarin from brown algae. Int J Food Sci Technol 2015;50(1):24–31.

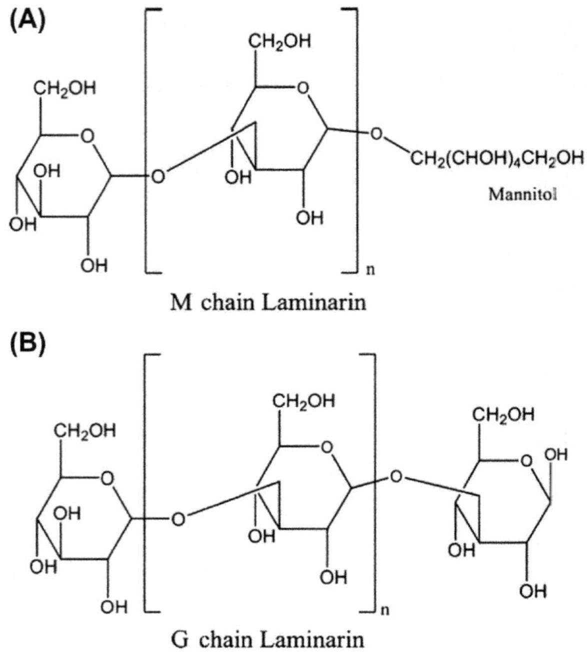

FIGURE 5.6

Chemical structure of one unit of (A) M and (B) G chain of laminarian.

Adapted from Kadam SU, Tiwari BK, O'Donnell CP. Extraction, structure and biofunctional activities of laminarin from brown algae. Int J Food Sci Technol 2015;50(1):24–31.

5.6.3 EXTRACTION METHODS

Laminarin can be extracted from its algal source using one of the two extraction methods developed by Black et al. [130]. The first method involves precipitation of laminarin from its aqueous solutions using ethanol or other organic solvents. This product is termed as cold water–soluble laminarin because of its solubility in cold water. The second method involves precipitation of laminarin from acidic solutions. The most commonly used acids are hydrochloric acid and sulfuric acid with pH up to 2.4. This product is termed as cold water–insoluble laminarin which is insoluble in cold water but readily dissolves in warm water. The solubility of laminarin depends on the degree of branching in its structure, with a higher level of branching associated with higher solubility [131].

In addition to the extraction methods described above, more efficient methods of extraction such as microwaves, ultrasonic waves, and supercritical fluid must also be investigated to improve the extraction rate and yield [132].

Unlike other common algal polysaccharides such as agar, alginate, and carrageenan, laminarin does not exhibit thickening and gelation properties [131]. The main utility of the compound lies in the food industry and biomedicine because of its various biofunctional properties [132]. Modification of laminarin using techniques such as irradiation, sulfation, oxidation, and reduction can lead to an enhanced bioactivity of the modified molecule. For example, laminarin has immunostimulatory effects that can activate macrophages leading to the healing of wounds and antitumor properties [133]. An enhanced antioxidant activity is also observed when this laminarin is irradiated with gamma rays [134].

5.6.4 BIOMEDICAL APPLICATIONS

5.6.4.1 Dietary Fiber/Functional Food

Use of laminarin as a functional dietary fiber was evaluated in an in vitro study by Deville et al. [135,136]. The in vitro digestibility of the fibers was assessed using hydrochloric acid and homogenates of the human digestive system. The authors observed that the fibers showed no hydrolysis in the upper digestive tract of humans. It was, however, observed that the endogenous intestinal microflora were able to ferment laminarin resulting in production of the short-chain fatty acids specifically butyrate. About 70% of energy consumption in the colon is associated with butyrate metabolism, and therefore fermentation of laminarin into butyrate can lead to modulation of intestinal metabolism. In addition to its role as an active substrate for the prebiotic bacteria, the molecule also has the ability to affect mucus composition of the intestine and its pH [136]. Furthermore, studies done on rats showed that it lowers low-density lipid (LDL) cholesterol, triglycerides, and phospholipids in serum and reduces blood pressure which further endorses its role as a functional dietary fiber [20].

5.6.4.2 Antioxidant Properties and Immunostimulatory Effect

According to Tsiapali et al., β-glucans show a weak ability to scavenge free radicals and stimulate macrophages [137]. Lee et al. reported an increase in intracellular calcium, H_2O_2, and NO in RAW 264.7 macrophages in mice. Furthermore, an increase in a number of transcription factors involved in immune reactions was also reported which points toward the immunostimulatory properties of laminarin [133]. Modification of the molecule can enhance these effects. Choi et al. studied the effect of gamma irradiation on the antioxidant ability of laminarins and concluded that a lower molecular weight, degraded product increased the antioxidant potential [138]. The antioxidant properties of laminarin were used by incorporation of laminarin-rich seaweed in porcine and fish feed. A reduction in lipid oxidation of muscles in laminarin-fed pigs was consistently observed even after the animals were processed for their meat and it increased the shelf life of pork meat, keeping it fresh for a longer time [139]. Similarly, it was observed that incorporation of a laminarin-rich diet in *Epinephelus* species, which is a popular farmed fish in China and South Asian countries, improves the immune responses and stimulates growth of fish [140].

5.6.4.3 Antiinflammatory Effect

In addition to stimulation of the immune system, laminarin also shows immunomodulatory properties that help prevent inflammation. Neyrinck et al. studied the tissue protective effects of laminarin in livers of laminarin-fed rats when injected with *E. coli* LPS. Laminarin was postulated to reduce the number of serum monocytes, nitrite (NO_2), and tumor necrosis factor alpha (TNF-α), and increase the recruitment of Kupffer cells. This reduction in secretion of inflammatory mediators can be linked to an overall lower systemic inflammation. Laminarin can therefore be considered as a potential therapeutic agent in case of systemic inflammatory syndrome and associated hepatic damage [141].

Laminarin also promotes wound healing by promoting the adhesion of human skin fibroblast cells and favors bone formation by inducing proliferation of human osteoblast cells [127]. Laminarin isolated from *Saccharina longicruris* was used as a substrate to promote matrix deposition of human skin fibroblasts. It was used along with an ascorbate, and the cells were cultured for 7 or 35 days to produce a dermis. An increase in the thickness of dermis due to a subsequent increase in deposition of collagen was observed, which suggests a potential application of this compound in cosmetic or therapeutic skin formulations [142].

5.6.4.4 Antibacterial and Antiviral Properties

The antibacterial properties of Laminarin were also explored recently. Kuda et al. studied the effect of laminarin and sodium alginate on the infections of *Salmonella typhimurium*, *Listeria monocytogenes*, and *Vibrio parahaemolyticus* in the human enterocyte-like HT-29-Luc cells. They concluded that laminarin prevents the adhesion of these bacteria on the surface of human enterocyte-like HT-29-Luc cells, thereby preventing infection. Similarly, colonization and infection of *S. typhimurium* in liver of BALB/c mice was also prevented [143]. Another study based on purified laminarin extracts from Irish brown seaweeds *L. hyperborea* and *A. nodosum* also showed significant antimicrobial activity against *S. aureus*, *E. coli*, *S. typhimurium*, and *L. monocytogenes* [144]. Laminarin was also shown to inhibit HIV adsorption on the surface of lymphocytes, and inhibits the activity of HIV reverse transcriptase which shows its potential to be developed as a novel antiviral drug with high efficiency and lower toxicity [145].

5.6.4.5 Anticancer Activity

Laminain also shows antitumorigenic and antiproliferative properties that can lead to its potential application as an anticancer drug.

Although the role of laminarin in prevention of colon cancer might as well be due to its functional properties as a dietary fiber, the compound also shows antitumorigenic activity through its ability to promote apoptosis in cancerous cells. While fermentation of laminarin in the intestine is thought to inhibit compounds such as indole, *p*-cresole, and sulfide which are risk indicators of colon cancer [146], it also exhibited the tendency to cause apoptosis in cancerous cells in experimental studies. A study by Ji et al. showed that laminarin is able to induce apoptosis in human colon cancer LOVO cells through the TRAIL/DR pathway and can be considered as an effective antitumor agent [147,148]. Ji et al. also studied the effect of sulfation on the antitumor activity of laminarin. The structural characterization revealed clear differences in the surface conformation of the sulfated laminarin, which showed an enhanced antitumor activity [149]. Another study by Park demonstrated its anticancer effects in HT-29 human colon cells, where laminarin induced apoptosis in a dose-dependent manner through its effect on the IGF-IR pathway [150].

In addition to colon cancer prevention, it is suggested that a diet based on *Laminaria* spp., which is a rich source of laminarin, may be responsible for observation of a lower rate of breast cancer in the Japanese population [151]. Sulfated derivatives of laminarin showed enhanced antitumor activities. Miao et al. demonstrated the potential clinical application of sulfated laminarin, which was able to inhibit heparanase activity and tumor metastasis in B16-BL6 mouse melanoma cells and 13762 MAT rat mammary adenocarcinoma cells [152]. Hoffman et al. reported that LAM S5 (a polysulfated derivative of laminarin) inhibits the binding of basic fibroblast growth factor (bFGF) and the bFGF-stimulated proliferation of fetal bovine heart endothelial (FBHE) cells. It was also observed that LAM S5 exhibited an antiangiogenic activity, as demonstrated by inhibition of tubule formation by endothelial cells cultured on Matrigel. It also inhibited vascularization of the chick chorioallantoic membrane [153].

5.6.4.6 Anticoagulant Activity

Laminarin exhibits anticoagulation activity upon sulfation. Moreover, its anticoagulant behavior increases with an increase in the degree of sulfation. For a modified laminarin sulfate molecule, up to two sulfate groups per glucose unit showed effective anticoagulant activity. In addition to the degree of sulfation, position of added sulfates also affects the behavior of laminarin, with preparations containing both O-sulfate and N-sulfate groups being more active as compared to those containing only O-groups [70]. Miao et al. showed that laminarin sulfate exhibited about one-third anticoagulant activity of heparin and shows potential to be therapeutically active in case of cerebrovascular diseases, wound healing, and atherosclerosis [152].

5.7 CHITIN/CHITOSAN

5.7.1 ISOLATION, SOURCE, AND OCCURRENCE

Chitin was first identified in 1884. It is the second most abundant biopolymer in the world which is synthesized by a wide variety of sea animals, insects, and microorganisms. Chitin is basically a structural polysaccharide that provides strength and reinforcement in the form of highly ordered crystalline microfibrils present in the exoskeleton or cell walls of its source organisms.

On a commercial level, main sources of chitin include the discarded wastes from marine food industry where processing of wastes from crabs, lobster, and shrimp can yield up to 15,000 tons of chitin per annum in the United States alone [154]. Table 5.9 shows the sources of chitin and chitosan.

5.7.2 CHEMICAL STRUCTURE

Chitosan is the most common deacetylated derivative of chitin with varying degrees of deacetylation (DA, expressed as molar percentage). When the average DA is lower than 0.5 the product is termed as chitosan and becomes soluble in acidic aqueous solutions [156].

Chitin is a linear polysaccharide containing chains of β-(1-4)-*N*-acetyl-D-glucosamine residues (Fig. 5.7).

Considering the crystalline nature of its molecule, two source-dependent allomorphs of chitin have been identified to date. These are termed as α-chitin and β-chitin both of which are distinguishable from each other by characterization techniques, such as solid-state NMR spectroscopy and X-ray diffraction [157,158]. Another allomorph of chitin termed γ-chitin was also described but detailed structural analysis classifies it as a variant of α-chitin family [159].

The most abundant form is α-chitin which occurs in yeast, fungi, insects, crabs, and shrimps. β-chitin is relatively rare and is found in the form of microtubules in squid pens [160], pogonophoran, and vestimentiferan worms [161]. The crystal structure of α-chitin consists of two antiparallel molecules per unit cell, whereas β-chitin

Table 5.9 Table Showing Occurrence and Commercial Sources of Chitin/Chitosan

Arthropods (Invertebrates)		
Sea Animals	**Insects**	**Microorganisms**
Annelida	Scorpions	Green algae
Mollusks	Spiders	Yeast (β-type)
Coelentrata	Brachiopods	Fungi (cell wall)[b]
Crustaceans	Ants	*Mycelia penicillium*[b]
Lobster[a]	Cockroaches	Green and brown algae
Crabs[a]	Beetles	Spores
Shrimp[a]		Chytridiaceae
Prawn[a]		Blastocladiaceae
Krill[b]		Ascomydes

[a] *Current sources.*
[b] *Future potential sources when demand increases.*
Adopted from Mathur NK, Narang CK. Chitin and chitosan, versatile polysaccharides from marine animals. J Chem Educ 1990;67(11):938.

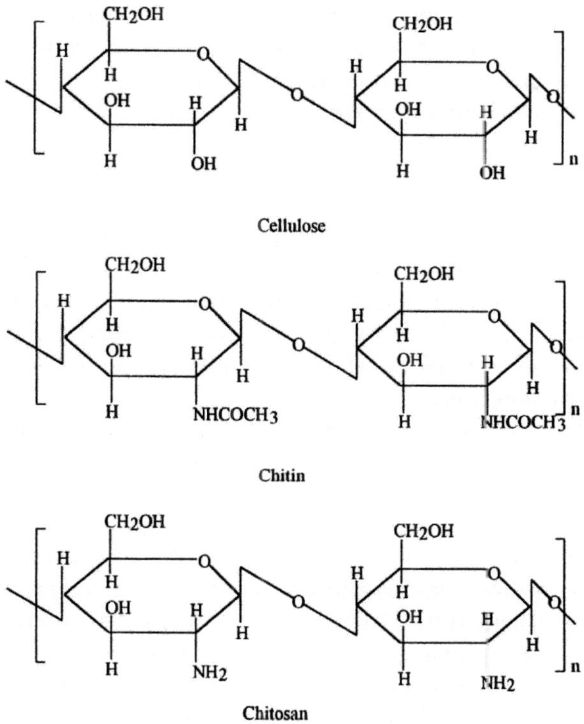

FIGURE 5.7

Chemical structure of chitin and chitosan.

Adapted from Kumar MNR. A review of chitin and chitosan applications. React Funct Polym 2000;46(1):1–27.

has only one molecule per unit cell, in a parallel arrangement. Hydrogen bonds are abundantly present in the molecular structure of both the allomorphs which governs the solubility, reactivity, and swelling behavior of chitin chains. Both inter and intra-sheet hydrogen bonds are present in α-chitin which prevent diffusion of small molecules into the crystalline phase whereas an absence of inter-sheet hydrogen bonds in β-chitin gives it a lose structure allowing diffusion of small molecules such as water, alcohol, and amines that result in its swelling in the presence of polar guest molecules. In addition to swelling behavior, the reactivity of β-chitin is also higher. Both α- and β-chitins are insoluble in all common solvents which limits the applications of chitin and demands development of more soluble derivatives [162].

5.7.3 EXTRACTION OF CHITIN

The steps involved in the extraction of chitin depend on the source materials. Since source materials for chitin are mostly the shells of crabs and shrimp, the first step involves selection of shells of same size and ideally the same species. These shells are then cleaned and ground into small pieces for further treatment. In case of crustaceans, chitin is found in form of a complex network of proteins and minerals such as calcium carbonate. Further processing involves deproteinization and demineralization through chemical or biological means [163].

Chemical extraction of chitin involves use of synthetic chemicals for chitin extraction. It involves two basic steps, chemical deproteinization and chemical demineralization. Chemical deproteinization uses a number of various deproteinization reagents such as NaOH, Na_2CO_3, $NaHCO_3$, KOH, K_2CO_3, etc. but all of them invariably result in the depolymerization of chitin during the process resulting in a decrease in the molecular weight of the biopolymer. Chemical demineralization involves removal of minerals such as calcium carbonate. It is performed by using acids such as dilute HCl. The use of acids results in formation of water-soluble salts that can be removed by filtration. However, demineralization reactions depend on the degree of mineralization of the shells, extraction time, particle size, and acid concentration, etc., and stronger acids or harsh chemical extraction conditions may be needed to complete the reactions. A number of chemical extraction methods were reviewed in detail by Yones et al. [164].

Biological extraction methods for deproteinization involve green synthesis routes that require the use of enzymes and fermentation processes that simultaneously result in deproteinization and demineralization of source material. These processes were also reviewed in detail by Yones et al. and help overcome drawbacks associated with chemical extraction methods such as decrease in molecular weight and undesirable deacetylation of purified chitin along with concerns for toxicity of effluent wastewater generated by these processes [164].

5.7.4 PREPARATION OF CHITOSAN

Chitosan is basically a family of polymers obtained from deacetylation of chitin, which results in the removal of acetyl groups and in turn changes the molecular weight of the product. The product is termed chitosan when degree of acetylation (DA) expressed as molar percentage is 50 mol%. Chitosan is soluble in acidic solutions.

Chitin can be prepared by chemical and biological or enzymatic deacetylation. Chemical deacetylation can be carried out using acids or bases but basic methods are preferable because glyosidic bonds are more susceptible to acids [165]. Enzymatic deacetylation using chitin deacetylases obtained from various fungal and insect species can be used for preparation of chitosan from chitin [166,167]. The enzymatic method is a better alternative to the chemical method, because it provides a means to overcome the associated drawbacks such as consumption of energy, environmental pollution, and a broad range of soluble and insoluble products. The main advantages of this process are its controllability, nondegradability, and production of a well-defined chitosan. Moreover, enzymatic method specifically produces chitosan oligomers [168].

5.7.5 CHITOSAN-BASED MATERIALS

Unlike alginic acid, agar, agarose, cargeenan, and other natural polysaccharides which are acidic or neutral, chitin and chitosan solutions are alkaline [169].

Acidic solutions of chitosan can be prepared in the form of a number of different conformations such as film, fibers, sponges, etc. These products can then be precipitated under alkaline conditions, washed, and dried. Chitin-based materials are stable whereas chitosan-based materials are easy to process. To introduce stability in chitosan-based materials, chitosan can be cross-linked using epichlorohydrin, diisocyanate, 1,4-butanediol diglycidyl ether, glutaraldehyde, oxalic acid, citric acid, etc., in the form of hydrogels [170–173].

Similarly, chitosan-based blends and composite materials can also be produced. Based on polycationic properties of chitosan, electrostatic complexes of chitosan can also be synthesized where properties of complexes depend on polymer concentration, ionic strength, temperature, and pH [174]. Chitosan can also form complexes with lipid vesicles with various applications in pharmaceutics and drug delivery [175]. Similarly, it can also form complexes with DNA, RNA, and other charged polysaccharides such as alginate with many biomedical applications [176–178].

Most of the biotechnological, biomedical, and pharmaceutical properties of chitin and chitosan are due to their intrinsic properties such as biocompatibility, biodegradability, and safety [179].

Other attractive properties include antimicrobial, antioxidant, and antitumor properties that, combined with the versatile ability to form different types of materials, lead to a broad range of applications in the food biotechnology, pharmaceutics, and healthcare.

5.7.6 APPLICATIONS IN FOOD BIOTECHNOLOGY AND FOOD PACKAGING

The antimicrobial activity of chitosan solution, powders, edible films, and coatings was studied by Friedman et al. [180]. They concluded that low molecular weight chitosans produced an antioxidative and antimicrobial effect against a number of different food-borne pathogens. Therefore these were suitable for producing a preservative effect in a variety of food stuff including fruit juices, eggs and dairy, cereal, meat, and seafood.

Similar preservative properties were reviewed by Kardas et al. [181] and Alishahi and Aider [182]. Alishahi and Aider pointed out that chitosan exhibits a selective gas permeable effect but its strong hydrophilic characteristics somewhat limits its use.

Some applications of chitosan are reviewed by Rinaudo et al. [183] in Table 5.10.

5.7.7 BIOMEDICAL AND PHARMACEUTICAL APPLICATIONS

A wide range of applications of chitin and chitosan in the pharmaceutics and biomedicine are compiled in Table 5.11.

Table 5.10 Principal Applications of Chitosan

Agriculture	Defensive mechanism in plants
	Stimulation of plant growth
	Seed coating, frost protection
	Time release of fertilizers and nutrients into the soil
Food and beverages	Not digestible by human (dietary fiber)
	Bind lipids (reduce cholesterol)
	Preservative
	Thickener and stabilizer for sauces
	Protective, fungistatic, antibacterial coating for fruit
Water and waste treatment	Flocculent to clarify water (drinking water, pools)
	Removal of metal ions
	Ecological polymer (eliminate synthetic polymers)
	Reduce odors
Biopharmaceutics	Immunologic, antitumoral
	Hemostatic and anticoagulant
	Healing, bacteriostatic
Cosmetics and toiletries	Maintain skin moisture
	Treat acne
	Improve suppleness of hair
	Reduce static electricity in hair
	Tone skin
	Oral care (toothpaste, chewing gum)

Adapted from Rinaudo M. Chitin and chitosan: properties and applications. Prog Polym Sci 2006;31(7): 603–32.

Table 5.11 Main Applications of Chitin and Chitosan in Pharmaceutical and Biomedical Domains

Forms	Application
Solutions	Cosmetics
	Bacteriostatic agent
	Hemostatic agent
	Anticoagulant
	Antitumor agent
	Gene delivery
	Spermicide
Beads	Drug delivery
Microspheres	Enzyme immobilization
Nanoparticles	Gene delivery vehicle
	Encapsulation of sensitive drugs
Fibers	Medical textiles
	Suture

Table 5.11 Main Applications of Chitin and Chitosan in Pharmaceutical and Biomedical Domains—cont'd

Forms	Application
Nanofibers	Guided bone regeneration
	Scaffold for nerve tissue regeneration
Nanowoven bioactive fibers	Wound healing
Films	Wound care
	Dialysis membrane
	Antitumoral
	Semipermeable film for wound dressing
Powder	Adsorbent for pharmaceutical and medical devices
	Surgical glove powder
	Enzyme immobilization
Sponge	Mucosomal hemostatic dressing
	Wound dressing
	Drug delivery
	Enzyme entrapment
	Artificial skin
Gels	Delivery vehicle
	Implants, coating
	Tissue engineering
	Wound dressing for wet treatment
Tablets	Compressed diluent
	Disintegrating agent
	Excipient
Coatings	Surface modification
	Textile finishes
Capsules	Delivery vehicle
Shaped objects	Orthopedics
	Contact lenses

Adapted from Younes I, Rinaudo M. Chitin and chitosan preparation from marine sources. Structure, properties and applications. Mar Drugs 2015;13(3):1133–74.

REFERENCES

[1] Chandini SK, Ganesan P, Bhaskar N. In vitro antioxidant activities of three selected brown seaweeds of India. Food Chem 2008;107(2):707–13.

[2] Murata M, Nakazoe J-I. Production and use of marine algae in Japan. Jpn Agric Res Q 2001;35(4):281–90.

[3] Kraan S. Algal polysaccharides, novel applications and outlook. In: Chang C-F, editor. Carbohydrates – comprehensive studies on glycobiology and glycotechnology. InTech; 2012.

[4] Holdt SL, Kraan S. Bioactive compounds in seaweed: functional food applications and legislation. J Appl Phycol 2011;23(3):543–97.

[5] Haug A, Jensen A. Seasonal variations in the chemical composition of *Alaria esculenta*, *Laminaria saccharina*, *Laminaria hyperborea* and *Laminaria digitata* from northern Norway. Akademisk Trykningssentral; 1954.
[6] Haug A, Jensen A. Geographical and seasonal variation in the chemical composition of *Laminaria hyperborea* and *Laminaria digitata* from the Norwegian Coast. Akademisk Trykningssentral; 1956.
[7] Arasaki S, Arasaki T. Low calorie, high nutrition vegetables from the sea to help you look and feel better. Japan Publications; 1983.
[8] Rasmussen RS, Morrissey MT. Marine biotechnology for production of food ingredients. Adv Food Nutr Res 2007;52:237–92.
[9] Gessner F. V. J. Chapman: seaweed and their uses. With 56 fig. — London: Methuen & Co 1970. 304pp. 4 £ (nur in U. K.). Int Rev Gesamten Hydrobiol Hydrogr 1971;56(4): 678–9.
[10] Gillespie RD, Critchley AT. Hydrobiologia 1999;398/399:201–10.
[11] Draget KI, Taylor C. Chemical, physical and biological properties of alginates and their biomedical implications. Food Hydrocoll 2011;25(2):251–6.
[12] Rehm BHA, Valla S. Bacterial alginates: biosynthesis and applications. Appl Microbiol Biotechnol 1997;48(3):281–8.
[13] Ertesvåg H, Høidal HK, Hals IK, Rian A, Doseth B, Valla S. A family of modular type mannuronan C-5-epimerase genes controls alginate structure in *Azotobacter vinelandii*. Mol Microbiol 1995;16(4):719–31.
[14] Remminghorst U, Rehm BH. Bacterial alginates: from biosynthesis to applications. Biotechnol Lett 2006;28(21):1701–12.
[15] Bixler HJ, Porse H. A decade of change in the seaweed hydrocolloids industry. J Appl Phycol 2010;23(3):321–35.
[16] Pawar SN, Edgar KJ. Alginate derivatization: a review of chemistry, properties and applications. Biomaterials 2012;33(11):3279–305.
[17] Kim IH, Lee JH. Antimicrobial activities against methicillin-resistant *Staphylococcus aureus* from macroalgae. J Ind Eng Chem 2008;14(5):568–72.
[18] Burtin P. Nutritional value of seaweeds. Electron J Environ Agric Food Chem 2003; 2(4):498–503.
[19] Pal A, Kamthania MC, Kumar A. Bioactive compounds and properties of seaweeds—a review. OALib 2014;01(04):1–17.
[20] Kiriyama S, Okazaki Y, Yoshida A. Hypocholesterolemic effect of polysaccharides and polysaccharide-rich foodstuffs in cholesterol-fed rats. J Nutr 1969;97(3): 382–8.
[21] Lamela M, Anca J, Villar R, Otero J, Calleja JM. Hypoglycemic activity of several seaweed extracts. J Ethnopharmacol 1989;27(1–2):35–43.
[22] Panlasigui LN, Baello OQ, Dimatangal JM, Dumelod BD. Blood cholesterol and lipid-lowering effects of carrageenan on human volunteers. Asia Pac J Clin Nutr 2003;12(2): 209–14.
[23] Dumelod D, Ramirez RPB, Tiangson CLP, Barrios EB, Panlasigui LN. Carbohydrate availability of arroz caldo with lambda-carrageenan. Int J Food Sci Nutr 1999;50(4): 283–9.
[24] Nishide E, Anzai H, Uchida N. Effects of alginates on the ingestion and excretion of cholesterol in the rat. J Appl Phycol 1993;5(2):207–11.
[25] Kimura Y, Watanabe K, Okuda H. Effects of soluble sodium alginate on cholesterol excretion and glucose tolerance in rats. J Ethnopharmacol 1996;54(1):47–54.

[26] Zec S. Body weight loss with the aid of alginic acid. Med Arh 1991;45(3–4):113–4.

[27] Torsdottir I, Alpsten M, Holm G, Sandberg AS, Tölli J. A small dose of soluble alginate-fiber affects postprandial glycemia and gastric emptying in humans with diabetes. J Nutr 1991;121(6):795–9.

[28] Vaugelade P, Hoebler C, Bernard F, Guillon F, Lahaye M, Duee P-H, et al. Non-starch polysaccharides extracted from seaweed can modulate intestinal absorption of glucose and insulin response in the pig. Reprod Nutr Dev 2000;40(1):33–47.

[29] Khotimchenko YS, Kovalev VV, Savchenko OV, Ziganshina OA. Physical–chemical properties, physiological activity, and usage of alginates, the polysaccharides of brown algae. Russ J Mar Biol 2001;27(Suppl. 1):53–64.

[30] Washington N, Denton G. Effect of alginate and alginate-cimetidine combination therapy on stimulated postprandial gastro-oesophageal reflux. J Pharm Pharmacol 1995; 47(11):879–82.

[31] Katayama H, Nishimura T, Ochi S, Tsuruta Y, Yamazaki Y, Shibata K, et al. Sustained release liquid preparation using sodium alginate for eradication of *Helicobacter pyroli*. Biol Pharm Bull 1999;22(1):55–60.

[32] Terada A, Hara H, Mitsuoka T. Effect of dietary alginate on the faecal microbiota and faecal metabolic activity in humans. Microb Ecol Health Dis 1995;8(6): 259–66.

[33] Wang Y, Han F, Hu B, Li J, Yu W. In vivo prebiotic properties of alginate oligosaccharides prepared through enzymatic hydrolysis of alginate. Nutr Res 2006;26(11): 597–603.

[34] Doyle JW, Roth TP, Smith RM, Li Y-Q, Dunn RM. Effect of calcium alginate on cellular wound healing processes modeled in vitro. J Biomed Mater Res 1996;32(4): 561–8.

[35] Glyantsev SP, Annaev AG, Savvina TV. Morphological substantiation of the choice of composition and structure of a sodium alginate-based biologically active composition for healing wounds. Bull Exp Biol Med 1993;115(1):69–72.

[36] Choi YS, Hong SR, Lee YM, Song KW, Park MH, Nam YS. Study on gelatin-containing artificial skin: I. Preparation and characteristics of novel gelatin-alginate sponge. Biomaterials 1999;20(5):409–17.

[37] Segal HC, Hunt BJ, Gilding K. The effects of alginate and non-alginate wound dressings on blood coagulation and platelet activation. J Biomater Appl 1998;12(3):249–57.

[38] Katayama S, Ohshita J, Sugaya K, Hirano M, Momose Y, Yamamura S. New medicinal treatment for severe gingivostomatitis. Int J Mol Med 1998;2(6):675–9.

[39] Rioux LE, Turgeon SL, Beaulieu M. Characterization of polysaccharides extracted from brown seaweeds. Carbohydr Polym 2007;69(3):530–7.

[40] Lahaye M. NMR spectroscopic characterisation of oligosaccharides from two *Ulva rigida* ulvan samples (Ulvales, Chlorophyta) degraded by a lyase. Carbohydr Res 1998;314(1–2):1–12.

[41] Lahaye M, Robic A. Structure and functional properties of ulvan, a polysaccharide from green seaweeds. Biomacromolecules 2007;8(6):1765–74.

[42] Lahaye M, Ray B. Cell-wall polysaccharides from the marine green alga Ulva "rigida" (Ulvales, Chlorophyta)—NMR analysis of ulvan oligosaccharides. Carbohydr Res 1996;283:161–73.

[43] Husemann E. Chemistry and enzymology of marine algal polysaccharides. Von E. Percival und R. H. McDowell. Academic Press, London-New York 1967. 1. Aufl., XII, 219 S., mehrere Abb. u. Tab., geb. 60s. Angew Chem 1968;80(20):856.

[44] Vera J, Castro J, Gonzalez A, Moenne A. Seaweed polysaccharides and derived oligosaccharides stimulate defense responses and protection against pathogens in plants. Mar Drugs 2011;9(12):2514–25.
[45] Lahaye M, Brunel M, Bonnin E. Fine chemical structure analysis of oligosaccharides produced by an ulvan-lyase degradation of the water-soluble cell-wall polysaccharides from *Ulva* sp., (Ulvales, Chlorophyta). Carbohydr Res 1997;304(3–4):325–33.
[46] Cunha L, Grenha A. Sulfated seaweed polysaccharides as multifunctional materials in drug delivery applications. Mar Drugs 2016;14(3).
[47] Percival E, McDowell RH. Algal polysaccharides. In: Dey PM, editor. Methods in plant biochemistry, vol. 2; 1990. p. 523–47.
[48] Nazia Auckloo B, Wu B. Structure, biological properties and applications of marine-derived polysaccharides. Curr Org Chem 2016;20(19):2002–12.
[49] Paradossi G, Cavalieri F, Chiessi E. A conformational study on the algal polysaccharide ulvan†. Macromolecules 2002;35(16):6404–11.
[50] Fujiwara-Arasaki T, Mino N, Kuroda M. The protein value in human nutrition of edible marine algae in Japan. In: Ragan CJBMA, editor. Eleventh international seaweed symposium; 1984. p. 513–6.
[51] Nagaoka (MT, JP), Shibata, Hideyuki (Tokyo, JP), Kimura, Itsuko (Tokyo, JP), Hashimoto, Shusuke (Tokyo, JP), inventor; Kabushiki Kaisha Yakult Honsha (Tokyo, JP), assignee. Oligosaccharide derivatives and process for producing the same. United States 2003.
[52] Nagaoka M, Shibata H, Kimura-Takagi I, Hashimoto S, Kimura K, Makino T, et al. Structural study of fucoidan from *Cladosiphon okamuranus* tokida. Glycoconj J 1999;16(1):19–26.
[53] Chopin T, Sharp G, Belyea E, Semple R, Jones D. Hydrobiologia 1999;398/399: 417–25.
[54] Rodrigueza MRC, Montaño MNE. Bioremediation potential of three carrageenophytes cultivated in tanks with seawater from fish farms. J Appl Phycol 2007;19(6):755–62.
[55] Percival E. The polysaccharides of green, red and brown seaweeds: their basic structure, biosynthesis and function. Br Phycol J 1979;14(2):103–17.
[56] Lahaye M. J Appl Phycol 2001;13(2):173–84.
[57] Estevez JM, Ciancia M, Cerezo AS. The system of low-molecular-weight carrageenans and agaroids from the room-temperature-extracted fraction of *Kappaphycus alvarezii*. Carbohydr Res 2000;325(4):287–99.
[58] De Ruiter GA, Rudolph B. Carrageenan biotechnology. Trends Food Sci Technol 1997;8(12):389–95.
[59] van de Velde F, Peppelman HA, Rollema HS, Tromp RH. On the structure of κ/ι-hybrid carrageenans. Carbohydr Res 2001;331(3):271–83.
[60] Funami T, Hiroe M, Noda S, Asai I, Ikeda S, Nishinari K. Influence of molecular structure imaged with atomic force microscopy on the rheological behavior of carrageenan aqueous systems in the presence or absence of cations. Food Hydrocoll 2007;21(4): 617–29.
[61] Zhou G, Sheng W, Yao W, Wang C. Effect of low molecular lambda-carrageenan from *Chondrus ocellatus* on antitumor H-22 activity of 5-Fu. Pharmacol Res 2006;53(2): 129–34.
[62] Skoler-Karpoff S, Ramjee G, Ahmed K, Altini L, Plagianos MG, Friedland B, et al. Efficacy of Carraguard for prevention of HIV infection in women in South Africa: a randomised, double-blind, placebo-controlled trial. Lancet 2008;372(9654):1977–87.

[63] Vlieghe P, Clerc T, Pannecouque C, Witvrouw M, De Clercq E, Salles J-P, et al. Synthesis of new covalently bound κ-Carrageenan–AZT conjugates with improved anti-HIV activities. J Med Chem 2002;45(6):1275–83.

[64] Yan X, Zheng L, Chen H, Lin W, Zhang W. Enriched accumulation and biotransformation of selenium in the edible seaweed *Laminaria japonica*. J Agric Food Chem 2004;52(21):6460–4.

[65] Zhou Y, Yang H, Hu H, Liu Y, Mao Y, Zhou H, et al. Bioremediation potential of the macroalga *Gracilaria lemaneiformis* (Rhodophyta) integrated into fed fish culture in coastal waters of north China. Aquaculture 2006;252(2–4):264–76.

[66] Morrissey J, Kraan S, Guiry MD. A guide to commercially important seaweeds on the Irish Coast. 2001.

[67] Cáceres PJ, Carlucci MJ, Damonte EB, Matsuhiro B, Zúñiga EA. Carrageenans from chilean samples of *Stenogramme interrupta* (Phyllophoraceae): structural analysis and biological activity. Phytochemistry 2000;53(1):81–6.

[68] Carlucci MJ, Pujol CA, Ciancia M, Noseda MD, Matulewicz MC, Damonte EB, et al. Antiherpetic and anticoagulant properties of carrageenans from the red seaweed *Gigartina skottsbergii* and their cyclized derivatives: correlation between structure and biological activity. Int J Biol Macromol 1997;20(2):97–105.

[69] Luescher-Mattli M. Algae, a possible source for new drugs in the treatment of HIV and other viral diseases. Curr Med Chem Anti-Infect Agents 2003;2(3):219–25.

[70] Shanmugam M, Mody KH. Heparinoid-active sulphated polysaccharides from marine algae as potential blood anticoagulant agents. Curr Sci 2000;79(12):1672–83.

[71] Witvrouw M, De Clercq E. Sulfated polysaccharides extracted from sea algae as potential antiviral drugs. General Pharmacol Vasc Syst 1997;29(4):497–511.

[72] Buck CB, Thompson CD, Roberts JN, Muller M, Lowy DR, Schiller JT. Carrageenan is a potent inhibitor of papillomavirus infection. PLoS Pathog 2006;2(7):e69.

[73] Zeitlin L, Whaley KJ, Hegarty TA, Moench TR, Cone RA. Tests of vaginal microbicides in the mouse genital herpes model. Contraception 1997;56(5):329–35.

[74] Spieler R. Seaweed compound's anti-HIV efficacy will be tested in southern Africa. Lancet 2002;359(9318):1675.

[75] Hawkins WW, Leonard VG, Maxwell JE, Rastogi KS. A study of the prolonged intake of small amounts of ethylenediaminetetraacetate on the utilization of low dietary levels of calcium and iron by the rat. Biochem Cell Biol 1962;40(3):391–5.

[76] Hawkins WW, Leonard VG. The antithrombic activity of carrageenin in human blood. Biochem Cell Biol 1963;41(5):1325–7.

[77] Kindness G, Williamson FB, Long WF. Effect of polyanetholesulphonic acid and xylan sulphate on antithrombin III activity. Biochem Biophys Res Commun 1979; 88(3):1062–8.

[78] Sen AK, Das AK, Banerji N, Siddhanta AK, Mody KH, Ramavat BK, et al. A new sulfated polysaccharide with potent blood anti-coagulant activity from the red seaweed *Grateloupia indica*. Int J Biol Macromol 1994;16(5):279–80.

[79] van de Velde F, Lourenço ND, Pinheiro HM, Bakker M. Carrageenan: a food-grade and biocompatible support for immobilisation techniques. Adv Synth Catal 2002; 344(8):815–35.

[80] Noa M, Mas R, Carbajal D, Valdes S. Effect of D-002 on acetic acid-induced colitis in rats at single and repeated doses. Pharmacol Res 2000;41(4):391–5.

[81] Dannhardt G, Kiefer W. Cyclooxygenase inhibitors – current status and future prospects. Eur J Med Chem 2001;36(2):109–26.

[82] Carthew P. Safety of carrageenan in foods. Environ Health Perspect 2002;110(4): A176–7.

[83] Tobacman JK. Review of harmful gastrointestinal effects of carrageenan in animal experiments. Environ Health Perspect 2001;109(10):983–94.

[84] Berteau O, Mulloy B. Sulfated fucans, fresh perspectives: structures, functions, and biological properties of sulfated fucans and an overview of enzymes active toward this class of polysaccharide. Glycobiology 2003;13(6):29R–40R.

[85] Kylin H. Zur Biochemie der Meeresalgen. Z für Physiol Chem 1913;83:171–97.

[86] Cumashi A, Ushakova NA, Preobrazhenskaya ME, D'Incecco A, Piccoli A, Totani L, et al. A comparative study of the anti-inflammatory, anticoagulant, antiangiogenic, and antiadhesive activities of nine different fucoidans from brown seaweeds. Glycobiology 2007;17(5):541–52.

[87] Ale MT, Meyer AS. Fucoidans from brown seaweeds: an update on structures, extraction techniques and use of enzymes as tools for structural elucidation. RSC Adv 2013; 3(22):8131–41.

[88] Chevolot L, Foucault A, Chaubet F, Kervarec N, Sinquin C, Fisher A-M, et al. Further data on the structure of brown seaweed fucans: relationships with anticoagulant activity. Carbohydr Res 1999;319(1–4):154–65.

[89] Zhuang C, Itoh H, Mizuno T, Ito H. Antitumor active fucoidan from the brown seaweed, umitoranoo (*Sargassum thunbergii*). Biosci Biotechnol Biochem 1995; 59(4):563–7.

[90] Li N, Zhang Q, Song J. Toxicological evaluation of fucoidan extracted from *Laminaria japonica* in Wistar rats. Food Chem Toxicol 2005;43(3):421–6.

[91] Zhao X, Xue C-H, Li B-F. Study of antioxidant activities of sulfated polysaccharides from *Laminaria japonica*. J Appl Phycol 2007;20(4):431–6.

[92] Li B, Lu F, Wei X, Zhao R. Fucoidan: structure and bioactivity. Molecules 2008;13(8): 1671–95.

[93] Usov AI, Smirnova GP, Klochkova NG. Russ J Bioorg Chem 2001;27(6):395–9.

[94] Hayashi L, Yokoya NS, Ostini S, Pereira RTL, Braga ES, Oliveira EC. Nutrients removed by *Kappaphycus alvarezii* (Rhodophyta, Solieriaceae) in integrated cultivation with fishes in re-circulating water. Aquaculture 2008;277(3–4):185–91.

[95] Wang WT, Zhou JH, Xing ST, Guan HS. Immunomodulating action of marine algae sulfated polysaccharides on normal and immunosuppressed mice. Chin J Pharmacol Toxicol 1994;8:199–202.

[96] Xiaomin W, Mingliang Y, Xiaolan H, Jun Y, Qiong L. Effect of fucoidan on splenic lymphocyte apoptosis induced by radiation. Chin J Radiol Med Prot 2003;23(6): 430–2.

[97] Xiaolin Y, Juyun S, Hannian X, Xiaohui L, Yishen Z, Shaolun Z. An experimental study on immunoregulatory effect of fucoidan. Chin J Mar Drugs 1995;(3):9–13.

[98] Mandal P, Mateu CG, Chattopadhyay K, Pujol CA, Damonte EB, Ray B. Structural features and antiviral activity of sulphated fucans from the brown seaweed *Cystoseira indica*. Antivir Chem Chemother 2007;18(3):153–62.

[99] Iqbal M, Flick-Smith H, McCauley JW. Interactions of bovine viral diarrhoea virus glycoprotein E(rns) with cell surface glycosaminoglycans. J Gen Virol 2000; 81(Pt 2):451–9.

[100] Li F, Tian T, Shi Y, Liu Y, Zhang S. Study on anti-virus effect of fucoidan in vitro. Baiqiuen Yike Daxue Xuebao 1995;21(3):255–7.

[101] Alekseyenko TV, Zhanayeva SY, Venediktova AA, Zvyagintseva TN, Kuznetsova TA, Besednova NN, et al. Antitumor and antimetastatic activity of fucoidan, a sulfated polysaccharide isolated from the Okhotsk sea *Fucus evanescens* brown alga. Bull Exp Biol Med 2007;143(6):730–2.

[102] Li Y, Qian Z-J, Le Q-T, Kim M-M, Kim S-K. Bioactive phloroglucinol derivatives isolated from an edible marine brown alga, *Ecklonia cava*. J Biotechnol 2008;136: S578.

[103] Ellouali M, Boisson-Vidal C, Durand P, Jozefonvicz J. Antitumor activity of low molecular weight fucans extracted from brown seaweed *Ascophyllum nodosum*. Anticancer Res 1993;13(6A):2011–9.

[104] Khotimchenko YS. Antitumor properties of nonstarch polysaccharides: fucoidans and chitosans. Russ J Mar Biol 2010;36(5):321–30.

[105] Liu JM, Bignon J, Haroun-Bouhedja F, Bittoun P, Vassy J, Fermandjian S, et al. Inhibitory effect of fucoidan on the adhesion of adenocarcinoma cells to fibronectin. Anticancer Res 2005;25(3B):2129–33.

[106] Aisa Y, Miyakawa Y, Nakazato T, Shibata H, Saito K, Ikeda Y, et al. Fucoidan induces apoptosis of human HS-sultan cells accompanied by activation of caspase-3 and down-regulation of ERK pathways. Am J Hematol 2005;78(1):7–14.

[107] Maruyama H, Tanaka M, Hashimoto M, Inoue M, Sasahara T. The suppressive effect of Mekabu fucoidan on an attachment of *Cryptosporidium parvum* oocysts to the intestinal epithelial cells in neonatal mice. Life Sci 2007;80(8):775–81.

[108] Verdrengh M, Erlandsson-Harris H, Tarkowski A. Role of selectins in experimental *Staphylococcus aureus*-induced arthritis. Eur J Immunol 2000;30(6):1606–13.

[109] Nasu T, Fukuda Y, Nagahira K, Kawashima H, Noguchi C, Nakanishi T. Fucoidin, a potent inhibitor of l-selectin function, reduces contact hypersensitivity reaction in mice. Immunol Lett 1997;59(1):47–51.

[110] Granert C, Raud J, Waage A, Lindquist L. Effects of polysaccharide fucoidin on cerebrospinal fluid interleukin-1 and tumor necrosis factor alpha in pneumococcal meningitis in the rabbit. Infect Immun 1999;67(5):2071–4.

[111] Ostergaard C, Yieng-Kow RV, Benfield T, Frimodt-Moller N, Espersen F, Lundgren JD. Inhibition of leukocyte entry into the brain by the selectin blocker fucoidin decreases interleukin-1 (IL-1) levels but increases IL-8 levels in cerebrospinal fluid during experimental pneumococcal meningitis in rabbits. Infect Immun 2000;68(6): 3153–7.

[112] Omata M, Matsui N, Inomata N, Ohno T. Protective effects of polysaccharide fucoidin on myocardial ischemia-reperfusion injury in rats. J Cardiovasc Pharmacol 1997; 30(6):717–24.

[113] Lasky LA. Selectin-carbohydrate interactions and the initiation of the inflammatory response. Annu Rev Biochem 1995;64:113–39.

[114] Qi H, Zhao T, Zhang Q, Li Z, Zhao Z, Xing R. Antioxidant activity of different molecular weight sulfated polysaccharides from *Ulva pertusa* Kjellm (Chlorophyta). J Appl Phycol 2006;17(6):527–34.

[115] Qi H, Zhang Q, Zhao T, Chen R, Zhang H, Niu X, et al. Antioxidant activity of different sulfate content derivatives of polysaccharide extracted from *Ulva pertusa* (Chlorophyta) in vitro. Int J Biol Macromol 2005;37(4):195–9.

[116] Xue C, Chen L, Li Z, Cai Y, Lin H, Fang Y. Antioxidative activities of low molecular fucoidans from kelp *Laminaria japonica*. Dev Food Sci 2004;42:139–45.

[117] Zhang Q, Li N, Zhou G, Lu X, Xu Z, Li Z. In vivo antioxidant activity of polysaccharide fraction from Porphyra haitanesis (Rhodephyta) in aging mice. Pharmacol Res 2003;48(2):151–5.

[118] Hu JF, Gen MY, Zhang JT, Jiang HD. An in vitro study of the structure-activity relationships of sulfated polysaccharide from brown algae to its antioxidant effect. J Asian Nat Prod Res 2001;3(4):353–8.

[119] Rupérez P, Ahrazem O, Leal JA. Potential antioxidant capacity of sulfated polysaccharides from the edible marine brown seaweed *Fucus vesiculosus*. J Agric Food Chem 2002;50(4):840–5.

[120] Rocha de Souza MC, Marques CT, Guerra Dore CM, Ferreira da Silva FR, Oliveira Rocha HA, Leite EL. Antioxidant activities of sulfated polysaccharides from brown and red seaweeds. J Appl Phycol 2007;19(2):153–60.

[121] Wang J, Zhang Q, Zhang Z, Li Z. Antioxidant activity of sulfated polysaccharide fractions extracted from *Laminaria japonica*. Int J Biol Macromol 2008;42(2):127–32.

[122] Sezer AD, Hatipoglu F, Cevher E, Ogurtan Z, Bas AL, Akbuga J. Chitosan film containing fucoidan as a wound dressing for dermal burn healing: preparation and in vitro/in vivo evaluation. AAPS PharmSciTech 2007;8(2), 39.

[123] Bull AT, Chesters CG. The biochemistry of laminarin and the nature of laminarinase. Adv Enzym Relat Areas Mol Biol 1966;28:325–64.

[124] Quillet M. Glucide metabolism of brown algae. Presence of small quantities of Laminarin in numerous new species, distributed over the entire group of Phaeophyceae. Compt rend hebd Séanc Acad Sci, Paris 1958;246:812–5.

[125] Nelson T, Lewis B. Separation and characterization of the soluble and insoluble components of insoluble laminaran. Carbohydr Res 1974;33(1):63–74.

[126] Alderkamp A-C, Van Rijssel M, Bolhuis H. Characterization of marine bacteria and the activity of their enzyme systems involved in degradation of the algal storage glucan laminarin. FEMS Microbiol Ecol 2007;59(1):108–17.

[127] Rioux L-E, Turgeon SL, Beaulieu M. Structural characterization of laminaran and galactofucan extracted from the brown seaweed *Saccharina longicruris*. Phytochemistry 2010;71(13):1586–95.

[128] Rinaudo M. 2.21-Seaweed polysaccharides A2-Kamerling, Hans. Comprehensive glycoscience. Oxford: Elsevier; 2007. p. 691–735.

[129] Chizhov AO, Dell A, Morris HR, Reason AJ, Haslam SM, McDowell RA, et al. Structural analysis of laminarans by MALDI and FAB mass spectrometry. Carbohydr Res 1998;310(3):203–10.

[130] Black W, Cornhill W, Dewar E, Woodward F. Manufacture of algal chemicals. VI.—laboratory-scale isolation of l-fucose from brown marine algae. J Sci Food Agric 1953; 4(2):85–91.

[131] Miao HQ, Ishai-Michaeli R, Peretz T, Vlodavsky I. Laminarin sulfate mimics the effects of heparin on smooth muscle cell proliferation and basic fibroblast growth factor-receptor binding and mitogenic activity. J Cell Physiol 1995;164(3):482–90.

[132] Kadam SU, Tiwari BK, O'Donnell CP. Extraction, structure and biofunctional activities of laminarin from brown algae. Int J Food Sci Technol 2015;50(1):24–31.

[133] Lee JY, Kim Y-J, Kim HJ, Kim Y-S, Park W. Immunostimulatory effect of laminarin on RAW 264.7 mouse macrophages. Molecules 2012;17(5):5404–11.

[134] Choi J-I, Kim H-J, Lee J-W. Structural feature and antioxidant activity of low molecular weight laminarin degraded by gamma irradiation. Food Chem 2011;129(2):520–3.

[135] Deville C, Damas J, Forget P, Dandrifosse G, Peulen O. Laminarin in the dietary fibre concept. J Sci Food Agric 2004;84(9):1030–8.

[136] Deville C, Gharbi M, Dandrifosse G, Peulen O. Study on the effects of laminarin, a polysaccharide from seaweed, on gut characteristics. J Sci Food Agric 2007;87(9): 1717–25.

[137] Tsiapali E, Whaley S, Kalbfleisch J, Ensley HE, Browder IW, Williams DL. Glucans exhibit weak antioxidant activity, but stimulate macrophage free radical activity. Free Radic Biol Med 2001;30(4):393–402.

[138] Choi JI, Kim HJ, Kim JH, Lee JW. Enhanced biological activities of laminarin degraded by gamma-ray irradiation. J Food Biochem 2012;36(4):465–9.

[139] Moroney N, O'Grady M, O'Doherty J, Kerry J. Addition of seaweed (*Laminaria digitata*) extracts containing laminarin and fucoidan to porcine diets: influence on the quality and shelf-life of fresh pork. Meat Sci 2012;92(4):423–9.

[140] Yin G, Li W, Lin Q, Lin X, Lin J, Zhu Q, et al. Dietary administration of laminarin improves the growth performance and immune responses in *Epinephelus coioides*. Fish Shellfish Immunol 2014;41(2):402–6.

[141] Neyrinck AM, Mouson A, Delzenne NM. Dietary supplementation with laminarin, a fermentable marine β (1–3) glucan, protects against hepatotoxicity induced by LPS in rat by modulating immune response in the hepatic tissue. Int Immunopharmacol 2007; 7(12):1497–506.

[142] Ayoub A, Pereira JM, Rioux L-E, Turgeon SL, Beaulieu M, Moulin VJ. Role of seaweed laminaran from *Saccharina longicruris* on matrix deposition during dermal tissue-engineered production. Int J Biol Macromol 2015;75:13–20.

[143] Kuda T, Kosaka M, Hirano S, Kawahara M, Sato M, Kaneshima T, et al. Effect of sodium-alginate and laminaran on *Salmonella typhimurium* infection in human enterocyte-like HT-29-Luc cells and BALB/c mice. Carbohydr Polym 2015;125: 113–9.

[144] Kadam SU, O'Donnell CP, Rai DK, Hossain MB, Burgess CM, Walsh D, et al. Laminarin from irish brown seaweeds *Ascophyllum nodosum* and *Laminaria hyperborea*: ultrasound assisted extraction, characterization and bioactivity. Mar drugs 2015; 13(7):4270–80.

[145] Wang W, Wang S-X, Guan H-S. The antiviral activities and mechanisms of marine polysaccharides: an overview. Mar drugs 2012;10(12):2795–816.

[146] Kuda T, Yano T, Matsuda N, Nishizawa M. Inhibitory effects of laminaran and low molecular alginate against the putrefactive compounds produced by intestinal microflora in vitro and in rats. Food Chem 2005;91(4):745–9.

[147] Ji YB, Ji CF, Zhang H. Laminarin induces apoptosis of human colon cancer LOVO cells through a mitochondrial pathway. Molecules 2012;17(8):9947–60.

[148] Ji CF, Ji YB. Laminarin-induced apoptosis in human colon cancer LoVo cells. Oncol Lett 2014;7(5):1728–32.

[149] Ji C-F, Ji Y-B, Meng D-Y. Sulfated modification and anti-tumor activity of laminarin. Exp Ther Med 2013;6(5):1259–64.

[150] Park H-K, Kim I-H, Kim J, Nam T-J. Induction of apoptosis by laminarin, regulating the insulin-like growth factor-IR signaling pathways in HT-29 human colon cells. Int J Mol Med 2012;30(4):734–8.

[151] Teas J. The dietary intake of *Laminaria*, a brown seaweed, and breast cancer prevention. Nutr cancer 1982;4(3):217–22.

[152] Miao HQ, Elkin M, Aingorn E, Ishai-Michaeli R, Stein CA, Vlodavsky I. Inhibition of heparanase activity and tumor metastasis by laminarin sulfate and synthetic phosphorothioate oligodeoxynucleotides. Int J Cancer 1999;83(3):424−31.

[153] Hoffman R, Paper D, Donaldson J, Vogl H. Inhibition of angiogenesis and murine tumour growth by laminarin sulphate. Br J cancer 1996;73(10):1183.

[154] Mathur NK, Narang CK. Chitin and chitosan, versatile polysaccharides from marine animals. J Chem Educ 1990;67(11):938.

[155] Kumar MNR. A review of chitin and chitosan applications. React Funct Polym 2000; 46(1):1−27.

[156] Roberts GA. Structure of chitin and chitosan. Chitin chemistry. Springer; 1992. p. 1−53.

[157] Rudall K, Kenchington W. The chitin system. Biol Rev 1973;48(4):597−633.

[158] Blackwell J. Chitin. New York (NY, USA): Academic Press; 1973. p. 474−89.

[159] Skjåk-Braek G, Anthonsen T, Sandford PA. Chitin and chitosan: sources, chemistry, biochemistry, physical properties, and applications. Kluwer Academic Pub; 1989.

[160] Jang MK, Kong BG, Jeong YI, Lee CH, Nah JW. Physicochemical characterization of α-chitin, β-chitin, and γ-chitin separated from natural resources. J Polym Sci Part A Polym Chem 2004;42(14):3423−32.

[161] Gaill F, Persson J, Sugiyama J, Vuong R, Chanzy H. The chitin system in the tubes of deep sea hydrothermal vent worms. J Struct Biol 1992;109(2):116−28.

[162] Kurita K, Tomita K, Ishii S, Nishimura SI, Shimoda K. β-chitin as a convenient starting material for acetolysis for efficient preparation of *N*-acetylchitooligosaccharides. J Polym Sci Part A Polym Chem 1993;31(9):2393−5.

[163] Horst MN, Walker AN, Klar E. The pathway of crustacean chitin synthesis. Crustac Integument Morphol Biochem 1993:113−49.

[164] Younes I, Rinaudo M. Chitin and chitosan preparation from marine sources. Structure, properties and applications. Mar Drugs 2015;13(3):1133−74.

[165] No HK, Meyers SP. Preparation and characterization of chitin and chitosan—a review. J Aquat Food Prod Technol 1995;4(2):27−52.

[166] Martinou A, Kafetzopoulos D, Bouriotis V. Chitin deacetylation by enzymatic means: monitoring of deacetylation processes. Carbohydr Res 1995;273(2):235−42.

[167] Tsigos I, Martinou A, Kafetzopoulos D, Bouriotis V. Chitin deacetylases: new, versatile tools in biotechnology. Trends Biotechnol 2000;18(7):305−12.

[168] Eaton P, Fernandes JC, Pereira E, Pintado ME, Malcata FX. Atomic force microscopy study of the antibacterial effects of chitosans on *Escherichia coli* and *Staphylococcus aureus*. Ultramicroscopy 2008;108(10):1128−34.

[169] Hench LL. Biomaterials: a forecast for the future. Biomaterials 1998;19(16):1419−23.

[170] Welsh ER, Schauer CL, Qadri SB, Price RR. Chitosan cross-linking with a water-soluble, blocked diisocyanate. 1. Solid state. Biomacromolecules 2002;3(6):1370−4.

[171] Argüelles-Monal W, Goycoolea F, Peniche C, Higuera-Ciapara I. Rheological study of the chitosan/glutaraldehyde chemical gel system. Polym Gels Netw 1998;6(6): 429−40.

[172] Hirano S, Yamaguchi R, Fukui N, Iwata M. A chitosan oxalate gel: its conversion to an *N*-acetylchitosan gel via a chitosan gel. Carbohydr Res 1990;201(1):145−9.

[173] Wei YC, Hudson SM, Mayer JM, Kaplan DL. The crosslinking of chitosan fibers. J Polym Sci Part A Polym Chem 1992;30(10):2187−93.

[174] Rinaudo M. Materials based on chitin and chitosan. Chichester (UK): John Wiley & Sons; 2014. p. 63−80.

[175] Rinaudo M, Quemeneur F, Pepin-Donat B. Stabilization of liposomes against stress using polyelectrolytes: interaction mechanisms, influence of pH, molecular weight, and polyelectrolyte structure. Int J Polym Anal Charact 2009;14(8):667–77.

[176] Lavertu M, Methot S, Tran-Khanh N, Buschmann MD. High efficiency gene transfer using chitosan/DNA nanoparticles with specific combinations of molecular weight and degree of deacetylation. Biomaterials 2006;27(27):4815–24.

[177] Alameh M, Jean M, DeJesus D, Buschmann MD, Merzouki A. Chitosanase-based method for RNA isolation from cells transfected with chitosan/siRNA nanocomplexes for real-time RT-PCR in gene silencing. Int J Nanomed 2010;5:473–81.

[178] Majima T, Funakosi T, Iwasaki N, Yamane S-T, Harada K, Nonaka S, et al. Alginate and chitosan polyion complex hybrid fibers for scaffolds in ligament and tendon tissue engineering. J Orthop Sci 2005;10(3):302–7.

[179] Farkaš V. Fungal cell walls: their structure, biosynthesis and biotechnological aspects. Acta Biotechnol 1990;10(3):225–38.

[180] Friedman M, Juneja VK. Review of antimicrobial and antioxidative activities of chitosans in food. J Food Prot 2010;73(9):1737–61.

[181] Kardas I, Struszczyk MH, Kucharska M, van den Broek LA, van Dam JE, Ciechańska D. Chitin and chitosan as functional biopolymers for industrial applications. The European Polysaccharide Network of Excellence (EPNOE). Springer; 2012. p. 329–73.

[182] Alishahi A, Aïder M. Applications of chitosan in the seafood industry and aquaculture: a review. Food Bioprocess Technol 2012;5(3):817–30.

[183] Rinaudo M. Chitin and chitosan: properties and applications. Prog Polym Sci 2006; 31(7):603–32.

CHAPTER

Algae-Based Biologically Active Compounds

6

Muhammad Ibrahim, Mahwish Salman, Shagufta Kamal, Saima Rehman, Aneeza Razzaq, Sajid Hamid Akash

Government College University Faisalabad, Faisalabad, Pakistan

6.1 INTRODUCTION

Algae are the group of plants known since ancient civilizations. Linnaeus introduced the term algae for the first time in 1753, and in 1789 de Jussieu classified the plants and separated algae from the rest of the plants. This classification is accepted till date [1]. The word "algae" is derived from the Latin word "alga" which means seaweeds [2]. The study of algae is called phycology (phyco means algae and logos means study of). Phycos is a Greek word which means seaweeds. The history of algae is mentioned in the early Chinese, Roman, and Greek literatures. Roman named it Fucus, whereas Chinese called it Tsao. The ancient Hawaiians used algae as food and called them Limu. The algae were used as manure on the north coast of France as early as 12th century [3,4]. They are eukaryotic and predominantly aquatic, photosynthetic organisms, which range in size from tiny flagellate micromonas that is 1 μm (0.000039 in.) in diameter to giant kelps that reach 60 m (200 ft) in length [5]. Biologists divided the living organisms into five kingdoms named as animals, plants, fungus, protists, and prokaryotes [6]. Algae belong to protista kingdom. Protists often share certain morphological and physical characteristics with animals, plants, and fungi. However, protists are neither animals nor plants, and nor fungi [5]. In earlier classifications, algae, which are photosynthetic like plants, were placed in the kingdom planta. Most of the classifications have now removed algae from the plant kingdom due to their simple construction (majority being unicellular), variety of accessory pigments used for photosynthesis, lack of protecting tissue around gametic cells, and their highly varied life histories [7]. Although, algae are photosynthetic organisms, they lack specialized reproductive systems of the plants, which always have multicellular reproductive structures that contain fertile gametes surrounded by sterile cells. Algae also lack true stems, leaves, and roots [5].

Algae are photosynthetic organisms, which contain chlorophyll and other photosynthetic pigments. These pigments trap light energy coming from the sun, convert it into chemical energy, and store it in the form of photosynthetic products such as starch, carbohydrates, and various complex sugars [8]. Their photosynthetic efficiency is higher than other plants, and some species of algae are considered among

Algae Based Polymers, Blends, and Composites. http://dx.doi.org/10.1016/B978-0-12-812360-7.00006-9

the fastest growing organisms. It is estimated that photosynthetic efficiency of algae ranges from 3% to 8%, compared to 0.5% of many other terrestrial crops [9]. As the algae cells are complete simple organisms capable of photosynthesis and synthesis of a multitude of other compounds that make up the cell, they have attracted special interest from the biological and phytochemical scientists [10].

The purpose of this chapter is to give brief information about algae and their biologically active constituents. The isolation and characterization of the biologically active components from seaweeds have gained much attention from various research groups across the world and the seaweeds have become a recognized potential natural product in pharmaceutical industries.

6.2 OCCURRENCE

Algae have universal occurrence and are found in almost all types of habitats. They are found in freshwater, seawater, on rocks, soil, stones, on other plants and even on animals, in deserts, and on permanent snowfields etc. [11,12]. They sometimes appear in such large quantity that the landscape becomes colored due to the presence of pigments in their cells. The once-mysterious "red snow" is now known to be due to the presence of algae rich in red pigments that mask the green chlorophyll. On the other hand, some algal species grow in hot springs even at a high temperature of 90°C [13].

6.3 GENERAL CHARACTERS OF ALGAE

Algae are chlorophyllous autotrophic organisms [14]. Most of the algae are aquatic; some are terrestrial [14]. They possess thalloid plant body without any variation of plant tissues [14]. They lack vascular tissues (xylem and phloem) and epidermis with stomata [14]. The sex organs are mostly unicellular [14]. Both gametophytic and sporophytic generations are independent in the life cycle of algae [14]. There is no formation of embryo after gametic union [14]. The germination of zygote can be of direct or indirect type [2]. All kinds of reproductions, sexual, asexual, and vegetative, are found in algae [2].

6.4 CLASSIFICATION OF ALGAE

Classification is a scientific categorization of organisms in hierarchical series of groups [11]. Harvey is considered as one of the first algologist who gave the first descriptive algal classification [1]. Although different scientists classified algae in a number of ways, according to the structure of algae, it can be classified into two main groups, i.e., microalgae and macroalgae. Microalgae are unicellular organisms, and that means they have only one cell with complex and robust cell walls, whereas macroalgae are multicellular, have complex structures, and look like plants. Based on the coloring pigments present in macroalgae, they are further divided into three large groups, i.e., red, green, and brown algae [9]. In the older classifications, algae actually were simply divided into four groups, i.e., Phaeophyceae or brown algae, Chlorophyceae or green algae, Rhodophyceae or red algae, and Cyanophyceae or blue-green algae. However, now there is more knowledge about simpler organisms, which were not used to be considered as algae, it has been realized that there is no real reason for such a distinction, and therefore the number of groups is increased. At present, it is most convenient to divide algae into 10–11 classes [15,16]. Generally, algae are categorized into different classes on the basis of nature and properties of pigments, nature of reserve and storage products, type, number, insertion and morphology of flagella, chemical composition and physical features of cell wall, and morphology and characteristics of cells and thalli [10,17].

6.4.1 CHLOROPHYCEAE (GREEN ALGAE)

Chlorophyceae or green algae are a large and diverse group of organisms consisting more than 700 species [17]. Green algae occur in wide range of habitants. They are found in aquatic, amphibious, terrestrial, as well as subaerial conditions. They are principally freshwater species but are also found in seawater. Green algae are eukaryotes characterized by chlorophyll *a* and *b* as major photosynthetic pigments. Majority of the unicellular green algae contain one chloroplast per cell. They store their food as true starch and have quite rigid cell walls composed of cellulose with pectic substances incorporated into the wall structure. Motile gametes possess two, or multiples of two, equal and terminated inserted flagella [11,13,14,17].

6.4.2 XANTHOPHYCEAE (YELLOW-GREEN ALGAE)

Xanthophyceae (yellow-green algae) are represented by 16 genera and 376 species [18]. Most of the members of xanthophyceae are found in free-floating freshwater conditions. Some are found attached to the walls or tree trunks, while others are soil inhabitants. A few representatives are also marine [11,18]. These yellow-green algae were once classified with the green algae. However, their pale green or yellow-green coloration indicates that they have a unique group of pigments. The main pigment in these algae is xanthophyll along with chlorophyll *a* and *e*. Pyrenoids are absent. Reserve food material is oil (glucan and lipids). Their cell walls are composed of cellulose and pectin. Motile gametes possess two or more anterior and unequal flagella, of which larger one is pantonematic-type and smaller is acronematic. In structure xanthophyceae shows similarity with Chlorophyceae [10,11,14,18].

6.4.3 CHRYSOPHYCEAE (GOLDEN ALGAE)

Members of this class contain phycochrysin pigment which imparts brown or orange color to the algae. Motile cells possess one or two equal or unequal and dissimilar flagella. One of the flagellum is anteriorly directed, long, hairy, and pantonematic-type, while the second one is posteriorly directed, short, smooth, and acronematic-type. Reserve food material is oil and chrysolaminarin. Cellulose is absent in the cell wall which has a tendency to become silicified [14].

6.4.4 BACILLARIOPHYCEAE (YELLOW OR GOLDEN-BROWN ALGAE)

Bacillariophyceae or golden-brown algae are unicellular microorganisms, popularly called as diatoms, containing about 190 genera and are classified into 5500 species [18]. It is the largest unicellular class of algae and most widely spread of all the unicellular algae. It occurs in all types of habitats except in hot waters and extremely dry areas. They are found in both freshwater and salt water and in the moist soil. Abundant in cold waters. The yellow or golden-brown color is due to higher proportion of carotenoids and xanthophylls, diatoxanthin, diadino, and fucoxanthin. Their reserve food material is fat; motile structures are almost absent; and pyrenoids are present. Their cell wall is composed of silica and partially of pectin substances [2,14,18].

6.4.5 CRYPTOPHYCEAE

The color of this class varies from brown, red, or olive green to even blue-green due to the presence of different types of pigments. Photosynthetic pigments include biloproteins—different from those of red algae and blue-green algae. They are unicellular, motile, and naked. Flagella are usually equal and located at the anterior end of the cell. Both are band-shaped [14].

6.4.6 DINOPHYCEAE

Members of this class, commonly known as dinoflagellates, are dark yellow or brown with characteristic combination of pigments, chlorophyll *a* and *c*, and carotenoids (dinoxanthin and peridinin). Members are unicellular and biflagellate. One of the flagellum is acronematic-type whereas the second is in helical form, and described as band shaped. Cells contain many discoid chloroplasts with pyrenoids [14].

6.4.7 CHLOROMONADINEAE

Members of this class are bright green in color, unicellular, and biflagellate. Chromatophores are numourous, discoid, and without pyrenoids. Sexual reproduction is absent. Reserve food is fat and oil. The flagella originate from the anterior depression. Of the two, one is anteriorly directed and bears hair, and the second is posteriorly directed and smooth [14].

6.4.8 EUGLENINEAE

Members of this class almost resemble to those of Chlorophyceae in having green color, which is due to excess of chlorophyll. The chloroplasts can be discoid or stellate. Their reserve food material is paramylum, with a number of pantonematic-type of flagella, may be 1, 2, or 3. They reproduce by cell division [14].

6.4.9 PHAEOPHYCEAE (BROWN ALGAE)

Phaeophyceae or brown algae are represented by about 240 genera and over 1500 species of which 99.7% are marine [11] and live attached to the rocks around sea coasts. They therefore occupy same kinds of habitat as the red algae. Brown algae grow as attached not only to the rocks but also to the dykes, quays-attached mollusks, eelgrass, or even other seaweeds [11,19]. These are multicellular algae and comprise a brown pigment which is responsible for their characteristic brown color and the common name of brown algae or brown seaweeds. They are structurally quite complex [10]. The brown algae owe their color to the pigment fucoxanthin. Fucoxanthin is present in large quantity and masks the green color of the chlorophylls. Other pigments include chlorophyll *a*, *c*, β-carotene, violaxanthin, flavoxanthin, etc. Their photosynthetic products are polysaccharides, fats, traces of simple sugars, and alcohols. Food is stored in the form of soluble carbohydrate called laminarin and an alcohol called mannitol. The cell wall is composed of cellulose, fuicinic acid, and alginic acid. The flagellated motile cells are pyriform and bear two laterally borne flagella, of which one is longer, anteriorly directed and pantonematic type and the second one is short, posteriorly directed and acronematic-type [14].

6.4.10 MYXOPHYCEAE (CYANOPHYCEAE)

The class Myxophyceae or Cyanophyceae is commonly called as blue-green algae because of the presence of a principal bluish-green pigment (c-phycocyanin) along with chlorophyll *a*, β-carotene, and some quantity of Myxoxanthin, Myxoxanthophyll as well as small quantities of carotene, flavacin, and c-phycoerythrin [11]. They can be unicellular or multicellular, free-living or colonial, unbranched filamentous or branched filamentous, and acquatic or terrestrial in nature [1]. The blue-green algae are also known as *Cyanobacteria* because they resemble to bacteria [19]. The photosynthetic reserve food of blue-green algae is glycogen. Motile flagellated cells are altogether absent, cellular organization is simple prokaryotic type lacking membrane-bound cell organelles such as nucleus, chromatophores, dictyosomes, ER, and true vacuoles. Cells lack definite chromosomes and divide by simple fission [14].

6.4.11 RHODOPHYCEAE (RED ALGAE)

The red algae are a distinct lineage of eukaryotic algae, containing about 5000–6000 species [20]. The majority of red algae live in tropical marine habitats, and most of the species are multicellular. Red algae have a complex life history, which means they go through several stages of independent organisms to complete their life cycles. Most undergo sexual reproduction. Like many other algae, rhodophytes also contain a phosynthetic pigment chlorophyll and are able to photosynthesize their own food. These algae contain reddish pigment phycoerythrin in a large amount; the red color of these algae is due to the presence of this pigment [21]. Other pigments include chlorophyll *a* and *d*, carotenoids (α- and β-carotenes), and xanthophylls (taraxanthin, lutein, zeaxanthin, etc.). Reserve food is stored in the form of floridean starch and galacto-fructosides. Their cell wall is mainly composed of cellulose and pectin, and in addition some components of polysulfate esters of carbohydrates also occur. Majority of the red algae exhibit triphasic life cycles but some also exhibit biphasic life cycles [14].

6.5 BIOLOGICALLY ACTIVE COMPOUNDS EXTRACTED FROM ALGAE

6.5.1 SULFATED POLYSACCHARIDES

Sulfated polysaccharides are the polysaccharides present in the cell wall of algae, which play storage and structural roles in seaweeds, and may exhibit many interesting biological properties. Seaweeds are the main source of sulfated polysaccharides [22,23]. The details of these compounds are briefly summarized in Table 6.1.

6.5.2 POLYPHENOLIC COMPOUNDS

Phenolic compounds comprise a wide variety of molecules that have a polyphenol structure. Polyphenols are divided into several classes according to the number of phenol rings they contain and the structural elements that bind these rings to one another. The main groups of polyphenols are flavonoids, phenolic acids, tannins (hydrolyzable, condensed, and phlorotannin), stilbenes, and lignans [32].

6.5.3 FLAVONOIDS

Flavonoids are the natural products that consist of C6−C3−C6 carbon framework, or more specifically phenylbenzopyran functionality. Depending on the position of the linkage of aromatic ring to the benzopyran moiety, flavonoids can be divided into three classes [33].

- Flavonoids (2-phenylbenzopyrans)
- Isoflavonoids (3-benzopyrans)
- Neoflavonoids (4-benzopyrans)

The details of these compounds are briefly summarized in Table 6.2.

6.5.4 PHLOROTANNINS

Phlorotannins are phenolic compounds formed by polymerization of phloroglucinol (1,3,5-trihydroxybenzene) monomer units and biosynthesized through the

Table 6.1 Sulfated Polysaccharides: Biologically Active Compounds Extracted From Algae

Sr. No.	Name	Molecular Formula	Biological Source	Biological Use	Structure	References
1	Heparin	$C_{12}H_{19}NO_{20}S_3$	Red alga *Delesseria sanguinea*	Anticoagulant, antiinflammatory, and antimetastastic effects	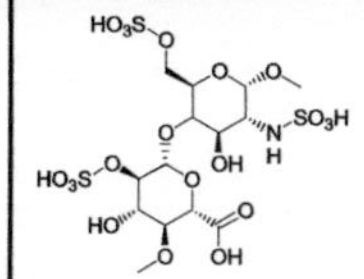	[24]
2	Agar		Extracted from red marine algae (Rhodophyceae); *Gelidiella acerosa*, *Gelidiella various* Gelidium spp. *Gracilaria confervoides*, *Pterocladia capillacea*, and *Pterocladia lucida*		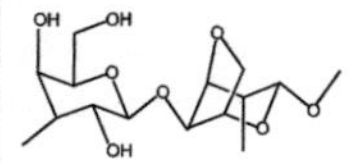	[25]
3	Chrysolaminarin		Isol. from diatoms and related algae	Reserve carbohydrate	N/A	[26]
4	Lambda (λ)-carrageenan	$C_{14}H_{26}O_{19}S_3$	Isol. from species of the *Gigartina* and *Chondrus* genera	Antiviral, antioxidant	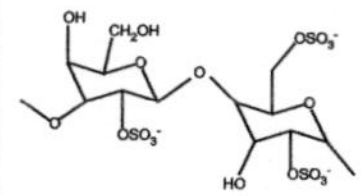	[27]
5	Iota (ι)-carrageenan	$C_{14}H_{24}O_{15}S_2$	Constit. of red seaweed *Eucheuma spinosum*	Antioxidant	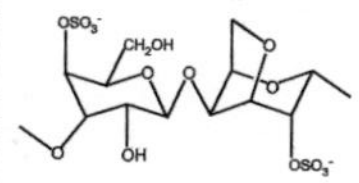	[27]
6	Kappa (κ)-carrageenan	$C_{14}H_{25}O_{12}S$	Constit. of red seaweed *Kappaphycus alvarezii*	Antioxidant	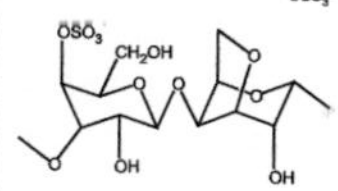	[27]
7	Fucans		Isol. from *Analipus japonicas*, *Ascophyllum nodosum*, *Chorda filum*, *Fucus evanescens*, and *Laminaria saccharina*	Anticoagulant	N/A	[27]

Continued

Table 6.1 Sulfated Polysaccharides: Biologically Active Compounds Extracted From Algae—cont'd

Sr. No.	Name	Molecular Formula	Biological Source	Biological Use	Structure	References
8	Alginate	$(C_6H_8O_6)_n$	Constit. of red and brown algae		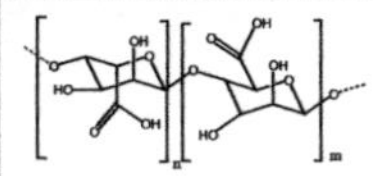	[28]
9	Ulvan		Isol. from cell walls of marine green algae *Ulva rigida*, *Ulva pertusa*, and *Ulva lactuca*	Shows cytotoxic props. Antioxidant	N/A	[29]
10	Sulfated mannans		Present in red and green seaweeds	Inhibit the propagation of HSV-1 in Vero cells	N/A	[30]
11	Rhamnan sulfate		Present in cell walls of *Monostroma latissimum* and *Monostroma nitidum*	Antiviral	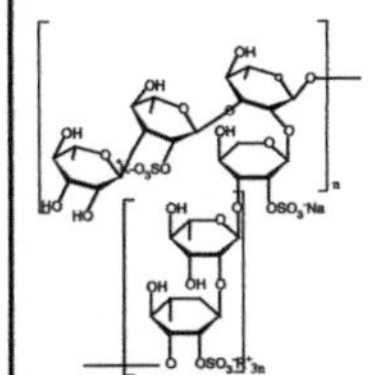	[31]
12	Laminarin	$(C_6H_{10}O_5)_n$	Found in brown algae especially the *Laminaria* subgroup; isol. from *Laminaria cloustoni*	Anticoagulant; antilipemic agent; plant activator	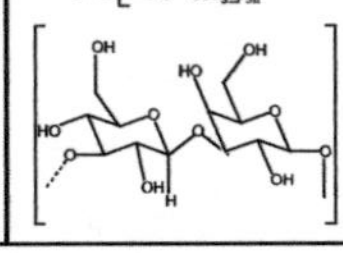	[26]

Constit., constituents; *Isol.*, isolated.

Table 6.2 Flavonoids: Biologically Active Constituents Extracted From Algae

Sr. No.	Name	Molecular Formula	Biological Source	Biological Use	Structure	References
1	4′-Chloro-2-hydroxyaurone	$C_{15}H_{11}O_3Cl$	Constit. of brown algae *Spatoglossum variabile*	Antioxidant	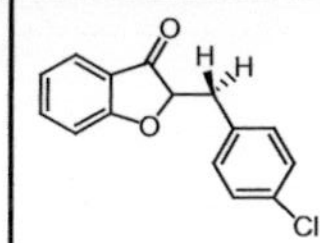	[34]
2	4′-Chloroaurone	$C_{15}H_9O_2Cl$	Constit. of brown algae *S. variabile*	Antioxidant	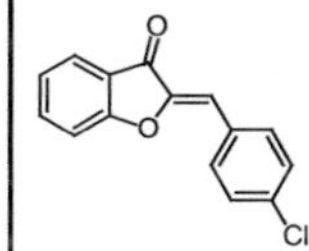	[34]
3	Morin	$C_{15}H_{10}O_7$	Constit. of green algae *Caulerpa serrulata*	Antioxidant	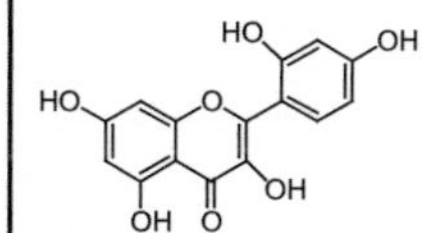	[35]
4	Hesperidin	$C_{28}H_{34}O_{15}$	Constit. of red algae	Antioxidant	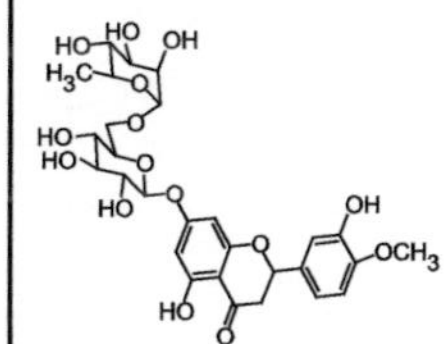	[35]
5	Myricetin	$C_{15}H_{10}O_8$	Constit. of *Tubinaria ornate*, *Chondrus verruscosus*	Antioxidant	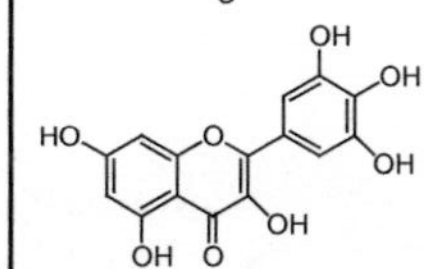	[35]

Continued

Table 6.2 Flavonoids: Biologically Active Constituents Extracted From Algae—cont'd

Sr. No.	Name	Molecular Formula	Biological Source	Biological Use	Structure	References
6	Quercetin	$C_{15}H_{10}O_7$	Constit. of *Undaria pinnatifida* and *Padina arborescens*	Antioxidant	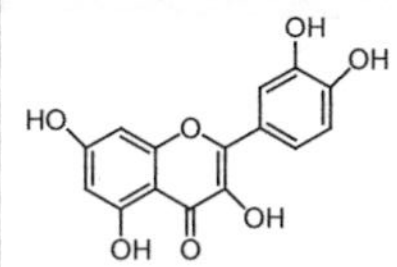	[35]
7	(−)-Catechin (2*S*,3*R*)	$C_{15}H_{14}O_6$	Constit. of red alga *Acanthophora spicijka*	Antioxidant	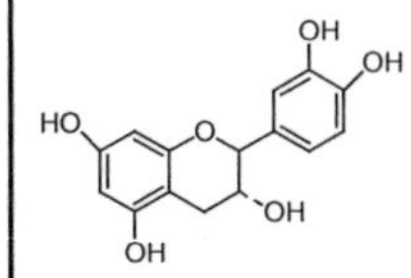	[36]
8	Apigenin	$C_{15}H_{10}O_5$	Constit. of *Acanthophora spicifera*	Analgesic and antiinflammatory	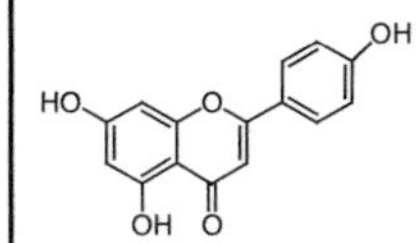	[37]
9	Scutellarein 4′-methyl ether	$C_{16}H_{12}O_6$	Constit. of *Osmundea pinnatifida*	Antileishmanial	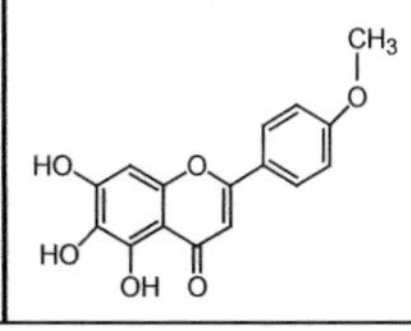	[38]

Constit., constituents.

acetate—malonate pathway, also known as polyketide pathway. Phlorotannins are highly hydrophilic components with a wide range of molecular sizes ranging between 126 Da and 650 kDa. Based on the types of linkage, phlorotannins can be classified into four subclasses such as [39]:

- Fuhalols and phlorethols (phlorotannins with an ether linkage)
- Fucols (with a phenyl linkage)
- Fucophloroethols (with an ether and phenyl linkage)
- Eckols (with a dibenzodioxin linkage)

The details of these compounds are briefly summarized in Table 6.3.

6.5.5 TERPENES AND TERPENOIDS

Terpenes are the hydrocarbons composed of carbon and hydrogen, and are the compounds whose carbon skeleton can be divided into two or more units identical with the carbon skeleton of isoprene. They are formed by bonding the tail of one isoprene unit to the head of another. Terpenes can be oxygenated into terpene—alcohols, terpene—esters, terpene—oxides, terpene—ketones, terpene—aldehydes, terpene—esters, and other varieties of oxygenated compounds [71,72]. The details of these compounds are briefly summarized in Table 6.4.

6.5.6 PHYCOBILIPROTEINS

Phycobiliproteins (PBPs) are the important photosynthetic accessory pigment biomolecules assembled into a supramolecular light-harvesting antenna complex, phycobilisomes (PBS), in cyanobacteria and red algae [145]. The details of these compounds are briefly summarized in Table 6.5.

6.5.7 STEROLS

Sterols are steroids carrying a hydroxyl group at C-3 [146]. Sterols are a type of lipids found in both plants and animals. Although sterols are classified as lipids, they differ significantly from triglycerides and phospholipids in structure and function. Sterol molecules consist of multiple rings made primarily of carbon and hydrogen atoms that are attached to each other [147]. The details of these compounds are briefly summarized in Table 6.6.

6.5.8 POLYHYDROXYALKONATES

Polyhydroxyalkonates (PHAs) are polyesters of various hydroxyalkanoates that are synthesized by many gram-positive and gram-negative bacteria in at least 75 different genera. These polymers are accumulated intracellularly to levels as high as 90% of the cell dry weight under conditions of nutrient stress and act as a carbon and energy reserve. The molecular mass of PHAs varies per PHA producer, but is

Table 6.3 Phlorotannins: Biologically Active Compounds Extracted From Algae

Sr. No.	Name	Molecular Formula	Biological Source	Biological Use	Structure	References
1	1,2,3,5-Benzenetetrol; 2,5-*O*-disulfate	$C_6H_6O_{10}S_2$	Constit. of brown algae *Ascophyllum nodosum*, *Petalonia fascia*, *Scytosiphon lomentaria*, *Chorda filum*, *Fucus distichus* spp. *anceps*, *Fucus vesiculosus*, *Pelvetia canaliculata,* and *Dictyota dichotoma*	Biogenetic precursor of the wide range of oligomeric phlorotannins which are a major component of brown algae	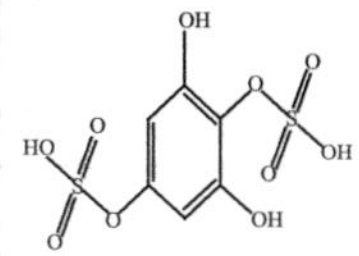	[40]
2	1,2,3,5-Benzenetetrol,2-*O*-sulfate	$C_6H_6O_7S$	Constit. of brown algae *P. fascia*, *S. lomentaria*, *D. dichotoma*, *A. nodosum*, *F. vesiculosus*, and *Himanthalia elongata*		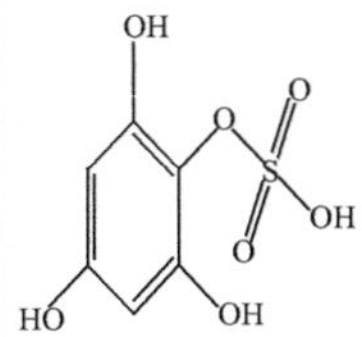	[41,42]
3	1,3,5-Benzenetriol	$C_6H_6O_3$	A minor constit. of brown algae, e.g., *Halidrys siliquosa*, *Cystophora retroflexa*, *Eisenia arborea*, *Laminaria ochroleuca*, *Analipus japonicus*, *Cystophora congesta*, and *Carpophyllum angustifolium*	Spasmolytic agent, hyaluronidase inhibitor	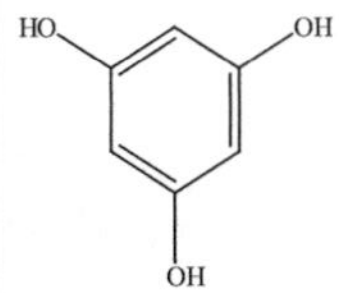	[43]
4	Benzo[1,2-b:3,4-b′]bis [1,4]benzodioxin-1,3,6,9,11-pentol	$C_{18}H_{10}O_9$	Constit. of brown algae *E. arborea*, *Ecklonia maxima*, and *Ecklonia stolonifera*		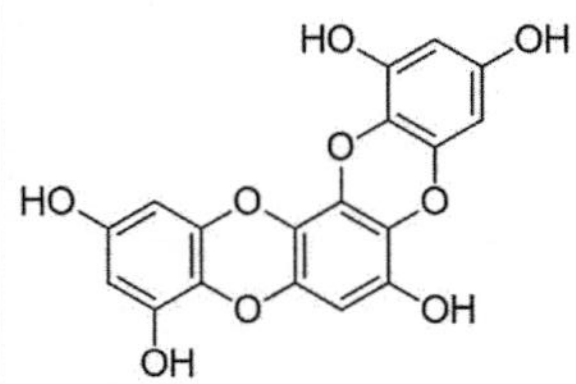	[44]

5	4‴,7-Bieckol	$C_{36}H_{22}O_{18}$	Constit. of brown algae *E. maxima*, *E. arborea*, and *Eisenia bicyclis*		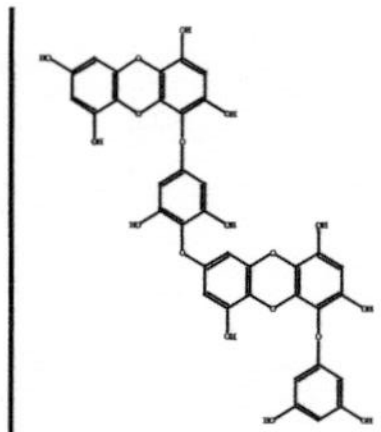	[45]
6	6,6′-Bieckol	$C_{36}H_{22}O_{18}$	Constit. of brown algae *Ecklonia kurome* and *E. arborea*	Antiplasmin inhibitor; α2-macroglobulin inhibitor; exhibits anti-HIV activity, antioxidant	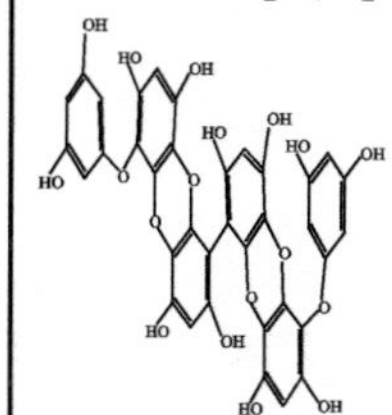	[44]
7	8,8′-Bieckol	$C_{36}H_{22}O_{18}$	Constit. of brown algae *E. kurome* and *E. arborea*	Antiplasmin inhibitor; α2-macroglobulin inhibitor, hyaluronidase inhibitor	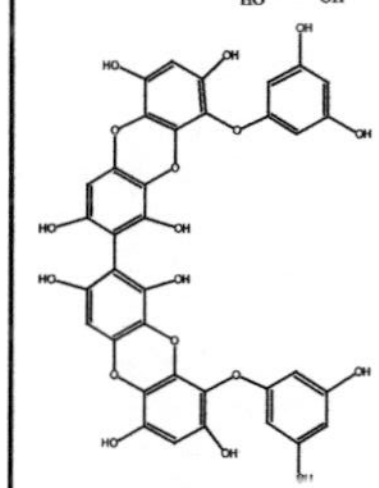	[44]
8	2,2′,4,4′,6,6′-Biphenylhexol	$C_{12}H_{10}O_6$	Occurs in several brown algae, e.g., *F. vesiculosus*, *C. retroflexa*, *C. angustifolium*		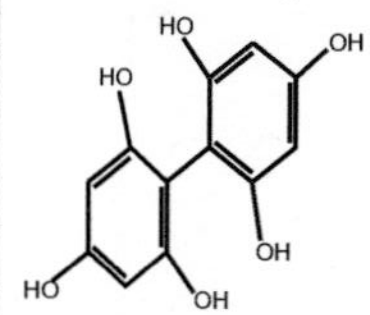	[46]
9	2-Chloro-1,3,5-benzenetriol	$C_6H_5ClO_3$	Constit. of *Rhabdonia verticillata* and from brown algae *E. arborea* and *C. angustifolium*		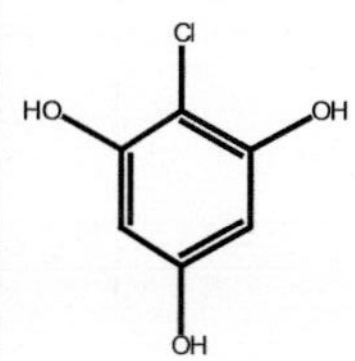	[47]

Continued

Table 6.3 Phlorotannins: Biologically Active Compounds Extracted From Algae—cont'd

Sr. No.	Name	Molecular Formula	Biological Source	Biological Use	Structure	References
10	Decafuhalol-A	$C_{60}H_{42}O_{35}$	Constit. of brown algae *Sargassum spinuligerum* and *C. angustifolium*		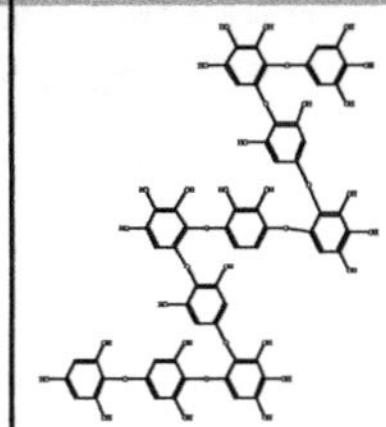	[48]
11	Dieckol	$C_{36}H_{22}O_{18}$	Constit. of brown algae *E. kurome* and *Ei. bicyclis*	Inhibitor of α2-macroglobulin, glycation and α-amylase	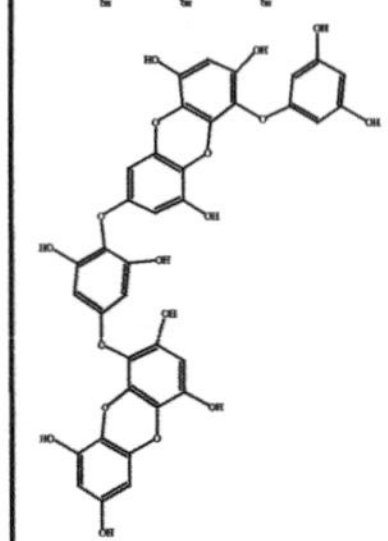	[49,50]
12	Difucophlorethol-A	$C_{24}H_{18}O_{12}$	Isol. from brown algae *H. elongata*, *C. retroflexa*, *Cystophora torulosa*, and *Xiphophora chondrophylla*		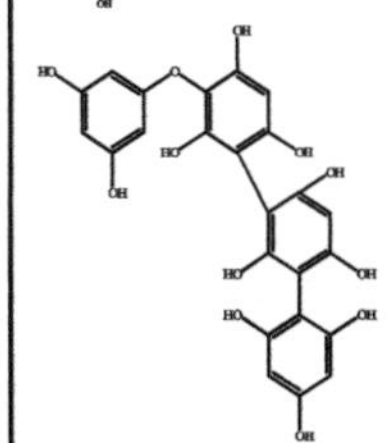	[51,52]
13	Dihydroxyfu-cotriphlorethol-B	$C_{30}H_{22}O_{17}$	Isol. from brown algae *S. spinuligerum* and *C. angustifolium*		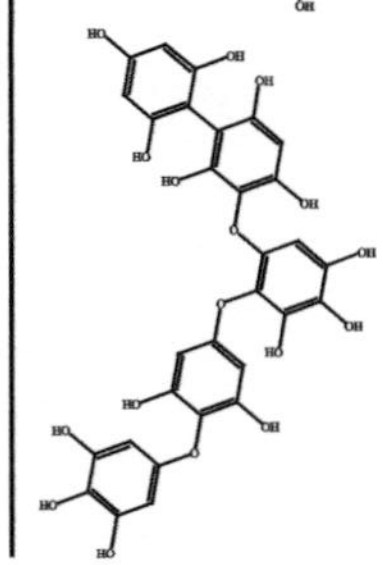	[53]

14	Eckol	$C_{18}H_{12}O_9$	Constit. of brown algae *E. kurome*, *E. maxima*, *E. arborea*, and *Ei. bicyclis*	Inhibitor of α2-macroglobulin, glycation, and α-amylase	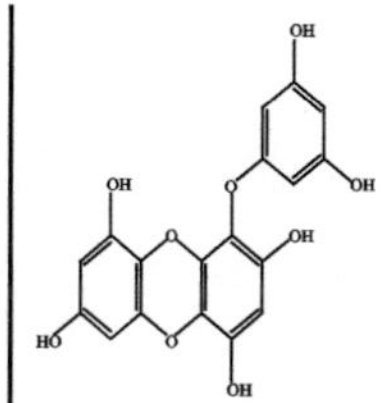	[49]
15	Fucodifucot-etraphlorethol-A	$C_{48}H_{34}O_{24}$	Isol. from brown algae *C. torulosa*, *S. spinuligerum*, and *C. angustifolium*		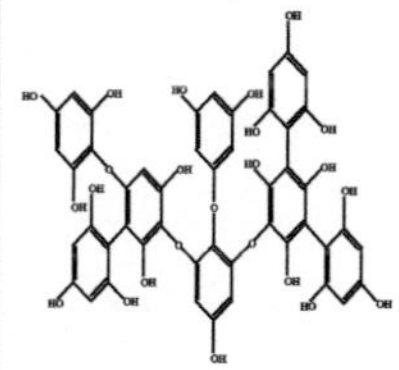	[53]
16	Fucodiphlorethol-D	$C_{24}H_{18}O_{12}$	Isol. from brown algae *Cystoseira baccata*, *C. congesta*, *C. angustifolium*, *S. spinuligerum*, and *C. retroflexa*		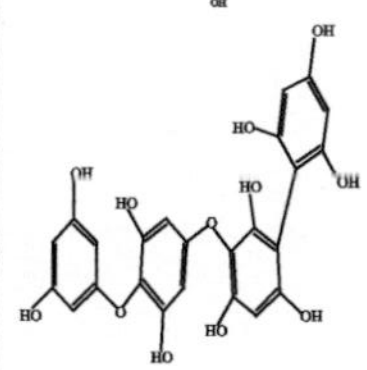	[54]
17	Fucodiphlorethol-D; 4‴-hydroxy	$C_{24}H_{18}O_{13}$	Isol. from brown algae *S. spinuligerum*, *Carpophyllum maschalocarpum*, and *C. torulosa*		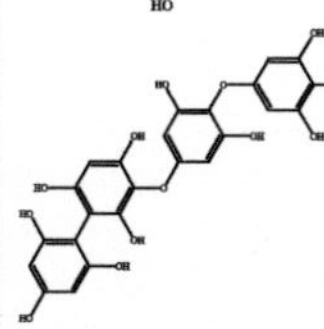	[55,56]
18	Fucodiphlorethol-E	$C_{24}H_{18}O_{12}$	Isol. from brown algae *Durvillaea antarctica*, *S. spinuligerum*, *C. torulosa*, and *F. vesiculosus*		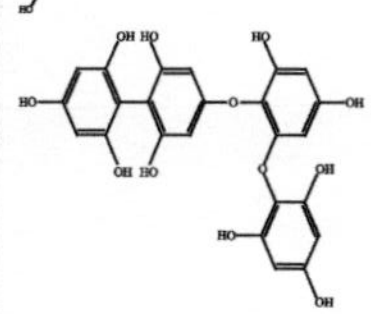	[55]

Continued

Table 6.3 Phlorotannins: Biologically Active Compounds Extracted From Algae—cont'd

Sr. No.	Name	Molecular Formula	Biological Source	Biological Use	Structure	References
19	Fucophlorethol-A	$C_{18}H_{14}O_9$	Constit. of *F. vesiculosus*	Hyaluronidase inhibitior	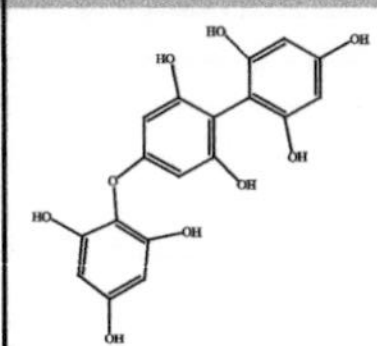	[57]
20	2,3′,4,4′,5′,6-Hexahydroxydiphenyl ether	$C_{12}H_{10}O_7$	Constit. of *Bifurcaria bifurcata*, *Halidrys* sp., *C. angustifolium*, and other brown algae		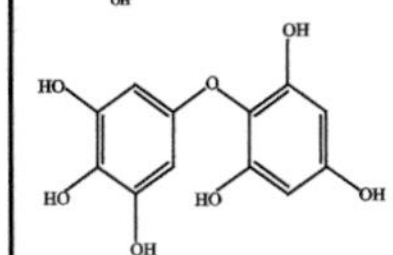	[58]
21	1-C-Methyl-scyllo-inositol	$C_7H_{14}O_6$	Isol. from red alga *Polysiphonia fastigiata* and brown algae		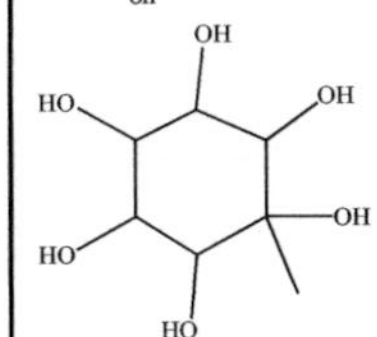	[59]
22	4-C-Methyl-myo-inositol; D-form	$C_7H_{14}O_6$	Isol. from red algae, e.g., *P. fastigiata* and brown alga, e.g., *Laminaria cloustoni*	Osmoregulator, especially in brown algae frequently exposed to both fresh water and salt water	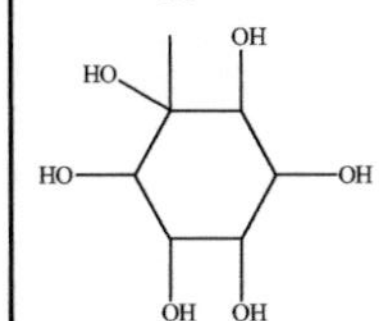	[59]

23	Nonafuhalol-A	$C_{54}H_{38}O_{31}$	Constit. of brown algae *Sargassum muticum*, *S. spinuligerum*, *C. maschalocarpum*, and *C. angustifolium*		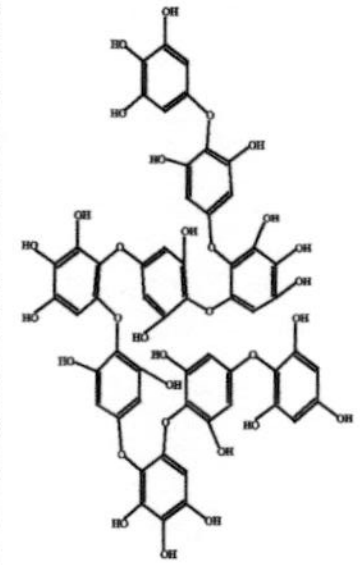	[60]
24	Octafuhalol-A	$C_{48}H_{34}O_{28}$	Constit. of brown algae *S. muticum*, *S. spinuligerum*, *C. maschalocarpum*, and *C. angustifolium*		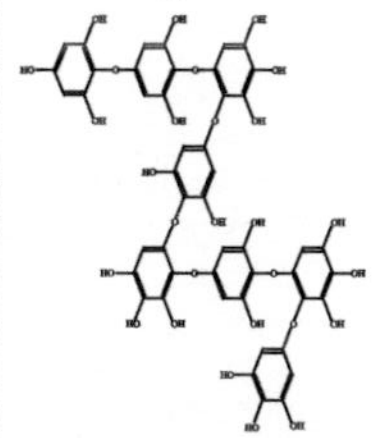	[61]
25	Pentafucol-A	$C_{30}H_{22}O_{15}$	Isol. from brown algae *Scytothamnus australis* and *A. japonicas*		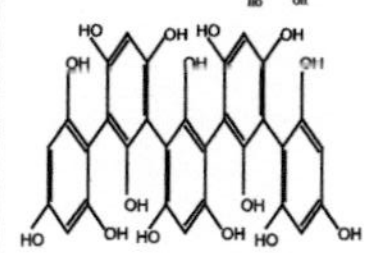	[62,63]
26	2,3′,4,5′,6-Pentahydroxydiphenyl ether	$C_{12}H_{10}O_{6}$	Constit. of brown algae including *A. japonicus*, *L. ochroleuca*, *Sargassum thunbergii*, *Ecklonia bicyclis*, *H. elongata*, *B. bifurcata*, *Cystoseira tamariscifolia*, *C. congesta*, and *S. muticum*		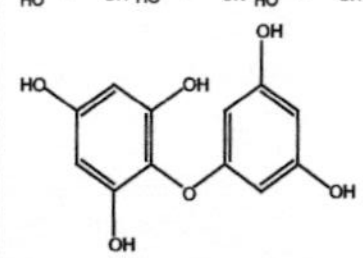	[64]

Continued

Table 6.3 Phlorotannins: Biologically Active Compounds Extracted From Algae—cont'd

Sr. No.	Name	Molecular Formula	Biological Source	Biological Use	Structure	References
27	Pseudoheptafuhalol-B	$C_{42}H_{30}O_{24}$	Constit. of brown algae *S. spinuligerum* and *C. angustifolium*		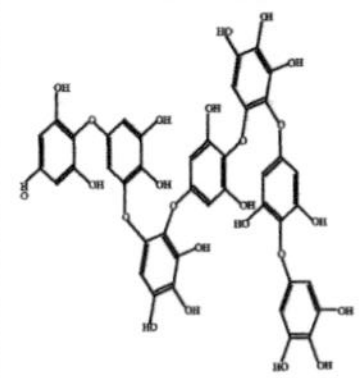	[65]
28	Pseudoheptafuhalol-C	$C_{42}H_{30}O_{24}$	Constit. of brown algae *S. spinuligerum* and *C. angustifolium*		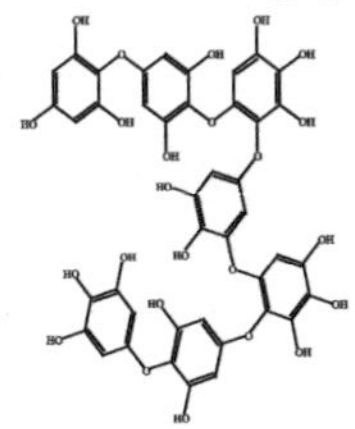	[65]
29	Pseudoheptafuhalol-D	$C_{42}H_{30}O_{24}$	Constit. of brown algae *S. spinuligerum* and *C. angustifolium*		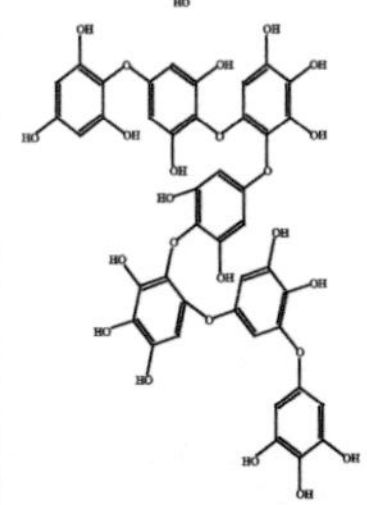	[65]
30	Pseudohexafuhalol-B	$C_{36}H_{26}O_{21}$	Constit. of brown algae *S. spinuligerum* and *C. angustifolium*		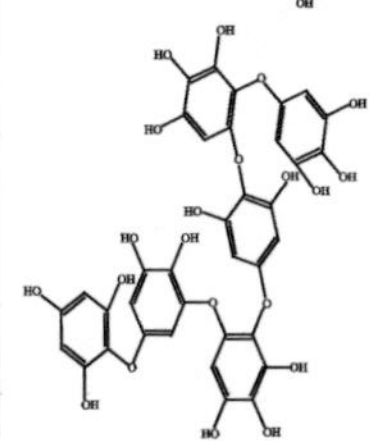	[65]

31	Pseudohexafuhalol-C	$C_{36}H_{26}O_2$	Constit. of brown algae *S. spinuligerum* and *C. angustifolium*		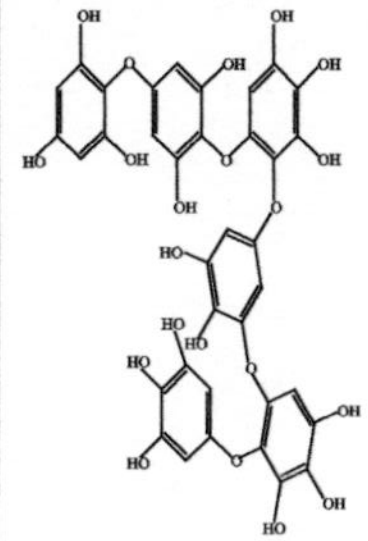	[65]
32	Pseudotetrafuhalol-A	$C_{24}H_{18}O_{14}$	Constit. of algae *C. maschalocarpum* and *S. spinuligerum*		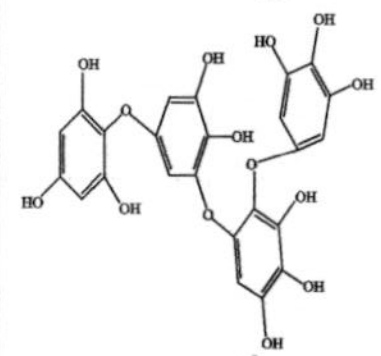	[65]
33	Terfucohex-aphlorethol-A	$C_{60}H_{42}O_{30}$	Isol. from brown algae *C. retroflexa*, *C. maschalocarpum*, and *C. angustifolium*		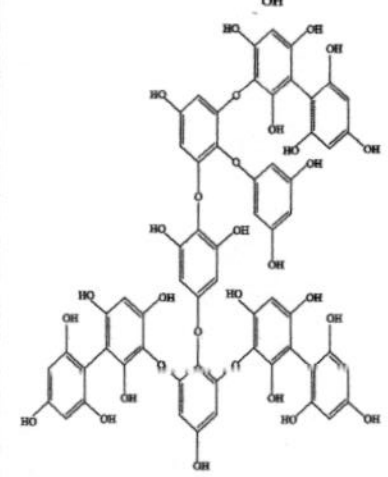	[52,56]
34	Terfucopent-aphlorethol-A	$C_{54}H_{38}O_{27}$	Isol. from brown algae *C. retroflexa* and *Carpophyllum maschalocarpum*		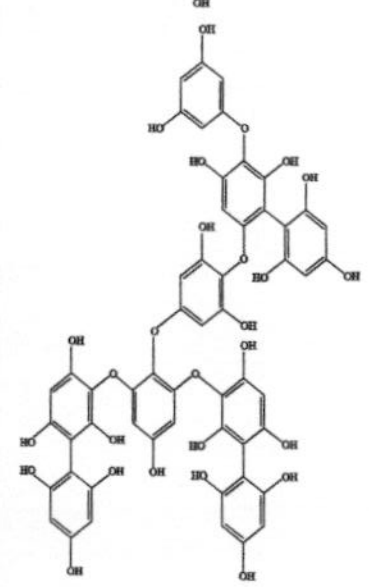	[56]

Continued

Table 6.3 Phlorotannins: Biologically Active Compounds Extracted From Algae—cont'd

Sr. No.	Name	Molecular Formula	Biological Source	Biological Use	Structure	References
35	[1,1′:3′,1″-Terphenyl]-2,2′,2″,4,4′,4″,6,6′,6″-nonol	$C_{18}H_{14}O_9$	Constit. of various brown algae incl. *A. japonicus*, *B. bifurcata*, *F. vesiculosus*, and *H. elongate*		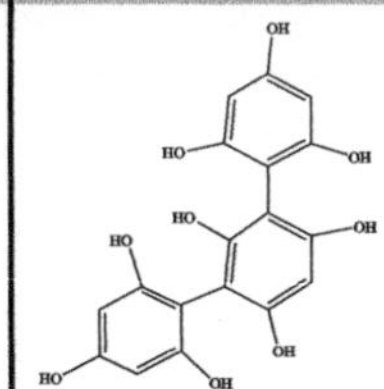	[45,66]
36	Tetrafucol-A	$C_{24}H_{18}O_{12}$	Constit. of brown algae *F. vesiculosus*, *A. japonicas*, and *C. angustifolium*		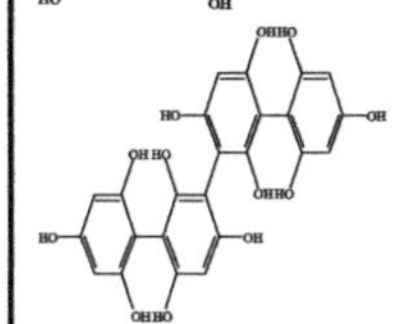	[62]
37	Tetraphlorethol-C	$C_{24}H_{18}O_{12}$	Constit. of algae *L. ochroleuca*, *C. congesta*, *E. maxima*, *Cystophora reflexa*, *S. spinuligerum*, and *C. angustifolium*		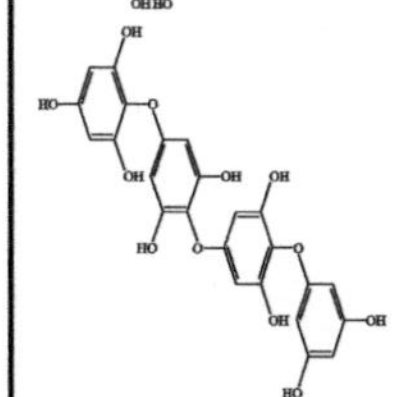	[54]
38	1-(2,4,6-Trihydroxyphenyl)-5,8,11,14,17-eicosapentaen-1-one; (all-*Z*)-form	$C_{26}H_{34}O_4$	Constit. of brown algae *Zonaria turneriana*, *Zonaria diesingiana*, *Zonaria farlowii*, *Zonaria tournefortii*, and *Distromium decumbens*		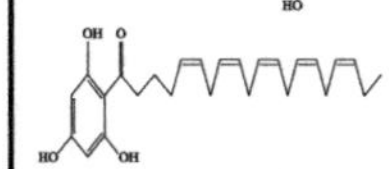	[67]

39	1-(2,4,6-Trihydroxyphenyl)-6,9,12,15-octadecatetraen-1-one; (all-*Z*)-form	$C_{24}H_{32}O_4$	Isol. from brown algae *Cystophora* spp.		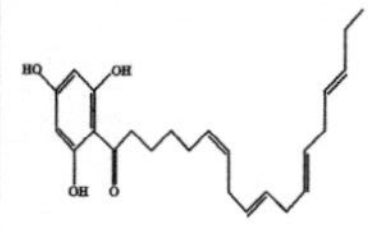	[68]
40	Triphlorethol-A	$C_{18}H_{14}O_9$	Constit. of *C. congesta* and other brown algae		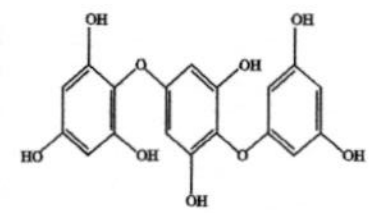	[69]
41	Triphlorethol-B	$C_{18}H_{14}O_9$	Constit. of brown algae *E. kurome*, *E. stolonifera*, *C. baccata*, and *C. angustifolium*	Shows antitrypsin and antiplasmin activities	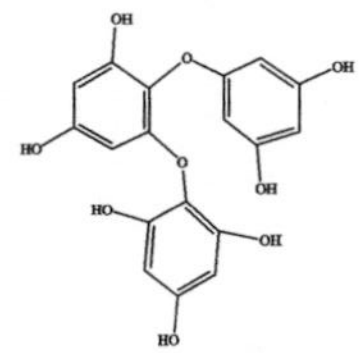	[47]
42	Undecafuhalol-A	$C_{66}H_{46}O_{38}$	Constit. of brown algae *B. bifurcata*, *Landsburgia quercifolia*, *C. angustifolium*, and *S. spinuligerum*		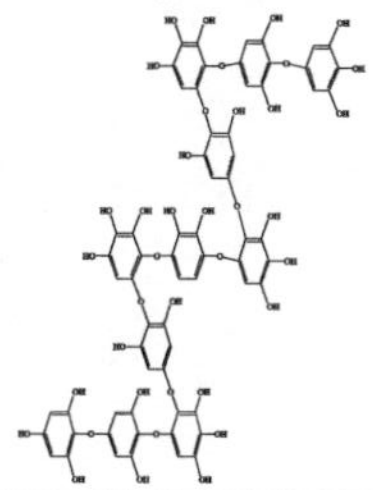	[70]

Constit., constituents; *Isol.*, isolated.

Table 6.4 Terpenes and Terpenoids: Biologically Active Constituents Extracted From Algae

Sr. No.	Name	Molecular Formula	Biological Source	Biological Use	Structure	References
1	2-[2-(Acetyloxy)ethenyl]-6,10-dimethyl-2,5,9-undecatrienal	$C_{17}H_{24}O_3$	*Penicillus capitatus* and *Udotea cyathiformis*	Antibacterial and ichthyotoxin	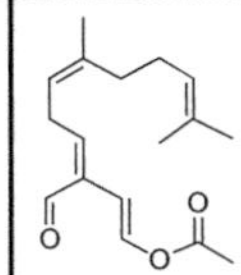	[73]
2	Adonixanthin	$C_{40}H_{54}O_3$	Isol. from green algae		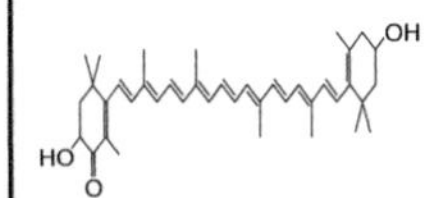	[74]
3	1(10)-Aristolen-9-ol; (ent-9α)-form	$C_{15}H_{24}O$	Constit. of marine algae		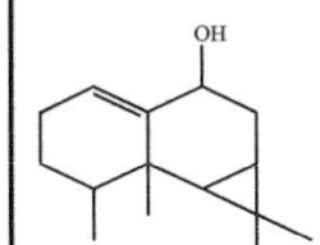	[75]
4	9-Aristolen-1-ol; (ent-1β)-form	$C_{15}H_{24}O$	Constit. of *Nardostachys chinensis*, *Nardostachys grandiflora*, and marine algae		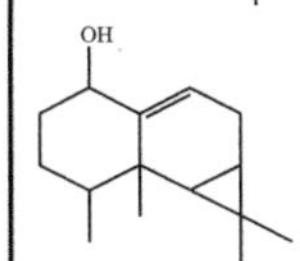	[76]
5	1,2-Benzenedicarboxylic acid; dibutyl ester	$C_{16}H_{22}O_4$	Reported from *Penicillium bilaii*. *Streptomyces nasri*, and *Streptomyces melanosporofaciens*. Also isol. from various red algae	Insect repellent; now superseded; glycosidase inhibitor	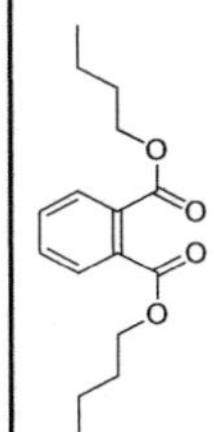	[77]
6	C34 Botryococcene; 1,2,6,7,21,22,24,29-Octahydro	$C_{34}H_{66}$	Constit. of *lacustrine sediments* attributed to algae		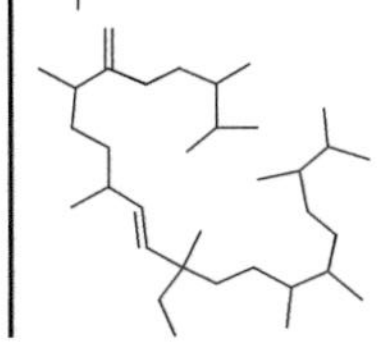	[78]

7	3-Bromo-4,5-dihydroxybenzaldehyde	$C_7H_5BrO_3$	Constit. of red algae *Polysiphonia* spp. and sponge *Psammaplysilla purpurea*	Antibiotic	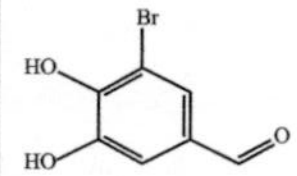	[79]
8	3-Bromo-4,5-dihydroxybenzoic acid	$C_7H_5BrO_4$	Constit. of Indian algae		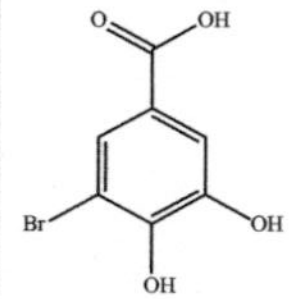	[80]
9	3-Bromo-4-hydroxybenzaldehyde	$C_7H_5BrO_2$	Isol. from *Asparagopsis taxiformis*, *P. purpurea,* and *Rhodomela larix;* constit. of Indian algae		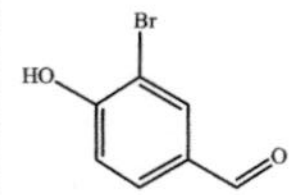	[81]
10	3-Bromo-4-hydroxybenzoic acid	$C_7H_5BrO_3$	Constit. of Indian algae		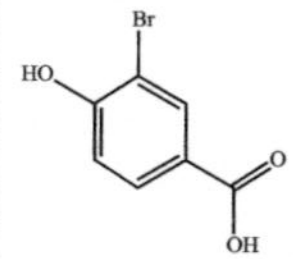	[82]
11	6-(1,3-Butadienyl)-1,4-cycloheptadiene; (*R*)-form	$C_{11}H_{14}$	Isol. from brown algae *Desmarestia aculeata* and *Desmarestia viridis*	Algal gamete sex attractant	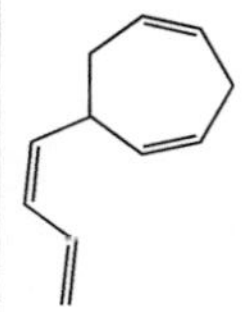	[83]
12	3-(1-Butenyl)-4-vinylcyclopentene; (3*S*,4*S*)-(*Z*)-form, 3′,4′-Didehydro	$C_{11}H_{14}$	Isol. from brown algae *D. aculeata*, *D. viridis,* and Syringoderma sp.	Algal gamete sex attractant	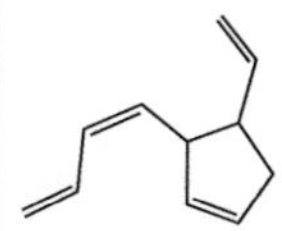	[84,85]
13	6-Butyl-1,4-cycloheptadiene; (*R*)-form	$C_{11}H_{18}$	Isol. from brown algae *Dictyopteris* spp. and *Dictyota dichotoma*		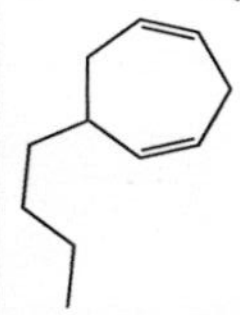	[86,87]

Continued

Table 6.4 Terpenes and Terpenoids: Biologically Active Constituents Extracted From Algae—cont'd

Sr. No.	Name	Molecular Formula	Biological Source	Biological Use	Structure	References
14	Colletodiol; 9,10-dihydro, 11,12-diketone	$C_{14}H_{18}O_6$	Constit. of *Colletotrichum capsici*, *Cytospora* sp.	Antibiotic, weakly active against gram-positive and gram-negative bacteria and blue-green algae; spore germination inhibitor	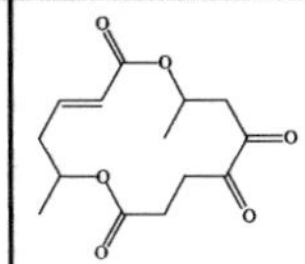	[88]
15	Crustaxanthin; tetraketone	$C_{40}H_{48}O_4$	Found in fish, crustaceans, bacteria (*Mycobacterium laticolum*), and algae (Euglena)		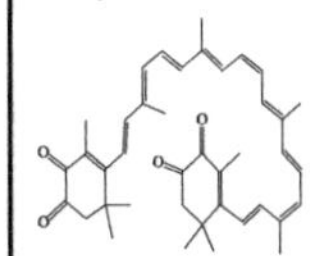	[89]
16	β-Cryptoxanthin	$C_{40}H_{56}O$	Isol. from papaya (*Carica papaya*) and many other higher plants, also from fungi, diatoms blue-green algae and fish eggs		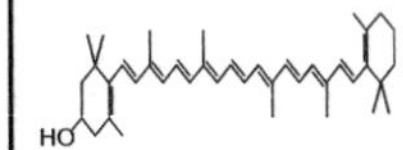	[90]
17	Diatoxanthin	$C_{40}H_{54}O_2$	Constit. of diatoms, algae, and fish		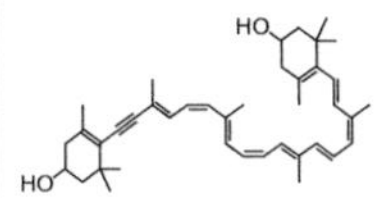	[91,92]
18	2,10-Dibromo-3-chloro-5,10-epoxy-8-chamigren-7-ol	$C_{15}H_{21}Br_2ClO_2$	Constit. of red algae *Laurencia pacifica*, *Laurencia majuscula*, *Laurencia nidifica*, *Laurencia claviformis*, and *Laurencia marianensis* and the mollusk *Aplysia dactilomela*	Feeding deterrent for aphids, insecticide, antimitotic agent	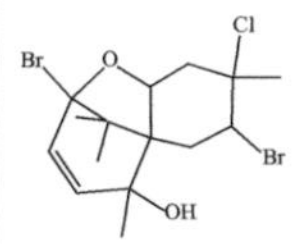	[93]
19	2,3-Dibromo-4,5-dihydroxybenzaldehyde	$C_7H_4Br_2O_3$	Isol. from algae *Polysiphonia lanosa* and *R. larix*		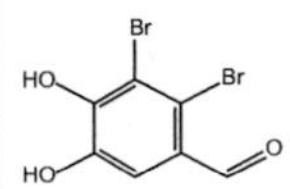	[94,95]
20	3,5-Dibromo-4-hydroxybenzyl alcohol	$C_7H_6Br_2O_2$	Isol. from red algae *Polysiphonia* spp. and *R. larix* and *Rhodomela confervoides*		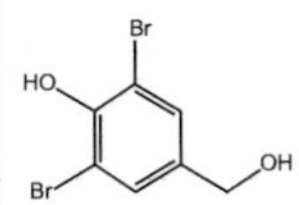	[96]

21	3,4-Dibromo-5-hydroxymethyl-1,2-benzenediol	$C_7H_6Br_2O_3$	Isol. from red algae *Lenormandia prolifera*, *Polysiphonia*, *Odonthalia*, and *Rhodomela* spp.	Glucosidase inhibitor	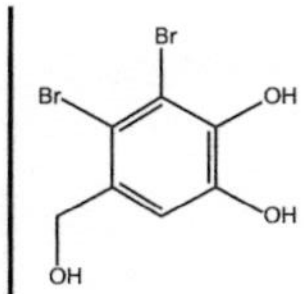	[79,97]
22	2-(Dibromomethylene) hexanoic acid	$C_7H_{10}Br_2O_2$	Isol. from red algae *Bonnemaisonia hamifera* and *Bonnemaisonia nootkana*		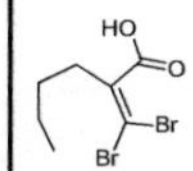	[98]
23	2,4-Dibromophenol	$C_6H_4Br_2O$	Isol. from the acorn worms *Balanoglossus carnosus* and *Ptychodera* sp. Commonly found in marine algae, e.g., *Grateloupia elliptica*, mollusks, and crustaceans		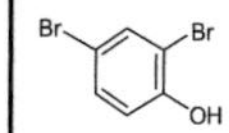	[99,100]
24	2,6-Dibromophenol	$C_6H_4Br_2O$	Widespread in marine algae, fish, mollusks and crustaceans, such as *Barantolla lepte*, *Marphysa sanguinea*, *Glycera americana*, *Lumbrineris latreilli*, *Nephtys australiensis*, *Ceratonereis aequisetis*, *Australonuphis teres*, *Scoloplos normalis*, *Penaeus plebejus*, *Penaeus latisulcatus*, *Penaeus merguiensis*, *Platycephalus caeruleopunctatus*, *Nemadactylus douglasii*, *Polysiphonia sphaerocarpa*, *Ulva lactuca*, *Ptychoderma flava* laysanica, and *Capitella* sp	Important flavor component of marine fish, mollusks and crustaceans	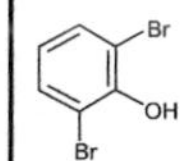	[101,102]
25	2,6-Di-tert-butyl-4-ethylphenol	$C_{16}H_{26}O$	Metabolites of blue-green algae	Antioxidant	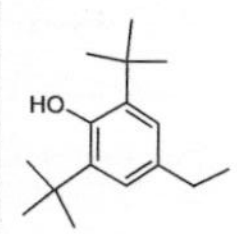	[103]

Continued

Table 6.4 Terpenes and Terpenoids: Biologically Active Constituents Extracted From Algae—cont'd

Sr. No.	Name	Molecular Formula	Biological Source	Biological Use	Structure	References
26	2,3-Didehydro-3,3′-dihydroxy-β,ε-caroten-4-one	$C_{40}H_{52}O_3$	Constit. of *Akategani estuarine* crab *Sesarma haematocheir*, green algae, and goldfish		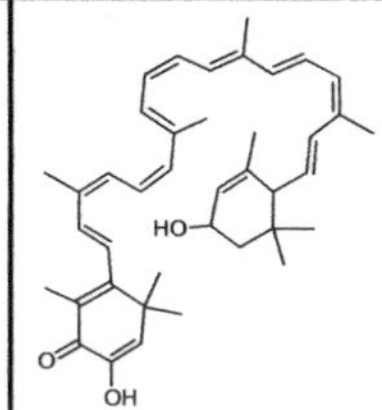	[104]
27	5,6-Dihydro-3-[2-(4-hydroxyphenyl)-2-oxoethyl]-2(1H)-pyridinone,	$C_{13}H_{13}NO_3$	Isol. from an algae-infested Caribbean sponge *Halichondria melanodocia*		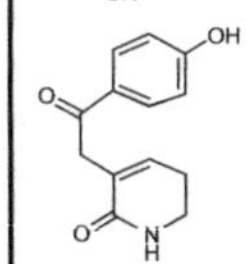	[105]
28	4,18-Dihydroxy-1(9),6,13-xenicatrien-19-al	$C_{20}H_{32}O_3$	Brown algae *D. dichotoma*		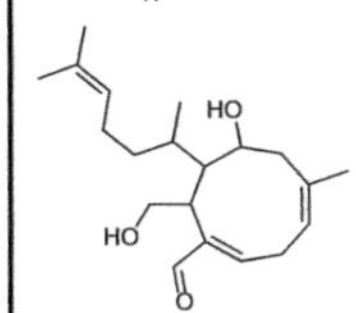	[106]
29	3,5-Diiodothyronine; (*S*)-form	$C_{15}H_{13}I_2NO_4$	Occurs in proteins of marine algae		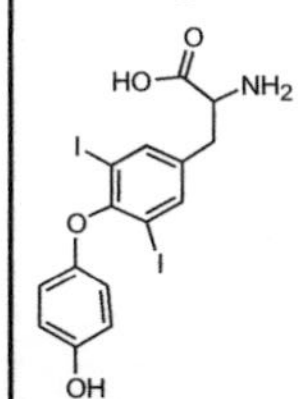	[107]
30	Dilophol	$C_{20}H_{34}O$	Constit. of algae *Dilophus ligulatus* and *D. dichotoma*		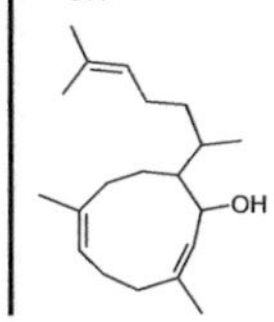	[108]

31	Echinenone	$C_{40}H_{54}O$	Found in echinoids, crustacean and green algae		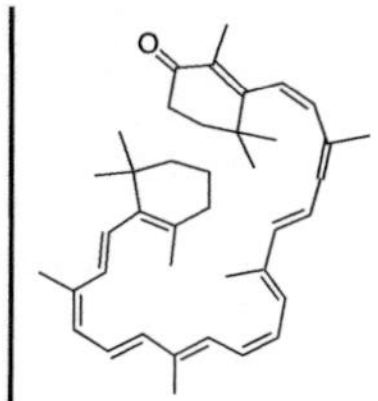	[109]
32	Echinenone; 4-Alcohol	$C_{40}H_{56}O$	Isol. from blue-green algae		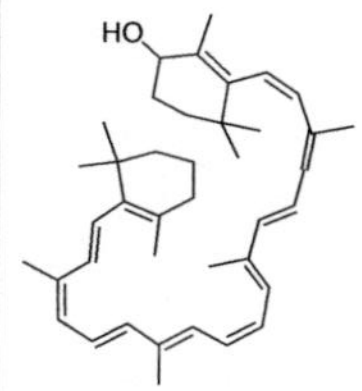	[110]
33	Echinenone; 3*S*-Hydroxy	$C_{40}H_{54}O_2$	Isol. from *Adonis annua*, *Rhizobium lupini*, fish eggs, algae		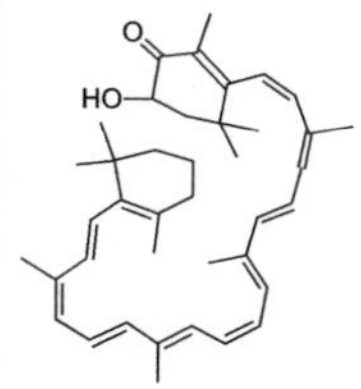	[111]
34	Echinenone; 4′-Hydroxy	$C_{40}H_{54}O_2$	Isol. from the spindle shell *Fusinus perplexus*, goldfish eggs, green algae, and other spp.		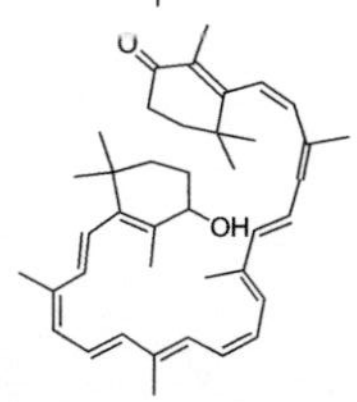	[111]
35	Galactaric acid	$C_6H_{10}O_8$	Isol. from brown algae, various fruits and fungi	Used as 2% aq. solution to form color complexes with Fe, Co, Cr(III), Mn, U(VI). (pH 3–10)	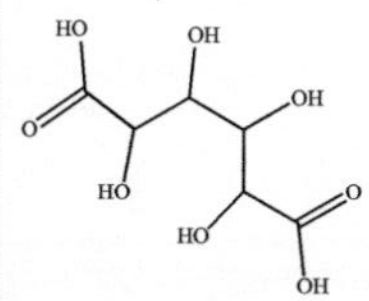	[112]

Continued

Table 6.4 Terpenes and Terpenoids: Biologically Active Constituents Extracted From Algae—cont'd

Sr. No.	Name	Molecular Formula	Biological Source	Biological Use	Structure	References
36	Grahamimycin-B	$C_{14}H_{20}O_7$	Isol. from *Cytospora* sp.	Weakly active against gram-positive and gram-negative bacteria, and blue-green algae	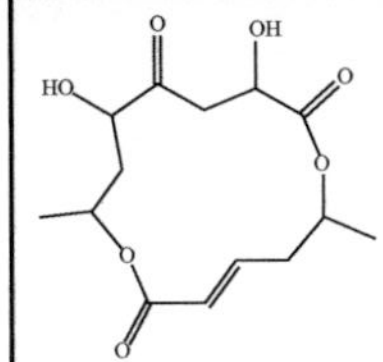	[113]
37	1-Heptadecene-4,6,8,10,12-pentol; (4*S*,6*S*,8*R*,10*R*,12*R*)-form, Penta-Me ether	$C_{22}H_{44}O_5$	Isol. from the blue-green algae *Scytonema burmanicum* and *Scytonema mirabile*		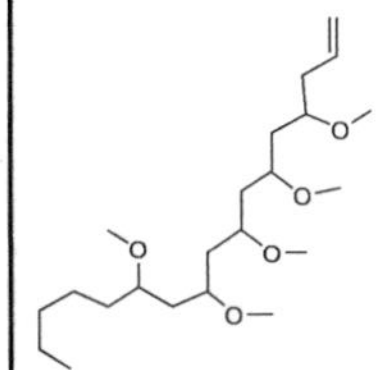	[114]
38	1-(1,3-Hexadienyl)-2-vinylcyclopropane; (1*R*,1′*E*,2*R*,3′*Z*)-form	$C_{11}H_{16}$	Constit. of algae *Dictyopteris* spp.		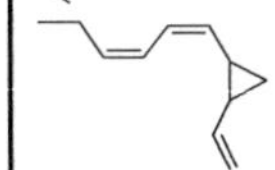	[115]
39	Hyatellaquinone; (+)-form	$C_{22}H_{30}O_4$	Constit. of *Hyatella intestinalis* and *Spongia* sp.; isol. from red algae	HIV reverse transcriptase (HIV-rt) inhibitor	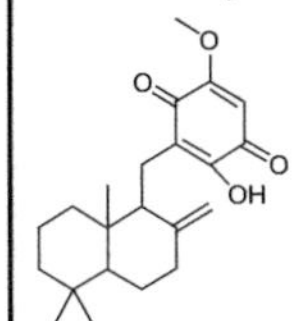	[116]
40	3-(2-Hydroxy-4,8-dimethyl-3,7-nonadienyl)benzaldehyde; Ac	$C_{20}H_{26}O_3$	Constit. of green algae *Halimeda scabra*, *Halimeda macroloba,* and *Halimeda discoidea*		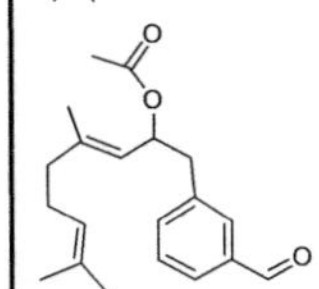	[117]
41	10-Hydroxy-4-oplopanone; (+)-form	$C_{15}H_{26}O_2$	Constit. of soft coral *Nephthea* sp. and the algae *Laurencia subopposita*		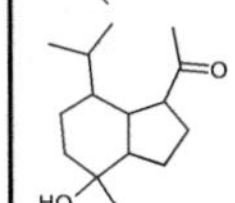	[118]
42	18-Hydroxy-1(9),6,13-xenicatrien-19-al; Ac	$C_{22}H_{34}O_3$	Constit. of marine algae *D. dichotoma* and *Pachydictyon coriaceum*		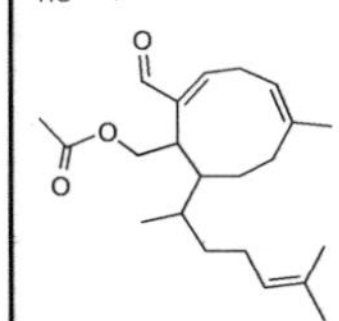	[106]

43	7-Isopropyl-1,4-dimethylazulene	$C_{15}H_{18}$	Obtained from essential oils, e.g., chamomile oil; found also in marine red algae and the gorgonians *Euplexaura erecta* and *Alcyonium* sp.	Antioxidant, inhibits lipid peroxidation; antiinflammatory agent, also used to treat gastrointestinal disorders; hepatoprotectant; immunodepressant; immunomodulatory and antiulcer agent	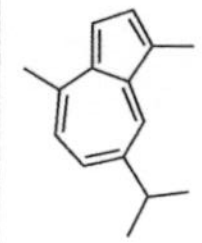	[119]
44	Isozeaxanthin; diketone	$C_{40}H_{52}O_2$	Constit. of edible mushroom *Cantharellus cinnabarinus*, sea trout, salmon and brine shrimp, *Corynebacterium michiganense* also in green algae	Food coloring	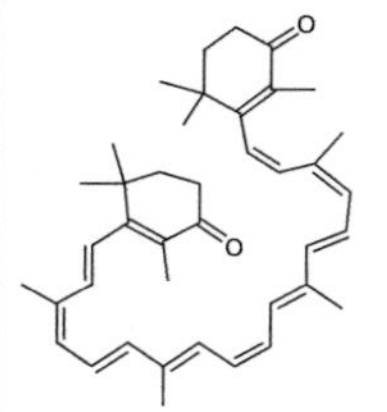	[120]
45	Lutein	$C_{40}H_{56}O_2$	Found in all higher plants, e.g., *Mimosa invasiva*, *Cosmos caudatus*, and also in microorganisms e.g., *Staphylococcus aureus*, green algae, *Porphyra* spp.	Shows antitumor, antimutagenic and a wide range of antimicrobial activity; potentially useful for treating macular degeneration, antioxidant	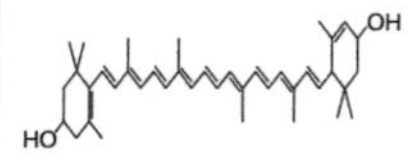	[121]
46	Lycopene	$C_{40}H_{56}$	Constit. of tomatoes and many other fruits; also occurs in bacteria and fungi; widely distributed in marine algae (red, green and brown)	Used as food coloring	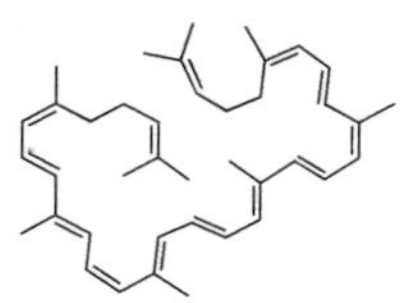	[122]
47	Cystoseirol-A	$C_{27}H_{36}O_4$	Constit. of brown algae *Cystoseira mediterranea*, *Cystoseira stricta*, and *Cystoseira tamariscifolia*		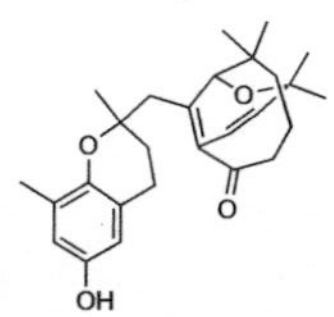	[123]
48	Mutatochrome	$C_{40}H_{56}O$	Constit. of orange peel and blue-green algae; also in *Calendula officinalis*, *Capsicum annuum* (paprika), *Delonix regia*, and others		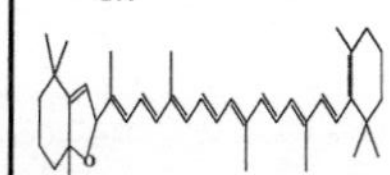	[124]

Continued

Table 6.4 Terpenes and Terpenoids: Biologically Active Constituents Extracted From Algae—cont'd

Sr. No.	Name	Molecular Formula	Biological Source	Biological Use	Structure	References
49	3,10(18)-Pachydictyadiene-6,14,15-triol; (1α,5β,6β,11*R*,14*R*)-form	$C_{20}H_{34}O_3$	Isol. from brown algae *D. dichotoma* and *Dictyota indica*		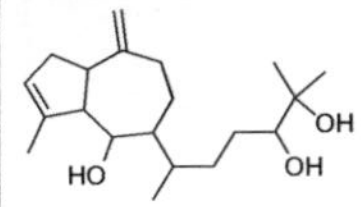	[125]
50	Polybromohydroxydiphenyl ethers; 2,3′,4-tribromo-4′-hydroxydiphenyl ether	$C_{12}H_7Br_3O_2$	Isol. from *Crustose coralline* red algae		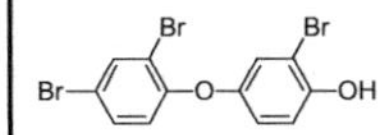	[126]
51	Prasinoxanthin	$C_{40}H_{56}O_4$	Constit. of algae within Prasinophyceae	Useful as a chemosystematic marker for algae	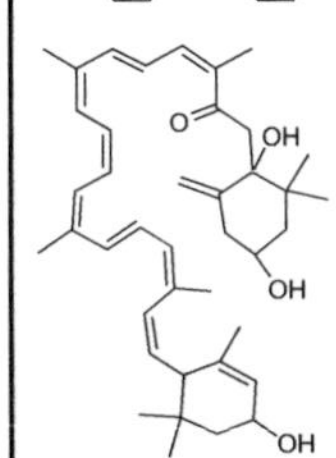	[127]
52	Sargassumketone	$C_{15}H_{18}O_{10}$	Isol. from brown algae *Sargassum kjellmanianum* and *Sargassum thunbergii*		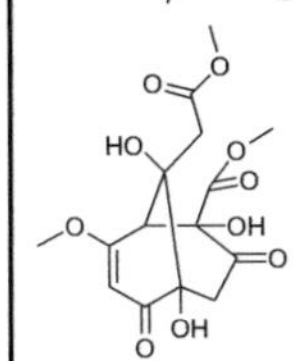	[128]
53	Siphonaxanthin	$C_{40}H_{56}O_4$	*Caulerpa prolifera* and other algae belonging to Prasinophyceae	Antiobesity	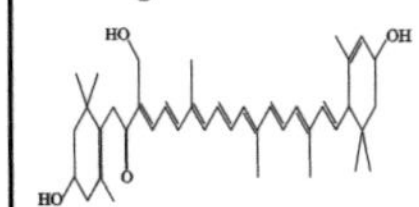	[129,130]
54	1,4,9,20-Tetraacetoxy-1,3(20),6,10,14-phytapentaen-19-al	$C_{28}H_{38}O_9$	*Halimeda* spp. of green algae	Feeding deterrent to fish	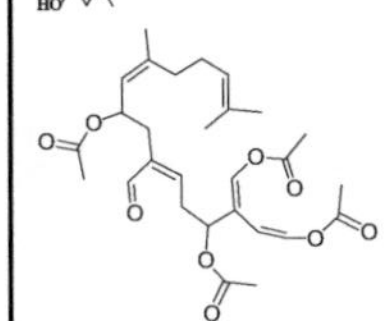	[131]
55	2,2′,3,3′-Tetrabromo-4,4′,5,5′-tetrahydroxydiphenylmethane	$C_{13}H_8Br_4O_4$	*R. larix*, *R. confervoides*, *Polysiphonia nigrescens*, and *Polysiphonia brodiaci*		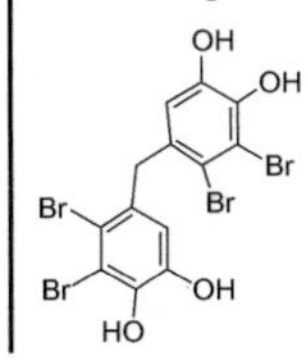	[132]

56	Thyroxine; (*S*)-form	$C_{15}H_{11}I_4NO_4$	*Heterochordaria abietina*, *Undaria pinnatifida*, *Sargassum thunbergii*, *Polysiphonia urceolata*, *Dendrodoa grossularia*, *Porphyra umbilicalis,* and *Enteromorpha intestinalis*	Thyromimetic, antihypercholesterolaemic	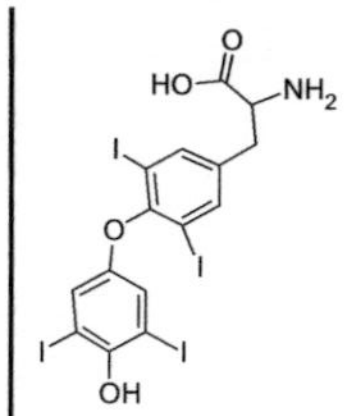	[133]
57	1,5,6-Trichloro-2-(dichloromethyl)-6-methyl-1,3,7-octatriene; (1*Z*,3*E*,5*S*,6*R*)-form	$C_{10}H_{11}Cl_5$	Isol. from red algae *Plocamium cartilagineum* and *Plocamium coccineum* and from mollusk *Aplysia limacina*		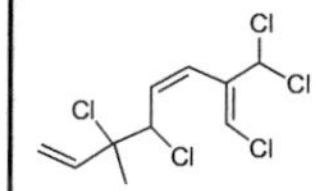	[134]
58	9,11,15-Trihydroxyprosta-5,13-dienoic acid; (5*Z*,8*R*,9*S*,11*R*,12*R*,13*E*,15*S*)-form	$C_{20}H_{34}O_5$	Produced by a variety of marine algae and invertebrates, such as *Phascolosoma japonica*, *Haliotis ovina*, *Crenomytilus grayanus*, *Modiolus difficilis*, *Stichopus japonicus*, *Distolasterias nipon*, and *Halocynthia aurantium*	Abortifacient, oxytocic and smooth muscle stimulant	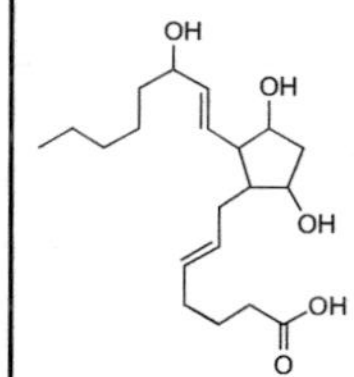	[135]
59	Zeaxanthin; (3*R*,3′*R*, all-*E*)-form	$C_{40}H_{56}O_2$	Found in cyanobacteria and most marine algae		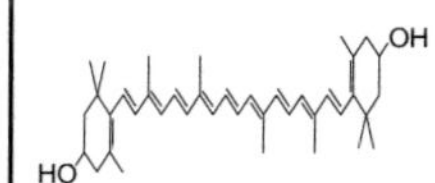	[136]
60	Brasilenyne	$C_{15}H_{19}ClO$	Isol. from sea hare *Aplysia brasiliana* and algae *L. nidifica*	Shows antifeedant activity	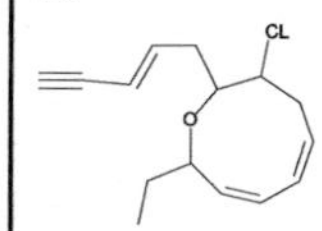	[137]
61	8-Bromo-1,5,6-trichloro-2-(dichloromethyl)-6-methyl-1,3,7-octatriene; (1*Z*,3*E*,5*R**,6*R**,7*E*)-form	$C_{10}H_{10}BrCl_5$	Constit. of red algae *P. cartilagineum* and *Plocamium suhrii*		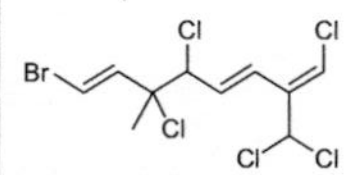	[138,139]
62	Telfairine	$C_{10}H_{14}$	Constit. of red alga *Plocamium telfairia*	Insecticidal	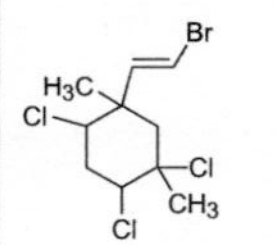	[140]

Continued

Table 6.4 Terpenes and Terpenoids: Biologically Active Constituents Extracted From Algae—cont'd

Sr. No.	Name	Molecular Formula	Biological Source	Biological Use	Structure	References
63	Sargatriol	$C_{27}H_{38}O_4$	Constit. of brown alga *Sargassum tortile*		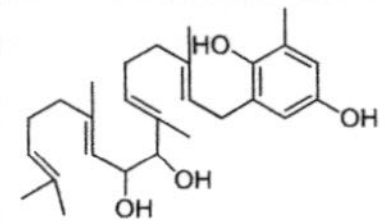	[141]
64	Methoxybifurcarenone	$C_{28}H_{40}O_5$	Constit. of brown alga *Cystoseira tamariscifolia*		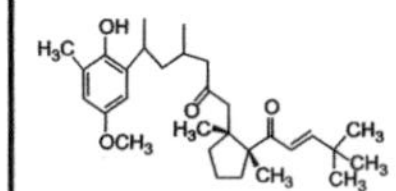	[142]
65	Astaxanthin	$C_{40}H_{52}O_4$	Constit. of *Haematococcus pluvialis*	Antioxidant	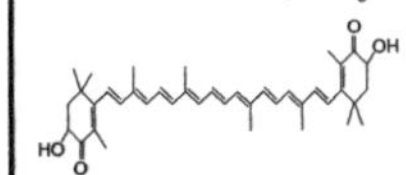	[143]
66	4-Bromophenol	C_6H_5BrO	Marine fish, mollusks, and algae		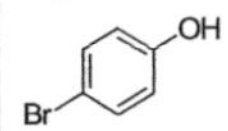	[101]
67	Geosmin	$C_{12}H_{22}O$	Produced by *Streptomyces* spp. and blue-green algae	Implicated in off-flavor of shellfish, freshwater fish, drinking water, and some vegetables	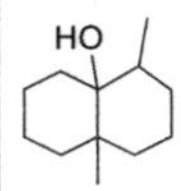	[144]

Constit., constituents; *Isol.*, isolated.

Table 6.5 Phycobiliproteins: Biologically Active Constituents Extracted From Algae

Sr. No.	Name	Molecular Formula	Biological Source	Biological Use	Structure	References
1	Phycocyanin		Constit. of blue-green algae	Antioxidative and antiinflammatory	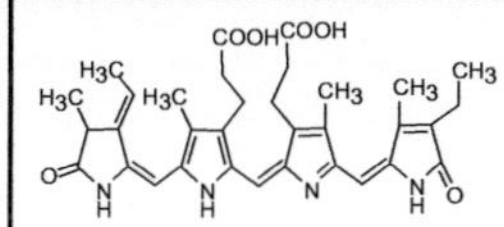	[145]
2	Phycoerythrin		Constit. of blue-green algae		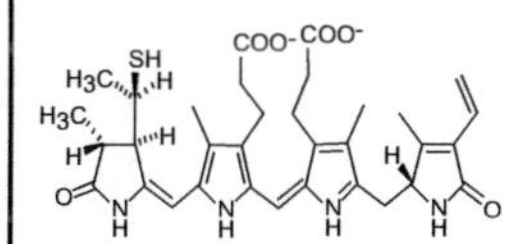	[145]
3	Allophycocyanin		Constit. of blue-green algae		N/A	[145]
4	Phycoerythrocyanin		Constit. of blue-green algae		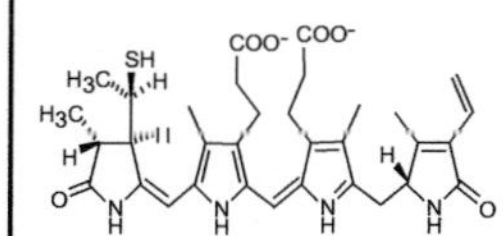	[145]

Constit., constituents.

Table 6.6 Sterols: Biologically Active Constituents Extracted From Algae

Sr. No.	Name	Molecular Formula	Biological Source	Biological Use	Structure	References
1	Cholesta-5,23-diene-3,25-diol; (3β,23*E*)-form	$C_{27}H_{44}O_2$	Constit. of red algae *Liagora distenta* and *Scinaia furcellata*		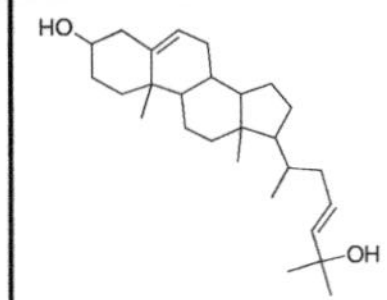	[148]
2	Cholesta-5,23-diene-3,25-diol; (3β,23*Z*)-form	$C_{27}H_{44}O_2$	Constit. of red algae *L. distenta* and *S. furcellata*		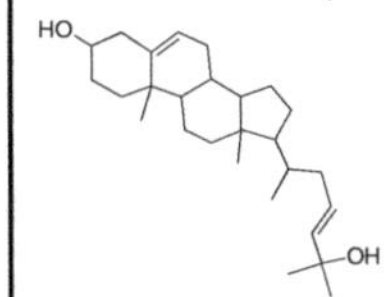	[148]
3	Cholesta-5,25-diene-3,24-diol; (3β,24*S*)-form	$C_{27}H_{44}O_2$	Isol. from brown alga *Rhodymenia palmata*, red algae *Asparagopsis armata*, *Rissoella verruculosa,* and the tetraspora *Falkenbergia rufolanosa*		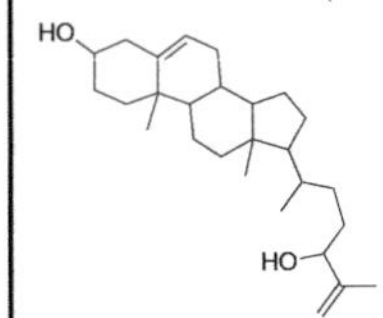	[149]
4	Cholesta-5,24-dien-3-ol; 3β-form	$C_{27}H_{44}O$	Constit. of *Funtumia latifolia*; found in red algae *R. palmata* and *Halosaccion ramentaceum* and *Patinopecten yessoensis*		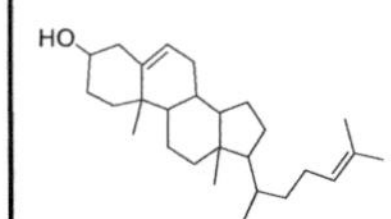	[150,151]
5	Cholesta-5,20,24-trien-3-ol; 3β-form	$C_{27}H_{42}O$	Constit. of *Mandevilla pentlandiana* and the marine algae *Bryopteris pennata* and *Scinaia fascicularis*		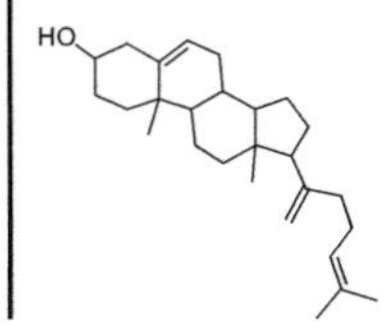	[152,153]

6	4,23-Dimethylergost-22-en-3-ol; (3β,4α,5α,22*E*,24*R*)-form	$C_{30}H_{52}O$	Constit. of *Gonyaulax tamarensis*, *Styela plicata*, *Sarcophyton glaucum*, *Holothuria scabra*, *Pseudostichopus trachus*, *Synapta maculate*, and dinoflagellate sediments	Occurs in dinoflagellates and rarely in other algae, and has been used as a marker of dinoflagellate contribution to organic sediments	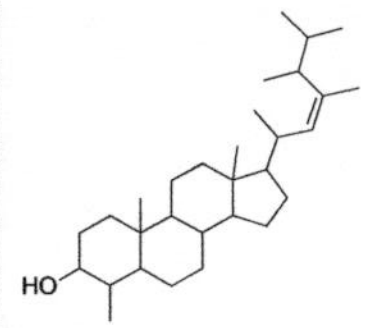	[154]
7	Ergosta-5,24(28)-dien-3-ol; 3β-form	$C_{28}H_{46}O$	Constit. of many marine organisms, such as *Eupentacta fraudatrix*, *Holothuria nobilis*, *S. maculata*, *Echinus esculentus*, *Echinocardium cordatum*, *Antedon bifida*, and *Ophiocomina nigra*; major sterol of green algae	Cytotoxic	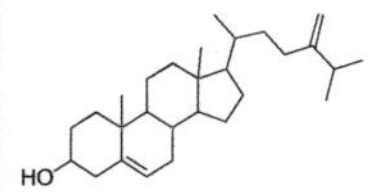	[155,156]
8	Ergosta-5,25-dien-3-ol; (3β,24*R*)-form	$C_{28}H_{46}O$	Constit. of *Verongia cauliformis*, *Jaspis stellifera*, and Chrysophyte unicellular algae		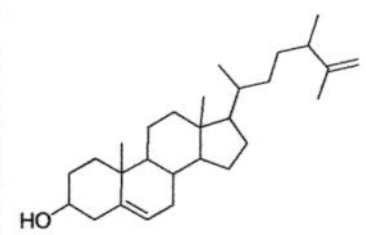	[157]
9	Ergosta-5,7,22,25-tetraen-3-ol; (3β,22*E*,24*S*)-form	$C_{28}H_{42}O$	Isol. from the algae *Dictylostelium discoideum*, *Prototheca wickerhamii*, and the sponge *Ciocalypta* sp.		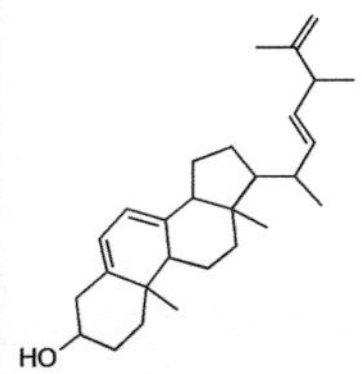	[158]
10	24-Hydroperoxystigmasta-5,28-dien-3-ol; (3β,24ξ)-form	$C_{29}H_{48}O_3$	Obtained from tunicates *Phallusia mamillata*, *Ciona intestinalis*, *Turbinaria conoides*, and *Padina pavonica* as well as brown algae	Shows cytotoxic activity	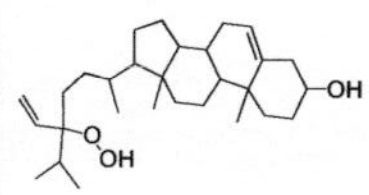	[159,160]

Continued

Table 6.6 Sterols: Biologically Active Constituents Extracted From Algae—cont'd

Sr. No.	Name	Molecular Formula	Biological Source	Biological Use	Structure	References
11	25-Methylergosta-5,7,22-trien-3-ol; (3β,22*E*,24ξ)-form	$C_{29}H_{46}O$	Constit. of red algae and *Haliclona longleyi*; also from green alga *Monostroma nitidum*		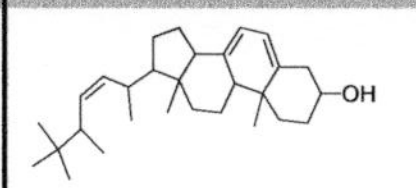	[161]
12	Stigmasta-5,28-diene-3,24-diol; (3β,24ξ)-form	$C_{29}H_{48}O_2$	Constit. of *Sargassum ringgoldianum* also from green algae *Ulva rigida* and *Ulva fasciata*		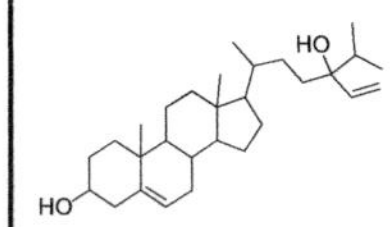	[162,163]
13	Stigmasta-5,22-dien-3-ol; (3β,22*E*,24*R*)-form	$C_{29}H_{48}O$	Constit. of sponges, e.g., *Cliona celata*, *Spheciospongia vesparia*, and *Haliclona variabilis*, green algae *Chlorella* spp., *Ochromonas malhamensis*, and clam *P. yessoensis*		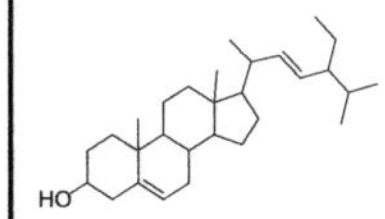	[164,165]
14	Stigmasta-5,24(28)-dien-3-ol; (3β,24(28)*E*)-form	$C_{29}H_{48}O$	Characteristic sterol of brown algae *Fucus vesiculosus*; also present in *Sargassum tortile*, *Cystoseira* sp., *Bifurcaria* sp., and a marine unicellular alga; also in sponges *Stelletta clarella*, *Tethya aurantia*, *Lissodendoryx noxiosa*, *Haliclona permollis*, and other *Haliclona* spp.		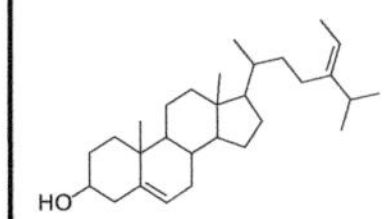	[161]
15	Stigmasta-5,24(28)-dien-3-ol; (3β,24(28)*Z*)-form	$C_{29}H_{48}O$	Major sterol of *Callyspongia diffusa* and other sponges and some green algae; also found in marine unicellular alga		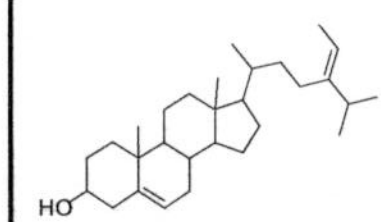	[161]

16	Stigmasta-5,7,22-trien-3-ol; (3β,24*R*)-form	$C_{29}H_{46}O$	Constit. of *Corbicula leana* and green algae and higher plants, also isol. from the *Phellodendron* sp.		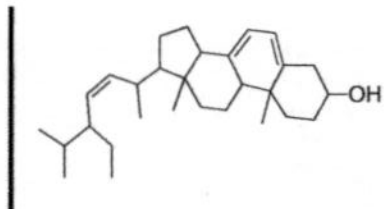	[166]
17	Stigmasta-7,9(11),22-trien-3-ol; (3β,22*E*,24*S*)-form	$C_{29}H_{46}O$	Constit. of *Haloxylon recurvum* and marine algae		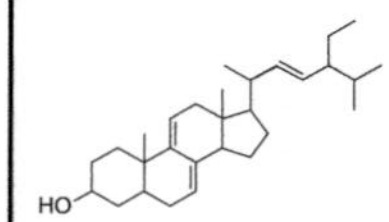	[167]
18	Stigmast-5-en-3-ol; (3β,24*S*)-form	$C_{29}H_{50}O$	Constit. of *S. vesparia*, *C. celata*, *T. aurantia*, liverworts, green algae *Chaetomorpha aurea*, *Caulerpa prolifera*, *Bryopsis plumosa*, and *Udotea petiolata*		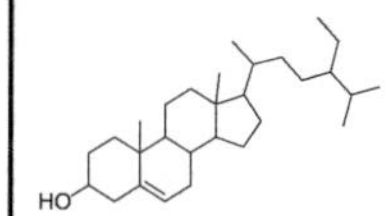	[168]

Constit., constituents; *Isol.*, isolated.

generally of the order of 50,000–1,000,000 Da [169]. The details of these compounds are briefly summarized in Table 6.7.

6.5.9 ALKALOIDS

The definition of the term alkaloid is not a simple one and is in many cases a source of academic controversy. They constitute a major class of natural products. Alkaloids are nitrogen-containing compounds derived from plants and animals. On the basis of structure, alkaloids are classified into three types [173].

- True alkaloids
- Protoalkaloids
- Pseudoalkaloids

The details of these compounds are briefly summarized in Table 6.8.

6.5.10 AMINO ACIDS

Amino Acids (AA) are defined as organic substances that contain both amino and acid groups. Some amino acid contain two carboxyl groups and some contain two amino groups [196]. The details of these compounds are briefly summarized in Table 6.9.

6.5.11 FATTY ACIDS

Fatty acids are the carboxylic acids derived from vegetable oils and animal fats. These are composed of a chain of alkyl groups usually containing an even number (4–22) of carbon atoms and are characterized by a terminal carboxylic group (–COOH). Their general formula is CH_3 $(CH_2)_x$ COOH. They can be saturated or unsaturated [218]. The details of these compounds have been briefly summarized in Table 6.10.

6.5.12 HYDROCARBONS

Hydrocarbons are the compounds composed of only carbon and hydrogen. Alkanes are saturated hyrocarbons, that is, they contain only carbon–carbon single bonds. Those containing one or more carbon–carbon double bonds, triple bonds, or benzene rings are classified as unsaturated hydrocarbon [71]. The details of these compounds are briefly summarized in Table 6.11.

6.5.13 OXYGEN HETEROCYCLES

Heterocycles are defined as cyclic molecules that contain one or more heteroatoms in a ring. A heteroatom is an atom other than carbon; common heteroatoms are oxygen, nitrogen, and sulfur. Oxygen heterocycles are compounds that contain

Table 6.7 Polyhydroxyalkonates: Biologically Active Constituents Extracted From Algae

Sr. No.	Name	Molecular Formula	Biological Source	Biological Use	Structure	References
1	Poly-3-hydroxybutyrate (PH3B)		Blue-green alga *Chlorogloea fritschii*		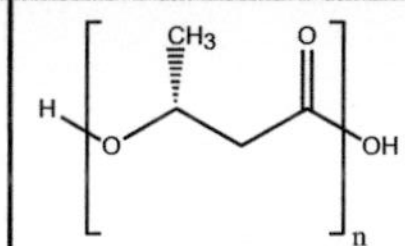	[170]
2	Poly(3-hydroxybutyrate-*co*-3-hydroxyvalerate) PHBV		Produced by the cyanobacterium *Nostoc muscorum Agardh*		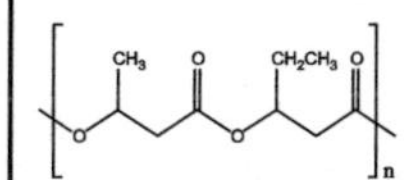	[171]
3	Polyhydroxy valerate (PHV)		Blue-green algae *Osxillatoris limosa*		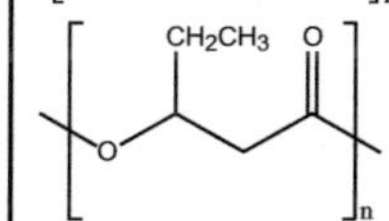	[172]

Table 6.8 Alkaloids: Biologically Active Constituents Extracted From Algae

Sr. No.	Name	Molecular Formula	Biological Source	Biological Use	Structure	References
1	Ambiguine-B isonitrile; Deoxy	$C_{26}H_{31}ClN_2$	Blue-green algae *Fischerella ambigua*, *Hapalosiphon hibernicus*, and *Westiellopsis prolifica*		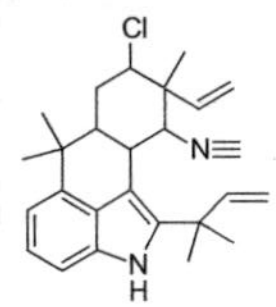	[174,175]
2	Ambiguine-D isonitrile	$C_{26}H_{29}ClN_2O_3$	Blue-green algae *F. ambigua* and *W. prolifica*		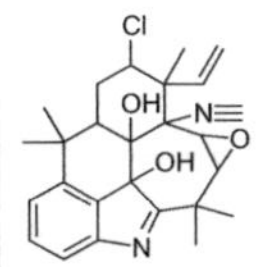	[174,175]
3	Ambiguine-E isonitrile	$C_{26}H_{29}ClN_2O_2$	*F. ambigua*, *H. hibernicus*, and *W. prolifica*		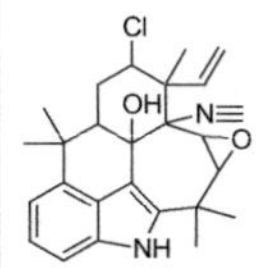	[174,175]
4	Ambiguine-G nitrile	$C_{26}H_{27}ClN_2$	*F. ambigua* and *Hapalosiphon delicatulus*		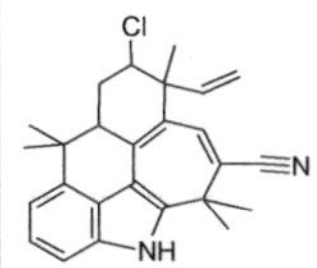	[176]
5	Biliverdin; 15,16-Dihydro	$C_{33}H_{36}N_4O_6$	Cyanobacteria and red algae, including *Cyanidium caldarium*		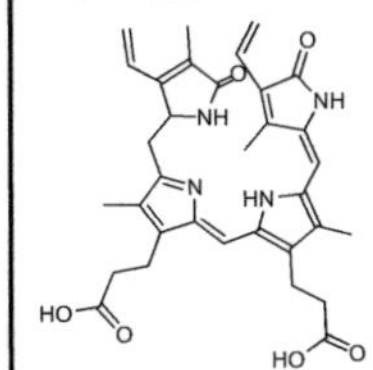	[177]
6	Chlorophyll *d*	$C_{54}H_{70}MgN_4O_6$	Red algae		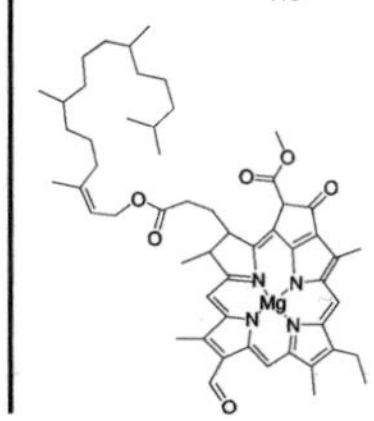	[14]

7	4,6-Dibromo-2,3-bis(methylthio)-1H-indole	$C_{10}H_9Br_2NS_2$	Isol. from red algae *Laurencia brongniartii* and *Laurencia grevilleana*		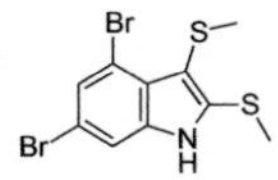	[178]
8	5,12-Dihydrocycloocta[1,2-b:5,6-b′]diindole-6,13-dicarboxylic acid, Di-Me ester	$C_{24}H_{18}N_2O_4$	Green algae *Caulerpa serrulata*, *Caulerpa racemosa*, *Caulerpa sertularioides*, *Caulerpa taxifolia*, *Caulerpa cuppresoides*, *Caulerpa scalpelliformis*, *Caloglossa leprieurii*, and *Laurencia majuscule*	Plant growth regulator, phycotoxin, vermifuge	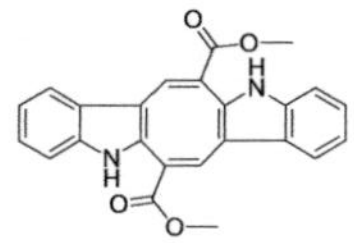	[179,180]
9	5,6-Dihydro-3-[2-(1H-indol-3-yl)-2-oxoethyl]-2(1H)-pyridinone,	$C_{15}H_{14}N_2O_2$	Isol. from an algae-infested Caribbean sponge *Halichondria melanodocia*		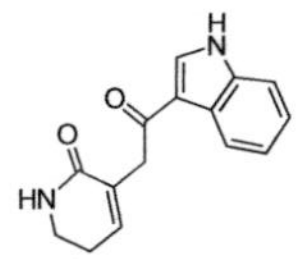	[181]
10	Hordenine	$C_{10}H_{15}NO$	*Anhalonium tissuratum*, *Hordeum vulgare*; also present in the Amaryllidaceae, Gramineae, Leguminosae, and algae and fungi including marine alga *Phyllophora nervosa*	Diuretic, disinfectant, antihypotensive (in large doses) agent; used for treatment of dysentery; feeding repellant for grasshoppers; shows similar actions to 2-(methylamino)-1-phenyl-1-propanol	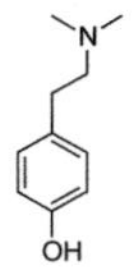	[182]
11	1H-Indole-3-acetic acid	$C_{10}H_9NO_2$	Isol. from the marine alga *Undaria pinnatifida*, also in bacteria, yeasts, and fungi	Plant growth hormone (auxin); involved in root development; phytotoxic agent	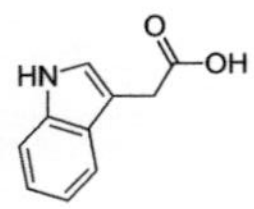	[183]

Continued

Table 6.8 Alkaloids: Biologically Active Constituents Extracted From Algae—cont'd

Sr. No.	Name	Molecular Formula	Biological Source	Biological Use	Structure	References
12	1H-Indole-3-carboxylic acid	$C_9H_7NO_2$	Present in apple, garden pea, *Brassica* spp., and the marine algae *U. pinnatifida* and *Botryocladia leptopoda*		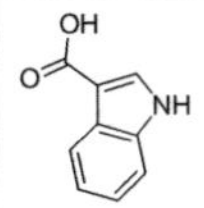	[183]
13	1H-Indole-3-ethanol; *O*-β-D-Glucopyranoside	$C_{16}H_{21}NO_6$	Isol. from numerous plant and algae		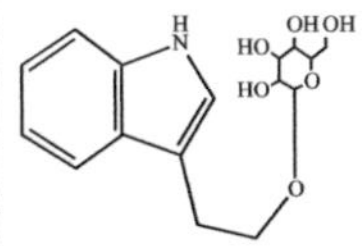	[184]
14	1′-Methylzeatin; (*R*)-form, 2-Hydroxy	$C_{11}H_{15}N_5O_2$	Produced by *Alternaria brassicae* and isol. from marine green algae	Shows cytokinin activity; plant growth stimulator	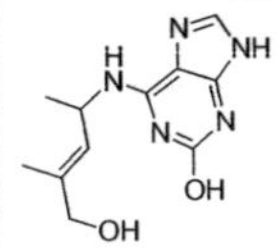	[185]
15	2-Phenylethylamine	$C_8H_{11}N$	*Acacia* spp., *Crataegus* spp.; also present in animal tissues, some algae, fungi and cacti (Leguminosae, Rosaceae, Cactaceae). Metab. of *Prosopis alba*	Shows DNA binding activity. Monoamine oxidase inhibitor	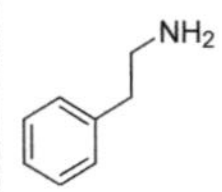	[186]
16	Phycocyanobilin	$C_{33}H_{38}N_4O_6$	*Phormidium luridum*, *Synechococcus lividus*, and *Plectonema boryanum*; also present in red algae		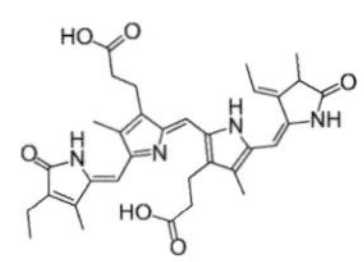	[187]

17	Phycoerythrobilin	$C_{33}H_{38}N_4O_6$	Blue-green algae as well as most red algae		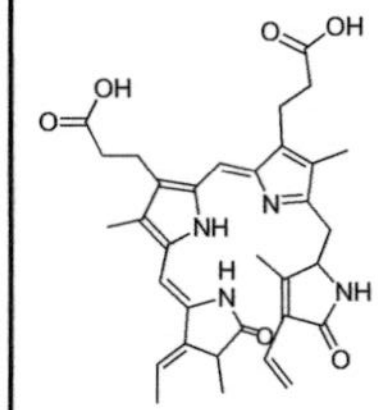	[188]
18	2,4-Pyrrolidinedicarboxylic acid; (2*S*,4*R*)-form	$C_6H_9NO_4$	Constit. of Mediterranean red algae *Chondria coerulescens*, *Chondria dasiphylla,* and *Ceramium rubrum*		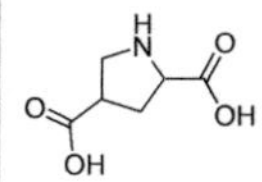	[189]
19	Saxitoxin; N^1-Hydroxy	$C_{10}H_{17}N_7O_5$	Found in shellfish, marine algae, and other organisms incl. *Gonyaulax tamarensis*, *Saxidomus giganteus*, *Vibrio* sp., *Pyrodinium* sp.	Neurotoxic neuromuscular blocker	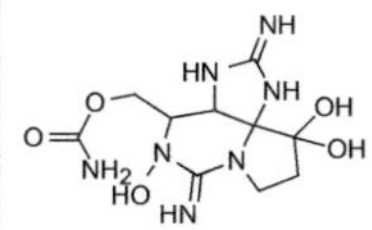	[190]
20	2,3,5,6-Tetrabromo-1H-indole	$C_8H_3Br_4N$	*L. brongniartii* and *Laurencia similis*		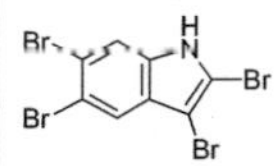	[191]
21	Thallusin	$C_{25}H_{31}NO_7$	Isol. from a marine bacterium *Cytophaga* sp. YM2-23 obt. from a *Monostroma* sp.	Morphogenesis inducer in algae	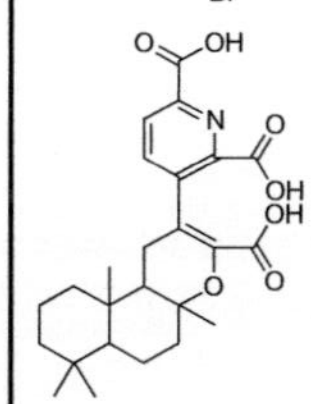	[192]
22	Welwitindolinone-B isothiocyanate	$C_{21}H_{21}ClN_2O_2S$	Blue-green algae *Hapalosiphon welwitschii* and *Westiella intricata*		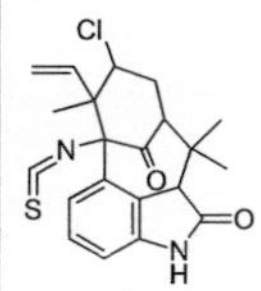	[193]

Continued

Table 6.8 Alkaloids: Biologically Active Constituents Extracted From Algae—cont'd

Sr. No.	Name	Molecular Formula	Biological Source	Biological Use	Structure	References
23	Zeatin; (*E*)-form	$C_{10}H_{13}N_5O$	Isol. from sweet corn Zea mays and numerous other plants; also from algae, bacteria, and basidiomycetes	Induces cell division; most effective of all known natural cytokinins; nematicide	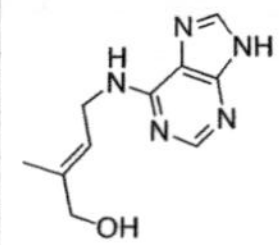	[194]
24	Adenosine diphosphate glucose	$C_{16}H_{25}N_5O_{15}P_2$	Present in ripening cereal grains; also obt. from algal cells	Serves as the glycosyl donor for formation of bacterial glycogen, amylose in green algae, and amylopectin in higher plants	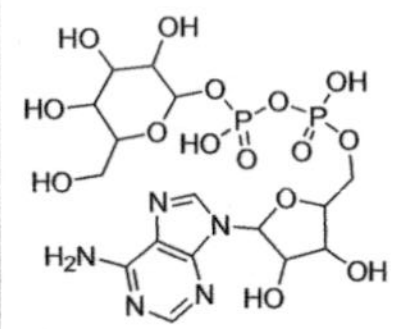	[195]

Constit., constituents; *Isol.*, isolated.

Table 6.9 Amino Acids: Biologically Active Constituents Extracted From Algae

Sr. No.	Name	Molecular Formula	Biological Source	Biological Use	Structure	References
1	3-Aminopropanoic acid	$C_3H_7NO_2$	Widely distributed in plants including algae, fungi, and many higher plants	Catalyst for Knoevenagel condensations. Used for synthesis of Pantothenic acid and derivatives	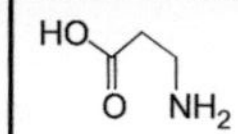	[197]
2	N2-Arginylglutamine; L-L-form	$C_{11}H_{22}N_6O_4$	Isol. from *Euglena gracilis* and from the green algae *Cladophora* spp., *Enteromorpha linza*, and *Ulva pertusa*		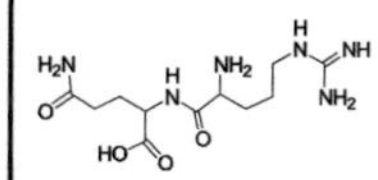	[198,199]
3	Chondrine; (1*S*,3*S*)-form	$C_5H_9NO_3S$	Red alga *Chondria crassicaulis*, brown alga *Undaria pinnatifida*		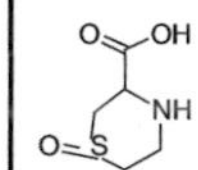	[200,201]
4	Gigartinine; (*S*)-form	$C_7H_{15}N_5O_3$	Constit. of many spp. of red algae, e.g., *Gymnogongrus flabelliformis*, *Gelidium amansii*, *Grateloupia livida*, *Grateloupia filicina*, *Polyopes polyideoides*, *Carpopeltis flabellata*, *Hypnea japonica*, and *Gracilaria textorii*		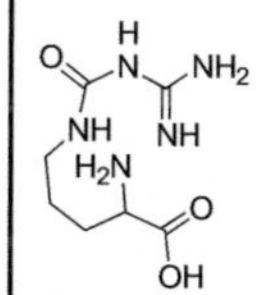	[202]
5	Gongrine	$C_6H_{12}N_4O_3$	Isol. from red algae *G. flabelliformis* and *G. filicina*		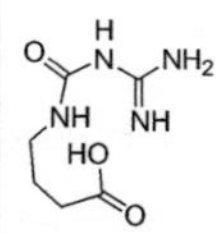	[203,204]
6	Methionine sulfoxide; (*S*)C(*R*)S-form	$C_5H_{11}NO_3S$	Isol. from various red algae incl. *Amphiroa beauvoisii* and *Grateloupia proteus*		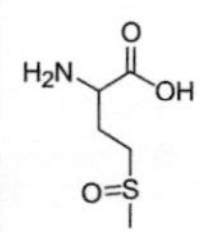	[205]

Continued

Table 6.9 Amino Acids: Biologically Active Constituents Extracted From Algae—cont'd

Sr. No.	Name	Molecular Formula	Biological Source	Biological Use	Structure	References
7	Methionine sulfoxide; (*S*)C(*R*)S-form, *N*-Me	$C_6H_{13}NO_3S$	Isol. from various red algae incl. *Centroceras clavulatum*, *Grateloupia doryphora*, and *Gelidium latifolium*		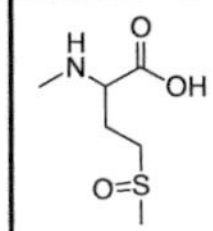	[205,206]
8	*N*-(2-Sulfoethyl)alanine; (*R*)-form	$C_5H_{11}NO_5S$	Isol. from algae *Chondrus ocellatus* and *Rhodoglossum japonicum*		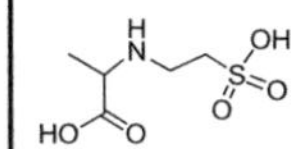	[207,208]
9	1,2,3,6-Tetrahydro-2-pyridinecarboxylic acid; (*S*)-form	$C_6H_9NO_2$	Present in *Baikiaea plurijuga*, *Caesalpinia tinctoria*, red algae also in *Russula subnigricans*	Neurotransmission inhibitor	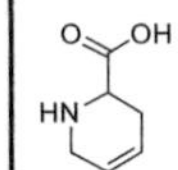	[205,209]
10	N-Trimethyllysine betaine; (*S*)-form	$C_9H_{20}N_2O_2$	Green alga *Enteromorpha intestinalis*, brown algae *Laminaria angustata*, and *Heterochordaria abietina*; also from the seaweed *Petalonia fascia*	Intermediate in biosynthesis of carnitine and γ-butyrobetaine; cell proliferation stimulant. Antihypertensive and hypocholesterolemic agent	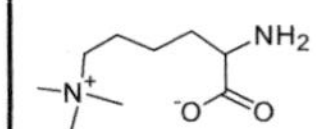	[210]
11	Citrulline; (*S*)-form	$C_6H_{13}N_3O_3$	Occurs in the watermelon *Citrullus vulgaris*, green alga *E. intestinalis*, and red alga *G. filicina*; widely distributed in Cucurbitaceae and fungi	Diuretic agent; used as arginine substitute in the treatment of inborn errors of urea synthesis, carbamyl phosphate synthetase and ornithine transcarbamylase deficiency	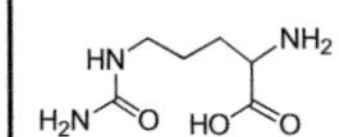	[211]
12	Domoic acid	$C_{15}H_{21}NO_6$	Constit. of red algae *Chondria armata* and *Alsidium corallinum*	Ionotropic glutamate (kainate) receptor agonist; neurotoxin. amnesic shellfish poison; vermifuge, insecticide	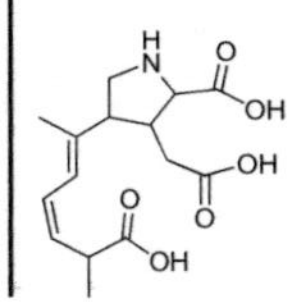	[205]

13	Kainic acid	$C_{10}H_{15}NO_4$	Constit. of red algae *Alsidium helminthochorton*, *C. clavulatum*, and *Digenea simplex*	Glutamate receptor agonist; neurotoxin, formerly used as an anthelmintic agent	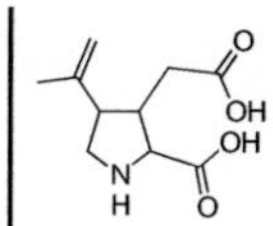	[205]
14	2,5-Pyrrolidinedicarboxylic acid; *S*,5*S*)-form	$C_6H_9NO_4$	Reported from red algae *Schizymenia dubyi* and *Haematocelis rubens*		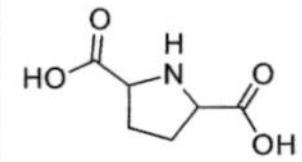	[205]
15	4-Hydroxy-2-pyrrolidinecarboxylic acid; (2*R*,4*R*)-form, *N*-Me	$C_6H_{11}NO_3$	Constit. of *Capsicum annuum*, *Copaifera* spp., *Croton gubougia*, *Dalbergia sympathetica*, *Erythroxylum argentinum*, *Toddalia aculeate*, and *Afrormosia elata*; also from red alga *Chondria coerulescens* and other red algae		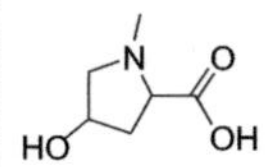	[212]
16	2-Aminoethanesulfonic acid	$C_2H_7NO_3S$	Occurs in animal tissues, bacteria, sponges, and red algae, e.g., isol. from *Macrocallista nimbosa*, *Turbo stenogyrus*, *Calyx nicaeensis*, *Geodia gigas*, *Mytilus edulis*. Also from green algae, e.g., *Caulerpa okamurai*, *Caulerpa racemosa*, *Chlorodesmis comosa*, *Codium adherens*, *Codium fragile*, and *E. linza*; and from higher plants, e.g., *leguminous seedlings*	Used as an adjunct in treatment of hypercholesterolaemia; metabolic regulator	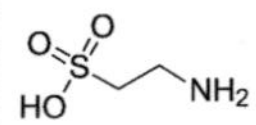	[213]

Continued

Table 6.9 Amino Acids: Biologically Active Constituents Extracted From Algae—cont'd

Sr. No.	Name	Molecular Formula	Biological Source	Biological Use	Structure	References
17	2-Aminoethanesulfonic acid; *N*,*N*-Di-Me	$C_4H_{11}NO_3S$	*Furcellaria fastigiata*		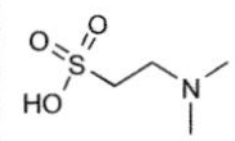	[214]
18	2-Aminoethanesulfonic acid; *N*-Me	$C_3H_9NO_3S$	Red alga *Ptilota pectinata* and green alga *C. comosa*.		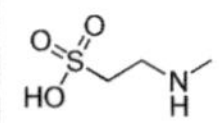	[215]
19	2-Amino-3-hydroxy-1-propanesulfonic acid; (*S*)-form	$C_3H_9NO_4S$	Isol. from brown (e.g., *Hijikia fusiforme*) and green (*U. pertusa*, *E. linza*) algae. Also from diatoms, e.g., *Navicula pelliculosa* and starfish *Asterina pectinifera*		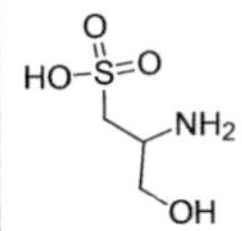	[59]
20	3-Amino-1-propanesulfonic acid	$C_3H_9NO_3S$	Constit. of marine red algae, e.g., *G. livida* and from green alga *Cladophora densa*	Inhibits amyloid A fibril formation and deposition; used in the treatment of Alzheimer disease and cerebral amyloid angiopathy	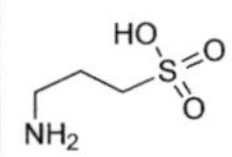	[216]
21	2,5-Diaminopentanoic acid; (*S*)-form	$C_5H_{12}N_2O_2$	Isol. from green algae *Codium decorticatum*, *E. intestinalis*, and *Ulva lactuca*	Used in treatment of hyperammonemia and liver disorders	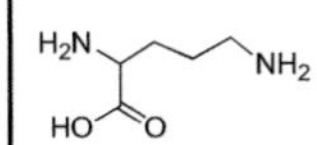	[217]

Constit., constituents; *Isol.*, isolated.

Table 6.10 Fatty Acids: Biologically Active Constituents Extracted From Algae

Sr. No.	Name	Molecular Formula	Biological Source	Biological Use	Structure	References
1	3-Bromo-2-heptenoic acid; (*E*)-form	$C_7H_{11}BrO_2$	Constit. of red algae *Bonnemaisonia hamifera* and *Bonnemaisonia nootkana*		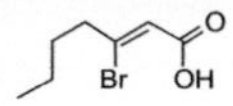	[219]
2	(2-Carboxyethyl) dimethylsulfonium(1+)	$C_5H_{11}O_2S^{1+}$	Isol. from green and red algae, e.g., *Enteromorpha intestinalis*, *Ulva lactuca*; also from *Spartina anglica*	Biological precursor of Dimethyl sulphide; fish feeding stimulant	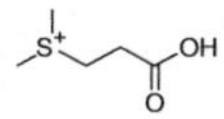	[220]
3	13-(2-Cyclopentenyl)tridecanoic acid; ξ-form, 2′,3′-dihydro	$C_{18}H_{34}O_2$	Constit. of *Hydnocarpus anthelmintica* and various red algae of Solieriaceae		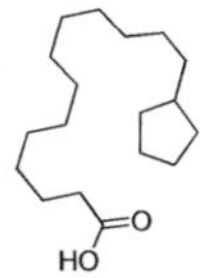	[221]
4	11-(2-Cyclopentenyl) undecanoic acid; (+)-form, 2′,3′-dihydro	$C_{16}H_{30}O_2$	Constit. of *H. anthelmintica* and various red algae of Solieriaceae		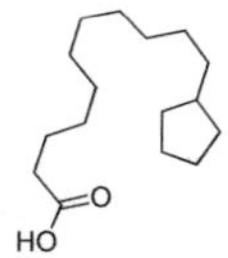	[221]
5	2-(Dibromomethylene)octanoic acid	$C_9H_{14}Br_2O_2$	Isol. from red algae *Bonnemaisonia* spp.		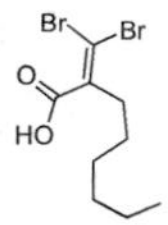	[222]
6	5,7,9,14,17-Eicosapentaenoic acid; (5*E*,7*E*,9*E*,14*Z*,17*Z*)-form	$C_{20}H_{30}O_2$	Isol. from red algae *Murraya periclados* and *Ptilota filicina*			[223,224]

Continued

Table 6.10 Fatty Acids: Biologically Active Constituents Extracted From Algae—cont'd

Sr. No.	Name	Molecular Formula	Biological Source	Biological Use	Structure	References
7	5,8,11,14,17-Eicosapentaenoic acid; (all-*Z*)-form	$C_{20}H_{30}O_2$	Present in fish oils, in animal phospholipids, and constit of various red algae	Nutriceutical with antioxidation props; precursor of PG3 series of prostaglandins; platelet aggregation inhibitor; allelopathic agent	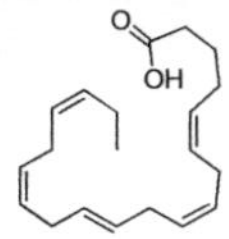	[225]
8	7,10,13-Hexadecatrienoic acid; (*Z*,*Z*,*Z*)-form	$C_{16}H_{26}O_2$	Present in a wide variety of angiosperm leaves and green algae; isol. from the sponge *Dysidea fragilis*		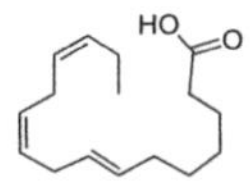	[226]
9	3-Hexadecenoic acid; (*E*)-form	$C_{16}H_{30}O_2$	Constit. of spinach leaves, red clover, *Grindelia oxylepis*, and *Helenium bigelovii*; also isol. from algae		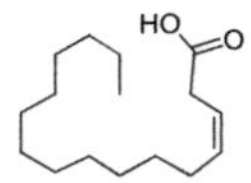	[227]
10	2-Oxohexadecanoic acid	$C_{16}H_{30}O_3$	Constit. of algae *Porphyra* sp. and *Ulva pertusa*		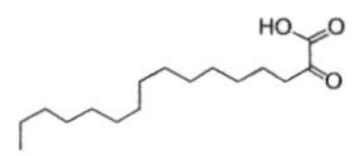	[228]
11	12,13-Dihydroxy-5,8,10,14,17-eicosapentaenoic acid; (5*Z*,8*Z*,10*E*,12*R*,13*S*,14*Z*,17*Z*)-form	$C_{20}H_{30}O_4$	Constit. of red algae *Farlowia mollis* and *Gracilariopsis lemaneiformis*		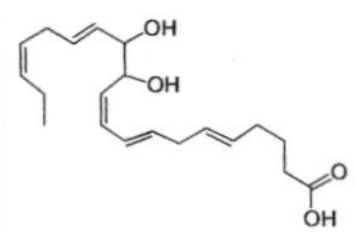	[229]
12	13-Hydroxy-5,8,11,14,17-eicosapentaenoic acid; (all-*Z*)-form, Et ester	$C_{22}H_{34}O_3$	Isol. from a mixture of algae *Lithothamnion calcareum* and *Lithothamnion corallioides*		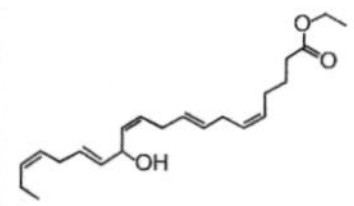	[230]

13	12-Hydroxy-5,8,10,14-eicosatetraenoic acid; (5*Z*,8*Z*,10*E*,12*S*,14*Z*)-form	$C_{20}H_{32}O_3$	Found in red algae *Murrayella periclados* and *Platysiphonia miniata*, and sponge *Echinochalina mollis*	Immunohormone; cell growth inhibitor; proinflammatory agent; toxic to brine shrimp	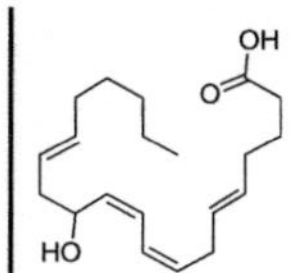	[224]
14	8-Hydroxy-13-oxo-5,9,11,14-eicosatetraenoic acid; (5*Z*,8ξ,9*E*,11*E*,14*E*)-form	$C_{20}H_{30}O_4$	Isol. from the algae *Lithothamnion* spp.		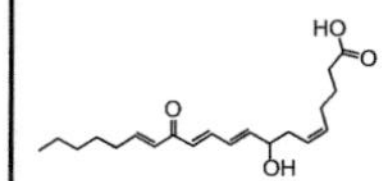	[230]
15	2-Propenoic acid	$C_3H_4O_2$	Found in green algae; produced by *Phaeocystis* sp., *Phaeocystis pouchetii*, *Enteromorpha* sp., *Ulva* sp., *Codium* sp., *Patinopecten yessonensis*, and *Protogonyaulax* sp.		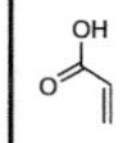	[226]
16	13-Octadecenoic acid; (*Z*)-form	$C_{18}H_{34}O_2$	Constit. of algae, bacteria, fish; also found in various plant spp.; isol. from *Haliclona cinerea*		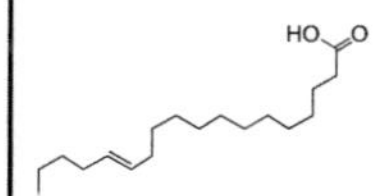	[231]

Constit., constituents; *Isol.*, isolated.

Table 6.11 Hydrocarbons: Biologically Active Constituents Extracted From Algae

Sr. No.	Name	Molecular Formula	Biological Source	Biological Use	Structure	References
1	Bromochloromethane	CH_2BrCl	Isol. from several marine algae	Intermediate for organic synthesis, fire extinguishant	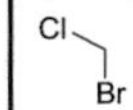	[232]
2	Bromodichloromethane	$CHBrCl_2$	Isol. from several marine algae	Source of dichlorocarbene on treatment with base	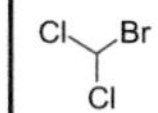	[232,233]
3	Bromoiodomethane	CH_2BrI	Isol. from several marine algae		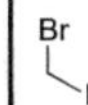	[232,234]
4	Bromomethane	CH_3Br	Isol. from various marine sources incl. giant kelp and ice algae.	Principally used as an insecticidal and nemacidal fumigant, especially for soil and agricultural produce	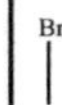	[232,234]
5	Carbon tetrabromide	CBr_4	Isol. from various marine algae including *Asparagopsis taxiformis*		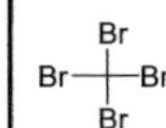	[232,234]
6	Carbon tetrachloride	CCl_4	Isol. from several plants and marine algae	Anthelmintic	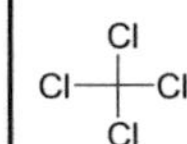	[232,234]
7	Chloroform	$CHCl_3$	Found in various plants and algae	Formerly used as anesthetic. Topical remedy for Herpes simplex sores	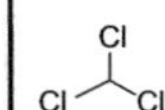	[232,234]
8	Chloroiodomethane	CH_2C_{11}	Isol. from phytoplankton and various marine algae	Versatile organic synthon. Commercially available	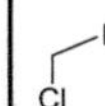	[232,234]
9	Dibromochloromethane	$CHBr_2Cl$	Isol. from several marine algae including *A. taxiformis*	Source of bromochlorocarbene in the presence of base and crown ether	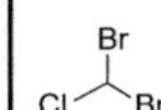	[232,234]
10	Dibromoiodomethane	$CHBr_2I$	Isol. from several marine algae including *A. taxiformis*		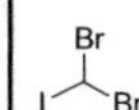	[232,234]

11	Dibromomethane	CH_2Br_2	Isol. from various marine algae including red alga *A. taxiformis*	Solvent, reagent for organic synthesis	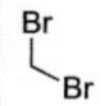	[232,234]
12	Dichloromethane	CH_2Cl_2	Found in various plants and algae and in volcanic gases	Widely used solvent used in coffee decaffeination. Component of paint removers. Blowing agent in foams. Can replace formaldehyde as reagent for Mannich reagents	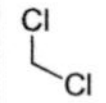	[232]
13	Diiodomethane	CH_2I_2	Isol. from various marine algae			[234]
14	Ethylamine	C_2H_7N	Produced by marine algae, *Clostridium* spp., *Candida albicans*, *Brevibacterium linens* and *Streptococcus lactis*	Used in manufacture of resins, rubber, herbicides, etc.	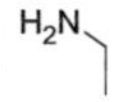	[235]
15	3,6,9,12,15,18-Heneicosahexaene; (all-*Z*)-form	$C_{21}H_{32}$	Isol. from marine algae and plankton, e.g., *Rhizosolenia setigera* and *Skeletonema costatum*		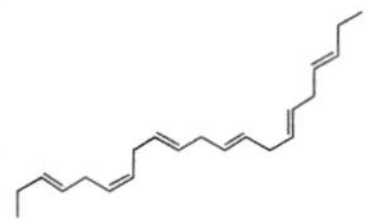	[236]
16	3,6,9,12,15-Heneicosapentaene; (all-*Z*)-form	$C_{21}H_{34}$	Constit. of *Polytrichum commune* and benthic green algae		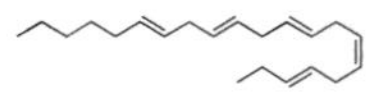	[237,238]
17	1-Hexacosene	$C_{26}H_{52}$	Constit. of *Acanthopanax giraldii*, *Aralia elata*, *Hippophae rhamnoides* and various algae incl. *Chlorella* sp.		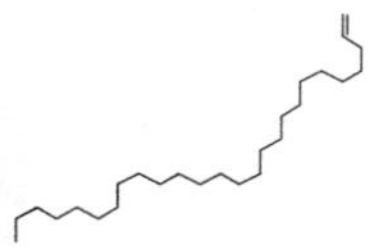	[239]
18	1-Iodobutane	C_4H_9I	Isol. from several marine algae	Used to esterify fatty acids for GC analysis		[232]

Continued

Table 6.11 Hydrocarbons: Biologically Active Constituents Extracted From Algae—cont'd

Sr. No.	Name	Molecular Formula	Biological Source	Biological Use	Structure	References
19	Iodoethane	C_2H_5I	Isol. from several marine algae	Used to esterify fatty acids for GC analysis		[232]
20	Iodomethane	CH_3I	Isol. from marine algae and giant kelp.	Methylating agent, Soil fumigant, insecticide, acaracide, rodenticide and fungicide		[232]
21	7-Methylheptadecane	$C_{18}H_{38}$	Isol. from various blue-green algae including *Anabaena* sp. and *Nostoc* sp.			[240]
22	3,6,9,12,15-Nonadecapentaene; (all-*Z*)-form	$C_{19}H_{30}$	Found in marine benthic algae		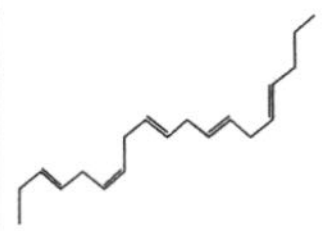	[238]
23	Pentadecane	$C_{15}H_{32}$	Blue-green algae *Oscillatoria splendida*, *Oscillatoria amoena*, *Oscillatoria geminate*, *Rumex japonicas*, *Vallisneria denseserrulate*, and *Aphanizomenon* sp.		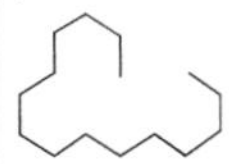	[238]
24	3,6,9,12-Pentadecatetraen-1-yne; (3*E*,6*Z*,9*Z*,12*Z*)-form	$C_{15}H_{20}$	Isol. from *Laurencia okamurai*	Possible precursor of a wide range of C-15-halogenated cyclic ethers found in marine red algae	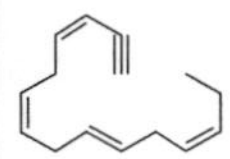	[241]
25	3,6,9-Pentadecatrien-1-yne; (3*E*,6*Z*,9*Z*)-form	$C_{15}H_{22}$	Constit. of *L. okamurai*	Possible precursor of a wide range of C-15-halogenated cyclic ethers found in marine red algae	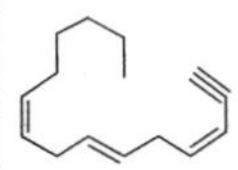	[242]

26	Tribromomethane	$CHBr_3$	Constit. of red algae *Asparagopsis* spp. and *Falkenbergia rufolanosa*	Antitussive; sedative, solvent for waxes, greases, and oils	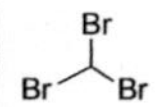	[243,244]
27	1,3,5,8-Undecatetraene; (3*E*,5*Z*,8*Z*)-form	$C_{11}H_{16}$	Constit. of the algae of the genera *Dictyopteris*, *Dictyosiphon* and *Spermatochnus*; also occurs in mandarin and tangerine oranges and pineapple	Smelling principle of the gametes of brown algae *Spermatochnus paradoxus*; sperm attractant of the marine brown alga *Ascophyllum nodosum*	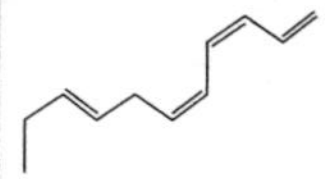	[245,246]
28	1,3,5-Undecatriene; (3*E*,5*Z*)-form	$C_{11}H_{18}$	Occurs in algae *Cystophora siliquosa* and *Dictyopteris* sp.; isol. from the fruit of *Ananas comosus* (pineapple)		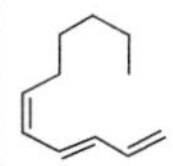	[247]

Constit., constituents; *Isol.*, isolated.

oxygen as a heteroatom in the ring [248]. The details of these compounds are briefly summarized in Table 6.12.

6.5.14 ALCOHOLS

Alcohol is a class of compound containing −OH (hydroxyl) group. Alcohols are important because they can be converted into other types of compounds, including alkenes, haloalkanes, aldehydes, ketones, carboxylic acids, and esters. Thus alcohols play a central role in the interconversion of organic functional groups [71]. The details of these compounds are briefly summarized in Table 6.13.

6.5.15 LIPIDS

Lipids are a heterogeneous group of naturally occurring organic compounds, classified together on the basis of their common solubility properties. Lipids are insoluble in water but soluble in nonpolar, aprotic organic solvents [71]. The details of these compounds are briefly summarized in Table 6.14.

6.5.16 CARBONYL COMPOUNDS

Carbonyl compounds contain a carbonyl group C=O in their molecules. Carbonyl group is the functional group of aldehyde and ketone and is one of the important functional groups in organic chemistry [71]. The details of these compounds are briefly summarized in Table 6.15.

6.5.17 OTHERS

Various other compounds have also been isolated from different species of algae. The details of these compounds are briefly summarized in Table 6.16.

6.6 THERAPEUTIC APPLICATIONS OF ALGAE

Scientists all across the world are trying to resolve the relationship between structure and therapeutic activities of algal-sulfated biopolymers. A spate intrigue has been observed to dab unexploited aquatic channels for the improvement of innovative therapeutics as algal cell wall polysaccharides and phytoconstituents have revealed a miraculous biological capital. The pharmacological properties of algae and their mode of action are reviewed as under.

6.6.1 ANTICOAGULANT ACTIVITY

Anticoagulant properties were first discovered in the extracts of marine algae over 50 years ago. The major active compounds are a variety of sulfated polysaccharides,

Table 6.12 Oxygen Heterocycles: Biologically Active Constituents Extracted From Algae

Sr. No.	Name	Molecular Formula	Biological Source	Biological Use	Structure	References
1	1,5-Anhydrofructose; D-form	$C_6H_{10}O_5$	Exists in rat liver, fungi, and algae	Antioxidant	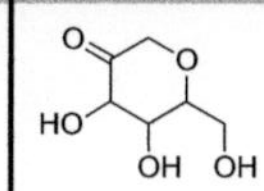	[249]
2	Antheraxanthin; (3*S*,3′*R*,5*S*,6*R*,9*Z*)-form	$C_{40}H_{56}O_3$	Constit. of *Lilium* spp., *Euonymus europaeus*, *Capsicum annuum*, and Chrysophyceae algae	Potential nutriceutical	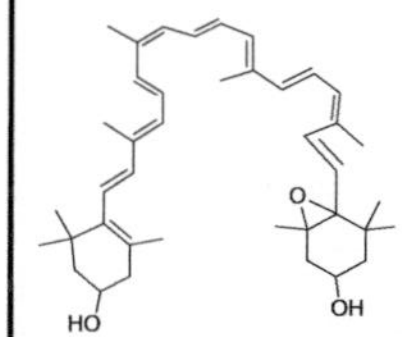	[250]
3	Auroxanthin	$C_{40}H_{56}O_4$	Isol. from *Viola tricolor*, *Lonicera japonica*, *Delonix regia*, and other plants; present in red algae		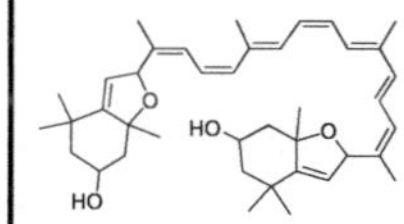	[251]
4	Auroxanthin; 3,3′-Dideoxy	$C_{40}H_{56}O_2$	Present in red algae		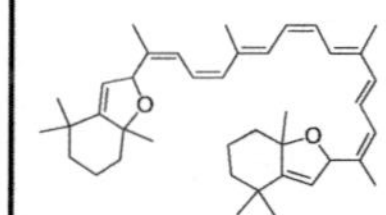	[251]
5	Cystoseirol-A	$C_{27}H_{36}O_4$	Constit. of brown algae *Cystoseira mediterranea*, *Cystoseira stricta*, and *Cystoseira tamariscifolia*		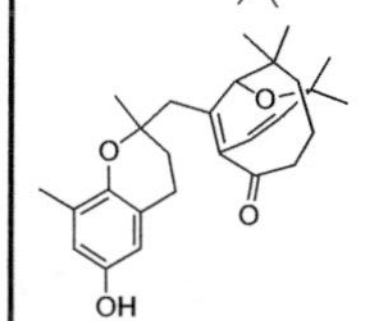	[123]
6	1-O-(6-Deoxy-6-sulfoglucopyranosyl)glycerol; α-D-form, 3-Hexadecanoyl	$C_{25}H_{48}O_{11}S$	Isol. from *Anthocidaris crassispina*, brown algae *Sargassum thunbergii*, and *Sargassum wightii*; and red alga *Caulacanthus ustulatus*		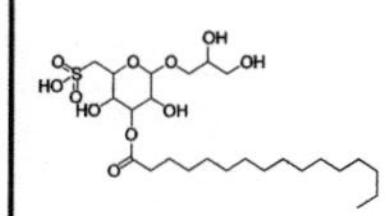	[252]

Continued

Table 6.12 Oxygen Heterocycles: Biologically Active Constituents Extracted From Algae—cont'd

Sr. No.	Name	Molecular Formula	Biological Source	Biological Use	Structure	References
7	3,4:7,8-Diepoxy-12-dolabellen-18-ol; (1*R**,3*R**,4*S**,7*R**,8*R**,11*R**)-form	$C_{20}H_{32}O_3$	Constit. of brown algae *Dictyota pardarlis*, *f. pseudohamata* A7749 and *Dictyota divaricate*	Antimalarial agent	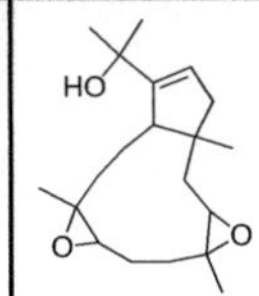	[253]
8	1,6-Di-O-β-D-glucopyranosyl-D-mannitol	$C_{18}H_{34}O_{16}$	Occurs in brown algae *Fucus vesiculosus*, *Fucus spiralis*, *Pelvetia canaliculata*, *Laminaria cloustoni*, and *Desmarestia aculeata*		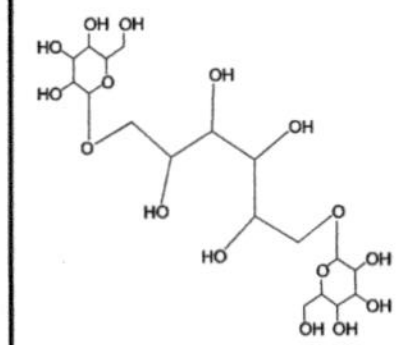	[254]
9	2,3-Dihydroxypropanoic acid; (*R*)-form, 2-*O*-α-D Mannopyranoside	$C_9H_{16}O_9$	Isol. from red algae *Alsidium*, *Chondria*, *Laurencia*, *Polysiphonia*, *Halopytis*, *Vidalia*, and *Digenea* spp.		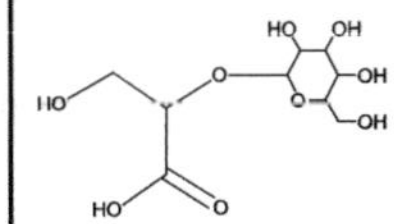	[255]
10	2,3-Dihydroxypropyl [5-deoxy-5-(dimethylarsino)] ribofuranoside; As-Oxide	$C_{10}H_{21}AsO_7$	Isol. from brown algae *Laminaria japonica* and *Ecklonia radiata*, giant clam *Tridacna maxima* and various freshwater mussels *Anodonta anatina*, *Dreissena polymorpha*, *Unio pictorum* and *Unio tumidas*		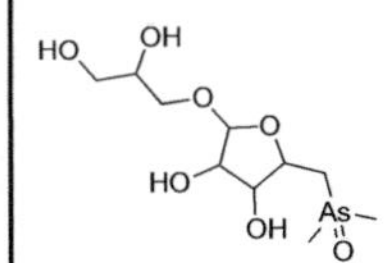	[256]

11	5,6-Epoxy-5,6-dihydro-β,ε-carotene-3,3′-diol	$C_{40}H_{56}O_3$	Isol. from a variety of higher plants and from algae		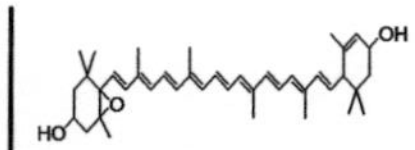	[251]
12	11,12-Epoxy-10-hydroxy-5,8,14-eicosatrienoic acid; (5*Z*,8*Z*,10*S*,11*S*,12*S*,14*Z*)-form	$C_{20}H_{32}O_4$	Constit. of *Murrayella periclados*, and algae *Platysiphonia miniata* and *Cottoniella filamentosa*	Insulin release enhancer; active against gram-positive bacteria; toxic to brine shrimp	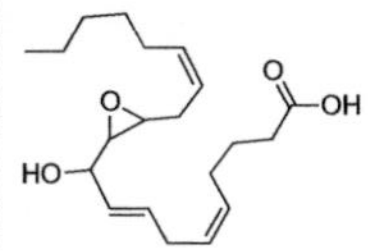	[257]
13	Flexixanthin; 2′-Hydroxy, 2′-*O*-α-L-rhamnopyranoside	$C_{46}H_{64}O_8$	Isol. from blue-green algae *Oscillatoria limosa* and *Phormidium faveolarum*		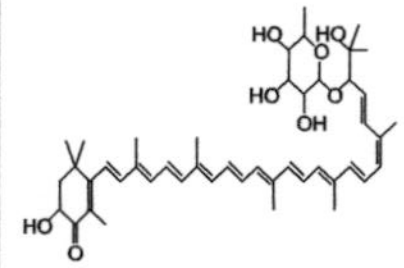	[258]
14	Fucoxanthinol; 3′-Ac	$C_{42}H_{58}O_6$	Constit. of algae, *Fucus virsoides*, *Polysiphonia nigrescens*, *Colpomenia peregrina*, *Phaeodactylum tricornutum*, *Ceramium rubrum*, and *Hijikia fusiformis*	Antioxidant, antifouling agent	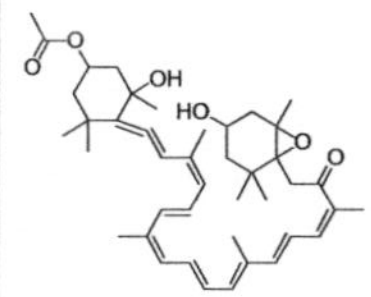	[136,259]
15	3-*O*-α-D-Galactopyranosyl-D-galactose	$C_{12}H_{22}O_{11}$	Isol. from *Aeodes ulvoidea* and *Pachymenia carnosa*		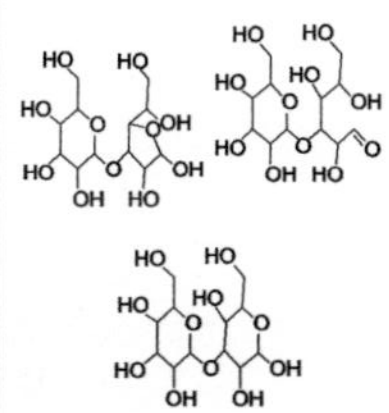	[260,261]
16	1-*O*-Galactopyranosylglycerol; α-D-(2*R*)-form	$C_9H_{18}O_8$	Constit. of various marine red algae		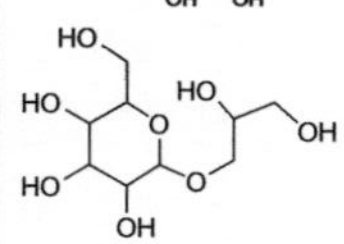	[205,262]

Continued

Table 6.12 Oxygen Heterocycles: Biologically Active Constituents Extracted From Algae—cont'd

Sr. No.	Name	Molecular Formula	Biological Source	Biological Use	Structure	References
17	2-*O*-α-D-Galactopyranosylglycerol	$C_9H_{18}O_8$	Constit. of many red algae, such as *Plocamium cartilagineum*, *Laurencia pinnatifida*, *Iridaea laminaroides*, and *Bacillus coagulans*	The main reserve carbohydrate in most red algae; prob. intracellular osmotic regulator	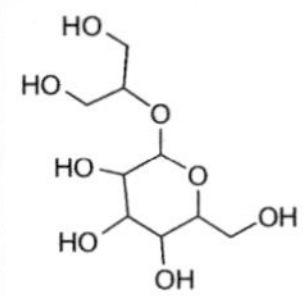	[205,263]
18	Mannuronic acid: D-form	$C_6H_{10}O_7$	*Macrocystis pyrifera*		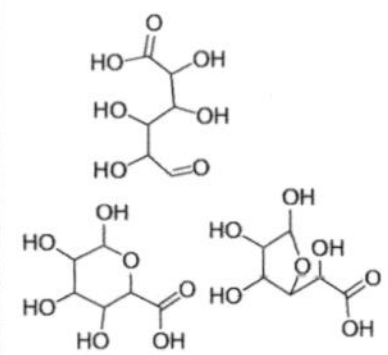	[264]
19	1,3,6-Tribromo-8-methoxydibenzofuran	$C_{13}H_7Br_3O_2$	Isol. from *crustose coralline* red algae		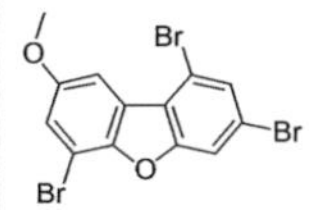	[126]
20	1-*O*-D-Glucopyranosyl-D-mannitol; β-form	$C_{12}H_{24}O_{11}$	Constit. of *F. vesiculosus* and other brown algae		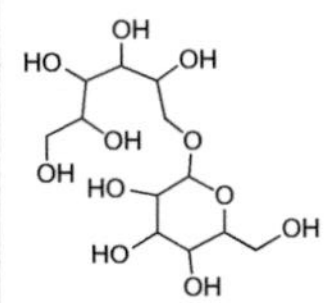	[265]
21	Stypoldione	$C_{27}H_{38}O_4$	Constit. of algae *Stypopodium zonale*, *Stypopodium flabelliforme*, and mollusk *Aplysia dactyomela*	Ichtyotoxin, algicide, toxic to sea urchin eggs; phospholipase A-2 inhibitor, microtubule inhibitor	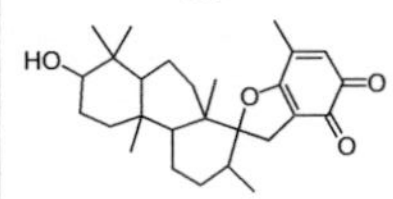	[266]

22	β-D-Xylopyranosyl-(1→3)-β-D-xylopyranosyl-(1→3)-D-xylose;	$C_{15}H_{26}O_{13}$	Constit. of the *Penicillus dumetosus*, *Rhodymenia palmate*, and several other green algae		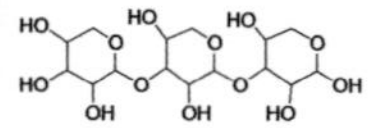	[267]
23	Myxol; 2′-*O*-6-Deoxy-α-L-glucoside	$C_{46}H_{66}O_{7}$	Isol. from *Oscillatoria rubescens*, *Oscillatoria agardhii*, *Arthrospira* spp., and *Phormidium luridum*		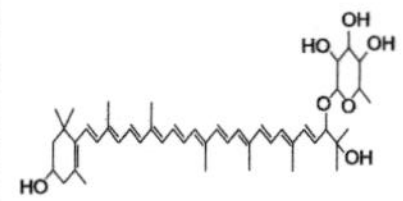	[109,268]
24	Neoxanthin; (all-*E*)-form	$C_{40}H_{56}O_{4}$	Constit. of paprika, lucerne, maple, Valencia orange; also in green algae		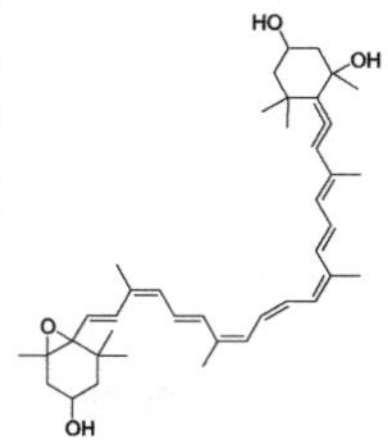	[269]
25	Violaxanthin; (all-*E*)-form	$C_{40}H_{56}O_{4}$	Constit. of many plants including *V. tricolor*; found in brown and green algae		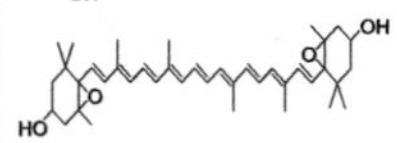	[269]
26	Glycerol 1,2-dialkanoates; glycerol 1,2-dioctanoate, 3-*O*-(6-Deoxy-6-sulfo-α-D-glucopyranoside)	$C_{41}H_{78}O_{12}S$	Constit. of the *Byrsonima crassifolia* and the algae *Gracilaria verrucosa* and *Sargassum parvivesiculosum*		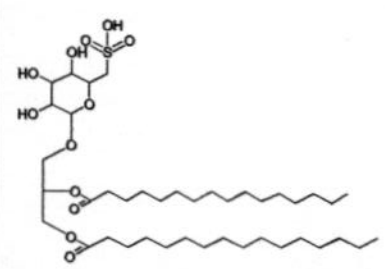	[270]

Continued

Table 6.12 Oxygen Heterocycles: Biologically Active Constituents Extracted From Algae—cont'd

Sr. No.	Name	Molecular Formula	Biological Source	Biological Use	Structure	References
27	Violaxanthin; (all-*E*)-form, Monodeoxy	$C_{40}H_{56}O_3$	Isol. from the algae of Xanthophyceae, *Eriobotrya japonica*, and *Prunus persica*		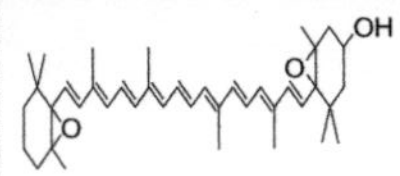	[269]
28	Chitin	$C_8H_{13}NO_5$	Occurs in crustacea, most fungi, mycelial yeast, green algae, and some brown and red algae	Antihemorrhagic; used as a wound healant	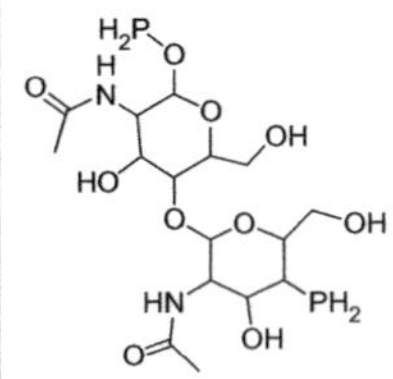	[271]
29	2,3-Dihydroxypropyl [5-deoxy-5-(dimethylarsino)] ribofuranoside; As-Oxide, *O3′*-sulfate	$C_{10}H_{21}AsO_{10}S$	Isol. from *T. maxima* and brown algae *Sargassum lacerifolium* and *Hizikia fusiforme*		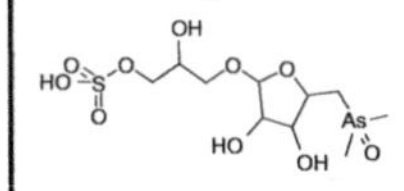	[272]
30	5-Methyl-2-furancarboxaldehyde	$C_6H_6O_2$	Isol. from brown algae and other plant sources	Flavoring ingredient	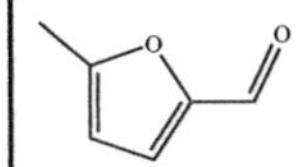	[273]
31	α,α-Trehalose	$C_{12}H_{22}O_{11}$	Occurs in fungi, molds, ergot, algae, yeast, and many insects	Probable energy reserve in many organisms and important cellular protectant in drought resistant organisms; associated with maintenance of the integrity of biological membranes in organisms subject to severe thermal stresses; marker for tubercuosis infection in vivo	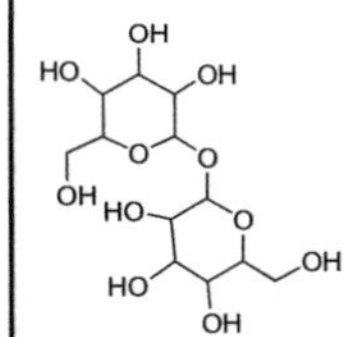	[274]

32	Galactose; L-form	$C_6H_{12}O_6$	Occurs in agar—agar, chagual gum, red algae, flax seed mucilage, and a snail galactan		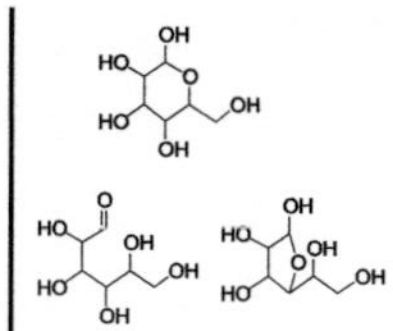	[275]
33	Mannose; D-form	$C_6H_{12}O_6$	*Phytelephas macrocarpa*, Orchidaceae, *Phoenix canariensis*, *Amorphophallus konjac*; they are proliferated by some red algae and yeasts; also occurs in trace amounts in apples and peaches	Inexpensive starting material for chiral synthesis	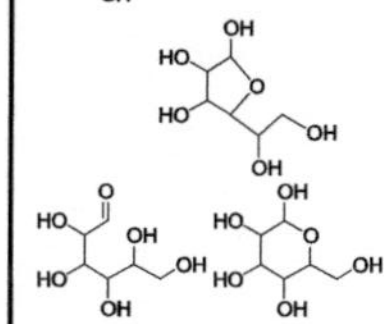	[275]

Constit., constituents; *Isol.*, isolated.

Table 6.13 Alcohols: Biologically Active Constituents Extracted From Algae

Sr. No.	Name	Molecular Formula	Biological Source	Biological Use	Structure	References
1	Erythritol	$C_4H_{10}O_4$	Found in a variety of algae, lichens and fungi; produced by *Protococcus vulgaris*, *Trentepohlia iolithus*, and *Aspergillus terreus*	Bulk sweetener with good taste props; also thickener, stabilizer, humectant, etc. in food, shows vasodilatory props	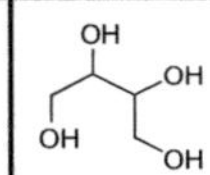	[276]
2	2-Methyl-1,2,3,4-butanetetrol; (2*R*,3*R*)-form, 4-phosphate	$C_5H_{13}O_7P$	Bacteria, algae, and plant chloroplasts	The first pathway-specific intermediate in the methylerythritol phosphate route for biosynthesis of isoprenoid compounds in bacteria, algae, and plant chloroplasts	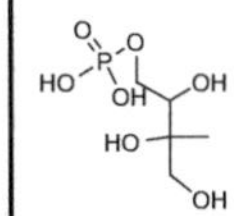	[277]
3	1-Deoxy-1-(dimethylarsinoyl) ribitol; 5-*O*-sulfate	$C_7H_{17}AsO_8S$	Isol. from *Chondria crassicaulis* and other red algae		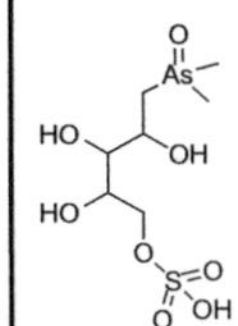	[278]
4	Talitol; D-form	$C_6H_{14}O_6$	Isol. from various brown algae including *Himanthalia elongata* and *Notheia anomala*		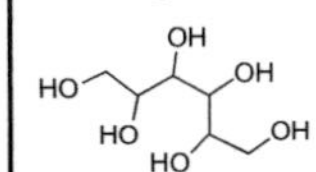	[279,280]
5	Mannitol; D-form	$C_6H_{14}O_6$	Olive and plane trees; obtained from manna and seaweeds; obtained industrially from fructose	Energy reserve in brown algae Diagnostic aid (renal function determination); diuretic; tablet and capsule diluent	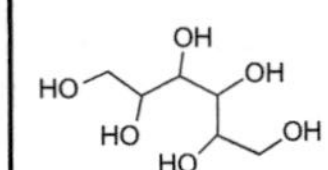	[276]

6	1,1,3,3-Tetrabromo-2-heptanol	$C_7H_{12}Br_4O$	Constit. of red algae *Bonnemaisonia* spp.		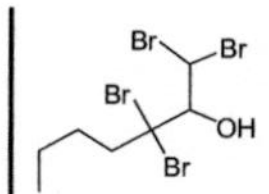	[222]
7	2-Nonen-1-ol	$C_9H_{18}O$	Present in melon, cucumber, brown algae, nectarine, and prickly pear	Used in food flavoring	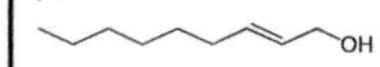	[281]
8	Glucitol; D-form	$C_6H_{14}O_6$	Present in algae, apples, pears, cherries, and apricots; also present in *Sorbus* and *Crataegus* spp.	Used in manufacture of sorbose, propylene glycol, ascorbic acid, resins, plasticizers and as antifreeze mixtures with glycerol or glycol; tablet diluent, sweetening agent and humectant, other food uses; used in photometric detn. of Ru(VI) and Ru(VIII); in acid—base titration of borate	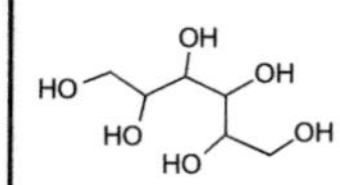	[280]
9	1-Bromo-3-iodo-2-propanol; (ξ)-form	C_3H_6BrIO	Present in red algae *Asparagopsis taxiformis*		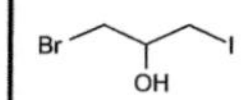	[282]
10	D-glycero-D-manno-heptitol	$C_7H_{16}O_7$	Occurs in mushroom *Lactarius volemus*, in primulae and in *Escherichia coli*; also in red algae		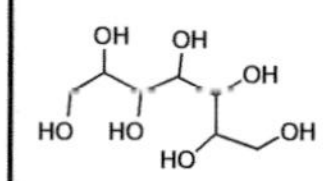	[276]

Constit., constituents; *Isol.*, isolated.

Table 6.14 Lipids: Biologically Active Constituents Extracted From Algae

Sr. No.	Name	Molecular Formula	Biological Source	Biological Use	Structure	References
1	1,2-Diacylglyceryl-3-(*O*-carboxyhydroxymethyl) choline		Betaine lipid present in some plants and algae		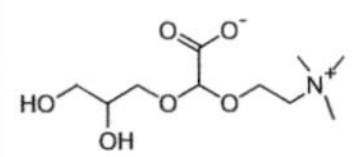	[283]
2	1,2-Diacylglyceryl-*O*-2′-(hydroxymethyl)-*N*,*N*,*N*-trimethyl-β-alanine		Isol. from various marine algae		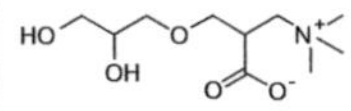	[283]
3	Ulvaline	$C_{10}H_{21}NO_5$	Isol. from the marine plant *Monostroma nitidum* and the mushroom *Lampteromyces japonicus*; isol. from numerous marine sources including algae and bryophytes	Shows hypocholesterolaemic activity	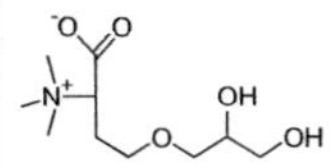	[284]
4	Lactyltrimethylarsonium betaine	$C_6H_{13}AsO_3$	Marine algae, e.g., *Dunaliella tertiolecta*; also in diatom *Chaetoceros concavicornis* and lobster *Homarus americanus*		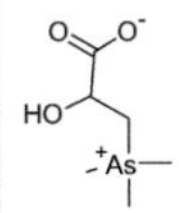	[285]

Isol., isolated.

Table 6.15 Carbonyl Compounds: Biologically Active Constituents Extracted From Algae

Sr. No.	Name	Molecular Formula	Biological Source	Biological Use	Structure	References
1	1,1-Dibromo-3-iodo-2-propanone	$C_3H_3Br_2IO$	Red algae, *Asparagopsis taxiformis*, and *Asparagopsis armata*		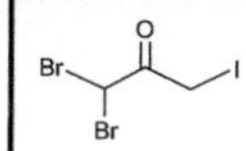	[234]
2	1,3-Dibromo-2-propanone	$C_3H_4Br_2O$	Constit. of the marine algae *A. armata*, *A. taxiformis*, and *Falkenbergia rufolanosa*		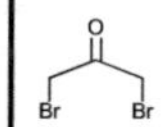	[234]
3	8,11,14-Heptadecatrienal; (all-*Z*)-form	$C_{17}H_{28}O$	Constit. of cucumber, tobacco, and wheat; also found in the algae *Enteromorpha* sp., *Scytosiphon lomentaria*, and *Ulva pertusa*		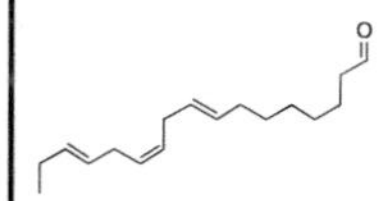	[286,287]
4	1-Bromo-2-propanone	C_3H_5BrO	Isol. from the algae *A. taxiformis* and *F. rufolanosa*		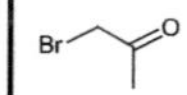	[234]
5	Dibromoacetaldehyde	$C_2H_2Br_2O$	Isol. from the algae *A. taxiformis* and *F. rufolanosa*		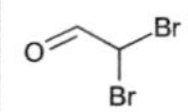	[234]
6	1,3-Dibromo-1-chloro-2-propanone; (±)-form	$C_3H_3Br_2ClO$	Component of red algae *A. taxiformis*, *A. armata*, and tetrasporophyte *F. rufolanosa*		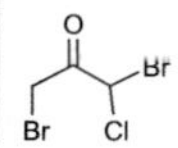	[234]
7	3-Hexyl-1,2-dithiepan-5-one; (−)-form	$C_{11}H_{20}OS_2$	Constit. of seaweed *Dictyopteris plagiogramma*, green algae *Bryopsis pennata*, *Caulerpa prolifera*, and *Cymopolia barbata*, and other marine algae		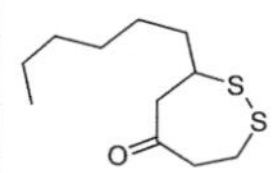	[288]

Continued

Table 6.15 Carbonyl Compounds: Biologically Active Constituents Extracted From Algae—cont'd

Sr. No.	Name	Molecular Formula	Biological Source	Biological Use	Structure	References
8	2-Tridecyl-2-heptadecenal; (*E*)-form	$C_{30}H_{58}O$	Constit. of red algae *Corallina mediterranea*, *Laurencia obtusa*, *Laurencia papillosa*, and *Laurencia undulata*; also prod. by a marine-derived *Cladosporium* sp.		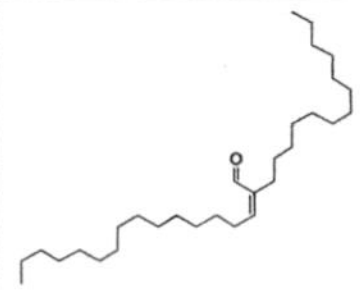	[289]
9	1,1,3-Tribromo-3-chloro-2-propanone	$C_3H_2Br_3ClO$	Red algae *A. taxiformis* and *F. rufolanosa*		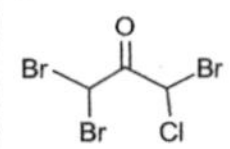	[234]
10	1,1,3-Tribromo-2-propanone	$C_3H_3Br_3O$	Constit. of red algae *Asparagopsis* spp. and *F. rufolanosa*		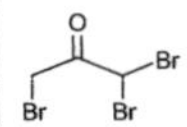	[234]
11	6,10,14-Trimethyl-2-pentadecanone; (ξ)-form	$C_{18}H_{36}O$	Present in marine organisms such as sponge *Spheciospongia vagabunda*, algae *Caulerpa taxifolia*, and *Padina tetrastromatica*, and crab *Carcinus maenas*		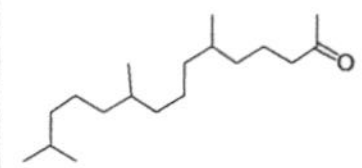	[290]

Constit., constituents; *Isol.*, isolated.

Table 6.16 Other Compounds Extracted From Algae

Sr. No.	Name	Molecular Formula	Biological Source	Biological Use	Structure	References
1	2-Amino-6-(1,2-dihydroxypropyl)-4(1H)-pteridinone; (1′*R*,2′*R*)-form	$C_9H_{11}N_5O_3$	Widely distributed in microorganisms, insects, algae, amphibian, and mammals; Found in urine	Growth factor	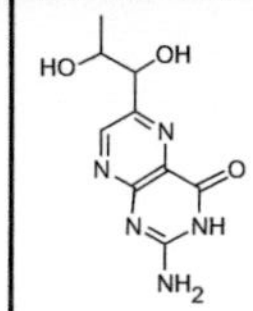	[291]
2	Arsenobetaine	$C_5H_{11}AsO_2$	Found in algae, lobsters, sharks, and dogfish		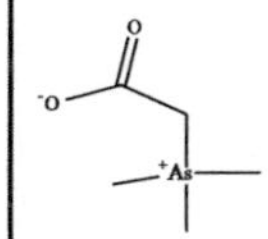	[292]
3	Carbon monoxide	CO	Produced in traces by algae and higher plants	Appears to be a hormonal messenger involved in cyclic GMP regulation; vasodilator. Used in the prophylaxis of graft dysfunction in patients receiving allogenic kidney transplants		[293]
4	Dimethyl sulfide	C_2H_6S	Isol. from green and red algae and higher plants	Selective reducing agent; reagent for ortho-alkylation of aromatic amines; used with diborane as effective reducing agent for esters, amides, etc.		[294]

Continued

Table 6.16 Other Compounds Extracted From Algae—cont'd

Sr. No.	Name	Molecular Formula	Biological Source	Biological Use	Structure	References
5	1-Pentacosene-4,6,8,10,12,14,16,18,20-nonol; (4*S*,6*S*,8*S*,10*S*,1214*R*,16*R*,18*R*)-form, Nona-Me ether	$C_{34}H_{68}O_9$	Isol. from the blue-green algae *Scytonema burmanicum* and *Scytonema mirabile*		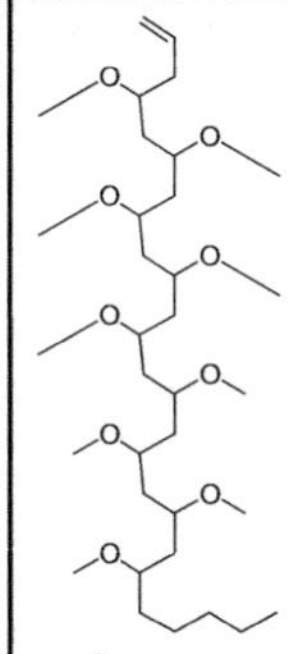	[114]
6	3-(Dimethylarsinyl)propanoic acid	$C_5H_{11}AsO_3$	Found in lobster, dogfish, oyster, blue mussel, horse mussel, scallop, cod, brown algae *Ascophyllum nodosum*, and *Fucus vesiculosus*		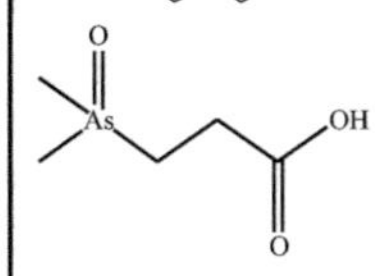	[295]

Isol., isolated.

some of which are characterized as proteoglycans [296]. Thrombin and plasmin are the proteolytic enzymes required for the activation of enzymes in fibrinolytic and coagulation cascade (Fig. 6.1). The balance between both pathways is obligatory, otherwise strokes and heart attacks are obvious due to excessive bleeding or thrombosis. Deep vein thrombosis is the major cause of mortality in posttraumatic cases and promulgates in operated patients [298]. Anticoagulant therapy is very beneficial in such situations. The first report of algal polysaccharide as anticoagulant agent was published in 1936 when Chargaff and colleagues succeeded to isolate the esters of sulfuric acid and galactan from *Iridae laminarioides* having anticoagulant activity equal to 30 U/mg of heparin. Consequent analysis revealed carrageenan and agar having similar anticoagulant properties. Up till now, approximately 150 species belonging to red-brown, green, and microalgae have been used to isolate this heterogeneous sulfated biopolymer (Table 6.17). Sulfated biopolymer heparin has been used as an anticoagulant since past 50 years in posttraumatic patients [317]. Its anticoagulating effect is expressed by heparin cofactor-II and antithrombin potentiation, a major anticoagulating enzyme (Fig. 6.1). However, algal-sulfated polysaccharides are most extensively studied as anticoagulants, more extensive studies are needed to overcome side effects, i.e., thrombocytopenia. [297]. This side effect was overcome by synthesizing low molecular weight algal-sulfated polysaccharides [318]. Similar to the algal-sulfated polysaccharides (heparin), dermatan sulfate, in combination with chondroitin sulfate, is also clinically employed, but unfortunately is little potent due to its high molecular weight [27]. Fucoidan from *Ecklonia kurome* is found to be more potent due to low molecular weight fractionation and low sulfation [319]. It was also observed that branched fucans isolated from brown algae have direct effect while linear fucans isolated from echinoderms require some antithrombin molecules for their action [320].

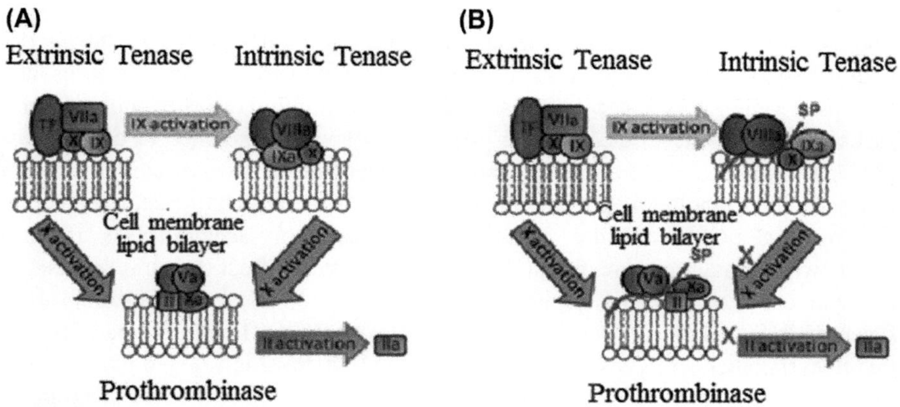

FIGURE 6.1

Schematic presentation of antithrombotic activity: antithrombotic effect of sulfated polysaccharides; (A) activation of prothrombin versus inactivation of prothrombinase by treating with sulfated polysaccharides [297].

Table 6.17 Major Sugar Molecules of Algal Polysaccharides (PSs) Exhibiting Anticoagulating and Antithrombotic Agents

Species	Major Sugar	PS	References
Grateloupia indica	Galactose	sPS	[299]
Gigartina skottsbergii	Galactose	sPS	[300]
Nothogenia fastigiata	Mannose	EsPS	[301]
Lomentaria catenata	Galactose	Galactan	[302]
Schizymenia binderi	Galactose	Galactan	[303]
Porphyra haitanensis	Galactose	sPS	[304]
Monostroma nitidum	Rhamnose	PS (sulfated)	[305]
Lessoni avadosa	Fucose	PS (sulfated)	[306]
Ecklonia cava	Fucose	PS (sulfated)	[307]
Ascophyllum nodosum	Fucose	PS (sulfated)	[308]
Ecklonia kurome	Fucose	PS (sulfated)	[309]
Monostroma latissimum	Rhamnose	sPS	[310]
Ulva conglobata	Rhamnose	sPS	[311]
Codium cylindricum	Galactose	Galactan	[312]
Codium pugniformis	Glucose	sPS	[302]
Codium fragile	Arabinose	Arabinans	[313]
Dictyota cervicornis	Mannose, fucose	Rhamnose	[314]
Caulerpa cupresoides	Galactose	Galactan	[315]
E. cava	Arabinose, galactose	Aribanogalactan	[316]

6.6.2 ANTIVIRAL ACTIVITIES OF ALGAE-BASED BIOLOGICALLY ACTIVE COMPOUNDS

The use of algal polysaccharides as antiviral agent is not new as it goes back more than 50 years when Gerber and his colleagues inhibited influenza-B virus and mumps in 1955 [321]. Later it was assumed that the antiviral effect of sulfated polysaccharides was due to electrostatic prevention of viral adherence to cell surface [322]. This gushed exploration of antiviral achievements of diverse-natured sulfated polysaccharides including polyanionic substances [323]. With the structural elucidation of algal polysaccharides, isolation of seaweed polysaccharides as antiviral agents was substantially increased due to their ability to affect the viral replication by a number of nonspecific mechanisms (Fig. 6.2) [324]. *Cladosiphon okamuranus* fucan inhibited DENV-2 (Dengue fever type-2) with BHK-21 cells, but had very little effect on other serotypes. Structural analysis of DENV-2 enveloped glycoprotein exhibits that arginine-323 close to heparin-binding site is censorious for fucan interaction [27]. In accordance with these findings, Talarico and colleagues published that sulfated biopolymer from two genera of red algae, i.e., carrageenan from *Gymnogongrus griffithsiae* and galactan from *Cryptonemia crenulata* completely impeded DENV-2, with less impedance on DENV-3 and DENV-4, while

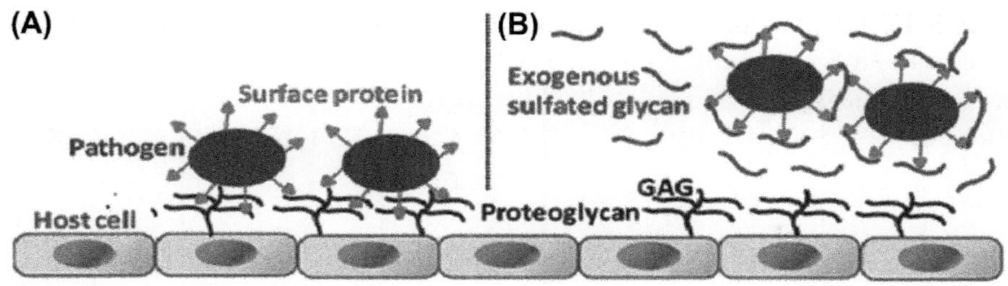

FIGURE 6.2

Antiviral mode of action of sulfated polysaccharides: antiviral mode of action of sulfated polysaccharides [297]; (A) binding of pathogen with host cell; (B) neutralization of pathogen.

remained ineffective against DENV-1 in cell lines [325,326]. These biopolymers use the phenomenon of adsorption, only when added soon after infection, as no inhibition was observed when viral entry was avoided by transfection into the cells [327]. Inhibition of herpes simplex virus-1 and -2 (HSV-1, HSV-2) by sulfated biopolymers has become a hot issue these days [328,329]. Sulfated xylomannan of *Sebdenia polydactyla* exhibited high potency against HSV-1 due to high sulfation, and the potency is further increased by increasing sulfation [330]. Inhibition of different viral species (Table 6.18) by biopolymers of different algal species indicated that

Table 6.18 Effect of Algal Sulfated Polysaccharides (PSs) on Different Virus Species

Species	Virus Strain	EC_{50}/ED_{50} (μg/mL)	Type of PS	References
Schizymenia pacifica	HIV-1	7–14	Carrageenan	[331]
Nothogenia Fastigiata	HIV-1	061–1.35	(Xylo)mannans	[332]
Digenea simplex	HIV	2.84	sp. non-characterized	[333]
Pterocladia capillacea	HSV-1	2.6–10	Sulfated agarans and hybrid DL-galactans	[334]
Gigartina skottsbergii	HSV-1, HSV-2	>1	Carrageenan	[335]
Cryptopleura ramosa	HSV-1, HSV-2	1.6–4.2	Sulfated agarans	[335]
Salvia interrupta	HSV-1, HSV-2	1.5–36	Carrageenans	[329]

Continued

Table 6.18 Effect of Algal Sulfated Polysaccharides (PSs) on Different Virus Species—cont'd

Species	Virus Strain	EC_{50}/ED_{50} (μg/mL)	Type of PS	References
		1 and 0.7 −6.6		
Asparagopsis armata	HIV	40.2	Sulfated agarans	[336]
Porphyridium cruentum	African swine fever virus (ASFV); vaccinia virus (VACV); vesicular stomatitis virus (VSV)	12–56; 20–45	Sulfated agarans	[242]
Gyrodinium impudicum	Encephalomyocarditis virus; influenza A virus (Flu-A)	0.19–0.48	Hybrid DL-galactans	[309]
Rhodella reticulata	HSV-2; varicela zoster virus (VZV); murine sarcoma virus (MuSV-124), and MuSV/MuLV (murine eukemia virus)	10–20; 8; 150 and 50	Sulfated agarans	[337]
Monostroma latissinum	HSV-1, HCMV, HIV-1	3.7	Sulfated Rhamnans	[338]
Porphyridium sp.	HSV-1, HSV-2	1–5 (in vivo, 100); 0.7; 10 and 5 (RT50)	Extract	[339]
Cochlodinium polykrikoides	HIV-1, RSV, Inf A, Inf B	0.45–1.1 and 7.1–8.3; 2.0–3.0 and 0.8; 1.7; 4.52–21.6; 0.8–25.3	Extract	[340]
Pennisetum purpureum	Vaccinia virus VACV and VACV-GFP; ectromelia virus (ECTV)	0.65	Fucans	[341]
Leathesia difformis	HSV-1	0.5–1.9	Fucans	[338]
Sargassum horneri	HSV-1, HCMV HIV-1	45.5	Fucans	[342]
Fucus vesiculosus	HIV-1	49	Fucans	[302]

antiviral activity is greatly affected by degree of sulfation, arrangement of sulfate, composition of sugar moieties, stereochemistry, or its molecular weight [343]. HSV-2 is inactivated by direct incubation with sulfated biopolymer as in vivo consequence of virucidal activity is related with amplification of antiviral activity [335].

6.6.3 ANTIOXIDANT ACTIVITY

Free radicals are highly reactive molecules with an unpaired electron and are produced by radiation or as by-products of metabolic processes. They initiate chain reactions which lead to disintegration of cell membranes and cell compounds, including lipids, proteins, and nucleic acids [327]. Antioxidant compounds scavenge free radicals such as peroxides and reduce the level of oxidative stress and slow/prevent the development of complications associated with oxidative stress—related diseases [344]. The use of algal biopolymers was totally ignored as an antioxidant agent but the recent studies revealed that algal compounds have considerable antioxidant competence [345]. As being photoautotrophs, micro/macroalgae possess a trenchant, antioxidant magpie complex for their self-defense against radical and oxidative stresses [346]. ROS (reactive oxygen species), i.e., superoxide anion, hydrogen peroxide, and hydroxyl radicals, dissimulate pharmaceutical security and diminish food quality by oxidation of lipids. Algal-sulfated biopolymers not only act as dietary fibers but also act as free-radical scavengers (Fig. 6.3). e.g., Tannin et al. [347] reported that *Porphyridium* spp. prevented autooxidation of linoleic acid and repressed oxidative damage to 3T3 cells caused by $FeSO_4$. Their further studies revealed that the antioxidant activity was dose dependent having positive correlation with sulfur contents conferring free-radical scavenger activity. Carrageenans and

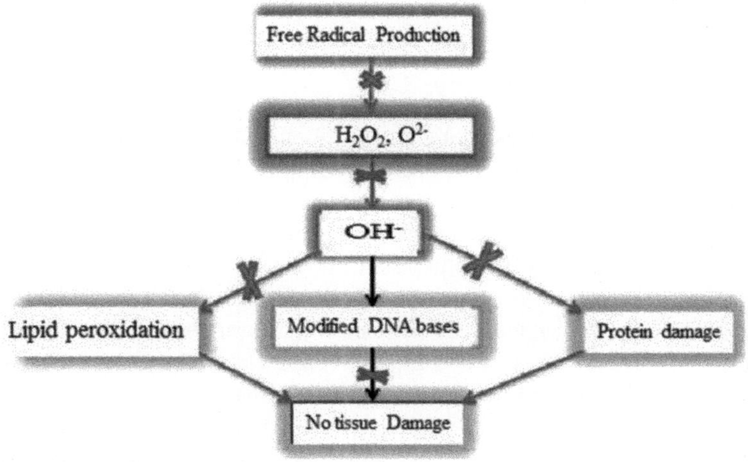

FIGURE 6.3 Antioxidant Activity of Algal Extract.

✕: Inhibition by algal extract. A step at which algal extracts prevents the tissue damage by inhibiting lipid peroxidation, superoxide, peroxide, hydroxyl ions formation.

ulvans bioactivities were found to be correlated with the contents of sulfate [348]. Fractions of fucan from *Laminaria japonica* and *Fucus vesiculosus* exhibited positive correlation between fucose and sulfate concentration to act as metal chelator and free-radical and insulin-dependent scavenger [349]. Sun et al. [350] reported that low molecular weight fractions (6.55 kDa) of sulfated biopolymers isolated from *Porphyridium cruentum* exhibited higher antioxidant activity against ascorbic acid than $FeSO_4$-induced peroxides having high molecular weight (256 kDa) in mouse cells. Scientists reported that sulfated biopolymers of *Arthrospira* were stronger superoxide anion radical scavenger than α-tocopherol. Crude extract was two times stronger than α-tocopherol. Seaweeds also reported recognized antioxidants such as terpenoids [351], carotenoids, mycosporine-like amino acids, phlorotannins, ascorbic acid, tocopherols [352], and algal-sulfated polysaccharides [353]. *Laminaria digitata* and *Himanthalia elongata* have been reported to exhibit antioxidant activity comparable with those of vitamin E and butylhydroxyl toluene (BHT) [347]. Antioxidant activity was also reported from *Sargassum micracanthum* [354], *Rhodomela confervoides*, *Symphocladia latiuscula* [355], *Kappaphycus alvarezzi* [356,357], *Gelidiella acerosa* [358], *F. vesiculosus* [359], diphlorethohydroxycarmalol and 6,6′-bieckol isolated from *Ishige okamurae* [360], *Undaria pinnatifida* [361], ethanolic extract of *Sargassum pallidum* [362], *Enteromorpha compressa*, *Capsosiphon fulvescens* [363], *Halimeda incrassate* [364], *Eisenia bicyclis*, *Kjellmaniella crassifolia*, *Alaria crassifolia*, *Sargassum horneri*, and *Cystoseira hakodatensis* [365]. Fucoidan derivatives, oversulphated, acetylated, and benzoylated fucoidan isolated from *L. japonica*, show potential antioxidant activity in vitro [345,353]. In vitro antioxidant activity of iota, kappa, and lambda carrageenans, homofucan from the edible seaweed *F. vesiculosus*, and heterofucans from the seaweed *Padina gymnospora* were also reported [366]. Astaxanthin, an antioxidant obtained from *Haematococcus pluvialis* is marketed [348].

6.6.4 ANTICANCER ACTIVITY

Cancer is a group of diseases characterized by uncontrolled growth and spread of abnormal cells. If the spread is not controlled (Fig. 6.4), it results in death. Isolation of cytotoxic antitumor substances from marine organisms was reported in several references during the last 40 years, while in recent years, hundreds of potential antitumor agents were isolated from marine origin especially from marine algae [367]. Biological properties of seaweeds such as reduction of plasma cholesterol, binding of biliary steroids, inhibition of carcinogenic fecal flora, binding of pollutants, stimulation of the immune system, and the protective effects of β-sitosterols seaweed suggested them as a potent therapeutic agent against breast cancer. The low rate of breast cancer in Japan is associated with high seaweeds consumption [368]. Certain algae have long been used in traditional Chinese herbal medicines for the treatment of cancer [369].

Sulfated polysaccharides, including fucoidans and carrageenans, inhibit tumor metastasis in rats by inhibiting the action of the tumor cell–derived heparanases

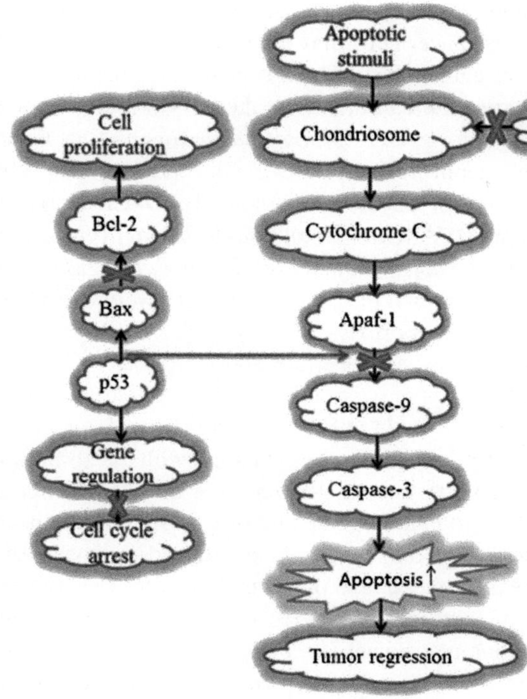

FIGURE 6.4 Anticancer Mode of Action of Algal Extract.

✖: Inhibition by algal extract. Algal polysaccharides inhibit the proliferation of various types of cancer and cell cycle arrest at G_0/G_1 S or G_2/M phase. *Apaf-1*, apoptotic protease activator factor; *Bax*, Bcl-2-like protein; *Bcl2*, B-cell lymphoma2; *P53*, tumor protein.

involved in membrane crossing [370–372]. Carrageenan from *Sargassum kjellmanianam* [373] significantly inhibited S-180 and leukemia-1210 ascites tumor growth effect of mitomycin in mice [374]. Carrageenan from rodophyta was also found to stimulate lectin-dependent, cell-mediated cytotoxicity against HEp-2 human epipharynx carcinoma cells [375]. Various brown algae, namely *Scytosiphon lomentaria, Lessonia nigrescens, L. japonica, Sargassum ringgoldianum*, red algae, *Porphyra yezoensis* and *Eucheuma gelatinae*, and the green alga, *Enteromorpha prolifera* have shown antitumor activity against Meth-A fibrosarcoma [376]. Fucoidans isolated from brown seaweed *Sargassum thunbergii* have proven antitumor activity [377,378]. Ulvan extracted from *Ulva lactuca* has shown cytotoxicity against human colon cell line [29]. Facosterols isolated from *Turbinaria conoides* also displayed cytotoxicity against human (HU) cell lines [379]. Caulerpenyne isolated from *Caulerpa taxifolia* exhibited antitumor activity against HU neuroblastoma cell line by inhibiting microtubule assembly and tubulin aggregation [380]. Compounds of dihydroxysargaquinone and sargatriol from *Sargassum tortile* [381] and diterpene from *Sargassum crispum* are known for their cytotoxic activities [382].

Cytotoxicity of fucodian extracted from *Sargassum swartzii* and *Sargassum denticaprum* against some human cancer cell lines [383], antitumor activity of *Gracilaria corticata* against Jurkat and molt-4 human cancer cell lines [384], *Spyridia filamentosa* against human prostate carcinoma epithelium-like cell lines DU-145 [385], *S. swartzii*, *Cystoseira myrica*, and *Colpomenia sinuosa* against Colon carcinoma (HT-29), colorectal adenocarcinoma (Caco-2), breast ductal carcinoma (T47D), tamoxifen-resistant breast ductal carcinoma (T47D-T.R), and estrogen-independent breast carcinoma (MDAMB468) cell lines [386], *Arthrospira platensis*, *Dunaliella salina* and *Aphanizomenon flosaquae*, and *H. pluvialis* against the growth of human leukemia cell line HL-60, and the biphenotypic B myelomonocytic leukemia cell line MV-4-11 [387] were reported.

6.6.5 ANTIINFLAMMATORY ACTIVITY

Algal-sulfated biopolymers are potent immunomodulators with the ability to control the activity of immune cell activity and to stimulate the immune response [388]. Algal cell wall polysaccharides jolt an outcome and progression of disease by complex targets [24]. Intravenous injection of fucoidan reduced leukocyte rolling in meningitis and similarly also reduced leukocyte enlisting to peritoneum [389]. Commercially available (Sigma−Aldrich Chemical Co.) fucoidan from *F. vesiculosus* exhibits antiinflammatory activity by inhibiting the cell adhesion process by binding with P- and L-selectins. Fucans from *Fucus* spp. *C. okamuranus*, *Ascophyllum nodosum*, and *Laminaria* spp. treat rats' acute peritonitis by inhibiting leukocyte recruitment toward abdominal cavity [390]. In addition to impairment of selection actions, these sulfated biopolymers have the ability to completely inhibit elastase and heparanase responsible for degradation of membrane integrity (Fig. 6.5). It was reported that sulfated biopolymers had the ability to inhibit the complement activation by binding with C-4, anticipate schism product required for the synthesis of C-3 convertase [391]. Tissot and colleagues [392] succeeded to isolate low molecular weight fractions of fucoidan that have shown their ability to inhibit the complement by binding with immune complexes and also influence the C-1q and activate C-1 to recognize IgG.

Structural elucidations using NMR revealed that branched fucoidan oligosaccharides are more potent than linear structures for inhibition of complement system [393,394]. Fucoidan influencing the complement system, also affect innate immunity rendering the pro-inflammatory state by reducing allergic reactions. Furthermore, it was also observed that sulfated algal cell wall polysaccharides also regulate innate immunity by binding with phagocytes and macrophages [395]. λ-carrageenan generates cytokine response by stimulation of T-cell cultures in a TLR4 (toll-like receptor-4) in mice [396]. However, TLR4-deficient splenocyte mice also produced γ-interferon after injection of λ-carrageenan indicating that PRRs (proline-rich polypeptides), in addition to TLR4, were also provoked. It was reported that λ-carrageenan might be applied to amend allergic reactions. *U. pinnatifida* fucoidan expressed similar results [397] in accordance with the

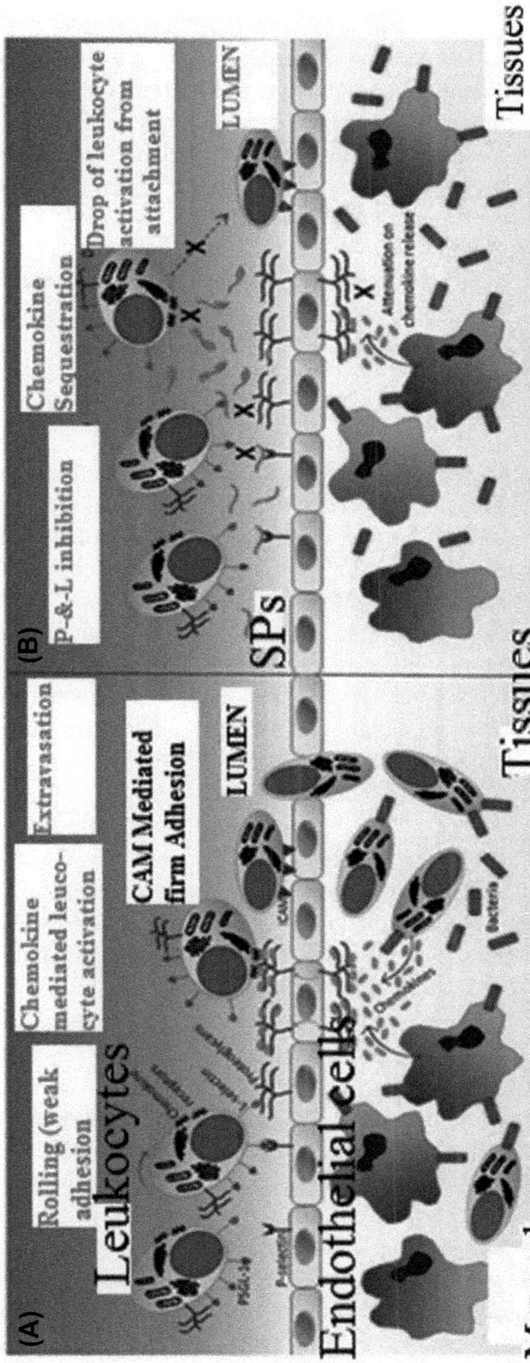

FIGURE 6.5 Antiinflammatory Role of Sulfated Polysaccharides.

(A) Untreated inflammation; (B) treated inflammtion with sulfated polysaccharides [297].

findings of Tsuji et al. [396]. Reduction in inflammation is due to its interaction with other factors such as inducible nitric oxide synthase (iNOS) and cytokines, e.g., fucoidan isolated from *F. vesiculosus* incurred NO by attracting macrophages, causing the reduction in inflammation beyond impairment of LPS (lipopolysaccharide) [398].

Desulfation of sulfated polysaccharides from *Ulva rigida* is associated with complete attenuation of immunostimulatory activity [309]. Moreover, sulfated biopolymers from various species of algae, i.e., *Phaeodactylum*, *Chlorella stigmatophora*, and *Porphyridium*, were found to be effective antiinflammatory agents. Raposo et al. [399] reported that carrageenans and fucoidans aggravated the cytotoxicity of macrophages, lymphocytes and natural-killer cells opposed to tumors. The antiinflammatory effect *of Sagarssum hemiphyllum* against Phorbol 12-myristate 13-acetate (PMA) and A23187-induced interleukin-8 (IL-8) and on tumor necrosis factor—α (TNF-α) secretion from human mast cells (HMC-1) [400], *Galaxaura marginata* against Croton oil induced mouse ear edema [401], *S. thunbergii* and *Sargassum fulvellum* against phorbol myristate acetate-induced ear edema, erythema, and blood flow [402], *Porphyra dentata* against LPS-induced RAW 264.7 macrophages [403], *Petalonia binghamiae* against ethyl acetate LPS-induced RAW 264.7 macrophages [404], *S. swartzii* and *Ulva reticulata* against Carrageenan-induce inflammation [405], *Dichotomaria obtusata*, against water-ear edema induced by TPA and writhing-induced hind paw edema in rats and peritonitis for acute and chronic inflammatory mode by acetic acid [406], and *T. conoides* against carrageenan-induced hind paw edema [407] were also reported. On the basis of these reports, sulfated biopolymers can be used for therapeutic purposes in allergy and autoimmune diseases.

6.6.6 ANTIULCER ACTIVITY

Chromene compound extracted from the seaweed *S. micracanthum* is reported to possess antiulcer property [408]. Methanolic extract of red algae *Gracilaria changii* shows antiulcer activity [409]. It is also exhibited by seaweeds *Turbinaria ornata*, *Gracillaria crassa*, and *Laurencia papillosa* [410].

6.6.7 ANTIDIABETIC ACTIVITY

Diabetes is one of the most serious, chronic diseases that is developing with the increase in obesity and aging in the general population. It is largely classified into insulin-dependent diabetes mellitus (type-1 diabetes) and non-insulin-dependent diabetes mellitus (type-2 diabetes) [411—413]. Phlorotannins derived from brown algae exhibited their various antidiabetic mechanisms such as α-glucosidase and α-amylase inhibitory effects (Fig. 6.6). Dieckol, Fucodiphloroethol G, 6,6′-Bieckol, 7-Phloroeckol, Phlorofucofuroeckol-A isolated from *Ecklonia cava*; Phloroglucinol, Eckol, and Dioxinodehydroeckol isolated from *Ecklonia Stolonifera* and *Ecklonia bicyclis*; and Octaphlorethol-A isolated from *Ishige foliacea* show antidiabetic

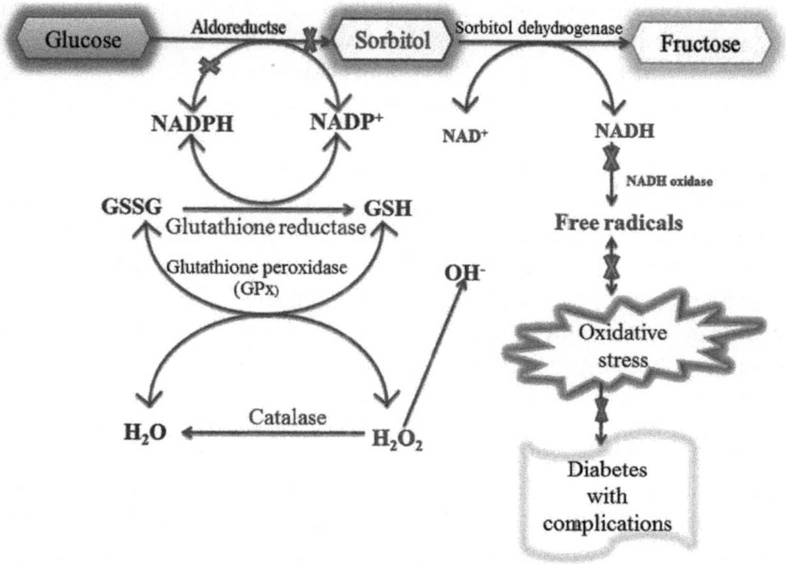

FIGURE 6.6 Schematic Presentation of Antioxidant Activity of Algal Extracts.

X: Inhibition by algal extract. Algal extracts prevents from diabetic complications by inhibiting the production of sorbitol, free radicals, and oxidative stress expressed by cross sign.

effects [395]. Octaphlorethol-A (OPA), a type of phlorotannin isolated from *Ishige foliacea*, is shown to have antidiabetic activities [414]. Diphlorethohydroxycarmalol (DPHC) isolated from *I. okamurae* is a potent inhibitor for α-glucosidase and α-amylase [415]. Seaweed lipids also show antidiabetic activity [416].

6.6.8 ANTITHROMBIN ACTIVITY

Strong antithrombin or antithrombotic activity is exhibited by fucoidans isolated from *Laminaria saccharina*, *L. digitata*, *Fucus serratus*, *Fucus distichus*, and *Fucus evanescens* [390]. A novel sulfated galactofucan isolated from *Spatoglossum schroederi* showed potent antithrombotic activity [417]. A fucan sulfate isolated from *E. kurome* [418] and *S. schroederi* exhibited antithrombic activity [419]. A bromophenol derivative named (+)-3-(2,3-dibromo-4,5-dihydroxy-phenyl)-4-bromo-5,6-dihydroxy-1,3-dihydroiso-benzofuran, isolated from brown alga *Leathesia nana*, also exhibited good in vivo antithrombotic activity [420].

6.6.9 ANTIOBESITY ACTIVITY

Obesity is a medical condition of excess body fat accumulation that impairs health. It is a major health problem throughout the world and an issue of growing concern in the 21st century. According to the World Health Organization, overweight and obesity are the fifth leading risk for global death, and at least 2.8 million adults die each year as a result of being overweight or obese. It is associated with the development of metabolic diseases such as type-2 diabetes as well as cardiovascular disorders, some types of cancer, and osteoarthritis [421]. Obesity is controlled by controlling appetite, blocking fat absorption, stimulating energy expenditure, suppressing adipose tissue growth, and by increasing body fat mobilization. Algal polysaccharides chitin and chitosan isolated from green algae, and alginate and fucoidan from brown algae show potential for weight loss with different mechanisms. Anti-obesity effect of phlorotannins 7-phloroeckol, fucofuroeckol-A and eckol isolated from *Ei. bicyclis* and *E. cava*; carotenoids fucoxanthin from *Ei. bicyclis* and *U. pinnatifida*; astaxanthin isolated from *H. pluvialis* have also been reported [422]. Green algae carotenoid siphonaxanthin shows potent antifat activity [423].

6.6.10 ANTIANGIOGENIC ACTIVITY

Angiogenesis results in new blood vessel formation from a preexisting vasculature that occurs under either physiological or pathological conditions [424]. In pathological conditions, such as inflammatory diseases, rheumatoid arthritis, and tumor metastasis, a chronic, unregulated angiogenic state often helps spreading of the diseases (Fig. 6.7). Hence preventing angiogenesis under pathological conditions is a promising approach in the prevention of cancer and other angiogenic-related diseases [425]. Algal carotenoid fucoxanthin effectively suppressed the differentiation of endothelial progenitor cells into endothelial cells involving new blood vessel formation (Fig. 6.7). Fucoxanthin and fucoxanthinol suppressed microvessel outgrowth in vivo and ex vivo angiogenesis assay using a rat aortic ring [426]. Antiangiogenic effect is also exhibited by siphonaxanthin derived from green algae *Codium fragile* [427].

6.6.11 HEPATOPROTECTIVE ACTIVITY

Phloroglucinol and phloroglucinol derivatives eckstolonol, phlorofucofuroeckol-A (Fig. 6.8) isolated from the brown alga *Ecklonia stolonifera* exhibit hepatoprotective activity [428]. Hepatoprotective activities of carotenoids, β-carotene, and xanthophyll from microalgae, *Spirulina platensis*, and *D. salina* [429], and low molecular weight sulfated polysaccharide isolated from *L. japonica* were also reported [430]. Phloroglucinol protects liver damage due to oxidative stress produced by tributyl hydroperoxide (T-BHP), carbon tetra chloride, etc.

6.6.12 RADIOPROTECTIVE EFFECT

Ionizing radiations are known to generate reactive oxygen species in irradiated tissue and cells. The primary damage caused by radiation is production of aqueous free radicals, generated by the action of radiation on water within the cell. These free radicals react with cellular macromolecules such as DNA, protein, and lipid membrane

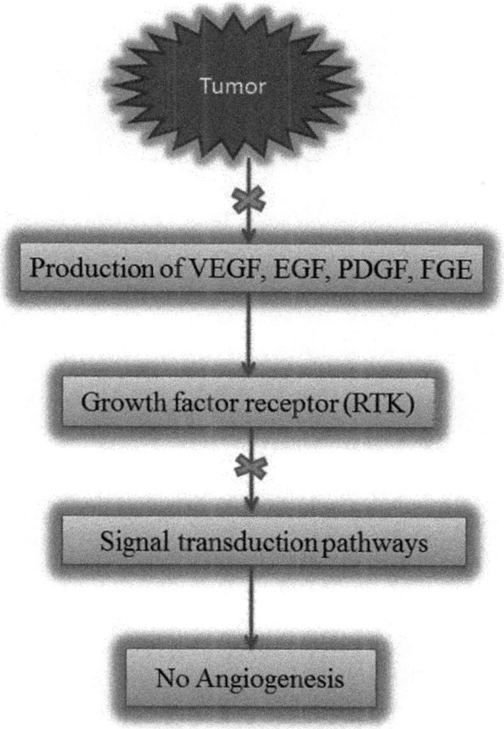

FIGURE 6.7 Antiangiogenic Effect of Algal Derived Fucoxanthin.

: Inhibition by algal extract. Algal extracts inhibits the synthesis of tumor by inhibiting the VEGF. *EGF*, epidermal growth factor; *PDGF*, platelet derived growth factor; *VEGF*, vascular endothelial growth factor.

causing cell dysfunction and mortality. Lungs are reported to be especially sensitive to oxidative stress. Eckol, a trimeric phloroglucinol with a dibenzeno-1,4-dioxin skeleton, one of the major phlorotannins isolated from brown alga *E. cava* is found to be a radioprorective agent. It acts by reducing the intracellular reactive oxygen species generated by γ-ray radiation (Fig. 6.9). Moreover, eckol also protected against radiation-induced cellular DNA damage and membrane lipid peroxidation [431].

6.6.13 ANTI-ALZHEIMER ACTIVITY

Alzheimer's disease (AD) is an irreversible, progressive, neurodegenerative disorder of the central nervous system associated with progressive cognitive memory loss [432]. Neuropathological studies demonstrated that AD was associated with deficiency of the brain neurotransmitter acetylcholine (ACh). The inhibition of

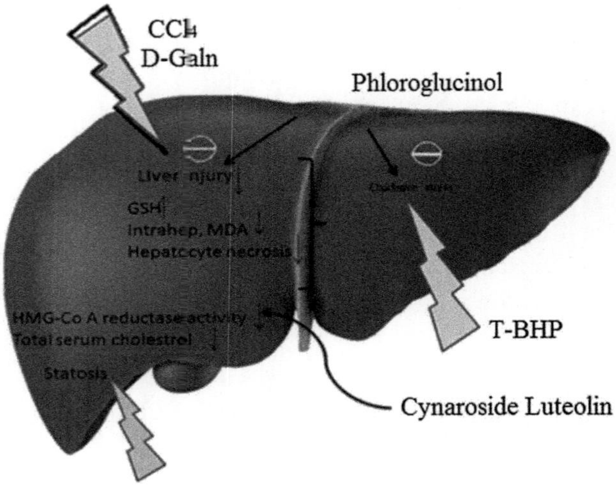

FIGURE 6.8 Hepatoprotective Effect of Algal Polymers.

Phloroglucinol protects liver damage due to oxidative stress produced due to tributyl hydroperoxide (T-BHP), carbon tetra chloride, etc.

acetylcholinesterase (AChE) enzyme, which catalyzes the breakdown of ACh, can be one of the most realistic approaches to the symptomatic treatment of AD [433]. Methanolic extracts of marine algae *Caulerpa racemosa*, *Codium capitatum*, *Ulva fasciata*, *Halimeda cuneata*, *Amphiora ephedraea*, *Amphiora bowerbankii*, *Dictyota humifusa* [434], *Hypnea valentiae*, *P. gymnospora*, *Ulva reticulate*, and

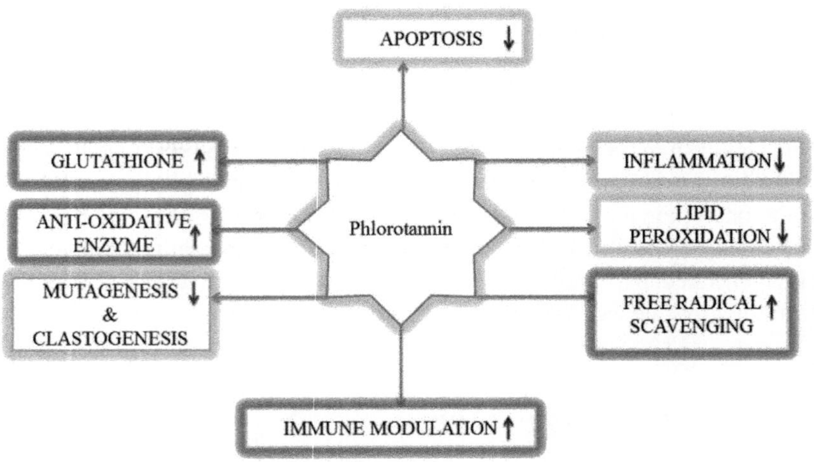

FIGURE 6.9 Radioprotective Mode of Action of Phlorotannin.

Phlorotannin from *Ecklonia cava* protects from radiation by activating immune system through increasing the levels of glutathione, oxidative enzymes, and free-radical scavengers whereas lowers apoptosis, mutation, and free-radical production.

Gracilaria edulis exhibit acetylcholinesterase inhibitory activity [435]. Fucosterol and 24-hydroperoxy 24-vinylcholestrol was isolated from *n*-hexane fraction of frozen *Hypomecis formosana*. Eckstolonol, eckol, phlorofucofluoroeckol−A, dieckol, 2-phloroeckol, and 7-phloroeckol were isolated from the EtOAc extracts of *Ecklonia stolonifera* [436]. MeOH extracts and EtOAc extracts of *I. okamurae* exhibit inhibitory potential against AChE. 6,6′-bieckol isolated from *I. okamurae* serve as a potential AChE inhibitor that could be used as a potential functional food ingredient or nutraceuticals for preventing Alzheimer's disease [437]. Sargaquinoic acid isolated from *Sargassum sagamianum* [438] and a glycoprotein from *U. pinnatifida* UPGP also exhibit acetylcholinesterase (AChE) inhibitory activity [439].

6.6.14 ACE INHIBITION ACTIVITY

Increased blood pressure or hypertension is one of the major chronic medical condition, and one of the major independent risk factors for cardiovascular diseases. Angiotensin-I converting enzyme (ACE) plays a significant physiological role in regulating blood pressure by converting angiotensin-I to angiotensin-II, a potent vasoconstrictor. Therefore the inhibition of ACE activity is a major target in the prevention of hypertension (Fig. 6.10). The ethanol extract of *E. cava*, *Ecklonia stolonifera*, Phlorotannins such as eckol, phlorofucofuroeckol-A, and dieckol derived from *E. stolonifera* have shown considerable inhibitory activity against ACE [440]. Aqueous extracts of red algae *Lomentaria catenata* and *Lithophyllum okamurae*, methanolic extracts of *Ahnfeltiopsis flabelliformis* show inhibitory activity against ACE [441].

6.6.15 ANTITUBERCULOSIS ACTIVITY

Antituberculosis activities of red algae *Gelidium* sp. were studied. Ether and alcoholic extracts of *Gelidium amansii* have shown a slight inhibition against *M. tuberculosis* while acetone extract of *Gypsophila capillaris* exhibited considerable inhibition toward *Mycobacterium avium* and *Mycobacterium tuberculosis* [442]. Diterpene, (1*E*,2*R*,3*R*,4*S*,6*E*,18*S*)-4,18-dihydroxydictyolactone isolated from the methanol extract of the brown alga *Dictyota* sp. exhibit weak antituberculosis activity [443]. Caulerpin a bis-indole alkaloid isolated from *C. racemosa* and *Caulerpa serrulata* exhibit excellent activity against *M. tuberculosis*. Biological results indicated that Caulerpin may be useful as a lead compound for the development of novel antituberculosis agents [444].

6.6.16 INSECTICIDAL ACTIVITY

Polyhalogenated monoterpenes isolated from the red alga *Plocamium cartilagineum* exhibit insecticidal activity against the aster leafhopper [445]. Telfairine a monoterpene isolated from the red alga *Plocamium telfairia* shows strong insecticidal activity

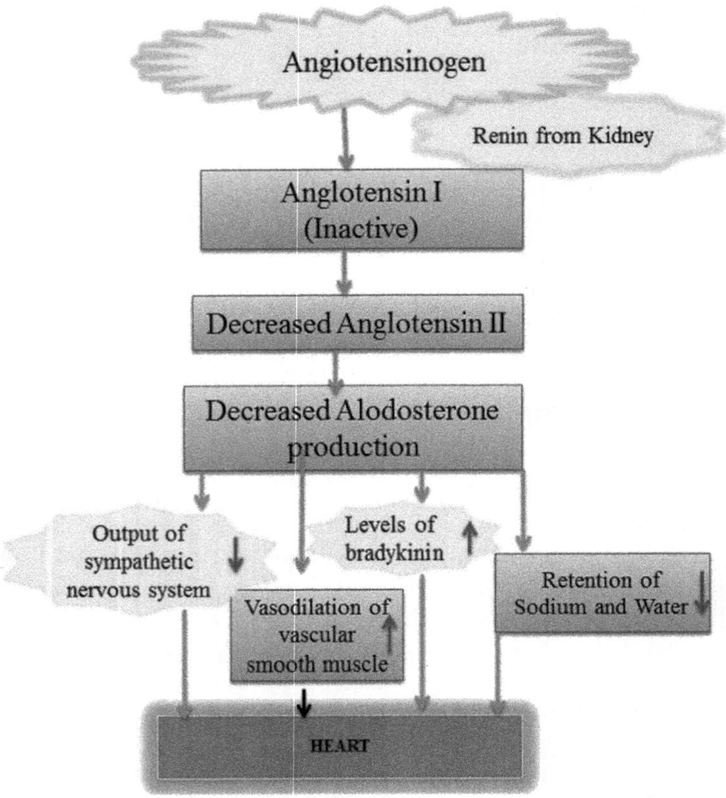

FIGURE 6.10 ACE (Angiotensin Converting Enzyme) Inhibition Role of Algal Extract.

: Inhibition by algal extract. Algal extract acts on angiotensin converting enzyme. Algal extract inhibits the enzyme responsible (ACE and angiotensin-II) for hypertension and heart failure.

against the mosquito larvae *Culex pipiens pallens* [140]. Isodomic acids isolated from the red alga *Chondria arnata* show significant insecticidal activity when they are injected subcutaneously into the abdomen of American cockroach [446].

6.6.17 HYALURONIDASE-INHIBITION ACTIVITY

Hyaluronidase is an enzyme that depolymerizes the polysaccharide hyaluronic acid in the extracellular matrix of connective tissue. The enzyme is known to be involved in allergic effects and migration of cancer and inflammation. Phlorotannins, phloroglucinol, such as eckol, phlorofucofuroeckol-A, dieckol, and 8,8-bieckol isolated from the brown algae *Ei. bicyclis* and *E. kurome* were reported to exhibit a stronger inhibition effect against hyaluronidase compared to well-known inhibitors such as catechin and sodium cromoglycate [447].

6.6.18 ANTIFUNGAL ACTIVITY

A meroterpenoid isolated from the brown alga *Cystoseira tamariscifolia*, characterized as methoxybifurcarenone, possesses antifungal activity against three tomato pathogenic fungi [142]. Hexane extracts of three algal species *Dictyota dichotoma*, *D. dichotoma* var. implexa, and *Dilophus spiralis* exhibited a wide spectrum of antifungal activities having ability from process of cell membrane permeability to metabolic disruption (Fig. 6.11) which varied during the seasons [448]. Terpenoids isolated from algal species *Stypopodium zonale*, *Laurencia dendroidea*, *A. nodosum*, *Sargassum muticum*, *Pelvetia canaliculata*, *Fucus spiralis*, *Sargassum filipendula*, *Sargassum stenophyllum*, *Laminaria hyperborea* and *G. edulis*. *S. zonale*, *L. dendroidea*, *P. canaliculata*, *S. muticum*, *A. nodosum*, and *F. spiralis* possess antifungal activity; their extracts significantly inhibited the *Colletotrichum lagenarium* growth [449].

6.6.19 ANTIMALARIAL ACTIVITY

Snyderol sesquiterpene derivative isolated from the red alga *Laurencia obtusa* was active against D6 and W2 clones of the malarial parasite *Plasmodium falciparum* [450]. Sphaerococcenol-A isolated from red alga *Sphaerococcus coronpifolius* is responsible for the antimalarial activity of the extract, against the chloroquine-resistant *P. falsciparum* FCBI strains with an IC_{50} of 1 μM [451].

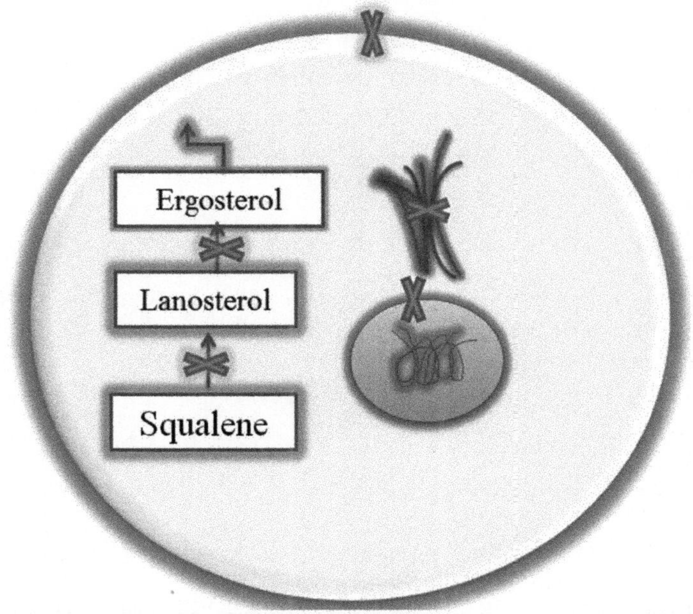

FIGURE 6.11 Antifungal Mode of Action of Algal Extract.

: Inhibition by algal extract. Different steps at which algal extracts inhibit the synthesis of ergosterol, microtubules and DNA replication.

6.6.20 ANTIGLYCEMIC AND ANTILIPIDEMIC EFFECTS

It was reported that macroalgal cell wall polysaccharides had an ability to decrease TG (triglycerides) and cholesterol [27], but unfortunately this area of research for microalgae is not appropriately investigated. In spite Ginzberg and colleagues [452] reported that modified fatty acid profile improved carotene contents in chicken egg yolks after the injection of *Porphyridium* sulfated biopolymer. Furthermore, considerably low levels of insulin and glucose, VLDL, and serum cholesterol, and improved levels of hepatic cholesterol without any toxic effect were observed when polysaccharides of both *Porphyridium* and *R. reticulate* were used in combination in rats [453,454]. Huang et al. [242] suggested that sulfated polysaccharides of *Porphyridium* as potent antilipidemic (Fig. 6.12) and antiglycemic agent have the ability to reduce blood-glucose levels without any modification of islets of Langerhans. Due to rich contents of fibers, algal polysaccharides acts as bile acids binder, cation exchangers, and hypocholesterolemic agent could be used as nutraceuticals [455]. Algal-sulfated polysaccharides affect the absorption of nutrients, lower the lipid absorption and micelles formation, and make internal contents more viscous, and are therefore known as potent fibers [456]. They are equally effective in coronary heart disease by increasing bile excretion [457]. Fucans isolated from *Sargassum polycystum* were reported to have a significant antilipidemic effect in

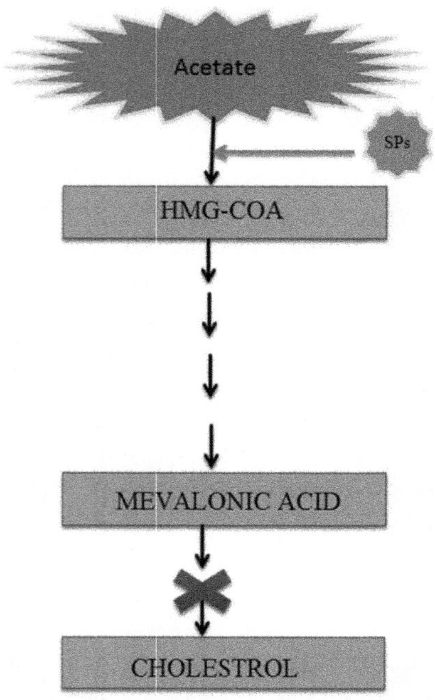

FIGURE 6.12 Antilipidemic Mode of Action of Algal Sulfated Polysaccharides.

✖: Inhibition by algal extract. Algal extract inhibits the synthesis of cholesterol via HMG 3-hydroxy-3-methylglutaryl-CoA (HMG), isopentenylpyrophosphate squalene synthesis.

liver tissues and serum in hepatitis caused by acetaminophen toxicity [458]. Sulfated *Sargassum wightii* polysaccharide caused the normalization of lecithin–cholesterol acyltransferase (LCAT) and lipoprotein lipase (LPL) of plasma in nephrotoxicity [459]. Sulfated polysaccharides acted as potent renoprotective agent by reducing the LDL oxidation. Fucoidan from *L. japonica* were suggested as antilipidemic, showed their potency by raising LPL, HL (hepatic lipase), and LCAT, thus rendering LDL and TG levels while elevating HDL levels.

Sulfated polysaccharides from *F. vesiculosus* affected the serum cholesterol and TG levels in dose-subsidiary-tone, e.g., smaller fractions of *Ulva pertusa* did not reduce the serum cholesterol, but only caused the normalization of TG by raising HDL [460]. It was observed that many sulfated biopolymers, by attracting the enzymes present on heparin-binding sites, influenced HL and LPL, e.g., fucoidan isolated from *F. vesiculosus* stabilized LPL activity by binding with heparin-binding sites [27]. Drug-toxicities-associated hyperlipidemia is effectively addressed by algal cell wall polysaccharides. Fucoidan polysaccharide sulfuric acid ester from *L. japonica* reduced the concentration of serum total cholesterol (TC), triglyceride (TG), and low-density lipoprotein cholesterol (LDL) of hyperlipidemic rats, and increased the concentration of high-density lipoprotein cholesterol (HDL). The treatment also increases the activities of lipoprotein lipase (LPL), hepatic lipoprotein (HL), and LCAT in serum [461]. Antilipidemic effect is also reported from the ulvan polysaccharide of *U. pertusa* [462].

6.6.21 ANTIALLERGY ACTIVITY

Allergic diseases are one of the major public health problems in the developed world. It was estimated that approximately one-third of the general population was affected by allergic diseases. These diseases include asthma, dermatitis, bee-sting allergy, food allergy, conjunctivitis, and severe systemic anaphylaxis [463]. A large number of antiallergic agents from marine algae were identified. Several bioactive phloroglucinol derivatives, fucodiphloroethol G, eckol, dieckol, 6,6-bieckol, phlorofucofuroeckol-A, and 1-(3′,5′-dihydroxyphenoxy)-7-(2″,4″,6-trihydroxyphenoxy)-2,4,9-trihydroxydibenzo-1,4-dioxin isolated from *E. cava* are evidenced to be efficient against A23187 or Fc RI-mediated histamine release from KU812 and RBL-2H3 cells [464,465]. The red algae *Carpopeltis affinis* was confirmed to be effective against atopic allergic reactions in vitro [466]. The ethanolic extracts of edible red alga *Laurencia undulata* and brown alga *E. cava* exhibit a significant inhibition of all asthmatic reactions in OVA-induced mice [467,468]. Alginic acid, a naturally occurring hydrophilic colloidal polysaccharide obtained from several species of brown seaweeds, exhibits inhibitory effects against hyaluronidase activity and histamine release from mast cells [469]. Phycocyanin, one of the major pigment constituents of blue-green microalgae *Spirulina*, has been found to be an inhibitor of different allergic responses such as histamine release from rat peritoneal mast cells (Fig. 6.13), ear swelling in mice induced by OVA, and skin reactions in rats caused by histamine [470].

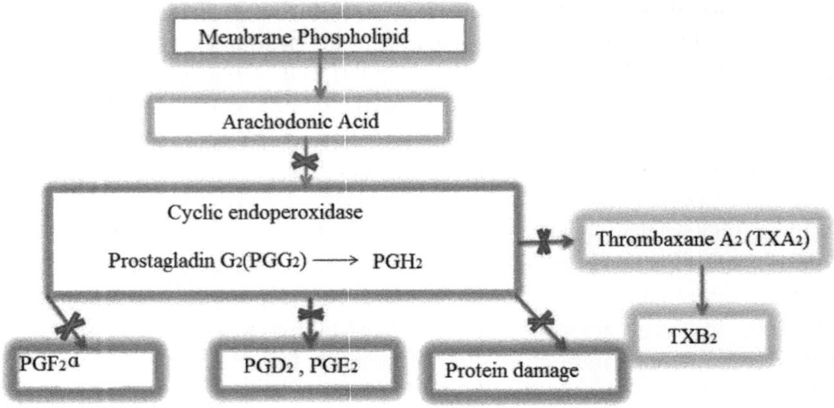

FIGURE 6.13 Antiallergy Mode of Action of Algal Extract.

: Inhibition by algal extract. Algal extract protects from allergic reactions by inhibiting the synthesis of prostaglandin E_2.

6.6.22 ANTIFEEDENT ACTIVITY

Antifeedent activity of two diterpenoids with a novel skeleton, dictyterepenoids A and B isolated from the brown algae *Dilophus okamurae* was reported against young abalone [471]. Crude organic extract of brown alga *Dictyota pfaffii* significantly exhibits antifeedent activity against sea urchin *Lytechinus variegatus* and generalist herbivore fishes [472].

6.6.23 ANTIADHESIVE EFFECT

Algal cell wall polysaccharides are potent antiadhesive agents as they inhibited the pathogenic adhesion. Currently, several algal polysaccharides were isolated which affect the adherence of fish pathogens to skin, gills, and gut, and *Helicobacter pylori* to HeLa S3 [473]. Consequential impersonation of carbohydrate moieties of sulfated polysaccharides in provision of recognition sites for infection attachment was already been reported, e.g., heparin sulfates [474]. Antiadhesive property of heparin sulfate is contributed due to its molecular stereochemistry and net charge [475]. However, due to certain risk associated with use of sulfated polysaccharides as antiadhesive agent, further studies to understand cytotoxicity are needed.

6.6.24 BIOLUBRICATING AGENT

Very little is known about the use of sulfated polysaccharides as biolubricants. However, Arad et al. [476] first time reported the lubricating properties of *Porphyridium* exopolysaccharides ascribable to its rheological idiosyncrasy [477]. Arad and coworkers suggested sulfated polysaccharides as better lubricating agent for joints than ordinary lubricating agents, i.e., hyaluronic acid, hydrogel, as they were thermostable [477]. In addition to these, they also reported that incubation of sulfated polysaccharide of *Porphyridium* with hyaluronidase did not show any significant change in viscosity while only 1% polysaccharide exhibited significant lubricating properties under high load. They concluded from their experiments that hyaluronic acid could be locum with sulfated biopolymers and could be used in the manufacturing of joint-lubricating solutions to alleviate the severity of arthritis.

6.6.25 DRAG-REDUCING AGENTS

Very little is known for algal-sulfated polysaccharides as drag-reducing agents. Exclusively a couple of studies were made to check the drag-reducing potential of sulfated polysaccharides to extend their functional characteristics for engineering purposes. Gasljevic et al. reported that exopolysaccharides of microalgae have potential to act as drag reducers. Exopolysaccharides of both *R. maculate* and *P. cruentum* were found to be very potent drag reducers in very low concentration followed by *Schizochlamydella* [478].

6.6.26 OTHER APPLICATIONS

It was reported that both micro- and macroalgae had their potential in nutraceuticals, functional foods, and health food industry [457]. Yet, due to their rheological characteristics and chemical composition, they also have diverse applications in other fields, e.g., cosmetics, prophylactic therapy. In conclusion to all applications of algal cell wall polysaccharides, adhesion is seemed to be more important as it is not only involved in algal movement but also in sand and soil particle aggregation, affecting cohesiveness and stability of sediments [479].

6.7 CONCLUSION

This chapter briefly described algal classification and their possible pharmacological and therapeutic applications with respect to phytochemical constituents. This is an attempt to compile and document comprehensive information on different aspects of algae to help scientific community for further research and investigative study.

REFERENCES

[1] Sahoo D, Seckbach J. The algae world, vol. 26. India: Springer; 2015. p. 3.
[2] Trivedi PS, Pandey SN. A textbook of botany. 11th ed., vol. 1. India: Vikas Publishing House; 2009. p. 367−9.
[3] Wahid A, Khan MA. Viruses, bacteria and thalloid organisms. Pakistan: Higher Education Commission; 2006. p. 25.
[4] Vashishta BR, Sinha AK, Singh VP. Botany for degree students algae. New Delhi: Neelab Printers; 1960. p. 1.

[5] Kara R. Fungi, algae and protists. New York: Britannica Educational Publishing; 2011. p. 89−142.
[6] Roberts M, Reiss MJ, Monger G. Advanced biology. UK: Nelson Thornes; 2000. p. 84.
[7] Dawes CJ. Marine botany. 2nd ed. New York: John Wiley & Sons; 1998. p. 2.
[8] Singh SP, Singh P. Effect of temperature and light on the growth of algae species: a review. Renew Sustain Energy Rev 2015;50:431−44.
[9] Rodriguez C, Alaswad A, Mooney J, Prescott T, Olabi AG. Pre-treatment techniques used for anaerobic digestion of algae. Fuel Process Technol 2015;138:765−79.
[10] Pelczar Jr MJ, Chan ECS, Kreig NR. Microbology. 5th ed. Islamabad: National Book Foundation; 1986. p. 365−77.
[11] Sharma OP. Textbook of algae. New Delhi: Tata Mcgraw-Hill Publishing Company Limited; 2007. p. 2−296.
[12] Bhargava M. The latest portfolio of theory & practice in algae. Dominant Publishers and Distributers; 2003. p. 1.
[13] Roger DR, Michael J, Pelczar JR. Microbology. Singapore: Mcgraw-Hill International Book Co.; 1972. p. 309−28.
[14] Awasthi DK. Cryptogams: algae, bryophyta and pterldophyta. New Delhi: Published by Krishna Praashan Media; 2009. p. 1−21.
[15] Chapman VJ. The algae. New York: St. Martin's Press; 1962. p. 1.
[16] Chapman VJ. An introduction to the study of the algae. UK: Cambridge University Press; 1941. p. 1.
[17] Pelczar Jr MJ, Chan ECS, Kreig NR. Microbiology: application based approach. New Delhi: Tata McGraw-Hill Education; 2010. p. 268.
[18] Reddy SM. University botany I: (algae, fungi, bryophyta and pteridophyta), vol. 1. New Delhi: New Age International; 2001. p. 60−9.
[19] Hoek C, Mann DG, Jahns HM. Algae: an introduction to phycology. UK: Cambridge University Press; 1995. p. 16. 166.
[20] Piganeau G. Genomic insights into the biology of algae. Advances in botanical research, vol. 64. London: Academic Press; 2012. p. 61.
[21] Hollar SA. Closer look at bacteria, algae, and protozoa. New York: Britannica Educational Publishing; 2011. p. 54.
[22] Hossain MA, Wani SH, Bhattacharjee S, Burritt DJ, Tran LSP. Drought stress tolerance in plants. Physiology and biochemistry, vol. 1. Switzerland: Springer; 2016. p. 240.
[23] Taylor S. Marine medicinal foods: implications and applications, macro and microalgae, vol. 64. UK: Academic Press; 2011. p. 392.
[24] Groth I, Grünewald N, Alban S. Pharmacological profiles of animal- and nonanimal-derived sulfated polysaccharides − comparison of unfractionated heparin, the semisynthetic glucan sulfate PS3, and the sulfated polysaccharide fraction isolated from *Delesseria sanguinea*. Glycobiology 2009;19(4):408−17.
[25] Duckworth M, Hong KC, Yaphe W. The agar polysaccharides of *Gracilaria* species. Carbohydr Res 1971;1(18):1−9.
[26] Beattie A, Hirst EL, Percival E. Studies on the metabolism of the Chrysophyceae. Comparative structural investigations on leucosin (chrysolaminarin) separated from diatoms and laminarin from the brown algae. Biochem J 1961;79(3):531−7.
[27] Jiao G, Yu G, Zhang J, Ewart S. Chemical structures and bioactivities of sulfated polysaccharides from marine algae. Mar Drugs 2011;9(2):196−223.

[28] Usov AI, Bilan MI, Klochkova NG. Polysaccharides of algae. 48. Polysaccharide composition of several calcareous red algae: isolation of alginate from *Corallina pilulifera* P. (Rhodophyta, Corallinaceae). Bot Mar 1995;1:43–51.

[29] Kaeffer B, Bénard C, Lahaye M, Blottière HM, Cherbut C. Biological properties of ulvan, a new source of green seaweed sulfated polysaccharides, on cultured normal and cancerous colonic epithelial cells. Planta Med 1999;65(6):527–31.

[30] Pérez RM, Noseda MD, Pujol CA, Carlucci MJ, Matulewicz MC. Sulfated mannans from the red seaweed *Nemalion helminthoides* of the South Atlantic. Phytochemistry 2009;70(8):1062–8.

[31] Lee JB, Hayashi K, Hayashi T, Sankawa U, Maeda M. Antiviral activities against HSV-1, HCMV, and HIV-1 of rhamnan sulfate from *Monostroma latissimum*. Planta Med 1999;65(5):439–41.

[32] Ignat I, Volf I, Popa VI. A critical review of methods for characterization of polyphenolic compounds in fruits and vegetables. Food Chem 2011;126:1821–35.

[33] Grotewold E. The science of flavonoids. USA: Springer Science & Business Media; 2007. p. 1–4.

[34] Atta-ur-Rahman, Choudhary MI, Hayat S, Khan AM, Ahmed A. Two new aurones from marine brown alga *Spatoglossum variabile*. Chem Pharm Bull 2001;49(1):105–7.

[35] Yumiko YS, Ya-Pei H, Takwshi S. Distribution of flavonoids and related compounds from seaweeds in Japan. J Tokyo Univ Fish 2003;89:1–6.

[36] Zeng LM, Wang CJ, Su JY, Li D, Owen NL, Lu Y, Lu N, Zheng QT. Flavonoids from the red alga *Acanthophora spicifera*. Chin J Chem 2001;11(19):1097–100.

[37] Shoubaky GA, Abdel-Daim MM, Mansour MH, Salem EA. Isolation and identification of a flavone apigenin from marine red alga *Acanthophora spicifera* with antinociceptive and anti-inflammatory activities. J Exp Neurosci 2016;10:21–9.

[38] Sabina H, Aliya R. Bioactive assessment of selected marine red algae against *Leishmania* major and chemical constituents of *Osmunde pinnatifida*. Pak J Bot 2011; 43(6):3053–6.

[39] Li YX, Wijesekara I, Li Y, Kim SK. Phlorotannins as bioactive agents from brown algae. Process Biochem 2011;46(12):2219–24.

[40] Jensen A, Ragan MA. 1,2,3,5-tetrahydroxybenzene 2,5-disulfate ester: the "phenolic precursor" in gelbstoff-forming exudates from the marine brown alga *Ascophyllum nodosum* (L.) Lejol. Tetrahedron Lett 1979;9(19):847–50.

[41] Ragan MA, Jensen A. Widespread distribution of sulfated polyphenols in brown algae. Phytochemistry 1979;2(18):261–2.

[42] Ragan MA. Phenol sulfate esters: ultraviolet, infrared, ^{1}H and ^{13}C nuclear magnetic resonance spectroscopic investigation. Can J Chem 1978;56(20):2681–5.

[43] Glombitza KW, Rosener HU, Vilter H, Rauwald W. Antibiotics from algae. 8. Phloroglucinol from phaeophyceae. Planta Med 1973;24(4):301–3.

[44] Glombitza KW, Gerstberger G. Phlorotannins with dibenzodioxin structural elements from the brown alga *Eisenia arborea*. Phytochemistry 1985;3(24):543–51.

[45] Glombitza KW, Rauwald HW, Eckhardt G. Fucole, polyhydroxyoligophenyle aus *Fucus vesiculosus*. Phytochemistry 1975;14:1403–5.

[46] Blackman AJ, Matthews DJ. Halogenated phloroglucinols from *Rhabdonia verticillata*. Phytochemistry 1982;8(21):2141–2.

[47] Glombitza KW, Schmidt A. Nonhalogenated and halogenated phlorotannins from the brown alga *Carpophyllum angustifolium*. J Nat Prod 1999;62(9):1238–40.

[48] Glombitza KW, Keusgen M. Fuhalols and deshydroxyfuhalols from the brown alga *Sargassum spinuligerum.* Phytochemistry 1995;4(38):987–95.

[49] Okada Y, Ishimaru A, Suzuki R, Okuyama T. A new phloroglucinol derivative from the brown alga *Eisenia bicyclis*: potential for the effective treatment of diabetic complications. J Nat Prod 2004;67(1):103–5.

[50] Kang HS, Chung HY, Jung JH, Son BW, Choi JS. A new phlorotannin from the brown alga *Ecklonia stolonifera.* Chem Pharm Bull 2003;8(51):1012–4.

[51] Glombitza KW, Hauperich S. Phlorotannins from the brown alga *Cystophora torulosa.* Phytochemistry 1997;4(46):735–40.

[52] Glombitza KW, Sailler B. Phlorethols and fucophlorethols from the brown alga *Cystophora retroflexa.* Phytochemistry 1999;5(50):869–81.

[53] Glombitza KW, Keusgen M, Hauperich S. Fucophlorethols from the brown algae *Sargassum spinuligerum* and *Cystophora torulosa.* Phytochemistry 1997;8(46): 1417–22.

[54] Koch M, Gregson RP. Brominated phlorethols and nonhalogenated phlorotannins from the brown alga *Cystophora congesta.* Phytochemistry 1984;11(23):2633–7.

[55] Glombitza KW, Hauperich S, Keusgen M. Phlorotannins from the brown algae *Cystophora torulosa* and *Sargassum spinuligerum.* Nat Toxins 1997;2(5):58–63.

[56] Glombitza KW, Li SM. Fucophlorethols from the brown alga *Carpophyllum maschalocarpum.* Phytochemistry 1991;10(30):3423–7.

[57] Ragan MA. Brown algal polyphenols: synthesis of "fucophlorethol A" octamethyl ether (2,2′,4,6,6′-pentamethoxy-4′-(2,4,6-trimethoxyphenoxy)biphenyl). Can J Chem 1985;63(2):291–3.

[58] Glombitza KW, Rösener HU. Bifuhalol: Ein diphenyläther aus *Bifurcaria bifurcate.* Phytochemistry 1974;7(13):1245–7.

[59] Wickberg B. Isolation of 2-L-Amino-3-hydroxy-l-propane sulphonic acid from *Polysiphonia fastigiata.* Acta Chem Scand 1957;3(11):506–11.

[60] Glombitza KW, Li SM. Hydroxyphlorethols from the brown alga *Carpophyllum maschalocarpum.* Phytochemistry 1991;8(30):2741–5.

[61] Glombitza KW, Forster M, Farnham FM. Antibiotics from algae—Part 25 polyhydroxyphenyl ethers from the brown alga *Sargassummuticum* (Yendo) Fensholt, Part II. Bot Mar 1982;9(25):249–54.

[62] Glombitza KW, Zieprath G. Phlorotannins from the brown alga *Analipus japonicus*1. Planta Med 1989;55(2):171–5.

[63] Glombitza KW, Pauli K. Fucols and phlorethols from the Brown Alga *Scytothamnus australis* Hook. et Harv. (Chnoosporaceae). Bot Mar 2003;3(46):315–20.

[64] Glombitza KW, Rosener HU, Müller D. Bifuhalol und diphlorethol aus *Cystoseira tamariscifolia.* Phytochemistry 1975;4(14):1115–6.

[65] Keusgen M, Glombitza KW. Pseudofuhalols from the brown alga *Sargassum spinuligerum.* Phytochemistry 1997;8(46):1403–15.

[66] Glombitza KW, Rauwald HW, Koch M. Polyhydroxyoligophenyle und phenyläther aus *Bifurcaria bifurcata.* Phytochemistry 1976;8(15):1279–81.

[67] Blackman AJ, Rogers GI, Volkman JK. Phloroglucinol derivative. Phloroglucinol derivatives from three Australian marine algae of the genus *Zonaria.* J Nat Prod 1988; 51(1):158–60.

[68] Kazlauskas R, King L, Murphy PT, Warren RG, Wells RJ. New metabolites from the brown algal genus *Cystophora.* Aust J Chem 1981;34(2):439–47.

[69] Gregson RP, Daly JJ. Polyhydroxy biphenyl ethers from the brown alga *Cystophora congesta*. Aust J Chem 1982;35(3):649–57.

[70] Li SM, Glombitza KW. Phlorotannins from the brown alga *Landsburgia quercifolia*. Bot Mar 1991;5(34):455.

[71] Brown W, Foote C, Iverson B, Anslyn E. Organic chemistry. 5th ed. USA: Cengage Learning; 2008. p. 194, 374, 560, 1010.

[72] Stewart D. The chemistry of essential oils made simple: god's love manifest in molecules. New York: Care Publications; 2005. p. 255–60.

[73] Paul VJ, Fenical V. Bioactive terpenoids from caribbean marine algae of the genera *penicillus* and *udotea* (chlorophyta). Tetrahedron 1984;15(40):2913–8.

[74] Yuan JP, Chen F, Liu X, Li XZ. Carotenoid composition in the green microalga *Chlorococcum*. Food Chem 2002;3(76):319–25.

[75] Mayer AMS, Paul VJ, Fenical W, Norris JN, Carvalho MS, Jacobs RS. Phospholipase A_2 inhibitors from marine algae. Hydrobiologia 1993;260:521–9.

[76] Ji NY, Li XM, Ding LP, Wang BG. Aristolane sesquiterpenes and highly brominated indoles from the marine red alga *Laurencia similis* (Rhodomelaceae). Helv Chim Acta 2007;2(90):385–91.

[77] Namikoshi M, Fujiwara T, Nishikawa T, Ukai K. Natural abundance ^{14}C content of dibutyl phthalate (DBP) from three marine algae. Mar Drugs 2006;4(4):290–7.

[78] Yongsong H, Murray M, Metzger P, Eglinton G. Novel unsaturated triterpenoid hydrocarbons from sediments of Sacred lake, Mt. Kenya, Kenya. Tetrahedron 1996;20(52): 6973–82.

[79] Weinstein B, Rold TL, Harrell Jr CE, Burns MW, Waaland JR. Reexamination of the bromophenols in the red alga *Rhodomela larix*. Phytochemistry 1975;12(14): 2667–70.

[80] Fan X, Xu N-J, Shi J-G. Bromophenols from the red alga *Rhodomela confervoides*. J Nat Prod 2003;66(3):455–8.

[81] Suzuki M, Kowata N, Kuroswa E. Bromophenols from the red alga *Rhodomela larix*. Bull Chem Soc Jpn 1980;7(53):2099–100.

[82] Xu XL, Fan X, Song FH, Zhao JL, Han LJ, Yang YC, Shi JG. Bromophenols from the brown alga *Leathesia nana*. J Asian Nat Prod Res 2004;3(6):217–21.

[83] Böcker SP, Pohnert G, Lui IF, Boland W, Peters AF. Synthesis and absolute configuration of desmarestene, the gamete-releasing and gamete-attracting pheromone of the brown algae *Desmarestia aculeata* and *D. firma* (Phaeophyceae). Tetrahedron 1995;29(51):7927–36.

[84] Boland W, Jakoby K, Jaenicke L. Synthesis of (±)-Desmarestene and (±)-Viridiene, the two sperm releasing and attracting pheromones from the brown algae *Desmarestia aculeata* and *Desmarestia viridis*. Helv Chim Acta 1982;7(65):2355–62.

[85] Boland W, Jakoby K, Jaenicke L, Müller DG, Fölster E. Absolute configuration of multifidene and viridiene, the sperm releasing and attracting pheromones of brown algae. Helv Chim Acta 1983;6(66):1905–13.

[86] Pickenhagen W, Näf F, Ohloff G, Müller P, Perlberger JC. Thermal and photochemical rearrangements of Divinylcyclopropanes to Cycloheptadienes. – A model for the biosynthesis of the cycloheptadiene derivatives found in a seaweed (*Dictyopteris*). Preliminary communication. Helv Chim Acta 1968;6(56):1868–74.

[87] Stratmann K, Boland W, Müller DG. Biosynthesis of pheromones in female gametes of marine brown algae (Phaeophyceae). Tetrahedron 1993;18(49):3755–66.

[88] Ronald RC, Gurusiddaiah S. Grahamimycin A_1: a novel dilactone antibiotic from *Cytospora*. Tetrahedron Lett 1980;8(21):681–4.

[89] Andrewes AG, Phaff HJ, Starr MP. Carotenoids of *Phaffia rhodozyma*, a red-pigmented fermenting yeast. Phytochemistry 1976;6(15):1003–7.

[90] Godoy HT, Rodriguez-Amaya DB, Connor AE, Britton G. Confirmation of the structure of papaya β-cryptoxanthin monoepoxide. Food Chem 1990;4(36):281–6.

[91] Chapman DJ. Three new carotenoids isolated from algae. Phytochemistry 1966;6(5): 1331–3.

[92] Bjornland T, Jensen SL, Throndsen J. Carotenoids of the marine chrysophyte *Pelagococcus subviridis*. Phytochemistry 1989;12(28):3347–53.

[93] Argandona VH, San-martín A, Rovirosa J. Halogenated sesquiterpenes pacifenol and pacifenol derivatives on the aphid *Schizaphis graminum*. Phytochemistry 1993;5(32): 1159–61.

[94] Pedersen M, Saenger P, Fries L. Simple brominated phenols in red algae. Phytochemistry 1974;10(13):2273–9.

[95] Shoeib NA, Bibby MC, Blunden G, Linley PA, Swaine DJ, Wheelhouse RT, Wright CW. In-vitro cytotoxic activities of the major bromophenols of the red alga *Polysiphonia lanosa* and some novel synthetic isomers. J Nat Prod 2004;67(9): 1445–9.

[96] Craigie JS, Gruenig DE. Bromophenols from red algae. Science 1967;3792(157): 1058–9.

[97] Saenger P, Pedersen M, Rowan KS. Bromo-compounds of the red alga *Lenormandia prolifera*. Phytochemistry 1976;12(15):1957–8.

[98] Jacobsen N, Madsen JO. Halogenated metabolites including brominated 2-heptanols and 2-heptyl acetates from the tetrasporophyte of the red alga *Bonnemaisonia hamifera*. Tetrahedron Lett 1978;33(19):3065–8.

[99] Kim KY, Choi KS, Kurihara H, Kim SM. β-glucuronidase inhibitory activity of bromophenols purified from *Grateloupia elliptica*. Food Sci Biotechnol 2008;5(17): 1110–4.

[100] Whitfield FB, Helidoniotis F, Shaw KJ, Svoronos D. Distribution of bromophenols in species of marine algae from Eastern Australia. J Agric Food Chem 1997;45(11): 4398–405.

[101] Whitfield FB, Helidoniotis F, Shaw KJ, Svoronos D. Distribution of bromophenols in species of marine algae from Eastern Australia. J Agric Food Chem 1999;47(6): 2367–73.

[102] Whitfield FB, Helidoniotis F, Shaw KJ, Svoronos D. Distribution of bromophenols in species of marine algae from Eastern Australia. J Agric Food Chem 1999;47(11): 4756–62.

[103] Stillson GH, Sawyer DW, Hunt CK. The hindered phenols. J Am Chem Soc 1945; 67(2):303–7.

[104] Burczyk J. Cell wall carotenoids in green algae which form sporopollenins. Phytochemistry 1986;1(26):121–8.

[105] Gopichand Y, Hirschmann H. D-Homosteroids. 6. Diverse bond shifts in the solvolyses of uranediol 3-acetate 17a-tosylate. J Org Chem 1979;44(2):185–92.

[106] Enoki N, Iahida R, Matsumoto T. Structures and confirmations of new nine-membered ring diterpenoids from the marine alga *Dictyota dicoma*. Chem Lett 1982;11(11): 1749–52.

[107] Cody V, Duax WL, Norton DA. Molecular conformation of the thyroxine analogue 3,5-diiodo-L-thyronine N-methylacetamide complex (1:1). Acta Crystallogr 1972; B28:2244–52. Available from: http://journals.iucr.org/b.

[108] Ishitsuka M, Kusumi T, Kakisawa H, Kawakami Y, Nagai Y, Sato T. Structural elucidation and conformational analysis of germacrane-type diterpenoids from the brown alga *Pachydictyan coriaceum*. Tetrahedron Lett 1974;51(15):4463–6.

[109] Hertzberg S, Liaaen-Jensen S, Siegelman HW. The carotenoids of blue-green algae. Phytochemistry 1971;12(10):3121–7.

[110] Francis GW, Halfen LN. γ-Carotene and lycopene *Oscillatoria princeps*. Phytochemistry 1972;7(11):2347–8.

[111] Egger K. Die ketocarotinoide in *Adonis annua* L. Phytochemistry 1965;2(2):609–18.

[112] Sut JC, Hassid WZ. Carbohydrates and nucleotides in the red alga *Porphyra perforate*. Biochemistry 1962;3(1):468–74.

[113] Gurusiddaiah S, Ronald RC. Grahamimycins: antibiotics from *Cytospora* sp. Ehrenb. W.F.P.L. 13A. Antimicrob Agents Chemother 1981;1(19):153–65.

[114] Mori Y, Kohchi Y, Suzuki M, Carmeli S, Moore RE, Patterson GML. Isotactic polymethoxy 1-alkenes from blue-green algae. Synthesis and absolute stereochemistry. J Org Chem 1991;56(2):631–7.

[115] Müller DG, Clayton MN, Gassmann G, Boland W, Marner FJ, Jaenicke L. The sperm attractant of *Hormosira banksii* (Phaeophyceae, Fucales), a seaweed common to Australia and New Zealand. Experientia 1984;2(40):211–2.

[116] Talpir R, Rudi A, Kashman Y. Three new sesquiterpene hydroquinones from marine origin. Tetrahedron 1994;14(50):4179–84.

[117] Paul VJ, Fenical W. Novel bioactive diterpenoid metabolites from tropical marine algae of the genus *Halimeda* (Chlorophyta). Tetrahedron 1984;16(40):3053–62.

[118] Wratten SJ, Faulkner DJ. Metabolites of the red alga *Laurencia subopposita*. J Org Chem 1977;42(21):3343–9.

[119] Fusetani N, Matsunaga S, Konosu S. Bioactive marine metabolites I. Isolation of guaiazulene from the gorgonian *Euplexaura erecta*. Experientia 1981;7(37):680–1.

[120] Accadia Di FD, Gribanovski-Sassu O, Romagnoli A, Tuttobello L. Isolation and identification of carotenoids produced by a green alga (*Dictyococcus cinnabarinus*) in submerged culture. Biochem J 1966;101(3):735–40.

[121] Frank HA, Young A, Britton G, Cogdell RJ. The Photochemistry of carotenoids, vol. 8. New York: Springer Science & Business Media; 2006. p. 31.

[122] El-Raey MA, Ibrahim GE, Eldahshan OA. Lycopene and lutein; a review for their chemistry and medicinal uses. J Pharmacogn Phytochem 2013;1(2):245–54.

[123] Francisco C, Banaigs B, Codomier L, Cave A, Cystoseirol A. A novel rearranged diterpene of mixed biosynthesis from the brown alga *Cystoseira mediterranea*. Tetrahedron Lett 1985;26(40):4919–22.

[124] Hertzberg S, Jensen SL. The carotenoids of blue-green algae—II.: The carotenoids of *Aphanizomenon flos-aquae*. Phytochemistry 1966;4(5):565–70.

[125] Enoki N, Ishida R, Urano S, Ochi M, Tokoroyama T, Matsumoto T. New hydroazulenoid diterpenes from the marine alga *Dictyota dichotoma*. Chem Lett 1982;11(11): 1837–40.

[126] Kitamura M, Koyama T, Nakano Y, Uemura D. Corallinafuran and Corallinaether, novel toxic compounds from crustose coralline red algae. Chem Lett 2005;9(34): 1272–3.

[127] Foss P, Guillard RRL, Liaaen-Jensen S. Prasinoxanthin—a chemosystematic marker for algae. Phytochemistry 1984;8(23):1629–33.

[128] Nozaki H, Ohira S, Takaoka D, Senda N, Nakayama M. Structure of Saragassumketone, a novel highly oxygenated ketone from *Sargassum kjellmanianum*. Chem Lett 1995;4:331.

[129] Walton TJ, Britton G, Goodwin TW. The structure of siphonaxanthin. Phytochemistry 1970;12(9):2545–52.

[130] Egeland ES, Guillard RRL, Liaaen-Jensen S. Additional carotenoid prototype representatives and a general chemosystematic evaluation of carotenoids in prasinophyceae (chlorophyta)Walton TJ, Britton G, Goodwin TW, editors. Struct siphonaxanthin. Phytochemistry 1997;6(44):1087–97.

[131] Tillekeratne LMV, Schmitz FJ. 4,9-diacetoxyudoteal: a linear diterpene aldehyde from the green alga *Halimeda opuntia*. Phytochemistry 1984;6(23):1331–3.

[132] Kurata K, Amiya T. Two new bromophenols from the red alga, *Rhodomela Larix*. Chem Lett 1977;12(6):1435–8.

[133] Chemburkar SR, Deming KC, Reddy RE. Chemistry of thyroxine: an historical perspective and recent progress on its synthesis. Tetrahedron 2010;11(66):1955–62.

[134] Napoli LD, Fattorusso E, Magno S, Mayol L. Acyclic polyhalogenated monoterpenes from four marine hydroids. Biochem Syst Ecol 1984;3(12):321–2.

[135] Korotchenko OD, Alekseev SM, Evstigneeva RP, Karpova GV, Isai SV. HPLC-fluorometric analysis of prostaglandins from marine organisms. Chem Nat Compd 1999;6(35):612–5.

[136] Palermo JA, Gros EG, Seldes AM. Carotenoids from three red algae of the Corallinaceae. Phytochemistry 1991;9(30):2983–6.

[137] Dabdoub MJ, Dabdoub VB, Pereira MA, Zukerman-Schpector J. Iodocyclization of (*Z*)-1-(Butyltelluro)-1,4-diorganylbut-1-en-3-ynes. Synthesis and reactions of 3-iodotellurophenes. J Org Chem 1996;61(26):9503–11.

[138] Antunes EM, Afolayan AF, Chiwakata MT, Fakee J, Knott MG, Whibley CE, Hendricks DT, Bolton JJ, Beukes DR. Identification and in vitro anti-esophageal cancer activity of a series of halogenated monoterpenes isolated from the South African seaweeds *Plocamium suhrii* and *Plocamium cornutum*. Phytochemistry 2011;8(72): 769–72.

[139] Mynderse JS, Faulkner DJ. Polyhalogenated monoterpenes from the red alga *Plocamium cartilagineum*. Tetrahedron 1975;16(31):1963–7.

[140] Watanabe K, Miyakado M, Ohno N, Okada A, Yanagi K, Umeda K, Okada A. A polyhalogenated isecticidal monoterepene from the red alga *Plocamium telfairiae*. Phytochemistry 1989;1(28):77–8.

[141] Kikuchi T, Mori Y, Yokoi T, Nakazawa S, Kuroda H, Masada Y, Kitamura K, Kuriyama K. Structure and absolute configuration of sargatriol, a new isoprenoid chromenol from a brown alga *Sargassum tortile* C. Agardh. Chem Pharm Bull 1983;31(1): 106–13.

[142] Bennamara A, Abourriche A, Berrada M, Charrouf M, Chaib N, Boudouma M, Garneau FX. Methoxybifurcarenone: an antifungal and antibacterial meroditerpenoid from the brown alga *Cystoseira tamariscifolia*. Phytochemistry 1999;52:37–40.

[143] Régnier P, Bastias J, Rodriguez-Ruiz V, Caballero-Casero N, Caballo C, Sicilia D, Fuentes A, Maire M, Crepin M, Letourneur D, Gueguen V, Rubio S, Pavon-Djavid G. Astaxanthin from *Haematococcus pluvialis* prevents oxidative stress on human endothelial cells without toxicity. Mar Drugs 2015;13(5):2857–74.

[144] Gerber NN. Geosmin, from microorganisms, is *trans*-1, 10-dimethyl-*trans*-9-decalol. Tetrahedron Lett 1968;25(9):2971–4.

[145] Rastogi RP, Sonani RR, Madamwar D. Physico-chemical factors affecting the in vitro stability of phycobiliproteins from *Phormidium rubidum* A09DM. Bioresour Technol 2015;190:219–26.

[146] Lamparczyk H. CRC handbook of chromatography: analysis and characterization of steroid. Handbook of chromatography, vol. 8. London: CRC Press; 1992. p. 119.

[147] Fink HH, Mikesky AE. Practical applications in sports nutrition. 2nd ed. USA: Jones & Bartlett Publishers; 2013. p. 105.

[148] Fattorusso E, Magno S, Santacroce C, Sica D, Impellizzeri G, Mangiafico S, Oriente G, Piattelli M, Sciuto S. Sterols of some red algae. Phytochemistry 1975; 7(14):1579–82.

[149] Sheu JH, Huang SY, Duh CY. Cytotoxic oxygenated desmosterols of the red alga *Galaxaura marginata*. J Nat Prod 1996;59(1):23–6.

[150] Kabore SA, Combaut G, Vidal JP, Codomier L, Passet J, Girard JP, Rossi JC. Sterols of the red alga *Rissoella verruculosa*. Phytochemistry 1983;5(22):1239–40.

[151] Idler DR, Saito A, Wiseman P. Sterols in red algae (Rhodophyceae). Steroids 1968; 4(11):465–73.

[152] Memon AH, Shameel M, Usman G, Ahmad M, Khan R, Ahmad VU. Phycochemical studies on *Scinaia Fascicularies*. Pak J Pharm Sci 1991;4(1):27–34.

[153] Siddiqui S, UsmanGhani K, Shameel M. Sterols and fatty acid compositions of a marine alga *Bryopsis pennsta* (Bryopsidophyceae Chlorophyta). Pak J Pharm Sci 1994; 7(1):73–82.

[154] Shimizu Y, Alam M, Kobayashi A. Dinosterol, the major sterol with a unique side chain in the toxic dinoflagellate, *Gonyaulax tamarensis*. J Am Chem Soc 1976; 98(4):1059–60.

[155] Volkman JK, Rijpstra WIC, de Leeuw JW, Mansour MP, Jackson AE, Blackburn SI. Sterols of four dinoflagellates from the genus *Prorocentrum*. Phytochemistry 1999; 4(52):659–68.

[156] Permeh P, Saeidnia S, Mashinchian-Moradi A, Gohari AR. Sterols from *Sargassum oligocystum*, a brown algae from the Persian Gulf, and their bioactivity. Nat Prod Res 2012;8(26):774–7.

[157] Kokke WCMC, Shoolery JN, Fenical W, Djerassi C. Biosynthetic studies of marine lipids. 4. mechanism of side chain alkylation in (E)-24-propylidenecholesterolby a chrysophyte alga. J Org Chem 1984;20(49):3742–52.

[158] Norton RA, Nes WD. Identification of ergosta-6(7),8(14),25(27)-trien-3β-ol and ergosta-5(6),7(8),25(27)-trien-3β-ol, two new steroidal trienes synthesized by *Prototheca wickerhamii*. Lipids 1991;3(26):247–9.

[159] Sheu JH, Sung PJ. Isolation of 24-hydroperoxy-24-vinylcholesterol and fucosterol from the brown alga *Turbinaria conoides*. J Chin Chem Soc 1991;5(38): 501–3.

[160] Guyot M, Davoust D, Belaud C. Hydroperoxy-24 vinyl-24 cholesterol, nouvel hydroperoxyde naturel isole de deux tuniciers: *Phallusia mamillata et Ciona intestinalis*. Tetrahedron Lett 1982;18(23):1905–6.

[161] Patterson GW. The distribution of sterols in algae. Lipids 1971;2(6):120–7.

[162] Ayyad SEN, Sowellim SZA, El-Hosini MS, Abo-Atia A. The structural determination of a new steroidal metabolite from the brown alga *Sargassum asperifolium*. Z Naturforsch 2003;58c:333–6.

[163] Popov SS, Marekov NL, Konaklieva ML, Panayotova MI, Dimitrova-Konaklieva S. Sterols from some black sea ulvaceae. Phytochemistry 1985;9(24):1987–90.

[164] Nicotra E, Ronchetti F, Russo G, Toma L, Gariboldi P, Ranzi BM. Mechanism of the transmethylation reaction by *S*-adenosylmethionine: stereochemistry of hydride migration from C-24 to C-25 in the biosynthesis of poriferasterol in the crysophyte *Ochromonas malharnensis*. J Chem Soc Perkin Trans 1 1985:521–4.

[165] Colombo D, Ronchetti F, Russo G, Toma L. Biosynthesis of poriferasterol in *Ochromonas malhamensis*: ^{13}C NMR assignment of the isopropyl methyl groups of 2-methylpentan-3-ol. J Chem Soc Perkin Trans 1 1991;1:962–4.

[166] Rezanka T, Vyhnálek O, Podojil M. Identification of sterols and alcohols produced by green algae of the genera *Chlorella* and *Scenedesmus* by means of gas chromatography—mass spectrometry. Folia Microbiol 1986;31:44.

[167] Neal AC, Prahl FG, Eglinton G, O'Hara SCM, Corner EDS. Lipid changes during a planktonic feeding sequence involving unicellular algae, *Elminius nauplii* and adult *Calanus*. J Mar Biol Assoc UK 1986;1(66):1–13.

[168] De Napoli L, Magno S, Mayol L, Novellino E. Sterol composition of some mediterranean green algae. Phytochemistry 1982;8(21):1993–4.

[169] Madison LL, Huisman GW. Metabolic engineering of poly (3-hydroxyalkanoates): from DNA to plastic. Microbiol Mol Biol Rev 1999;63(1):21–53.

[170] Jensen TE, Sicko LM. Fine structure of poly-3-hydroxybutyric acid granules in a blue-green alga, *Chlorogloea fritschii*. J Bacteriol 1971;2(106):683–6.

[171] Bhati R, Mallick N. Poly (3-hydroxybutyrate-co-3-hydroxyvalerate) copolymer production by the diazotrophic cyanobacterium *Nostoc muscorum* Agardh: process optimization and polymer characterization. Algal Res 2015;7:78–85.

[172] Stal LJ. Polyhydroxyalkanoate in cyanobacteria: an overview. FEMS Microbiol Rev 1992;103:169–80.

[173] Aniszewski T. Alkaloids – secrets of life: aklaloid chemistry, biological significance, applications and ecological role. UK: Elsevier; 2007. p. 1–5.

[174] Smitka TA, Bonjouklian R, Doolin L, Jone ND, Deetar JB. Ambiguine isonitriles, fungicidal hapalindole-type alkaloids from three gernera of blue green algae belong to the stingonemataceae. J Org Chem 1992;57:857–61.

[175] Raveh A, Carmeli C. Antimicrobial ambiguines from the cyanobacterium *Fischerella* sp. collected in Israel. J Nat Prod 2007;70(2):196–201.

[176] Huber U, Moore RE, Patterson GML. Isolation of a nitrile-containing indole alkaloid from the terrestrial blue-green alga *Hapalosiphon delicatulus*. J Nat Prod 1998;61: 1304–6.

[177] Beale SI, Cornejo J. Biosynthesis of phycobilins. 15,16-dihydrobiliverdin IX alpha is a partially reduced intermediate in the formation of phycobilins from biliverdin IX alpha. J Biol Chem 1991;266:22341–5.

[178] El-Gamal AA, Wang WL, Duh CY. Sulfur-containing polybromoindoles from the formosan red alga *Laurencia brongniartii*. J Nat Prod 2005;68(5):815–7.

[179] Schwede JG, Cardellina JH, Grode SH, James Jr TR, Blackman AJ. Distribution of the pigment caulerpin in species of the green alga *Caulerpa*. Phytochemistry 1986;1(26): 155–8.

[180] Anjaneyulu ASR, Prakash CVS, Mallavadhani UV. Two caulerpin analogues and a sesquiterpene from *Caulerpa racemosa*. Phytochemistry 1991;9(30):3041–2.

[181] Johnson AL, Johnson J. Synthetic approaches towards an indole alkaloid isolated from the marine sponge *Halichondria melanodocia*. Tetrahedron 2006;47(62):10815–20.

[182] Güven KC, Bora A, Sunam G. Hordenine from the alga *phyllophora nervosa*. Phytochemistry 1970;8(9):189.

[183] Abe H, Uchiyama M, Sato R. Isolation and identification of native auxins in marine algae. Agric Biol Chem 1972;12(36):2259–60.

[184] Lacan G, Magnus V, Jericevic B, Kunst L, Iskrić S. Formation of tryptophol galactoside and an unknown tryptophol ester in *Euglena gracilis*. Plant Physiol 1984;4(76): 889–93.

[185] Fujit T, Ohba M, Haneishi T, Matsubara S, Abad Farooqi AH, Shukla YN. Synthesis and absolute configuration of the green alga cytokinin 2-hydroxy-1′-methylzeatin. Heterocycles 1992;1(34):21–4.

[186] Rolle I, Hobucher HE, Kneifel H, Paschold B, Riepe W, Soeder CJ. Amines in unicellular green algae: 2. Amines in *Scenedesmus acutus*. Anal Biochem 1977;1(77): 103–9.

[187] Gossauer A, Hinze RP. Synthesis of bile pigments. 7. An improved chemical synthesis of racemic phycocyanobilin dimethyl ester. J Org Chem 1978;43(2):283–5.

[188] O'Carra P, O'Heocha C, Carrol DM. Spectral properties of the phycobilins. 11. Phycoerythrobilin Biochem 1964;9(3):1343–50.

[189] Impellizzeri G, Piattelli M, Sciuto S, Fattorusso E. Pyrrolidine-2,4-dicarboxylic acid, a new naturally occurring imino acid. Phytochemistry 1977;10(16):1601–2.

[190] Onodera H, Satake M, Oshima Y, Yasumoto T, Carmichael WW. New saxitoxin analogues from the freshwater filamentous cyanobacterium *Lyngbya wollei*. Nat Toxins 1997;4(5):146–51.

[191] Carter GT, Rinehart Jr KL, Li LH, Kuentzel SL, Connor JL. Brominated indoles from *Laurencia brongniartii*. Tetrahedron Lett 1978;46(19):4479–82.

[192] Matsuo Y, Imagawa H, Nishizawa M, Shizuri Y. Isolation of an algal morphogenesis inducer from a marine bacterium. Science 2005;5715(307):1598.

[193] Stratmann K, Moore RE, Bonjouklian R, Deeter JB, Patterson GML, Shaffer S, Smith CD, Smitkat TA. Welwitindolinones, unusual alkaloids from the blue-green algae *Hapalosiphon welwitschii* and *Westiella intricata*. Relationship to fischerindoles and hapalindoles. J Am Chem Soc 1994;116:9935–42.

[194] Tarakhovskaya ER, Maslov YI, Shishova MF. Phytohormones in algae. Russ J Plant Physiol 2007;2(54):163–70.

[195] Shen LC, Atkinson DE. Regulation of adenosine diphosphate glucose synthase from *Escherichia coli*. Interactions of adenylate energy charge and modifier concerations. J Biol Chem 1970;245:3996–4000.

[196] Wu G. Amino acids: biochemistry and nutrition. UK: CRC Press; 2013. p. 1.

[197] Inouye K, Tanaka A, Otsuka H. Synthesis of corticotropin peptides. XI synthesis and biological properties of [1-β-Alanine]-ACTH (1-18)-octadecapeptide amide. Bull Chem Soc Jpn 1970;43:1163–72.

[198] Makisumi S. Occurrence of arginylglutamine in green alga, *Cladophora* species. J Biochem 1995;1(46):63–71.

[199] Kempner ES, Miller JH. The molecular biology of *Euglena gracilis* IX. Amino acid pool composition. J Protozool 1974;2(21):363–7.

[200] Tominaga F, Oka K. On the isolation and identification of 1,4-thriazane-3-carboxylic acid S-oxide from the brown alga *Undaria pinnatifida*. J Biochem 1963;3(54):222–4.

[201] Palmer KJ, Lee KS, Wong RY, Carson JF. Crystal and molecular structure of chondrine, C5H9NO3S. Acta Crystallogr 1972;B28:2789–93. Available from: http://journals.iucr.org/bhttp://journals.iucr.org/b.

[202] Ito K, Hashimoto Y. Gigartinine: a new amino-acid in red algae. Nature 1966;211:417.
[203] Ito K, Hashimoto Y. Occurrence of γ-(Guanylureido)butyric acid in a red alga *Gymnogongrus flabelliformis*. Agric Biol Chem 1965;9(29):832–5.
[204] Ito K, Hashimoto Y. Syntheses of DL-gigartinine and gongirine. Agric Biol Chem 1965;2(13):237–41.
[205] Impellizzeri G, Mangiafico S, Oriente G, Piattelli M, Sciuto S, Fattorusso E, Magno S, Santacroce C, Sica D. Amino acids and low-molecular-weight carbohydrates of some marine red algae. Phytochemistry 1975;7(14):1549–57.
[206] Simon-Colin S, Kervarec N, Pichon R, Bessières MA, Deslandes E. Characterization of N-methyl-L-methionine sulfoxide and isethionic acid from the red alga *Grateloupia doryphora*. Phycol Res 2002;2(50):125–8.
[207] Sato M, Kanno N, Sato Y. Studies on the distribution and metabolism of D-rhodoic acid in algae. Hydrobiologia 1987;151–152:457–62.
[208] Kuriyama M. New ninhydrin-reactive substance from red algae. Nature 1961;192:969.
[209] Sciuto S, Chillemi R, Morrone R, Patti A, Piattelli M. Dragendorff-positive compounds in some Mediterranean red algae. Biochem Syst Ecol 1989;1(17): 5–10.
[210] Larsen PO. N6-trimethyl-L-lysine betamine from seeds of *Reseda luteola* L. Acta Chem Scand 1968;22(4):1369–70.
[211] Wrkamiya T, Kobayashi Y, Shiba T, Setogawa K, Matsutani H. Isolations and structures of new ureido amino acids, lividine and grateloupine, from red algae *Ograteloupia* c. Agardh genus. Tetrahedron 1984;1(40):235–40.
[212] Sciuto S, Chillemi R, Piattelli M, Impellizzeri G. The identification of 4-hydroxy-*N*-methylproline in the red alga *Chondria coerulescens*—spectral information. Phytochemistry 1983;10(22):2311–2.
[213] Wickberg B. Isolation of N-[D-2, 3-Dihydroxy-n-propyl]-taurine from *Gigartina leptorhynchos*. Acta Chem Scand 1956;7(10):1097–9.
[214] Lindberg B. Low-molecular carbohydrates in algae. 10. Investigation of *Furcellaria fastigiata*. Acta Chem Scand 1955;7(9):1093–6.
[215] Lindberg B. Methylated taurines and choline sulphate in red algae. Acta Chem Scand 1955;7(9):1323–6.
[216] Wu S, Yue Y, Tian H, Tao L, Wang Y, Xiang J, Wang S, Ding H. Tramiprosate protects neurons against ischemic stroke by disrupting the interaction between PSD95 and NOS. Neuropharmacology 2014;83:107–17.
[217] Albertson NF, Archer S. A synthesis of DL-ornithine hydrochloride. J Am Chem Soc 1945;67(11):2043–4.
[218] Mukhopadhyay M. Natural extracts using supercritical carbon dioxide. New York: CRC Press; 2000. p. 321.
[219] McConnell OJ, Fenical M. Halogenated metabolites including favorsxy rearrangement products from the red seweed *Bonnemaisonia nooteana*. Tetrahedron Lett 1977; 48(18):4159–62.
[220] Greene RC. Biosynthesis of dimethyl-β-propiothetin. J Biol Chem 1962;7(237): 251–4.
[221] Miralles J, Aknin M, Micouin L, Gaydou EM, Kornprobst JM. Cyclopentyl and ω-5 monounsaturated fatty acids from red algae of the Solieriaceae. Phytochemistry 1990;7(29):2161–3.
[222] McConnell OJ, Fenical W. Halogen chemistry of the red alga *Bonnemaisonia*. Phytochemistry 1980;2(19):233–47.

[223] Lopez A, Gerwick WH. Two new icosapentaenoic acids from the temperate red seaweed *Ptilota filicina* J. Agardh. Lipids 1987;3(22):190–4.

[224] Bernart MW, Gerwick WH. Eicosanoids from the tropical red alga *Murrayella periclados*. Phytochemistry 1994;5(36):1233–40.

[225] Suzuki M, Wakana I, Denboh T, Tatewaki M. An allelopathic polyunsaturated fatty acid from red algae. Phytochemistry 1996;1(43):63–5.

[226] Rasoul-Amini S, Ghasemi Y, Morowvat MH, Mohagheghzadeh A. PCR amplification of 18S rRNA, single cell protein production and fatty acid evaluation of some naturally isolated microalgae. Food Chem 2009;116:129–36.

[227] Radunz R. Über die lipide der pteridophyten—I.: Die isolierung und identifizierung der polyensäuren. Phytochemistry 1967;3(6):399–406.

[228] Kajiwara T, Kashibe M, Matsui K, Hatanaka A. Enzymatic formation of (2*R*)-hydroxy- and 2-oxo-hexadecanoic acid in *Ulva pertusa* and *Porphyra* sp. Phytochemistry 1991; 1(30):193–5.

[229] Solem ML, Jiang ZD, Gerwick WH. Three new and bioactive icosanoids from the temperate red marine alga *Farlowia mollis*. Lipids 1989;4(24):256–60.

[230] Guerriero A, D'Ambrosio M, Pietra F. Novel hydroxyicosatetraenoic and hydroxyicosapentaenoic acids and a 13-oxo analog. Isolation from a mixture of the calcareous red algae *Lithothamnion corallioides* and *Lithothamnion calcareum* of brittany waters. Helv Chim Acta 1990;8(73):2183–9.

[231] Mohy El-Din SM, El-Ahwany AMD. Bioactivity and phytochemical constituents of marine red seaweeds (*Jania rubens*, *Corallina mediterranea* and *Pterocladia capillacea*). J Taibah Univ Sci 2016;10:471–84.

[232] Gribble GW. Naturally occurring organohalogen compounds – a comprehensive update, vol. 91. New York: Springer Science & Business Media; 2009.

[233] Fedoryński M, Popławska M, Nitschke K, Kowalski W, Mâkosza M. A simple method for preparation of dibromochloromethane and bromodichloromethane. Synth Commun 1977;4(7):287–92.

[234] Burreson BJ, Moore RE, Roller PP. Volatile halogen compounds in the alga *Asparagopsis taxiformis* (Rhodophyta). J Agric Food Chem 1976;24(4):856–61.

[235] Steiner M, Hartmann T. The occurence and distribution of volatile amines in marine algae. Planta 1968;79(2):113–21.

[236] Blumer M, Mullin MM, Guillard RRL. A polyunsaturated hydrocarbon (3, 6, 9, 12, 15, 18-heneicosahexaene) in the marine food web. Mar Biol 1970;3(6):226–35.

[237] Karunen P. Polyunsaturated hydrocarbons from *Polytrichum commune*spores. Phytochemistry 1974;10(13):ss.

[238] Youngblood WW, Blumer M, Guillard RL, Fiore F. Saturated and unsaturated hydrocarbons in marine benthic algae. Mar Biol 1971;3(8):190–201.

[239] Yao L, Gerde JA, Lee SL, Wang T, Harrata TA. Microalgae lipid characterization. J Agric Food Chem 2015;63(6):1773–87.

[240] Tsuchiya Y, Matsumoto A. Identification of volatile metabolites produced by blue-green algae. Water Sci Technol 1988;20(8–9):149–55.

[241] Kigoshi H, Shizuri Y, Niwa H, Yamada K. Four new C_{15} acetylenic polyenes of biogenetic significance from the red alga: *Laurencia okamurai*: structures and synthesis. Tetrahedron 1986;14(42):3781–7.

[242] Huang J, Liu L, Yu Y, Lin W, Chen B, Li M. Reduction in the blood glucose level of exopolysaccharide of *Porphyridium cruentum* in alloxan-induced diabetic mice (in Chinese). J Fujian Normal Univ 2006;22:77–80.

[243] Combaut G, Bruneau Y, Teste J, Codomier L. Composes halogenes d'une algue rouge, *Falkenbergia rufolanosa* tetrasporophyte d' *Asparagopsis armata*. Phytochemistry 1978;9(17):1661–3.

[244] McConnell O, Fenical W. Halogen chemistry of the red alga *Asparagopsis*. Phytochemistry 1977;3(16):367–74.

[245] Jaenicke L, Boland W. Signal substances and their reception in the sexual cycle of marine brown algae. Angew Chem 1982;9(21):643–53.

[246] Muller DG, Gassmnn G, Marner FJ, Boland W, Jaenicke L. The sperm attractant of the marine brown alga *Ascophyllum nodosum* (phaeophyceae). Science 1982;4577(218): 1119–20.

[247] Muller DG, Clayton MN, Gassmnn G, Boland W, Marner FJ, Schotten T, Jaenicke L. Cystophorene and hormosirene, sperm attractants in Australian brown algae. Naturwissenschaften 1985;2(72):97–9.

[248] Lemke TL. Review of organic functional groups: introduction to medicinal organic chemistry. UK: Lippincott Williams & Wilkins; 2003. p. 71.

[249] Deffieux G, Baute R, Baute MA, Atfani M, Carpy A. 1,5-D-anhydrofructose, the precursorof the pyrone microthecin in *Morchella vulgaris*. Phytochemistry 1987;5(26): 1391–3.

[250] Alberte RS, Andersen RA. Antheraxanthin, a light harvesting carotenoid found in a chromophyte alga. Plant Physiol 1986;2(80):583–7.

[251] Czeczuga B, Taylor FJ. Carotenoid content in some species of the brown and red algae from the coastal area of New Zealand. Biochem Syst Ecol 1987;1(15):5–8.

[252] Arunkumar K, Selvapalam N, Rengasamy R. The antibacterial compound sulphoglycerolipid 1-0 palmitoyl-3-0(6′-sulpho-α-quinovopyranosyl)-glycerol from *Sargassum wightii* Greville (Phaeophyceae). Bot Mar 2005;48:441–5.

[253] Wright AD, König GM, Sticher O. Two new dolabellane derivatives from the brown alga *dictyota pardarlis*. Tetrahedron 1990;11(46):3851–3858c.

[254] Lindberg B. Low molecular carbohydrates in algae. Investigation of *Fucus vesiculosus* 1953;7:1119–22.

[255] Kirst GO. Low MW carbohydrates and ions in rhodophyceae: quantitative measurement of floridoside and digeneaside. Phytochemistry 1980;6(19):1107–10.

[256] Shibata Y, Morita M, Edmonds JS. Purification and identification of arsenic-containing ribofuranosides from the edible brown seaweed, *Laminaria japonica (MAKONBU)*. Agric Biol Chem 2014;2(51):391–8.

[257] Moghaddam MF, Gerwick WH, Ballantine DL. Discovery of the mammalian insulin release modulator, hepoxilin B3, from the tropical red algae *Platysiphonia miniata* and *Cottoniella filamentosa*. J Biol Chem 1990;265:6126–30.

[258] Francis GW, Hertzberg S, Andersen K, Jensen SL. New carotenoid glycosides from *Oscillatoria limosa*. Phytochemistry 1970;3(8):629–35.

[259] Mori K, Ooi T, Hiraoka M, Oka N, Hamada H, Tamura M, Kusumi T. Fucoxanthin and its metabolites in edible brown algae cultivated in deep seawater. Mar Drugs 2004; 2(2):63–72.

[260] Farrant AS, Nunn JR, Parolis H. Sulphated polysaccharides of the *grateloupiaceae* family: Part VII. Investigation of the acetolysis products of a partially desulphated sample of the polysaccharide of *pachymenia carnosa*. Carbohydr Res 1972;2(25): 283–92.

[261] Allsobrook AJR, Nunn JR, Parolis H. Investigation of the acetolysis products of the sulphated polysaccharide of *Aeodes ulvoidea*. Carbohydr Res 1975;2(40):337–44.

[262] Meng J, Rosell KG, Srivastava LM. Chemical characterization of floridosides from *Porphyra perforata*. Carbohydr Res 1987;2(161):1804171−4.

[263] Abreu PM, Galindro JM, Relva AM, Ramos AM. Non-terpenoid compounds from *Plocamium cartilaginuem*. Phytochemistry 1997;8(45):1601−3.

[264] Nelson WL, Cretcher LH. The properties of d-mannuronic acid lactone. J Am Chem Soc 1932;54(8):3409−12.

[265] Lindberg B. Low-molecular carbohydrates in algae. Synthesis of 1-D-mannitol β glucoside. Acta Chem Scand 1953;7:1218−9.

[266] Gerwick WH, Fenical W. Ichthyotoxic and cytotoxic metabolites of the tropical brown alga *Stypopodium zonale* (Lamouroux) Papenfuss. J Org Chem 1981;46(1):22−7.

[267] Precival EGV, Chanda SK. The xylan of *Rhodymenia palmate*. Nature 1950;166: 787−8.

[268] Hertzberg S, Jensen SL. The carotenoids of blue-green algae—I.: the carotenoids of *Oscillatoria rubescens* and an *Athrospira* sp. Phytochemistry 1966;4(5):557−63.

[269] Haugan JA, Liaaen-Jensen S. Algal carotenoids 54. Carotenoids of brown algae (Phaeophyceae). Biochem Syst Ecol 1994;22:31−41.

[270] Qi SH, Zhang S, Huang JS, Xiao ZH, Wu J, Long LJ. Glycerol derivatives and sterols from *Sargassum parvivesiculosum*. Chem Pharm Bull 2004;8(52):986−8.

[271] Durkin CA, Mock T, Armbrust EV. Chitin in diatoms and its association with the cell wall. Eukaryot Cell 2009;7(8):1038−50.

[272] Francesconi KA, Edmonds JS, Stick RV, Skelton BW, White AH. Arsenic-containing ribosides from the brown alga *Sargassum lacerifolium*: X-ray molecular structure of 2-amino-3-[5′-deoxy-5′-(dimethylarsinoyl) ribosyloxy] propane-1-sulphonic acid. J Chem Soc Perkin Trans 1 1991;1:2707−16.

[273] Anastasakis K, Ross AB, Jones JM. Pyrolysis behaviour of the main carbohydrates of brown macro-algae. Fuel 2011;2(90):598−607.

[274] Nagashima H, Ozaki H, Nakamura S, Nisizwa K. Physiological studies on floridean starch, floridoside and trehalose in a red alga, *Serraticardia maxima*. Botan Mag Tokyo 1969;82:462−73.

[275] Blumreisinger M, Meindl D, Loos E. Cell wall composition of chlorococcal algae. Phytochemistry 1983;7(22):1603−4.

[276] Feige GB, Kremer BP. Unusual carbohydrate pattern in *Trentepohlia* species. Phyochemistry 1980;8(19):1844−5.

[277] Koppisch AT, Blagg BSJ, Poulter CD. Synthesis of 2-*C*-Methyl-d-erythritol 4-phosphate: the first pathway-specific intermediate in the methylerythritol phosphate route to isoprenoids. Org Lett 2000;2(2):215−7.

[278] Edmonds JS, Shibata Y, Yang F, Morita M. Isolation and synthesis of 1-Deoxy-1-dimetylarsinoylribitol-5-sulfate, a neutral constituent of *Chondria crassicaulis* and other red algae. Tetrahedron Lett 1977;33(38):5819−20.

[279] Chudek JA, Foster R, Davison IR, Reed RH. Altritol in the brown alga *Himanthalia elongate*. Phytochemistry 1984;5(23):1081.

[280] Raven JA, Beardall J, Chudek JA, Scrimgeour CM, Clayton MN, McInroy SG. Altritol synthesis by *Notheia anomala*. Phytochemistry 2001;3(58):389−94.

[281] Manilal A, Sujith S, Selvin J, Shakir C, Kiran GS. Antibacterial activity of *Falkenbergia hillebrandii* (born) from the Indian coast against human pathogens. Int J Exp Bot 2009;78:161−6.

[282] Woolard FX, Moore RE. Halogenated acetamides, but-3-en-2-ols, and isopropanols from *asparagopsis taxiformis* (delile) trev. Tetrahedron 1976;22(32):2843−6.

[283] Kato M, Sakai M, Adachi K, Ikemoto H, Sano H. Distribution of betaine lipids in marine algae. Phytochemistry 1996;5(42):1341−5.

[284] Fushiya S, Komato Y, Nozoe S. Two betaine type amino acid derivatives of *Lampteromyces joponicus*. Nat Med 1997;51(6):558.

[285] Cooney RV, Mumma RO, Benson AA. Arsoniumphospholipid in algae. Proc Natl Acad Sci USA 1978;9(75):4262−4.

[286] Kajiwara T, Yoshikawa H, Matsui K, Hatanaka A, Kawai T. Specificity of the enzyme system producing long chain aldehydes in the green alga, *Ulva pertusa*. Phytochemistry 1989;2(28):407−9.

[287] Fujimura T, Kawai T, Shiga M, Kajiwara T, Hatanaka A. Long-chain aldehyde production in thalli culture of the marine green alga *Ulva pertusa*. Phytochemistry 1990; 3(29):745−7.

[288] Schnitzler I, Boland W, Hay ME. Organic suilhu compounds from *Dictyopteris* ssp. Dater feeding by an herbivorous amphipod (*Ampithoe longimana*) but not by an herbivorous sea urchin (*Arbacia punctulata*). J Chem Ecol 1998;10(24):1715−32.

[289] De Rosa S, De Giulio A, Iodice C, Alcaraz MJ, Paya M. Long-chain aldehydes from the red alga, *Corallina mediterranea*. Phytochemistry 1995;3(40):995−6.

[290] Abdel-Aal FI, Haroon AM, Mofeed J. Successive solvent extraction and GC−MS analysis for the evaluation of the phytochemical constituents of the filamentous green alga *Spirogyra longata*. Egypt J Aquat Res 2015;3(41):233−46.

[291] Taylor EC, Jacobi PA. Pteridines. XXXVII. A total synthesis of *L-erythro*-biopterin and some related 6-(polyhydroxyalky1) pterins. J Am Chem Soc 1976;98:2301−7.

[292] Nischwitz V, Pergantis SA. First report on the detection and quantification of arsenobetaine in extracts of marine algae using HPLC-ES-MS/MS. Analyst 2005;130:1348−50.

[293] Wilks SS. Carbon monoxide in green plants. Science 1959;3354(129):964−6.

[294] Malin G, Erst GO. Algal production of dimethyl sulfide and its atmospheric role. J Physiol 1997;6(33):889−96.

[295] Sloth JJ, Larsen EH, Julshamn K. Report on three aliphatic dimethylarsinoyl compounds as common minor constituents in marine samples. An investigation using high-performance liquid chromatography/inductively coupled plasma mass spectrometry and electrospray ionisation tandem mass spectrometry. Rapid Commun Mass Spectrom 2005;2(13):227−35.

[296] Mclellan DS, Jurd KM. Anticoagulants from marine algae. Blood Coagul Fibrinolysis 1992;1(3):6−17.

[297] Pomin VH. Structural and functional insights into sulfated galactans: a systematic review. Glycoconj J 2010;27:1−12.

[298] Toida T, Amornrut C, Robert L. Structure and bioactivity of sulfated polysaccharides. Trends Glycosci Glycotechnol 2003;81(15):29−46.

[299] Lee JB, Hayashi K, Maeda M, Hayashi T. Antiherpetic activities of sulfated polysaccharides from green algae. Planta Med 2004;70:813−7.

[300] Geresh S, Adin I, Yarmolinsky E, Karpasas M. Characterization of the extracellular polysaccharide of *Porphyridium* sp.: molecular weight determination and rheological properties. Carbohydr Polym 2002;50:183−9.

[301] Ginzberg A, Korin E, Arad SM. Effect of drying on the biological activities of a red microalga polysaccharide. Biotechnol Bioeng 2008;99:411−20.

[302] Kim M, Yim JH, Kim S-Y, Kim HS, Lee WG, Kim SJ, Kang P-S, Lee C-K. In vitro inhibition of influenza A virus infection by marine microalga-derived sulphated polysaccharide p-KG03. Antivir Res 2012;93:253−9.

[303] Badel S, Callet F, Laroche C, Gardarin C, Petit E, El Alaoui H, Bernardi T, Michaud P. A new tool to detect high viscous exopolymers from microalgae. J Ind Microbiol Biotechnol 2011;38:319–26.

[304] Proksch E, Holleran WM, Menon GK, Elias PM, Feingold KR. Barrier function regulates epidermal lipid and DNA synthesis. Br J Dermatol 1993;128:473–82.

[305] Heaney-Kieras J, Roden L, Chapman DJ. The covalent linkage of protein to carbohydrate in the extracellular protein-polysaccharide from the red alga *Porphyridium cruentum*. Biochem J 1977;165:1–9.

[306] Yim JH, Son E, Pyo S, Lee HK. Novel sulfated polysaccharide derived from red-tide microalga *Gyrodinium impudicum* strain KG03 with immunostimulating activity in vivo. Mar Biotechnol 2005;7:331–8.

[307] Namikoshi M. Bioactive compounds produced by cyanobacteria. J Ind Microbiol Biotechnol 1996;17:373–84.

[308] Bae SY, Yim JH, Lee HK, Pyo S. Activation of murine peritoneal macrophages by sulphated exopolysaccharide from marine microalga *Gyrodinium impudicum* (strain KG03): involvement of the NF-kappa B and JNK pathway. Int Immunopharmacol 2006;6:473–84.

[309] Leiro JM, Castro R, Arranz JA, Lamas J. Immunomodulating activities of acidic sulphated polysaccharides obtained from the seaweed *Ulva rigida* C. Agardh. Int Immunopharmacol 2007;7(7):879–88.

[310] Xing RE, Yu HH, Liu S. Antioxidant activity of differently regioselective chitosan sulfates in vitro. Bioorg Med Chem Lett 2005;13:1387–92.

[311] Mao W, Zang X, Li Y, Zhang H. Sulfated polysaccharides from marine green algae *Ulva conglobata* and their anticoagulant activity. J Appl Phycol 2006;18(1):9–14.

[312] Eteshola E, Karpasas M, Arad SM, Gottlieb M. Red microalga exopolysaccharides: 2. Study of the rheology, morphology and thermal gelation of aqueous preparations. Acta Polym 1998;49:549–56.

[313] Li H, Mao W, Zhang X, Qi X, Chen Y, Chen Y, Xu J, Zhao C, Hou Y, Yang Y, Li N, Wang C. Structural characterization of an anticoagulant-active sulfated polysaccharide isolated from green alga *Monostroma latissimum*. Carbohydr Polym 2011;85(2):394–400.

[314] Costa LS, Fidelis GP, Cordeiro SL, Oliveira RM, Sabry DA, Câmara RB, Nobre LT, Costa MS, Almeida-Lima J, Farias EH, Leite EL. Biological activities of sulfated polysaccharides from tropical seaweeds. Biomed Pharmacother 2010;64(1):21–8.

[315] Ciancia M, Quintana I, Vizcarguenaga MI, Kasulin L, de Dios A, Estevez JM, Cerezo AS. Polysaccharides from the green seaweeds *Codium fragile* and *C. vermilara* with controversial effects on hemostasis. Int J Biol Macromol 2007;41(5):641–9.

[316] Wijesinghe WAJP, Athukorala Y, Jeon YJ. Effect of anticoagulative sulfated polysaccharide purified from enzyme-assistant extract of a brown seaweed *Ecklonia cava* on Wistar rats. Carbohydr Polym 2011;86(2):917–21.

[317] Tan ML, Choong PFM, Dass CR. Cancer, chitosan nanoparticles and catalytic nucleic acids. J Pharm Pharmacol 2009;2009(61):3–12.

[318] Weitz JI. Low-molecular-weight heparins. N Engl J Med 1997;337(10):688–99.

[319] Kusaykin M, Bakunina I, Sova V, Ermakova S, Kuznetsova T, Besednova N, Zaporozhets T, Zvyagintseva T. Structure, biological activity, and enzymatic transformation of fucoidans from the brown seaweeds. Biotechnol J 2008;3(7):904–15.

[320] Pomin VH, Mourao PAS. Structure, biology, evolution, and medical importance of sulfated fucans and galactans. Glycobiology 2008;18:1016–27.

[321] Witvrouw M, De Clercq E. Sulfated polysaccharides extracted from sea algae as potential antiviral drugs. Vasc Syst 1997;29(4):497–511.

[322] Hidari KIPJ, Takahashi N, Arihara M, Nagaoka M, Morita K, Suzuki T. Structure and anti-dengue virus activity of sulfated polysaccharide from a marine alga. Biochem Biophys Res Commun 2008;376(1):91–5.

[323] Ghosh T, Chattopadhyay K, Marschall M, Karmakar P, Mandal P, Ray B. Focus on antivirally active sulfated polysaccharides: from structure-activity analysis to clinical evaluation. Glycobiology 2009;19:2–15.

[324] Damonte EB, Matulewicz MC, Cerezo AS. Sulfated seaweed polysaccharides as antiviral agents. Curr Med Chem 2004;11:2399–419.

[325] Talarico LB, Pujol CA, Zibetti RGM, Farea PCS, Noseda MD, Duarte MER, Damonte EB. The antiviral activity of sulfated polysaccharides against dengue virus is dependent on virus serotype and host cell. Antivir Res 2005;66:103–10.

[326] Talarico LB, Duarte MER, Zibetti RGM, Noseda MD, Damonte EB. An algal-derived DL-galactan hybrid is an efficient preventing agent for in vitro dengue virus infection. Planta Med 2007;73:1464–8.

[327] Talarico LB, Damonte EB. Interference in dengue virus adsorption and uncoating by carrageenans. Virology 2007;363:473–85.

[328] Ghosh T, Pujol CA, Damonte EB, Sinha S, Ray B. Sulfated xylomannans from the red seaweed *Sebdenia polydactyla*: structural features, chemical modification and antiviral activity. Antivir Chem Chemother 2009;19:235–42.

[329] Harden EA, Falshaw R, Carnachan SM, Kern ER, Prichard MN. Virucidal activity of polysaccharide extracts from four algal species against herpes simplex virus. Antivir Res 2009;83:282–9.

[330] Mohsen MSA, Mohamed SF, Ali FM, El-Sayed OH. Chemical structure and antiviral activity of water-soluble sulfated polysaccharides from *Sargassum latifolium*. J Appl Sci Res 2007;3:1178–85.

[331] Ahmadi A, Moghadamtousi SZ, Abubakar S, Zandi K. Antiviral potential of algae polysaccharides isolated from marine sources: a review. Synthesis 2015;41:2.

[332] Damonte E, Neyts J, Pujol CA, Snoeck R, Andrei G, Ikeda S, Witvrouw M, Reymen D, Haines H, Matulewicz MC, Cerezo A, Coto CE, Clerco E. Antiviral activity of a sulphated polysaccharide from the red seaweed *Nothogenia fastigiata*. Biochem Pharmacol 1994;47(12):2187–92.

[333] Pakker H, Klerk H, Campen JH, Olsen JL, Breeman AM. Evolutionary and ecological differentiation in the pantropical to warm-temperate seaweed *Digenea simplex* (Rhodophyta) 1. J Phycol 1996;32(2):250–7.

[334] Machado LP, Carvalho LR, Young MCM, Cardoso-Lopes EM, Centeno DC, Zambotti-Villela L, Colepicolo P, Yokoya NS. Evaluation of acetylcholinesterase inhibitory activity of Brazilian red macroalgae organic extracts. Rev Bras Farmacogn 2015;25(6):657–62.

[335] Carlucci MJ, Scolaro LA, Noseda MD, Cerezo AS, Damonte EB. Protective effect of a natural carrageenan on genital herpes simplex virus infection in mice. Antivir Res 2004;64:137–41.

[336] Pinteus S, Alves C, Monteiro H, Araújo E, Horta A, Pedrosa R. *Asparagopsis armata* and *Sphaerococcus coronopifolius* as a natural source of antimicrobial compounds. World J Microbiol Biotechnol 2015;31(3):445–51.

[337] Talyshinsky MM, Souprun YY, Huleihel MM. Anti-viral activity of red microalgal polysaccharides against retroviruses. Cancer Cell Int 2002;2(1):8.

[338] Carlucci MJ, Mateu CG, Artuso MC, Scolaro LA. Polysaccharides from red algae: genesis of a renaissance. Life 2012;7:8.

[339] Huheihel M, Ishanu V, Tal J, Arad SM. Activity of *Porphyridium* sp. polysaccharide against herpes simplex viruses in vitro and in vivo. J Biochem Biophys Method 2002;50(2):189–200.

[340] Hasui M, Matsuda M, Okutani K, Shigeta S. In vitro antiviral activities of sulfated polysaccharides from a marine microalga (*Cochlodinium polykrikoides*) against human immunodeficiency virus and other enveloped viruses. Int J Biol Macromol 1995;17(5):293–7.

[341] Raposo MFDJ, de Morais RMSC, Bernardo de Morais AMM. Bioactivity and applications of sulphated polysaccharides from marine microalgae. Mar Drugs 2013; 11(1):233–52.

[342] Bazes A, Silkina A, Douzenel P, Faÿ F, Kervarec N, Morin D, Berge JP, Bourgougnon N. Investigation of the antifouling constituents from the brown alga *Sargassum muticum* (Yendo) Fensholt. J Appl Phycol 2009;21(4):395–403.

[343] Nakashima H, Kido Y, Kobayashi N, Motoki Y, Neushul M, Yamamoto N. Purification and characterization of an avian myeloblastosis and human immunodeficiency virus reverse transcriptase inhibitor, sulfated polysaccharides extracted from sea algae. Antimicrob Agents Chemother 1987;10(31):1524–8.

[344] Wu XJ, Hansen C. Antioxidant capacity, phenolic content, polysaccharide content of *Lentinus edodes* grown in whey permeate-based submerged culture. J Food Sci 2008; 1(73):M1–8.

[345] Wang J, Liu L, Zhang Q, Zhang Z, Qi H, Li P. Synthesized oversulphated, acetylated and benzoylated derivatives of fucoidan extracted from *Laminaria japonica* and their potential antioxidant activity in vitro. Food Chem 2009;114(4):1285–90.

[346] Pulz O, Gross W. Valuable products from biotechnology of microalgae. Appl Microbiol Biotechnol 2004;65:635–48.

[347] Le Tutour B. Antioxidative activities of algal extracts, synergistic effect with vitamin E. Phytochemistry 1990;12(29):3759–65.

[348] Ambat RR, Moi PS, Ravi S, Aswathanarayana RG. Astaxanthin: sources, extraction, stability, biological activities and its commercial applications—a review. Mar Drugs 2014;12(1):128–52.

[349] Wang J, Zhang Q, Zhang Z, Song H, Li P. Potential antioxidant and anticoagulant capacity of low molecular weight fucoidan fractions extracted from *Laminaria japonica*. Int J Biol Macromol 2010;46:6–12.

[350] Sun L, Wang C, Shi Q, Ma C. Preparation of different molecular weight polysaccharides from *Porphyridium cruentum* and their antioxidant activities. Int J Biol Macromol 2009;45:42–7.

[351] Fisch KM, Böhm V, Wright AD, Konig GM. Antioxidative meroterpenoids from the brown alga *Cystoseira crinite*. J Nat Prod 2003;66(7):968–75.

[352] Zubia M, Fabre MS, Kerjean V, Le Lann K, Stiger-Pouvreau V, Fauchon M, Deslandes E. Antioxidant and antitumoural activities of some Phaeophyta from Brittany coasts. Food Chem 2009;3(116):693–701.

[353] Tannin-Spitz T, Bergman M, van Moppes D, Grossman S, Arad SM. Antioxidant activity of the polysaccharide of the red microalga *Porphyridium* sp. J Appl Phycol 2005; 3(17):215–22.

[354] Iwashima M, Mori J, Ting X, Matsunaga T, Hayashi K, Shinoda D, Saito H, Sankawa U, Hayashi T. Antioxidant and antiviral activities of plastoquinones from the brown alga *Sargassum micracanthum*, and a new chromene derivative converted from the plastoquinones. Biol Pharm Bull 2005;2(28):374–7.

[355] Huang H, Wang BG. Antioxidant capacity & lipophilic content of seaweeds collected from the Qingdao coastline. J Agric Food Chem 2004;52(16):4993–7.

[356] Fayaz M, Namitha KK, Murthy KNC, Swamy MM, Sarada R, Khanam S, Subbarao PV, Ravishankar GA. Chemical composition, iron bioavailability and antioxidant activity of *Kappaphycus alvarezzi* (Doty). J Agric Food Chem 2005;53(3): 792–7.

[357] Kumar KS, Ganesan K, Subba Rao PV. Antioxidant potential of solvent extracts of *Kappaphycus alvarezii* (Doty) Doty – an edible seaweed. Food Chem 2008;1(107):289–95.

[358] Devi KP, Suganthy N, Kesika P, Pandian SK. Bioprotective properties of seaweeds: *in vitro* evaluation of antioxidant activity and antimicrobial activity against food borne bacteria in relation to polyphenolic content. BMC Complement Altern Med 2008; 8(38):1–11.

[359] Zaragoz MC, Lopez D, Saiz MP, Poquet M, Perez J, Puig-Parellada P, Marmol F, Simonetti P, Gardana C, Lerat Y, Burtin P, Inisan C, Rousseau I, Besnard M, Mitjavila MT. J Agric Food Chem 2008;56(17):7773–80.

[360] Zou Y, Qian ZJ, Li Y, Kim MM, Lee SH, Kim SK. Antioxidant effects of phlorotannins isolated from *Ishige okamurae* in free radical mediated oxidative systems. J Agric Food Chem 2008;56(16):7001–9.

[361] Je Jae Y, Park P-J, Kim E-K, Park J-S, Yoon Ho-D, Kim K-R, Ahn C-B. Antioxidant activity of enzymatic extracts from the brown seaweed *Undaria pinnatifida* by electron spin resonance spectroscopy. LWT–Food Sci Technol 2009;4(42):874–8.

[362] Ye H, Zhou C, Sun Y, Zhang X, Liu J, Hu Q, Zeng X. Antioxidant activities in vitro of ethanol extract from brown seaweed *Sargassum pallidum*. Eur Food Res Technol 2009;230:101–9.

[363] Cho ML, Kang IJ, Won MH, Le HS, You SG. The antioxidant properties of ethanol extracts and their solvent-partitioned fractions from various green seaweeds. J Med Food 2010;5(13):1232–9.

[364] Novoa AV, Andrade-Wartha ER, Linares AF, Genovese MI, González AE, Vuorela P, Costa A, Mancini-Filho J. Antioxidant activity and possible bioactive components in hydrophilic and lipophilic fractions from the seaweed *Halimeda incrassata*. Rev Bras Farmacogn 2011;21(1):53–7.

[365] Airanthi WAMK, Hosokawa M, Miyashita K. Comparative antioxidant activity of edible Japanese brown seaweeds. J Food Sci 2011;1(76):104–11.

[366] Rocha de Souza MC, Marques CT, Dore CMG, Ferreira da Silva FR, Rocha HAO, Leite EL. Antioxidant activities of sulfated polysaccharides from brown and red seaweeds. J Appl Phycol 2007;2(19):153–60.

[367] Azhaguraj R, Lenin AE, Viswanathan, Sangeetha B, Selvanayagam M. Prediciation of biological activity of algal antitumor drugs using pass. Pharmacologyonline 2010;3: 22–34.

[368] Teas J. The consumption of seaweed as a protective factor in the etiology of breast cancer. Med Hypotheses 1981;5(7):601–13.

[369] Yamamoto I, Takahashi M, Tamura E, Maruyama H, Mori H. Antitumor activity of edible marine algae: effect of crude fucoidan fractions prepared from edible brown seaweeds against L-1210 leukemia. Hydrobiologia 1984;116:145–8.

[370] Parish CR, Coombe DR, Jakobsen KB, Bennett FA, Underwood PA. Evidence that sulphated polysaccharides inhibit tumour metastasis by blocking tumour-cell-derived heparanases. Int J Cancer 1987;4(40):511–8.

[371] Coombe DR, Parish CR, Ramshaw IE, Snowden JM. Analysis of the inhibition of tumour metastasis by sulphated polysaccharides. Int J Cancer 1987;1(39):82–8.

[372] Parish CR, Snowden JM. Lymphocytes express a diverse array of specific receptors for sulfated polysaccharides. Cell Immunol 1985;1(91):201–14.

[373] Nagumo T, Mizui LN, Fujihara M, Himeno J, Komiyama K, Umezawa I. Separation of sulfated, fucose-containing polysaccharides from the brown seaweed *Sargassum kjellmanianum* and their heterogeneity and antitumor activity. Kitasato Arch Exp Med 1988;61(1):59–67.

[374] Matsumoto T, Kitano A, Oshitami N, Obata A, Hiki M, Hashimura H, Okawa K, Nagura H, Kobayashi K. Immunoglobulin-containing cells in the colonic mucosa of rabbits with carrageenan induced colitis. Dis Colon Rectum 1988; 31(9):723–9.

[375] Perl A, Gonzalez-Cabello R, Gergely P. Stimulation of lectin-dependent cell-mediated cytotoxicity against adherent HEp-2 cells by carrageenan. Clin Exp Immunol 1983; 54(2):567–72.

[376] Hiroyuki N, Amano H, Arashima K, Nisizawa K. Antitumor activity of marine algae. Hydrobiologia 1990;1(204):577–84.

[377] Zhuang C, Itoh H, Mizuno T, Ito H. Antitumor active fucoidan from the brown seaweed, umitoranoo (*Sargassum thunbergii*). Biosci Biotechnol Biochem 1995; 4(59):563–7.

[378] Itoh H, Noda H, Amano H, Ito H. Immunological analysis of inhibitor of lung metastases by fucoidan (GIV-A) prepared from brown seaweed *Sargassum thunbergii*. Anticancer Res 1995;15(5B):1937–47.

[379] Sheu JH, Wang GH, Sung PJ, Duh CY. New cytotoxic oxygenated fucosterols from the brown alga *Turbinaria conoides*. J Nat Prod 1999;62(2):224–7.

[380] Barbier P, Guise S, Huitorel P. Caulerpenyne from *Caulerpa taxifolia* has an antiproliferative activity on tumor cell line SK–N–SH and modifies the microtubule network. Life Sci 2001;4(70):415–29.

[381] Numata A, Kanbara S, Takahashi C, Fujuki R, Yoneda M, Fujuta E, Nabashima Y. Cytotoxic activity of marine algae and a cytotoxic principle of the brown algae *Sargassum tortile*. Chem Pharm Bull 1991;8(39):2129–31.

[382] Ayyad SE, Slama M, Mokhtar AH, Anter AF. Cytotoxic bicyclic diterpene from the brown algae *Sargassum crispum*. Boll Chim Farm 2001;140(3):155–9.

[383] Bui LM, Ngo BO, Nguyen ND, Pham TD, Tran VTT. Studies on fucoidan and its production from Vietnamese brown sea weeds. ASEAN J Sci Technol Dev 2005;4(22): 371–80.

[384] Zandi K, Tajbakhsh S, Nabipour I, Rastian Z, Yousefi F, Sharafian S, Sartavi K. *In vitro* antitumor activity of *Gracilaria corticata* (a red alga) against Jurkat and molt-4 human cancer cell lines. Afr J Biotechnol 2010;9(40):6787–90.

[385] Taskin E, Caki Z, Ozturk M, Taskin E. Assessment of in vitro antitumoral and antimicrobial activities of marine algae harvested from the eastern Mediterranean sea. Afr J Biotechnol 2010;27(9):4272–7.

[386] Khanavi M, Nabavi M, Sadati N, Ardekani S, Sohrabipour J, Nabavi SMB, Ghaeli P, Ostad SN. Cytotoxic activity of some marine brown algae against cancer cell lines. Biol Res 2010;1(43):31–7.

[387] Bechelli J, Coppage M, Rosell K, Liesveld J. Cytotoxicity of algae extracts on normal and malignant cells. Leuk Res Treat 2011;2011.

[388] Chen D, Wu XZ, Wen ZY. Sulfated polysaccharides and immune response: promoter or inhibitor? Panminerva Med 2008;50(2):177–83.

[389] Granert C, Raud J, Xie X, Lindquist L, Lindbom L. Inhibition of leukocyte rolling with polysaccharide fucoidin prevents pleocytosis in experimental meningitis in the rabbit. J Clin Investig 1994;93(3):929–36.

[390] Cumashi A, Ushakova NA, Preobrazhenskaya ME, D'Incecco A, Piccoli A, Totani L, Tinari N, Morozevich GE, Berman AE, Bilan MI, Usov AI, Ustyuzhanina NE, Grachev AA, Sanderson CJ, Kelly M, Rabinovich GA, Iacobelli S, Nifantiev NE. A comparative study of the anti-inflammatory, anticoagulant, antiangiogenic, and anti-adhesive activities of nine different fucoidans from brown seaweeds. Glycobiology 2007;17:541–52.

[391] Pomin H. Marine medicinal glycomics. Front Cell Infect Microbiol 2014;5:1–13.

[392] Tissot B, Montdargent B, Chevolot L, Varenne A, Descroix S, Gareil P, Daniel R. Interaction of fucoidan with the proteins of the complement classical pathway. Biochim Biophys Acta Proteins Proteom 2003;1651(1):5–16.

[393] Tissot B, Gonnet F, Iborra A, Berthou C, Thielens N, Arlaud GJ, Daniel R. Mass spectrometry analysis of the oligomeric C1q protein reveals the B chain as the target of trypsin cleavage and interaction with fucoidan. Biochemistry 2005;44(7): 2602–9.

[394] Clement MJ, Tissot B, Chevolot L, Adjadj E, Du Y, Curmi PA, Daniel R. NMR characterization and molecular modeling of fucoidan showing the importance of oligosaccharide branching in its anticomplementary activity. Glycobiology 2010;20: 883–94.

[395] Lee SH, Jeon YJ. Anti-diabetic effects of brown algae derived phlorotannins marine polyphenols through diverse mechanisms. Fitoterapia 2013;86:129–36.

[396] Tsuji RF, Hoshino K, Noro Y, Tsuji NM, Kurokawa T, Masuda AS, Nowak B. Suppression of allergic reaction by lambda-carrageenan: toll-like receptor 4/MyD88-dependent and independent modulation of immunity. Clin Exp Allergy 2003;33(2): 249–58.

[397] Maruyama H, Tamauchi H, Hashimoto M, Nakano T. Suppression of Th2 immune responses by mekabu fucoidan from *Undaria pinnatifida* sporophylls. Int Arch Allergy Immunol 2005;137(4):289–94.

[398] Yang JW, Yoon SY, Oh SJ, Kim SK, Kang KW. Bifunctional effects of fucoidan on the expression of inducible nitric oxide synthase. Biochem Biophys Res Commun 2006; 346:345–50.

[399] Raposo MFDJ, de Morais AMMB, de Morais RMSC. Emergent sources of prebiotics: seaweeds and microalgae. Mar Drugs 2016;14(2):27.

[400] Na HJ, Moon PD, Ko SG, Lee HJ, Jung HA, Hong SH, Seo Y, Oh JM, Lee BH, Choi BW, Kim HM. *Sargassum hemiphyllum* inhibits atopic allergic reaction via the regulation of inflammatory mediators. J Pharmacol Sci 2005;2(97):219–26.

[401] Rozas E, Freitas JC. Anti inflammatory activity of the apolar extract from the seaweed *Galaxaura marginata* (Rhodophyta, Nemaliales). J Venom Anim Toxins Incl Trop Dis 2007;2(13):545.

[402] Kang JY, Khan MNA, Park NH, Cho JY, Lee MC, Fujii H, Hong YK. Antipyretic, analgesic, and anti-inflammatory activities of the seaweed *Sargassum fulvellum* and *Sargassum thunbergii* in mice. J Ethnopharmacol 2008;1(116):187–90.

[403] Kazlowska K, Hsu T, Hou CC, Yang WC, Tsai GJ. Antiinflammatory properties of phenolic compounds and crude extract from *Porphyra dentate*. J Ethnopharmacol 2010;128:123–30.

[404] Yang EJ, Moon JY, Kim MJ, Kim DS, Lee WJ, Lee NH, Hyun CG. Anti-inflammatory effect of *Petalonia binghamiae* in LPS-induced macrophages is mediated by suppression of iNOS and COX-2. Int J Agric Biol 2010;5(12):754–8.

[405] Hong DD, Hien HM, Anh HTL. Studies on the analgesic and anti-inflammatory activities of *Sargassum swartzii* (Turner) C. Agardh (Phaeophyta) and *Ulva reticulata* Forsskal (Chlorophyta) in experiment animal models. Afr J Biotechnol 2011;12(10): 2308–14.

[406] Vazquez AIF, Sánchez CMD, Delgado NG, Alfonso AMS, Ortega YS, Sánchez HC. Anti-inflammatory and analgesic activities of red seaweed *Dichotomaria obtusata*. Braz J Pharm Sci 2011;47:111–8.

[407] Boonchum W, Peerapornpisal Y, Kanjanapothi D, Pekkoh J, Amornlerdpison D, Pumas C, Sangpaiboon P, Vacharapiyasophon P. Antimicrobial and anti-inflammatory properties of various seaweeds from the gulf of Thailand. Int J Agric Biol 2011;1(13):100–4.

[408] Mori J, Hayashi T, Iwashima M, Matsunaga T, Saito H. Effects of plastoquinones from the brown alga *Sargassum micracanthum* and a new chromene derivative converted from the plastoquinones on acute gastric lesions in rats. Biol Pharm Bull 2006; 6(29):1197–201.

[409] Shu MH, Appleton D, Zandi K, AbuBakar S. Anti-inflammatory, gastroprotective and anti-ulcerogenic effects of red algae *Gracilaria changii* (Gracilariales, Rhodophyta) extract. Complement Altern Med 2013;13:61.

[410] Senthil KA, Murugan A. Antiulcer, wound healing and hepatoprotective activities of the seaweeds *Gracilaria crassa, Turbinaria ornata* and *Laurencia papillosa* from the southeast coast of India. Braz J Pharm Sci 2013;4(49):670–8.

[411] Akash MSH, Rehman K, Chen S. Role of inflammatory mechanisms in pathogenesis of type 2 diabetes mellitus. J Cell Biochem 2013;114(3):525–31.

[412] Akash MSH, Shen Q, Rehman K, Chen S. Interleukin-1 receptor antagonist: a new therapy for type 2 diabetes mellitus. J Pharm Sci 2012;101(5):1647–58.

[413] Akash MSH, Rehman K, Tariq M, Chen S. *Zingiber officinale* and type 2 diabetes mellitus: evidence from experimental studies. Crit Rev Eukaryot Gene Expr 2015;25(2): 91–112.

[414] Lee SH, Ko SC, Kang MC, Lee DH, Jeon YJ, Octaphlorethol A. A marine algae product, exhibits antidiabetic effects in type 2 diabetic mice by activating AMP-activated protein kinase and upregulating the express ion of glucose transporter 4. Food Chem Toxicol 2016;91:58–64.

[415] Heo SJ, Hwang JY, Choi JI, Han JS, Kim HJ, Jeon YJ. Diphlorethohydroxycarmalol isolated from *Ishige okamurae*, a brown algae, a potent α-glucosidase and α-amylase inhibitor, alleviates postprandial hyperglycemia in diabetic. Eur J Pharmacol 2009; 615:252–6.

[416] Miyashita K, Mikami N, Hosokawa M. Chemical and nutritional characteristics of brown seaweed lipids: a review. J Funct Foods 2013;5:1507–17.

[417] Rocha HA, Moraes FA, Trindade ES, Franco CR, Torquato RJ, Veiga SS, Valente AP, Mourao PA, Leite EL, Nader HB, Dietrich CP. Structural and hemostatic activities of a sulfated galactofucan from the brown alga *Spatoglossum schröederi*, an ideal antithrombotic agent? J Biol Chem 2005;280(50):41278–88.

[418] Nishino T, Aizu Y, Nagumo T. Antithrombin activity of a fucan sulfate from the brown seaweed *Ecklonia kurome*. Thrombosis Res 1991;6(62):765–73.

[419] Lima JA, Santos ND, Gomes LD, Cordeiro LS, Sabry DA, Costa SL, Freitas DML, Silva BN, Carlos MEB, Telma MA, Lemos M, Leite LE, Rocha HAO. Evaluation of acute and subchronic toxicity of a non-anticoagulant, but antithrombotic algal heterofucan from the *Spatoglossum schröederi* in Wistar rats. Braz J Pharmacogn 2011; 4(21):674–9.

[420] Shi D, Li J, Guo S, Han L. Antithrombotic effect of bromophenol, the alga-derived thrombin inhibitor. J Biotechnol 2008;136:577–S588.

[421] Kang MC, Kang N, Ko SC, Kim YB, Jeon YJ. Anti-obesity effects of seaweeds of Jeju Island on the differentiation of 3T3-L1 preadipocytes and obese mice fed a high-fat diet. Food Chem Toxicol 2016;90:36–44.

[422] Hu X, Tao N, Wang X, Xiao J, Wang M. Marine-derived bioactive compounds with anti-obesity effect: a review. J Funct Foods 2016;21:372–87.

[423] Li ZS, Noda K, Fujita E, Manabe Y, Hirata T, Sugawara T. The green algal carotenoid siphonaxanthin inhibits adipogenesis in 3T3-L1 preadipocytes and the accumulation of lipids in white adipose tissue of KK-Ay mice. J Nutr 2014;145(3):490–8.

[424] Carmeliet P. Angiogenesis in health and disease. Nat Med 2003;6(9):653–60.

[425] Kirk S, Frank JA, Karlik S. Angiogenesis in multiple sclerosis: is it good, bad or an epiphenomenon? J Neurol Sci 2004;2(217):125–30.

[426] Sugawara T, Matsubara K, Akagi R, Mori M, Hirata T. Antiangiogenic activity of brown algae fucoxanthin and its deacetylated product, fucoxanthinol. J Agric Food Chem 2006;54(26):9805–10.

[427] Ganesan P, Matsubara K, Ohkubo T, TanakaY NK, Sugawara T. Anti-angiogenic effect of siphonaxanthin from green alga, *Codium fragile*. Phytomedicine 2010;14(17):1140–4.

[428] Kim YC, An RB, Yoon NY, Nam TJ. Hepatoprotective constituents of the edible brown alga *Ecklonia stolonifera* on tacrine-induced cytotoxicity in hep G2 cells. Arch Pharmacal Res 2005;12(28):1376–80.

[429] Chidambara Murthy KN, Rajesha J, Swamy MM, Ravishankar GA. Comparative evaluation of hepatoprotective activity of carotenoids of microalgae. J Med Food 2005; 8(4):523–8.

[430] Zhao X, Xue CH, Li ZJ, Cai YP, Liu HY, Qi HT. Antioxidant and hepatoprotective activities of low molecular weight sulfated polysaccharide from *Laminaria japonica*. J Appl Phycol 2004;2(16):111–5.

[431] Zhang R, Kang KA, Piao MJ, Ko DO, Wang ZH, Lee IK, Kim BJ, Jeong IY, Shin T, Park JW, Lee NH. Eckol protects V79-4 lung fibroblast cells against '-ray radiation-induced apoptosis via the scavenging of reactive oxygen species and inhibiting of the c-Jun NH2-terminal kinase pathway. Eur J Pharmacol 2008;591(1):114–23.

[432] Parihar MS, Hemnani T. Alzheimer's disease pathogenesis and therapeutic interventions. J Clin Neurosci 2004;5(11):456–67.

[433] Pangestuti R, Kim SK. Neuroprotective properties of chitosan and its derivatives. Mar Drugs 2010;8(7):2117–28.

[434] Stirk W, Reinecke D, Staden JV. Seasonal variation in antifungal, antibacterial and acetylcholinesterase activity in seven South African seaweeds. J Appl Phycol 2007; 3(19):271–6.

[435] Suganthy N, Pandian SK, Pandima Devi K. Neuroprotective effect of seaweeds inhabiting South Indian coastal area (Hare Island, Gulf of Mannar marine biosphere reserve): Cholinesterase inhibitory effect of *Hypnea valentiae* and *Ulva reticulata*. Neurosci Lett 2010;3(468):216–9.

[436] Yoon N, Chung H, Kim H, Choi J. Acetyl and butyrylcholinesterase inhibitory activities of sterols and phlorotannins from *Ecklonia stolonifera*. Fish Sci 2008;1(74): 200−7.

[437] Yoon NY, Lee SH, Yong L, Kim SK. Phlorotannins from *Ishige okamurae* and their acetyl- and butyrylcholinesterase inhibitory effects. J Funct Foods 2009;4(1):331−5.

[438] Choi BW, Ryu G, Park SH, Kim ES, Shin J, Roh SS, Shin HC, Lee BH. Anticholinesterase activity of plastoquinones from *Sargassum sagamianum*: lead compounds for Alzheimer's disease therapy. Phytother Res 2007;5(21):423−6.

[439] Rafiquzzaman SM, Kim EY, Lee JM, Mohibbullah M, Alam B, Moon S, Kim JM, Kong IS. Anti-Alzheimers and anti-inflammatory activities of a glycoprotein purified from the edible brown alga *Undaria pinnatifida*. Food Res Int 2015;77:118−24.

[440] Wijesekara I, Kim SK. Angiotensin-I-converting enzyme (ACE) inhibitors from marine resources: prospects in the pharmaceutical industry. Mar Drugs 2010;8(4): 1080−93.

[441] Cha SH, Lee KW, Jeon YJ. Screening of extracts from red algae in Jeju for potentials marine angiotensin-I converting enzyme (ACE) inhibitory activity. Algae 2006;21(3): 343−8.

[442] Levring T, Tanaka Y. Marine algae in pharmaceutical science, vol. 1. New York: Walter de Gruyter; 1979. p. 694.

[443] Jongaramruong J, Kongkam N. Novel diterpenes with cytotoxic, anti-malarial and anti-tuberculosis activities from a brown alga *Dictyota* sp. J Asian Nat Prod Res 2007;8(9):743−51.

[444] Chay CIC, Cansino RG, Espitia Pinzón CI, Torres-Ochoa RO, Martínez R. Synthesis and anti-tuberculosis activity of the marine natural product caulerpin and its analogues. Mar Drugs 2014;12:1757−72.

[445] San-Martin A, Negrete R, Rovirosa J. Insecticide and acaricide activities of polyhalogenated monoterpenes from Chilean *Plocamium cartilagineum*. Phytochemistry 1991; 7(30):2165−9.

[446] Maeda M, Kodama T, Tanaka T, Yoshizumi H, Takemoto T,Nomoto K, Fujita T. Structures of isodomic acids A, B and C novel insecticidal amino acids from the red alga Chonriaarmata. Chem Pharm Bull 1986;11(34):4892−5.

[447] Shibata T, Fujimoto K, Nagayama K, Yamaguchi K, Nakamura T. Inhibitory activity of brown algal phlorotannins against hyaluronidase. Int J Food Sci Technol 2002;37: 703−9.

[448] Moreau J, Pesando D, Bernard P, Caram B, Pionnat JC. Seasonal variations in the production of antifungal substances by some dictyotales (broouwn algae) from the french mediterranean coast. Hydrobiologia 1988;2(162):157−62.

[449] Peres JCF, Retz de Carvalho L, Gonçalez E, Berian LOS, Dárc Felicio J. Evaluation of antifungal activity of seaweed extracts. Ciência e Agrotecnologia 2012;3(36):294−9.

[450] Topcu G, Aydogmus Z, Imre S, Gören AC, Pezzuto JM, Clement JA, Kingston DGI. Brominated sesquiterpenes from the red alga *Laurencia obtuse*. J Nat Prod 2003; 66(11):1505−8.

[451] Etahiri S, Poncé VB, Caux C, Guyot M. New bromoditerpenes from the red alga *Sphaerococcus coronopifolius*. J Nat Prod 2001;64(8):1024−7.

[452] Ginzberg A, Cohen M, Sod-Moriah UA, Shany S, Rosenshtrauch A, Arad SM. Chickens fed with biomass of the red microalga *Porphyridium* sp. have reduced blood cholesterol levels and modified fatty acids composition in egg yolk. J Appl Phycol 2000;12:325−30.

[453] Dvir I, Chayoth R, Sod-Moriah U, Shany S, Nyska A, Stark AH, Madar Z, Arad SM. Soluble polysaccharide of red microalga *Porphyridium* sp. alters intestinal morphology and reduces serum cholesterol in rats. Br J Nutr 2000;84:469–76.

[454] Dvir I, Stark AH, Chayoth R, Madar Z, Arad SM. Hycholesterolemic effects of nutraceuticals produced from the red microalga *Porphyridium* sp. in rats. Nutrients 2009;1:156–67.

[455] Chen B, You B, Huang J, Yu Y, Chen W. Isolation and antioxidant property of the extracellular polysaccharide from *Rhodella reticulata*. World J Microbiol Biotechnol 2010;26:833–40.

[456] Oakenfull D. Physicochemical properties of dietary fiber: overview. In: Cho SS, Dreher MD, editors. Handbook of dietary fibers. New York (NY, USA): Marcel Dekker Inc.; 2001. p. 195–206.

[457] Wijesekara I, Pangestuti R, Kim S-K. Biological activities and potential health benefits of sulfated polysaccharides derived from marine algae. Carbohydr Polym 2011;84:14–21.

[458] Raghavendran HR, Sathivel A, Devaki T. Effect of *Sargassum polycystum* (Phaeophyceae) sulphated polysaccharide extract against acetaminophen-induced hyperlipidemia during toxic hepatitis in experimental rats. Mol Cell Biochem 2005;276:89–96.

[459] Josephine A, Veena CK, Amudha G, Preetha SP, Varalakshmi P. Protective role of sulphated polysaccharides in abating the hyperlipidemic nephropathy provoked by cyclosporine A. Arch Toxicol 2007;81(5):371–9.

[460] Vázquez-Freire MJ, Lamela M, Calleja JM. Hypolipidaemic activity of a polysaccharide extract from *Fucus vesiculosus* L. Phytother Res 1996;8(10):647–50.

[461] Huang L, Wen K, Gao X, Liu Y. Hypolipidemic effect of fucoidan from *Laminaria japonica* in hyperlipidemic rats. Pharm Biol 2010;4(48):422–6.

[462] Pengzhan Y, Ning L, Xiguang L, Gefei Z, Quanbin Z, Pengcheng L. Antihyperlipidemic effects of different molecular weight sulfated polysaccharides from *Ulva pertusa* (Chlorophyta). Pharmacol Res 2003;6(48):543–9.

[463] Ono SJ. Molecular genetics of allergic diseases. Annu Rev Immunol 2000;18:347–66.

[464] Li Y, Lee SH, Le QT, Kim MM, Kim SK. Anti-allergic effects of phlorotannins on histamine release via binding inhibition between IgE and Fc RI. J Agric Food Chem 2008;56(24):12073–80.

[465] Le QT, Li Y, Qian ZJ, Kim MM, Kim SK. Inhibitory effects of polyphenols isolated from marine alga *Ecklonia cava* on histamine release. Process Biochem 2009;2(44):168–76.

[466] Na HJ, Moon PD, Lee HJ, Kim HR, Chae HJ, Shin T, Seo Y, Hong SH, Kim HM. Regulatory effect of atopic allergic reaction by *Carpopeltis affinis*. J Ethnopharmacol 2005;101(1):43–8.

[467] Jung WK, Choi I, Oh S, Park SG, Seo SK, Lee SW, Lee DS, Heo SJ, Jeon YJ, Je JY, Ahn CB. Anti-asthmatic effect of marine red alga (*Laurencia undulata*) polyphenolic extracts in a murine model of asthma. Food Chem Toxicol 2009;47(2):293–7.

[468] Kim SK, Lee DY, Jung WK, Kim JH, Choi I, Park SG, Seo SK, Lee SW, Lee CM, Yea SS, Choi YH. Effects of Ecklonia cava ethanolic extracts on airway hyperresponsiveness and inflammation in a murine asthma model: role of suppressor of cytokine signaling. Biomed Pharmacother 2008;5(62):289–96.

[469] Asada M, Sugie M, Inoue M, Nakagomi K, Hongo S, Murata K, Irie S, Takeuchi T, Tomizuka N, Oka S. Inhibitory effect of alginic acids on hyaluronidase and on histamine release from mast cells. Biosci Biotechnol Biochem 1997;6(61):1030–2.

[470] Remirez D, Ledón N, González R. Role of histamine in the inhibitory effects of phycocyanin in experimental models of allergic inflammatory response. Mediat Inflamm 2002;11:81–5.

[471] Suzuki M, Yamada H, Kurata K. Dictyterpenoids A and B two novel diterpenoids with feeding-deterrent activity from the brown alga *Dilophus okamurae*. J Nat Prod 2002; 65(2):121–5.

[472] Barbosa JP, Teixeira VL, Pereira RC. A dolabellane diterpene from the brown alga *Dictyota pfaffii* as chemical defense against herbivores. Bot Mar 2004;2(47):147–51.

[473] Guzmán-Murillo MA, Ascencio F. Anti-adhesive activity of sulphated exopolysaccharides of microalgae on attachment of the red sore disease-associated bacteria and *Helicobacter pylori* to tissue culture cells. Lett Appl Microbiol 2000;30:473–8.

[474] Ofek L, Beachery EH, Sharon N. Surface sugars recognition in bacterial adherence. Trends Biochem Sci 1978;3(3):159–60.

[475] Ascencio F, Fransson LA, Wadstrom T. Affinity of the gastric pathogen *Helicobacter pylori* for the N-sulphated glycosaminoglycan heparin sulphate. J Med Microbiol 1993;38:240–4.

[476] Arad SM, Weinstein J. Novel lubricants from red microalgae: interplay between genes and products. J Biomed (Isr) 2003;1:32–7.

[477] Arad SM, Rapoport L, Moshkovich A, van Moppes D, Karpasan M, Golan R, Golan Y. Superior biolubricant from a species of red microalga. Langmuir 2006;2:7313–7.

[478] Gasljevic K, Hall K, Chapman D, Matthys EF. Drag-reducing polysaccharides from marine microalgae: species productivity and drag reduction effectiveness. J Appl Phycol 2008;20:299–310.

[479] Wetherbee R, Lind JL, Burke J, Quatrano RS. The first kiss: establishment and control of initial adhesion by raphid diatoms. J Phycol 1998;34:9–15.

CHAPTER

Production and Processing of Algal Biomass

7

Ayesha Shahid[1], Aqib Zafar Khan[1], Tianzhong Liu[2], Sana Malik[1], Ifrah Afzal[1], Muhammad A. Mehmood[1]

Government College University Faisalabad, Faisalabad, Pakistan[1]; Microalgae Biotechnology Centre, Qingdao Institute of Bio-energy & Bioprocess Technology, Chinese Academy of Sciences, Qingdao, People's Republic of China[2]

7.1 NUTS AND BOLTS OF ALGAL CULTIVATION

Microalgae are believed to have approximately 350,000 species [1], and different strains of the same species can behave differently under the same growth conditions [2]. Therefore, to utilize the full potential of this bioresource for biotechnological applications, the most important aspect is to explore its vast natural biodiversity and isolate the potential strains for any desired industrial or environmental applications. For any application, we need to establish selection criteria to isolate potential strains. For instance, from biodiesel production point of view, the ideal algal strain should have higher lipid productivity, higher CO_2 fixation capacity, should require limited supply of nutrients, have higher photosynthetic efficiency, should provide resistance to predators [3], and should be easy for harvesting and extraction of lipids. Moreover, the strain of choice should be able to tolerate the shear stresses of medium flow with an ability to resist the inversion of wild strains in open pond systems [4,5]. Moreover, it should have the capability to tolerate a range of temperatures and irradiations resulting from the diurnal rhythms and seasonal variations. Moreover, it must provide valuable coproducts and have a fast and efficient productivity cycle [6].

Any strain which appears with aforementioned features can be declared as "ideal strain" but unfortunately, till date, almost all the oleaginous microalgae isolated and used are single cellular algal species. No such unicellular algal strain has been discovered which can meet all the criteria. However, Wang [7] and Guo [8] recently reported a novel filamentous *Tribonema* sp. which was able to accumulate more than 50% lipid oil. Its filamentous structure, with an average length of 3 mm, makes it resistant to predators and an easy harvest by simple filtration, sedimentation, or flotation. However, for future applications, the concept of metabolic engineering of microalgae strains to make them possess the abovementioned properties seems really attractive [3,6]. Microalgae are being engineered to develop a technically efficient and straightforward process of algal biomass utilization in terms of economic and environmental feasibility [3].

Algae Based Polymers, Blends, and Composites. http://dx.doi.org/10.1016/B978-0-12-812360-7.00007-0

The most important factor that influences biofuel production using microalgae is the lipid productivity. Therefore, extensive algal biofuel research is focused on developing strains with higher lipid content [9]. The average lipid content of various microalgal species ranges from 1% to 70% but some species, e.g., *Nannochloropsis* spp. and *Botryococcus braunii* can accumulate up to 80% of lipid as their dry cell mass under different conditions [10–12]. Lipid content of microalgae can be increased by optimizing the growth conditions, such as temperature, light irradiance, and most importantly nutrient requirements, and together these factors affect both lipid content and composition in many algae [13–16]. For instance, nonpolar triacylglycerols (TAGs) tend to accumulate under high radiation, while low radiations stimulate mainly polar lipids (glycolipids or phospholipids). Lipid content is very often connected to a decrease in biomass production [17].

The highest lipid productivity of microalgae is associated with optimum CO_2 concentrations. At lower concentrations of CO_2, the growth of microalgae will be inhibited due to limited carbon supply. It is shown that various microalgae display maximum growth and high intracellular lipid content with increasing temperatures [18]. On the other hand, certain microalgae have shown to hyper-accumulate lipids under stress, such as nitrogen and nutrient deficiencies [19]. For instance, lipid content of diatom *Navioua pelliculosa* was increased up to 60% under silicon deficiency. Salt content may also influence the lipid content of certain microalgae, and some algal species grown in saltwater have shown to accumulate higher lipid content than those grown in low salt content [2]. However, more efficient and favorable TAGs are produced under nitrogen deficiency, which may influence the fatty acid chain length and saturation for biofuel transformation [18] to a favorable extent. *Chlorella pyrenoidosa* had shown to accumulate up to 85% lipid content of its biomass under nitrogen deficiency [20]. An increase in lipid and carbohydrate content was also observed in *Isochrysis* during nitrogen deficiency [21]. It is found that lipid content of *Spirulina* species increased threefold at low nitrogen concentrations [22] compared with the one grown in standard growth media.

Interestingly, different microalgae and diatoms respond differently to nitrogen deficiency. Unlike green microalgae, diatoms [23] and *Tribonema* sp. [8] do not respond to nitrogen deficiency in the similar fashion. Green microalgae show different responses from high lipid content (as in the case of *C. pyrenoidosa*) to no distinct change (as in the case of *Dunaliella* spp.) [24]. Even some species of the same genus were shown to accumulate neutral lipids while others accumulated starch in response to nitrogen deficiency [14]. Therefore, detailed genomic and transcriptomic analyses of such microalgae are required to completely understand the factors affecting their response to exploit these strains to their full potential. In addition to nitrogen, the effect of carbon on biodiesel production under stress was also studied. It was revealed that the biomass content can be increased when provided with glucose as organic carbon source in lower amounts during mixotrophic conditions. Whereas, higher lipid content production and accumulation is observed when glucose supplement is provided during nutrient deprivation conditions [25].

Moreover, nutrients are not the only factors that influence lipid content and composition, rather cultivation methods such as fed batch, batch and semicontinuous culture [26,27] may also influence the biomass composition. The lipid productivity and high growth rates were achieved by using semicontinuous culture strategy through controlled pH and limited nitrogen, which was four times higher than the batch culture strategy [28]. In addition to cultivation methods, pH is another factor affecting the lipid content. For instance, microalgae cultures (e.g., *Chlorella* sp. and *Scenedesmus* sp.) showed maximum growth and high TAG accumulation rate when grown at high pH [11,29–31].

7.2 ALGAL BIOMASS PRODUCTION SYSTEMS

There are various methods established recently to enhance the growth rate and biomass production of the microalgae. Few of them are discussed here.

7.2.1 OPEN POND CULTIVATION

It is the most common method for microalgae cultivation [32], which is based on the photosynthetic fixation of atmospheric carbon. If CO_2 is the only carbon source, this method is feasible for algal cultivation near power plants or factories where it can get larger CO_2 supply [9], and remove nutrients from the industrial wastewater prior to sending it to wastewater treatment facility. Although, oil productivity is lower, it is still a method of choice among outdoor cultivation systems due to ease of use [5]. Open pond cultivation is in use since 1950 [4] and natural open ponds are often referred to as cultivation systems, however, advanced forms of it also exist namely raceway ponds, tanks, shallow bio-ponds, and circular ponds [6].

Purpose-oriented open ponds are known as "Raceway Ponds" and are commonly used for commercial microalgae cultivation (Fig. 7.1). These are man-made, shallow, oval-shaped ponds, internally lined with cement, PVC, or clay, having depth of approximately 0.25–0.4 m. Baffles are used for dividing the pond [33]. Mechanical paddle wheels are used for efficient aeration and thorough mixing of nutrients for uniform biomass growth [34]. Raceway ponds at least require an area of 1.0 ha at least, while extensive ponds require approximately 200 ha area [35]. Additional nutrient supply is required for the growth of algae in the form of suspension [36]. These are generally operated by natural light source, but some efforts were also made to use artificial light as energy source for enhanced biomass production. However, artificial light source do not contribute to enhance the growth [5] significantly, instead it raises the cost of production. Environmental conditions and algae strains affect the choice of system. It is widely reported that raceway ponds can maintain cell concentration of 0.5 g/L with productivity of 25 $g^{-2}day^{-1}$ [37]. Interestingly, raceway ponds were shown to be more promising when used for large-scale algal cultivation [4]. Microalgae such as *Dunaliella salina*, *Nannochloropsis*, and *Chlorella* sp. are the most suitable choices for such system [38].

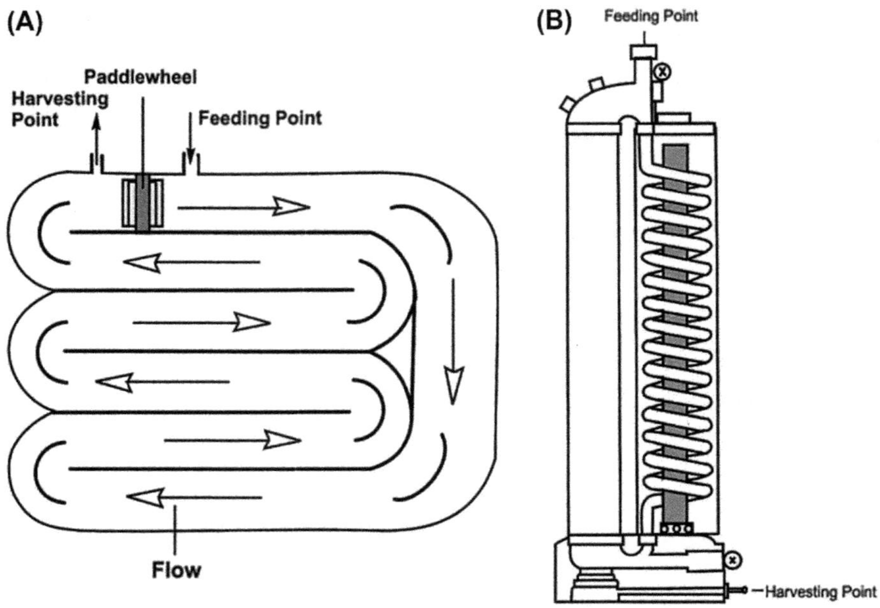

FIGURE 7.1

Schematic diagram of commonly used photobioreactors. (A) Raceway pond, an example of open production system and (B) vertical photobioreactor, an example of closed production system.

Open systems have a number of advantages, e.g., ease of use, cost-effectiveness (both in production and operation), easy to construct (on flat surface) [34], where a nonarable land can be productively used for the pond, and easy to maintain and clean after cultivation [39]. Despite being cheap, these ponds face several challenges including less environmental control, inefficient CO_2 and light dispersion, higher evaporation losses due to larger surface area and lower depth [40], and higher contamination and cell mortality risks due to presence of microbes and in case of bright days, respectively, followed by lower biomass production [34]. However, these limitations can be overcome by using mass cultivation systems [41]. Moreover, advancements in the design of open ponds can be implied to make it more productive to address the challenges regarding contamination risks and lower biomass and lipid productivity.

7.2.2 CLOSED PHOTOBIOREACTOR SYSTEMS

To overcome the drawbacks of open pond system, focus turned toward the closed production systems [10]. This system uses transparent recipient enclosed cultures in which water circulates with the help of a pumping system. Common types of closed photobioreactors (PBRs) include flat panel, tubular system, vertical columns and internally illuminated system, or fermenter-like systems. These are available in varied shapes and sizes [41]. Most preferable and popular photobioreactor is vertical due to its advantages over other types [42] (Fig. 7.1). PBRs can be operated in both outdoor and indoor environments by using natural, artificial, or combination of both types of lights as energy source [43]. Maintenance of light wavelength and intensity is highly important due to its impact on lipid profile and on lipid productivity and biomass production [44]. Closed PBRs have been used for cultivation of various microalgae such as *Chlorella vulgaris*, *Nannochloropsis* sp., *Phaeodactylum tricornutum*, *Haematococcus pluvialis*, and *Chlorella sorokiniana* [38]. Closed PBRs have shown to be more efficient than open ponds in terms of biomass production owing to their higher ability to capture radiant energy, optimal cultivation area use, variable values of energy consumption for mixing, and mass transfer of gas or liquid. However, these values vary depending on the PBR type [45].

In addition to high efficiency, the advantages of closed PBRs include more control on environment and growth conditions, such as CO_2, light intensity, temperature, and pH; least contamination risks; and continuous operation [41]. However, the disadvantages include difficulty in scaling up, maintenance of optimal conditions, mixing of algae with water, bio-fouling, overheating, oxygen accumulation, and cell damage due to shear stress [9]. Solving these problems may increase the cost of operation which makes it less attractive for use at commercial scale [35]. However, closed production systems are more popular for microalgae cultivation as compared to natural or open cultivation systems (Fig. 7.2) worldwide [46].

However, some advancement is still required in design and growth conditions. Optical fibers are used to achieve better lighting system, which carry sunlight from outdoor to indoor system [43]. For artificial light, light emitting diodes (LEDs) are preferred over regular fluorescent or tungsten bulbs as they reduce power consumption and heat generation while emitting specific light wavelength to

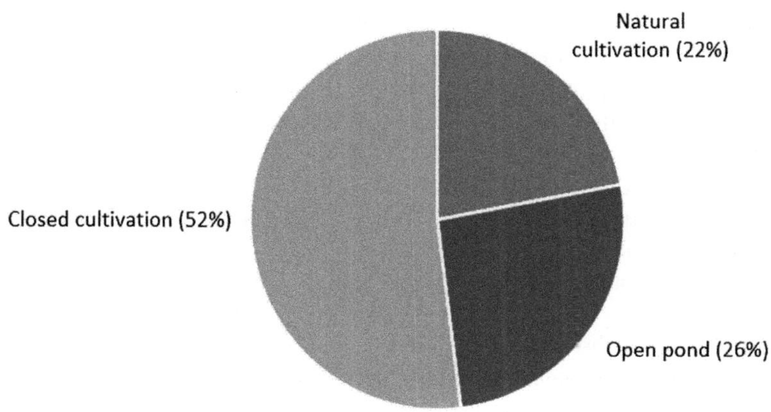

FIGURE 7.2

Comparison of worldwide usage of microalgae cultivation technologies [46].

increase the efficiency of process [41]. To achieve large algae culture production, a hybrid system was suggested that combined PBRs with open cultivation system. In this system, growth is done in two stages. In first stage, cells present in PBRs are allowed to maximize in number under nitrogen deficiency, and are then transferred to open raceway pond under stress during second stage. In the second stage, nitrogen deficiency allows lipids to be accumulated in the cells [17,40].

7.2.3 HYBRID CULTIVATION SYSTEM

To overcome the problems of open and closed cultivation systems, a two-stage hybrid system was introduced for microalgae [6]. This system combines the growth stages from open ponds and closed PBRs. In the first stage, microalgae are allowed to grow under controlled conditions in PBRs to achieve maximum biomass while minimizing the contamination risks. In the second stage, microalgae are directed to raceway ponds and subjected to nutrient stress conditions to enhance lipid production [47].

Generally, hybrid systems are of two typologies: the *covered open pond* and *inflated and widened tubular design*. In *covered open pond* gaseous space above the culture channel is protected from surrounding environment by a cover to avoid contamination by dust, and prevent rain and animals from entering into culture. The *inflated and widened tubular design* is partially filled and is approximately the size of the covered open pond [48]. However, various designs of hybrid systems have been developed by different companies to reach the optimum yield. These designs provide additional benefits to open and closed production systems such as conversion of organic matter up to 99% and production of direct ethanol from biomass [48].

7.2.4 HETEROTROPHIC CULTIVATION

In the absence of light, microalgae also utilize organic carbon as source of energy for their growth, known as *heterotrophic growth*. Heterotrophic cultivation can address the problems related to light quality, light intensity, and absorption inefficiency which limit biomass production. Higher lipid productivity and biomass production can be achieved by using this approach [49]. It is reported that this approach can produce 20 times higher amount of lipid than that obtained by phototrophic cultivation [50]. A variety of carbon sources such as glucose, acetate, glycerol, fructose, sucrose, lactose, galactose, and mannose can be used for heterotrophic cultivation. However, cheaper organic sources such as corn powder hydrolyzate can be used to obtain higher biomass and lipid production [5]. Food waste hydrolyzate is also suitable for this type of cultivation; for example, *Schizochytrium mangrovei* and *C. pyrenoidosa* have shown best growth and high lipid content on food waste media. Use of food and other related industrial wastages also helps to reduce the cost of production during mass cultivation [51]. Owing to its contained cultivation method, minimum risk of contamination is another advantage of heterotrophic cultivation over phototrophic mode [52].

However, several problems may occur during heterotrophic mode of cultivation. For example, usage of carbon sources may incur additional cost, which can be avoided by using industrial and municipal wastewater. Glycerol is a major by-product of biodiesel industry, which can be recycled to produce algal biomass, if the selected strains are capable of the efficient use of glycerol as a carbon source. Sometimes, dark periods during heterotrophic growth influence the production of some valuable phytochemicals in algal biomass, subsequently lowering the economic value of biomass produced [9]. In spite of closed cultivation, heterotrophic culture can be contaminated due to the presence of sugars. Moreover, high amount of carbon dioxide may be produced during large-scale cultivation [53]. However, despite these challenges, heterotrophic cultivation is being used in various areas of research for production of high value compounds, in wastewater cleanup, and for biodiesel production [53,54].

7.2.5 MIXOTROPHIC CULTIVATION

Mixotrophic cultivation is a combination of both phototrophic and heterotrophic cultivations, in which the growth is supplemented with organic carbon sources along with the phototrophic conditions [55]. This cultivation fixes the produced carbon dioxide, which overcomes the problem of heterotrophic cultivation, and increases the growth rate of microalgae [56]. However, mixotrophic cultivation is a complex process, where light intensity and temperature are critical and strongly control the content and composition of lipids and fatty acid. This mode of cultivation shortens the growth cycle and helps to minimize the biomass loss due to dark periods. Microalgae such as *Nannochloropsis oculata*, *C. sorokiniana*, and *C. vulgaris* have shown higher lipid content under mixotrophic conditions [57,58]. Moreover, this cultivation system also lowers the chlorophyll content in algal biomass, which reduces the downstream processing of biomass for biodiesel production [59]. Interestingly, mixotrophic cultivation also addresses the light-induced problems, such as photoinhibition and photooxidation damages, which commonly occur during phototrophic cultivation in closed system [60]. However, low energy conversion efficiency and contamination risks are the major drawbacks of mixotrophic cultivation. In spite of these problems, this mode of cultivation still improves energy economics as compared to autotrophic growth [53,61].

7.3 ALGAL BIOMASS HARVESTING

Harvesting process accounts for approximately 20%–30% of total cost involved in algal biomass production [62]. Almost, 90% of equipment cost is due to harvesting and dewatering procedures during open system algae growth [63]. Therefore, efficient and cost-effective harvesting methods are required to be established. Harvesting methods such as centrifugation, filtration, sedimentation, and ultrafiltration in addition to flocculation are commonly used [12]. These are mostly used in combination to increase the efficiency of the process. To date, no single harvesting method is developed that can be applied universally; therefore, it is an active area of research [9]. The following criteria should be considered for designing an effective and efficient harvesting technique:

1. Characteristics of selected microalgae and type of desired end product
2. Shortcomings of various harvesting methods (Table 7.1)
3. Recycling of nutrients and water, along with the dilute suspension with complete cell separation
4. Minimum interferences with the downstream processing, such as extraction of lipid [64]

Various harvesting methods are implemented nowadays depending upon the algal species being cultivated, pH of the media, nutrients, environmental conditions, choice of medium, and available resources. Method of choice will be the one which provides maximum efficiency in a cost-effective and environmental-friendly manner. The most common methods are described in the following sections (see Table 7.2 for comparative analyses of various methods).

7.3.1 FLOCCULATION

Flocculation is a cost-effective and efficient harvesting method, which can be used for large-scale harvesting of various microalgae [71,72]. In this method, microalgae cells collide and adhere to form an aggregate (*floc*) due to the addition of various flocculants (synthetic or natural) into the growth media [73,74]. As a result, a microalgae paste containing almost 20% of dry matter is produced. It is suggested that the use of two-step harvesting, i.e., flocculation pre-concentration at first step followed by dewatering in the second step (Fig. 7.3), can lower the cost [72].

Table 7.1 Efficiency of Different Combined Harvesting Methods [64]

Combination	Performance Efficiency	References
Centrifugation and filtration	Can save 90% of broth by using submerged filtration for pre-concentration	[65]
Centrifugation and bio-flocculation	Can save 90% of energy during centrifugation using bio-flocculation	[66]
Filtration and inorganic flocculation	Use of inorganic flocculent polyaluminum chloride (PAC) improved the solid concentration up to 34% w/w by reducing the moisture content	[67]
Flotation and inorganic flocculation	Flotation efficiency of cells and allogenic organic matter was enhanced by metallic coagulants; flotation efficiency may be enhanced up to 99%	[68,69]
Ultrasound and inorganic flocculation	*Microcystis aeruginosa* removal efficiency increased from 35% to 67% by using 0.5 mg of PAC for 1–5 s of ultrasonic irradiation	[70]

Table 7.2 Comparison of Different Harvesting Methods

Harvesting Method	Solid Output Concentration (%)	Advantages	Disadvantages
Centrifugation	10–22	Common, reliable, compatible to many species, time-efficient, higher cell recovery (90%)	Energy-intensive, cell damage due to extensive forces, higher cost
Sedimentation	0.5–3.0	Low cost, energy efficient, can separate larger size particles with high density	Cannot be used for smaller size algae, slow rate
Filtration	2–27	Simple, continuous harvesting, maximum separation efficiency (70%–80%), no chemicals required	Membrane fouling/clogging, intensive maintenance, higher operational cost
Flocculation	3–8	Efficient, low cost, applicable to different microalgae species and for mass scale	Require additional methods, require flocculants
Flotation	7	Use of natural properties, flexibility, suitable for unicellular microalgae, less reactive time, versatility	May be effectively (depending on variant), additional chemical may require, contamination risks in some cases
Ultrasound	–	No shear or stress damage and stress, non-fouling, applicable to various algae species	Quite expensive, not suitable for mass scale

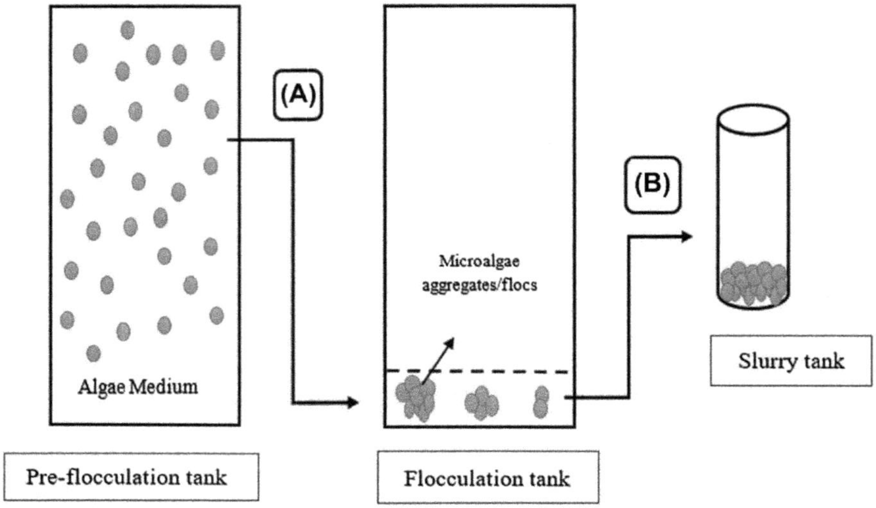

FIGURE 7.3

Schematic diagram of harvesting via flocculation method. (A) Flocculation occurs to aggregate microalgae cells and (B) microalgae paste obtained.

There are two flocculation approaches, namely *combined flocculation* and *auto-flocculation*. In combined flocculation, different flocculants, including inorganic and polymeric, are used in combination. Moreover, microalgae are treated with oxidants prior to flocculation to improve flocculation efficiency and reduce the requirement of inorganic flocculants [75]. In some microalgae species, cells can produce floc naturally in response to pH changes, environmental stress, and changes in dissolved oxygen levels and nitrogen. This type of flocculation is termed as auto-flocculation, but it is not reliable and is a very slow process [62]. It is widely reported in micro-organisms. *C. vulgaris* JSC-7 [76], *Ettlia texensis* (SAG79.80), *Ankistrodesmus falcatus* (SAG202-9) [66], and *Scenedesmus obliquus* AS-6-1 [77] are some self-flocculating microalgae.

It is necessary to understand the genetic basis of flocculation. This process is poorly understood in microalgae [78], however, biochemical basis have shown that polysaccharides biosynthesized by *C. vulgaris* JSC and *S. obliquus* AS-6-1 are responsible for self-flocculation of these strains [76,77]. Another study in *E. texensis* (SAG79.80) proposed the involvement of glycoprotein in self-flocculation [79]. Self-flocculation of microalgae can be achieved by overexpression of flocculating agent biosynthesis genes. In past few years, genetic manipulation methods and systems have been developed for different microalgae strains [80].

Spontaneous flocculation in natural microalgae blooms is termed as *bio-flocculation*. It is assumed that extracellular polymeric substances are responsible for this activity [3]. Fungi and bacteria may also induce bio-flocculation. In case of fungi, interaction of positively charged fungus hyphae with negatively charged microalgae cell surface results in flocculation [81]. It can also be induced by specific bacterial consortia [64]. This process helps to avoid the chemical contamination during harvesting but may cause microbiological contamination which can interfere with food and feed applications of microalgae [82]. *Chlorella* can be flocculated by bio-flocculation in which bacterial culture is used to induce flocculation [62,83,84]. *Bacillus cereus* and *Pseudomonas stutzeri* containing microbial mixture had been successfully used for harvesting of *Pleurochrysis carterae* (CCMP647) [64].

However, none of the flocculants or flocculation methods is applicable to all types of microalgae. Marine microalgae flocculation is still unresolved at industrial scale. Combination of flocculation with sedimentation requires lowest energy input [62]. Magnetite nanoparticles (Fe_3O_4) were used for induction of flocculation and separation of microalgae cells present in media [71]. These nanoparticles are used in either nonfunctionalized form [85] or functionalized form (surface coating of nanoparticles with cationic functional group). Functional nanoparticles can be removed and reused by adjusting pH [86].

7.3.2 ULTRASONIC AGGREGATION

Microalgae can also be harvested via ultrasound-induced aggregation. It can also be used for oil extraction in downstream processing [64]. Combination of ultrasound with enhanced sedimentation is considered as separation technique. Advantages include smaller footprint, non-fouling, continuous operation, absence of mechanical failure, and no shear and stress damage due to low sound pressure and higher frequency of ultrasound [70]. It is performed in a resonating chamber which also consists of reflector and transducer [87]. Microalgae culture is continuously pumped into the resonating chamber. Fields of maximum and minimum potential energy are created when the apparatus is turned on [88] which drive the cells into a zone of minimum acoustic potential energy called node plane. Particle–particle and acoustic interactions cause accumulation of cells. After this, ultrasonic wave field is nullified; as a result large aggregates of microalgae go to sedimentation due to gravitation, and the process is repeated until maximum yield is reached [87]. This method can be used to harvest 92% algal biomass with 90% efficiency having concentration factor of 11 and flow rates of 4–6 L/day. Low particle density and small size of cells result in unfruitful attempts of high harvesting efficiency. Concentration factor is greatly influenced by biomass concentration, feed flow rate, and ratio between feed flow and harvesting [87]. It is suggested that this method can be applied to various algal species; however, ultrasound aggregation–sedimentation method may not be feasible for large-scale harvesting due to higher energy requirements [89].

7.3.3 FLOTATION

Flotation is an effective approach to harvest microalgae by gravity, in which attachment of gas or air bubbles with solid particles carry the cells to liquid surface

[90]. This approach has several advantages including the use of natural microalgae properties such as self-floating and low density, small footprint, moderate operational cost [91], less reaction time, flexibility of usage, higher efficiency than sedimentation, and suitability to harvest smaller unicellular microalgae with an average diameter up to 500 μm [92]. In some cases, the process is facilitated by using chemical or natural collectors. For instance, *N*-acetyl-*N*-*N*-*N*-tri-methyl ammonium bromide (CTAB) and sodium dodecyl sulfate (SDS) are frequently employed, while naturally methylated egg ovalbumin is considered good as natural inducer of flotation [90,93].

Flotation methods are often divided into three categories based on the bubble size, namely *electrolytic flotation*, *dispersed air flotation*, and *dissolved air flotation* [5]. Electrolytic flotation is effective for different microalgae but is energy−intensive, and therefore, it is not considered as method of choice. However, it is considered as effective method for salted or marine water species [72] due to its safety, environmental compatibility, cost and energy efficiency, selectivity, and versatility. On the other hand, system damages due to high temperature and fouling of cathodes are the major challenges [92].

Another variation of flotation method is dispersed air flotation in which bubbles of approximately 700−1500 mm size are formed by using a high-speed mechanical agitator and air injection system. These bubbles interact with negatively charged algal cell surface [5] to induce the flotation process. Overall effectiveness of the process depends on, and may be enhanced by, creating net positive charge on bubbles with the addition of chemicals or cationic surfactants [92]. This system cannot be applied to open pond system due its higher cost and contamination risks [92,94].

Flotation through dissolved air is the third approach. In this process 10−100 μm bubbles are produced by reducing water stream pressure which is already presaturated with air [92]. These bubbles carry suspended solids when they rise to the surface from where they can be separated. Particles floating rate, pressure of tank, hydraulic retention time, and rate of recycling are some important factors which determine microalgae harvesting by dissolved air flotation method [5,72]. This method is more suitable for the treatment of industrial effluents. Zhou [95] reported electro-flotation of *Chlorella* sp. assisted with first two steps of flocculation by chitosan and then sequential bubbling flotation by nonsacrificial graphite electrodes. This method was shown to be cost-effective in terms of power saving and the algal biomass collected is free of any residual metal ions such as Al^{3+} or Fe^{3+}, emitted by sacrificial aluminum and steel anodes, respectively. Around 90% of total recovery efficiency and 50 times concentration rate was reached, and only 23.7 g chitosan and 0.43 kWh of electricity was required to harvest 1 kg biomass.

7.3.4 CENTRIFUGAL SEDIMENTATION

It is the most commonly used and reliable harvesting method applicable to majority of microalgae species to obtain algae concentrates with extended shelf lives [83] quickly compared to other methods. But the process is energy-intensive with higher operational and establishment costs, also the cell structures are damaged due to higher gravitational and shear forces [5]. Based on various centrifugation conditions, the extent of cell recovery and impact on cell viability can be evaluated. Different types of centrifuges, such as disc stack centrifuge and spiral plate centrifuge, can be used depending upon choice, each of which has its own advantages and disadvantages [62]. Centrifugation method can be made energy-positive by using entire algae biomass [96], as it is shown that 1 kg of dry algae, containing 20% oil content, may yield 1.9 kWh biodiesel with calorific value of approximately 6 kWh of entire biomass [97].

7.3.5 GRAVITY SEDIMENTATION

This method is used to separate solid and liquid particles in suspension by using gravitational forces. It is cost-effective, energy efficient [92], and can separate microalgae having large size (>70 μm) and high density, but its disadvantage is the slow rate, especially in case of smaller particle size or low density difference [62]. Rate of this process depends on induced sedimentation velocity calculated by Stoke's law. In addition, it determines suspended solid characteristics using density and radius of microalgae cells [98]. Settling velocity is determined by the type of microalgae, light intensity, nutrients, and water quality. Settlement varies among and within species; for instance, diatoms, cyanobacteria, and green microalgae have settlement rate of 0.2, 0.05, and 0.1 m/day, respectively [62]. However, sedimentation rate may be enhanced using tailor-made sedimentation tanks, along with lamella separators and flocculation methods [92].

7.3.6 BIOMASS FILTRATION

This technique is highly applicable due to simplicity, maximum separation efficiency, no chemical requirement, and continuous operation [99]. However, this method is suitable for large colony-forming or elongated microalgae [6]. Though feasible for many reasons, its operational cost is high due to intensive maintenance involved in the cleaning of filters on regular intervals to avoid fouling and clogging [94]. A number of variants of this technology such as *microfiltration*, *vacuum filtration*, *ultrafiltration*, *dead-end filtration*, and *pressure filtration* are available which provide enhanced efficiency and address the common problems [100].

Common problems of conventional filtration can be minimized through ultrafiltration and membrane microfiltration [83]. Membrane microfiltration is more suitable for fragile cells and is most commonly used by small aquaculture farms to recover algal cells for shellfish larvae feedstock [101]. Large volumes of algae recovery (approximately 70%–80% of algae biomass) is possible by using cross-flow or tangential-flow filtration harvesting process [5] where culture is tangentially passed through microporous or ultrafiltration membranes [102]. Dynamic filtration is an improved harvesting method involving turbulence to generate high membrane shear stress. Microfiltration has shown threefold plateau flux while using comparatively low energy as compared to nonrotational filtration system [103].

There is a need to develop a membrane with not only high flux but which can surmount fouling and clogging and are compatible with different microalgae species. Development of continuous integrated system, with cultivation and harvesting processes combined, is also believed to improve significance of this method as large amounts of microalgae can be harvested and water can be recycled for further batches [64].

7.4 PROCESSING OF ALGAL BIOMASS

Algal biomass produced cannot be used immediately for any purpose, unless it undergoes extensive downstream processing, such as dewatering, extraction of desired biomolecules, and purification, depending upon the species selected, purpose of biomass production, or products to be obtained. Overall, the downstreaming involved in the processing of algal biomass to produce the desired products is tedious and energy-intensive, raising the cost of algae-based products. Hence, efficient and cost-effective methods of algal biomass processing are required to be developed. We discuss some common approaches for the processing algal biomass along with their merits and demerits in the following sections.

7.4.1 DEHYDRATION

After recovery, the algal biomass has to be dried for its stability and for the extraction of oil or other important molecules. Dehydrated biomass is more suitable for oil or metabolite extraction which is further processed for biodiesel [89]. A number of techniques have been used to dry out the microalgae without posing any risk to the quality of oil or metabolites. Method of dehydration depends on the purpose the biomass is used for. Common dehydration methods developed from conventional dehydration are as follows:

- *Rotary drying*: In this method, a rotary dryer is used in which algal biomass is moved downward by gravity while the supernatant contains water or medium contents. This method not only involves dehydration of cell but also sterilizes the samples and breaks down the cell wall. About 25%–30% dry biomass can be obtained through this method. But it is an energy–intensive process, where 1.4 kWh of energy is consumed to obtain 1000 g of algal biomass [104,105].
- *Spray drying*: Atomized droplets of media are sprayed vertically through hot gas which helps drying the biomass within seconds. However, the waste gases are exhausted separately. Algal biomass used for human consumption is dried through this method, considering it to be an appropriate method. However, it ruptures the intact cells due to high pressure inside the tank which may deteriorate the product quality, and as a result, dried cells obtained through this

method have low digestibility [5,106]. However, this method is suitable for a few thermosensitive products, as the biomass is exposed to heat for a very short time, ranging from seconds to minutes.

- *Solar drying*: Sunlight can be used for drying biomass, in areas which receive more solar radiation. However, intense solar radiations can dehydrate algal chlorophyll, converting it into tiny particles which affect the texture and color of the final product [104]. Moreover, the process is time-consuming and may not dry the biomass completely.
- *Cross-flow air drying*: To produce a biomass with 4%–8% moisture contents, wet slurry algal biomass is dried at 62°C for 14 h via passing dry air through it. However, the method is not very efficient, because the biomass near the source of dried air is over-dried, while that away from the source is left under-dried. As a result, the moisture gradient exists after completion of drying. This method is cheaper compared to rotary drying and spray drying, but may not be economically viable due to higher energy input [89,107].
- *Vacuum-shelf drying*: This method can produce biomass containing 4% moisture content, at a temperature ranging from 50 to 65°C and a pressure 0.06 atm [89]. Biomass is placed inside the dryer where the water molecules are allowed to escape by heat and pressure and then it is moved to the condensers. Heat transfer takes place by the mechanism of conduction. Use of pressure makes it faster and cheaper, but the entire process requires higher installation and operational costs [104].
- *Incinerator drying*: Incineration is used to obtain dry biomass using lower input heat. It is useful to process the heat-sensitive biomass; however, the process requires higher capital cost [104].

7.4.2 EXTRACTION AND PURIFICATION OF LIPIDS

Different microalgae can accumulate lipids in the form of various fatty acids (Table 7.3) intracellularly. After drying, the algal cells should be disrupted to obtain the desired product. However, algal cells generally resist to cell disruption; therefore a suitable cell disruption method needs to be established for efficient recovery of cellular content, i.e., the required products or feedstock of various products. Mechanical and chemical methods are used for lipid extraction from microalgae [111]. Most commonly used mechanical extraction methods are *oil expeller*, *microwave-assisted*, and *ultrasound-assisted* methods. Chemical methods are *solvent extraction*, *ionic liquid extraction*, and *supercritical CO_2 extraction* [112].

- *Oil expeller*: This method is based on mechanical pressing of cells to obtain lipids contents and it can release up to 75% of lipids from dried algal biomass [113]. This method is slow and is only suitable for large scale [100].
- *Microwave-assisted extraction*: In this method, algal biomass is subjected to electromagnetic radiations of 0.3–300 GHz which increase its temperature and interact with polar molecules present in it, such as water, releasing oil from the

Table 7.3 Fatty Acid Content in Different Microalgae Species [108–110]

Microalgae	Fatty Acids						
	Polyunsaturated Fatty Acids			Omega 3 mg/100 g			Omega 6
	14:0	16:0	18:0	ALA (18:3)	EPA (20:5)	DHA (22:6)	GHA (18:3)
Spirulina maxima	9	1078	32	10	–	–	452
Chlorella vulgaris (green)	124	1016	25	661	19	16	112
C. vulgaris (orange)	210	5606	406	3665	39	80	23
Haematococcus pluvialis	154	5977	603	3981	579	–	472
Diacronema vlkianum	2081	1413	14	14	3212	836	112
Isochrysis galbana	3272	2711	50	421	4875	1156	–
Nannochloropsis gaditana	6	23.8	0.5	–	27.5	–	–
	3	16	–	9	20.5	–	–

biomass. Better quality oil can be produced in less time using this approach [112]. Microwave-assisted extraction with organic solvent (hexane) has shown up to 77% lipid extraction efficiency from *S. obliquus* in 30 min [114].

- *Ultrasound-assisted extraction*: Ultrasound frequency of 20 kHz produces high and low pressure cycle which collapses and produces a phenomenon called *cavitation*. During cavitation high pressure forces act on algal cell wall to break it and enhance lipid extraction about 50–500 times with 10-fold decrease in extraction time [42].
- *Solvent extraction*: It is based on the *like-dissolves-like* concept, where lipids are extracted from microalgae using lipid-dissolving solvents. Both polar and nonpolar solvents can be used. Interactions between neutral lipids and nonpolar lipids are disrupted by nonpolar solvents [112]. A [2:1 v/v] mixture of chloroform:methanol [115] is a commonly used solvent. The solvents used for extracting lipid from microalgae biomass are *n*-hexane, ethanol, 1-butanol, DBU (1,8-diazabicyclo-[5.4.0]-undec-7-ene), dimethyl ether, and mixtures of chloroform/methanol, *n*-hexane/ethanol, *n*-hexane/isopropanol (*n*-hexane/2-propanol), methanol/1-ethyl-3-methylimidazolium methyl sulfate, DBU/ethanol, DBU/octanol, methylene chloride/methanol, dichloroethane/methanol, dichloroethane/ethanol, and acetone/dichloromethane [116].
- *Supercritical CO_2 extraction*: Ethanol in combination with CO_2 has solvent power which can be varied by changing the extraction pressure and temperature to produce solvent-free crude lipids [117]. Biomass of *N. oculata* has shown

90% of recovery efficiency through supercritical CO_2 along with air-flow drying at 35°C without any phospholipids [118]. In another study, through supercritical CO_2 along with ethanol as cosolvent at 35 MPa and 50°C, *N. oculata* produced 239.7 mg triglycerides per gram of extracted oil [119].

- *Ionic liquid extraction*: Ionic liquid extraction has advantages of synthetic flexibility, thermal stability, and nonvolatility, over other methods. Ionic mixtures of [Bmim][CF_3SO_3] and [$MeSO_4$]. [Bmim][$MeSO_4$] is used for lipid extraction from *C. vulgaris* [120,121] and it is reported [120] that [Bmim][CF_3SO_3] along with methanol yielded 19% of extract oil from *C. vulgaris* by using Bligh and Dyer's method.

Other than the method of choice, algal biomass, temperature, and enzymes are also the factors affecting the extraction of lipids for biofuel production. Enzyme cocktail used for cell disruption generally contains cellulase, pectinase, and hemicellulase (in 1:1:1 w/w).

7.4.3 EXTRACTION AND PURIFICATION OF METABOLITES

Microalgae are an abundant source of natural compounds, which could be used as functional ingredients [122]. There is a large number of available microalgae species, and knowledge of their chemical composition is mandatory as a first step in screening methodology, since it will help to target valuable compounds, pigments, antioxidants, and polyunsaturated fatty acids (PUFAs) to extract and purify (Table 7.4). The first step in extraction of metabolites is cell harvesting and then breaking of cell wall by either mechanical or chemical methods [108].

Conventional methods involve the use electric field, i.e., pulse electric field (PEF) and high-voltage electric discharge (HVED) [128]. When *PEF* (3 kV/cm) is applied to *Nannochloropsis salina* and *C. vulgaris*, it creates pores in cell membrane which can be observed through microscope [129]. Increasing yield-results of proteins, chlorophyll, and carotenoids are obtained by treating *Chlorella* and *Spirulina* species with *PEF* (15 kV/cm and 100 kJ/kg) [130]. In HVED, a pulse streamer of 20 kV/cm is discharged in water which also causes production of localized high electric fields, hydroxide radicals, ozone, shock waves, cavitation, etc. This phenomena damages cell wall, leading to release of intracellular compounds [131].

High-performance liquid chromatography (HPLC) is an advanced technique which is used for purification of metabolites from microalgae. In HPLC, mostly C-18 or RP-18 columns with column size 4.6 mm × 250 mm to 4.6 mm × 300 mm, particle size 5–10 μm, and PDA or UV/VIS detectors are used. Absorbance is usually checked at different wavelengths (361, 546, 460, 480, and 660 nm). Eluent mixture consists of methanol and acetonitrile, however, instead of acetonitrile, acetic acid and other compounds can also be used at the flow rate of 1.0 mL/min [108,124,127,132,133].

Table 7.4 Treatment Conditions and Yield of Different Pigments From Microalgae

Microalgae	Pigments	Temperature (°C)	Pressure (MPa)	Time (min)	pH	Yield	References
Monoraphidium sp.	Astaxanthin	60	20	60	–	2.02–2.45 mg/g	[123]
Haematococcus pluvialis	Astaxanthin	45, 55	7, 20	120	–	53.48, 62.57 mg/g	[124]
Limnothrix sp.	C-phycocyanin	4	Atmospheric	Overnight	7.0	18% (w/w)	[125]
Spirulina platensis	Allophycocyanin, C-phycocyanin	Room	Atmospheric	4 h	7.0		[126]
	Vitamin-B_{12}	42–121	Atmospheric	10–60	4.8–6.0	0.11–0.40 mg 100 g^{-1}	[127]
Nannochloropsis gaditana	Neoxanthin	30–40	–	1–24 h	1.0–6.5	36.8% w/w	[110]
	Violaxanthin					33.1% w/w	
	Zeaxanthin					79.8% w/w	
	β-carotene					96.1% w/w	
Spirulina maxima	Lutein/Zeaxanthin	80–220	–	60–100	–	0.9% DW	[108]
	β-carotene chlorophyll *a*						
Chlorella vulgaris (green)	Astaxanthin Lutein/Zeaxanthin Canthaxanthin mono-, di-esters of Canthaxanthin and/	80–220	–	60–100	–	1.2% DW	[108]

	or astaxanthin Chlorophyll *a/b*						
C. vulgaris (orange)	Astaxanthin Lutein/ Zeaxanthin Canthaxanthin mono-, di-esters of Canthaxanthin and/ or astaxanthin Chlorophyll *a/b*	80–220	–	60–100	–	1.3% DW	[108]
H. pluvialis	Astaxanthin Lutein/ Zeaxanthin Canthaxanthin mono-, di-esters of Canthaxanthin and/ or astaxanthin Chlorophyll *a/b*	80–220	–	60–100	–	3.0% DW	[108]
Diacronema vlkianum	Fucoxanthin Lutein/ Zeaxanthin β-carotene Chlorophyll *a* Chlorophyll *c*	80–220	–	60–100	–	2.4% DW	[108]
Isochrysis galbana	Fucoxanthin Lutein/ Zeaxanthin β-carotene Chlorophyll *a* Chlorophyll *c*	80–220	–	60–100	–	3.2% DW	[108]

Thin layer chromatography is also used for separation of metabolites. Instruments containing plates of silica gel—alumina and an eluent mixture of petroleum ether and acetone (72:28 v/v) are generally used [108]. PUFAs which mostly contain omega molecules may be obtained through Soxhlet method using ether as a solvent after digestion of cells with concentrated acid (4N), while acetone extraction can be used other pigments. Supercritical CO_2 along with/without ethanol as cosolvent has been used to extract astaxanthin, which is a high value pigment, from GK12 cells of *Monoraphidium* sp. [123]. Activated carbon, distilled water, and 0.1 g/L of chitosan are used to extract C-phycocyanin from *Limnothrix* sp. After extraction, the metabolites can be purified. For instance, ammonium sulfate can be used to precipitate the phycocyanin, followed by resuspension in phosphate-buffer saline and tangential-flow filtration [125]. Allophycocyanin and C-phycocyanin have also been extracted by continues stirrer shaking for 4 h in 100 mM PBS [126].

REFERENCES

[1] Brodie J, Zuccarello G, Hodkinson T, Parnell J, Waldren S. Systematics of the species-rich algae: red algal classification, phylogeny and speciation. 2007.

[2] Rawat I, Kumar RR, Mutanda T, Bux F. Biodiesel from microalgae: a critical evaluation from laboratory to large scale production. Appl Energy 2013;103:444–67.

[3] Larkum AW, Ross IL, Kruse O, Hankamer B. Selection, breeding and engineering of microalgae for bioenergy and biofuel production. Trends Biotechnol 2012;30(4):198–205.

[4] Borowitzka MA. Commercial production of microalgae: ponds, tanks, tubes and fermenters. J Biotechnol 1999;70(1):313–21.

[5] Chen C-Y, Yeh K-L, Aisyah R, Lee D-J, Chang J-S. Cultivation, photobioreactor design and harvesting of microalgae for biodiesel production: a critical review. Bioresour Technol 2011;102(1):71–81.

[6] Brennan L, Owende P. Biofuels from microalgae—a review of technologies for production, processing, and extractions of biofuels and co-products. Renew Sustain Energy Rev 2010;14(2):557–77.

[7] Wang H, Gao L, Chen L, Guo F, Liu T. Integration process of biodiesel production from filamentous oleaginous microalgae *Tribonema* minus. Bioresour Technol 2013;142:39–44.

[8] Guo F, Wang H, Wang J, Zhou W, Gao L, Chen L, et al. Special biochemical responses to nitrogen deprivation of filamentous oleaginous microalgae *Tribonema* sp. Bioresour Technol 2014;158:19–24.

[9] Mata TM, Martins AA, Caetano NS. Microalgae for biodiesel production and other applications: a review. Renew Sustain Energy Rev 2010;14(1):217–32.

[10] Chisti Y. Biodiesel from microalgae. Biotechnol Adv 2007;25(3):294–306.

[11] Li Y, Horsman M, Wang B, Wu N, Lan CQ. Effects of nitrogen sources on cell growth and lipid accumulation of green alga *Neochloris oleoabundans*. Appl Microbiol Biotechnol 2008;81(4):629–36.

[12] Li Y, Horsman M, Wu N, Lan CQ, Dubois-Calero N. Biofuels from microalgae. Biotechnol Prog 2008;24(4):815–20.

[13] Guschina IA, Harwood JL. Lipids and lipid metabolism in eukaryotic algae. Prog Lipid Res 2006;45(2):160−86.

[14] Hu Q. 5. Environmental effects on cell composition. Handbook of microalgal culture: biotechnology and applied phycology. 2004. p. 83.

[15] Hu Q, Sommerfeld M, Jarvis E, Ghirardi M, Posewitz M, Seibert M, et al. Microalgal triacylglycerols as feedstocks for biofuel production: perspectives and advances. Plant J 2008;54(4):621−39.

[16] Roessler PG. Environmental control of glycerolipid metabolism in microalgae: commercial implications and future research directions. J Phycol 1990;26(3):393−9.

[17] Rodolfi L, Chini Zittelli G, Bassi N, Padovani G, Biondi N, Bonini G, et al. Microalgae for oil: strain selection, induction of lipid synthesis and outdoor mass cultivation in a low-cost photobioreactor. Biotechnol Bioeng 2009;102(1):100−12.

[18] Sibi G, Shetty V, Mokashi K. Enhanced lipid productivity approaches in microalgae as an alternate for fossil fuels − a review. J Energy Inst August 2016;89(3):330−4.

[19] Pruvost J, Van Vooren G, Le Gouic B, Couzinet-Mossion A, Legrand J. Systematic investigation of biomass and lipid productivity by microalgae in photobioreactors for biodiesel application. Bioresour Technol 2011;102(1):150−8.

[20] Leonardi PI, Popovich CA, Damiani MC. Feedstocks for second-generation biodiesel: microalgae's biology and oil composition. INTECH Open Access Publisher; 2011.

[21] Mata TM, Almeidab R, Caetanoa N. Effect of the culture nutrients on the biomass and lipid productivities of microalgae *Dunaliella tertiolecta*. Chem Eng J 2013;32:973.

[22] Macedo RVTD, Alegre RM. Influence of the nitrogen level in the cultivation of spirulina maxima in two levels of temperatures: part II: "production of lipids". Food Sci Technol (Camp) 2001;21(2):183−6.

[23] Benemann JR, Oswald WJ. Systems and economic analysis of microalgae ponds for conversion of CO {sub 2} to biomass. Final report. Berkeley (CA, United States): California Univ., Dept. of Civil Engineering; 1996.

[24] Borowitzka MA, Borowitzka LJ. Micro-algal biotechnology. Cambridge University Press; 1988.

[25] Chandra R, Rohit M, Swamy Y, Mohan SV. Regulatory function of organic carbon supplementation on biodiesel production during growth and nutrient stress phases of mixotrophic microalgae cultivation. Bioresour Technol 2014;165:279−87.

[26] Rodrigues MS, Ferreira LS, Converti A, Sato S, De Carvalho JCM. Influence of ammonium sulphate feeding time on fed-batch *Arthrospira* (*Spirulina*) platensis cultivation and biomass composition with and without pH control. Bioresour Technol 2011; 102(11):6587−92.

[27] Miron AS, Garcia M-CC, Camacho FG, Grima EM, Chisti Y. Growth and biochemical characterization of microalgal biomass produced in bubble column and airlift photobioreactors: studies in fed-batch culture. Enzyme Microb Technol 2002;31(7):1015−23.

[28] Han F, Huang J, Li Y, Wang W, Wan M, Shen G, et al. Enhanced lipid productivity of *Chlorella pyrenoidosa* through the culture strategy of semi-continuous cultivation with nitrogen limitation and pH control by CO_2. Bioresour Technol 2013;136:418−24.

[29] Takagi M, Watanabe K, Yamaberi K, Yoshida T. Limited feeding of potassium nitrate for intracellular lipid and triglyceride accumulation of *Nannochloris* sp. UTEX LB1999. Appl Microbiol Biotechnol 2000;54(1):112−7.

[30] Gardner R, Peters P, Peyton B, Cooksey KE. Medium pH and nitrate concentration effects on accumulation of triacylglycerol in two members of the chlorophyta. J Appl Phycol 2011;23(6):1005−16.

[31] Illman A, Scragg A, Shales S. Increase in *Chlorella* strains calorific values when grown in low nitrogen medium. Enzyme Microb Technol 2000;27(8):631–5.

[32] Borowitzka MA. Microalgae for aquaculture: opportunities and constraints. J Appl Phycol 1997;9(5):393–401.

[33] Handler RM, Canter CE, Kalnes TN, Lupton FS, Kholiqov O, Shonnard DR, et al. Evaluation of environmental impacts from microalgae cultivation in open-air raceway ponds: analysis of the prior literature and investigation of wide variance in predicted impacts. Algal Res 2012;1(1):83–92.

[34] Mehmood MA, Rashid U, Ibrahim M, Abbas F, Taufiq-Yap YH. Algal biomass production using waste water. In: Biomass bioenergy. Springer; 2014. p. 307–27.

[35] Slade R, Bauen A. Micro-algae cultivation for biofuels: cost, energy balance, environmental impacts and future prospects. Biomass Bioenergy 2013;53:29–38.

[36] Singh A, Nigam PS, Murphy JD. Mechanism and challenges in commercialisation of algal biofuels. Bioresour Technol 2011;102(1):26–34.

[37] Lee Y-K. Microalgal mass culture systems and methods: their limitation and potential. J Appl Phycol 2001;13(4):307–15.

[38] Jorquera O, Kiperstok A, Sales EA, Embiruçu M, Ghirardi ML. Comparative energy life-cycle analyses of microalgal biomass production in open ponds and photobioreactors. Bioresour Technol 2010;101(4):1406–13.

[39] Ugwu C, Aoyagi H, Uchiyama H. Photobioreactors for mass cultivation of algae. Bioresour Technol 2008;99(10):4021–8.

[40] McGinn PJ, Dickinson KE, Bhatti S, Frigon J-C, Guiot SR, O'Leary SJ. Integration of microalgae cultivation with industrial waste remediation for biofuel and bioenergy production: opportunities and limitations. Photosynth Res 2011;109(1–3):231–47.

[41] Leite GB, Abdelaziz AE, Hallenbeck PC. Algal biofuels: challenges and opportunities. Bioresour Technol 2013;145:134–41.

[42] Suali E, Sarbatly R. Conversion of microalgae to biofuel. Renew Sustain Energy Rev 2012;16(6):4316–42.

[43] Chen C-Y, Saratale GD, Lee C-M, Chen P-C, Chang J-S. Phototrophic hydrogen production in photobioreactors coupled with solar-energy-excited optical fibers. Int J Hydrog Energy 2008;33(23):6886–95.

[44] Das P, Lei W, Aziz SS, Obbard JP. Enhanced algae growth in both phototrophic and mixotrophic culture under blue light. Bioresour Technol 2011;102(4):3883–7.

[45] Sierra E, Acién F, Fernández J, García J, González C, Molina E. Characterization of a flat plate photobioreactor for the production of microalgae. Chem Eng J 2008;138(1): 136–47.

[46] Singh J, Gu S. Commercialization potential of microalgae for biofuels production. Renew Sustain Energy Rev 2010;14(9):2596–610.

[47] Martín LA, Popovich CA, Martinez AM, Damiani MC, Leonardi PI. Oil assessment of *Halamphora coffeaeformis* diatom growing in a hybrid two-stage system for biodiesel production. Renew Energy 2016;92:127–35.

[48] Zittelli GC, Biondi N, Rodolfi L, Tredici MR. Photobioreactors for mass production of microalgae. In: Handbook of microalgal culture: applied phycology and biotechnology. 2nd ed. 2013. p. 225–66.

[49] Chojnacka K, Marquez-Rocha F-J. Kinetic and stoichiometric relationships of the energy and carbon metabolism in the culture of microalgae. Biotechnology 2004;3(1): 21–34.

[50] Xu H, Miao X, Wu Q. High quality biodiesel production from a microalga *Chlorella protothecoides* by heterotrophic growth in fermenters. J Biotechnol 2006;126(4): 499−507.

[51] Pleissner D, Lam WC, Sun Z, Lin CSK. Food waste as nutrient source in heterotrophic microalgae cultivation. Bioresour Technol 2013;137:139−46.

[52] Devi MP, Subhash GV, Mohan SV. Heterotrophic cultivation of mixed microalgae for lipid accumulation and wastewater treatment during sequential growth and starvation phases: effect of nutrient supplementation. Renew Energy 2012;43:276−83.

[53] Lowrey J, Brooks MS, McGinn PJ. Heterotrophic and mixotrophic cultivation of microalgae for biodiesel production in agricultural wastewaters and associated challenges—a critical review. J Appl Phycol 2015;27(4):1485−98.

[54] Gouveia L, Oliveira AC. Microalgae as a raw material for biofuels production. J Ind Microbiol Biotechnol 2009;36(2):269−74.

[55] Chojnacka K, Noworyta A. Evaluation of *Spirulina* sp. growth in photoautotrophic, heterotrophic and mixotrophic cultures. Enzyme Microb Technol 2004;34(5):461−5.

[56] Bhatnagar A, Chinnasamy S, Singh M, Das K. Renewable biomass production by mixotrophic algae in the presence of various carbon sources and wastewaters. Appl Energy 2007;88(10):3425−31.

[57] Ogbonna JC, Tanaka H. Cyclic autotrophic/heterotrophic cultivation of photosynthetic cells: a method of achieving continuous cell growth under light/dark cycles. Bioresour Technol 1998;65(1):65−72.

[58] Wan M, Liu P, Xia J, Rosenberg JN, Oyler GA, Betenbaugh MJ, et al. The effect of mixotrophy on microalgal growth, lipid content, and expression levels of three pathway genes in *Chlorella sorokiniana*. Appl Microbiol Biotechnol 2011;91(3): 835−44.

[59] Wang J, Yang H, Wang F. Mixotrophic cultivation of microalgae for biodiesel production: status and prospects. Appl Biochem Biotechnol 2014;172(7):3307−29.

[60] Li T, Zheng Y, Yu L, Chen S. Mixotrophic cultivation of a *Chlorella sorokiniana* strain for enhanced biomass and lipid production. Biomass Bioenergy 2014;66:204−13.

[61] Kim S, Park J-E, Cho Y-B, Hwang S-J. Growth rate, organic carbon and nutrient removal rates of *Chlorella sorokiniana* in autotrophic, heterotrophic and mixotrophic conditions. Bioresour Technol 2013;144:8−13.

[62] Milledge JJ, Heaven S. A review of the harvesting of micro-algae for biofuel production. Rev Environ Sci Biotechnol 2013;12(2):165−78.

[63] Amer L, Adhikari B, Pellegrino J. Technoeconomic analysis of five microalgae-to-biofuels processes of varying complexity. Bioresour Technol 2011;102(20):9350−9.

[64] Kim J, Yoo G, Lee H, Lim J, Kim K, Kim CW, et al. Methods of downstream processing for the production of biodiesel from microalgae. Biotechnol Adv 2013;31(6): 862−76.

[65] Bilad MR, Vandamme D, Foubert I, Muylaert K, Vankelecom IF. Harvesting microalgal biomass using submerged microfiltration membranes. Bioresour Technol 2012; 111:343−52.

[66] Salim S, Vermuë M, Wijffels R. Ratio between autoflocculating and target microalgae affects the energy-efficient harvesting by bio-flocculation. Bioresour Technol 2012; 118:49−55.

[67] Lee D-J, Liao G-Y, Chang Y-R, Chang J-S. Coagulation-membrane filtration of *Chlorella vulgaris*. Bioresour Technol 2012;108:184−9.

[68] Hanotu J, Bandulasena H, Zimmerman WB. Microflotation performance for algal separation. Biotechnol Bioeng 2012;109(7):1663–73.

[69] Henderson RK, Parsons SA, Jefferson B. The impact of differing cell and algogenic organic matter (AOM) characteristics on the coagulation and flotation of algae. Water Res 2010;44(12):3617–24.

[70] Zhang G, Zhang P, Fan M. Ultrasound-enhanced coagulation for *Microcystis aeruginosa* removal. Ultrason Sonochem 2009;16(3):334–8.

[71] Muylaert K, Vandamme D, Foubert I, Brady PV. Harvesting of microalgae by means of flocculation. Biomass and biofuels from microalgae. Springer; 2015. p. 251–73.

[72] Uduman N, Qi Y, Danquah MK, Forde GM, Hoadley A. Dewatering of microalgal cultures: a major bottleneck to algae-based fuels. J Renew Sustain Energy 2010;2(1). 012701.

[73] Pieterse A, Cloot A. Algal cells and coagulation, flocculation and sedimentation processes. Water Sci Technol 1997;36(4):111–8.

[74] König RB, Sales R, Roselet F, Abreu PC. Harvesting of the marine microalga *Conticribra weissflogii* (Bacillariophyceae) by cationic polymeric flocculants. Biomass Bioenergy 2014;68:1–6.

[75] Sukenik A, Bilanovic D, Shelef G. Flocculation of microalgae in brackish and sea waters. Biomass 1988;15(3):187–99.

[76] Alam MA, Wan C, Guo S-L, Zhao X-Q, Huang Z-Y, Yang Y-L, et al. Characterization of the flocculating agent from the spontaneously flocculating microalga *Chlorella vulgaris* JSC-7. J Biosci Bioeng 2014;118(1):29–33.

[77] Guo S-L, Zhao X-Q, Wan C, Huang Z-Y, Yang Y-L, Alam MA, et al. Characterization of flocculating agent from the self-flocculating microalga *Scenedesmus obliquus* AS-6-1 for efficient biomass harvest. Bioresour Technol 2013;145:285–9.

[78] Wan C, Alam MA, Zhao X-Q, Zhang X-Y, Guo S-L, Ho S-H, et al. Current progress and future prospect of microalgal biomass harvest using various flocculation technologies. Bioresour Technol 2015;184:251–7.

[79] Salim S, Kosterink N, Wacka NT, Vermuë M, Wijffels R. Mechanism behind autoflocculation of unicellular green microalgae *Ettlia texensis*. J Biotechnol 2014;174:34–8.

[80] Trentacoste EM, Shrestha RP, Smith SR, Glé C, Hartmann AC, Hildebrand M, et al. Metabolic engineering of lipid catabolism increases microalgal lipid accumulation without compromising growth. Proc Natl Acad Sci 2013;110(49):19748–53.

[81] Zhou W, Cheng Y, Li Y, Wan Y, Liu Y, Lin X, et al. Novel fungal pelletization-assisted technology for algae harvesting and wastewater treatment. Appl Biochem Biotechnol 2012;167(2):214–28.

[82] Vandamme D, Foubert I, Muylaert K. Flocculation as a low-cost method for harvesting microalgae for bulk biomass production. Trends Biotechnol 2013;31(4):233–9.

[83] Grima EM, Belarbi E-H, Fernãndez FA, Medina AR, Chisti Y. Recovery of microalgal biomass and metabolites: process options and economics. Biotechnol Adv 2003;20(7): 491–515.

[84] Shelef G, Sukenik A, Green M. Microalgae harvesting and processing: a literature review. Haifa (Israel): Technion Research and Development Foundation Ltd.; 1984.

[85] Procházková G, Šafařík I, Brányik T. Surface modification of *Chlorella vulgaris* cells using magnetite particles. Procedia Eng 2012;42:1778–87.

[86] Seo JY, Lee K, Lee SY, Jeon SG, Na J-G, Oh Y-K, et al. Effect of barium ferrite particle size on detachment efficiency in magnetophoretic harvesting of oleaginous *Chlorella* sp. Bioresour Technol 2014;152:562–6.

[87] Bosma R, van Spronsen WA, Tramper J, Wijffels RH. Ultrasound, a new separation technique to harvest microalgae. J Appl Phycol 2003;15(2−3):143−53.

[88] Hawkes JJ, Limaye M, Coakley WT. Filtration of bacteria and yeast by ultrasound enhanced sedimentation. J Appl Microbiol 1997;82(1):39−47.

[89] Show K-Y, Lee D-J, Chang J-S. Algal biomass dehydration. Bioresour Technol 2013; 135:720−9.

[90] Chen Y, Liu J, Ju Y-H. Flotation removal of algae from water. Colloids Surf B Biointerfaces 1998;12(1):49−55.

[91] Garg S, Li Y, Wang L, Schenk PM. Flotation of marine microalgae: effect of algal hydrophobicity. Bioresour Technol 2012;121:471−4.

[92] Pragya N, Pandey KK, Sahoo P. A review on harvesting, oil extraction and biofuels production technologies from microalgae. Renew Sustain Energy Rev 2013;24:159−71.

[93] Kurniawati HA, Ismadji S, Liu J. Microalgae harvesting by flotation using natural saponin and chitosan. Bioresour Technol 2014;166:429−34.

[94] Rawat I, Kumar RR, Mutanda T, Bux F. Dual role of microalgae: phycoremediation of domestic wastewater and biomass production for sustainable biofuels production. Appl Energy 2011;88(10):3411−24.

[95] Zhou W, Gao L, Cheng W, Chen L, Wang J, Wang H, et al. Electro-flotation of *Chlorella* sp. assisted with flocculation by chitosan. Algal Res 2016;18:7−14.

[96] Milledge JJ. The challenge of algal fuel: economic processing of the entire algal biomass. Condens Matter Mater Eng Newsl 2010;1(6):4−6.

[97] Milledge J. The potential yield of microalgae oil. Biofuels Int 2010;4(1):44−5.

[98] Schenk PM, Thomas-Hall SR, Stephens E, Marx UC, Mussgnug JH, Posten C, et al. Second generation biofuels: high-efficiency microalgae for biodiesel production. Bioenergy Res 2008;1(1):20−43.

[99] Barros AI, Gonçalves AL, Simões M, Pires JC. Harvesting techniques applied to microalgae: a review. Renew Sustain Energy Rev 2015;41:1489−500.

[100] Harun R, Singh M, Forde GM, Danquah MK. Bioprocess engineering of microalgae to produce a variety of consumer products. Renew Sustain Energy Rev 2010;14(3): 1037−47.

[101] Petrusevski B, Bolier G, Van Breemen A, Alaerts G. Tangential flow filtration: a method to concentrate freshwater algae. Water Res 1995;29(5):1419−24.

[102] Chen X, Huang C, Liu T. Harvesting of microalgae *Scenedesmus* sp. using polyvinylidene fluoride microfiltration membrane. Desalin Water Treat 2012;45(1−3):177−81.

[103] Rios SD, Clavero E, SalvadÃ J, Farriol X, Torras C. Dynamic microfiltration in microalgae harvesting for biodiesel production. Ind Eng Chem Res 2010;50(4):2455−60.

[104] Delrue F, Setier P-A, Sahut C, Cournac L, Roubaud A, Peltier G, et al. An economic, sustainability, and energetic model of biodiesel production from microalgae. Bioresour Technol 2012;111:191−200.

[105] Singh A, Olsen SI. A critical review of biochemical conversion, sustainability and life cycle assessment of algal biofuels. Appl Energy 2011;88(10):3548−55.

[106] Show K-Y, Lee D-J, Tay J-H, Lee T-M, Chang J-S. Microalgal drying and cell disruption − recent advances. Bioresour Technol 2015;184:258−66.

[107] Liu X, Clarens AF, Colosi LM. Algae biodiesel has potential despite inconclusive results to date. Bioresour Technol 2012;104:803−6.

[108] Batista AP, Gouveia L, Bandarra NM, Franco JM, Raymundo A. Comparison of microalgal biomass profiles as novel functional ingredient for food products. Algal Res 2013;2(2):164−73.

[109] López BC, Cerdán LE, Medina AR, López EN, Valverde LM, Peña EH, et al. Production of biodiesel from vegetable oil and microalgae by fatty acid extraction and enzymatic esterification. J Biosci Bioeng 2015;119(6):706–11.

[110] Peña EH, Medina AR, Callejón MJJ, Sánchez MDM, Cerdán LE, Moreno PAG, et al. Extraction of free fatty acids from wet *Nannochloropsis gaditana* biomass for biodiesel production. Renew Energy 2015;75:366–73.

[111] Donot F, Strub C, Fontana A, Jouy N, Delbes C, Gunata Z, et al. Rapid analysis and quantification of major neutral lipid species and free fatty acids by HPLC-ELSD from microalgae. Eur J Lipid Sci Technol 2016.

[112] Mubarak M, Shaija A, Suchithra T. A review on the extraction of lipid from microalgae for biodiesel production. Algal Res 2015;7:117–23.

[113] Topare NS, Raut SJ, Renge V, Khedkar SV, Chavanand Y, Bhagat S. Extraction of oil from algae by solvent extraction and oil expeller method. Int J Chem Sci 2011;9(4): 1746–50.

[114] Balasubramanian S, Allen JD, Kanitkar A, Boldor D. Oil extraction from *Scenedesmus obliquus* using a continuous microwave system – design, optimization, and quality characterization. Bioresour Technol 2011;102(3):3396–403.

[115] Wang G, Wang T. Characterization of lipid components in two microalgae for biofuel application. J Am Oil Chem Soc 2012;89(1):135–43.

[116] Neto AMP, de Souza RAS, Leon-Nino AD, da Costa JDAA, Tiburcio RS, Nunes TA, et al. Improvement in microalgae lipid extraction using a sonication-assisted method. Renew Energy 2013;55:525–31.

[117] Santana A, Jesus S, Larrayoz M. Supercritical carbon dioxide extraction of algal lipids for the biodiesel production. Procedia Eng 2012;42:1755–61.

[118] Crampon C, Mouahid A, Toudji S-AA, Lépine O, Badens E. Influence of pretreatment on supercritical CO_2 extraction from *Nannochloropsis oculata*. J Supercrit Fluids 2013;79:337–44.

[119] Liau B-C, Shen C-T, Liang F-P, Hong S-E, Hsu S-L, Jong T-T, et al. Supercritical fluids extraction and anti-solvent purification of carotenoids from microalgae and associated bioactivity. J Supercrit Fluids 2010;55(1):169–75.

[120] Kim Y-H, Choi Y-K, Park J, Lee S, Yang Y-H, Kim HJ, et al. Ionic liquid-mediated extraction of lipids from algal biomass. Bioresour Technol 2012;109:312–5.

[121] Kim Y-H, Park S, Kim MH, Choi Y-K, Yang Y-H, Kim HJ, et al. Ultrasound-assisted extraction of lipids from *Chlorella vulgaris* using [Bmim][$MeSO_4$]. Biomass Bioenergy 2013;56:99–103.

[122] Gouveia L, Marques AE, Sousa JM, Moura P, Bandarra NM. Microalgae – source of natural bioactive molecules as functional ingredients. Food Sci Technol Bull Funct Foods 2010;7(2):21.

[123] Fujii K. Process integration of supercritical carbon dioxide extraction and acid treatment for astaxanthin extraction from a vegetative microalga. Food Bioprod Process 2012;90(4):762–6.

[124] Reyes FA, Mendiola JA, Ibañez E, del Valle JM. Astaxanthin extraction from *Haematococcus pluvialis* using CO_2-expanded ethanol. J Supercrit Fluids 2014;92: 75–83.

[125] Gantar M, Simović D, Djilas S, Gonzalez WW, Miksovska J. Isolation, characterization and antioxidative activity of C-phycocyanin from *Limnothrix* sp. strain 37-2-1. J Biotechnol 2012;159(1):21–6.

[126] Chaiklahan R, Chirasuwan N, Bunnag B. Stability of phycocyanin extracted from *Spirulina* sp.: influence of temperature, pH and preservatives. Process Biochem 2012;47(4):659–64.

[127] Kumudha A, Sarada R. Effect of different extraction methods on vitamin B12 from blue green algae, spirulina platensis. Pharm Anal Acta 2015.

[128] Barba FJ, Grimi N, Vorobiev E. New approaches for the use of non-conventional cell disruption technologies to extract potential food additives and nutraceuticals from microalgae. Food Eng Rev 2015;7(1):45–62.

[129] Coustets M, Al-Karablieh N, Thomsen C, Teissié J. Flow process for electroextraction of total proteins from microalgae. J Membr Biol 2013;246(10):751–60.

[130] Töpfl S. Pulsed electric fields (PEF) for permeabilization of cell membranes in food- and bioprocessing – applications, process and equipment design and cost analysis. 2006.

[131] Grimi N, Dubois A, Marchal L, Jubeau S, Lebovka N, Vorobiev E. Selective extraction from microalgae *Nannochloropsis* sp. using different methods of cell disruption. Bioresour Technol 2014;153:254–9.

[132] Campenni L, Nobre BP, Santos CA, Oliveira A, Aires-Barros M, Palavra A, et al. Carotenoid and lipid production by the autotrophic microalga *Chlorella protothecoides* under nutritional, salinity, and luminosity stress conditions. Appl Microbiol Biotechnol 2013;97(3):1383–93.

[133] Casal C, Cuaresma M, Vega JM, Vilchez C. Enhanced productivity of a lutein-enriched novel acidophile microalga grown on urea. Mar Drugs 2010;9(1):29–42.

CHAPTER

8

Algae Biotechnology: A Green Light for Engineered Algae

Ijaz Rasul[1], Farrukh Azeem[1], Muhammad H. Siddique[1], Saima Muzammil[1], Azhar Rasul[1,2], Anam Munawar[3], Muhammad Afzal[1], Muhammad A. Ali[4], Habibullah Nadeem[1]

Government College University Faisalabad, Faisalabad, Pakistan[1]; RIKEN Center for Sustainable Resource Science (CSRS), Saitama, Japan[2]; National Institute for Biotechnology and Genetic Engineering, Faisalabad, Pakistan[3]; University of Agriculture Faisalabad, Faisalabad, Pakistan[4]

Algae are chlorophyll-containing photosynthetic organisms found everywhere on the earth, such as in the sea, rivers, lakes, soil, in animal and plants (as symbiotic partners collaborating together). They vary from small, single-celled types to complex multicellular ones. There are two main types of algae: (1) the macroalgae (multicellular and plant-like algae with roots) are present in littoral zone, which includes red algae, green algae, and brown algae; and (2) the microalgae (unicellular species with no roots) which occupy both benthic and littoral habitats as well as the ocean waters as phytoplankton. Phytoplankton includes organisms such as green and yellow-brown flagellates (chlorophyta, cryptophyta, chrysophyta, prasinophyta, prymnesiophyta, and rhaphidophyta), dinoflagellates (dinophyta), diatoms (bacillariophyta), and blue-green algae (cyanobacteria) [1,2].

It was estimated that there are around 1−10 million species of algae, most of which are microalgae recognized as one of the oldest residing microorganisms on earth [3]. Culturable microalgae can be classified into four categories: cyanobacteria, green algae, chrysophyte, and red algae. Cyanobacteria (blue-green algae) are distinguished from other bacteria because they perform oxygenic photosynthesis [4]. Microalgae are currently used as an important source of valuable, natural, biologically active molecules, such as carotenoids, chlorophyll, long-chain polyunsaturated fatty acids (PUFAs), phycobiliproteins, enzymes, carbohydrates, lipids, proteins, starch, cellulose, pigments, antioxidants, and pharmaceuticals. Significant advances were achieved in microalgae biotechnology over the last decade. Products such as vaccines, antibodies, enzymes, blood-clotting factors, immunoregulators, growth factors, hormones, and taxol (valuable anticancer agent) can also be synthesized via the genetic engineering of microalgae [5].

Photosynthetic marine algae are also attractive targets for the production of bio-products because of their ability to capture and fix carbon dioxide using solar energy and to grow in seawater, thereby minimizing freshwater usage. Marine algae and their bio-products are also getting greater attention in the treatment of

Algae Based Polymers, Blends, and Composites. http://dx.doi.org/10.1016/B978-0-12-812360-7.00008-2

diabetes. There is a great interest in screening marine algae for therapeutic drugs, as their ability to produce secondary metabolites is extensively documented. Marine algae contain considerable amounts of polyphenols, which are effective antioxidants and have particular biological activities [6]. In this chapter, we outline the potential of this chemically rich group of organisms to obtain various useful products as shown in Fig. 8.1.

8.1 MAKE YOUR OWN ALGAL BIOPLASTIC/GREEN PLASTIC

Plastics are carbon-based polymers made from petroleum. Life became much more convenient with the discovery of plastics, because it is used in a wide array of materials. The annual consumption of plastic is approximately 140 million tons across the world, which requires processing of about 150 million tons of

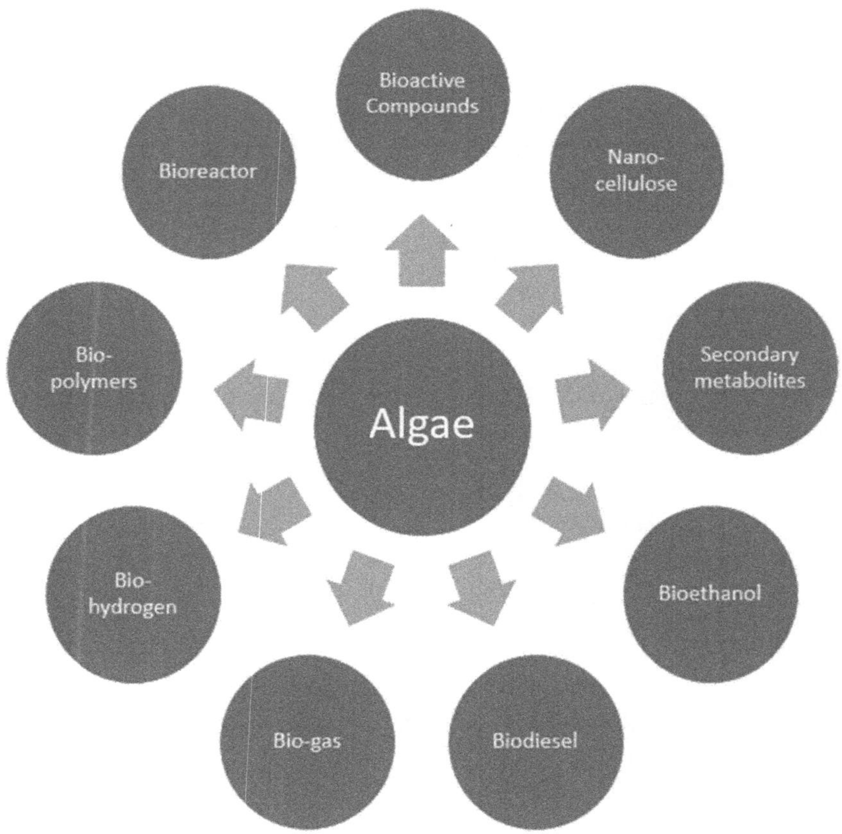

FIGURE 8.1

Multidimensional uses of algae.

petrochemicals. Such a huge plastic production generates massive amount of waste materials [7]. Moreover, these plastics are so durable that it will take many centuries for these plastics to degrade completely.

Getting rid of petroleum-based plastics is extremely difficult, their burning gives off harmful chemicals such as dioxins that contribute to global warming. Recycling these plastics is also difficult because there are numerous kinds of plastics and each has to be recycled by a different process. Though the plastics are considered to be one of the greatest innovations ever, they are also wreaking a great havoc on the environment, the wildlife, and the general public. At present, a major share of organic chemicals and energy in the world is derived from the fossil fuels alone. The consumption of organic chemicals and energy is increasing continuously throughout the world due to increasing population, leading to inflated crude oil prices and constant dwindling of fossil fuels. Fossil fuel resources are projected to diminish approximately by 2050 if the current rate of utilization persists [8]. Therefore, it is significant to explore renewable resources for sustainable production of organic chemicals, polymers, heat, fuels, and electricity.

Plastics are referred as "Green" if they exhibit one or more of the following properties: source renewability, biodegradability/compostability after life, and environment-friendliness [9]. Therefore, bioplastics derived from renewable sources, which can be easily biodegraded, are a good substitute of fossil fuel—based plastics. Biomass including trees and crops, aquatic plants, agricultural food and feed crop residue, wood and wood residue, animal wastes, and other waste materials are the most important renewable resources to produce value-added products and energy [10]. Not only corn and soybean, other food crops such as cassava, wheat, potato, and sago are also now turned into plastics to replace the use of petroleum-based plastics. However, continued usage of crops for plastics will interfere with human food supply, creating additional pressure on world's already diminishing food resources. There was a need to develop sustainable alternate sources of raw materials other than the food crops to ensure the food security of the world and to continue using crops as a raw material for plastics.

In this respect, algae represent a potential biomass to be explored as a source to develop bioplastics, because algal biomass is abundant, fast growing, and unexploited resource often left to decompose on the shores posing waste problems. Low percentage of lignin and high percentage of carbohydrates make algae an excellent candidate for the synthesis of bioplastics [1,2]. The history of bioplastics is given in Table 8.1.

8.1.1 TYPES OF ALGAL BIOPLASTICS

Various forms of plastics, including hybrid plastics, cellulose-based plastics, polylactic acid (PLA), bio-polyethylene, can be produced from algal feedstock (Table 8.2).

Table 8.1 History of Bioplastics [11]

Year	Development
1941	Henry Ford experimented with plastics made from soybeans and produced a plastic car. World War II played an important role in the development of bioplastics.
1992	Metabolix (bioscience company) provided solutions for world needs for plastics, chemicals, and energy.
2000	Metabolix initiated the research programs to develop the engineered industrial crops for the production of bioplastics.
2005	Toyota began a pilot plant at Hirose in Toyota city (Japan) to test the ease of producing bioplastics.
2006	LONDON-NEC corporation and Unitika Ltd. developed a bioplastic material reinforced with fiber from the kenaf plant to reduce the environmental impact of mobile phones.
2010	Cardia Bioplastics Malaysia Manufacturing (CBMM) was developed to manufacture bioplastic products.

Table 8.2 Algae Strains Used for the Synthesis of Bioplastics

Sr. No.	Strain	References
1	*Ulva armoricana*	[12]
2	*Gracilaria verrucosa*	[13]
3	*Eucheuma cottonii*	[14]
4	*Spirulina dregs*	[15]
5	*Spirulina platensis*	[16]
6	*Chlorella vulgaris* and *S. platensis*	[17]
7	*Cladophora rupestris*	[18]
8	*Plocamium cartilagineum*	[19]
9	*Gracilariopsis longissima*	[20]

8.1.1.1 Hybrid Plastics

These are made by adding denatured algal biomass to petroleum-based plastics, such as polyurethane and polyethylene, as fillers. It thus decreases the amount of petroleum used per unit of plastic and often provides these plastics with very desirable properties including biodegradability. Filamentous green algae of the order *Cladophorales* are claimed to be well-suited for use in hybrid plastics.

8.1.1.2 Cellulose-Based Plastics

The oldest forms of bioplastics, first made over a century ago, are cheap and low-quality plastics derived from cellulose (naturally occurring biopolymer of glucose). For some strains of algae, 30% of the biomass produced after extraction of algal oil is known to contain cellulose. These strains are thus an ideal feedstock for cellulose-based plastics [21].

8.1.1.3 Polylactic Acid

Lactic acid is usually produced by fermentation of natural feedstock and is polymerized to produce PLA. Lactic acid and its polymer (PLA) are already used as a biodegradable alternative and are believed to be economically viable alternatives on a large scale in the future. Lactic acid can be produced by bacterial fermentation of algal biomass [22].

8.1.1.4 Bio-Polyethylene

Ethylene (monomer used in the production of polyethylene) is easily produced by cracking and dehydration of ethanol. Ethanol is presently derived from natural gas or petroleum; however, it can also be produced from bacterial digestion of algal biomass, or directly from algae [23].

While the above types of plastics from algae are technically feasible, their economic (cost) viability is still being worked out.

8.1.2 ALGAE AS HETEROLOGOUS EXPRESSION SYSTEM FOR THE PRODUCTION OF BIOPLASTICS

Microalgae provide a platform for the expression of foreign genes due to their quick growth under photoautotrophic conditions and their low cost of maintenance, as compared to terrestrial plants, mammalian cells, yeast, and bacteria. In addition, many algal species have developed unique metabolic pathways that produce compounds of commercial value. Genetic manipulations of microalgae can lead to major changes in microalgal biotechnology—based industries. Such manipulations include the transfer of biosynthetic cascades into organisms that are suitable for growth in open ponds, or alternatively, adapting organisms that produce compounds of interest to grow under extreme conditions, thus reducing the risk of contamination. Recent developments in genome sequencing, combined with old and new systems for genetic manipulation, act synergistically to advance all aspects of microalgal research, both basic and applied [24].

Microalgae have great potential as novel low-cost expression systems, especially if aiming at the biosynthesis of recombinant proteins needed in numerous industrial, therapeutic, or diagnostic applications. Poly(3-hydroxyalkanoate) also known as PHA is recommended as a green substitute of conventional plastics, produced from sustainable resources and biodegraded by the enzymatic action. Another polymer, poly(3-hydroxybutyrate) also known as PHB has a melting point close to polypropylene with similar mechanical, and better oxygen barrier properties. PHB is synthesized from acetyl coenzyme A (acetyl coA) by three enzymatic reactions. It is produced by several bacteria (e.g., *Ralstonia eutropha* and *Bacillus megaterium*) as intracellular carbon and storage material, and is 100% biodegradable. However, cost of PHB production by bacterial fermentation is still very high. PHB production is feasible in a microalgal system when bacterial PHB pathway of *R. eutropha* is introduced into the diatom *Phaeodactylum tricornutum* [25]. Cyanobacteria (blue-green algae), as the only group of PHB-accumulating oxygenic photoautotrophs, can contribute to decrease costs of carbon source and oxygen supply. Cyanobacteria provide an excellent feedstock for plastic production due to its high yield and ability to grow in a variety of environments. The utilization of cyanobacterial plastics kicks off the opportunity of exploiting carbon source and neutralizing greenhouse gas emissions from factories and power plants. Genetically engineered cyanobacteria that accumulate PHB were transformed with the genes encoding PHB synthesis (3-ketothiolase, acetoacetyl-CoA reductase, and PHB synthase). Metabolic engineering is extremely investigated to establish novel metabolic pathways to augment bioplastic synthesis and to expand the utilizable substrate range [23,24].

8.1.3 CHALLENGES FOR BIOPLASTICS

Even though bioplastics are anticipated to be good for the environment, they can damage the nature in certain ways, e.g., emanation of greenhouse gases such as methane and carbon dioxide, while they are degrading. This can be controlled by designing plastics so that they can disintegrate and break down slowly, or by accumulating the released methane and use it elsewhere as fuel. Some bioplastics require specific conditions to degrade, and these requirements may not be available at all landfills or consumers may not have access to landfills. In such cases it is essential to design bioplastics that are biodegradable in ordinary soil-composting conditions.

Starch-based bioplastics are manufactured generally from plants such as corn and potatoes. This sets huge pressure on the agricultural crops as they have to gratify the needs of the ever-growing populace. To craft out bioplastics, crops have to be grown and this could lead to deforestation. Similarly, bioplastics derived from bacterial sources offer a least cost-effective solution. Use of algae as bioreactor to express bioplastic-producing pathways of bacteria or other organisms is an attractive solution. However, such algal strains are to be genetically modified to advance their resistance to ecological stresses and increase the yield. This exercise, however, involves a high risk not only for environment but also for humans and animals.

Bioplastics are a novel technology and entail still more research and development to get ascertained. Bioplastics are not thus equivalent to conventional plastics with respect to cost issues.

8.1.4 FUTURE PROSPECTS OF BIOPLASTICS

The prompt development of plastic production was a 20th century phenomenon of a historical scale. The low cost of plastics and its resourcefulness flagged a way for an extensive range of applications. As the plastics are nonbiodegradable and found to have toxic effects on human, animals, and environment, the biodegradable bioplastics came into being which can be stemmed from renewable biological sources. Bioplastics have identical applications as plastics. Though, there are various sources of bioplastics such as plants, animals, and microbes, they have some boundaries such as unavailability of biomass in bulk and complications involved in refinement. Algae can be used as an alternate source for the production of bioplastics. It can be used for wastewater remediation for the removal of nitrogenous waste, which would only increase their protein production capacity. In addition, using wastewater as algae growth media makes these algae production models more suitable than conventional food crops because they remediate water for further use, and can be grown in urban environments or on nonarable land not suitable for conventional food crop production. Bioplastics from seaweeds are stated to be more resilient to microwave radiation and are less brittle. The technology expansions for the seaweeds-based bioplastics are still in the research stage and it is anticipated that suggestive developments would be made in the bioplastic industries, which will make seaweed bioplastics a reality in future. Fermentation and genetic engineering can take the lead in employing novel systems to produce bioplastics from seaweeds which would make them a sustainable alternate. Cassava, corn, sago, and other food yields are frequently used as raw materials to prepare bioplastics. However, plastics fabricated from such crops cannot be personalized to suit particular requirements owing to their weak water resistance and mechanical properties. A large amount of green plastics shortcomings are overcome by synthetic blending, such as water resistance, strength, and elasticity.

Researchers are currently considering alternative raw materials from other sustainable resources to design plastics. Current studies have showed that marine red algae, extensively used as a raw material for producing biofuels in the past, are one of the prospective algae crop for bioplastics.

8.2 MICROALGAE AS A SOURCE OF USEFUL COMPOUNDS OTHER THAN BIOFUELS

Algae are found in nearly every environment on earth, with many of them possessing the ability to shape the ecosystem they inhabit. With current challenges to climate stability, understanding how these important organisms interact with their environment as well as one another might afford better insight into how they respond to a changing climate. It is estimated that there are 1–10 million species of algae, most of which are microalgae. Culturable microalgae can be classified into four categories: cyanobacteria, green algae, chrysophyte, and red algae.

8.2.1 MICROALGAE AS SOURCE OF BIOACTIVE COMPOUNDS

Many of the molecules used for therapeutic purposes have their origins in nature, more specifically in plants. Many molecules of natural origin also serve as structural models for creating synthetic analogs. Falkowski and Raven [26] estimated that there are approximately 10^{25} cells of blue-green algae in the oceans, therefore the fixation of photosynthetic carbon and nitrogen in modern oceans is carried out by blue-green algae. Blue-green algae are known to contain a huge variety of toxic and bioactive substances. These metabolites have been explored as potential anticancer, antiviral, antibiotics, antioxidant, and antiinflammatory drugs. This unique species is known to cause the dermatitis called "swimmers itch," and is shown to produce a wide range of metabolites which include polyketides, lipopeptides, potent toxins, and many others [27]. Analysis of 424 natural products obtained from blue-green algae showed that 40.2% are lipopeptides, 5.6% are pure amino acids, 4.2% are fatty acids, 4.2% are macrolides, and 9% are amides [28]. Also, blue-green algae remain a largely unexplored resource of new bioactive compounds. Dinoflagellates also demonstrate an extraordinary chemical diversity. However, in addition to producing bioactive molecules of interest, such as macrolides and polyethers, they also yield potent toxins. For example, the dinoflagellate species *Gambierdiscus toxicus*, which is very common near reefs, produces the gambiertoxin. This toxin enters the food chain through fish that have ingested toxic microalgae [29]. The dinoflagellate marine genus *Amphidinium* also produces a wide range of polyketides and cytotoxic macrolides that may lead to new anticancer drugs. The biological activities of the compounds obtained from several species of microalgae and blue-green algae are shown in Table 8.3.

8.2.2 SECONDARY METABOLITES AS BIOACTIVE COMPOUNDS

Secondary metabolites are present in all organisms including algae and may perform antibiotic, anticancer, antiviral, and other activities.

8.2.2.1 Antibiotics

Fatty acids, bromophenols, tannins, polysaccharides, alcohols, halogenated compounds, peptides, lipopeptides, alkaloids, amides, tertiary sulfoniums, etc. are the substances that can act as antibiotics. Noscomin (8-[(5-carboxy-2-hydroxy)-benzyl]-2-hydroxy-1,1,4a,7,8-pentamethyl-1,2,3,4, 4a,6,7,8, 8a,9,10,10*a*-dodecahydrophenanthrene), an extracellular metabolite isolated from the *Nostoc commune*, showed antibacterial activity against *Bacillus cereus*, *Staphylococcus epidermidis*, and *Escherichia coli* [63,95,96]. Two exometabolites norharmane (9*H*-pyrido(3,4-*b*)indole) and 4,4′-dihydroxybiphenyl were extracted from *Nodularia harveyana* and *Nostoc insulare*, respectively. These exhibited antibacterial and antifungal activities. They perform an allelopathic function which can be regarded as a chemical defense [62]. Hapalindole-T, a methanolic extract of *Fischerella* sp., showed activity against *Mycobacterium tuberculosis*, *Enterobacter aerogenes*, *Staphylococcus aureus*, *Pseudomonas aeruginosa*, *Salmonella typhi*, and *E. coli* [44]. Luteophanol-D isolated from the cultured marine dinoflagellate *Amphidinium* sp. exhibited antibacterial activity against *Micrococcus luteus* [30].

8.2.2.2 Anticancer

Lipopeptides, a family of secondary metabolites of blue-green algae, are used to develop new and more effective drugs against cancer. These molecules exhibited their cytotoxic activities against cancer cells via induction of apoptosis.

Table 8.3 Products/Compounds Obtained From Various Species of Microalgae/Blue-Green Algae and Their Biological Activity

Sr. No.	Algae (Genus/Species)	Products	Biological Activity	References
1	*Amphidinium operculatum*	Amphidinolide N (macrolide)	Cytotoxic against human cancer cell lines	[30]
2	*Anabaena flos-aquae*	Saxitoxins	Neurotoxic alkaloid, potent blocker of nerve conduction that can produce prolonged anaesthesia	[31]
3	*Botryococcus braunii*	Carotenoids	Antioxidant activity	[32]
4	*Calothrix* sp.	Calothrixin	Antimalarial and anticancer activities	[33]
5	*Chlorella stigmatophora*	Sulfated polysaccharides	Antiviral, antiinflammatory and immunomodulatory activities. Aqueous extracts showed effects on the CNS	[34,35]
6	*Cochlodinium polykrikoides*	A1A2	Anti-HIV-1 activity and antiviral activity against parainfluenza virus type-2 and HSV-1	[36]
7	*Dunaliella primolecta*	Pheophorbides	Anti-herpes simplex virus (HSV) activity	[37]
8	*Dunaliella salina*	β-carotene	Free-radical scavenger, antitumor and cancer preventative activity, food additive, antioxidant, pro-vitamin A, health food supplement, pharmaceuticals	[27,38–41]
9	*Dunaliella tertiolecta*	No references about the compounds	Aqueous extract showed effects on the CNS, activity as a CNS depressant, and potential muscle relaxant	[42]
10	*Fischerella* sp.	Isonitrile-containing alkaloids Hapalindole-T	Antibacterial and antimycotic activity	[43,44]
11	*Gymnodinium catenatum*	Saxitoxin, saxitoxin analogues	Sodium channel blocking activity, aesthetic, effects on the cardiovascular system, and hypotensive effect	[45,46]

Continued

Table 8.3 Products/Compounds Obtained From Various Species of Microalgae/Blue-Green Algae and Their Biological Activity—cont'd

Sr. No.	Algae (Genus/Species)	Products	Biological Activity	References
12	*Gyrodinium impudicum*	p-KG03	Anti-influenza virus activity and anti-encephalomyocarditis virus activity	[47,48]
13	*Haematococcus pluvialis*	Astaxanthin	Antiinflammatory, antioxidant, treatment of muscular lesions, natural pigment, and dietary supplement	[49–52]
14	*Isochrysis galbana*	No references about the compounds	Aqueous extract showed effects on the CNS	[53]
15	*Lyngbya majuscula*	Lyngbyatoxin-a	Potent activator of protein kinase-C	[46]
16	*Lyngbya majuscula*	Apratoxin A (cyclodepsipeptide)	Cancer cell cytotoxicity	[54]
17	*Lyngbya majuscula/Schizothrix* sp. Assemblage of marine cyanobacteria	Somocystinamide A (lipopeptide)	Cytotoxicity against mouse neuro-2a neuroblastoma cells	[55]
18	*Lyngbya* sp.	Bisebromoamide	Cytotoxicity against HeLa cells and inhibitory activity against protein kinase	[56]
19	*Microcystis aeruginosa*	Microvirin	Antiviral activity	[57]
20	*Microcystis* sp.	Micropeptins	Inhibitory activity for the serine protease trypsin	[58]
21	*Nannochloropsis*, *Phaeodactylum tricornutum*, *Porphyridium cruentum*, *Crypthecodinium*, *Odontella*, *Pavlova lutheri*	Eicosapentaenoic acid (EPA)	Arteriosclerosis, anticancer, rheumatoid arthritis, Alzheimer disease, macular degeneration, cardiovascular development and protection of human CNS, nutraceutical, antimicrobial, and antiinflammatory	[59–62]
22	*Nodularia harveyana*	9*H*-pyrido (3,4-*b*) indole	Antibacterial and antifungal activities	[62]
23	*Nostoc commune*	Noscomin	Antibacterial activity against *Bacillus cereus*, *Staphylococcus epidermidis*, and *Escherichia coli*	[63]
24	*Nostoc ellipsosporum*	Cyanovirin-N	Anti-HSV-1 activity, ebola, hepatitis C, and anti-influenza activity	[64]

25	*Nostoc insulare*	4,4′-dihydroxybiphenyl	Antibacterial and antifungal activities	[62]
26	*Nostoc muscorum*	C-phycocyanin	Antimalarial activity	[65]
27	*Oscillatoria agardhii*	*O. agardhii* agglutinin (OAA)	Antiviral activity	[66]
28	*Ph. tricornutum*	Sulfated polysaccharides	Antiinflammatory effects, immunomodulatory activity, aqueous extract showed effects on the CNS. Activity as a CNS depressant, and potential muscle relaxant	[32,67]
29	*Skeletonema costatum*	No references about the compounds	Aqueous extract showed antidopaminergic activity and muscle relaxing activity	[68]
30	*S. platensis*	Allophycocyanin	Antiinflammatory, antioxidant, and antiviral activities	[69]
31	*Symploca* sp.	Largazole, Dolastatin 10	Cancer cell cytotoxicity	[70,71]
32	*Tetraselmis suecica*	No references about the compounds	Aqueous extract showed effects on the CNS	[53,72]
33	*Arthrospira*, *Porphyridium*	γ-Linolenic acid (GLA)	Integrity of tissues, delay of aging, and immunomodulatory activity	
34	*Po. cruentum*	Arachidonic acid (ARA)	Aggregative and vasoconstrictive for platelets, and health ingredients	[72,73]
35	*Crypthecodinium*, *Schizochytrium*, *I. galbana*	Docosahexaenoic acid (DHA)	*P. lutheri* nutraceutical, brain development, health ingredients, baby food, and poultry feed additive (eggs with ω-3 fatty acids)	[52,74,75]
36	*I. galbana*, *Chaetoceros*, *Skeletonema*, *P. lutheri*	Brassicasterol, Stigmasterol	Hypocholesterolemic and oyster food	[76,77]
37	*Arthrospira/Spirulina*	Phycocyanin	Natural pigment, antiinflammatory and antioxidant, health food supplement, and nutrient ingredient	[52,78]

Continued

Table 8.3 Products/Compounds Obtained From Various Species of Microalgae/Blue-Green Algae and Their Biological Activity—cont'd

Sr. No.	Algae (Genus/Species)	Products	Biological Activity	References
38	*Pavlova cruentum*	Phycoerythrin	Natural pigment and food marker	[52]
39	*Chlorella protothecoides*	Lutein, zeaxanthin, canthaxanthin	Antioxidant activity	[52,79]
40	*Dunaliella*, *Ph. tricornutum*, *Arthrospira platensis*, *Pa. cruentum*	Proteins	Health food supplement	[80–82]
41	*Pavlova tricornutum*, *Porphyridium*, *Anabaena*, *Synechococcus*	Superoxide dismutase (SOD)	Antioxidant, antiinflammatory, health food supplement, and therapeutics	[83–86]
42	*I. galbana*, *Amphidinium carterae*, *Prorocentrum minimum*	Carbonic anhydrase	Responsible for the conversion of CO_2 into H_2CO_3 and HCO_3^-	[87,88]
43	*Arthrospira*, *I. galbana*, *Po. cruentum*, *Pavlova*, *Tetraselmis*	Vitamin C, vitamin K, vitamins B12, A and E, α-tocopherol	Antioxidant, blood cell formation, blood-clotting mechanism, and immune system	[52,76]
44	*Porphyridium*	γ-Aminobutyric acid (GABA)	Neurotransmitter, antioxidant, antiinflammatory, development of abalone larvae, regulation of CNS in mammals, and immune system	[89–91]
45	*Gambierdiscus toxicus*, *Prorocentrum lima*	Okadaic acid	Antifungal, promotion of the secretion of nerve growth factor (NGF), and therapeutics	[92,93]
46	*Lyngbya majuscule*	Microcolin-A	Immunosuppressive activity	[94]

Somocystinamide-A is the lipopeptide isolated from *Lyngbya majuscule*, *Schizothrix* sp. which possesses exceptional biological properties and shows significant cytotoxicity against mouse neuro-2a neuroblastoma cells [55]. Apratoxin-A isolated from *L. majuscule* exhibits the cytotoxic activity against cancer cell lines. However, in vivo activity was reduced against a colon tumor and was ineffective against a mammary tumor [54]. The dolastatins (a family of peptides) isolated from the marine blue-green algae *Symploca* sp. are antiproliferative agents, inhibiting cancer cell growth and inducing apoptosis. Mitotic cell division is ceased when dolastatin interacts with the tubulin (microtubular subunit protein), which is an important component of the cell structure and kinetics [97]. Bisebromoamide (a cytotoxic peptide), isolated from *Lyngbya* sp., exhibited cytotoxicity against HeLa cells, together with potent protein kinase inhibition [56].

8.2.2.3 Antivirals

One of the antivirals isolated from algae and blue-green algae is cyanovirin-N (CV-N), which is highly effective virucidal protein from extracts of the cultured *Nostoc ellipsosporum*. CV-N irreversibly inactivates a broad spectrum of human immunodeficiency virus (HIV) strains by binding with high affinity to high mannose glycans on the viral envelop glycoprotein gp120, and interferes with viral interactions with target cell receptors essential for viral entry and cell-to-cell fusion. CV-N is also active against a number of other enveloped viruses including the herpes virus [64].

Microvirin (MVN) is the cyanobacterial lectin isolated from *Microcystis aeruginosa*, which inhibits the infection of a wide variety of HIV-1 strains. MVN binds to multiple glycans present on the envelope of HIV and subsequently inhibits the viral entry process. However, in contrast to CV-N, MVN does not enhance viral replication in pretreated peripheral blood mononuclear cells, and is more than 50-fold less cytotoxic than the former [57]. Sulfated exopolysaccharides are known to interfere with the absorption and penetration of some enveloped viruses into host cells. The marine microalgae *Gyrodinium impudicum* produce an extracellular sulfated polysaccharide p-KG03, which exhibited extraordinary in vitro antiviral activity against the encephalomyocarditis virus (EMCV). The p-KG03 also exhibits antiviral activity against the influenza virus type-A and type-B and inhibits the binding of the influenza virus to host cells, by capturing viral particles, and prevented cellular internalization of the virus and viral replication [47,48].

8.2.3 MICROALGAE FOR HUMAN HEALTH

Microalgae are marketed in different forms such as tablets, capsules, liquids, and can also be incorporated into pastas, snack foods, candy bars, gums, and beverages [98,99]. Owing to their diverse chemical properties, they can act as a nutritional supplement or represent a source of natural food colorants [60,61,100]. Four strains (*Arthrospira*, *Chlorella*, *Dunaliella salina*, and *Aphanizomenon flos-aquae*) are more dominant for commercial applications. The alleviation of hyperlipidemia, suppression of hypertension, protection against renal failure, growth promotion of intestinal *Lactobacillus*, and suppression of elevated serum glucose level are the health promoting effects of *Arthrospira* [98,99,101]. Many other companies sell a wide variety of *Arthrospira*-based nutraceuticals. For example, the Myanmar Spirulina Factory (Yangon, Myanmar) sells tablets, chips, pasta, and liquid extract; and Cyanotech (Kona, Hawaii, USA) produces products ranging from pure powder to packaged bottles under the name Spirulina pacifica. Tablets, extracts, and products in powdered form produced from *Arthrospira* are distributed in over 20 countries around the world.

Chlorella is produced by more than 70 companies over the world; Taiwan Chlorella Manufacturing and Co. (Taipei, Taiwan) is the largest producer with 400 t of dried biomass per year, and 130–150 t dry biomass per year is produced by Klötze, Germany. The world annual sales of *Chlorella* are in excess of US$38 billion. β-1,3-glucan which is an active immune-stimulator, a free-radical scavenger, and a reducer of blood lipids is produced by *Chlorella*. However, efficacy on gastric ulcers, wounds, constipation, and preventive action against atherosclerosis, hypercholesterolemia, and antitumor action is also performed by *Chlorella* [98,102]. *Chlorella* can also be used as a food additive owing to the taste and flavor-adjusting actions of its coloring agent [98,103].

D. salina is important for its β-carotene content that can reach 14% of dry weight [104]. For human consumption, Cognis Nutrition and Health, the world's largest producer of this strain, offers *Dunaliella* powder as an ingredient of dietary supplements and functional foods. The last major commercial strain application is *A. flos-aquae*. According to many research studies, used alone or in combination with other nutraceuticals and natural food products, *Anabaena flos-aquae* promotes overall good health [105–107].

8.3 ENGINEERING ALGAE TO MAKE WONDER MATERIAL NANO-CELLULOSE FOR BIOFUELS

Modern world is facing many major challenges such as resource depletion, climate change, energy crises, and increasing demand for oil. One of the promising solutions of these global issues is algal biofuels. Compared to different terrestrial plants, algal biomass has become more popular for the production of biofuels. Algal biomass is processed by three steps: alkalization of fibers, treatment with bleach, and finally acid hydrolysis. Major advantages of biofuel production from algae are: potentially high yield, no harsh and low-cost treatment required, and high biomass production without requiring large land area.

This part is aimed to summarize the state of the art of biofuel production from nano-cellulose material by engineering algae.

Energy security, increase of oil worth, resources depletion, and changes in temperature are the greatest demanding challenges facing our society in coming years. It has been predicted that biofuels and biorefineries will, at least partially, mitigate these issues and build a lot of balanced and sustainable economies. Thus far three generations of energy crops for biofuel production are developed. First-generation biofuels is made up of edible feedstock such as corn, oilseed, soybean, and sugarcane. Production of biofuel from these resources may contribute in increasing the food costs. Second-generation biofuels from waste and dedicated nonstandard plastic feed stocks such as *Miscanthus*, switch grass, or poplar have already got benefits over first generation. The primary and main benefits are higher yields and lower land demand (in each quality and quantity). Third generation biofuel feedstock, micro- and macroalgae will have a foothold over the previous two generations [108,109].

Microalgae are recognized as one of the oldest residing microorganisms on earth [3]. They develop at an extraordinary quick rate: 100 times quicker than terrestrial plants and they can double their biomass within 24 h [110]. Microalgae appear to be the only source of renewable biodiesel that can meet the global demand for transport fuels. Like plants microalgae also use sunlight to produce oils, but with more efficiency [111].

According to recent studies, a representative production of microalgae biomass lies between 15 and 25 tons/ha year. It is hypothesized that 4.5−7.5 tons/ha year lipids can be produced from 30% lipid contents present in microalgae cells [112]. This quantity is highly comparable to the assembly of oil from soybean (0.4 tons/ha year), oilseed (0.68 tons/ha year), feather palm (3.62 tons/ha year), and dicot genus (4.14 tons/ha year) [111,113]. In other words, culturing microalgae for biodiesel production needs the least acreage and holds a vital key feature for effective and powerful land usage. Moreover, microalgae are superior feedstock for bioethanol production. In addition to their high macromolecule contents, some microalgae contain carbohydrates (generally not cellulose) that can be used as carbon supply or substrate for fermentation [114].

The promise of maintainable energy production from algae has gained remarkable interest in recent years. Crude oil shortage and, therefore, the climate implications have driven research, analysis, and business ventures toward algae-based biofuels. That attention stems from many of apparently fascinating characteristics contained by algae that set it with the exception of alternative biomass sources. First, compared to terrestrial plants algae tends to supply additional biomass per unit area and can be cultivated on otherwise marginal land exploiting fresh or seawater. Moreover, their fast growing and invasive properties compared to traditional crops are appealing, since most developed nations consume additional energy that might offset exploiting slow growing terrestrial crops. Second, algae do not contend directly with food crops. Third, algae by virtue of their quick growth rates and aquatic surroundings may be cultivated in systems designed for concurrent biomass production, uptake of phylogeny dioxide, and removal of water pollutants [115]. There is a stable interest in algae-to-energy systems over the past many decades [109].

8.3.1 NANO-CELLULOSE

Cellulose is the most abundant natural polymer on the earth, and is widely used for many industries such as pharmaceutical applications, paper making, biofuel production, food packaging, and green composites with different innovative applications. Due to its different and unique characteristics such as high surface area to volume ratio, less weight, high strength and modulus, high aspect ratio, low cost, high tensile strength and stiffness, recyclability and biodegradability (environmental benefits), the market demand for this natural polymer has increased exponentially. During past few years, several studies accomplished the preparation of nanostructured-type cellulose materials from microalgae biomass resources by using top-down approach [109,116].

New nano-material sources of solid nano-cellulose polymer were recently isolated successfully from microalgae [116]. Compared to terrestrial plants, microalgae biomass is going to become more popular as a promising source of nano-cellulose for biofuel production in near future [117]. This section is aimed to comprise state of the art of biofuel (bioethanol, biodiesel, biohydrogen, and biomethane etc.) production from algal feedstock as described below.

8.3.2 BIOETHANOL

Bioethanol is long perceived to be the foremost possible short-term replacement of fossil fuels due to a well-established production technology. First-generation bioethanol suffered from supposed food versus fuel quandary [118,119]. Second-generation bioethanol, in spite of successes in recent years [120–122], still lacks economic viability at an outsized scale principally due to high investment prices [123–125]. The biggest downside related to conversion of lignocellulose to biofuels is its unexpected resistance to degradation typically represented as unmanageable biomass that makes saccharification expensive [126]. It is believed that fermentation could be another way to produce bioethanol from algal feedstock. Due to lack of lignin in algal biomass, it is feasible to convert it into simple sugar as compared to plant biomass. In addition to cellulose and hemicellulose, several microalgal species, e.g., *Chlorella vulgaris*, genus *Chlamydomonas reinhardtii* [127,128] accumulate more content of starch as their backup material.

Production of bioethanol by fermentation of algal feedstock depends upon the growth of the specific strain. Utilization and harvesting of algal feedstock, typically invasive, from the natural surroundings or growth of fittingly chosen algal strain in artificial conditions, is basically species-dependent. Macro algal species used for the production of ethanol are: red (*Kappaphycus alvarezii*, *Gelidium amansii*, *Gelidium elegans*, and *Gracilaria salicornia*); brown (*Laminaria japonica*, *Laminaria hyperborea*, *Saccharina latissima*, *Sargassum fulvellum*, *Undaria pinnatifida*, and *Alaria crassifolia*); and green (*Ulva lactuca* and *Ulva pertusa*). All these macro algal species have been used as a source of fermentable carbohydrate for the production of ethanol using yeast and bacteria [127,129,130].

8.3.3 ETHANOL-PRODUCING SUGARS IN MACROALGAE

Composition of algal cell wall differs from those of terrestrial plants. The key distinction is low content or absence of polymer in macro- and microalgal feedstock.

Polymer could be a heterogeneous, aromatic type that forms a physical seal and supports polysaccharide fibrils in plant cells [131]. The organic process which helped plants to adapt land system, made plant cell walls resistant to microbial degradation. For biomass conversion purpose it is a problematic attribute because it needs expensive pretreatment steps to liberate polysaccharide fibers for reaction. Despite a longer period of analysis and development, it is still a lively area of analysis for optimization and improvement [132,133].

The crystalline structure of algal cellulose is advantageous while choosing an acceptable mixture of enzymes. Unlike plants that contain largely polyose Iβ (monoclinic crystalline form), algal cells comprise polyose Iα (triclinic crystalline form) [134,135]. The latter type consists of weaker chemical bonding ensuing from spatial arrangement of individual cellulose chains with respect to one another. This loose chemical bonding would cause easier access to endo-cellulases throughout protein reaction ultimately exposing extra sites for subsequent exo-cellulase action on liberated fibrils. It should be noted that cellulose–cellulase interactions are advanced and complex while numerous cellulases usually exhibit preference toward completely different kinds of cellulose to confirm economical reaction. Polyose Iβ is additionally recalcitrant to conversion than polyose Iα [134,136–139]. A recent close study of substrate-specificity of cellulases proved that there conjointly exist enzymes that preferentially hydrolyze polyose Iβ [140]. Thus it is essential to select appropriate enzymes and methodology for specific protein feedstock [141].

Carbohydrate profiles of macroalgae and terrestrial plants conjointly dissent considerably. Both teams contain hemicelluloses, heterogeneous carbohydrate composed of pentoses, in the main wood sugar, which would be utilized for bioethanol production. Red seaweeds, for instance, are mainly composed of polymers of modified galactose: gum and agar. The major cell wall portion of algae *K. alvarezii* is κ-carrageenan [141,142], a linear, sulfated carbohydrate composed of sucrose that cannot be directly metabolized to plant product. Another rhodophyte, *G. amansii* is preponderantly composed of agar [143], a carbohydrate containing D- and L-galactose derivatives. On the other hand, cell walls of a macroalgae, *Laminaria* sp. are rich in diuretic and contain giant quantities of laminarin, a carbohydrate composed of 1,3-linked and 1,6-linked gluco pyranose units terminated with D-mannitol [143–145]. These sugars and sugar alcohols might be a further pool of carbohydrates when combined with applicable conversion theme. In addition to these hetero-polysaccharides each micro- and macroalgae store their reserves as starch, which ranges from 35% to 45% of dry cellular mass [127].

8.3.4 PREPROCESSING AND SACCHARIFICATION OF ALGAL BIOMASS

To address the giant diversity of algal feedstock composition, an array of strategies for cell disruption, pretreatment, and saccharification are used. These strategies are divided into physical, chemical, and catalytic approaches. Macroalgae as giant cellular organisms need initial size reduction to extend surface area for consequent reactions. Common techniques such as cutting, edging, crushing, and blending are employed to reduce the size [143–148]. Dried macroalgae are liquefied in distilled water to unharness cellular elements [125]. Hydrothermal fractionation is used to separate completely different cellular fractions of microalgae. Cells are treated at 150°C to yield super molecule, super molecule fractions, and later on carbohydrates are used successfully for bioethanol production [143].

Utilization of waste macroalgal biomass for production of alcohol by fermentation requires special attention. This shifts algal biofuel production toward a biorefinery construct due to the long-run sustainable answer for biomass energy [149,150]. Waste biomass of Rhodophyta *K. alvarezii* generated through extraction of sap, a natural biofertilizer, produced κ-carrageenan that was used for bioethanol production. Floating residues, a by-product of alginate extraction method, are also used as a feedstock for alcohol production [142]. Floating residues are alkali-insoluble fractions containing algae residue and are shaped by natural process and flotation of scrap throughout alginate production. Residues are subsequently collected, milled, dried, and used for biofuel production.

The most applied methodology of chemical pretreatments is treatment with dilute acid (0.1%–3% v/v H_2SO_4) carried out at high temperatures (120–130°C) for 15–120 min [120,141,146]. The acid treatment of polysaccharides has a further advantage over catalytic reaction because it yields both chain depolymerization and chemical reaction of salt moiety. Resultant D-galactose molecules would be later on utilized by most of the yeast strains. The acid chemical reaction of agar yields D- and L-galactose units. D-galactose would be solely soured by yeast whereas L-galactose acts as a fermentation matter and might significantly decrease overall potency of the method [151]. Saccharification is usually increased by introduction of ballroom dance process of biomass treatment; cells are initially treated with acid and then with an acceptable mixture of catalysts. The composition of catalysts combined, for the most part, depends on biomass origin. Catalyst treatment of algae is best reported for genus *Laminaria* sp. It is successfully saccharified by employing a combination of different catalysts cocktails after acid pretreatment. Additional aldohexose was liberated from laminarin by adding laminarase in fermentation mixture to extend overall potency [145,152].

To address heterogeneity of algal carbohydrates such as multienzymatic preparations containing preponderantly cellulase and cellobiase were applied successfully [143,147,153]. Two-step hydrolysis reaction of algae is used to saccharify predominant cellulose fraction with combination of a dilute acid methodology and cellulase treatment [148]. Starch-rich microalgae, *C. reinhardtii*, were transformed successfully to yield alcohol in simultaneous saccharification and solid state fermentation process, followed by successive treatment with α-amylase and glucoamylase [127]. Alkaline pretreatment with 0.75% (w/v) NaOH for 15 min at 120°C is solely used for saccharified algae to unleash possible sugars [154].

8.3.5 FERMENTATION

Although variations in sugar composition of various algae resulted in utilization of many nonstandard organisms for ethanol production, the foremost common organism for bioethanol production from algae is still yeast. Efficiency of plant product fermentation is stricken by two factors. First, the potency differs considerably, however, not because different algae species are additionally season- and site-dependent. Second, small molecular-weight compounds such as 5-hydroxymethylfurfural and levulinic acid (generated through acid reaction of biological feedstock) might have a profound impact on the potency of production from plant products [155].

Though acid reaction of biomass has been the most efficient methodology, the unsought compounds might influence overall potency of the method. This problem is additionally encountered in fermentative production of plant product from algal feedstock [141]. Fermentation for alcohol production from microalgae has higher conversion rates than that of macroalgae. Variety of factors such as cell disruption potency, fermentable sugar content, and cell membrane structure contribute to this distinction [146]. Microalgae are less studied as compared to macroalgae due to a diluent cell membrane. Microalgae designated for bioethanol production are renowned to supply significant content of starch that is a wonderful substrate for saccharification and fermentation by alcohol-producing organisms. Moreover, industrial strains for fermenting alcohol were designated based on the idea of efficient transformation of aldohexose instead of nonstandard sugars obtained from the reaction of macroalgal polysaccharides. Due to these factors, genetic engineering of ethanol-producing species is required to increase their specificity for their substrate, an acceptable way to increase bioethanol production.

8.3.6 BIODIESEL

Most work in the area of liquid biofuels was done in two basic fields: bioethanol production from macro- and microalgae, and biodiesel from microalgae oils. To date macroalgal biodiesel is barely reported and yields are much lower than those of microalgae. This scenario changes radically if microalgae are used to produce biodiesel.

Biodiesel is a monoalkyl ester of long-chain fatty acids currently derived from renewable feedstock, such as vegetable oil or animal fats, but not from microalgae. Unfortunately, biodiesel from oil crops, waste cooking oils, and animal fat is insufficient to satisfy the existing demand for fuels. This is likely to change as several companies are endeavoring to commercialize microalgal biodiesel. Microalgae appear to be the only source of renewable biodiesel that is capable of meeting the global demand. Technology for producing and using biodiesel is known for more than 50 years. The primary advantage of biodiesel is that it is one of the most renewable fuels, nontoxic, and biodegradable [111,149].

Biodiesel is usually defined as methyl (or ethyl) esters of fatty acids obtained by transesterification (alcoholysis) of triglycerides. Biodiesel is one of the biofuels with the most compatible characteristics with the fossil fuel (petroleum diesel). For example, the higher heating value (HHV) of biodiesel (39–41 MJ/kg) is comparable with petrodiesel (43 MJ/kg) and other important parameters, for instance flash point, cetane number, and kinematic viscosity, are similar to its fossil alternative. Thus, biodiesel can be directly used in the existing diesel engines [57]. The idea of producing diesel from microalgae is not new but is being taken seriously now because of the escalating price of petroleum and, more importantly, the emerging concern about global warming associated to burning of fossil fuels [111].

Microalgal biomass as feedstock for biofuel production is an attracting alternative to terrestrial plant utilization for biofuel production. Microalgae accumulate considerable amounts (up to 60%–65% of dry weight) of lipids or carbohydrates along with several secondary metabolites. Some of the latter are especially valuable compounds with an enormous range of industrial applications. The simultaneous production of lipids, and carbohydrates for biofuel production, and secondary metabolites in a biorefinery concept might allow the microalgal production to be economically feasible. Like bioethanol, biodiesel also suffers from food versus fuel dilemma. Increased demand for the production of edible oils has put limitations on using the same feedstock for biodiesel and vice versa raising concerns about long-term sustainability of this fuel.

In the United States, biodiesel is produced mainly from soybeans. Other sources of commercial biodiesel include canola oil, animal fat, palm oil, corn oil, and waste cooking oil. Obviously, oil crops cannot significantly contribute to replace petroleum-derived liquid fuels in the foreseeable future. Moreover, typical feedstock for biodiesel production, i.e., rapeseed and soybean, are very energy-intensive to cultivate or such as oil palm are likely to cause deforestation of tropical forests. For this reason attention turned to waste and nonedible sources of oil for biodiesel production. High theoretical productivities and oil accumulation exhibited by certain algal species under cellular stress made biodiesel production by algal oil transesterification a logical choice for research and development after relative success of biodiesel production from other feedstock. Biodiesel has now come to mean a very specific chemical modification of natural oils. One of the biggest advantages of biodiesel compared to many other alternative transportation fuels is that it can be used in existing diesel engines without modification, and its suitability for blending at any ratio with petroleum diesel. Any biofuel production process, which can successfully replace an equivalent conventional fuel, needs to fulfill three basic requirements. First, a sufficient feedstock to produce fuel at commercial scale should be grown; second it should cost less than conventional fossil fuel; and third it should match standard specifications of fuel quality. The oilseed crops mentioned above as well as other identified nonfood crops, e.g., *Jatropha curcas*, switch grass, and Karanj (*Pongamia pinnata*), are known to interfere with the terrestrial food chain crops up to some extent. Microalgae have potential to be used as a raw material for biodiesel production as they meet all those requirements. Microalgal lipids are mostly neutral lipids with lower degree of unsaturation which make these lipids a potential replacement for fossil fuel [149].

Microalgae possess high growth rate and commonly double their biomass within 24 h. Oil content in microalgae can exceed 80% by weight of dry biomass.

Microalgal biomass for biodiesel synthesis was produced using either pure or mixed cultures in phototrophic or heterotrophic growth systems.

In addition to microalgae, oil-producing heterotrophic microorganisms grown on a natural organic carbon source such as sugar can also be used to make biodiesel; however, heterotrophic production is not as efficient as using photosynthetic microalgae. This is because the renewable organic carbon sources required for growing heterotrophic microorganisms are produced ultimately by photosynthesis, usually in crop plants [111]. The fast growing photosynthetic microalgae are widely used in aquaculture, food, health, and biofuels.

Microalgal oils differ from most vegetable oils in being quite rich in polyunsaturated fatty acids with four or more double bonds. For example, eicosapentaenoic acid (EPA) having five double bonds and docosahexaenoic acid (DHA) having six double bonds occur commonly in algal oils. Fatty acids and fatty acid methyl esters (FAME) with four and more double bonds are susceptible to oxidation during storage, which reduces their acceptability for use in biodiesel. Some vegetable oils also face this problem. For example, vegetable oils such as high oleic canola oil contain large quantities of linoleic acid (with two double bonds) and linolenic acid (with three double bonds). Although these fatty acids have much higher oxidative stability compared to DHA and EPA, the European Standard EN 14214 limits the linolenic acid methyl ester content in biodiesel for vehicle use to 12%. No such limitation exists for biodiesel intended for use as heating oil, but acceptable biodiesel must meet other criteria related to the extent of total unsaturation of the oil. Total unsaturation of oil is indicated by its iodine value. Standards EN 14214 and EN 14213 require the iodine value of biodiesel to not exceed 120 and 130 g iodine/100 g biodiesel, respectively. Furthermore, both the European biodiesel standards limit the contents of FAME with four and more double bonds to a maximum of 1% mol. In view of the composition of many microalgal oils, most of them are unlikely to comply with the European biodiesel standards, but this need not be a significant limitation. The extent of unsaturation of microalgal oil and its content of fatty acids with more than four double bonds can be reduced easily by partial catalytic hydrogenation of the oil, the same technology that is commonly used in making margarine from vegetable oils [111].

Microalgae are a diverse group of organisms with considerable commercial potential as sources of various biochemicals, storage molecules, and metabolites such as lipids, sugars, amino acids, pigments, and toxins. Algal lipids can be processed into biooils and biodiesel. Most common methods for extraction of lipids from microalgae are expeller/oil press, liquid–liquid extraction (solvent extraction), supercritical fluid extraction (SFE), and ultrasound techniques. Oil presses or expellers are the most common method for extraction of oil from nuts and seeds. Same equipment and process would be appropriate for extraction of oil from microalgae. To ensure the efficacy of this process, algae must first need to be dried. Press uses pressure to break cells and compress out the oil. Though this method extracts almost 75% of oil without any special skill required, this conventional method has been reported to be less effective due to relatively longer extraction time. Another method for lipid extraction from microalgae is solvent extraction. In this approach, algae paste in water is extracted by adding organic solvents, such as benzene, cyclohexane, hexane, acetone, chloroform, etc. Solvent destroys the algal cell wall, and extracts oil from aqueous medium because of their higher solubility in organic solvents than water. The oil may be separated from the solvent extract by distillation process, and can be recycled for further use. Hexane is reported to be the most efficient solvent in extraction based on its highest extraction capability and low cost. A two-step process to extract improved lipids is used. In first step ethanol is used for extraction of the lipids and in the second step, hexane is used for purifying the extracted lipids. Around 80% lipids are extracted by this two-stage extraction. The fatty acid contents in microalgae varied between different species when extracted using chloroform: methanol (2:1, v/v) as a solvent. However, this particular method is unsuitable due to use of environmentally destructive solvents.

Another commonly employed method of extraction is supercritical fluid extraction (SFE), which is extremely time efficient. It makes use of high pressures and temperatures to rupture the cells. Another competent mechanism to be used in extraction of microalgae is via ultrasound. This method involves the exposure of algae to a high-intensity ultrasonic wave, which creates tiny cavitation bubbles around cells. Collapse of bubbles emits shockwaves, shattering the cell wall thus disrupting it and releasing the desired compounds into solution. By using this method, 90% extraction of fatty acids and pigments, and complete extraction of lipids is reported.

Conversion of algal oils to biodiesel is largely obtained with acid catalysis. Typical for the biodiesel industry, homogenous alkali catalysis was reported only occasionally; most likely because of the high content of free fatty acids in many algal feedstock that largely disqualify this method due to soap formation in competitive reaction between base and free fatty acids. Efficient acid catalysis has been reported for many strains as well as mixed cultures using both sequential (oil extraction followed by alcoholysis) and in situ transesterification. The major drawback of this method is the requirement for methanol to drive the reaction forward that impacts cost-effectiveness, sustainability, and applicability of glycerol, a coproduct for many industries. There are few reports on the biodiesel synthesis with the biological catalyst lipase. This most environment-friendly method of biofuel production is long seen as a potential breakthrough in biodiesel production, however, high catalyst cost still hinders the development of lipase-catalyzed biodiesel production in a large scale. Significant advances are made in biofuel production through metabolic engineering of blue-green algae.

A lot of research was done on microalgae and blue-green algae with the aim to produce numerous biotechnological products. However, native strains have a few shortcomings, such as limitations in cultivation, harvesting, and product extraction, which prevent reaching optimal production value at lowest costs. Such limitations require the intervention of genetic engineering to produce strains with superior properties. Promising advancements in the cultivation of cyanobacteria and microalgae have been achieved by improving photosynthetic efficiency through increasing

RuBisCO activity and truncation of light-harvesting antennae. Genetic engineering has also contributed to final product extraction by inducing autolysis and product secretory systems, to enable direct product recovery without going through costly extraction steps.

8.3.7 BIOGAS

Numerous works are done on biodiesel production technologies from algae. In contrast, the use of microalgae and macroalgae as potential substrates in processes of biogas production is addressed recently. The use of algal biomass for biogas production processes has proved a promising alternative and prospective source of organic substrate. Algae have many advantages over typical higher energetic crops. The algae, mainly the marine ones, were shown to contain high quantities of polysaccharides and lipid substances and to be free of sparingly degradable lignocellulose compounds. The algal species, e.g., *Scenedesmus* sp., *Spirulina* sp., *Euglena* sp., *C. vulgaris*, *Melosira* sp., *Oscillatoria* sp., *Laminaria* sp., *Macrocystis* sp., *Gracilaria ceae*, and *Ulva* sp. could be used as substrates for biogas production. The up-to-date studies confirm methane fermentation to be an effective technology for algae biomass conversion for energy purposes [156].

Anaerobic digestion (AD) or methane fermentation consists of organic carbon degradation into organic acids and biogas through a complex microbial community of bacteria and archaea. Biogas mainly consists of methane (around 65%), the most reduced state of carbon, and carbon dioxide (around 35%), the most oxidized state. Other gases (normally less than 1%), such as nitrogen, nitrogen oxides, hydrogen, ammonia, and hydrogen sulfide are also formed. AD comprises four stages, i.e., hydrolysis, acidogenesis, acetogenesis, and methanogenesis [157]. Each stage plays a different role during AD:

Hydrolysis involves the degradation of complex high-molecular-weight organic macromolecules into low-molecular-weight monomers. The active genera in hydrolysis include *Lactobacillus*, *Sporobacterium*, *Bacteroides*, etc.

Acidogenesis is the conversion of soluble monomers into various metabolic products, including (1) gaseous metabolic products such as hydrogen and carbon dioxide, and (2) soluble metabolic products such as volatile fatty acids and alcohols. The active genera in acidogenesis include *Clostridium*, *Ruminococcus*, *Paenibacillus*, etc.

Acetogenesis is the conversion of products of acidogenesis to acetic acid and hydrogen. The active genera in acetogenesis include *Aminobacterium*, *Acidaminococcus*, *Desulfovibrio*, along with many more.

Methanogenesis involves the production of methane by (1) hydrogenotrophic methanogenic archaea via consumption of hydrogen and carbon dioxide, and (2) aceticlastic methanogenic archaea via consumption of acetic acid. The active genera in methanogenesis include *Methanobacterium*, *Methanosphaera*, *Methanosarcina*, *Methanosaeta*, and *Methanobrevibacter* [158].

Pretreatment techniques were pointed out as a necessary step for algal cell disruption and biogas production. The effectiveness of pretreatment methods on biogas production depends on the characteristics of algae, i.e., the toughness and structure of the cell wall and the macromolecular composition of cells. In fact, species without cell wall (e.g., *Dunaliella* sp. and *Pavlova* sp.) or containing a glycoprotein cell wall (e.g., *Euglena* sp. and *Tetraselmis* sp.) produce higher methane yields than those with a more complex cell wall containing recalcitrant compounds (e.g. *Scenedesmus* sp. and *Chlorella* sp.). Pretreatment methods can be divided into four categories: thermal, mechanical, chemical, and biological processes. Thermal and mechanical pretreatments are regarded as the most effective for microalgae cell disruption. Mechanical pretreatments were less dependent on microalgae species, but required a higher energy input compared with chemical, thermal, and biological methods. Chemical pretreatments have been proved successful, particularly when combined with heat. However, the use of chemicals may contaminate downstream products. Enzymatic pretreatment seems to improve microalgae hydrolysis, which is promising due to its low-energy input [157]. The mechanical pretreatment seems to positively affect the methane production of macroalgae, though the result may be dependent on the species used. Indeed, the main effect is to reduce the particle size of the substrate making the complex organic matter more available to the attack of the hydrolytic enzymes [159].

8.3.8 BIOHYDROGEN

Algae are also a potential source of commercial biohydrogen used as a gas fuel or for electricity generation. Macroalgae are a potential source of biomass for the production of this gas due to their fast growth rates, ability to grow in oceanic environments, and their lack of the structural lignin which is especially difficult to digest. Many species of macroalgae are known for having high levels of carbohydrate, although in many cases these carbohydrates consist of nonglucose monosaccharides such as galactose. The red algae *G. amansii* and the brown algae *L. japonica* are both potential biomass sources for biohydrogen production through anaerobic fermentation. Thus, as bioprospecting of macroalgae for their fermentative future continues, it will be necessary to optimize pretreatment methods for maximum biohydrogen production. Photosynthetic microalgae are also able to produce biohydrogen directly through photofermentation in an anaerobic process involving oxidation of ferredoxin by hydrogenase enzyme. However, hydrogenases directly compete with many other metabolic processes for the partitioning of electrons, not all activities of hydrogenases function equally. Although hydrogen production from algae still seems far away from commercial viability, continued progress in this area shows its ultimate potential [160].

8.3.9 ECONOMICS OF BIODIESEL PRODUCTION

Recovery of oil from microalgal biomass and conversion of oil to biodiesel are not affected by whether the biomass is produced in raceways or photobioreactors. Hence, the cost of producing the biomass is the only relevant factor for a comparative assessment of photobioreactors and raceways for producing microalgal biodiesel. Moreover, microalgal biomass contains significant quantities of proteins, carbohydrates, and other nutrients. Therefore, the residual biomass from biodiesel production processes can be potentially used as animal feed. Some of the residual biomass can be used to produce methane by anaerobic digestion, for generating the electrical power necessary for running the microalgal biomass production facility.

8.3.10 ENHANCING ALGAL BIOLOGY

Genetic and metabolic engineering are likely to have the greatest impact on improving the economics of production of microalgal diesel. Genetic modification of microalgae has received little attention. Molecular level engineering can be used to potentially increase: photosynthetic efficiency to enable increased biomass yield on light, biomass growth rate, oil content in biomass, temperature tolerance to reduce the expense of cooling, and to reduce photoinhibition that actually reduces growth rate at midday light intensities in temperate and tropical zones, and susceptibility to photooxidation that damages cells. In addition, it is required to identify possible biochemical triggers and environmental factors that might favor accumulation of oil. Stability of engineered strains and methods for achieving stable production in industrial microbial processes are also important issues but are hardly examined for microalgae [111].

8.3.11 FUTURE PROSPECTS

As demonstrated here, microalgal biodiesel is technically feasible. It is the only renewable biodiesel that can completely displace petroleum-based fuels. Economics of producing microalgal biodiesel need to improve substantially to make it competitive with fossil-fuel diesel, but the level of necessary improvement appears to be attainable. Producing low-cost microalgal biodiesel primarily requires improvements to algal biology through genetic and metabolic engineering. Use of the biorefinery concept and advances in photobioreactor engineering will further lower the cost of production. In view of their much greater productivity than raceways, tubular photobioreactors are likely to be used in producing much of the microalgal biomass required for making biodiesel. Photobioreactors provide a controlled environment that can be tailored to the specific demands of highly productive microalgae to attain a consistently good annual yield of oil [108,111,161].

8.3.12 CHALLENGES AND HURDLES

Microalgal systems could contribute to a sustainable bioenergy production (bioethanol, biodiesel, and biogas), however, different biotechnical, environmental, and economic challenges have to be overcome before energy products from these systems can enter the market.

8.3.12.1 Biotechnical Challenges

The main biotechnical challenges are cultivation (large-scale production), harvesting (recovery and extraction), and genetic engineering of microalgae.

8.3.12.2 Ecological Challenges

A major advantage of microalgae is their ability to provide additional environmental benefits (CO_2 recycling and wastewater treatment). However, to realize these benefits some hurdles (nutrients requirement, recycling of CO_2, and suitability of land) should overcome.

8.3.12.3 Economic Challenges

The development of microalgae for different types of mass energy production is in its beginning stage. Because of this it seems critical to assume the cost requirements for the state of the art techniques used for small-scale production of high-value products. Cultivating and processing algae consumes energy, both in organization and operation. Depending on the method of growing, the process of harvesting, and yield, the energetic inputs of microalgae production could also increase the energetic output. However, ongoing research in the reactor design is promising and will lead to cheaper and more energy-effective designs [162].

REFERENCES

[1] Kuruppalil Z. Green plastics: an emerging alternative for petroleum-based plastics. Int J Eng Res Innov 2011;3:59–64.

[2] Pathak S, Sneha CLR, Mathew BB. Bioplastics: its timeline based scenario & challenges. J Polym Biopolym Phys Chem 2014;2:84–90.

[3] Song D, Fu J, Shi D. Exploitation of oil-bearing microalgae for biodiesel. Chin J Biotechnol 2008;24:341–8.

[4] Chu W-L. Biotechnological applications of microalgae. IeJSME 2012:24–37.

[5] Yan N, Fan C, Chen Y, et al. The potential for microalgae as bioreactors to produce pharmaceuticals. Int J Mol Sci 2016;17:962.

[6] Lordan S, Smyth TJ, Soler-Vila A, et al. The α-amylase and α-glucosidase inhibitory effects of Irish seaweed extracts. Food Chem 2013;141:2170–6.

[7] Suriyamongkol P, Weselake R, Narine S, et al. Biotechnological approaches for the production of polyhydroxyalkanoates in microorganisms and plants — a review. Biotechnol Adv 2007;25:148–75.

[8] Corma A, Iborra S, Velty A. Chemical routes for the transformation of biomass into chemicals. Chem Rev 2007;107:2411–502.

[9] Stevens ES. What makes green plastics green? Biocycle 2003;44:24–7.

[10] Ohara H. Biorefinery. Appl Microbiol Biotechnol 2003;62:474–7.

[11] Rajendran N, Puppala S, SR M, et al. Seaweeds can be a new source for bioplastics. J Pharm Res 2012;5:1476–9.

[12] Chiellini E, Cinelli P, Ilieva VI, et al. Biodegradable thermoplastic composites based on polyvinyl alcohol and algae. Biomacromolecules 2008;9:1007–13.

[13] Amel I, Wahiba H, Fethi M, et al. Bioplastic from agar: hydrophilic and thermomechanical properties. Bull Inst Natn Scien Tech Mer Salammbô 2015;42.

[14] Machmud MN, Fahmi R, Abdullah R, et al. Characteristics of red algae bioplastics/latex blends under tension. Int J Sci Eng 2013;5. http://dx.doi.org/10.12777/ijse.5.2.81-88. Epub ahead of print October 15, 2013.
[15] Thurmund W. Seaweed : a new wave of investment in macro-algae. Algae 2010:65−6.
[16] Maheswari N, Ahilandeswari K. Production of bioplastic using *Spirulina* plate and comparison with commercial plastic. Res Environ Life Sci 2011;4:133−6.
[17] Zeller MA, Hunt R, Jones A, et al. Bioplastics and their thermoplastic blends from *Spirulina* and *Chlorella* microalgae. J Appl Polym Sci 2013;130:3263−75.
[18] Stabili L, Acquaviva MI, Biandolino F, et al. Biotechnological potential of the seaweed *Cladophora rupestris* (Chlorophyta, Cladophorales) lipidic extract. N Biotechnol 2014;31:436−44.
[19] Abreu PM, Galindro JM, Relva AM, et al. Non-terpenoid compounds from *Plocamium cartilagineum*. Phytochemistry 1997;45:1601−3.
[20] Stabili L, Acquaviva MI, Biandolino F, et al. The lipidic extract of the seaweed *Gracilariopsis longissima* (Rhodophyta, Gracilariales): a potential resource for biotechnological purposes? N Biotechnol 2012;29:443−50.
[21] Mishra AK, Mishra SB. Cellulose based green bioplastics for biomedical engineering. In: Pilla S, editor. Handbook of bioplastics and biocomposites engineering applications. Hoboken, NJ, USA: John Wiley & Sons, Inc.; 2011. http://dx.doi.org/10.1002/9781118203699.ch12.
[22] Rhim JW. Preparation and characterization of vacuum sputter silver coated PLA film. LWT Food Sci Technol 2013;54:477.
[23] Noreen A, Zia KM, Zuber M, et al. A critical review of algal biomass: a versatile platform of bio-based polyesters from renewable resources. Int J Biol Macromol 2016;86: 937−49.
[24] Doron L, Segal N, Shapira M. Transgene expression in microalgae—from tools to applications. Front Plant Sci 2016;7:1−24.
[25] Hempel F, Bozarth AS, Lindenkamp N, et al. Microalgae as bioreactors for bioplastic production. Microb Cell Fact 2011;10:81.
[26] Falkowski PG, Raven JA. Aquatic photosynthesis. Oxford: Blackwell Scientific Publishers; 1997.
[27] Singh S, Kate BN, Banerjee UC. Bioactive compounds from cyanobacteria and microalgae: an overview. Crit Rev Biotechnol 2005;25:73−95.
[28] Burja AM, Banaigs B, Abou-Mansour E, et al. Marine cyanobacteria—a prolific source of natural products. Tetrahedron 2001;57:9347−77.
[29] Klaassen CD, editor. Casarett and Do ull's Toxicology: the basic science of poisons. 7th ed. New York: McGrawHill; 2008.
[30] Kobayashi J, Kubota T. Bioactive macrolides and polyketides from marine dinoflagellates of the genus *Amphidinium*$^{\perp}$. J Nat Prod 2007;70:451−60.
[31] Wiese M, D'Agostino PM, Mihali TK, et al. Neurotoxic alkaloids: saxitoxin and its analogs. Mar Drugs 2010;8:2185−211.
[32] Rao AR, Sarada R, Baskaran V, et al. Antioxidant activity of *Botryococcus braunii* extract elucidated in vitro models. J Agric Food Chem 2006;54:4593−9.
[33] Rickards RW, Rothschild JM, Willis AC, et al. Calothrixins A and B, novel pentacyclic metabolites from *Calothrix* cyanobacteria with potent activity against malaria parasites and human cancer cells. Tetrahedron 1999;55:13513−20.
[34] Guzmán S, Gato A, Lamela M, et al. Anti-inflammatory and immunomodulatory activities of polysaccharide from *Chlorella stigmatophora* and *Phaeodactylum tricornutum*. Phyther Res 2003;17:665−70.

[35] Laguna MR, Villar R, Calleja JM, et al. Effects of *Chlorella stigmatophora* extract on the central nervous system. Planta Med 1993;59:125–30.

[36] Hasui M, Matsuda M, Okutani K, et al. In vitro antiviral activities of sulfated polysaccharides from a marine microalga (*Cochlodinium polykrikoides*) against human immunodeficiency virus and other enveloped viruses. Int J Biol Macromol 1995;17:293–7.

[37] Ohta S, Ono F, Shiomi Y, et al. Anti-herpes simplex virus substances produced by the marine green alga, *Dunaliella primolecta*. J Appl Phycol 1998;10:349–55.

[38] Borowitzka MA. Microalgae as sources of pharmaceuticals and other biologically active compounds. J Appl Phycol 1995;7:3–15.

[39] Murthy KNC, Vanitha A, Rajesha J, et al. In vivo antioxidant activity of carotenoids from *Dunaliella salina* — a green microalga. Life Sci 2005;76:1381–90.

[40] El Baz FK, Aboul-Enein AM, El-Baroty GS, et al. Accumulation of antioxidant vitamins in *Dunaliella salina*. J Biol Sci 2002;2:220–3.

[41] Plaza M, Herrero M, Cifuentes A, et al. Innovative natural functional ingredients from microalgae. J Agric Food Chem 2009;57:7159–70.

[42] Laguna MR, Villar RM, Cadavid I, et al. Effect of *Phaeodactylum tricornutum* and *Dunaliella tertiolecta* extracts on the central nervous system. Planta Med 1990;56: 152–7.

[43] Mo S, Krunic A, Santarsiero BD, et al. Hapalindole-related alkaloids from the cultured cyanobacterium *Fischerella ambigua*. Phytochemistry 2010;71:2116–23.

[44] Asthana RK, Srivastava A, Singh AP, et al. Identification of an antimicrobial entity from the cyanobacterium *Fischerella* sp. isolated from bark of *Azadirachta indica* (Neem) tree. J Appl Phycol 2006;18:33–9.

[45] Negri A, Stirling D, Quilliam M, et al. Three novel hydroxybenzoate saxitoxin analogues isolated from the dinoflagellate *Gymnodinium catenatum*. Chem Res Toxicol 2003;16:1029–33.

[46] Bhakuni DS, Rawat DS. Bioactive marine natural products. Berlin/Heidelberg: Springer-Verlag; 2005. http://dx.doi.org/10.1007/1-4020-3484-9. Epub ahead of print 2005.

[47] Kim M, Yim JH, Kim SY, et al. In vitro inhibition of influenza A virus infection by marine microalga-derived sulfated polysaccharide p-KG03. Antivir Res 2012;93: 253–9.

[48] Yim JH, Kim SJ, Ahn SH, et al. Antiviral effects of sulfated exopolysaccharide from the marine microalga *Gyrodinium impudicum* strain KG03. Mar Biotechnol (NY) 2004;6:17–25.

[49] Todd LR, Ysewski G. Commercial potential for *Haematococcus* microalgae as a natural source of astaxanthin. Trends Biotechnol 2000;18:160–7.

[50] Guerin M, Huntley ME, Olaizola M. Haematococcus astaxanthin: applications for human health and nutrition. Trends Biotechnol 2003;21:210–6.

[51] Ohr LM. Riding the nutraceuticals wave. Food Technol 2005;59:95–6.

[52] Pulz O, Gross W. Valuable products from biotechnology of microalgae. Appl Microbiol Biotechnol 2004;65:635–48.

[53] Laguna MR, Villar R, Cadavid I, et al. Effects of extracts of *Tetraselmis suecica* and *Isochrysis galbana* on the central nervous system. Planta Med 1993;59:207–14.

[54] Luesch H, Yoshida WY, Moore RE, et al. Total structure determination of apratoxin A, a potent novel cytotoxin from the marine cyanobacterium *Lyngbya majuscula*. J Am Chem Soc 2001;123:5418–23.

[55] Nogle LM, Gerwick WH. Somocystinamide A, a novel cytotoxic disulfide dimer from a Fijian marine cyanobacterial mixed assemblage. Org Lett 2002;4:1095–8.

[56] Teruya T, Sasaki H, Fukazawa H, et al. Bisebromoamide, a potent cytotoxic peptide from the marine cyanobacterium *lyngbya* sp.: isolation, stereostructure, and biological activity. Org Lett 2009;11:5062–5.

[57] Huskens D, Férir G, Vermeire K, et al. Microvirin, a novel alpha(1,2)-mannose-specific lectin isolated from *Microcystis aeruginosa*, has anti-HIV-1 activity comparable with that of cyanovirin-N but a much higher safety profile. J Biol Chem 2010;285: 24845–54.

[58] Zafrir E, Carmeli S. Micropeptins from an Israeli fishpond water bloom of the cyanobacterium *Microcystis* sp. J Nat Prod 2010;73:352–8.

[59] Lukiw WJ, Bazan NG. Docosahexaenoic acid and the aging brain. J Nutr 2008;138: 2510–4.

[60] Apt KE, Behrens PW. Commercial developments in microalgal biotechnology. J Phycol 1999;35:215–26.

[61] Borowitzka MA. Commercial production of microalgae: ponds, tanks, tubes and fermenters. J Biotechnol 1999;70:313–21.

[62] Volk R-B, Furkert FH. Antialgal, antibacterial and antifungal activity of two metabolites produced and excreted by cyanobacteria during growth. Microbiol Res 2006;161: 180–6.

[63] Jaki B, Orjala J, Sticher O. A novel extracellular diterpenoid with antibacterial activity from the cyanobacterium *Nostoc commune*. J Nat Prod 1999;62:502–3.

[64] Tiwari V, Shukla SY, Shukla D. A sugar binding protein cyanovirin-N blocks herpes simplex virus type-1 entry and cell fusion. Antivir Res 2009;84:67–75.

[65] Pankaj PP, Mallick N, Biswas S, et al. In vitro antimalarial activity of c-phycocyanin from nostoc muscorum. Bioscan 2010;1:69–78.

[66] Koharudin LMI, Furey W, Gronenborn AM. Novel fold and carbohydrate specificity of the potent anti-HIV cyanobacterial lectin from *Oscillatoria agardhii*. J Biol Chem 2011;286:1588–97.

[67] Villar R, Laguna MR, Calleja JM, et al. Effects of *Phaeodactylum tricornutum* and *Dunaliella tertiolecta* extracts on the central nervous system. Planta Med 1992;58: 405–9.

[68] Villar R, Laguna MR, Calleja JM, et al. Effects of *Skeletonema costatum* extracts on the central nervous system. Planta Med 1992;58:398–404.

[69] Shih S-R, Tsai K-N, Li Y-S, et al. Inhibition of enterovirus 71-induced apoptosis by allophycocyanin isolated from a blue-green alga *Spirulina platensis*. J Med Virol 2003;70:119–25.

[70] Taori K, Liu Y, Paul VJ, et al. Combinatorial strategies by marine cyanobacteria: symplostatin 4, an antimitotic natural dolastatin 10/15 hybrid that synergizes with the coproduced HDAC inhibitor largazole. Chembiochem 2009;10:1634–9.

[71] Luesch H, Moore RE, Paul VJ, et al. Isolation of dolastatin 10 from the marine cyanobacterium *Symploca* species VP642 and total stereochemistry and biological evaluation of its analogue symplostatin 1. J Nat Prod 2001;64:907–10.

[72] Cohen Z, Heimer Y. Production of polyunsaturated fatty acids (EPA, ARA and GLA) by the microalgae *Porphyridium* and *Spirulina*. In: Industrial applications of single cell oils. AOCS Publishing; 1992. http://dx.doi.org/10.1201/9781439821855.ch14. Epub ahead of print June 1992.

[73] Gordon PVR. The implication of omega−3 fatty acids in human health. Lancaster, USA: Technomic Publishing; 1992.
[74] Barclay W, Zeller S. Nutritional enhancement of n-3 and n-6 fatty acids in rotifers and *Artemia* Nauplii by feeding spray-dried *Schizochytrium* sp. J World Aquac Soc 1996; 27:314−22.
[75] Sijtsma L, de Swaaf ME. Biotechnological production and applications of the ω-3 polyunsaturated fatty acid docosahexaenoic acid. Appl Microbiol Biotechnol 2004; 64:146−53.
[76] Bandarra NM, Pereira PA, Batista I, et al. Fatty acids, sterols and α-tocopherol in *Isochrysis galbana*. J Food Lipids 2003;10:25−34.
[77] Tsitsa-Tzardis E, Patterson GW, Wikfors GH, et al. Sterols of *Chaetoceros* and *Skeletonema*. Lipids 1993;28:465−7.
[78] Mao TK, Van de Water J, Gershwin ME. Effects of a *Spirulina*-based dietary supplement on Cytokine production from allergic Rhinitis patients. J Med Food 2005;8: 27−30.
[79] Shi XFC. High yield production of lutein by heterotrophic *Chlorella protothecoides* in fed-batch systems. The Netherlands: Kluwer Academic Publishers; 2001.
[80] Becker E. Microalgae biotechnology and microbiology. Cambridge: Cambridge University Press; 1994.
[81] Becker E. Microalgae in human and animal nutrition. Oxford: Blackwell Publishing; 2004.
[82] Rasmussen RS, Morrissey MT. Marine biotechnology for production of food ingredients. Adv Food Nutr Res 2007;52:237−92.
[83] García-González A, Ochoa JL. Anti-inflammatory activity of *Debaryomyces hansenii* Cu,Zn-SOD. Arch Med Res 1999;30:69−73.
[84] Guzmán-Murillo MA, López-Bolaños CC, Ledesma-Verdejo T, et al. Effects of fertilizer-based culture media on the production of exocellular polysaccharides and cellular superoxide dismutase by *Phaeodactylum tricornutum* (Bohlin). J Appl Phycol 2007;19:33−41.
[85] Priya B, Premanandh J, Dhanalakshmi RT, et al. Comparative analysis of cyanobacterial superoxide dismutases to discriminate canonical forms. BMC Genomics 2007;8: 435.
[86] Thépenier C, Chaumont D, Gudin C. Mass culture of *Porphyridium cruentum*: a multiproduct strategy for the biomass valorisation. London: Elsevier Applied Science Publishers; 1988.
[87] Nimer NA, Merrett MJ. Calcification and utilization of inorganic carbon by the coccolithophorid *Emiliania huxleyi* Lohmann. New Phytol 1992;121:173−7.
[88] Sun Y, Wang C. The optimal growth conditions for the biomass production of *Isochrysis galbana* and the effects that phosphorus, Zn^{2+}, CO_2, and light intensity have on the biochemical composition of *Isochrysis galbana* and the activity of extracellular CA. Biotechnol Bioprocess Eng 2009;14:225−31.
[89] Li K, Xu E. The role and the mechanism of gamma-aminobutyric acid during central nervous system development. Neurosci Bull 2008;24:195−200.
[90] Morse DE, Hooker N, Duncan H, et al. γ-aminobutyric acid, a neurotransmitter, induces planktonic abalone larvae to settle and begin metamorphosis. Science 1979; 204:407−10.
[91] Obrietan K, Gao X-B, Van Den Pol AN. Excitatory actions of GABA increase BDNF expression via a MAPK-CREB-dependent mechanism−a positive feedback circuit in developing neurons. J Neurophysiol 2002;88:1005−15.

[92] Nagai H, Murata M, Torigoe K, et al. Gambieric acids, new potent antifungal substances with unprecedented polyether structures from a marine dinoflagellate *Gambierdiscus toxicus*. J Org Chem 1992;57:5448–53.

[93] Pshenichkin SP, Wise BC. Okadaic acid increases nerve growth factor secretion, mRNA stability, and gene transcription in primary cultures of cortical astrocytes. J Biol Chem 1995;270:5994–9.

[94] Vikrant Arya VKG. A review on marine immunomodulators. Int J Pharm life Sci 2011; 2:751–8.

[95] Hannon M, Gimpel J, Tran M, et al. Biofuels from algae: challenges and potential. Biofuels 2010;1:763–84.

[96] Cohen Z. Chemicals from microalgae. Oxford: CRC Press; 1999.

[97] Zheng L-H, Wang Y-J, Sheng J, et al. Antitumor peptides from marine organisms. Mar Drugs 2011;9:1840–59.

[98] Yamaguchi K. Recent advances in microalgal bioscience in Japan, with special reference to utilization of biomass and metabolites: a review. J Appl Phycol 1996;8: 487–502.

[99] Liang S, Liu X, Chen F, et al. Current microalgal health food R & D activities in China. In: Asian Pacific Phycology in the 21st Century: Prospects and Challenges. Dordrecht: Springer Netherlands; 2004. p. 45–8.

[100] Soletto D, Binaghi L, Lodi A, et al. Batch and fed-batch cultivations of *Spirulina platensis* using ammonium sulphate and urea as nitrogen sources. Aquaculture 2005;243: 217–24.

[101] Vílchez C, Garbayo I, Lobato MV, et al. Microalgae-mediated chemicals production and wastes removal. Enzyme Microb Technol 1997;20:562–72.

[102] Jong-Yuh C, Mei-Fen S. Potential hypoglycemic effects of *Chlorella* in streptozotocin-induced diabetic mice. Life Sci 2005;77:980–90.

[103] Gouveia L. Evolution of pigment composition in *Chlorella vulgaris*. Bioresour Technol 1996;57:157–9.

[104] Metting FB. Biodiversity and application of microalgae. J Ind Microbiol Biotechnol 1996;17:477–89.

[105] Jensen GS. Blue-Green algae as an immuno-enhancer and biomodulator. Jana 2001;3: 24–30.

[106] Benedetti S, Benvenuti F, Pagliarani S, et al. Antioxidant properties of a novel phycocyanin extract from the blue-green alga *Aphanizomenon flos-aquae*. Life Sci 2004;75: 2353–62.

[107] Pugh N. Characterization of human monocyte activation by a water soluble preparation of *Aphanizomenon flos-aquae*. Phytomedicine 2001;8:445–53.

[108] Daroch M, Geng S, Wang G. Recent advances in liquid biofuel production from algal feedstocks. Appl Energy 2013;102:1371–81.

[109] Lam E, Male KB, Chong JH, et al. Applications of functionalized and nanoparticle-modified nanocrystalline cellulose. Trends Biotechnol 2012;30:283–90.

[110] Tredici MR. Photobiology of microalgae mass cultures: understanding the tools for the next green revolution. Biofuels 2010;1:143–62.

[111] Chisti Y. Biodiesel from microalgae beats bioethanol. Trends Biotechnol 2008;26: 126–31.

[112] Tsukahara K, Sawayama S. Liquid fuel production using microalgae. J Jpn Pet Inst 2005;48:251–9.

[113] Lam MK, Lee KT. Renewable and sustainable bioenergies production from palm oil mill effluent (POME): win-win strategies toward better environmental protection. Biotechnol Adv 2011;29:124−41.

[114] Harun R, Danquah MK, Forde GM. Microalgal biomass as a fermentation feedstock for bioethanol production. J Chem Technol Biotechnol 2010;85:199−203.

[115] Parmar A, Singh NK, Pandey A, et al. Cyanobacteria and microalgae: a positive prospect for biofuels. Bioresour Technol 2011;102:10163−72.

[116] Bettaieb F, Khiari R, Dufresne A, et al. Mechanical and thermal properties of *Posidonia oceanica* cellulose nanocrystal reinforced polymer. Carbohydr Polym 2015;123: 99−104.

[117] Korzen L, Pulidindi N, Israel A, et al. RSC advances single step production of bioethanol from the seaweed *Ulva rigida* using sonication †. RSC Adv 2015;5:16223−9.

[118] Fargione J, Hill J, Tilman D, et al. Land clearing and the biofuel carbon debt. Science (80) 2008;319:1235−8.

[119] Cassman KG, Liska AJ. Food and fuel for all: realistic or foolish? Biofuels, Bioprod Biorefin 2007;1:18−23.

[120] Heinzelman P, Snow CD, Wu I, et al. A family of thermostable fungal cellulases created by structure-guided recombination. Proc Natl Acad Sci USA 2009;106: 5610−5.

[121] Talebnia F, Karakashev D, Angelidaki I. Production of bioethanol from wheat straw: an overview on pretreatment, hydrolysis and fermentation. Bioresour Technol 2010; 101:4744−53.

[122] Sørensen A, Teller PJ, Hilstrøm T, et al. Hydrolysis of *Miscanthus* for bioethanol production using dilute acid presoaking combined with wet explosion pre-treatment and enzymatic treatment. Bioresour Technol 2008;99:6602−7.

[123] Wingren A, Galbe M, Zacchi G. Energy considerations for a SSF-based softwood ethanol plant. Bioresour Technol 2008;99:2121−31.

[124] Balat M, Balat H. Recent trends in global production and utilization of bio-ethanol fuel. Appl Energy 2009;86:2273−82.

[125] He J, Zhang W. Techno-economic evaluation of thermo-chemical biomass-to-ethanol. Appl Energy 2011;88:1224−32.

[126] Himmel ME, Ding S-Y, Johnson DK, et al. Biomass recalcitrance: engineering plants and enzymes for biofuels production. Science 2007;315:804−7.

[127] Choi SP, Nguyen MT, Sim SJ. Enzymatic pretreatment of *Chlamydomonas reinhardtii* biomass for ethanol production. Bioresour Technol 2010;101:5330−6.

[128] Brányiková I, Maršálková B, Doucha J, et al. Microalgae-novel highly efficient starch producers. Biotechnol Bioeng 2011;108:766−76.

[129] Uduman N, Qi Y, Danquah MK, et al. Dewatering of microalgal cultures: a major bottleneck to algae-based fuels. J Renew Sustain Energy 2010;2. http://dx.doi.org/10.1063/1.3294480. Epub ahead of print 2010.

[130] Christenson L, Sims R. Production and harvesting of microalgae for wastewater treatment, biofuels, and bioproducts. Biotechnol Adv 2011;29:686−702.

[131] Jung SK, Parisutham V, Jeong SH, et al. Heterologous expression of plant cell wall degrading enzymes for effective production of cellulosic biofuels. J Biomed Biotechnol 2012. http://dx.doi.org/10.1155/2012/405842. Epub ahead of print 2012.

[132] Hendriks ATWM, Zeeman G. Pretreatments to enhance the digestibility of lignocellulosic biomass. Bioresour Technol 2009;100:10−8.

[133] Agbor VB, Cicek N, Sparling R, et al. Biomass pretreatment: fundamentals toward application. Biotechnol Adv 2011;29:675–85.

[134] Hayashi N, Sugiyama J, Okano T, et al. Selective degradation of the cellulose I(α) component in *Cladophora* cellulose with *Trichoderma viride* cellulase. Carbohydr Res 1997;305:109–16.

[135] VanderHart DL, Atalla RH. Studies of microstructure in native celluloses using solid-state carbon-13 NMR. Macromolecules 1984;17:1465–72.

[136] Nishiyama Y, Sugiyama J, Chanzy H, et al. Crystal structure and hydrogen bonding system in cellulose I α from synchrotron X-ray and neutron fiber diffraction. J Am Chem Soc 2003;125:14300–6.

[137] Wada M, Chanzy H, Nishiyama Y, et al. Cellulose III I crystal structure and hydrogen bonding by synchrotron X-ray and neutron fiber diffraction. Macromolecules 2004;37: 8548–55.

[138] Qian X, Ding S-Y, Nimlos MR, et al. Atomic and electronic structures of molecular crystalline cellulose Iβ: a first-principles investigation. Macromolecules 2005;38: 10580–9.

[139] Hayashi N, Kondo T. Enzymatically produced nano-ordered elements containing cellulose Iβ crystalline domains of *Cladophora* cellulose. Handb Polym Nanocomp Process Perform Appl Vol C Polym Nanocomp Cellul Nanoparticles 2015;61:1–14.

[140] Corgié SC, Smith HM, Walker LP. Enzymatic transformations of cellulose assessed by quantitative high-throughput fourier transform infrared spectroscopy (QHT-FTIR). Biotechnol Bioeng 2011;108:1509–20.

[141] Meinita MDN, Kang JY, Jeong GT, et al. Bioethanol production from the acid hydrolysate of the carrageenophyte *Kappaphycus alvarezii* (cottonii). J Appl Phycol 2012; 24:857–62.

[142] Khambhaty Y, Mody K, Gandhi MR, et al. *Kappaphycus alvarezii* as a source of bioethanol. Bioresour Technol 2012;103:180–5.

[143] Kim NJ, Li H, Jung K, et al. Ethanol production from marine algal hydrolysates using *Escherichia coli* KO11. Bioresour Technol 2011;102:7466–9.

[144] Horn SJ, Aasen IM, Stgaard K. Ethanol production from seaweed extract. J Ind Microbiol Biotechnol 2000;25:249–54.

[145] Adams JM, Gallagher JA, Donnison IS. Fermentation study on *Saccharina latissima* for bioethanol production considering variable pre-treatments. J Appl Phycol 2009; 21:569–74.

[146] Lee S, Oh Y, Kim D, et al. Converting carbohydrates extracted from marine algae into ethanol using various ethanolic *Escherichia coli* strains. Appl Biochem Biotechnol 2011;164:878–88.

[147] Yanagisawa M, Nakamura K, Ariga O, et al. Production of high concentrations of bioethanol from seaweeds that contain easily hydrolyzable polysaccharides. Process Biochem 2011;46:2111–6.

[148] Wang X, Liu X, Wang G. Two-stage hydrolysis of invasive algal feedstock for ethanol fermentation. J Integr Plant Biol 2011;53:246–52.

[149] Singh J, Gu S. Commercialization potential of microalgae for biofuels production. Renew Sustain Energy Rev 2010;14:2596–610.

[150] Greenwell HC, Laurens LML, Shields RJ, et al. Placing microalgae on the biofuels priority list: a review of the technological challenges. J R Soc Interface 2010;7: 703–26.

[151] Yun EJ, Shin MH, Yoon JJ, et al. Production of 3,6-anhydro-*l*-galactose from agarose by agarolytic enzymes of *Saccharophagus degradans* 2-40. Process Biochem 2011;46: 88–93.

[152] Adams JMM, Toop TA, Donnison IS, et al. Seasonal variation in *Laminaria digitata* and its impact on biochemical conversion routes to biofuels. Bioresour Technol 2011;102:9976–84.

[153] Ge L, Wang P, Mou H. Study on saccharification techniques of seaweed wastes for the transformation of ethanol. Renew Energy 2011;36:84–9.

[154] Harun R, Jason WSY, Cherrington T, et al. Exploring alkaline pre-treatment of microalgal biomass for bioethanol production. Appl Energy 2011;88:3464–7.

[155] Klinke HB, Thomsen AB, Ahring BK. Inhibition of ethanol-producing yeast and bacteria by degradation products produced during pre-treatment of biomass. Appl Microbiol Biotechnol 2004;66:10–26.

[156] D□bowski M, Zieliński M, Grala A, et al. Algae biomass as an alternative substrate in biogas production technologies – review. Renew Sustain Energy Rev 2013;27: 596–604.

[157] Passos F, Uggetti E, Carrère H, et al. Pretreatment of microalgae to improve biogas production: a review. Bioresour Technol 2014;172:403–12.

[158] Xia A, Cheng J, Murphy JD. Innovation in biological production and upgrading of methane and hydrogen for use as gaseous transport biofuel. Biotechnol Adv 2015; 34:451–72.

[159] Montingelli ME, Tedesco S, Olabi AG. Biogas production from algal biomass: a review. Renew Sustain Energy Rev 2015;43:961–72.

[160] Jones CS, Mayfield SP. Algae biofuels: versatility for the future of bioenergy. Curr Opin Biotechnol 2012;23:346–51.

[161] Xue J, Wang L, Zhang L, et al. The pivotal role of malic enzyme in enhancing oil accumulation in green microalga *Chlorella pyrenoidosa*. Microb Cell Fact 2016;15:120.

[162] Posten C, Schaub G. Microalgae and terrestrial biomass as source for fuels-A process view. J Biotechnol 2009;142:64–9.

FURTHER READING

1. Patil V, Källqvist T, Olsen E, et al. Fatty acid composition of 12 microalgae for possible use in aquaculture feed. Aquac Int 2007;15:1–9.

CHAPTER

A Biorefinery Processing Perspective for the Production of Polymers

9

Aqdas Noreen[1], Khalid M. Zia[1], Mudassir Jabeen[1], Shazia Tabasum[1], Fazal-ur-Rehman[1], Saima Rehman[1], Nadia Akram[1], Qun Wang[2]

Government College University Faisalabad, Faisalabad, Pakistan[1]; Iowa State University, Ames, IA, United States[2]

9.1 INTRODUCTION

Nature reveals that long-lasting stability depends on an effective use of resources in closed circuits. Enduring perspective is not only an issue of environmental research in the current eras of rising petrochemical prices but also has an economical value [1]. Petrochemicals, most commonly used raw materials for industrial production of chemicals and fuels, are neither sustainable nor ecofriendly. It is gradually recognized worldwide that plant-derived raw materials (biomass) [2–8] have the prospective to substitute a huge part of fossil resources to produce energy and nonenergy materials on industrial scale [9,10]. Production of bio-based chemicals is an essential approach for the sustainable development of biorefining techniques owing to the lower material demands and high value of this industry. Numerous commodity products such as citric acid, acetic acid, lactic acid, and methylene succinic acid were produced through fermentation in the early 20th century [11,12]. More than 65% of *n*-butanol and 10% of acetone were manufactured by the fermentation of starch and molasses, in the United States between 1945 and 1950 [11]. Henry Ford gave the idea of manufacturing cars from plant materials such as fuel from vegetable oils, car body from soybean meal, resin, and flex, while the tires made up from the goldenrod-based latex [13]. The development in employment of raw materials based on biomass for the manufacturing of automobile parts continued and for the fabrication of several inner automobile parts such as seat cushions, sun shades, structural foams, energy absorbance, seat cushions, carpet backing, and arm rests, Wood Bridge Group utilized bio-derived polyol foams [14]. The progress of biorefineries indicates the key for the access to an integrated manufacturing of food, chemicals, materials, and fuels for the future [15]. Combination of agroenergy crops and biorefinery technologies offers the possibility for the growth of viable biomaterials and biopower that may lead to an innovative manufacturing paradigm [16].

Algae Based Polymers, Blends, and Composites. http://dx.doi.org/10.1016/B978-0-12-812360-7.00009-4

9.2 THE BIOREFINERY CONCEPT: DEFINITION AND PERSPECTIVES

The biorefinery concept, analogous to the petroleum refinery, considered an approach that helps in the reduction of carbon dioxide (greenhouse gases), reduced dependence on the rapidly depleting crude oil, and the uncertainty in energy supply [17,18]. A biorefinery combines the biomass conversion process with the equipment for the production of power, fuel, and value-added chemicals from biomass with minimum waste and emissions [19–21]. This concept is illustrated in Fig. 9.1 [22]. A biorefinery process has various stages in which the first step, after feedstock selection, is pretreatment of biomass to make it more suitable for further processing. Subsequently, the biomass constituents are subjected to biological, chemical, thermochemical, and/or mechanical treatments. Bio-based chemicals produced from this step can be further transformed to chemical building blocks for production of novel materials such as specialty polymers, fuels, composites. Biomass has a complicated composition similar to petroleum, and its primary fractionation into

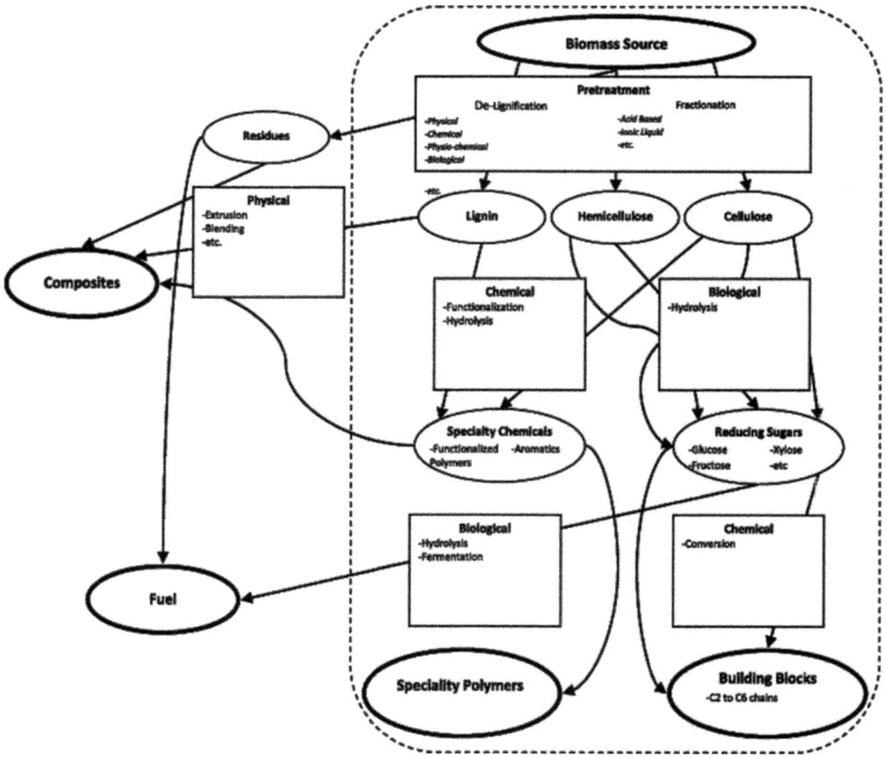

FIGURE 9.1

Biorefinery concept [22].

simple constituents permits the processing of various products. Unlike petroleum, biomass usually exhibits higher degree of functionality and lower thermostability. Therefore, biomass-based raw material needs specific reaction conditions as compared to petroleum-based raw material [23]. A comparison is often made between traditional petrochemical refineries and biorefineries. Table 9.1 gives an overview of the main similarities and dissimilarities between biorefineries and petroleum refineries [24].

Table 9.1 Comparison of Biorefineries and Petroleum Refineries [24]

	Biorefinery	Petroleum Refinery
Feedstock	Feedstock heterogeneous regarding bulk components, e.g., carbohydrates, lignin, proteins, oils, extractives, and/or ash	Feedstock relatively homogenous
	Most of the starting material present in polymeric form (cellulose, starch, proteins, lignin)	
	High in oxygen content	Low in oxygen content
	The weight of the product generally decreases with processing	The weight of the product generally increase with processing
	It is important to perceive the functionality in the starting material	
	Low sulfur content	Some sulfur present, sometimes high in sulfur
	Sometimes high in inorganics, especially silica	
Building-block composition	Main building blocks: glucose, xylose, fatty acids (e.g., oleic, stearic, sebacic)	Main building blocks: ethylene, propylene, methane, benzene, toluene, xylene isomers
Biochemical processes	Combination of chemical and biotechnological processes removal of oxygen	Almost exclusively chemical processes of heteroatoms (O,N,S)
	Relative heterogeneous processes to arrive to building blocks	Relative heterogeneous processes to arrive to building blocks, steam cracking, catalytic reforming
	Smaller range of conversion chemistries: dehydration, hydrogenation, fermentation	Wide range of conversion chemistries
Chemical intermediates produced at commercial scale	Few but increasing (e.g., ethanol, furfural, biodiesel, monoethanol glycol, lactic acid, succinic acid, etc.)	Many

9.2.1 BIOMASS AS MULTIPLE FEEDSTOCK FOR BIOREFINERY

Photosynthetic organisms have ability to use sunlight, water, and carbon dioxide to generate primary and secondary metabolites. These biomolecules could be used to produce biomass [25]. Primary metabolites are carbohydrates and lignin present in high amount in biomass while secondary metabolite, such as triglycerides, alkaloids, gums, waxes, resins, rubber, steroids, tannin, terpenes, terpenoids, and plant acids, are found in less quantity in the plant biomass [2]. The secondary metabolites could be used for synthesis of value-added chemicals such as flavors used in food industry, nutraceuticals, cosmaceuticals, and pharmaceuticals using integrated processing system. Renewable bio-based raw material for biorefinery comes from the following four areas:

- Agriculture (crops and residues)
- Forestry
- Industrial and household (solid waste and wastewater)
- Aquaculture (algae and seaweeds)

Biomass derived from plants, aquatic plants, crops, trees, grasses, and agroforest residues is versatile and main renewable raw material for biorefinery as shown in Fig. 9.2 [26]. The biomass feedstock is classified into four wide groups: (1) lignin

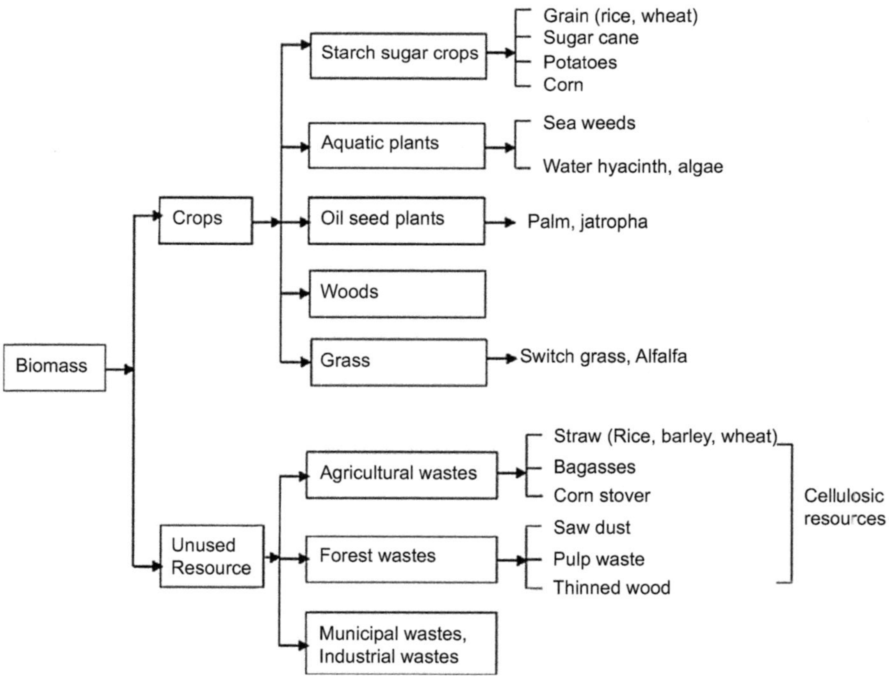

FIGURE 9.2

Biomass as renewable raw material for biorefinery [26].

and carbohydrates, (2) triglycerides, (3) mixed organic residues, and (4) chitin and chitosan from seafood waste.

9.2.1.1 Carbohydrates and Lignin

Carbohydrates are the most common component present in plant biomass. Most common six-carbon monosaccharide sugars are glucose, galactose, and mannose, whereas five-carbon sugars are xylose and arabinose. Sugarcane and sugar beet are two important sugar crops which along with maize starch yield nearly 100% ethanol [27]. Cellulose, hemicellulose, and lignin are three main constituents of lignocellulosic biomass. Cellulose and hemicellulose can be hydrolyzed to simple oligosaccharides and monosaccharides, which upon fermentation result in the formation of alcohol and other products, while lignin which constitutes 15%–25% of lignocellulosic feedstock is composed of phenolic polymers and is not suitable for the fermentation process. However, it is highly suitable for the energy generation or chemical extraction. Crops as well as the residues are major source of lignocellulosic biomass. Massive quantities of lignocellulosic biomass can be obtained from certain crops which can be exclusively grown for this purpose, e.g., perennial herbaceous plants or woody crops. Waste and residues are other sources of lignocellulosic biomass, for instance, woody residue of paper and pulp industry, straw from agriculture, and forestry waste. Exploitation of waste/residual biomass provides a path of generating worth for the humanity, substitute fossil fuels with normally decomposing materials without the requirement of additional land usage [28].

9.2.1.2 Triglycerides

Oils and fats are triglycerides generally composed of saturated/unsaturated fatty acids and glycerol. The chain length of fatty acids ranges from 8 to 20 carbon atoms; however, C:16, C:18, and C:20 are among the most commonly existing fatty acids. Vegetable and animal fats are the main sources of triglycerides. In terms of global oil production, sunflower, palm, soybean, and rapeseed are the most common [29,30]. The current process of producing biodiesel is to carry out a reaction between vegetable oils and an alcohol, typically methanol. There are two reactive centers in these oils for various chemical reactions, which yield various commodity monomers and polymers, (1) pi-bonds of unsaturated fatty acids and (2) acidic moiety of the fatty acid [31]. In the imminent times, nonedible crops such as *Jatropha curcas* and *Pongamia pinnata*, which need minor inputs and are suitable for minimal lands, might become the main source of oils for biorefinery purpose, particularly in dry and semi-arid areas [32]. The other major source of waste oil is the waste streams of food industry, where the waste oil mainly comes from the food processing plants, households and commercial services such as fast-food chains and restaurants [33].

9.2.1.3 Mixed Organic Residues

Municipal solid waste (MSW), containing mixed organic residues coming from wild crops, manure, proteins, and fresh fruits and vegetables residues, has high prospective for energy recovery. The chemical and physical properties of this broad-spectrum biomass alter largely; therefore biomass waste involves different conversion methods. Certain streams, such as sewage sludge, residues from food processing, and manure from swine and dairy farms, contain very high moisture content (above 70%). Hence, these feedstocks are more appropriate to undergo anaerobic digestion for the production of biomass instead of other chemicals [34].

9.2.1.4 Chitin and Chitosan From Seafood Waste

Marine biomass is considered as another important feedstock for biorefinery as terrestrial biorefineries involve certain problems related to the land usage, which can be overcome by using aquatic feedstock [35]. In addition to aquatic plants, aquatic animals are also attractive biorefinery feedstock having high amounts of chitin (biopolymer). Approximately 45% of processed seafood waste comprises shrimp exoskeleton and cephalothoraxes which has become a problem for the environment [36]. This waste represents 50%–70% of the weight of the raw material; however, it contains valuable components such as protein and chitin. Chitin, the second most abundant biopolymer next to cellulose and its derivatives such as chitosan, is widely recognized to have immense applications in many fields [37]. Chitin and chitosan are extensively used in the food industry, chemical industries, medicinal fields, textiles, water treatment plants, etc. [38,39]. The reasons for larger use of these biopolymers in numerous industries are cost of the manufacturing method and the technical advantages. The commercial method of preparation of chitin from shrimp shell involves strong acid and alkali treatment to remove the minerals and proteins, respectively [40]. Potential applications of chitin and its derivatives, mainly chitosan, are estimated to be more than 200. In addition to being biodegradable and biocompatible, they also have antimicrobial activity [41]. They have a variety of applications in several fields, such as cosmetics, biomedical, pharmacy, paper industry, agriculture, food, and also as absorbent materials for wastewater treatment [42–44]. Chitosan is used to modify the surface of nonwoven fabrics and polypropylene films to improve antimicrobial properties [45,46]. Chitin and chitosan can be obtained as a feedstock from seafood processing industry. Chitin can be converted into several value-added chemicals such as polyols, amine sugars, amide alcohols, and pyrrole [47].

Inherent properties of polymers can be improved for specific applications by the modification of chitosan such as biodegradability, biocompatibility, chemical versatility, and lower toxicity [48]. Chitosan as well as chitosan-based polymers are employed as flocculent and metal chelator in water treatment [49]. Owing to antimicrobial activity, chitosan is used for pathogen removal from drinking water [50]. Moreover, chitosan has several applications in cosmetics, catalysis, and biomedication [51].

9.2.2 PRETREATMENT AND FRACTIONATION OF BIOMASS

Pretreatment of biomass is mostly carried out to increase processing, surface area, and reactivity. Variety of chemical, physical, or thermal pretreatment systems were studied widely [52–54]. Common pretreatment techniques include steam explosion, lime, ammonia, dilute acid, and hot water (liquid). Prior to gasification, hydrocracking, or fast pyrolysis, physical pretreatment (steam explosion or ball milling) is necessary for the water-insoluble biomass requiring complicated processing [52,55]. During the pretreatment process, breakdown of carbon–oxygen network present in lignin and the loss of crystallinity of cellulose are helpful in overcoming the integrity of sustainable resources [55–57]. Utilizing dilute acid for the pretreatment of lignocellulosic biomass enhances the surface area, modifies the structure of lignin, and hydrolyzes hemicelluloses to xylose [53]. As hemicelluloses surrounds the cellulose fibers (Fig. 9.3) [58], pretreatment with dilute acids assist to enhance successive reactions on the cellulose. Acid-catalyzed hydrolysis of biomass in the presence of dilute sulfuric acid yields levulinic acid and furfural. Levulinic acid is one of the DOE's top-12 sustainable chemicals, and DuPont have been using it for synthesis of lactones and pyrrolidones [59].

Starch-based biomass, obtained from wheat, corn, sago palm, sorghum, and cassava, requires a pretreatment with amylases before fermentation for the production of glucose [60]. Additionally, initial degradation step is required for lignocellulosic materials to obtain ethanol by fermentation [61]. Plant oils are recovered by solvent extraction, grinding, and pressing. Subsequently, monounsaturated as well as polyunsaturated alkenes are outstanding contenders for a wide range of rearrangement

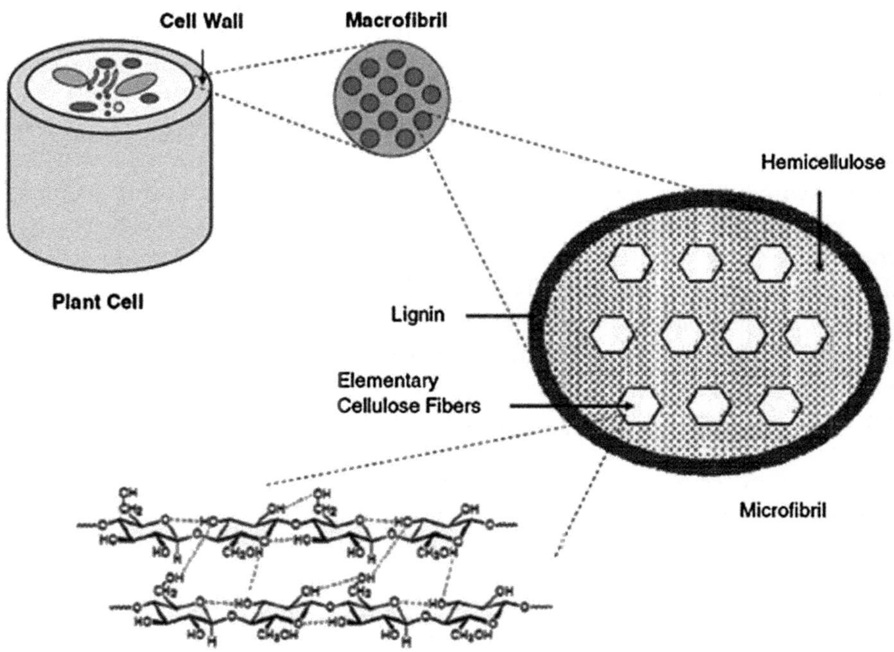

FIGURE 9.3

Lignocellulose structure containing cellulose, hemicellulose, and lignin [58].

reactions in the presence of metathesis catalysts [62] and several addition reactions, such as those involving acrylates [63,64], carboxylic acids [65], enones [66,67], or epoxides [68–71]. Plant oils are quite suitable for the synthesis of polyurethanes [69,72]. Moreover, Diels–Alder reactions are usually used for the isomerization of nonconjugated alkenes to conjugated alkenes in plant oils [73,74].

Biomass pretreatment through fractionation results in increased hydrolysis as well as the separation of basic constituents. Fractionation is utilized in a biorefinery for the separation of primary refined products such as conversion of plant or wood into lignin, cellulose, and hemicellulose [16,75]. Fractionation methods contain steam explosion, hot water systems, and aqueous separation. Basic fractionation products from plant or wood biomass are as follows:

- Breakdown of biomass constituents → Lignin + Cellulose + Oligosaccharides
- Saccharification (cellulose hydrolysis) → Glucose
- Fermentation of glucose → Lactic acid + Ethanol
- Decomposition of cellulose → Xylitol + Levulinic acid
- Chemical decomposition of lignin → Phenolics

9.3 TYPES OF BIOREFINERIES

9.3.1 GREEN BIOREFINERY

A green biorefinery is a system in which refinery products are in accordance with the physiology of the corresponding plant material as described by Kamm and Kamm [76] and Fernando et al. [77]. Natural wet raw material obtained from natural products, such as green crops, plants, or grass, are used as inputs in green biorefinery (Fig. 9.4). The first step is to treat biomass by applying wet fractionation to make a fiber-rich press cake and a nutrient-rich green juice. The constituents of press cake include starch, cellulose, crude drugs, pigments, and valuable dyes, etc., whereas the green juice contains free amino acids, proteins, organic acids, hormones, enzymes, dyes, and minerals, etc. The press cake can be used as a feedstock for manufacturing value-added chemicals, such as levulinic acid, for transformations to syngas and fuel, and green feed pellets production.

9.3.2 THE FOREST AND LIGNOCELLULOSIC-BASED BIOREFINERY

Lignocellulosic feedstock has two varieties of polysaccharides, cellulose and hemicellulose, bounded together by a third constituent, lignin [78]. A summary of possible products of lignocellulosic-based biorefinery (LCB) is shown in Fig. 9.5 [26]. More commonly, rough fibrous plant substances produced through lumber or municipal waste are employed in LCB. Plant fibers are first washed and degraded into three basic constituents by using enzymatic hydrolysis or chemical digestion. Hemicellulose and cellulose may also be synthesized with the aid of alkali (caustic soda) and sulfite (acidic, bisulfite, alkaline, etc.). The cellulose and hemicellulose,

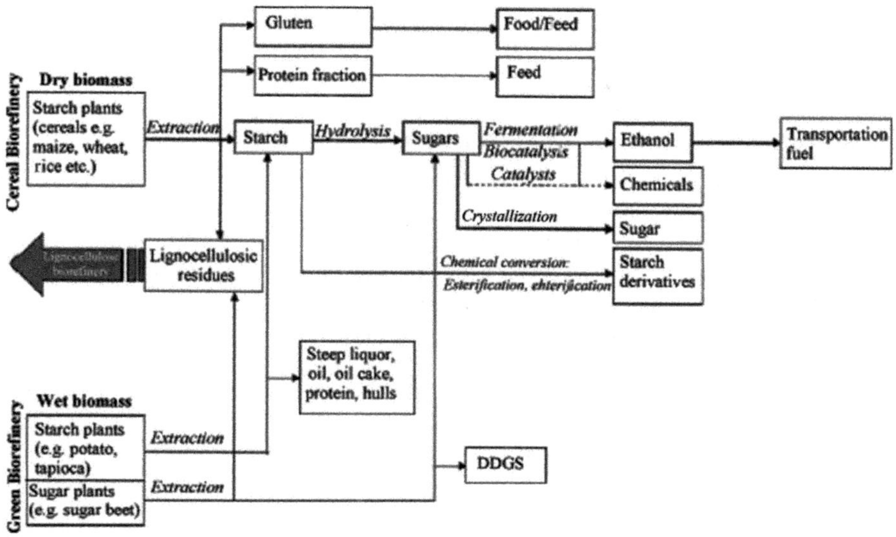

FIGURE 9.4

Green biorefinery [26].

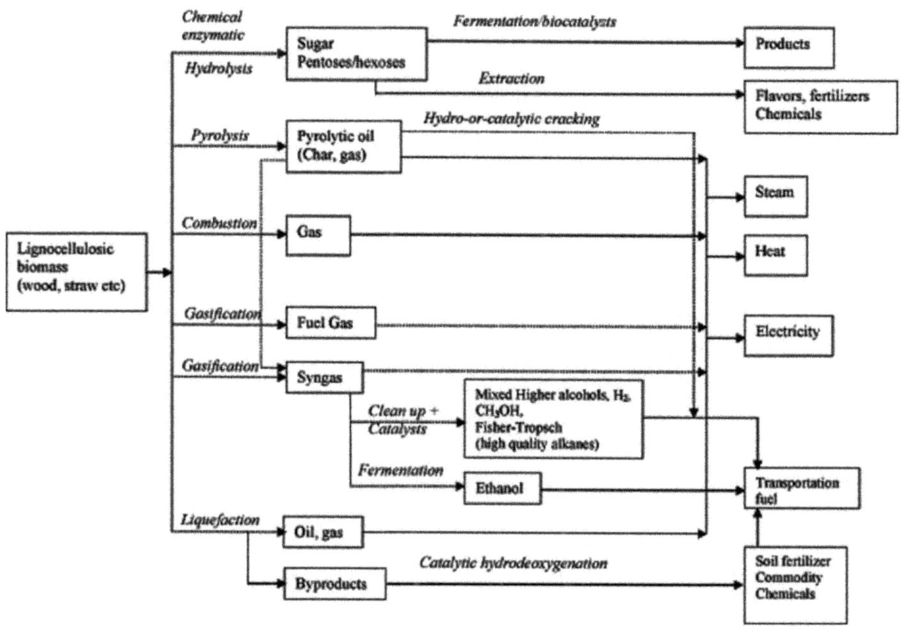

FIGURE 9.5

Forest-based and lignocellulosic biorefinery [26].

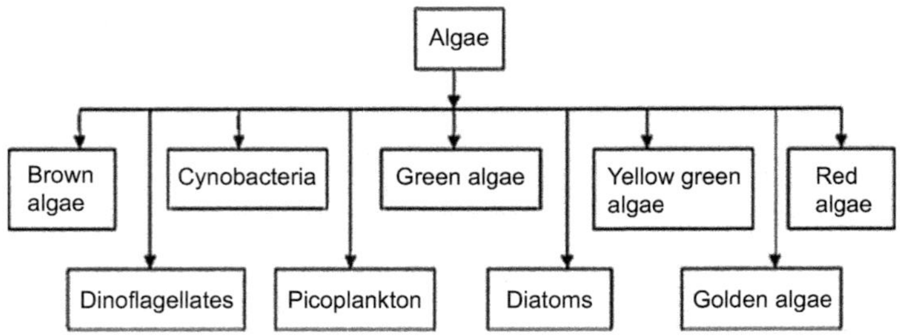

FIGURE 9.6

Classification of algal species [26].

sugar polymers, are transformed to their component sugars by hydrolysis. The enzymatic or chemical hydrolysis of cellulose yields glucose which is used to produce valuable chemicals such as acetone, ethanol, butanol, acetic acid, and different fermentation products. Lignin is used only for fuel, adhesive, or binder purposes.

9.3.3 ALGAE-BASED BIOREFINERY

Primary biomass production of the world is equally divided between terrestrial and aquatic systems. Till now strategies have primarily focused their attention to terrestrial biomass, while marine sources such as algae and their derived products might provide a prospective that is still not fully known [79]. Marine crops are known for their greenhouse gas-reduction potential and their capability to absorb CO_2 probably exceeding that of terrestrial species. There are above 40,000 well-known algal species and a few others yet to be recognized. Algae are categorized in the lots of most important groups (Fig. 9.6). They are able to live and reproduce in low-quality, excessively saline water [80,81]. Algae can accumulate considerable quantities of carbohydrates, starch, oils, and vitamins depending on species and growing conditions [82–84]. Table 9.2 gives the general composition of special algal strains.

The potential merits of algae as raw material for biorefineries are as follows:

1. Algae produce and store high amount of neutral oils.
2. High growth rates.
3. Grow in saline sea water.
4. Occupy marginal lands (e.g., desert, arid, and semiarid land) that are not suitable for usual agriculture.
5. Devour developmental nutrients such as nitrogen and phosphorus from different wastewater sources (e.g., agricultural runoff, municipal and industrial wastewater).
6. Fix CO_2 from fuel gases released from fossil-fuel-fired power plants and other sources, and consequently slash emissions of principal greenhouse gas [87,88].
7. Produce excessive value coproducts or by-products (e.g., proteins, polysaccharides, pigments, fertilizer, animal feed, and H_2).
8. Algae can be grown in a proper culture vessel (photobioreactor) all year long with yearly biomass production.

Table 9.2 General Composition of Different Algae (Percentage of Dry Matter) [85,86]

Alga	Protein	Carbohydrates	Lipids
Anabaena cylindrical	43–56	25–30	4–7
Aphanizomenon flos-aquae	62	23	3
Chlamydomonas reinhardtii	48	17	21
Chlorella pyrenoidosa	57	26	2
Chlorella vulgaris	51–58	12–17	14–22
Dunaliella salina	57	32	6
Dunaliella bioculata	49	4	8
Euglena gracilis	39–61	14–18	14–20
Porphyridium cruentum	28–39	40–57	9–14
Scenedesmus obliquus	50–56	10–17	12–14
Scenedesmus quadricauda	47	–	1.9
Scenedesmus dimorphus	8–18	21–52	16–40
Spirogyra sp.	6–20	33–64	11–21
Arthrospira maxima	60–71	13–16	6–7
Spirulina platensis	46–63	8–14	4–9
Spirulina maxima	60–71	13–16	6–7
Synechococcus sp.	63	15	11
C. vulgaris	51–58	12–17	14–22
Prymnesium parvum	28–45	25–33	22–38
Tetraselmis maculata	52	15	3
P. cruentum	8–39	40–57	9–14

Algae are considered to be an innovative feedstock for a biorefinery due to their prospective to form multiple products [19,89]. These products can be categorized as energy and nonenergy based on their prospective function. Fig. 9.7 shows the graphical flow sheet of the algae-based biorefinery [90].

9.3.3.1 Energy Products From Algae

Algal *biodiesel* is a carbon-neutral fuel, as it assimilates CO_2 during algal development and releases it upon fuel combustion [91,92]. However, algae-based fuels can be the most proficient and sustainable solution to climate changes [93]. Pyrolysis, catalytic cracking, and microemulsification are very expensive processes and generate a low-quality biodiesel. Transesterification is the commonly used method to convert oil into biodiesel [94–97]. It is a process that transfers algal lipids to low molecular weight fatty acid alkyl esters [98]. This algal biodiesel meets the International Biodiesel Standard for Vehicles (EN14214). The selection of algal species for biodiesel development depends on properties of fuel, amount of oil, engine performance, and emission characteristics [99]. The bio-based oil from microalgae

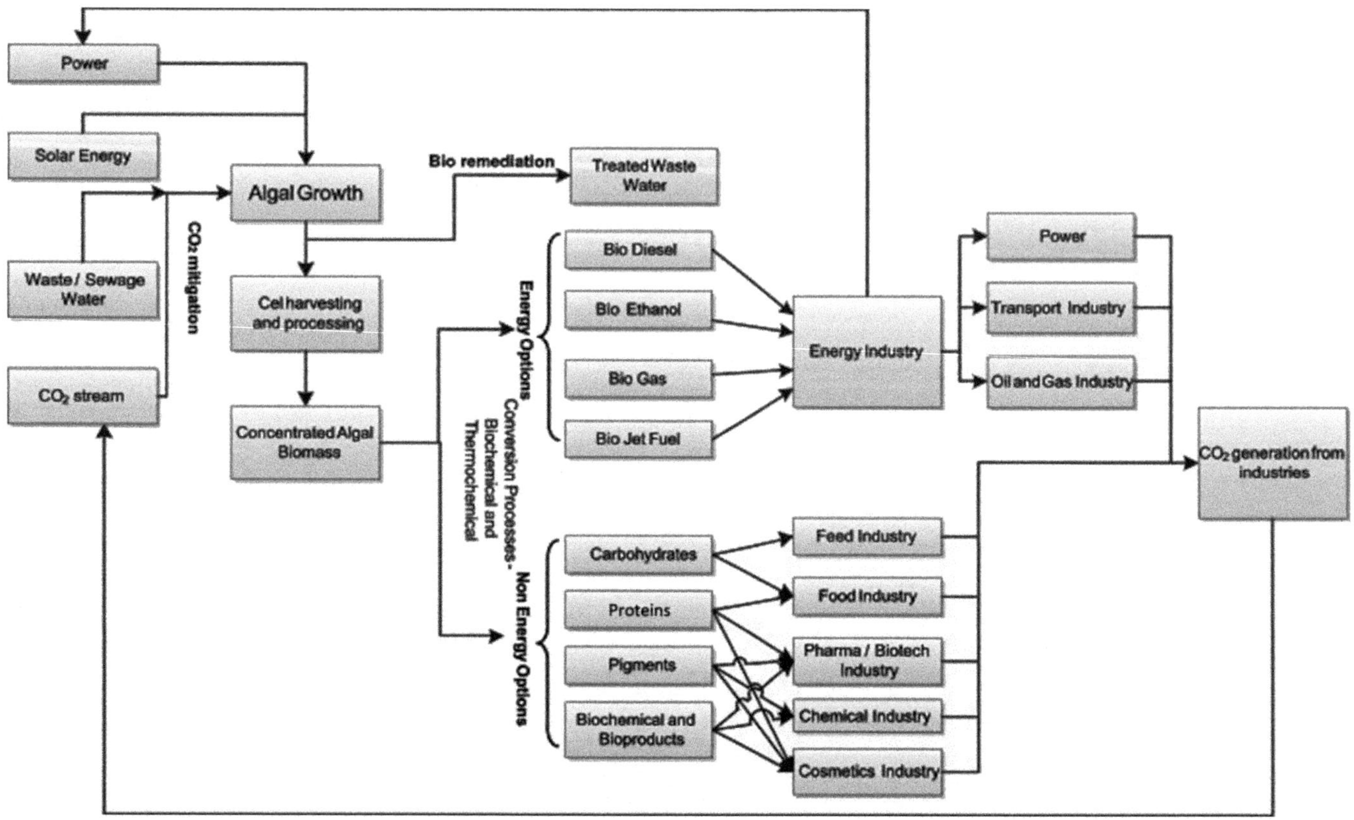

FIGURE 9.7

Schematic flow sheet for an algae biorefinery [90].

has higher density, lower viscosity, and lower heating values in contrast to fossil oil [100]. Glycerol obtained as a by-product in the transesterification method can be utilized as a carbon source. It can be converted into valuable chemicals, such as organic acids, single cell oil, microbial biomass, and mannitol, by using fungi or yeast.

The macroalgae demonstrate high methane production rates compared to terrestrial biomass. *Biogas* production from macroalgae is technically more viable than different fuels, even when it is not yet economically practicable due to the high price of macroalgae biomass [101]. Once the lipid is extracted, the microalgal biomass containing proteins and carbohydrates can be processed to synthesize biogas, a renewable fuel, by anaerobic means. Biogas is a mixture of methane and carbon dioxide. Hydrolysis, acetogenesis, acidogenesis, and methanogenesis are four basic steps in biogas production [102,103]. Another technique to produce biogas is gasification. It consists of partial oxidation of algal biomass at high temperatures (800–1000°C) [104]. When biomass reacted with steam and oxygen, it generated a mixture of gases (methane, carbon dioxide, nitrogen, and hydrogen) known as *syngas*. It can be utilized to produce energy, fuel, and chemicals (e.g., methane) [105,106,141,153]. Harmful algal blooms in lakes, ponds, or oceans produce toxic secondary metabolites that have severe effects on ecosystems; hence, biogas production from algal biomass plays a vital role in bioremediation [107,108].

Bioethanol from algae has great potential because of low percentage of hemicellulose and lignin as compared to other lignocellulosic plants [109]. Macroalgae have many carbohydrates (starch, agar, cellulose, mannitol, and laminarin) which are converted to sugars [110], and fermentation of these simple sugars by using suitable microorganisms produce bioethanol. *Cholorococcum*, *Chlorella*, and *Chlamydomonas* are a few species used for bioethanol production. Brown alga is a main feedstock for manufacturing of bioethanol due to significant carbohydrate content and can be readily mass cultivated with the current farming methods [108,111–115]. Biobutanol could also be prepared from macroalgae by the acetone–butanol fermentation method through anaerobic bacteria such as *Clostridium* sp. [116].

The aircraft fuel upon combustion produces carbon monoxide (CO), carbon dioxide (CO_2), sulfur oxides (SO_x), nitrogen oxides (NO_x), water vapors (H_2O), unburned or partly combusted hydrocarbons, particulates, and other trace compounds. These elements together pose a challenge for the aviation industry to confirm the safety of the fuels and to abate the undesirable hazard to the atmosphere. Aviation changes the composition of the environment worldwide and can thus drive climate change and ozone depletion [117,118]. The aviation industry is concerned to reduce its carbon foot print by employing environment-friendly fuel for air transport.

Renewable jet fuel or *bio-jet fuel* can decrease the greenhouse gas emissions by 60%–80% as compared to fossil-fuel-derived jet fuel. Bio-jet fuel is synthesized by blending microalgae biofuel with petroleum-based jet fuel that provides the mandatory specification characteristics [119]. Microalgae oil is transformed into jet fuel by hydro-treatment or by Fischer–Tropsch method. Liquid fuels can be produced from algal biomass by gasification, by the formation of synthesis gas (CO and H_2) and its transformation to liquid hydrocarbon fuel via Fischer–Tropsch process [120].

Table 9.3 Composition of Microalgal Cell Wall and Storage Products [121]

Division	Cell Wall	Storage
Cyanophyta	Lipopolysaccharides, peptidoglycan	Cyanophycean starch
Chlorophyta	Cellulose, hemicelluloses	Starch/lipid
Dinophyta	Absent or contains little cellulose	Starch
Cryptophyta	Periplast	Starch
Euglenophyta	Absent	Paramylum/lipid
Rhodophyta	Agar, carrageenan, cellulose, and calcium carbonate	Floridean starch
Heterokontophyta	Naked or covered by scales or with large quantities of silica	Leucosin/lipid

9.3.3.2 Nonenergy Products From Algae

The accumulation of carbohydrates in algae is due to CO_2-fixation during photosynthesis [121]. These carbohydrates can either be stored in the plastids as reserve materials (e.g., starch), or become the key component of cell walls. Composition of cell wall of microalgae and storage products is given in Table 9.3 [122]. Glucose, starch, and cellulose/hemicellulose are the most common algal carbohydrates. Among these, algal starch/glucose is utilized for biofuel production, mostly in the production of bioethanol [123] hydrogen and building-block chemicals. Except starch, other carbohydrates could also be converted to biofuel and biochemicals. Galactans such as agar and carrageenan are the chief polysaccharide constituents of red algae [124,125]. Carrageenan is obtained by extraction from red algae or by dissolving them into an aqueous solution. Major sugars of brown seaweeds are alginate, glucan, and mannitol. Alginate (alginic acid) accounts for up to 40% dry weight of the cell wall [126].

At present, algal polysaccharides represent a group of valuable materials with numerous applications, i.e., in food, textiles, cosmetics, and as thickening agents, stabilizers, emulsifiers, lubricants, and clinical drugs. Algal sulfated polysaccharides illustrate different pharmacological activities, such as antioxidant, antiinflammatory, antitumor, anticoagulant, antiviral, and immunomodulating activities. The sulfated polysaccharides obtained from *Porphyridium* sp. have capability to slow down the migration and adhesion of polymorphonuclear leukocytes. Therefore, they have an enormous prospective for antiinflammatory skin treatments [127]. *Nannochloropsis* sp. can be used for manufacturing oil, valuable pigments, and biohydrogen, while production of oil, pigments and H_2 by supercritical fluid extraction is an inexpensive biorefinery approach [128]. *Chlorella protothecoides*, grown autotrophically in high salinity and luminosity stress environment, could be used as a source of lipids and carotenoids [129]. The residual biomass could be exploited for H_2 or bioethanol production [130]. *Spirogyra* sp., being a sugar-rich microalga, could be used for H_2 and

Table 9.4 Few Value-Added Bioproducts Extracted From Microalgae [133]

Product Group	Applications	Examples (Producer)
Phycobiliproteins carotenoids	Pigments, cosmetics, provitamins, pigments	Phycocyanin (*Spirulina platensis*)
		β carotene (*Dunaliella salina*)
		Astaxanthin and leutin (*Haematococcus pluvialis*)
Polyunsaturated fatty acids (PUFAs)	Food additive, nutraceutics	Eicosapentaenoic acid (EPA) (*Chlorella minutissima*)
		Docosahexaenoic acid (DHA) (*Schizochytrium* sp.)
		Arachidonic acid (AA) (*Parietochloris incisa*)
Vitamins	Nutrition	Biotin (*Euglena gracilis*)
		α-tocopherol (vitamin E) (*E. gracilis*)
		Ascorbic acid (vitamin C) (*Prototheca moriformis*, *Chlorella* sp.)

pigment production [131,132]. A summary of value-added bioproducts extracted from algae is given in Table 9.4 [133].

9.3.4 INTEGRATED BIOREFINERY

Only one conversion process is used to generate a variety of chemicals in previously discussed biorefineries. A biorefinery is a capital-intensive plan and when it involves only one conversion method, it raises the price of products manufactured by them. Consequently, various conversion technologies, such as thermochemical and biochemical, can be combined together to decrease the total cost with more flexibility in product generation and to supply its own power. Fig. 9.8 presents a scheme of an integrated biorefinery [77]. Three different platforms, namely sugar, thermochemical, and non-platform or existing technologies, are integrated. An integrated biorefinery generates different products such as electricity (from thermochemical process) and bioproducts (obtained from the combination of sugar and other existing conversion platforms).

A promising scheme in biorefinery area is the transformation of bio-based oil, the product from biomass pyrolysis, which can be routed through petrochemical refinery to generate a variety of chemicals (Fig. 9.9). All required infrastructures for the separation and purification of products are already in place for this method. This idea gives an ideal sense as the majority of petroleum refineries are well equipped to handle variable feedstock [134,135]. Integration of the algal part with dairy industry produces bio-based methanol for biodiesel production. Integration of the algal fuel with aquaculture presents a novel inland-based animal production system to meet increasing protein demand of the world [102,136–139]. Amalgamation with the lignocellulosic industry synthesizes cellulase or hemicellulase enzyme for hydrolysis, and thus enhanced the commercial

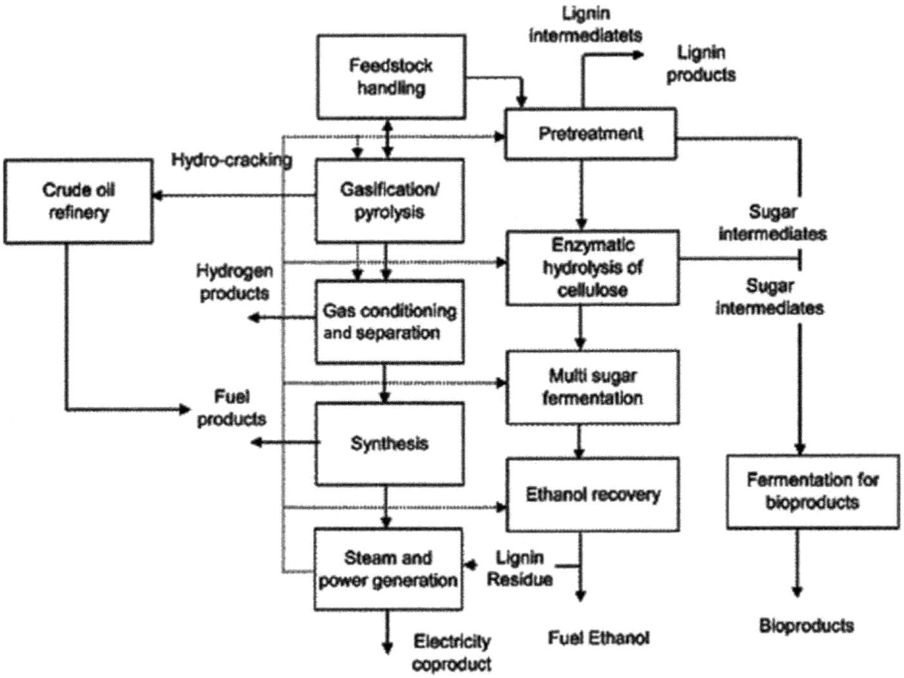

FIGURE 9.8

Schematic of an integrated biorefinery [77].

viability of both parts. Several algal strains, *Chlamydomonas* and *Dunaliella*, are genetically modified to express cellulases and hemicellulases which has opened the doors for integrating production of enzyme as a by-product from the algal biofuel area. They can be subsequently supplied to enzymatic hydrolysis step in cellulose-based raw material [136,140]. Preferred species of microalgae (saltwater algae, freshwater algae, and cyanobacteria) were used as a substrate for fermentative biogas production in a combined biorefinery. Anaerobic fermentation was considered as the final step in a future microalgae-based biorefinery concept [110].

9.4 TECHNOLOGICAL CONVERSION PROCESSES IN A BIOREFINERY

Depolymerization and deoxygenation of the biomass constituents is the aim of technological process in a biorefinery. Numerous technological conversion processes should be applied jointly for the conversion of biomass into important products in a biorefinery. Such processes have been classified into four groups as shown in Fig. 9.10 [26].

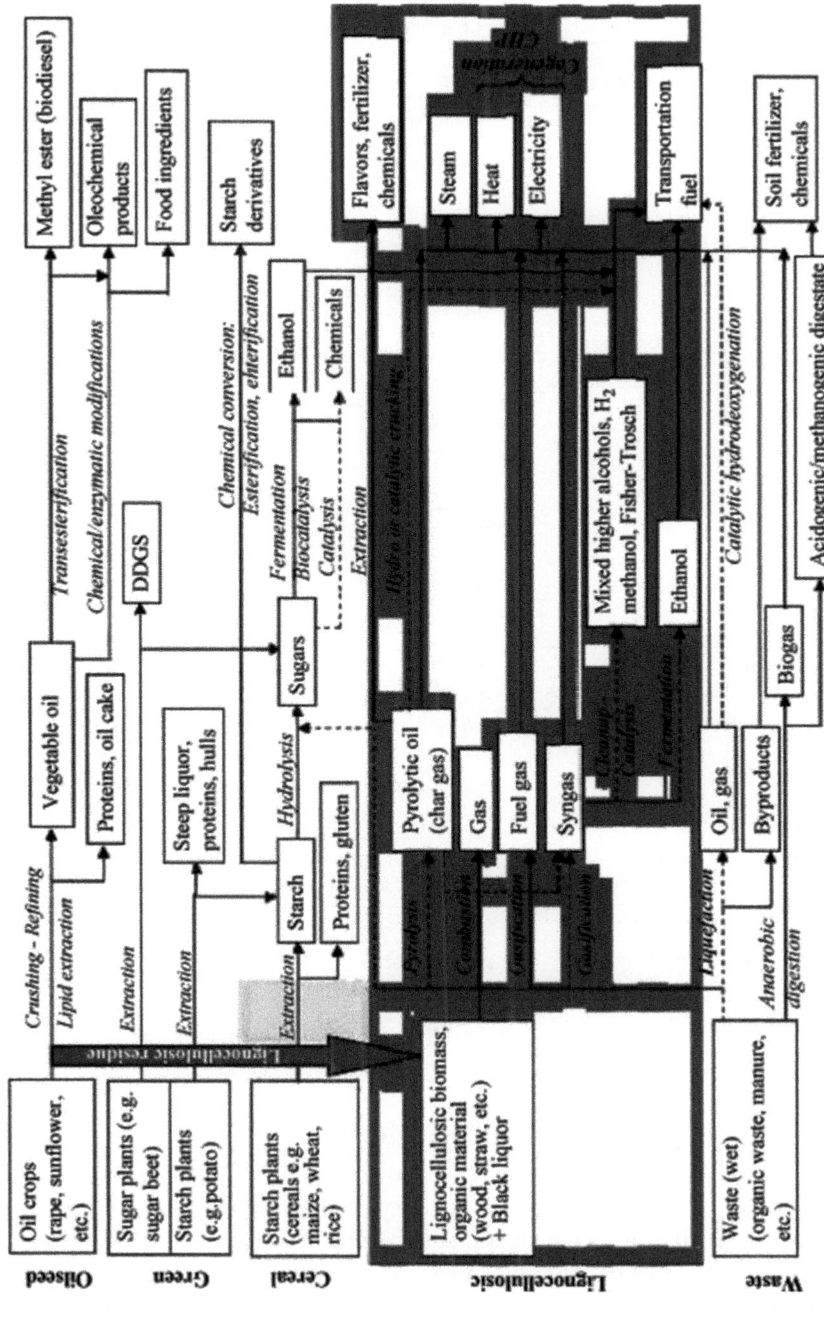

FIGURE 9.9

Integrated biorefinery [26].

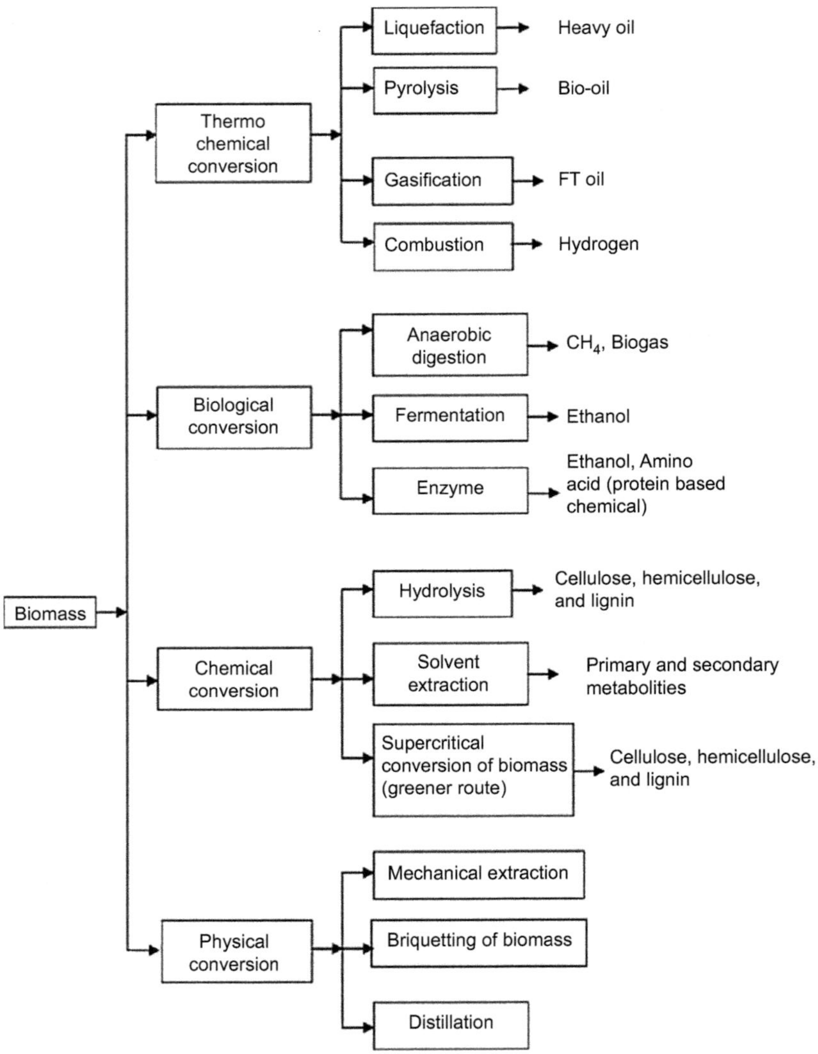

FIGURE 9.10

Biomass conversion processes [26].

9.4.1 THERMOCHEMICAL CONVERSION PROCESSES

Major thermochemical conversion techniques include pyrolysis, liquefaction, and gasification. These methods convert the biorenewable feedstock into gaseous or liquid state for the electricity, heat, value-added chemicals, and gaseous or liquid fuels purposes [141–143]. Main processes of conversion of biomass are indirect and direct liquefaction, physical extraction, thermochemical, electrochemical, and biochemical conversions [144–146].

Pyrolysis involves the heating of biomass/fuel in the absence of oxygen. Pyrolysis is a primary process used for the gasification and burning of fuels in the solid state.

Gasification of biomass offers an alternate energy resource which can be used for power generation in the internal combustion engines. In gasification process, the biomass is partially ignited resulting in the formation of a gas along with some char at the first step, followed by reduction of product gases such as H_2O, CO_2, CO, and H_2. In this process, low amount of methane and some other hydrocarbons are also generated depending on the operating conditions and design of the reactor [147,148].

Variety of chemicals, such as alcohol, aldehydes, ketones, acids, esters, phenolics, steroids, and hydrocarbons, are obtained through the fast pyrolysis of bio-oil/biomass. Cyclopentanone, phenol, methoxyphenol, acetone, methanol, formic acid, furfural, levoglucosan, alkylated phenols, and guaiocol are the major constituents of bio-oils. Thermal decomposition of all three major biomass components results in the formation of acetic acid through the removal of acetyl groups linked to xylose units. Dehydration of xylose results in the formation of furfural, methanol comes from the methoxyl group of uronic acid, water through dehydration, and carboxylic groups of uronic acid result in formic acid [149].

Appell et al. reported the liquefaction of biomass such as civic and agriculture waste [146]. In this process, biomass is reacted with water, sodium carbonate, and carbon monoxide/hydrogen, and converted into oil-like product.

The gasification process of biomass involves its thermal conversion into gaseous products along with small amounts of ash and char. This process is done at high temperatures to optimize the gas production. The product gas is known as producer gas, which is a mixture of H_2, CO, and methane together with N_2 and CO_2. Tars, chars, gaseous hydrocarbons, inorganic constituents, and ash are also produced. For the oxidation and combustion of biomass, usually oxygen or air is used. The composition of gas product depends mainly on the gasifying agent, gasification process, and composition of feedstock [150,151].

Initial step of biomass gasification involves thermochemical breakdown of cellulose, hemicellulose, and lignin compounds with char and volatiles production. Further gasification of these products is done in the next steps. Possible gasification products are represented in Fig. 9.11 [152].

Biomass liquefaction method yields a liquefied product. In this process, biomass is usually decomposed into smaller size molecules. These molecules being reactive and unstable, form oily compounds when repolymerize. The hydrothermal or direct liquefaction (HTL) is a highly promising technique in which treatment of waste streams from different sources generates valuable bioproducts [150,151,153].

In hydrothermal upgrading technique (HTU), biomass is treated at high pressure and temperature in the presence of water. HTU involves highly complicated phase equilibria due to the presence of several components such as water, alcohols, biocrude, and supercritical CO_2. The biocrude is normally a mixture consisting different types of molecules with broad molecular weight distributions. Biocrude

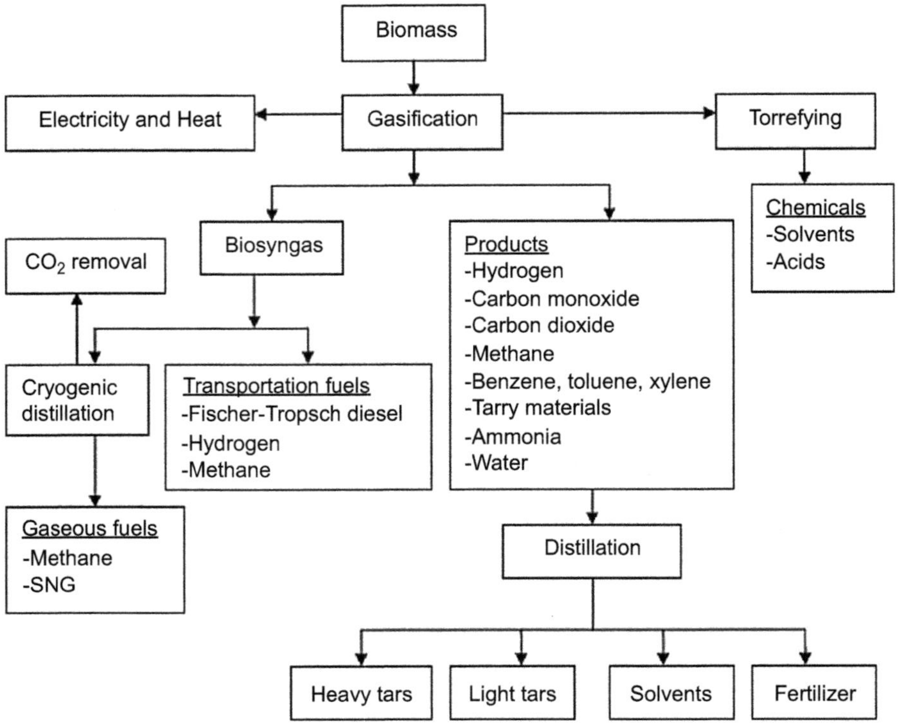

FIGURE 9.11

Products from gasification process [152].

is composed of 10%–13% O_2. It is upgraded by the catalytic hydrodeoxygenation. Earlier studies have showed that HTU process is a more attractive method than other processes such as pyrolysis or gasification. In HTU process the biomass (25% slurry in H_2O) is treated in liquid water at 575–625K temperature and 12–18 MPa for about 5–20 min to form a liquid biocrude mixture, CO_2 gas, and H_2O. Further processing is used to upgrade this biocrude into useable biofuel [141].

9.4.2 BIOCHEMICAL CONVERSION PROCESSES

Biochemical conversion process offers great selectivity for products. It proceeds by using low temperature and low rate of reaction. The production of bioethanol is an example of biochemical conversion technique for energy generation from a variety of biomass. For the production of ethanol, acid hydrolysis of hemicelluloses and enzymatic hydrolysis of cellulose has been mostly taken into account. Biodiesel formation has been successfully employed to generate energy from the oilseed crops [154–162].

Bioethanol is an imperative and renewable biofuel especially for motor vehicles. It can reduce the environmental pollution and consumption of crude oil. Bioethanol

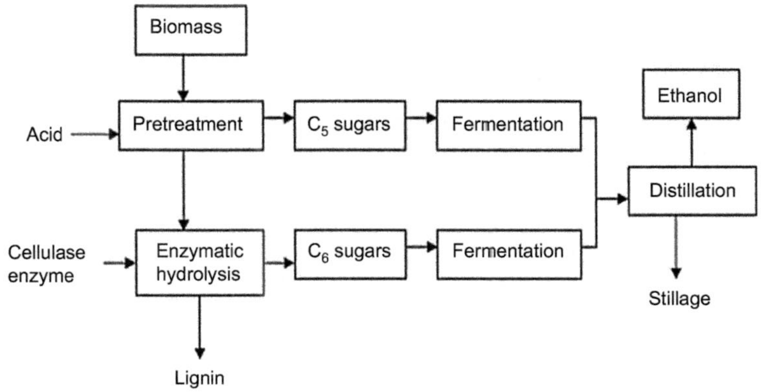

FIGURE 9.12

Enzymatic hydrolysis process [152].

production from cellulose requires pretreatment for the reduction of sample size, opening of cellulose structure, and conversion of hemicelluloses into simple sugars. The cellulose and hemicellulose is hydrolyzed into glucose by enzymes and acids, respectively, and is further fermented to generate bioethanol [163].

Pretreatment required for the fermentation of feedstock is usually referred as hydrolysis. Such pretreatments can be chemical, physical, or biological and are required for the conversion of complex carbohydrates to simple sugars and for opening the biomass structure. Fermentation of these obtained sugars is done in the presence of bacteria and yeast. Feedstock containing high amount of sugar and starch can be easily hydrolyzed. However, cellulosic feedstocks are not easy to hydrolyze and require extensive pretreatment methods. Fermentation process is usually employed industrially to convert the substrates, e.g., glucose to ethanol which is used in beverage, chemical, and fuel applications. Fermentation is anaerobic and enzymatically controlled method although this term is occasionally related to aerobic processing. Fig. 9.12 represents a flow diagram of enzymatic hydrolysis process.

9.4.3 MECHANICAL CONVERSION PROCESSES

In mechanical conversion processes, the composition and state of the biomass remains unaltered and only the biomass components are separated and reduced in size. Size reduction of biomass is a mechanical method that consists of either commuting/cutting process which changes the size or shape of biomass particles and its bulk density. Separation process separates the biomass into its simple components, while in extraction process valuable compounds are extracted and also concentrated from bulk. Pretreatment of lignocellulosic biomass (e.g., the opening up of lignocellulose into cellulose, lignin, and hemicellulose) fall in this category [164].

9.4.4 CHEMICAL CONVERSION PROCESSES

Chemical conversion processes involve the chemical modification of biomass feedstock by reacting it with other substances. The most common chemical methods for the substrate conversion are *trans*esterification and hydrolysis. In hydrolysis, mostly alkalis, acids, or enzymes are used to depolymerize the proteins and polysaccharides into sugars, cellulose to glucose or derivative chemicals, and glucose to levulinic acid [164]. Transesterification process is most common to produce biodiesel. During this process glycerin is coproduced which can be used in several commercial applications [29].

9.5 BIOREFINERY PRODUCTS

Biorefinery products are classified into two major groups: energy products and material products. The essential material products of biorefineries are chemicals such as organic acids (lactic, itaconic, succinic acid), polymers and resins, food, and fertilizers; while the most significant energy products include gaseous biofuels (biomethane, syngas, hydrogen, biogas), liquid biofuels (biodiesel, bioethanol, bio-oil FT-fuels), and solid biofuels (charcoal, pellets, lignin). These products substitute the ones obtained from fossil-fuel refineries. Instead of using fossil fuels, the same chemicals are synthesized from biomass in a biorefinery. Moreover, a molecule of the same function but different chemical formula can also be synthesized. The updated top-12 building blocks derived from biomass by chemical or biochemical manufacturing techniques are shown in Fig. 9.13 [85–87,165–167].

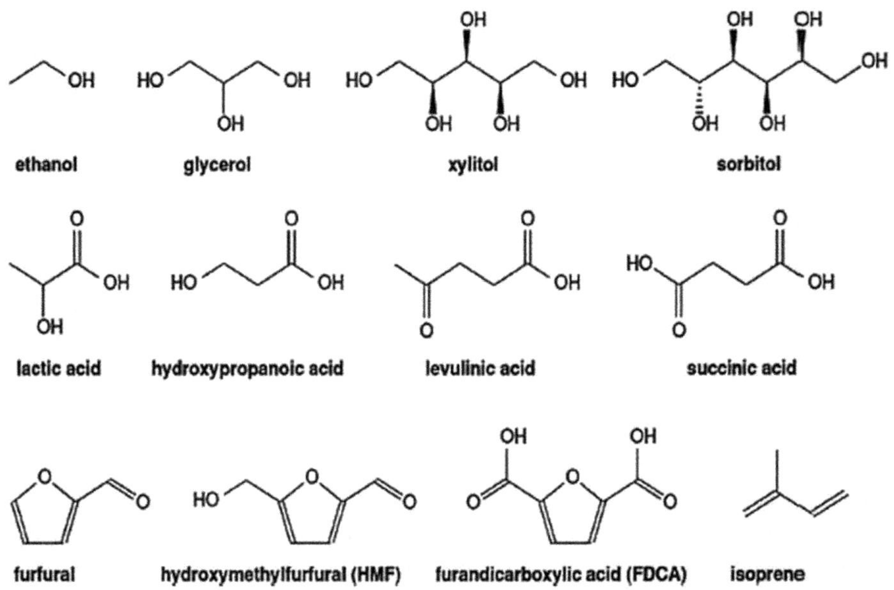

FIGURE 9.13

Top-12 bio-based platform molecules [165–167].

9.5.1 PRODUCTS OBTAINED FROM CONVENTIONAL CHEMICAL METHODS

Classical chemical methods are successfully used for the synthesis of wide range of building block of polymers from biomass. The most common traditional chemical method for the development of bio-based polymers involves the transformation of bio-based fatty acids into polymer building blocks. Carbon—carbon double bonds of triglycerides are chemically converted to methoxy and alcohol groups, which leads to a bio-based polyol (BiOH). These polyols are later utilized for the synthesis of polyurethane products. Industrially, hexose sugars of wood processing and agricultural wastes are converted into levulinic acid [168,169] which is a short chain (C_5) acid with two very reactive functional moieties, a carbonyl moiety (R—CO—R) and a carboxyl (—COOH) moiety. Levulinic acid could be used as a building block for certain specialty chemicals or directly in various products such as resins, plasticizers, and textiles [169]. Owing to its aromatic structure, lignin is converted into xylene, benzene, toluene, or other aromatic compounds [170].

9.5.1.1 Catalysis

Catalysts play a vital role in converting biomass to several value-added chemicals and fuels. One of the well-known methods involves the utilization of Fischer—Tropsch chemistry in pyrolyzed biomass. Catalysts were exploited in the manufacturing of biofuels from palm [171]. Moreover, platinum-catalyzed, aqueous-phase reforming of glycerol produces high quantity of hydrogen fuel with low CO level [172]. For the production of value-added products (pharmaceutical and fine chemicals), biocatalysts have also been extensively used [9,173]. Biocatalysts have the ability to selectively catalyze the reactions to confirm the formation of required products, to decrease the consumption of energy and waste generation, and to make products which are not feasible by chemical reactions alone [9].

9.5.1.2 Condensation Polymerization

Bio-derived monomers can be polymerized through condensation polymerization by using immobilized enzyme catalysts. For instance, chemically or biologically derived diacids are reacted with sorbitol or glycerol in presence of lipase by condensation polymerization [94,174]. Condensation polymerization reduces the reaction temperature and energy utilization and controls branching during polymerization [11]. Increase in control and decrease in temperature are particularly essential in the growth of biorefining technologies to compete with conventional petroleum refineries.

9.5.2 PRODUCTS OBTAINED FROM FERMENTATION

Fermentation is extensively used to produce highly desired building blocks, for instance, succinic acid, one of the DOE top-12 building-block chemical, is made by fermentation. As most of the microorganisms employed in fermentation cannot tolerate acidic conditions, the process is neutralized by preparing salts of acids.

Salts of succinic acid are produced by the fermentation of glucose, which fixes CO_2 from atmosphere and is therefore a green method [169,175]. Chemical processing, such as separation and recovery, of these salts is easier. Succinic acid is obtained by separation followed by dissolving these salts in acidic medium [175]. Glycerol, a by-product of biodiesel, has become an industrial commodity molecule and source of a variety of value-added chemicals [176,177]. 1,3-Propanediol, formed through glycerol fermentation [178,179], is a main component for the synthesis of polypropylene terephthalate (PPT) that is being employed as a fiber in the carpet and apparel industries. Another top-12 building block is itaconic acid that is produced by carbohydrate fermentation. Polymerized esters such as vinyl, ethyl, and methyl are used in adhesives and coatings. Itaconic acid is usually present in emulsions, enhances the polymer adhesion of emulsions, also act as a hardening agent for the organo-siloxanes used in the contact lenses. Owing to the two reactive carboxyl groups, itaconic acid can be combined with polymers. It is currently being evaluated as a substitute for methacrylic and acrylic acid in styrene–butadiene systems as well as in polymers [169].

Lactic acid, formed by fermentation, is converted to various significant chemicals, such as lactide, methyl lactate, and polylactic acid (a biodegradable substitute for polyethylene terephthalate) [180,181]. Lactic acid is a building block for wide range of high-value chemicals. Biomass can be transformed to acrylic acid through fermentation [182]. Acrylic acid, along with its ester and amide derivatives, is a basic constituent in the polymer synthesis and these polymers are used in surface coatings, absorbent, textiles, and detergents.

Ethylene can be synthesized from biomass hydrolysates by fermentation. Fermentation of sugars yields bioethanol which upon dehydration produces bioethylene. Dimerization of ethylene gives normal butane which reacts with bioethylene by metathesis to make propylene [184–186]. Bioethylene became a substitute of ethylene obtained from steam cracking of petroleum fractions, natural gas, or shale gas as the point of origin for the C_2 product tree (Fig. 9.14) [187].

9.5.3 PRODUCTS OBTAINED FROM IONIC LIQUID PHASE REACTION

Ionic liquids (ILs) phase biomass reaction involves direct incorporation of functional additives through dispersion or dissolution, before or after dissolving cellulose [188], resulting in the decrease of processing steps, power, and cost requirements. Ionic liquids can be mixed with catalytic amounts of acid to combine hydrolysis and pretreatment efficiently in a single step which increases reducing sugar yield from cellulose [189–191]. Stability of catalyst during reaction is also enhanced in the presence of ionic liquids. For instance, chromium chloride ($CrCl_2$) can be stabilized in 1-alkyl-3-methylimidazolium chloride (AMIM Cl) as well as ethyl-3-methylimidazolium chloride ([EMIM]Cl) to catalyze the synthesis of 5-hydroxymethyl furfural (HMF) from biomass, a highly valuable chemical [192,193]. The ILs are used for the preparation of cellulose-based initiator used for the atom-transfer radical polymerization, a medium for cellulose polymerization reactions and as a polymerizable composite in

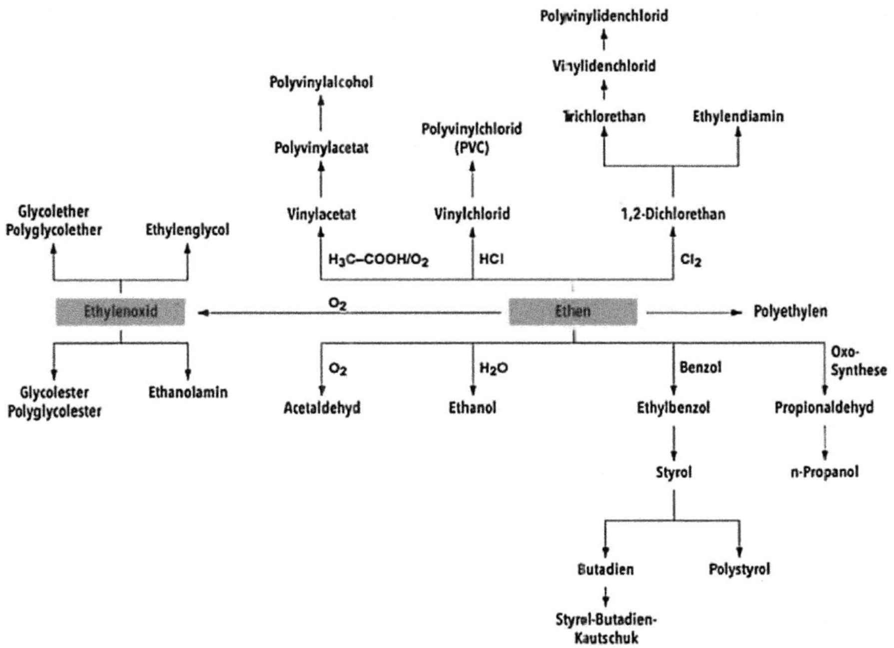

FIGURE 9.14

Most important product trees derived from ethylene [187].

radical polymerization [194,195]. Novel ionic liquids, such as switchable ionic liquids, can be used for the separation of products and activities assays of microbial enzymes specially obtained from the extremophiles as well as transformation of reducing sugars in solution [196,197].

9.5.4 PRODUCTS OBTAINED FROM DIRECT BIOLOGICAL CONVERSION

9.5.4.1 Extraction

Certain value-added chemicals can be synthesized in vivo, i.e., within the microorganism and plants. Efficient extraction technologies are required prior to advanced processing. Conventional extraction techniques can be employed for the direct extraction of commodity chemicals from the biomass, e.g., ferulic acid, used in the synthesis of valuable chemicals (guaiacol and vanillin) is directly extracted from corn fiber in high yields. Tulipalin A monomer, extracted from tulips, can be polymerized in a way analogous to methyl methacrylate with favorable durability and refractive index [169,198].

9.5.4.2 Enzymatic Transformation

Polymers, such as polyhydroxyalkanoates (PHAs), are produced completely within microbial cells. PHAs consist of more than 150 hydroxyalkanoates which are made by a variety of bacterial species as intracellular granules (90% of dry cell weight) [175,199–201]. They are extensively used in plastic industry due to their wide range of properties. Several investigations have been carried out to genetically modify the plants for direct PHA production. Various carbon sources can be used to make both medium and short-chained PHAs. In recent times, PHAs are synthesized by employing a forestry-based biorefinery with lignocellulosic streams, containing levulinic acid and hemicelluloses hydrolysates obtained from cellulose, and tall oil fatty acids obtained from kraft pulping, used as the sources of carbon for the bacteria *Burkholderia cepacia* [202,203]. Fermented municipal primary solids, industrial wastewaters from methanol-enriched paper and pulp mill foul condensate, and biodiesel upon passing through batch bioreactors consisting of microbial consortium (municipal activated sludge) give PHAs [204].

9.5.5 NEW BIOREFINERY TECHNOLOGIES AND PRODUCTS

Integration of several biomass conversion processes for the generation of energy, power, and value-added chemicals is the basic idea behind a biorefinery. As previously discussed, most commonly used biomass feedstocks in biorefineries include lignocelluloses, mono- or oligosaccharides, triglycerides, chitin, etc. Mostly, the optimization of substrate is not done resulting in the unutilized biomass such as large quantities of proteins. Protein obtained from the oilseed cakes is mostly utilized in animal feed [205]. Moreover, high-value protein is produced from dairy and meat processing industry. Considering the higher number of biorefineries being established, protein is a promising and cost-effective starting material for bioenergy and chemical production in a biorefinery [206,207]. Protein can be purified for the food and feed purpose along with nonfood applications. Moreover, proteins can be converted into biofuels and several chemical binders, adhesives, and building blocks, etc. A new area for the protein utilization is the application in pharmaceutics by the conversion into antiaging products, antibodies, hormones, and immunoglobulin, etc.

9.6 CONCLUSION

Biomass, versatile and main renewable raw material for biorefinery, has potential to substitute fossil resources to produce energy and nonenergy materials. Biomass is pretreated before processing to increase processing, surface area, and reactivity. Different technological conversion processes, such as thermochemical, biochemical, chemical, or mechanical processes, are used to convert biomass into important products in a biorefinery technique. The important material products obtained from biorefineries are chemicals, organic acids (lactic, succinic, itaconic acid), polymers and resins, food, and fertilizers; while the energy products are: gaseous biofuels, solid biofuels, and liquid biofuels. In a green biorefinery, natural wet raw material derived from green plants, green crops, or grass can be used as inputs. In lignocellulosic biorefinery, cellulose and hemicelluloses are converted to produce valuable chemicals such as ethanol, acetone, acetic acid, butanol, and other fermentation products. Lignin is used only for fuel, adhesive, or binder purposes. Algae, a marine crop, are known for their greenhouse gas-reduction potential and their capability to absorb CO_2 probably exceeding that of terrestrial species. Algae are considered to be novel feedstock for a biorefinery due to their prospective to form multiple products. Many conversion technologies, such as thermochemical and biochemical, can be combined together in an integrated biorefinery to decrease the total cost with more flexibility in product generation and to supply its own power.

REFERENCES

[1] Sedlmeyer FB. Xylan as by-product of biorefineries: characteristics and potential use for food applications. Food Hydrocoll 2011;25:1891−8.

[2] Clark JH. Green chemistry for the second generation biorefinery − sustainable chemical manufacturing based on biomass. J Chem Technol Biotechnol 2007;82:603−9.

[3] Shuangning X, Bo Z, Abolghasem S. Biorefinery processes for biomass conversion to liquid fuel. INTECH Open Access Publisher; 2011. p. 167−90.

[4] Choi S, Song CW, Shin JH, Lee SY. Biorefineries for the production of top building block chemicals and their derivatives. Metab Eng 2015;28:223−39.

[5] Mathers RT, Meier MAR. Green polymerization methods: renewable starting materials, catalysis and waste reduction. Weinheim: Wiley-VCH; 2011.

[6] Belgacem MN, Gandini A. Monomers, polymers and composites from renewable resources. New York: Elsevier; 2008.

[7] Bozell JJ, Patel MK. Feedstocks for the future: using technology development as a guide to product identification. (Washington, DC): ACS Symposium Series; 2006. p. 921.

[8] Anastas P, Eghbali N. Green chemistry: principles and practice. Chem Soc Rev 2010; 39:301−12.

[9] Hatti-Kaul R, Törnvall U, Gustafsson L, Börjesson P. Industrial biotechnology for the production of bio-based chemicals − a cradle-to-grave perspective. Trends Biotechnol 2007;25(3):119−24.

[10] Francesco C. The biorefinery concept: using biomass instead of oil for producing energy and chemicals. Energy Convers Manag 2010;51:1412−21.

[11] Dodds DR, Gross RA. Chemicals from biomass. Chem Sci 2007;318:1250−1.

[12] Lipinsky ES. Chemicals from biomass: petrochemical substitution options. Science 1981;212:1465−71.

[13] McLaughlin M. Bioproducts: significant stakes for the 21st century. In: The Canadian conference on industrial bioproduct innovation, November 5−6, Montreal; 2008.

[14] Khalil H. The use of biomaterials in automotive parts manufacturing. In: The Canadian conference on industrial bioproduct innovation, November 5−6, Montreal; 2008.

[15] Kamm B, Kamm M. Biorefinery systems. Chem Biochem Eng 2004;18(1):1−6.

[16] Ragauskas AJ, Williams CK, Davison BH, Britovsek G, Cairney J, Eckert CA. The path forward for biofuels and biomaterials. Science 2006;311:484−9.

[17] Adam K, Prodromos D. Process synthesis of biorefineries: optimization of biomass conversion to fuels and chemicals. Ind Eng Chem Res 2014;53:5261−73.

[18] Mariano M, Ignacio EG. On the systematic synthesis of sustainable biorefineries. Ind Eng Chem Res 2013;52:3044–64.

[19] Demirbas A. Biorefineries: current activities and future developments. Energy Convers Manag 2009;50:2782–801.

[20] Sharmaa P, Sarkerb BR, Romagnoli JA. A decision support tool for strategic planning of sustainable biorefineries. Comput Chem Eng 2011;35:1767–81.

[21] Stephane O, Daniel T. Mini-review biorefinery: toward an industrial metabolism. Biochimie 2009;91:659–64.

[22] Michael FP, Pascale C, Michael FC, Ralph AW. Review: a biorefinery processing perspective: treatment of lignocellulosic materials for the production of value-added products. Bioresour Technol 2010;101:8915–22.

[23] Huber GW, Dumesic JA. An overview of aqueous-phase catalytic processes for production of hydrogen and alkanes in a biorefinery. Catal Today 2006;111:119–32.

[24] de Jong E, Jungmeier G. Biorefinery concepts in comparison to petrochemical refineries. Ind biorefin white biotechnol. 2016.

[25] Carla SJ, Stephen PM. Algae biofuels: versatility for the future of bioenergy. Curr Opin Biotechnol 2012;23:346–51.

[26] Naik SN, Vaibhav VG, Prasant KR, Ajay KD. Production of first and second generation biofuels: a comprehensive review. Renew Sustain Energy Rev 2010;14:578–97.

[27] Rajagopal D, Zilberman D. Review of environmental, economic and policy aspects of biofuels. In: Policy research working paper of the World Bank development research group; September, 2007.

[28] Hoogwijk M, Faaij A, van den BR, Berndes G, Gielen D, Turkenburg W. Exploration of the ranges of the global potential of biomass for energy. Biomass Bioenergy 2003;25(2):119–33.

[29] Demirabas A. Biodiesel fuels from vegetable oils via catalytic and non-catalytic supercritical alcohol transesterifications and other methods: a survey. Energy Convers Manag 2003;44(13):2093–109.

[30] Hill K. Industrial development and application of biobased oleochemicals. In: Biorefienries – industrial processes and products (status quo and future directions), vol. 2. Wiley-VCH; 2006.

[31] Biermann U, Fürmeier S, Metzger JO. New chemistry of oils and fats, oleochemical manufacture and applications. Sheffield Academic Press and CRC Press; 2001.

[32] Achten WMJ, Mathijs E, Verchot L, Singh VP, Aerts R, Muys B. Jatropha biodiesel fueling sustainability. Biofuels Bioprod Biorefin 2007;1:283–91.

[33] Tsai WT, Lin CC, Yeh CW. An analysis of biodiesel fuel from waste edible oil in Taiwan. Renew Sustain Energy Rev 2007;11:838–57.

[34] Cherubini F, Bargigli S, Ulgiati S. Life cycle assessment of urban waste management: energy performances and environmental impacts. The case of Rome, Italy. J Waste Manag 2008;28:2552–64.

[35] Clark J, Luque R, Matharu A. Green chemistry, biofuels, and biorefinery. Annu Rev Chem Biomol Eng 2012;3:183–207.

[36] Naznin R. Extraction of chitin and chitosan from shrimp (*Metapenaeus monoceros*) shell by chemical method. Pak J Biol Sci 2005;8(7):1051–4.

[37] Sila A, Mlaik N, Sayari N, Balti R, Bougatef A. Chitin and chitosan extracted from shrimp waste using fish proteases aided process: efficiency of chitosan in the treatment of unhairing effluents. J Polym Environ 2013;22(1):78–87.

[38] Santhosh S, Sini TK, Anandan R, Mathew PT. Effect of chitosan supplementation on antitubercular drugs-induced hepatotoxicity in rats. Toxicology 2006;219:53–9.

[39] Santhosh S, Anandan R, Sini TK, Mathew PT. Protective effect of glucosamine against ibuprofen-induced peptic ulcer in rats. J Gastroenterol Hepatol 2007;22:949–53.

[40] Bautista J, Jover M, Gutierrez JF, Corpas R, Cremades O, Fontiveros E, Iglesias F, Vega J. Preparation of crayfish chitin by in situ lactic acid production. Process Biochem 2001;37:229–34.

[41] Ravi Kumar MNV. A review of chitin and chitosan applications. React Funct Polym 2000;46:1–273.

[42] Bautista-Baños S, Hernández-Lauzardo AN, Velázquez-del Valle MG, Hernández-López M, Ait Barka E, Bosquez-Molina E, Wilson CL. Chitosan as a potential natural compound to control pre and postharvest disease of horticultural commodities. Crop Prot 2006;25:108–18.

[43] Rashidova SS, Milusheva RY, Voropaeva NL, Pulatova SR, Nikonovich GV, Ruban IN. Isolation of chitin from a variety of raw materials, modification of the material, and interaction its derivatives with metal ions. Chromatographia 2004;59:783–6.

[44] Sashiwa H, Aiba S. Chemistry modified chitin and chitosan as biomaterials. Prog Polym Sci 2004;29:887–908.

[45] Abdou ES, Elkholy SS, Elsabee MZ, Mohamed E. Improved antimicrobial activity of polypropylene and cotton nonwoven fabrics by surface treatment and modification with chitosan. J Appl Polym Sci 2008;108:2290–6.

[46] Elsabee MZ, Abdou ES, Nagy KSA, Eweis M. Surface modification of polypropylene films by chitosan and chitosan/pectin multilayer. Carbohydr Polym 2008;71:187–95.

[47] Chen X, Yang H. Shell biorefinery: dream or reality? Chemistry September 12, 2016; 22(38):13402–21.

[48] Muzzarelli R, Muzzarelli C. Chitosan chemistry: relevance to the biomedical sciences. Adv Polym Sci 2005;186:151–209.

[49] Dai J, Ren F, Tao C, Bai Y. Synthesis of cross-linked chitosan and application to adsorption and speciation of Se (VI) and Se (IV) in environmental water samples by inductively coupled plasma optical emission spectrometry. Int J Mol Sci 2011;12: 4009–20.

[50] Elwakeel K. Environmental application of chitosan resins for the treatment of water and wastewater: a review. J Dispers Sci Technol 2010;31:273–88.

[51] Sayed S, Jardine A. Chitosan derivatives as important biorefinery intermediates. Quaternary tetraalkylammonium chitosan derivatives utilized in anion exchange chromatography for perchlorate removal. Int J Mol Sci 2015;16:9064–77.

[52] Huber GW, Iborra S, Corma A. Chem Rev 2006;106:4044–98.

[53] Mosier N, Wyman C, Dale B, Elander R, Lee YY, Holtzapple M, Ladisch M. Bioresour Technol 2005;96:673–86.

[54] Wyman CE, Dale BE, Elander RT, Holtzapple M, Ladisch MR, Lee YY. Bioresour Technol 2005;96:1959–66.

[55] Garcia PM, Wang SSJ, Rhodes M, Lee WJ, Li CZ. Energy Fuels 2008;22:2022–32.

[56] Han J, Kim H. Renew Sustain Energy Rev 2008;12:397–416.

[57] Matsumuraa Y, Minowab T, Poticc B, Kerstenc SRA, Prinsc W, Van Swaaijc WPM, Van de Beldd B, Elliotte DC, Neuenschwandere GG, Krusef A, Antal MJ. Biomass Bioenergy 2005;29:269–92.

[58] Zhang Y-HP. Reviving the carbohydrate economy via multi-product lignocellulose biorefineries. J Ind Microbiol Biotechnol 2008;35:367–75.

[59] Manzer LE. Feedstocks for the future: using technology development as a guide to product identification. In: Bozell JJ, Patel MK, editors. ACS symposium series 921. (Washington, D.C.): American Chemical Society; 2006. p. 40−51.
[60] Sanchez OJ, Cardona CA. Bioresour Technol 2008;99:5270−95.
[61] Olsson L, Hahn HB. Enzyme Microb Technol 1996;18:312−31.
[62] Mutlu H, De ELM, Meier MAR. Chem Soc Rev 2011;40:1404−45.
[63] De Espinosa LM, Ronda JC, Galia M, Cadiz V. J Polym Sci Part A Polym Chem 2009; 47:4051−63.
[64] De Espinosa LM, Ronda JC, Galia M, Cadiz V. J Polym Sci Part A Polym Chem 2009; 47:1159−67.
[65] Mathers RT, LeBlond C, Damodaran K, Kushner DI, Schram VA. Macromolecules 2008;41:524−6.
[66] Montero L, De Espinosa LM, Ronda JC, Galia M, Cadiz V. J Polym Sci Part A Polym Chem 2010;48:869−78.
[67] De Espinosa LM, Ronda JC, Galia M, Cadiz V. J Polym Sci Part A Polym Chem 2008; 46:6843−50.
[68] Del Rio E, Galia M, Cadiz V, Lligadas G, Ronda JC. J Polym Sci Part A Polym Chem 2010;48:4995−5008.
[69] Del RE, Lligadas G, Ronda JC, Galia M, Cadiz V. J Polym Sci Part A Polym Chem 2010;48:5009−17.
[70] Stemmelen M, Pessel F, Lapinte V, Caillol S, Habas JP, Robin JJ. J Polym Sci Part A Polym Chem 2011;49:2434−44.
[71] Lligadas G, Ronda JC, Galia M, Biermann U, Metzger JO. J Polym Sci Part A Polym Chem 2006;44:634−45.
[72] Petrovic ZS. Polym Rev 2008;48:109−55.
[73] Larock RC, Dong X, Chung S, Reddy CK, Ehlers LE. J Am Oil Chem Soc 2001;78: 447−53.
[74] Mathers RT. How well can renewable resources mimic commodity monomers and polymers? J Polym Sci Part A Polym Chem 2012;50:1−15.
[75] Huang HJ, Ramaswamy S, Tschirner UW, Ramarao BV. A review of separation technologies in current and future biorefineries. Sep Purif Technol 2008;62:1−21.
[76] Kamm B, Kamm M. Principles of biorefineries. Appl Microbiol Biotechnol 2004;64: 137−45.
[77] Fernando S, Adhikari S, Chandrapal C, Murali N. Biorefineries: current status challenges and future direction. Energy Fuels 2006:1727−37.
[78] Tyson KS, Bozell J, Wallace R, Petersen E, Moens L. Biomass oil analysis: research needs and recommendations. NREL Technical Report. 2005. Available from:http://www.eere.energy.gov/biomass/pdfs/34796.pdf.
[79] de Jong E, van Ree R, Sanders JPM, Langeveld H. Biorefineries: giving value to sustainable biomass use. In: Langeveld H, Meeusen M, Sanders J, editors. The biobased economy: biofuels, materials and chemicals in the post-oil era. London: Earthscan Publishers; 2010. p. 111−30 [Chapter 7].
[80] Osamu K, Carl HW. Biomass handbook. Gordon Breach Science Publisher; 1989.
[81] Brown LM, Zeiler KG. Aquatic biomass and carbon dioxide trapping. Energy Convers Manag 1993;34:1005−13.
[82] Raja A, Vipin C, Aiyappan A. Biological importance of marine algae − an overview. Int J Curr Microbiol App Sci 2013;2(5):222−7.
[83] Hu Q, Sommerfeld M, Jarvis E, Ghirardi M, Posewitz M, Seibert M, Darzins A. Microalgal triacylglycerols as feedstocks for biofuel production: perspectives and advances. Plant J 2008;54:621−39.

[84] Chisti Y. Biodiesel from microalgae beats bioethanol. Trends Biotechnol 2008;26(3): 126−31.

[85] Sahu A, Pancha I, Jain D, Paliwal C, Ghosh T, Patidar S. Fatty acids as biomarkers of microalgae. Phytochemistry 2013;89:53−8.

[86] Becker EW. Micro-algae as a source of protein. Biotechnol Adv 2007;25:207−10.

[87] Schmid-Straiger U. Algae biorefinery—concept. In: National German Workshop on Biorefineries; 2009. Worms.

[88] Harun R, Singh M, Forde GM, Danquah MK. Bioprocess engineering of microalgae to produce a variety of consumer products. Renew Sustain Energy Rev 2010;14: 1037−47.

[89] Jasvinder S, Gu S. Commercialization potential of microalgae for biofuels production. Renew Sustain Energy Rev 2010;14:2596−610.

[90] Jayati T, Mounika A, Bangwal DP, Savita K, Garg MO. Algae based biorefinery how to make sense? Renew Sustain Energy Rev 2015;47:295−307.

[91] Najafi G, Ghobadiana B, Yusaf TF. Algae as a sustainable energy source for biofuel production in Iran: a case study. Renew Sustain Energy Rev 2011;15:3870−6.

[92] Griffiths MJ, Harrison STL. Lipid productivity as a key characteristic for choosing algal species for biodiesel production. J Appl Phycol 2009;21:493−507.

[93] Ziolkowska JR, Simon L. Recent developments and prospects for algae-based fuels in the US. Renew Sustain Energy Rev 2014;29:847−53.

[94] Robles MA, González MPA, Esteban CL, Molina GE. Biocatalysis: towards ever greener biodiesel production. Biotechnol Adv 2009;27(4):398−408.

[95] Li P, Miao X, Li R, Zhong J. In situ biodiesel production from fast-growing and high oil content *Chlorella pyrenoidosa* in rice straw hydrolysate. J Biomed Biotechnol 2011. http://dx.doi.org/10.1155/2011/141207.

[96] Araujo GS, Matos LJBL, Goncalves LRB, Fernandes FAN, Farias WRL. Bio-prospecting for oil producing microalgal strains: evaluation of oil and biomass production for ten microalgal strains. Bioresour Technol 2011;102:5248−50.

[97] Johnson MB, Wen Z. Production of biodiesel fuel from the microalga *Schizochytrium limacinum* by direct transesterification of algal biomass. Energ Fuels 2009;23: 5179−83.

[98] Rawat I, Kumar RR, Mutanda T, Bux F. Dual role of microalgae: phyco remediation of domestic wastewater and biomass production for sustainable biofuels production. Appl Energy 2011;88:3411−24.

[99] Islam MA, Ayoko GA, Brown R, Stuart D, Heimann K. Influence of fatty acid structure on fuel properties of algae derived biodiesel. Proced Eng 2013;56:591−6.

[100] Talebi AF, Mohtashami SK, Tabatabaei M, Tohidfar M, Bagheri A, Zeinalabe-dini M, et al. Fatty acids profiling: a selective criterion for screening microalgae strains for biodiesel production. Algal Res 2013;2:258−67.

[101] Amaro HM, Macedo AC, Malcata FX. Microalgae: an alternative as sustainable source of biofuels? Energy 2012;44:158−66.

[102] Suominen KEH, Ojanen S, Ahtila P. A biorefinery concept for energy intensive industries focusing on microalgae and anaerobic digestion. J Mech Eng Autom 2014;4: 242−51.

[103] Singh B, Guldhe A, Rawat I, Bux F. Towards a sustainable approach for development of biodiesel from plant and microalgae. Renew Sustain Energy Rev 2014;29: 216−45.

[104] Clark J, Deswarte F. Introduction to chemicals from biomass. In: Clark J, Deswarte F, editors. Wiley series in renewable rescurces. John Wiley & Sons; 2008.

[105] Searchinger T, Heimlich R, Houghton RA, Dong F, Elobeid A, Fabiosa J, et al. Use of U.S. Croplands for biofuels increases greenhouse gases through emissions from land-use change. Science 2008;319:1238–40.

[106] Yuan X, Shi X, Zhang D, Qiu Y, Guo R, Wang L. Biogas production and microcystin biodegradation in anaerobic digestion of blue green algae. Energ Environ Sci 2011;4: 1511–5.

[107] Collet P, Hélias A, Lardon L, Ras M, Goy RA, Steyer JP. Life-cycle assessment of microalgae culture coupled to biogas production. Bioresour Technol 2011;102: 207–14.

[108] Harun R, Danquah MK, Forde GM. Microalgal biomass as a fermentation feedstock for bioethanol production. J Chem Technol Biotechnol 2010;85:199–203.

[109] Wi SG, Kim HJ, Mahadevan SA, Yang DJ, Bae HJ. The potential value of the seaweed Ceylon moss (*Gelidium amansii*) as an alternative bioenergy resource. Bioresour Technol 2009;100:6658–60.

[110] Mussgnug JH, Klassen V, Schluter A, Kruse O. Microalgae as substrates for fermentative biogas production in a combined biorefinery concept. J Biotechnol 2010;150: 51–6.

[111] Harun R, Jason WSY, Cherrington T, Danquah MK. Exploring alkaline pre-treatment of microalgae biomass for bioethanol production. Appl Energy 2011;88:3464–7.

[112] Nguyen MT, Choi SP, Lee J, Lee JH, Sim SJ. Hydrothermal acid pretreatment of *Chlamydomonas reinhardtii* biomass for ethanol production. J Microbiol Biotechnol 2011; 154:878–88.

[113] Choi SP, Nguyen MT, Sim SJ. Enzymatic pretreatment of *Chlamydomonas reinhardtii* biomass for ethanol production. Bioresour Technol 2010;101:5330–6.

[114] Lee S, Oh Y, Kim D, Kwon D, Lee C, Lee J. Converting carbohydrates extracted from marine algae into ethanol using various ethanolic *Escherichia coli* strains. Appl Biochem Biotechnol 2011;164:878–88.

[115] Maurych D, Shu G, Guangyi W. Recent advances in liquid biofuel production from algal feedstocks. Appl Energy 2013:1371–81.

[116] Jung KA, Lim SR, Kim Y, Park JM. Potentials of macroalgae as feedstocks for biorefinery. Bioresour Technol 2013;135:182–90.

[117] Campbell CJ. Oil & gas liquid 2004 scenario, uppsala hydrocarbon depletion study group. 2004.

[118] Lee DS, Pitari G, Grewe V, Gierens K, Penner JE, Petzold A. Transport impacts on atmosphere and climate: aviation. Atmos Environ 2010;44:4678–734.

[119] Marian E, Ihab HF. Bio-jet fuel from microalgae: reducing water and energy requirements for algae growth. Int J Eng Sci 2012;1(2):22–30.

[120] Judit S, Berta MG. Overview of biofuels for aviation. Chem Eng Trans 2012;29: 1147–52.

[121] Lehninger A, Nelson D, Cox M. Lehninger principles of biochemistry. 4th ed. New York (NY): W.H. Freeman; 2005.

[122] Chen CY, Zhao XQ, Yen HW, Ho SH, Cheng CL, Lee DJ, et al. Microalgae-based carbohydrates for biofuel production. Biochem Eng J 2013;78:1–10.

[123] John RP, Anisha GS, Nampoothiri KM, Pandey A. Micro and macroalgal biomass: a renewable source for bioethanol. Bioresour Technol 2011;102:186–93.

[124] Lobban CS, Wynne MJ. The biology of seaweeds. 1st ed. Blackwell Scientific Publications; 1981.

[125] McHugh DJ. A guide to the seaweed industry. FAO fish tech, vol. 441; 2003. p. 1–105.

[126] Draget KI, Smidsrod O, Skjak BG. Alginates from algae. In: Biopolymers online. Wiley-VCH Verlag GmbH & Co. KGa; 2005.

[127] Yen HW, Hu IC, Chen CY, Ho SH, Lee DJ, Chang JS. Microalgae-based biorefinery—from biofuels to natural products. Bioresour Technol 2013;135:166–74.

[128] Gouveia L. From tiny microalgae to huge biorefineries. Oceanography 2014;2: 2332–632.

[129] Campenni L, Nobre BP, Santos CA, Oliveira AC, Aires-Barros AR, et al. Carotenoids and lipids production of autotrophic microalga *Chlorella protothecoides* under nutritional, salinity and luminosity stress conditions. Appl Microbiol Biotechnol 2013; 97:1383–93.

[130] Olguín EJ. Dual purpose microalgae–bacteria based systems that treat waste-water and produce biodiesel and chemical products within a biorefinery. Biotechnol Adv 2012;30:1031–46.

[131] Pacheco R, Ferreira AF, Pinto T, Nobre BP, Loureiro D. Life cycle assessment of a *Spirogyra* sp. biorefinery for the production of pigments, hydrogen and leftovers energy valorisation. Appl Energy 2014.

[132] Budarin V, Ross AB, Biller P, Riley R, Clark JH, Jones JM. Microalgae biorefinery concept based on hydrothermal microwave pyrolysis. Green Chem 2012;14:3251.

[133] Yanqun L, Horsman M, Wu N, Lan CQ, Dubois-Calero N. Biofuels from microalgae. Biotechnol Prog 2008;24:815–20.

[134] Bridgewater A, Czernik C, Diebold J, Mekr D, Radlein P. Fast pyrolysis of biomass: a handbook. Newbury (UK): CPL Scientific Publishing Services, Ltd.; 1999.

[135] Bridgewater AV, Cottam ML. Opportunities for biomass pyrolysis liquids production and upgrading. Energy Fuels 1992;6:113–20.

[136] Bobban S, Grinson G. Algal biorefinery-based industry: an approach to address fuel and food insecurity for a carbon-smart world. J Sci Food Agric 2011;91(1):2–13.

[137] Stephens E, Ross IL, King Z, Mussgnug JH, Kruse O, Posten C. An economic and technical evaluation of microalgal biofuels. Nat Biotechnol 2010;28:126–8.

[138] Van HJ, Scott EL, Sanders J. Bulk chemical from biomass. Biofuels Bioprod Biorefin 2008;2:41–57.

[139] Bobban S. Sustainability of algal biofuel production using integrated renewable energy park (IREP) and algal biorefinery approach. Energy Policy 2010;38: 5892–901.

[140] Delgadoa ADG, Kafarov V. Microalgae based biorefinery: evaluation of several routes for joint production of biodiesel, chlorophylls, phycobiliproteins, crude oil and reducing sugars. Chem Eng Trans 2012;29:607–12.

[141] Demirbas A. Liquefaction of biomass using glycerol. Energy Sources Part A 2008;30: 1120–6.

[142] Demirbas A. Products from lignocellulosic materials via degradation processes. Energy Sources Part A 2008;30:27–37.

[143] Gercel HF, Gercel O. Bio-oil production from an oilseed by-product: fixed-bed pyrolysis of olive cake. Energy Sources Part A 2007;29:695–704.

[144] Demirbas A. Producing bio-oil from olive cake by fast pyrolysis. Energy Sources Part A 2008;30:38–44.

[145] Bakis R. Alternative electricity generation opportunities. Energy Sources Part A 2008; 30:141–8.

[146] Appel HR, Fu YC, Friedman S, Yavorsky PM, Wender I. Converting organic wastes to oil. US burea of mines report of investigation no. 7560. 1971.

[147] Marbe Å, Harvey S. Opportunities for integration of biofuel gasifiers in natural gas combined heat-and-power plants in district-heating systems. Appl Energy 2006;83: 723–48.

[148] López JM, Gómez Á, Aparicio F, Sánchez FJ. Comparison of GHG emissions from diesel, biodiesel and natural gas refuse trucks of the city of Madrid. Appl Energy 2009;86:610–5.

[149] Molten PM, Demmitt TF, Donovan JM, Miller RK. Mechanism of conversion of cellulose wastes to liquid in alkaline solution. In: Klass DL, editor. Energy from biomass and wastes III. Chicago (IL): Institute of Gas Technology; 1983. p. 293.

[150] Balat M. Possible methods for hydrogen production. Energy Sources Part A 2009;31: 39–50.

[151] Balat M. Mechanisms of thermochemical biomass conversion processes. Part 3: reactions of liquefaction. Energy Sources Part A 2008;30:649–59.

[152] Fatih Demirbas M. Biorefineries for biofuel upgrading: a critical review. Appl Energy 2009;89:151–61.

[153] Balat M. Mechanisms of thermochemical biomass conversion processes. Part 1: reactions of pyrolysis. Energy Sources Part A 2008;30:620–35.

[154] Sigar CP, Soni SL, Mathur J, Sharma D. Performance and emission characteristics of vegetable oil as diesel fuel extender. Energy Sources Part A 2009;31:139–48.

[155] Dincer K. Lower emissions from biodiesel combustion. Energy Sources Part A 2008; 30:963–8.

[156] Shieh CJ, Liao HF, Lee CC. Optimization of lipase-catalyzed biodiesel by response surface methodology. Bioresour Technol 2003;88:103–6.

[157] Demirbas A. Production of biodiesel from algae oils. Energy Sources Part A 2009;31: 163–8.

[158] Demirbas A. Oils from hazelnut shell and hazelnut kernel husk for biodiesel production. Energy Sources Part A 2008;30:1870–5.

[159] Ilkilic C, Yucesu HS. The use of cottonseed oil methyl ester on a diesel engine. Energy Sources Part A 2008;30:742–53.

[160] Sastry GSR, Murthy ASRK, Prasad PR, Bhuvaneswari K, Ravi PV. Identification and determination of bio-diesel in diesel. Energy Sources Part A 2006;28: 1337–42.

[161] Demirbas A. Biodiesel production via rapid transesterification. Energy Sources Part A 2008;30:1830–4.

[162] Lv P, Wang X, Yuan Z, Tan T. Conversion of soybean oil to biodiesel fuel with immobilized Candida lipase on textile cloth. Energy Sources Part A 2008;30:872–9.

[163] Demirbas A. Bioethanol from cellulosic materials: a renewable motor fuel from biomass. Energy Sources 2005;27:327–37.

[164] Sun Y, Cheng J. Hydrolysis of lignocellulosic materials for ethanol production: a review. Bioresour Technol 2002;83(1):1–11.

[165] Werpy T, Petersen G. Top value added chemicals from biomass, volume 1 results of screening for potential candidates from sugars and synthesis gas. 2004 [Online 2004]. Available at: http://www1.eere.energy.gov/bio mass/pdfs/35523.pdf.

[166] Bozell JJ, Petersen GR. Technology development for the production of biobased products from biorefinery carbohydrates – the US Department of Energy's "Top 10" revisited. Green Chem 2010;12:539–54.

[167] Werpy T, Pedersen G. Top value added chemicals from biomass, vol. 1. U.S. Department of Energy; 2005.

[168] Farone WA, Cuzens JE. Method for the production of levulinic acid and its derivatives. 2000. U.S. Patent: 6054,611.

[169] Lucia LA, Argyropoulos DS, Adamopoulos L, Gaspar AR. Chemicals and energy from biomass. Can J Chem 2006;84:960–70.

[170] Van HJ, Scott EL, Sanders J. Bulk chemicals from biomass. Biofuels Bioprod Biorefin 2007;2:41–57.

[171] Chew TL, Bhatia S. Catalytic processes towards the production of biofuels in a palm oil and oil palm biomass-based biorefinery. Bioresour Technol 2008;99:7911–22.

[172] Pagliaro M, Rossi M. The future of glycerol: new uses of a versatile raw material. Cambridge: Royal Society of Chemistry; 2008.

[173] Thomas SM, DiCosimo R, Nagarajan V. Biocatalysis: applications and potentials for the chemical industry. Trends Biotechnol 2002;20:238–42.

[174] Kumar A, Kulshrestha AS, Gao W, Gross RA. Versatile route to polyol polyesters by lipase catalysis. Macromolecules 2003;36:8219–21.

[175] Carole TM, Pellegrino J, Paster MD. Opportunities in the industrial biobased products industry. Appl Biochem Biotechnol 2004;115:871–85.

[176] Abraham TW, Höfer R. Lipid-based polymer building blocks and polymers. In: Matyjaszewski K, Möller M, editors. Polymer science: a comprehensive reference. McGrath JE, Hickner MA, Höfer R, editors. Polymers for a sustainable environment and green energy, vol. 10. Amsterdam: Elsevier; 2012. p. 15–58.

[177] Pagliaro M, Rossi M. The future of glycerol. 2nd ed. Cambridge: RSC Green Chemistry No. 8, RSC Publ.; 2010.

[178] Nakamura CE, Whited GM. Metabolic engineering for the microbial production of 1,3-propanediol. Curr Opin Biotechnol 2003;14:454–9.

[179] Chen L, Ren S, Re XP. Lactic acid production from glycerol using CaO as solid base catalyst. Fuel Process Technol 2014;120:40–7.

[180] Haas T, Kircher M, Köhler T, Wich G, Schörken U, Hagen R. White biotechnology. In: Höfer R, editor. Sustainable solutions for modern economies. Cambridge: RSC Green Chemistry, No. 4, RSC Publ.; 2010. p. 436–74.

[181] Xu X, Lin J, Cen P. Advances in the research and development of acrylic acid production from biomass. Chin J Chem Eng 2006;14:419–27.

[182] Danner H, Braun R. Biotechnology for the production of commodity chemicals from biomass. Chem Soc Rev 1999;28:395–405.

[183] Deleted in review.

[184] Keim W. C1 chemistry: potential and developments. Pure Appl Chem 1986;58: 825–32.

[185] Bare SR. Methanol to olefins (MTO): development of a commercial catalytic process. In: Modern methods in heterogeneous catalysis research, FHI lecture Nov. 30; 2007. http://www.fhi-berlin.mpg.de/acnew/department/pages/teaching/pages/teaching__wintersemester__2007_2008/bare_mto_301107.pdf.

[186] Chayot R. Direct fermentation for isobutene, butadiene and propylene production: a highway to renewable plastics, synthetic rubber and fuels. In: AIChE proceedings metabolic engineering X; 2014. http://www3.aiche.org/proceedings/Abstract.aspx?PaperID=354187.

[187] Bader HJ, Horn S, Fehrenbacher U, Grosshardt O, Kowollik K, Pohsner U, et al. Informationsserie Nachwachsende Rohstoffe. 2009. 54 p. https://www.vci.de/vci/Downloads-vci/textheft_farbig.pdf.

[188] Zhu S, Wu Y, Chen Q, Yu Z, Jin S. Dissolution of cellulose with ionic liquids and its application: a mini-review. Green Chem 2006;8:325–7.

[189] Li C, Wang Q, Zhao ZK. Acid in ionic liquid: an efficient system for hydrolysis of lignocellulose. Green Chem 2008;10:177–82.

[190] Zhang Y, Du H, Qian X, Chen EYX. Ionic liquid water mixtures: enhanced Kw for efficient cellulosic biomass conversion. Energy Fuels 2010;4:2410–7.

[191] Turner MB, Spear SK, Huddleston JG, Holbrey JD, Rogers RD. Ionic liquid salt-induced inactivation and unfolding of cellulase from *Trichoderma reesei*. Green Chem 2003;5:443–7.

[192] Zhao H, Holladay JE, Brown H, Zhang ZC. Metal chlorides in ionic liquid solvents convert sugars to 5-hydroxymethylfurfural. Science 2007;316:1597–600.

[193] Meng T, Gao X, Zhang J, Yuan J, Zhang Y, He J. Graft copolymers prepared by atom transfer radical polymerization (ATRP) from cellulose. Polymer 2009;50:447–54.

[194] Murakami M, Kaneko Y, Kadokawa J. Preparation of cellulose-polymerized ionic liquid composite by in-situ polymerization of polymerizable ionic liquid in cellulose-dissolving solution. Carbohydr Polym 2007;69:378–81.

[195] Oren A. Diversity of halophilic microorganisms: environments, phylogeny, physiology, and applications. J Ind Microbiol Biotechnol 2002;28:56–63.

[196] Phan L, Brown H, White J, Hodgson A, Jessop PG. Soybean oil extraction and separation using switchable or expanded solvents. Green Chem 2009;11:53–9.

[197] Mullin R. Sustainable specialties. Chem Eng News 2004;82:29–37.

[198] Munoz LEA, Riley MR. Utilization of cellulosic waste from tequila bagasse and production of polyhydroxyalkanoate (PHA) bioplastics by *Saccharophagus degradans*. Biotechnol Bioeng 2008;100:882–8.

[199] Sun Z, Ramsay JA, Guay M, Ramsay BA. Fermentation process development for the production of medium-chain-length poly-3-hyroxyalkanoates. Appl Microbiol Biotechnol 2007;75:475–85.

[200] Lee SY. Deciphering bioplastic production. Nat Biotechnol 2006;24:1227–9.

[201] Sun Z, Ramsay JA, Guay M, Ramsay BA. Carbon-limited fed-batch production of medium-chain-length polyhydroxyalkanoates from nonanoic acid by *Pseudomonas putida* KT2440. Appl Microbiol Biotechnol 2007;74:69–77.

[202] Suriyamongkol P, Weselake R, Narine S, Moloney M, Shah S. Biotechnological approaches for the production of polyhydroxyalkanoates in microorganisms and plants – a review. Biotechnol Adv 2007;25:148–75.

[203] Keenan TM, Nakas JP, Tanenbaum SW. Polyhydroxyalkanoate copolymers from forest biomass. J Ind Microbiol Biotechnol 2006;33:616–26.

[204] Coats ER, Loge FJ, Smith WA, Thompson DN, Wolcott MP. Functional stability of a mixed microbial consortium producing PHA from waste carbon sources. Appl Biochem Biotechnol 2007;137:909–25.

[205] Wernick DG, Liao JC. Protein-based biorefining: metabolic engineering for production of chemicals and fuel with regeneration of nitrogen fertilizers. Appl Microbiol Biotechnol 2013;97:1397–406.

[206] Huo YX, Cho KM, Rivera JGL, Monte E, Shen CR, Yan Y, Liao JC. Conversion of proteins into biofuels by engineering nitrogen flux. Nat Biotechnol 2011;29(4):346–51.

[207] Choi KY, Wernick DG, Tat CA, Liao JC. Consolidated conversion of protein waste into biofuels and ammonia using *Bacillus subtilis*. Metab Eng 2014;23:53–61.

CHAPTER

10 Blends of Algae With Natural Polymers

Shagufta Kamal[1], Maryam Rehman[1], Saima Rehman[1], Zill-i-Huma Nazli[2], Nazia Yaqoob[2], Razia Noreen[1], Saiqa Ikram[4], Ho S. Min[3]

Government College University Faisalabad, Faisalabad, Pakistan[1]; Government College Women University Faisalabad, Faisalabad, Pakistan[2]; INTI International University, Nilai, Malaysia[3]; Jamia Millia Islamia, New Delhi, India[4]

10.1 INTRODUCTION

The existing robust consciousness of sustainability concerns in society and the looming scarceness of oil sources are the two drivers behind the interest in the use of biopolymers (polymers extracted from living organisms) at both academic and industrial levels [1]. A polymer blend or polymer mixture (Fig. 10.1) belongs to a class of materials analogous to metal alloys; two polymers are blended together to create a new class of materials with different physical properties [2]. The flaw in one component can be camouflaged by the strength of the other component of polyblend [3].

The arithmetical average of blend components commonly known as "mixing rule" is usually followed by miscible blends, whereas phase-separated blends have the properties of all components of the blend. Ultimately, the performance

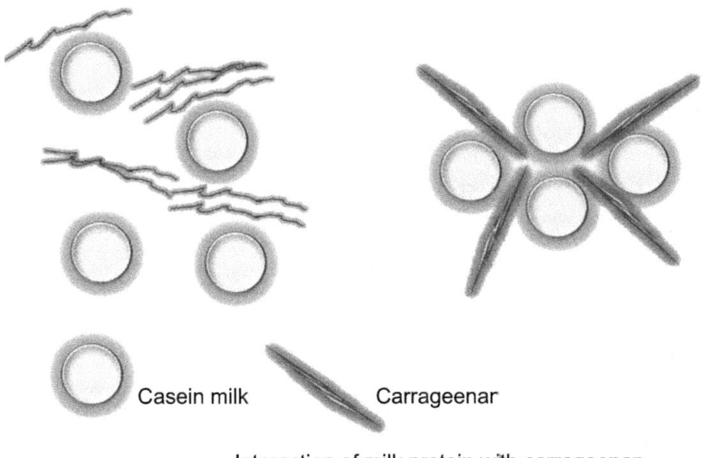

FIGURE 10.1

Casein and carrageenan blend.

Algae Based Polymers, Blends, and Composites. http://dx.doi.org/10.1016/B978-0-12-812360-7.00010-0

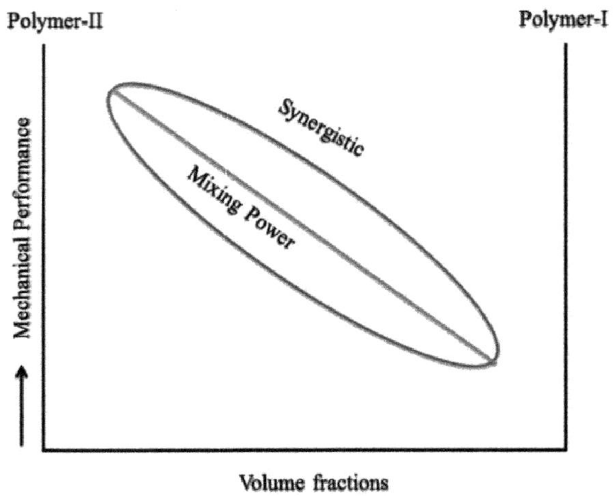

FIGURE 10.2

Mechanical performance of blends.

of blend depends on the size of structural elements and their interface adhesion. Generally, due to low interfacial adhesion and gross phase morphology, the immiscible blends are mostly incompatible and exhibit negative deviation from the "mixing rule" [3]. In few exceptional cases, incompatible blends remain futile, whereas the properties of compatible blends can be improved compared to individual blending components. Some synergistic effects are observed in these types of blends which are sometimes challenging to guess. Fig. 10.2 represents the mechanical performance of polymer blends.

The term polymer blend appeared first time in history when Thomas Hancock modified the mixture of natural rubber with gutta-percha, later on called as a new class of polymers or polymer blends [4]. Currently there are three main classes of polymer blends [5–7].

1. *Heterogeneous polymer blends*: These are commonly known as immiscible polymer blends due to presence of two phases. These are the most populous group of phase-separated polymer blends having two transition temperatures. This is represented as $\Delta G_m \approx \Delta H_m \geq 0$, where G_m is free energy of mixing and H_m is enthalpy of mixing.
2. *Compatible polymer blends*: It is a type of immiscible blend in which inhomogeneity is so small that it cannot be visualized. These blends most probably present homogeneous physical properties due to strong interactions of individual components.
3. *Homogeneous polymer blend*: It is a miscible polymer blend having a single-phase structure. This can be presented as $\Delta G_m \approx \Delta H_m \leq 0$. Polystyrene (PS) and polyphenylene oxide (PPO) are the examples of homogenous polymer

blend in which both polymers are held together by repeating units of aromatic rings. Poly(methyl methacrylate) (PMMA) and polyethylene terephthalate (PET)–polybutylene terephthalate (PBT) is another example of homo, homopolymer whereas acrylonitrile butadiene styrene (ABS)–polycarbonate (PC) is an example of homo, copolymer complex. The identity of blend components is lost in homogenous blends and they usually follow the mixing rule [4,8]. Endless efforts were started to develop new monomers for the production of novel polymers since 1940s. It was observed that improvements in the modification techniques of existing polymers might be economically feasible [9]. Being geographically wide spread, biopolymers are very cheap however, they individually exhibit insufficient mechanical properties (e.g., tensile properties), and erraticism at molecular level (e.g., compositions) remains a solemn concern [10]. Therefore blending of biopolymers provides entities of aggrandized beneficial characteristics beyond the array that can be achieved from only biopolymer counterparts [11].

10.2 METHODS OF BLENDING

Most of the blends are not formed spontaneously due to immiscible nature of polymer pairs. Furthermore, blending of polymers depends upon the preparation process as their phase structure is not in equilibrium [12]. Generally, five different methods (graft copolymerization, melt mixing, latex mixing, solution blending, and preparation of interpenetrating polymer networks (IPN)) are commonly used for blend preparation [13,14].

10.2.1 INTERPENETRATING POLYMER NETWORK

When two or more polymers are present in the form of a network in which one member is cross-linked or being synthesized in the presence of other member then resulting blend will be known as IPN [15]. It is different from grafts, blocks, and polymer blends as it swells without dissolving in the solvents [16]. It was discovered in 1914 for the first time and is most abundantly used in dentistry [16].

10.2.2 MELT PROCESSED BLENDS

Most commonly used and wide-spread method for the preparation of blend is "Melt mixing" which includes the mixing of blend components in a molten state [17]. Arvanitoyannis et al. [18] reported the preparation of biodegradable blends of gutta-percha (1,4-transpolyisoprene) and gelatinized starch for biomedical and food packaging applications. Carvalho et al. [19] heated corn starch and natural latex at 150°C to prepare the thermoplastic blend of natural rubber and starch.

10.2.3 AQUEOUS BLENDS

Most of the biopolymers degrade below 20°C; therefore melting is not appropriate for their processing. Due to this reason, aqueous blending is considered as the best technology for the modification of natural polymers [20]. Pereira et al. [21] polymerized acrylic acid with methyl methacrylate monomers by free-radical polymerization method to produce hydrogel of cornstarch/cellulose acetate blends for bone cements. Arvanitoyannis and Biliaderis [22] reported on aqueous blends of plasticized glycerol with sugars to produce soluble starch and methyl cellulose blends by hot pressing, extrusion, or casting. Ikada et al. [23] reported aqueous blends of poly-(D-lactide) enantiomeric PLAs [i.e., poly-(D-lactic acid) (PDLA) and poly-(L-lactic acid) (PLLA)] by casting. Suyatma et al. [24] prepared biodegradable blends of PLA with chitosan by film casting and solution mixing.

10.2.4 SOLUTION BLENDING

Solution blending is the most appropriate method for laboratory-scale preparation of polymer blends which involves intensive stirring after dissolution of blend components in solvent [24]. Precipitation and evaporation are required for separation of blend from solvent. It is a preferred method because of no energy expenditure for rapid mixing, and unfavorable chemical reactions can be avoided. During last few years, novel methods, such as cryogenic mechanical alloying and shear pulverization, have been established for effectual mixing of polymer blends. Nanoscale blend morphologies are achieved when polymers are disintegrated at cryogenic temperatures in pulverizers. Generally, blending methods include cheapest mechanical mixing, film casting by spray drying or freezing after dissolving in solvent, fine powder mixing, latex blending, and interpenetrating network [25].

10.3 ADVANTAGES OF BLENDING

The blending of natural polymers offers advantages in two ways, i.e., (1) improvement in product performance and (2) processability [26].

1. Blending has the ability to improve product performance of resin by fabricating materials having desired characteristics at nethermost rate. It incorporates cheaper polymers by extending engineered resin's performance. It improves brittleness without using low molecular weight additives (e.g., plasticizer) and allows the fabrication of dimension-stable modulus, and improvement in solvent, chemical, and flame resistance (e.g., blends of carrageenan with chitosan, styrene's or acrylics with PVC). Blending facilitates the production of enduringly antistatic, biodegradable, multilayered blends (e.g., carrageenan and Konjac glucom blend) and provides the means for recycling (rebuild high molecular weight, partially degraded polymers).
2. Blending is a way to improve the processability by mixing resins having low T_g (glass transition temperature) with resins having high T_g (e.g., alginate/

polyethylene blends). Blending increases the productivity of resins by reducing the drop across runners or dies (e.g., agar/starch blends). Their strain hardening (SH) becomes controllable (e.g., calcium alginate/starch) and they also gain the ability to control the foaming process by reducing foam density and bubble size. Blending technology produces advanced esthetic-value materials, ascertains rapid designs, mends product uniformity, and provides better mechanical performance.

10.4 ALGAL POLYMERS AND THEIR GEOGRAPHICAL DISTRIBUTION

Production of bioplastics is the most common route for the commercial manufacturing of biopolymer blend. These blends overcome all disadvantages by exploiting extrusion technology [27]. In 1998 European commission reported that possible markets for bioplastics include mainly packaging applications and their use as plastic bags [28]. Bioplastic industry is considered as one of the largest industry with 1,145,000 ton annual production in the first decade of the 20th century, yet it is in the state of infancy with foreseen application areas [29]. Literature reveals that several scientific/technical problems regarding the use of natural polymers in bioplastics have been only partially addressed and solved. Among these, the selection of a given natural polymer for a specific project is still a chief issue in the scheme of new chemical merchandises. Extensive studies of commercially important polymer blends with alginate [30], gelatin [31], starch [32], soy protein [33], natural fibers [34], and wood flour [35] are reported. Plasticizers such as extruders must be supplemented to biopolymer for processing in conventional equipment [36,37]. The situation of biopolymers (within every class) becomes further intricate due to molecular diversity as a result of geographical origin.

In this chapter all possible blends of algal polymers (alginate, carrageenan, and agar) with other natural polymers such as starch (St), chitosan (Cht), collagen, lignin, and gelatin are discussed, because blending of a natural polymer with a synthetic polymer is a chief detriment to compatibility [38]. Generally, mixing of natural polymer with synthetic ones results in immiscible blends because natural polymers are hydrophilic whereas synthetic polymers are hydrophobic in nature [39]. Moreover, stabilization of melt blends by modification of interfacial properties, additives, or compatibilizers is obligatory [40]. Consequently, blending of algal polymers with natural polymers is chosen carefully as they have wide-ranging applications reflected as excellent illustrations of the benefits and drawbacks related to their use in environmental, biomedical, agriculture materials, pharmaceutical, and bioplastics, and can be extracted or manufactured in huge volumes [41].

Biopolymers of algal origin are omnipresent in surface waters and have striking potential (Table 10.1 of applications of polymers). The seaweed extractives of commercial importance of red and brown algae fall into three main groups, which include agar and carrageenans are from red algae and alginates from brown algae.

Table 10.1 Applications of Algal Polymers and Their Functions

Application	Function	Type of Algal Polymer	References
Drugs	Emulsions, laxatives	Agar	[42]
Plant propagation	Fertilizer, land conditioner	Agar	[43]
Beverages	Clarification and refining of juices	Agar	[44]
Medical application	Anti-herpective	Alginate	[45]
Water-based paints	Suspension, flow control, emulsion stabilization	Kappa + galactomannans, iota	[44]
Lotions	Bodying, emollient	Sodium kappa, lambda, iota	[44]
Toothpastes	Binder	Sodium kappa, lambda, iota	[44]
Syrups	Suspension, bodying	Kappa + lambda	[44,46]
Low-calorie jellies	Gelation	Kappa + iota	[46]
Dessert gels	Gelation	Kappa + iota + locust bean gum	[47]
Acidified milk	Bodying, fruit suspension	Kappa + locust bean gum	[46,48]
Cold prepared milk	Suspension, bodying, stabilize overrun	Lambda	[46]
Whipped products	Stabilize overrun, suspension, bodying	Lambda	[46,49]
Milk gels	Gelation, thickening, syneresis control, level-search gelatinization	Kappa, lambda + iota	[50]
Sterilized milk products	Suspension, bodying, emulsion, fat and protein stabilization	Kappa	
Pasteurized milk products	Suspension, emulsion stabilization, bodying, cling	Kappa iota	[46]
Frozen desserts	Whey prevention	Kappa	[44]

All these three types of extractives are related to the algal cell walls and take after cellulose in elementary molecular organization. Red algae (Table 10.2 cultivation of algae) are considered as the most important source of many biologically active metabolites compared to other algal classes [53,54].

Table 10.2 Sea Weed and Other Worldwide Aquatic Plants Cultivation

Sr. No.	Country	Algal Source	Total Harvested (Weight per Year)
1	North America	*Macrocystis*	150,000 t
2	China	*Laminaria japonica*	1,000,000 t
3	Malaysia	*Laminaria digitata*	207,892 F[a]
4	Japan	*Porphyra*	Not known
5	Republic of Korea	*Undaria pinnatifida*	80,000 t
6	Vietnam	Sea weeds	35,000 F[a]
7	Zanzibar	Sea weeds	1,257 mt
8	Australia	Sea weeds	1,923 F[a]
9	Brazil	Sea weeds and aquatic plants	730 t
11	Canada	Sea weeds	37,632 t
12	Chile	Sea weeds	380,759 t
13	World[b]	Sea weeds and aquatic plants	19,892,703 t

[a] *Estimation by FAO or calculation based on specific assumptions.*
[b] *World data comprise on countries not mentioned in table.*
Data shown in metric ton from FAO, 2014, Brendal D, Lisa K, Christine A, Ward P, Demian C, Michael RH, Steven TK, Samuel HG. Global catches, exploitation rates, and rebuilding options for sharks. Mar Policy 2013;40:194–204.

10.4.1 ALGINATE

Brown seaweeds are considered as a potential source of alginate, and therefore their cultivation depends on the availability and properties of alginate, because these properties vary from species to species. Generally, *Turbinaria*, *Ascophyllum*, *Lessonia*, *Sargassum*, *Durvillaea*, *Macrocystis*, *Laminaria*, and *Ecklonia* are the sources of alginates, among which *Ascophyllum*, *Macrocystis*, and *Laminaria* are the most important species. Most of the alginate fabricators harvest the entire brown seaweed; therefore there its trading is restricted [46].

10.4.1.1 Structure of Alginic Acid

Alginate is the generic term used for alginic acids and their salts. Alginic acid is a rectilinear polymer composed of β-D-mannuronic acid and α-L-guluronic acid (Fig. 10.3), and is therefore considered as a polyuronide having tetrahydropyran ring with two possible chair forms [55].

Alginate exists in the form of M-blocks (sequences of poly mannuronic acids), G-blocks (sequences of poly guluronic acid), and random blocks of MG (Fig. 10.4). The number of blocks may vary from species to species of seaweeds [56]. The segments of M-blocks are flexible and linear having β (1−4) bonds whereas the segments of G-blocks have stiff, rigid, and folded conformation due of α (1−4) bonds, and as a result carboxylate groups are sterically hindered [57,58]. These G-blocks sequester the divalent metal ions for providing the aqueous environment for the coordination with other molecules to form ionotropic gels.

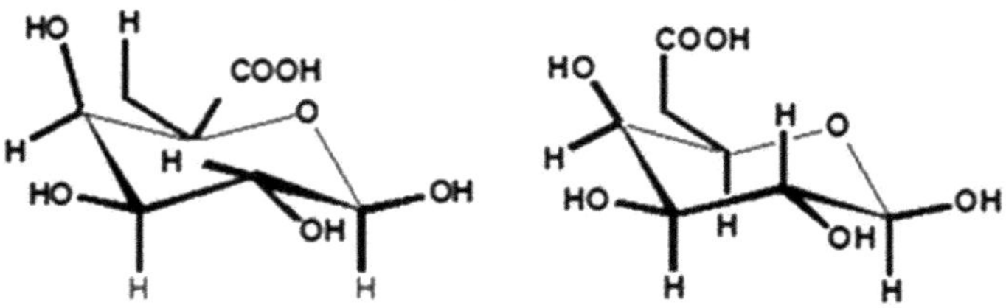

FIGURE 10.3

α-L-guluronic acid and β–D-mannuronic acid.

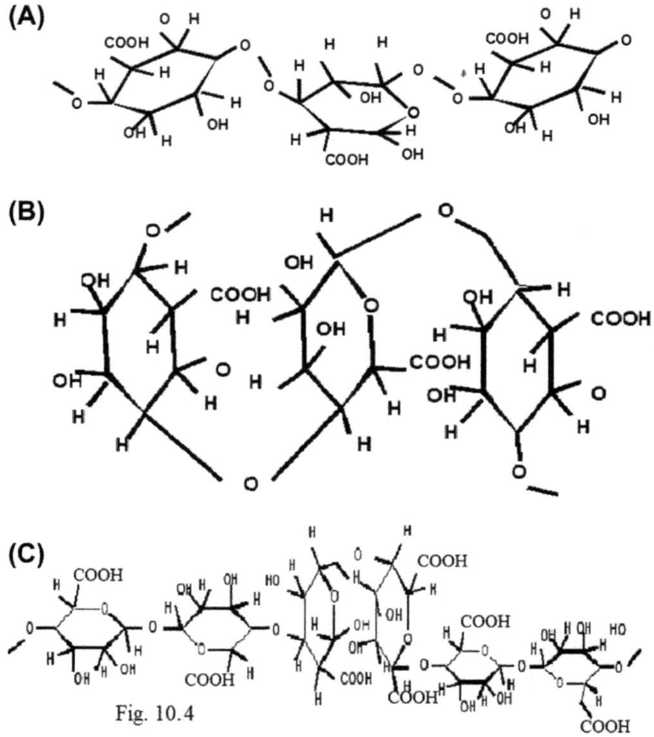

FIGURE 10.4

Structure of M-block (A), G-block (B), and MG-block (C).

That is why the proportional length of G-blocks and contents of M-blocks are the most valuable and most important gauges for gel formation and other physical properties of alginates. Many other properties, e.g., biological activities of alginate, are the function of these blocks [59]. The pH of medium is an important factor in the solubility of alginate because they are water soluble due to binding with monovalent

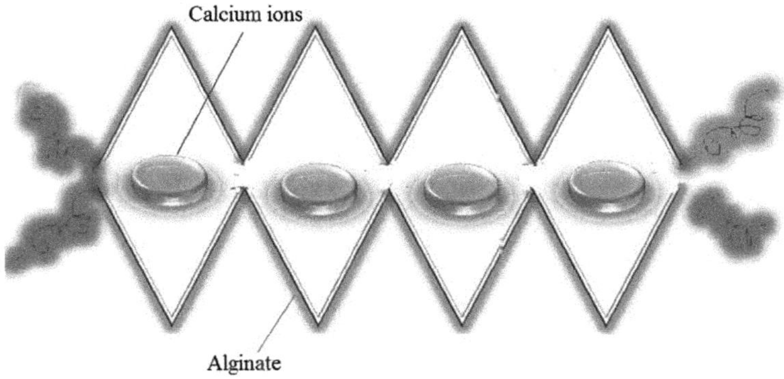

FIGURE 10.5

Unique egg-box structure of calcium and alginate.

ions and salted out after binding with divalents such as Ba^{2+}, Ca^{2+}, and Sr^{2+}, etc. Binding of G-blocks (Fig. 10.5) with Ca^{2+} is of particular importance as it gives unique egg-box structure [60]. Alginates have the ability to salt out in acidic medium due to transformation into alginic acid which is insoluble in water.

Alginate structure plays an important role in fabricating smart materials for superior utilization. It is a supreme material for chemical modification due to presence of large number of free carboxyl and hydroxyl groups. Biological and physicochemical properties can be exploited by chemical modifications such as oxidation, reductive amination, sulfation, grafting, esterification, and Ugi reaction [61,62]. Unique characteristics such as nontoxicity, immunogenicity (cytokine production stimulation), biodegradability, and biocompatibility are due to its hydrocolloid property which make alginate a supreme candidate for pharmaceutical and food industries [63,64].

10.4.1.2 Extraction of Alginate

Stanford extracted alginate for the first time, more than 100 years ago, from brown seaweed. It was later reported that alginate is present in all types of brown algae, with no exception, constituting 40% of their dry weight depending on growth conditions [59]. Numerous scientists reported that alginic acid could be more readily extracted with alkali treatment than Ca-alginate [65]. Fig. 10.6 presents the commercial extraction process of alginate from seaweeds.

First step of the extraction of alginate is to convert the salts of calcium or magnesium into sodium as it is soluble in water. For this purpose ion-exchange method is used [66]. Calcium ions influence the viscosity of the alginate depending on the composition of M- and G-blocks and must be present in residual amounts, e.g., only 1.2% calcium is present in sodium alginate extracted from Kelco [67]. The prolong storage of highly viscous alginate, for 6–12 months, alters its physical properties as compared to medium viscous alginate; therefore most of the manufacturers fabricate medium- and low-viscosity alginate for commercial applications [68].

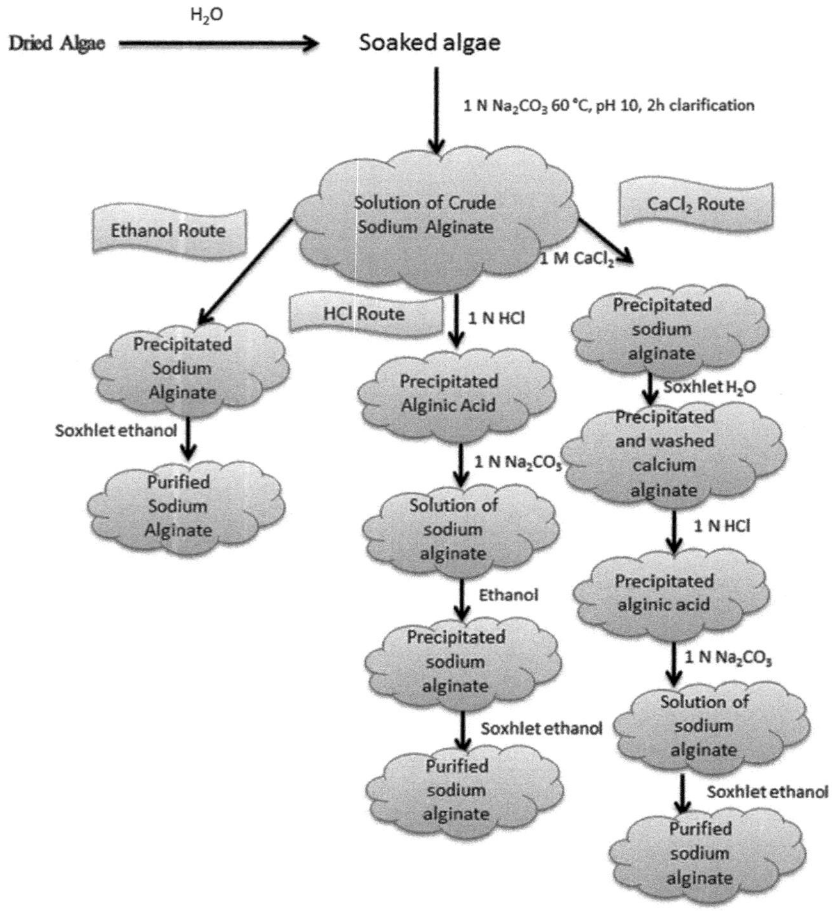

FIGURE 10.6

Schematic presentation of extraction of alginate from seaweeds.

10.4.1.3 Blends of Alginate

Blends of alginate exhibits advantage over other synthetic polymer blends in two ways (1) they are highly compatible and (2) homogenous [69]. Water-soluble blends of alginate/chitosan were prepared after the succinic acid-induced amide modification of chitosan [70,71]. Chitosan is a copolymer formed by deacetylation of naturally occurring chitin. Chitosan in alginate/chitosan blends has synergistic effect which results in hydrogels with improved mechanical properties. These blends with additional $CaSO_4$ as osteoinductive material are ideal candidates for cartilage engineering in medical applications [72]. Agar, collagen, and alginate ternary blend was reported by Wang and Whim [73]. This ternary blend with incorporation of silver nanoparticles possessed high water retention and antibacterial activity against food-borne gram-positive and gram-negative bacteria. These ternary blends present highly antifogging films for preservation of agriculture products. Blending of alginate with synthetic polymer polyvinyl alcohol in the presence of glycerol as natural plasticizer can improve the mechanical properties and compatibility [74]. These blends gained a lot of attentions for biofilms formation [75]. Generally, dehydrated alginate with more than 25% solids are suitable for blending with pectin in heavy pastes with solid alkali (sodium carbonate). The presence of small amounts of water in the solution does not affect the heavy paste formation, and sugar moieties do not exhibit significant affect on the gel formation. However less than 25% solid material result in a liquid paste with low viscosity [76]. Morris and Chilver [77] reported that blending of pectin with alginate involves the binding of methyl groups of pectin with G-blocks of alginate without assessing the influence of Ca^{2+}. Contrary to this, no effort for alginate/CMC blends is reported in the literature till date.

Starch is one of the low cost, abundantly present biopolymer having amylose (linear) and amylopectin (branched) structures [78]. The main objective of starch blending with algal polymer is to reduce the cost of blends and conferring/conserving biodegradable character [79]. It can also act as a filler. A ternary blend of alginate with starch and chitosan was prepared by adding stigmasterol as stabilizer by utilizing external ionic gelation method. Chitosan and starch acted as filler materials to provide strength and to lessen the permeability, whereas alginate formed the porous gels. Very easy method with respect to performance, external ionic gel method results in uniform, rounded blends having 1.4 mm diameter and 94.8% yield with wide-spread application [80].

Protein is considered as one of the most important macromolecule of the universe and natural resources (both animals and plants) are the main reservoirs of proteins [81]. Although, proteins are suitable due to their high molecular weights, in the range of 1.661019 g/mol, their lower decomposition temperature than other natural polymer limits their usage in blends. Soybean (soy protein) is a rich and cost-effective source of saccharides (15% starch and soluble saccharides), oil (18%), protein (38%), and moisture (4%) used in most of the countries, such as China and the United States, in the form of soy protein isolate (SPI) with P90% protein (13.7% N), defatted soy flour (DSF) with 40%–50% protein (including 8.5% N), and soy protein concentrate (SPC) with P70% protein (11.5% N) [82]. Jane and Wang [83] reported injection molding and extrusion for the preparation of soy protein thermoplastic. High-moisture absorption and low strength delimited the use of soy protein plastics [84]. However, it can be blended with other algal polysaccharides through film-forming dispersions using enzymatic, physical, or chemical methods [85]. Soybean-based materials were used for making car bodies in 1940 [86]. Liang et al. [87] observed that both SPI and SPC resins have characteristic rigid and brittle polymer properties suitable for molded specimen manufacturing. Blends of sodium alginate/soy protein were prepared using different concentrations ranging from 10% to 50% by weight. Naturally available SA was used to prepare blend membranes by adding different amounts of soy protein (10, 30, 40, and 50 wt%). Membranes were cross-linked with glutaraldehyde/hydrochloric acid, which not only acted as cross-linking agents but also provided the flux and strength to the membranes. FTIR and DSC techniques were used to measure compatibility and cross-linking [88].

Alginate consumers fall into two categories. First group of consumers include all those who exactly knows their needs but they do not have their own resources while second group includes those who have their own resources that is why do not require overhauling. This class includes small-scale gum companies having ability to manufacture their own blends from seaweeds and other colloids according to demand of customers.

10.4.1.4 Carrageenan

Commercially important water-soluble gum of Rhodophyta is called "carrageenan" which is similar in function to the cellulose of terrestrial plants [89]. Chemically, they are different from agars and alginates as they are sturdy, anionic polymers due to the presence of half-ester sulfate moieties of sulfated galactans [90]. Stable harvests of carrageenan in any climate depend on the consistent and adequate light and temperature. Unlimited cluster of *Chondrus crispus*, rich source of carrageenan, can be cultivated across maritime nations. Carrageenans with extremely flexible, large helical molecule are found to prevent global warming. Carrageenan being a copolymer of D- and L-galactose is widely used as a thickening and stabilizing agent in food and other industries [91]. Numerous isomers of carrageenan, depending on the number and position of sulfate groups on galactose moieties, are reported among which three, i.e., lambda (λ or LC), iota (ι or IC), and kappa (κ, or KC) are commercially important [92]. KC possesses a characteristic film-forming property and forms firm gels with K^+ ions. Different ranges of gel textures were achieved when hot KC solution was cooled between 30 and 70°C below gel point [93]. Fig. 10.7 depicts the two-step gel-formation mechanism, where stage B is elastic (iota) and stage C is brittle (kappa).

Ionotropic hydrogel is formed when two oppositely charged polyelectrolytes are mixed together [3]. The reversible ionotropic hydrogels are stabilized by

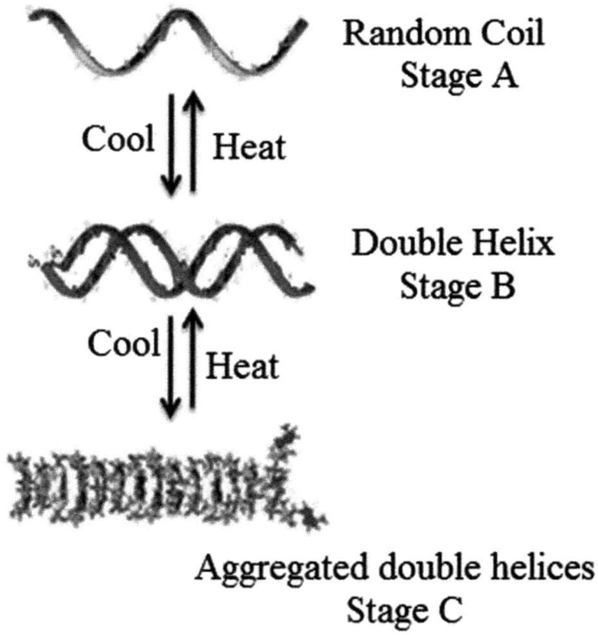

FIGURE 10.7

Gel-formation mechanism of carrageenan.

hydrophobic forces, ionic forces, or H-bonding, and these interactions can be easily interrupted by temperature, pH, and ionic strength [94]. Different structures of hydrogels such as nanoparticles, beads, and microparticles can be prepared by varying the concentration of polyelectrolyte [95]. Piyakulawat et al. [96] reported bead and tablet forms of chitosan and carrageenan hydrogels for efficient delivery of sodium diclofenac and diltiazem hydrochloride. These formulations, due to small size, surface functionality, and volume to surface ratio etc. would be very promising vehicles for drug delivery.

10.4.1.5 Sources and Chemical Properties

Many algal species, for example, *C. crispus* and *Gigartina stellate*, are considered as important sources of carrageenan [42]. The word carrageenan was coined in 1862 when Stanford extracted gelatinous material from *C. crispus* [97]. Commercially, Rhabdoniaceae, Solieriaceae, Hypnaceae, Phyllophoraceae, Gigartinaceae, Furcellariaceae, and most recently, Rhodophyllidaceae are exploited in Massachusetts, Philippines, France, Spain, Morocco, and Canada [98]. It is a well-known fact that both lambda and kappa do not occur together in the same plant, but they can be extracted from different stages of the reproductive cycle (Fig. 10.8), such as diploid tetrasporophytes possess lambda whereas haploid gametophytes have kappa carrageenan (KC) [99]. Lambda carrageenan (LC) is predominantly present in *Gigartina acicularis* and *Gigartina pistillata*, whereas *Eucheuma spinosum* provides nearly ideal iota carrageenan (IC) [92].

It is reported that above gel-melting temperature all forms of carrageenan were soluble in water between 40 and 70°C temperature depending on the concentration of solution and cations [100]. LC is soluble in cold water whereas other two types IC and KC are only soluble in cold water in the presence of sodium [99]. Presence of

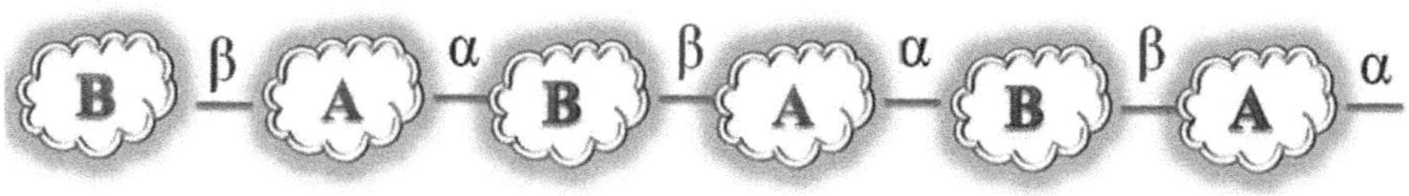

B-Units	Types of carrageenan
D-Galactose	λ, θ
D-Galactose 2- sulfate	λ, ζ, θ
D-Galactose 4- sulfate	ν, μ, κ, ι
A-Units	
D-Galactose 2- sulfate	ζ
D-Galactose 6- sulfate	μ
D-Galactose 2, 6- disulfate	λ, ν
3, 6- anhydro-D-Galactose	κ
3, 6- anhydro-D-Galactose 2- sulfate	ι, θ

FIGURE 10.8

Graph showing proportion of different types of carrageenan.

3,6-anhydro bridges in KC and IC results in different rheological behaviors in water [42]. Absence of 3,6-anhydro bridge in LC prevents from helical structure formation [100]. IC and KC remain insoluble while LC is soluble in cold milk; whereas all types remain soluble in hot milk (245). Upon cooling they form gels whose consistencies depend on the concentration of carrageenan and its sensitivity to calcium ions [99]. It is observed that different factors such as pH, gel strength, concentration, and droplet size influence the carrageenan and milk protein interaction [100]. Stronger the bond between milk protein and carrageenan, less hydrocolloidal the nature will be. Carrageenan forms highly viscous solutions due to linear nature. Viscosity of carrageenan depends on its type, concentration, temperature, and molecular weight [101].

10.4.1.6 Extraction of Carrageenan

Annual production of carrageenan has increased rapidly and exceeded that of agar owing to its superior characteristics in dairy products, food industry, and biotechnology [101]. The current annual production of carrageenan amounts to 15,000 tons worldwide, to which China alone contributes 600 tons [42]. Thirteen types of carrageenan are reported depending on the presence of sulfate groups on D-galactose residues with $(1 \rightarrow 3)$ and $(1 \rightarrow 4)$ bonding, out of which only three (kappa, iota, and lambda) are commercially significant [92]. These three carrageenans differ from each other as KC is a well-known gelling agent; LC is an emulsifier and stabilizer; whereas IL has peculiar properties applicable in food industries [42]. With difference in properties, these carrageenans have unique extraction processes which differ from each other.

10.4.1.7 Kappa, Lambda, and Iota Carrageenans

This family is more hydrophobic due to presence of 25% sulfate and 35% 3,6-AG residues. Gel-press and freeze–thaw are the most success techniques for the extraction of KC as expressed in Figs. 10.9 and 10.10. Some algal species such as *Hypnea* sp., *Eucheuma* sp., and *Furcellaria* sp. require the treatment of 5%–10% NaOH solution at 80–90°C and the resultant gel is washed with KCl solution [102].

10.4.1.8 Blends of Carrageenan

Blends of pure carrageenan with agar (natural polymer) are primarily considered as food additives, but blends of carrageenan with synthetic polymers are the possible candidates for some biomedical applications. For example, the synthesis of KC-graft-PVP blends and agar-graft-polyvinyl pyrrolidone (PVP) by microwave irradiation technique is testified [3] and the new blend hydrogels were found to be not as robust, but presented improved water-holding ability and spreadability with a potential to be used as active drug carriers and moisturizer formulations. Hydrogel dressing with the use of agar and PVP blends was also reported.

Ng and Camacho [103] prepared a blend of KC polymer with electrolytes such as poly(3,4-ethylene dioxythiophene) dye sensitized solar cells (DSSC). Blends of polymer with different KC concentrations and molecular weights were also

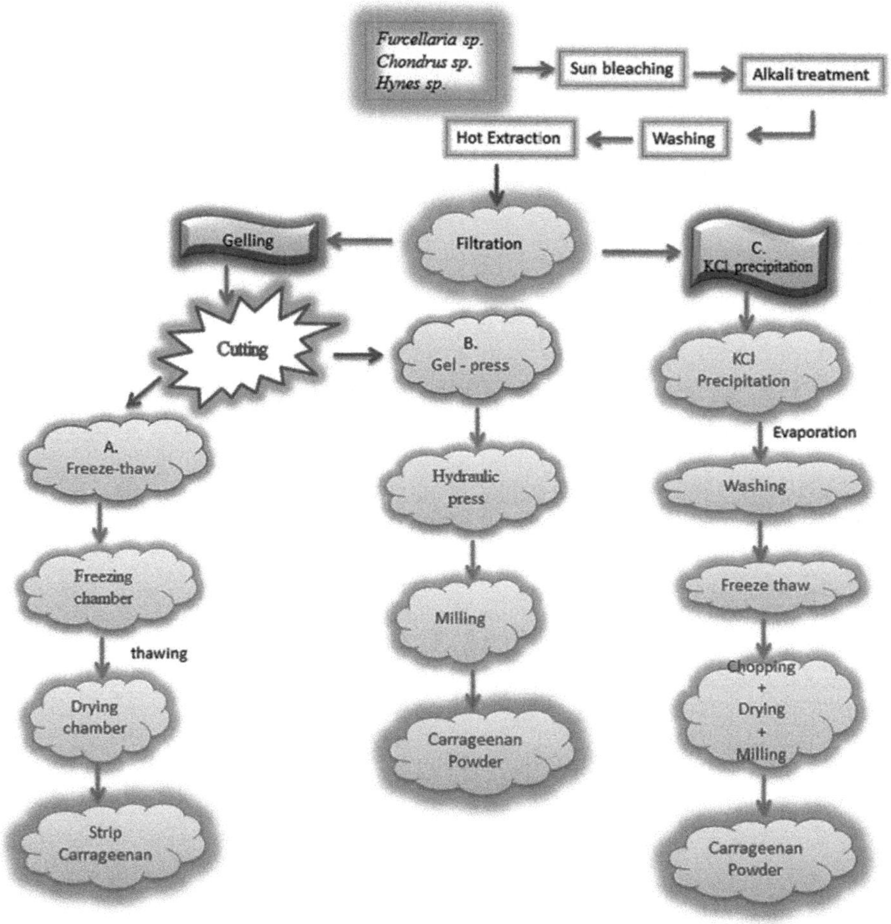

FIGURE 10.9

Schematic presentation of KC extraction from seaweed.

explored. It was establish that the conductivity of the polymer blends is enhanced with higher KC concentrations, and lowered with degraded KC. These blends when used in DSSC provide a solar cell with efficiency of 0.421%.

To improve the functional stability of agar/kappa-carrageenan (food hydrocolloid blend), it was treated with genipin; a natural crosslinker, in an aqueous medium.

In case of solvent-treated blends, total N contents lower significantly to 0.028% from 0.29% in the non modified blend after solvent treatment, only a slight blue color arises after 20 h during the cross-linking reaction, while the parent blend became dark blue after 30 min of reaction time. It was observed that genipin has an ability to react with minute amount of blood proteinaceous matter [104].

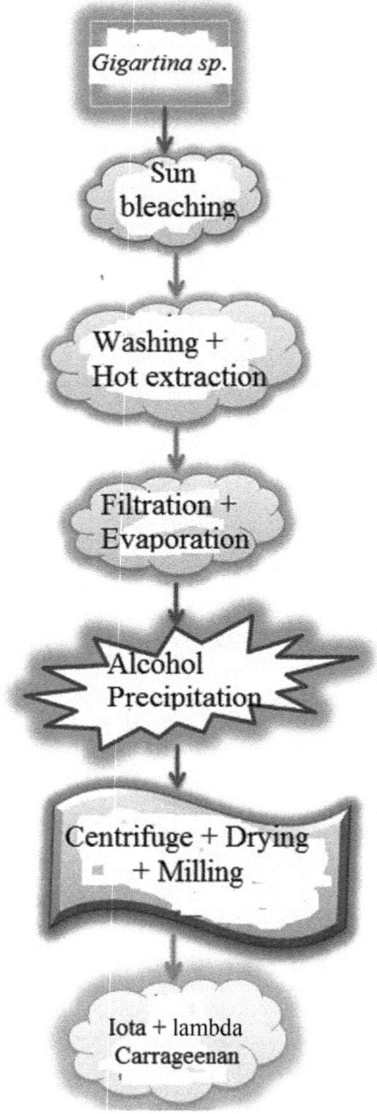

FIGURE 10.10

Schematic presentation of iota or lambda carrageenan extraction from seaweeds.

Arof et al. [105] reported the blend of chitosan/iota IC using H_3PO_4 as ionic dopant and PEG (polyethylene glycol) as plasticizer for EDLC (electrical double layer capacitor). They suggested that the blends followed the Arrhenius behavior which might increase with temperature, having the highest conductivity (6.29×10^{-4} S/cm) with stable capacitance (35 F/g for 30 cycles). Blends of biopolymer with KC and cellulose derivatives were produced using solution casting method. A derivative of cellulose, carboxymethyl cellulose, was prepared from cellulose taken out from Kenaf fibers. It was blended with different weight percentage of KC derivative to achieve free-standing films. The produced blend films were subjected to Fourier transform infrared characterization, tensile test, scanning electron microscopy, and linear sweep voltammetry to investigate their mechanical, structural, chemical, and physical properties. The FTIR outcome explained that both polymers were consistent with each other. The conductivity of the films was found to be influenced by the charge-carrier concentration, polymer-segmental motion, and polymer–salt interaction. The blending of carboxymethyl cellulose into carboxymethyl KC was found to be a promising strategy to improve the material properties such as conductivity [106].

Hydrocolloid blend foods such as agar and KC were blended with the natural cross-linker genipin in aqueous solution to relay functional stability. The fixed blend of genipin showed remarkable stability over a large range of pH 1–12 in Ringer's solution. Genipin and hydrocolloids ratios in the blend as well as cross-linked product were measured on the basis of swelling properties [104]. Soy protein and KC (GEL-PRO, 3:1 ratio) blends were prepared to improve the quality of scalded sausage [107].

A large number of blends, such as natural fiber–reinforced, are generally used as low-cost materials having chemical and structural properties. In fact, the petrochemical-based materials are diminishing day by day, and renewable sources are the best way to conserve petrochemical-based material.

10.4.2 AGAR

Intracellular matrix of red marine algae "Rhodophyta" contains a characteristic class of polysaccharides termed as agar. These galactan polysaccharides play a structural role similar to that of cellulose in terrestrial plants. Due to presence of cellulose, terrestrial plants have a rigid structure to face the constant gravity pull, whereas presence of agar gives flexibility to aquatic plants to accommodate wave motion and fluctuating stresses of currents [108]. Consequently, marine plants adapt the necessary pliability by evolving hydrophilic, gelatinous structural constituents [109]. The use of agar in food and other commercial application for gel preparation is very old, goes back to approximately 350 years ago and endures to date [110]. Families of *Gelidiaceae* and *Gracilariaceae* (agarophytes) are found to be rich sources of agar. Well-known species of algae producing agar are discussed in Table 10.3. Although mariculture of carrageenophytes achieved considerable commercial importance, unfortunately it is not the case for agarophytes. Commercial cultivation of agarophytes was approximately 15,700 dry tons as reported in Chile for the first time. Water, temperature, and other environmental factors critically influence the cultivation and transplantation [127].

10.4.2.1 Structure of Agar

Agar was first discovered in 1960 by a Japanese Mino Tarozaemon and it was called as Kanten. Derivatives of agarose play a structural role in certain algae [110]. Being a

Table 10.3 Annual Production of Agar From Different Seaweeds

Algal Species	Country	Production (MT)	References
Acanthopeltis japonica; *Gelidium amansii*; *Gelidium crinale*; *Gelidium divaricatum*; *Gelidium japonicum*; *Gelidium liatulum*	Japan	2440	[111,112]
Chara flaccida; *Ganoderma capense*; *Gelidium cartilagineum*; *Gelidium corneum*	South Africa	100	[113,114]
Gelidiella acerosa	Japan, India	Very limited	[115]
Gelidium arborescens	Southern California	100	[116]
G. cartilagineum	The United States, Mexico	80	[117]
Gelidium caulacanthum; *Pterocladia lucida*; *Pterocladia capillacea*	New Zealand	26	[110,118,119]
G. corneum; *Gelidium sesquipedale*	Portugal, Spain	320, 890	[120,121]
G. corneum; *Gelidium spinulosum*	Morocco	550	[122,123]
Gelidium coulteri	Mexico	52	[124]
Gelidium lingulatum	Chile	820	[110,125]
Gracilaria compressa	Canada	1	[126]

polysaccharide, agarose (the gelling fraction) has a neutral linear structure composed of 3,6-anhydro-L-galactopyranose and D-galactose units, free from sulfates. Agarose can be purified from agar by confiscating agaropectin [128]. The non-gelling portion or agaropectin contains different concentrations of D-glucuronic acid, esters of sulfate, and pyruvic acid in very small amounts along with agarose. The percentage of the two portions varies from species to species of seaweeds, while two-third portions of most of the seaweeds are agarose [129]. Fractionation, enzymatic, chemical hydrolysis, and different spectrophotometric methods are used to study the complex structure of agar and for separation of agarose from agaropectin [130].

Knutsen et al. [131] proposed a modified nomenclature and reported agaran and agarose, the two main polysaccharide of agar, depending on the presence of "anhydride bridge." If the bridge is absent then it will contain β-D- and α-L-galactose residues termed as agaran. Irrespective the selection of nomenclature, both of these contents with diverse arrangements should give agars with distinct properties. Formation of thermo-reversible hydrogels by huge thermal hysteresis, the most important property of agar, makes them a potential candidate for different industrial applications [132]. Agar quality is measured from gel strength (GS). The term GS was first time used in 1957 and has now become a standard test to measure the agar quality in industry [133].

10.4.2.2 Agar Gelation

The complexity of agar hydrogels remained in the trench for many past years [134]. It is believed that below the gelation point, agar shifted from random coil state to ordered structure and then interhelical association developed to attain minimum free-energy in the form of double helix (Fig. 10.11). This gelation route is strongly influenced by physicochemical properties, where the anhydride bridge plays a major role in the stability of helix. On the other hand, kinks of sulfate cause the softening of hydrogels [135]. Rodriguez et al. [136] reported that gelation process is also influenced by molecular masses of agars.

10.4.2.3 Extraction of Agars

Traditionally agar is extracted using ionizing liquid solutions (ILs) following the scheme presented in Fig. 10.12. The extraction process is started with dried *Gelidiaceae* which is a rich source of agar [137].

After the successful completion of agar extraction, IL was recovered from methanol filtrate and recycled for further extraction.

10.4.2.4 Blends of Agar

Blends of agar with other natural or synthetic polymers are very appealing and gaining interest due to their reversible gel-formation ability. Usually these blends improve water-holding capacity of other polymer components and the gelation property of agar because pure agar gels are brittle [132]. This characteristic makes it a potential candidate for biomedical industry. Agar and alginate blends for fresh fruits dehydration are superfluous and springy and can be easily managed allowing water permeability [138].

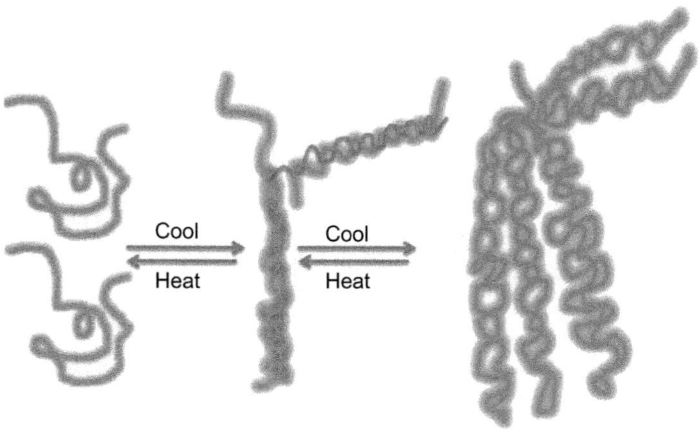

FIGURE 10.11

Gelling mechanism of agar.

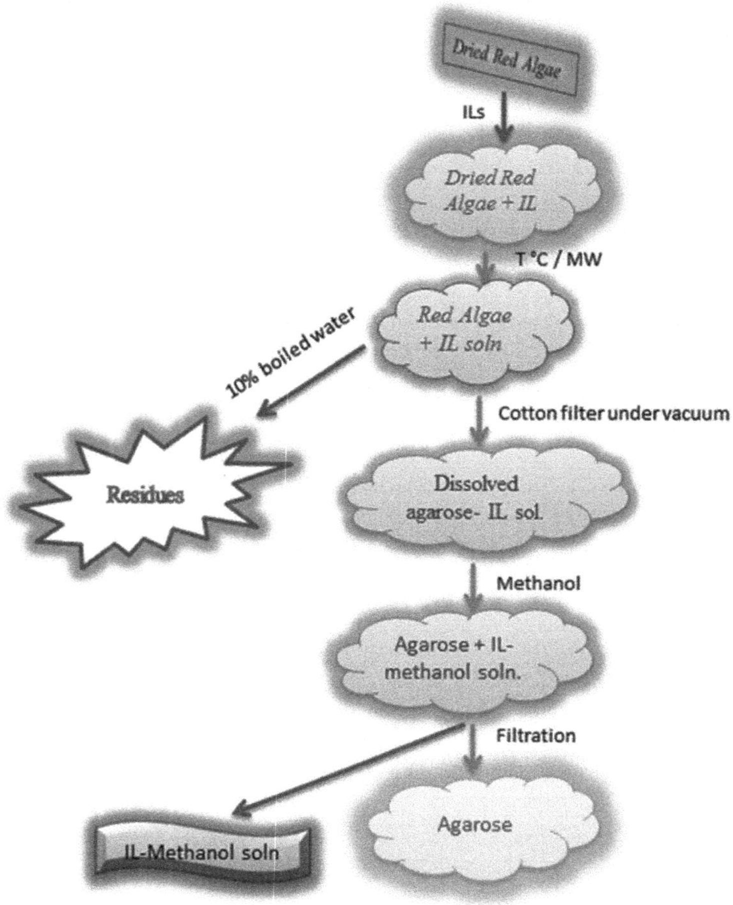

FIGURE 10.12

Extraction of agar from red algae (Gelidiaceae) using ionic liquids (ILs).

10.5 BIOMEDICAL AND PHARMACEUTICAL APPLICATIONS OF ALGAL BLENDS

Algal blends have been extensively used in human health system. These blends have been employed as drug-delivery recipients (DDS) in dentistry, wound dressing, and stomach reflux–preventing formulations. In stomach aid and oral medicines, alginates have been found superior to other formulations owing to their preservation of gel structure (alginic acid structure) in highly acidic conditions. This characteristic of alginate-based formulations prevents the denaturation of delicate compounds in acidic conditions of gastric juice. For the treatment of epidermal and dermal wounds, and diabetic foot, Ca-alginate dressings are found appropriate both in terms of healing as well as tolerance. Ca-alginate continuously releases Ca^{2+} that plays a significant role in healing, acting as a modulator in keratinocyte proliferation and differentiation [139]. Alginate anti-reflux (antacid) preparations are employed in the treatment of gastro-oesophageal decay disease due to formation of neutral floating gel or "raft" in the stomach. The formulations consist of high sodium alginate contents that undergo ionic gelation in stomach. Under acidic stomach conditions, soluble sodium alginate is converted into insoluble gel of alginic acid resulting in the formation of rafts. Ca^{2+} from the calcium carbonate in the formulation results in a better mechanical strength, while the bicarbonate release CO_2 under the action of acid, which gives aeration and elasticity to the alginic gel [140].

In vivo visualization of tagged cells is now possible by using MRI, giving new observations into the biodynamics of cell departure. The power of stem cell remedial treatment is also recorded, mostly by using cells tagged with superparamagnetic metal—contrast agents that need adaptable microscale capsules infixed with nano-sized elements capable of improving real-time visualization with MRI, and transport a powerful therapeutic to the diseases in animal models. Therapeutic zones of insulin for up to 4 weeks were produced in swine by injecting human insulin islets encaptured in magnetic microcapsules composed of alginate and ferric oxide [141]. Alginate-based microcapsules are not confined to diabetes but were also proved effective for the treatment of Alzheimer disease, Parkinson disease, and multiple sclerosis, providing a promising remedial treatment for the resistible damage of the central nervous system [142,143]. Administration of homologized mesenchymal stem cells is analytically practicable and quite safe procedure that provokes immediate immunomodulatory effects [144]. Alginate-poly(L-lysine)-alginate (APA) capsules are preferred for delivery of therapeutic medicines using perflurocarbon emulsion, barium sulfate, or bismuth. Alginate-based microcapsules have the potential application for other visualization techniques, including X-ray, computerized tomography, ultrasound, and fluorine imaging, etc. [145].

10.5.1 CONTROL OF TYPE-II DIABETES AND OBESITY

Alginic acid is a nonabsorbable, nutritional polysaccharide. Consumption of dietary alginic acid is associated with the improvement of the gastrointestinal function and a probable decline in the damage of the luminal contents by altering the colonic micro flora. Moreover, the intake of dietary alginic acid lessens nutrient absorption in intestines and enhances satiety, resulting in the control of obesity and Type-II diabetes [146].

Under the acidic conditions in stomach, alginates undergo ionic gelation and produce alginate gel that remains in stomach and supresses the feeling of emptiness. Moreover, by reducing the intestinal nutrient absorption and influencing the glycemic response, alginate gel significantly reduces the appetite and energy consumption in acute feeding and obesity models. By employing appropriate alginate, i.e., higher molar mass and higher G contents, ionic gelation can be obtained independent of gastric acid and pH. The alteration of pH enhances the solubility of $CaCO_3$ which regulates the gelling process [147]. The consumption of appropriate alginate-based formulation has been clinically effective in reducing the hunger, blood cholesterol, and glucose level hence helpful in the reduction of obesity and related issues [147].

10.5.2 DENTAL IMPRESSION

The reaction of soluble alginate with semi-soluble calcium salt gives an irreversible hydrocolloid used as dental impression. Such materials are used for the preparation of dental molds which give the patients oral information. Clinical characteristics of impressions, such as working and setting time, are dependent on the elasticity and viscosity. The clinicians can compensate the slight variations in working time and mixing temperatures by utilizing the chromogenic system of irreversible alginate hydrocolloids [148].

10.5.3 DURA MATER REPAIR

Minor dural defects can be treated by using adhesive materials as sealant or sutures whereas patches are required for large defects. Sutures can be quick but unable to seal the approximated tissue. Insertions of several neural devices result in small and large dural defects [149]. Thus there is a need for the development of biocompatible materials for efficient dural replacement patches. Hydrogel are applied to animal models for application as sealant as well as dural replacement patches. Owing to superior biocompatibility, quick gel formation, and mild reaction conditions, alginates have become promising candidates capable of direct cross-linking to neural tissues. Alginates undergo rapid cross-linking in the presence of polyvalent ions, enabling to seal any dural defect of any size, without any seizure or inflammation [150].

The success of alginate hydrogels as sealant and replaceable patch depends on the mechanism of gelation. Internal and diffusion gelation are two important methods for the formation of alginate hydrogels. The former utilizes $CaCO_3$ while the later involves $CaCl_2$ as a source of Ca^{2+} during the gelation process [151]. The rate of gel formation should be fast enough to seal the dural defects without any loss of neural fluid. After applying the sealant, it comes under cranial pressure and brain pulsation resulting in the compression of gel between skull and brain. Internal gelation process is slower as compared to diffusion gelation. However, it can be used by premixing the gel components and then applying the viscous material to the dural defect. Internal gelation process that occurs in the presence of $CaCO_3$ results in a more homogenous gel compared to diffusion gelation and hence is more suitable for dural patch [152]. For an ideal dural patch, the solid elastic characteristic of hydrogel is necessary for the retention of shape and orientation during dural movement. For the cross-linking of alginate used as dural patch $CaCO_3$ is more suitable due to long-term stability [152].

10.5.4 MUCOADHESION

Mucoadhesion is the property of a polymeric material to attach with the layer of mucus gel. It is an important characteristic for the designation of polymeric material for specific applications. Mucus gel consists of water (95%) and mucus glycoprotein known as mucin (5%) along with a number of secondary metabolites [153]. Initially, the mucoadhesive polymers were based on the interaction between polymer and mucin via physical or noncovalent chemical bonds. In later attempts, covalent bonding, such as disulfide, was considered to improve the polymer and mucin interaction. By using this approach thiolated polymers were developed. In these polymers, a thiol-containing molecule is attached [154]. However, alginate thiol mucoadhesives did not express good adhesive characteristic due to the development of intramolecular disulfide linkage in the hydrated environment [155].

Alginate is an anionic polymer, and it has a strong ability to adhere with the mucin glycoprotein through the formation of hydrogen bond between carboxyl and hydroxyl groups. Polyethylene glycol (PEG) is also known for its mucoadhesive ability. Based on its molecular weight, PEG has high water solubility and a quick in vivo clearance. The mucoadhesive property of PEG is due to formation of hydrogen bonding with glycoproteins [156].

G-residues of alginates can induce modification in the mucin network. Such mucin network modifications can be used for the treatment of respiratory disorders and mucosal surface. In cystic fibrosis, thick mucus along with hyperinflammation results in the failure of mucociliary clearance. For the treatment of this condition, oligo granulators can be utilized to maintain the viscosity of mucus by reducing the interaction between macromolecules and mucin. Such interactions can be diminished by the induction of competitive electrostatic inhibition achieved by small charged oligomers unable to establish cross-linking. For this purpose, nonimmunogenic alginate G-blocks, with degree of polymerization in the range of 10–20 G-blocks, are considered as suitable candidates [157].

Alginate-PEGAc, i.e., a mucoadhesive polymer, is synthesized by a novel two-step mechanism. In the first step alginate thiol is synthesized, and in the second step PEG-DA is incorporated in the alginate backbone [158]. Biomaterials with important biomedical and biotechnological applications can be synthesized by incorporating PEG into variety of biodegradable polymers. Tissue engineering through biosynthetic materials helps to support the biological and degradable parts, and allows the cell growth along with some synthetic materials for the provision of controlled properties and mechanical stability. While delivering ophthalmic medicines via conventional drug-delivery systems, a common and frequent problem of quick precorneal damage due to tear loss is encountered. Replacement of such delivery systems with mucoadhesives is an attractive step for the treatment of external ophthalmic diseases. Cornea and conjunctiva are anionic in nature, and therefore polymeric mucoadhesives that can attach with these eye structures for longer time are attractive candidates for efficient drug delivery. Owing to the superior bio affinity and biodegradability, chitosan, that is cationic in nature, and anionic alginates are preferred over other polymeric materials. Chitosan solution prolongs the retention time of antibiotics; hence chitosan-coated nanocapsules for the ocular drug delivery through diffusion are found to be superior. Due to hemocompatibility of alginates, they are easily degraded and excreted from the biological systems [159].

Colloidal particles used for the drug delivery were observed to be quickly picked up by mononuclear phagocytes (MPS) from the blood, especially from the liver cells. Such colloidal particles can be used for the targeted delivery of drugs to the specific organs. Alginate-loaded nanoparticles are found to be promising for treatment of liver metastasis in mice. This is a great approach for delivering drug to liver; however, disadvantageous when considered for targeted drug delivery to the other organs. To overcome this problem, scientists focused on developing hydrophilic colloidal matrix, such as sodium alginate, due to its gel characteristics. Moreover, the ability of alginate to form stable polyelectrolytes in polyamine environment is attractive for the aforesaid purpose. Chitosan alginate nanoparticle system is successfully developed for the delivery of broad spectrum antibiotics to cure the eye infections. This formulation is able to sustain the drug delivery for long time periods and hence is a better alternative to conventional eye drops in terms of easy administration and minimized dose [160].

10.5.5 WOUND DRESSING

Algal polymer blend–based wound dressings are a frequently used product, specifically for the treatment of bleeding wounds. The wound healing process is dynamic and requires altered dressing performance as healing progress. The ability of alginate to absorb the wound excretions results in a gel formation around the wound hence provides a better humid environment for the healing process. For the treatment of burn wounds, fluid balance plays a vital role, as heavy water loss during wound exudation and evaporation results in decreased water concentration and a fall in the body temperature. Alginates with high gel strength and flexibility can be employed for such wounds. Being transparent, moist healing, and able to absorb fluid, hydrogels are highly suitable for this purpose [161].

Some commercial dressings based on alginates include AlgiSite TM (nonwoven Ca-alginate fibers, from Smith & Nephew, Inc.), Algosteril (Ca-alginate, Johnson & Johnson), Kaltocarb (Ca-alginate fibers, ConvaTec), Kaltogel (Ca/Na-alginate gelling fiber, ConvaTec), Kaltostat (Ca-alginate fibers in nonwoven pads, ConvaTec), Melgisorb (Ca/Na-alginate gelling fibers, Molnlycke), Restore (Ca-alginate and carboxymethyl cellulose nonwoven pad, Hollister Wound Care LLC), SeasorbTM (Ca/Na-alginate gelling fibers, Coloplast Sween Corp.), Sorbalon (Ca-alginate, Hatrman), Sorbsan (Ca-alginate fibers in nonwoven pads, Dow Hickam), and TegagenTM HG (Ca-alginate fibers in nonwoven pads, 3 M Health Care) [162]. Some alginate dressings, e.g., Kaltostat, have the ability to induce excitation of monocytes resulting in improved healing. Such high bioactivities can be related to the endotoxin present in alginate dressings [161].

In addition to conventional alginate dressings, dressing with improved characteristics are also obtained [163]. By the absorption of hydrolyzed and unhydrolyzed chitosan on the key alginate fiber, improved alginate dressings have been developed [163]. In such dressings chitosan provides antibacterial, wound healing, and homeostatic characteristics while alginate absorbs the exudate. Different concentrations of hydrolyzed chitosan (7%–25%) are incorporated with alginate to improve wound healing. Lower molar mass improves the ability of chitosan to perforate into the alginate fibers resulting in enhanced tensile strength as compared to unhydrolyzed chitosan/alginate fibers. These fibers are able to provide sustained antibacterial compounds/activities [163]. Improved wound healing of oxidized alginate/gelatin hydrogels were also studied [164]. Water loss is much more from the wounds (280 g/m^2 day from normal injury; 5140 g/m^2 day from first degree burn) compared to the normal skin (200 g/m^2 day). Hence for a better wound healing control of water loss is necessary. Wound dressing should balance the water evaporation neither to dehydrate the wound nor to development of exudates [164]. A value of 2000–2500 g/m^2 day provides an appropriate wetness for improved wound healing [165]. Oxidized alginate/gelatin–based dressings are observed to have a permeability of 2700 g/m^2 day closed to appropriate level for fluid balance, and promote re-epithelization and cellular migration for better wound healing [164]. The wound healing ability of Ca-alginate fibers in nonwoven pads or ropes as well as in situ formed hydrogels can be improved by incorporation of growth factors and drugs into these fibers.

10.5.6 METAL-LOADED NANOPARTICLES

10.5.6.1 Silver and Gold Nanoparticles

Preparation of metallic nanoparticles can be improved by utilization of γ radiations. Several advantages are associated with this technique. Free electrons generated in the phenomenon of radiolysis of water are hydrated in the solution. These electrons are capable of reducing the metal without special reducing agents. By varying the irradiation time, the quantity of uncharged nuclei can be controlled. Homogeneous development results in an evenly dispersed nanomaterial. The reduction of Ag^+ to Ag is stepwise. Ag atoms principally reduced by hydrated electrons quickly combine with silver cations to develop dimmer clusters. Continual reduction of the Ag cations results in accumulation of tetramer clusters of nano-sized particles. Nano-sized clusters of silver prepared via gamma irradiation are then capped by sodium alginate [166].

Irradiation of $AgNO_3$ solution results in yellow dispersion representing the formation of even and homogenized Ag-nanoparticles. The stability of sodium alginate–capped Ag-nanoparticles is so good that they can be stored up to 6 months without any sedimentation [166]. The synthesis of gold nanoparticles with controllable size is also a striking area of research due to their valuable application in biological systems such as drug delivery, DNA sensors, and diagnostics and treatment of melanoma. Gold nanoparticles with different size have specific characteristics suitable for utilizations in biomedicine and cosmetics. However, the methods of preparation of gold nanoparticles by using chemicals, such as citrate, borohydride, or other organic compounds, as reducing and/or stabilizing agents are environmentally poisonous or biologically hazardous. The utilization of gamma radiation for the synthesis of gold nanoparticles can overcome these problems. The process involves getting gold nanoparticles, from trivalent gold solutions having alginic acid as a stabilizer, with predecided size in the range of 5–40 nm. Larger (40 nm) and more monodispersed gold nanoparticles were attained by enlargement of 20 nm seed particles at $[Au^{3+}]/[Au^{0}] = 6$. Alginates-stabilized gold nanoparticles with different sizes from 5 to 40 nm can be valuable for biological pertinence due to biocompatibility of sodium alginate [167].

10.5.6.2 Iron Nanoparticles

Biomedical applications, such as hyperthermia, cell-sorting systems, clinical diagnostic techniques, and magnetic drug delivery, directed the attention of scientists toward the biocompatible nanostructured magnetic blends. MRI due to its noninvasive nature, high resolution, and radiation-free exposure of soft tissues revolutionized the area of diagnostics. Minimum dose of superparamagnetic iron oxide nanoparticles generates high rest rates in specific organs due to their smaller magnetic moment [115]. Phagocytic action distributes these iron oxide supermagnetic nanoparticles in rough endoplasmic reticulum cells, where they have the ability to decline the intensity of magnetic resonance in lymph nodes, liver, spleen, and bone marrow, etc. and in all those tissues where they plinth up [168]. Transportation of drugs to specific tissue is an attractive application in drug delivery. Controlled drug liberation is only possible with high vulnerability and high dispersed magnetization after magnetic stimuli. Magnetic characteristics and biocompatibility of maghemite (c-Fe_2O_3) and magnetite (Fe_3O_4) made them ideal for these pertinences [169]. Llanes et al. [170] synthesized c-Fe_2O_3 nanoparticles with alginate network by filtration and washing ferrous alginate beads with methanol/water solution in 1:1 ratio. These ferrous alginate beads are converted to maghemite (c-Fe_2O_3) by treating with alkaline H_2O_2. Maghemite (c-Fe_2O_3) are hardened as compared to alginate beads which facilitated their integrity [170]. Nanoparticles of Fe^{2+} can be synthesized from Fe^{3+} at 60°C in the presence of ammonium hydroxide [171]. These iron particles are utilized for the synthesis of two different types of alginate beads. Superparamagnetic ferumoxides having 5 nm layer of iron oxide with 80–150 nm layers of dextran are utilized in spleen, myocardial perfusion, and liver MRI whereas ferumoxtran with 4–6 nm layer of iron oxide coated with 20–40 nm hydrodynamic dextran employed in late stage of lymph nodes and in early intravenous administration as blood pool agent. On the other hand, 10 nm layer of iron oxide with nonrecyclable 300 nm hydrodynamic starch or cellulose was used for bowl MRI [172]. Ma et al. [172] successfully synthesized alginate-stabilized supermagnetic particles of iron oxide by dissolving 2:1.5 M ratio of ferric chloride and ferric in distilled water [172]. Tissue division and pharmacokinetics of these alginate-stabilized supermagnetic nanoparticles of iron oxide were also investigated. Two types of tumor models, primary liver tumor in rats and VX2 liver cancer in rabbits, were utilized to check their worth for liver cancer detection [173].

10.5.7 BIOSENSORS

The improvement of alginic acid, generally by means of chemical modification via cross-linking of functional COOH groups on the backbone of alginic acid, raises the stability with decreasing the swelling capacity of alginate-based aqua-gels. This strategy is utilized for Staudinger covalent ligation. Staudinger covalent ligation developed biologically applicable, compatible, catalyst-free and chemoselective alginate-based aqua-gels with azide and phosphine functional groups. To further enhance the compatibility of azide-functionalized alginic acid polymer, it was blended with a high molecular weight poly(ethylene glycol) (PEG), terminated with 1-methyl-2-diphenylphosphino-terephthalate (MDT). XAlg PEG microbeads were found to be extremely compatible with insulinoma cell lines [174].

Alginic acids via lectins with avidin—biotin blends are used in drug-delivery systems [175]. Alginate avidin—biotin blends trap and connect the reporter cells onto optical surface of the spherical geometry triggering over a wide range of noxious substances. Bioilluminated alginate biosensor was applied for of *Escherichia coli* against different concentrations of mitomycin-C [175]. The strain of *E. coli* responded to the mitomycin-C-induced DNA damage at 490 nm [175]. Deposition is a technique in which biomolecules could be firmed and developed on the surface via entrapment. Cross-linking and covalent binding is then utilized in manufacturing the fabric of biosensor. Algal polymers are promising candidates due to their prick structure and biocompatibility. Alginic acid is a promising option because its biocompatibility and prick structure which permits for high diffusion rates.

Alg-Py matrix with enzymes improved the characteristics of hydrogels. Enzymes play a major role in the fabrication of sensible and selective biosensors [176]. Na-alginate, polyvinyl butyral (PVB), and horseradish peroxidase (HRP)-based biosensor of H_2O_2 are fabricated. The enzyme-based biosensor exhibited a strong catalytic activity against H_2O_2 [118]. The activation of luciferase gene for air virulence determination can be achieved through different genetic manipulation techniques. To establish a relation between immobilization technique and bioreporter sensitive bacteria, Eltzov et al. [177] employed genotoxic (DPD2794) and cytotoxic sensitive cells (TV1061) in a biosensor. This biosensor is capable of determining indoor toxins or endocrine distressing possessions [177].

10.5.8 TISSUE ENGINEERING

Stimulus-response polymeric blends composed of polyvinyl ether and algal polymers are tempting interest in a variety of fields. Algal polymer blends, in addition to calcium, are used in the formation of different temperature-sensitive devices [178]. Formation of nonthrombogenic porous surface, stabilization, active transportation of macromolecules, vasoconstriction, and provision of nonsticky surface for white blood cells are the vital roles of endothelial cells (EC). Synthetic polymers and peptides including extracellular proteins (ECM) are the substrates of EC. To ensure the porosity of endothelial cells in the prevascularization perception, scaffolds are planted in inert or dynamic (e.g., perfusion) conditions within endothelial cells. Vasculogenesis results in the outgrowth of endothelial cells to connect vasculature. Arg-Gly-Asp (modified macroperforate) is now used as endothelial ancestor for angiogenesis and cell enlargement. This is proved to be a potent therapeutic agent for heart failure patients [179]. Chemically modified or antiviral associated carrageenan of parenteral formulation has also proved to be effective against HPV transmission and can be very promising agent against HIV. Blends of carrageenan have also played a role of probes to examine the mechanism of powerful diseases, such as cancer and AIDS. KC and cellulose blends can also be employed as prototypes for the structure of novel therapeutics agents, more powerful and less toxic than current chemotherapeutics agents [180]. Integration of fibrous proteins with β-tri-calcium phosphate, $CaCO_3$ and hydroxyapatite causes an improvement in rigidness and strength [180].

Neural engineering, an emerging discipline of science, relies on the information from clinical neurology, computer engineering, nanotechnology, neural tissue engineering, robotics, computational neuroscience, and investigational neuroscience to analyze and manipulate the functional behavior of PNS or CNS. Neural interfaces such as brain pacemakers with ability to control and explain the voluntary processes of brains are generated which are very useful in Parkinson disease, dystonia, recent times depression, and epilepsy, etc. Algal polymers, particularly electrical-conducting biopolymers with antiinflammatory drug-loaded nanofibrils, were reported to have same functions similar to pace markers [181]. Controlled release of loaded drugs also lessens the side effects, i.e., skin atrophy diabetes and hemorrhagic ulcer. [181]. Algal polymer blends are also used in the reconstruction of intervertebral disc (IVD) whose degeneration results in severe health consequences starting from pain in lower back to paraplegia [182]. These algal polymer blends are also examined for gene countenance of ECM accrual [182]. Recombinant bone morphogenetic protein-2 (rhBMP-2) having ability to support and repair spinal fusion is developed which showed ability to overcome all the challenges of bone flaws [183].

10.5.9 DELIVERY SYSTEMS

10.5.9.1 Drug Delivery

Controlled formulations utilizing water-soluble polymers in oral route drug delivery is proved to be more potent than conventional delivery system. To reduce the flaws and facilitate the site-specific drug delivery having low half-lives is a characteristic of polymeric delivery system. Kulkarni et al. [184] reported controlled release of alginate cross-linked beads having low half-lives to protect from the side effects such as perforation, ulcer, and release of diclofenac sodium. They further reported that entrapment efficiency of sodium alginate and release of diclofenac sodium was dependent on pH, temperature, and properties of beads whereas cross-linking property depends on the time of contact with glutaraldehyde [184]. Similarly, IPN of alginates with gelatin were used for cefadroxil-loaded bead preparation with little eruption discharge rates and up to 88% encapsulation efficiencies [185]. Heparin, a potent anticoagulant, was encapsulated with alginate hydroxyl propyl cellulose for controlled release [186]. Chan et al. [187] noticed that the efficiency and discharge profile of encapsulated sulfaguanidine had been found to be robustly dependent on the formulations' viscosity and hydrophobicity of the cellulose derivatives [187]. In spite the advancement and development in understanding the mechanisms of neurodegenerative diseases and actions of neuroactive agents, drug delivery to central nervous system remains a challenge due to restriction of blood–brain barriers. However, for maintenance of neuronal regeneration, systemic administration of neuroactive biomolecules is toxic and less stable [188]. Ciofani et al. [189] reported that Ca-alginate microgels with biotinylated Wisteria floribunda agglutinin (bio-WFA) were the best treatment for CNS. These blended hydrogels have high efficacy, less invasion, and easy to handle in surgery. They followed the quick burst phase kinetics initially and then very slow phase for sustained release. This sort of behavior could be helpful for the treatment of lupus cerebritis, multiple sclerosis with experimental treatment of Alzheimer disease, and vacuities because these diseases require a strong therapeutic initial dose followed by minimal dose maintenance therapy [189]. Xu et al. [190] also developed alginic acids and chitosan cross-linked beads. Simulated gastric fluid (SGF), simulated colonic fluid (SCF), and simulated intestinal fluid (SIF) were used to analyze the BSA (bovine serum albumin) from both dual as well as single beads [190]. Less than 3% BSA release was achieved after 4 h in simulated colonic fluid, and dual cross-linked beads were stable after 10 days in simulated colonic fluid. BSA release after 4 h in simulated intestinal fluid is controlled by alginate/chitosan content in the beads; however, BSA total release after 4 h varies between 20% and 40%, and between 60% and 80% after 8 h. In simulated colonic fluid, the release of BSA increased compared to SIF since Ca-cross-linked alginate is less stable at pH above than 7 [190]. In vitro salicylic acid discharge from Na-alginate and starch blends with ionic strength of 0.145 M at pH 7.4 was investigated. It was observed that concentration of starch, pH, and ionic strength influenced the discharge drug from blend matrix [78]. Ca-alginate and chitosan blends were used for the preparation of nanospheres in which concentration of chitosan is very critical [191]. Alginic acids and gelatin blends cross-linked with calcium ions were used for the ophthalmic delivery of ciprofloxacin hydrochloride for antibacterial inflammation [192]. Mladenovska et al. [136] fabricated chitosan-Ca-alginate microparticles for 5-aminosalicylic acid (5-ASA), a widespread colon-specific drug employed for inflammatory bowel diseases (IBD) treatment. Irradiation plays a double role as a (1) sterilizing agent and (2) an initiator or cross-linking agent. Yang et al. [62] fabricated alginate and folic acid modified chitosan blends for photodynamic diagnosis of bowel cancer by bowel-specific delivery of 5-aminolevulinic acid [62]. Algal polymers and their blends are widely used as stabilizers, thickeners, and emulsifiers in food industry. These blends can also be chemically modified by the addition of nonpolar groups such as cholesteryl, alkyl chain (ranging from C_{12} to C_{18}), and *n*-octyl groups to improve the stabilizing properties [193]. It has also been revealed that cholesteryl-grafted Na-alginate has the ability to encapsulate pyrene [62]. Unfortunately, the synthesis involves organic solvents causing harmful effects to the human body as well as to the environment. To overcome these side effects, chemical modifications with nonpolar solvents such as oleoyl chloride to release vitamin D3 as a test-model nutraceutical was performed. It was observed that chemically modified, nonpolar polymers remain active in SGF as well as in SIF.

Vitamin D is of enormous significance for both disease and health playing a vital role in the fetal formation and development of osteoblasts. Since it is a fat-soluble vitamin, its insufficiency is a common health problem. Li et al. [194] investigated the discharge of vitamin D3 in GIT fluid from alginate nanoparticles. The nanocomposites of quinine (QUI), an antimalarial drug with alginic acids solution MSH-QUI/alginate were prepared in $CaCl_2$ solution to deliver the drug in the gastric environment [195].

10.5.9.2 Cells and Enzymes Delivery

Cells are most potent repository of drugs responsible for the discharge of active therapeutic molecules for extended time period and are considered as an alternate drug-delivery system (DDS). Cells help to maintain homeostasis in diseases such as Parkinson and diabetes by discharging drugs, cytokines, or proteins in response to the stimulus. Grafted, undifferentiated progenitor cells are considered as a potent source of drug-delivery vehicles and also have the ability to cure diabetes, ischemia, dementia, and hypertension by the liberation of endothelial growth factor and nerve growth factors. The efficacy and viability of grafted, undifferentiated progenitor cells can be further enhanced by lowering the immune response. Algal polymers are usually processed for the penetration into the membranes of grafted cells to prevent their accessibility into the immune cells [196]. Furthermore, release kinetics, degradation rate, and pore size could also be controlled by electing the type of algal polymer and coating agent.

Algal polymers are highly compatible and resistant to mammalian enzymes, but they can be thawed and eliminated in vivo via kidneys. On the contrary, vulnerability of alginates has been observed against in vivo hydrolytic degradation due to partial oxidation of uronic units [197]. For noninvasive algal polymer blend delivery formation for pharmaceutical formulations, self-gel formations are utilized [198]. Stable gels are obtained by blending Ca reservoirs with alginates. These stable gels of alginate are potent immunotherapeutic agents, which incorporate interleukin-2 (IL-2) into the matrix and deliver CpG oligonucleotides [198]. Algal polymers with poly-L-lysine capsules have been reported to reverse the diabetes in large animals and humans too. The mechanical stability of algal polymers can be further improved by semi-interpenetrating network gels [199]. The therapeutic effectiveness of the cells was consequently decreased by the adsorption of proteins due to surface polycations of islets encapsulating aqua-gels. These surface cations were beneficent for diabetic treatment. Biocompatibility and compositional constituents of algal polymer microcapsules are a perilous problem for the persistent efficacy [200].

Sakai et al. [201] developed perforate silicate layered and alginate-based microcapsules for the diffusion of lower molecular weight molecules such as glucose or insulin. Between the alginate core gel and the outer alginic acid layer were assimilated alginate/aminopropyl silicate/alginate particles [146]. These alginate/aminopropyl silicate/alginic acid microcapsules can further extended to design artificial pancreas [201]. Orive et al. [202] explained the development of highly standardized transplantable objects and biocompatible, permeable, and constant-sized microcapsules by varying the M/G ratio. Bunger et al. [203] reported the transient release of co-encapsulated steroids against alginate-PLL capsules. The beads formed by microfluidic devise could be applicable in bioengineering, pharmaceutical, and medical areas [204]. Ribeiro et al. [205] fabricated Ca-titanium phosphate-alginate and hydroxyapatite-alginate-based microspheres for bone revival templates and enzyme delivery matrices.

Enzymes are universal green biocatalysts exhibiting high degrees of regiospecificity and stereospecificity. These specificities elevate their applications in food processing, bioremediation, pharmaceutical synthesis, biosensors fabrication, and protein digestion in proteomic study. Polymer alteration strategy could be used to fabricate a multienzyme system. Among the various methods of polymer alteration, graft copolymerization is reported for the introduction of innovative functional groups to a polymer. These immobilized polymer systems have found extensive range of applications in food industries, agriculture, and medics such as lessening urea content in delivery treatment, blood detoxification in artificial kidneys, and removal of urea from beverages [206].

10.6 ENVIRONMENTAL APPLICATIONS

10.6.1 ACTIVE PACKAGING

Active packaging or wrapping is very important to sustain the nutritional quality of the product; moreover, it may enhance the shelf life and also improve the sensory characteristics. Numerous studies of edible or biodegradable films or coatings have been performed to enhance the shelf life and to amplify the quality of food in the past 10 years, e.g., alginate, as other biopolymers, acts as a food preservative and also as a particular barrier for humidity, gas, and solute migration. Antimicrobial consequences of alginate-based films holding essential oils for the maintenance of total beef muscle are reported [207]. Oregano-based films exhibited the highest antiradical properties, while cross-linking by calcium chloride (20%) reduced their solubility and enhanced their mechanical characteristics [208]. Na-alginate and lemon extract with sodium chloride exhibited an increase in the shelf life of all packaged Mozzarella cheese; confirming that the examined substance may apply an inhibitory effect on the microbes responsible for deterioration process [209].

10.6.2 BIODIESEL AND BIOETHANOL PRODUCTION

Biodiesel, a clean burning fuel, is generated by transesterification of vegetable oils or animal fats with short-chain ROH or by the esterification of fatty acids. Transesterification of algae oil into fatty acid methyl ester (biodiesel) is one of the well-known mechanism and general practice to get biofuel from algae. Different strains of micro- or macroalgae, rich in oil constituents, can be used for generating biodiesel used as fuels. These types of algae are also commonly used for the manufacturing of CH_3OH or EtOH, hydrogen, and other hydrocarbon fuels, and most commonly used fuels can be obtained from these via variety of methods. Algal polymers blends with natural polymers are considered a future substrate for bioenergy. Magnetic nanoparticles, i.e., superparamagnetic Fe_2O_3 nanoparticle and cross-magnetic silica nanoparticle were used for complete immobilization of recombinant alginate lyase. Immobilized alginate lyase was used to produce a dimer or trimmer mixture of alginate oligosaccharide. It is advantageous as immobilized alginates were reusable, frequently more than 10 times after magnetic separation [210]. Different molecular techniques are used to improve the strain for better extraction of polymeric materials as EtOH technology makes a direct use of algae. It was reported that improved algal strain are prone to bioenergy. The bioethanol production from algae is an encouraging approach that resolves dilemmas related to land biomass, such as bioethanol-food disagreement and the indirect land use [211].

10.6.3 ULTRA- AND NANOFILTRATION

Alginate-based nanofiltration and hybrid (organic/inorganic) tubular ultrafiltration eliminates Cd from wastewater with high throughput rates and at low pressure. Physical (filtration/cross-linking) and chemical (grafting) methods are applied for the stabilization and safe discharge of alginate from macroporous alkoxide tubes which are advantageous due to polymeric ligands. The interaction of heavy metals with blends could be improved by a permeable structure [212]. Temperature—sensitive, permeable aqua-gels made of poly (N-isopropilacrilamide) (PNIPAAm) and interpenetrated networks (IPN) of alginate-Ca ions are reported by Moura et al. [213]. These aqua-gels were fabricated by cross-linking PNIPAAm networks and Na-alginate with calcium ions. The crumpling of the PNIPAAm chains was done by heating IPN aqua-gels, resulting in intramolecular interactions due to nonpolarity. The empty spaces were occupied with water by rearrangement, thus lessening the pore size of aqua-gel. Among other, these temperature-sensitive objects could be more suitable as separation membranes [213].

10.7 CONCLUDING REMARKS

Algal polymers and their blends are regarded for medical, packaging, agriculture, dairy, and food pertinences such as icings, drug delivery, glazes, wound-coverage material, low-fat sausages, and cell encapsulation/grating based on their exceptional properties, such as thickeners, stabilizers, emulsifiers, gelling, and film—making, resulting in several pertinences. Moreover, alginate, carrageenan, and agar configurations can also be chemically improved to devise smart materials for special pertinences. The reactivity of algal polymer functional groups can be exploited as a potential tool for the modification of exciting characteristics such as nonpolarity, solubility, and physicochemical and biological properties. Alginate, carrageenan, and agar blends are considered as outstanding polysaccharides for the food and pharmaceutical industries because of their exceptional attributes due to nontoxic, immunogenic, biocompatibile, and biodegradable nature.

REFERENCES

[1] Mantia FP, Morreale M. Green composites: a brief review. Compos Part A Appl Sci Manuf 2011;42:579–88.

[2] Lecomte F, Siepmann J, Walther M, Macrae RJ, Bodmeier R. Blends of enteric and GIT-insoluble polymers used for film coating: physicochemical characterization and drug release patterns. J Control Release 2003;89:457–71.

[3] Prasad K, Mehta G, Meena R, Siddhanta AK. Hydrogel-forming agar-graft-PVP and k-carrageenan-graft-PVP blends: rapid synthesis and characterization. J Appl Polym Sci 2006;102:3654–63.

[4] Strobl GR. The physics of polymers concepts for understanding their structures and behavior. Springer Verlag; 1996. Section 3.2 Polymer Mixtures.

[5] Robeson LM, Solc K, editors. Polymer compatibility and incompatibility: principles and practices. New York: Harwood Academic Pub.; 1982.

[6] Hammer CF. In: Paul DR, Newman S, editors. "Polymer blends,", vol. 2. New York: Academic Press; 1978.

[7] Zini E, Scandola M. Green composites: an overview. Polym Compos 2011;32:1905–15.

[8] Mohanty AK, Misra M, Drzal LT. Sustainable bio-composites from renewable resources: opportunities and challenges in the green materials world. J Polym Environ 2002;10:19–26.

[9] Mohanty AK, Drzal LT, Misra M. Engineered natural fiber reinforced polypropylene composites: influence of surface modifications and novel powder impregnation processing. J Adhes Sci Technol 2002;16:999–1015.

[10] Dicker MPM, Duckworth PF, Baker AB, Francois G, Hazzard MK, Weaver PM. Green composites: a review of material attributes and complementary applications. Compos Part A Appl Sci Manuf 2014;56:280–9.

[11] Ibrahim BA, Karrer MK. Influence of polymer blending on mechanical and thermal properties. Mod Appl Sci 2010;4(9):157–61.

[12] Maniruzzaman M, Boateng JS, Bonnefille M, Aranyos A, Mitchell JC, Douroumis D. Taste masking of paracetamol by hot melt extrusion: an in vitro and in vivo evaluation. Eur J Pharm Biopharm 2012;80:433–42.

[13] Alexander P, Nikolay O, Alexander K, Galina S. Some aspects of preparation methods and properties of polyaniline blends and composites with organic polymers. Prog Poly Sci 2003;28:1701–53.

[14] Genies EM, Lapkowski M, Tsintavis C. Preparation, properties and applications of polyaniline. J Chem 1988;12(4):181–96.

[15] Sperling LH, Klemper D, Sperling LH, Utracki LA. Interpenetrating polymer networks. Washington, DC: American Chemical Society; 1994.

[16] Vallittu PK. Interpenetrating polymer networks (IPNs) in dental polymers and composites. J Adhes Sci Technol 2009;23:961–72.

[17] Ankit P, Deepak S, Ashok D, Rahul G, Piyush A, Pratik P, Parth P, Gaurang P. A review of hot melt extrusion technique. Int J Innov Res Sci Eng Technol 2013;2.

[18] Arvanitoyannis I, Kolokuris I, Nakayama A, Aiba SI. Preparation and study of novel biodegradable blends based on gelatinized starch and 1, 4-trans-polyisoprene (gutta percha) for food packaging or biomedical applications. Carbohydr Polym 1997;34(4):291–302.

[19] Carvalho AJFA, Job E, Alves N, Curvelo AA, Gandini A. Thermoplastic starch/natural rubber blends. Carbohydr Polym 2003;53:95–9.

[20] Huda MS, Drzal LT, Mohanty AK, Misra M. Effect of chemical modifications of the pineapple leaf fiber surfaces on the interfacial and mechanical properties of laminated biocomposites. Compos Interfaces 2008;15:169–91.

[21] Pereira RA, Mano EB, Dias ML, Acordi EB. Comparative study on the lamellar crystal structure of high and low density polyethylenes. Polym 1997;38:707–14.

[22] Arvanitoyannis I, Biliaderis CG. Physical properties of polyol-plasticized edible blends made of methyl cellulose and soluble starch. Carbohydr Polym 1999;38(1):47–58.

[23] Ikada Y, Tsuji H. Biodegradable polyesters for medical and ecological applications. Macromol Rapid Commun 2000:117–32.

[24] Suyatma NE, Copinet A, Coma V, Fricoteaux F. Compatibilization method applied to the chitosan-acid poly (L-lactide) solution. J Appl Polym Sci 2010;117:3083–91.

[25] Averous L. Biodegradable polymer blends and composites from renewable resources. Macromol Chem Phys 2009:210–890.

[26] Li QZ, Xiao YL, Su FZ, Xin FL, Zhi GT. Study on konjac glucom annan-carrageenan edible blend films. Adv Mat Res 2013;641–642:521–4.

[27] Satyanarayana KG, Arizaga GGC, Wypych F. Biodegradable composites based on lignocellulosic fibers – an overview. Prog Polym Sci 2009;34:982–1021.

[28] Laner D, Crest M, Scharff H, Morris JW, Barlaz MA. A review of approaches for the long-term management of municipal solid waste landfills. Waste Manag 2012;32: 498–512.

[29] Fishbein B. Building for the future: strategies to reduce construction and demolition waste in municipal projects. 1998.

[30] Chèze-Lange H, Beunard D, Dhulster P, Guillochon D, Cazé AM, Saude N, Morcellet M, Junter GA. Production of microbial alginate in a membrane bioreactor. Enzyme Microb Technol 2002;30:656–61.

[31] Emo C, Patrizia C, Elizabeth GF, El-Refaie SK, Andrea L. Gelatin-based blends and composites. Morphological and thermal mechanical characterization. Biomacromolecules 2001;2:806–11.

[32] Ma X, Yu J, Kennedy JF. Studies on the properties of natural fibers-reinforced thermoplastic starch composites. Carbohydr Polym 2005;62:19–24.

[33] Cláudia MV, Patrick FNM, Van D, Yilmaz G, Leontine A, Rui LR, António MC. Processing and characterization of biodegradable soy plastics: effects of crosslinking with glyoxal and thermal treatment 2005;97:604–10.

[34] Mather RR, Wardman RH. The chemistry of textile fibres. Cambridge: RSC Publishing; 2011.

[35] Matsumoto Y, Teramoto Y, Nishio Y. Preparation of thermoplastic magnetic wood via etherification and in-situ synthesis of iron oxide. J Wood Chem Tech 2010;30:373–81.

[36] Wu C, Liao H. The mechanical properties, biocompatibility and biodegradability of chestnut shell fibre and polyhydroxyalkanoate composites. Polym Degrad Stab 2014;99:274–82.

[37] Yu T, Jiang N, Li Y. Study on short ramie fiber/poly(lactic acid) composites compatibilized by maleic anhydride. Compos Part A Appl Sci Manuf 2014;64:139–46.

[38] Jawaid M, Abdul KHPS. Cellulosic/synthetic fibre reinforced polymer hybrid composites: a review. Carbohydr Polym 2011;86:1–18.

[39] Marais S, Gouanvé F, Bonnesoeur A, Grenet J, Poncin-Epaillard F, Morvan C, et al. Unsaturated polyester composites reinforced with flax fibers: effect of cold plasma

and autoclave treatments on mechanical and permeation properties. Compos Part A Appl Sci Manuf 2005;36:975–86.

[40] Venkateshwaran N, Perumal AE, Arunsundaranayagam D. Fiber surface treatment and its effect on mechanical and visco-elastic behaviour of banana/epoxy composite. Mater Des 2013;47:151–9.

[41] Sreekala MS, Kumaran MG, Joseph S, Jacob M, Thomas S. Oil palm fiber reinforced phenol formaldehyde composites influence of fiber surface modifications on the mechanical performance. Appl Compos Mater 2000;7:295–329.

[42] Bixler HJ, Porse H. A decade of change in the seaweed hydrocolloids industry. J Appl Phycol 2011;23:321–35.

[43] Cao Z, Gilbert RJ, He W. Simple agarose-chitosan gel composite system for enhanced neuronal growth in three dimensions. Biomacromolecules 2009;10:2954–9.

[44] Davidson RL. Handbook of water-soluble gums. New York: McGraw-Hill Book Co.; 1980.

[45] Lee JB, Hayashi K, Maeda M, Hayashi T. Antiherpetic activities of sulfated polysaccharides from green algae. Planta Med 2004;70:813–7.

[46] Graham HD. Food colloids. Cambridge. (Connecticut): AVI Publishing Co., Inc. Westport; 1977.

[47] Belarbi EH, Molina E, Chisti Y. A process for high yield and scaleable recovery of high purity eicosapentaenoic acid esters from microalgae and fish oil. Process Biochem 2000;35:951–69.

[48] Dhargalkar VK, Verlecar XN. Southern Ocean Seaweeds: a resource for exploration in food and drugs. Aquaculture 2009;287:229–42.

[49] Kumar M, Sharma MK, Kumar A. Spirulina fusiformis: a food supplement against mercury induced hepatic toxicity. J Health Sci 2005;51:424–30.

[50] Besada V, Andrade JM, Schultze F, Gonzalez JJ. Heavy metals in edible seaweeds commercialized for human consumption. J Mar Syst 2009;75:305–13.

[51] Deleted in review.

[52] Brendal D, Lisa K, Christine A, Ward P, Demian C, Michael RH, Steven TK, Samuel HG. Global catches, exploitation rates, and rebuilding options for sharks. Mar Policy 2013;40:194–204.

[53] Li X, Tabil LG, Panigrahi S. Chemical treatment of natural fibre for use in natural fibre-reinforced composites: a review. Polym Environ 2007;15:25–33.

[54] Saenghirunwattana P, Noomhorm A, Rungsardthong V. Mechanical properties of soy protein based "green" composites reinforced with surface modified cornhusk fiber. Ind Crop Prod 2014;60:144–50.

[55] Klöck G, Pfefferman A, Ryser C, Gröhn P, Kuttler B, Hahn HJ, Zimmermann U. Biocompatibility of mannuronic acid-rich alginates. Biomaterials 1997;18:707–13.

[56] Rowley JA, Madlambayan G, Mooney DJ. Alginate hydrogels as synthetic extracellular matrix materials. Biomaterials 1999;20:45–53.

[57] Avella M, Di PE, Immirzi B, Impallomeni G, Malinconico M, Santagata G. Addition of glycerol plasticizer to seaweeds derived alginates: influence of microstructure on chemical-physical properties. Carbohydr Polym 2007;69:503–11.

[58] Chopra A, Shan L, Eckelman WC. Molecular imaging and contrast agent database (MICAD): evolution and progress. Mol Imaging Biol 2012;14:4–13.

[59] Usov A. Alginic acids and alginates: analytical methods used for their estimation and characterization of composition and primary structure. Russ Chem Rev 1999;68: 957–66.

[60] Nakamura K, Nishimura Y, Hatakeyama T, Hatakeyama H. Thermal properties of water insoluble alginate films containing di- and trivalent cations. Thermochim Acta 1995;267:343–53.

[61] Gómez CG, Pérez LMV, Lozano JE, Rinaudo M, Villar MA. Influence of the extraction-purification conditions on final properties of alginates obtained from brown algae (*Macrocystis pyrifera*). Int J Biol Macromol 2009;44:365–71.

[62] Yang JS, Xie YJ, He W. Research progress on chemical modification of alginate: a review. Carbohydr Polym 2011;201(84):33–9.

[63] Mancini M, Moresi M, Rancini R. Uniaxial compression and stress relaxation tests on alginate gels. J Texture Stud 1999;30:639–57.

[64] Moe ST, Draget KI, Skjåk BG, Smidsrød O. Alginate. In: Stephen AM, editor. Food polysaccharides and their applications. New York: Marcel Dekker, Inc.; 1995.

[65] Higuera DL, Hernández CG, Rodríguez-Montesinos YE. Efecto del tipo de precipitación en el procesos de obtención de alginato de sodio: Método de alginato de calcio y método de ácido algínico. Cienc Mar 1997;23:195–207.

[66] Navarro SA, García CCH. In: Elnashar M, editor. Biopolymers by *Azotobacter vinelandii*. InTech Open Access Publisher; 2010.

[67] Ertesvåg H, Valla S. Biosynthesis and applications of alginates. Polym Degrad Stab 1998;59:85–91.

[68] McHugh DJ. Production, properties and uses of alginates. In: McHugh DJ, editor. Production and utilization of products from commercial seaweedsvol. 288. FAO Fish Aquaculture Technol Pap; 1987. p. 58–115.

[69] Laurienzo P. Marine polysaccharides in pharmaceutical applications: an overview. Mar Drugs 2010;8:2435–65.

[70] Daia YN, Lia P, Zhangc JP, Wangc AQ, Wei Q. A novel pH sensitive N-succinyl chitosan/alginate hydrogel bead for nifedipine delivery. Biopharm Drug Dispos 2008;29:173–84.

[71] Nobile MR, Pirozzi V, Somma E, Gomez AG, Laurienzo P. Development and rheological investigation of novel alginate/N-succinylchitosan hydrogels. J Polym Sci B 2008; 46:1167–82.

[72] Gomez AG, De RA, Laurienzo P, Malinconico M. Development of a new calcium sulfate-based composite using alginate and chemically modified chitosan for bone regeneration. J Biomed Mater Res A 2007;81:811–20.

[73] Wnag LF, Rahim JW. Preparation and application of agar/alginate/collagen ternary blend functional food packaging films. Int J Biol Macromol 2015;80:460–8.

[74] Lee YL, Shin DS, Kwon OW, Park WH, Choi HG, Lee YR, Han SS, Noh SK, Lyoo WS. Preparation of atactic poly(vinyl alcohol)/sodium alginate blend nanowebs by electrospinning. J Appl Polym Sci 2007;106:1337–42.

[75] Cho SH, Oh SH, Lee JH. Fabrication and characterization of porous alginate/polyvinyl alcohol hybrid scaffolds for 3D cell culture. J Biomater Sci Polym Ed 2005;16: 933–47.

[76] Toft K. Prog Food Nutr Sci 1982;6:89–96.

[77] Morris VL, Chilvers GR. J Sci Food Agric 1984;35:1370–6.

[78] Wang Q, Dua XHY, Kennedy JF. Alginate/starch blend fibers and their properties for drug controlled release. Carbohydr Polym 2010;82:842–7.

[79] Pohja S, Suihko E, Vidgren M, Paronen P, Ketolainen J. Starch acetate as a tablet matrix for sustained drug release. J Control Release 2004;94:293–302.

[80] Fujiwara GM, Campos R, Costa CK, Dias JFG, Miguel OG, Miguel MD, Marques F, Zanin W. Production and characterization of alginate-starch-chitosan microparticles containing stigmasterol through the external ionic gelation technique. Braz J Pharm Sci 2013;49.
[81] Santin M, Ambrosio L. Soybean-based biomaterials: preparation, properties and tissue regeneration potential. Expert Rev Med Devices 2008;5:349–58.
[82] Silva GA, Vaz CM, Coutinho OP, Cunha AM, Reis RL. In vitro degradation and cyto-compatibility evaluation of novel soy and sodium caseinate-based membrane. J Mater Sci 2003;14:1055–66.
[83] Jane JL, Wang S. Soy protein-based thermoplastic composition for preparing molded articles. 1996. p. 293. U.S. Patent 5, 523.
[84] Gennadios A. Protein based films and coatings. 1st ed. New York: CRC Press; 2002.
[85] Mo X, Sun XS, Wang Y. Effects of molding temperature and pressure on properties of soy protein polymers. J Appl Polym Sci 1999;73:2595–602.
[86] Jerez A, Partal P, Martinez I, Gallegos C, Guerrero A. Protein-based bioplastics: effect of thermo-mechanical processing. Rheol Acta 2007;46:711–20.
[87] Liang F, Wang Y, Sun XS. Curing process and mechanical properties of protein based polymers. J Polym Eng 1999;19:383–93.
[88] Nishinari K, Fang Y, Guo S, Phillips GO. Soy proteins: a review on composition, aggregation and emulsification. Food Hydrocoll 2014;39:301–18.
[89] Tojo E, Prado J. Chemical composition of carrageenan blends determined by IR spectroscopy combined with a PLS multivariate calibration method. Carbohydr Res 2003;338:1309–12.
[90] Roberts MA, Quemener B. Trends Food Sci Technol 1999;10:169–81.
[91] Therkelsen GH. Carrageenan. In: Whistler RL, BeMiller JN, editors. Industrial gums. San Diego, CA: Academic Press; 1993.
[92] Viana AG, Noseda MD, Duarte MER, Cerezo AS. Alkali modification of carrageenans. Part V. The iota–nu hybrid carrageenan from *Eucheuma denticulatum* and its cyclization to iota-carrageenan. Carbohydr Polym 2004;58:455–60.
[93] Karbowiak T, Debeaufort F, Champoin D, Voilley A. Wetting properties at the surface of iota-carrageenan edible films. J Colloid Interface Sci 2006;294:400–10.
[94] Goosen MFA, Shea GMO, Gharapetian HM, Chou S, Sun AM. Optimization of microencapsulation parameters: semipermeable microcapsules as a bioartificial pancreas. Biotechnol Bioeng 1985;27:146–50.
[95] Li L, Ni R, Shao Y, Mao S. Carrageenan and its applications in drug delivery. Carbohydr Polym 2014;103. 1–1.
[96] Bixler P, Praphairaksit N, Chantarasiri N, Muangsin N. Preparation and evaluation of chitosan/carrageenan beads for controlled release of sodium diclofenac. AAPS Pharm Sci Technol 2007;8:120–30.
[97] Bixler HJ. Recent developments in manufacturing and marketing carrageenan. Hydrobiologia 1996;326:35–57.
[98] Floc JY, Deslandes E, Le GY. Evidence for vegetative propagation of the carrageeno-phyte *Solieria chordalis* (Solieriaceae, Rhodophyceae) on the coast of Brittany (France) and in culture. Bot Mar 1987;30:315–22.
[99] Chen LC, McLachlan J, Neish AC, Shacklock PF. The ratio of kappa-to lambda-carrageenan in nuclear phases of the rhodophycean algae, *Chondrus crispus* and *Gigartina stellata*. J Mar Biol Assoc UK 1973;53:11–6.

[100] Ruiter GA, Rudolph B. Carrageenan biotechnology. Trend Food Sci Technol 1997;8: 389–95.
[101] Gu YS, Decker EA, McClements DJ. Influence of pH and ι-carrageenan concentration on physicochemical properties and stability of β-lactoglobulin-stabilized oil-in-water emulsions. J Agric Food Chem 2004;52:3626–32.
[102] Pelegrín FY, Robledo D, Azamar JA. Carrageenan of *Eucheuma isiforme* (Solieriaceae, Rhodophyta) from Yucatán, Mexico. I. Effect of extraction conditions. Bot Mar 2006;49:65–71.
[103] Ng CA, Camacho DH. Polymer electrolyte system based on carrageenan-poly (3, 4-ethylenedioxythiophene) (PEDOT) composite for dye sensitized solar cell. In: IOP Conference Series: Mat Sci Eng, vol. 79; 2015. p. 012–20.
[104] Meena R, Prasad K, Siddhanta AK. Development of a stable hydrogel network based on agar–kappa-carrageenan blend cross-linked with genipin. Food Hydrocoll 2009; 23:497–509.
[105] Arof AK, Shuhaimi NE, Alias NA, Kufian MZ, Majid SR. Application of chitosan/ iota-carrageenan polymer electrolytes in electrical double layer capacitor (EDLC). J Solid State Electrochem 2010;14:2145–52.
[106] Rudhziah S, Ahmad A, Ahmad I, Mohamed NS. Biopolymer electrolytes based on blend of kappa-carrageenan and cellulose derivatives for potential application in dye sensitized solar cell. Electrochimica Acta 2015;175:162–8.
[107] Pietrasik Z, Duda Z. Effect of fat content and soy protein/carrageenan mix on the quality characteristics of comminuted, scalded sausages. Meat Sci 2000;56:181–8.
[108] Oliviera EC, Alveal K. The mariculture of *Gracilaria* (Rhodophyta) in the production of agar, introduction to applied phycology. The Hague. 1991.
[109] Suzuki Y. Usage of agar in processed food. Chem Abstr 1999;131:156–991.
[110] Soriano EM. Agar polysaccharides from *Gracilaria* species (Rhodophyta, Gracilariaceae). J Biotechnol 2001;89:81–4.
[111] Li Z, Shi S, Huang J, Xu Z, Lee W, Guo Y, Zhang X. Agar. Chem Abstr 1993;119:231–92.
[112] Furneaux RH, Miller IJ. Chemical structure and physico-chemical properties of agar. Bot Mar 1989;68:137–48.
[113] Pontevida HB, Hurtado-Ponce AQ. Bot Mar 1996;39:123.
[114] Chirapart A, Ohno M, Ukeda H, Sawamura M, Kusunose HJ. Chemical composition of agars from a newly reported Japanese agarophyte, *Gracilariopsis lemaneiformis*. Appl Phycol 1995;7:359.
[115] Montano NE, Villanueva RD, Romero JB, Romero J. Chemical characteristics and gelling properties of agar from two Philippine *Gracilaria* spp. (Gracilariales, Rhodophyta). Appl Phycol 1999;137:541–8.
[116] Bautista LJ, Kapraun DF. Agar Analysis, nuclear genome quantification and characterization of four agarophytes (*Gracilaria*) from the Mexican Gulf Coast. J Appl Phycol 1995;7:351.
[117] Lewis RJ, Hanisak D. Effects of phosphate and nitrate supply on productivity, agar content and physical properties of agar of *Gracilaria* strain G-16S. J Appl Phycol 1996;8:41.
[118] Rebello J, Ohno M, Critchley AT, Sawamura M. Growth rates and agar quality of *Gracilaria gracilis* (Stackhouse) Steentoft from Namibia, Southern Africa 1996;39:273.
[119] Kapraun DF, Ganzon FE, Bird KT, Trono G, Breden C. Karyology and agar analysis of the agarophyte *Gelidiella acerosa* (Forsskål) Feldmann*et* Hamel from the Philippines. J Appl Phycol 1994;6:545–50.

[120] Villanueva RD, Montano NE, Romero JB, Aliganga AK, Enriquez EP. Seasonal variations in the yield, gelling properties, and chemical composition of agars from *Gracilaria eucheumoides* and *Gelidiella acerosa* (Rhodophyta) from the Philippines. Bot Mar 1999;2:175–82.

[121] Gorostiaga JM. Growth and production of the red alga *Gelidium sesquipedale* off the Basque coast (northern Spain). Mar Biol 1994;120:311–22.

[122] Freile PY, Robledo DR, Garcia RG. Seasonal agar yield and quality in *Gelidium canariensis* (Grunow) Seoane-Camba (Gelidiales, Rhodophyta) from Gran Canaria, Spain. J Appl Phycol 1995;7:141–4.

[123] Takano R, Hayashi K, Hara S, Hirase S. Carbohydr Polym 1995;27:305.

[124] Murano E, Jellus V, Piras A, Toffanin R. Food polysaccharides and their applications. J Appl Phycol 1998;10:315.

[125] Goff LJ, Moon DA, Coleman AW. Molecular delineation of species and species relationships in the red algal agarophytes *Gracilariopsis* and *Gracilaria* (Gracilariales). J Phycol 1994;30:521–37.

[126] Mazumder S, Ghosel PK, Pujol CA, Carbucci MJ, Damonte EB, Ray B. Isolation, chemical investigation and antiviral activity of polysaccharides from *Gracilaria corticata* (Gracilariaceae, Rhodophyta). Int J Biol Macromol 2002;31:87–95.

[127] Yokoya NS, Oliviera EC. Temperature responses of economically important red algae and their potential for mariculture in Brazilian waters. J Appl Phycol 1992;4:339–45.

[128] Minghou J, Lahaye M, Yaphe W. Structure of agar from *Gracilaria* spp. (Rhodophyta) collected in the People's Republic of China. Bot Mar 1985;28:521–8.

[129] Lahaye M, Yaphe W. (C. Agardh) Greville, *G. cervicornis* (Turner) J. Agardh, *G. damaecornis* J. Agardh and *G. domingensis Sonder ex* Kützing (Gigartinales, Rhodophyta). The chemical structure of agar from *Gracilaria compressa*. Bot Mar 1989;32:369–78.

[130] Armisen R, Galatas F. Production, properties and uses of agar. Production and utilization of products from commercial seaweeds, vol. 288. FAO Fish Technol Pap; 1987. p. 1–57.

[131] Knutsen SH, Myslabodski DE, Larsen B, Usov AI. A modified system of nomenclature for red algal galactans. Bot Mar 1994;37:163–70.

[132] Boral S, Saxena A, Bohidar HB. Universal growth of microdomains and gelation transition in agar hydrogels. J Phys Chem B 2008;112:3625–32.

[133] Matsuhashi T. Agar in food gels. Netherlands: Springer; 1990. p. 1–51.

[134] Xiong JY, Narayanan J, Liu XY, Chong TK, Chen SB, Chung TS. Topology evolution and gelation mechanism of agarose gel. J Phys Chem B 2005;109:5638–43.

[135] Ayyad O, Muñoz RD, Agulló N, Borrós S, Gómez RP. High-concentration compact agar gels from hydrothermal synthesis. Soft Matter 2010;6:2389–91.

[136] Llansola RF, Escuder B, Miravet JF. Remarkable increase in basicity associated with supramolecular gelation. Org Biomol Chem 2009;7:3091–4.

[137] Trivedi TJ, Kumar A. Efficient extraction of agarose from red algae using ionic liquids. Green Sustain Chem 2014;4:190.

[138] Meena R, Chhatbar M, Prasad K, Siddhanta AK. Development of a robust hydrogel system based on agar and sodium alginate blend. Polym Int 2008;57:329–36.

[139] Draget KI, Taylor C. Chemical, physical and biological properties of alginates and their biomedical implications. Food Hydrocoll 2011;25:251–6.

[140] Hampson FC, Farndale A, Strugala V, Sykes J, Jolliffe IG, Dettmar PW. Alginate rafts and their characterisation. Int J Pharm 2005;294:137–47.

[141] Herranz F, Almarza E, Rodríguez I, Salinas B, Rosell Y, Desco M, Bulte JW, Ruiz-Cabello J. The application of nanoparticles in gene therapy and magnetic resonance imaging. Microsc Res Tech 2011;74:577−91.

[142] Berman SM, Walczak P, Bulte JW. MRI of transplanted neural stem cells. Methods Mol Biol 2011;711:435−49.

[143] Kraitchman DL, Kedziorek DA, Bulte JW. MR imaging of transplanted stem cells in myocardial infarction. Methods Mol Biol Part 2 2011;680:141−52.

[144] Barnett BP, Kraitchman DL, Lauzon C, Magee CA, Walczak P, Gilson WD, Arepally A, Bulte JW. Radiopaque alginate microcapsules for X-ray visualization and immunoprotection of cellular therapeutics. Mol Pharm 2006;3:531−8.

[145] Karussis D, Karageorgiou C, Vaknin DA, Gowda KB, Gomori JM, Kassis I, Bulte JW, Petrou P, Ben HT, Abramsky O, Slavin S. Safety and immunological effects of mesenchymal stem cell transplantation in patients with multiple sclerosis and amyotrophic lateral sclerosis. Arch Neurol 2010;67:1187−94.

[146] Brownlee IA, Allen A, Pearson JP, Dettmar PW, Havler E, Atherton MR, Onsøyen E. Alginate as a source of dietary fiber. Crit Rev Food Sci Nutr 2005;45:497−510.

[147] Dettmar PW, Strugala V, Richardson JC. The key role alginates play in health. Food Hydrocoll 2011;25:263−6.

[148] King S, See H, Thomas G, Swain M. Determining the complex modulus of alginate irreversible hydrocolloid dental material. Dent Mater 2008;24:1545−8.

[149] Maynard EM, Fernández E, Normann RA. A technique to prevent dural adhesions to chronically implanted microelectrode arrays. J Neurosci Methods 2000;97:93−101.

[150] Becker TA, Preul MC, Bichard WD, Kipke DR, McDougall CG. Calcium alginate gel as a biocompatible material for endovascular arteriovenous malformation embolization: six-month results in an animal model. Neurosurgery 2005;56:793−803.

[151] Mammarella EM, Rubiolo AC. Crosslinking kinetics of cation-hydrocolloid gels. Chem Eng J 2003;94:73−7.

[152] Nunamaker EA, Otto KJ, Kipke DR. Investigation of the material properties of alginate for the development of hydrogel repair of dura mater. J Mech Behav Biomed Mater 2011;4:16−33.

[153] Bernkop SA. Mucoadhesive polymers. In: Dumitriu S, editor. Polymer biomaterial. New York: Marcel Dekker; 2002.

[154] Leitner VM, Walker GF, Bernkop SA. Thiolated polymers: evidence for the formation of disulfide bonds with mucus glycoproteins. Eur J Pharm Biopharm 2003;56:207−14.

[155] Davidovich PM, Harari O, Bianco PH. Evaluating the mucoadhesive properties of drug delivery systems based on hydrated thiolated alginate. J Control Release 2009;136:38−44.

[156] Wang YY, Lai SK, Suk J, Race A, Cone R, Hanes J. Addressing the PEG mucoadhesivity paradox to engineering nanoparticles that "slip" through the human mucus barrier. Angew Chem Int 2008;47:9726−9.

[157] Taylor C, Pearson JP, Draget KI, Dettmarc PW, Smidsrød O. Rheological characterisation of mixed gels of mucin and alginate. Carbohydr Polym 2005;59:189−95.

[158] Pinhas DM, Bianco PH. Alginate-PEGAc. A new mucoadhesive polymer. Acta Biomater 2011;7:625−33.

[159] Rajaonarivony M, Vauthier C, Couarraze A, Puisieux F, Couvreur P. Development of a new drug carrier made from alginate. J Pharm Sci 1993;2:912−7.

[160] Motwani SK, Chopra S, Talegaonkar S, Kohli K, Ahmad FJ, Khar RK. Chitosan-sodium alginate nanoparticles as submicroscopic reservoirs for ocular delivery: formulation, optimization and in vitro characterization. Eur J Pharm Biopharm 2008;68:513−25.

[161] Thomas A, Harding KG, Moore K. Alginates from wound dressings activate human macrophages to secret tumour necrosis factor-a. Biomaterials 2000;21:1797−802.

[162] Paul W, Sharma CP. Chitosan and alginate wound dressings: a short review. Trends Biomater Artif Organs 2004;18:18−23.

[163] Knill CJ, Kennedy JF, Mistry J, Miraftab M, Smart G, Groocock MR, Williams HJ. Alginate fibres modified with unhydrolysed and hydrolysed chitosans for wound dressing. Carbohydr Polym 2004;55:65−76.

[164] Balakrishnan B, Mohanty M, Umashankar PR, Jayakrishnan A. Evaluation of an in situ forming hydrogel wound dressing based on oxidized alginate and gelatine. Biomaterials 2005;26:6335−42.

[165] Queen D, Gaylor JDS, Evans JH, Courtney JM, Reid WH. The preclinical evaluation of the water vapour transmission rate through burn wound dressings. Biomaterials 1987;8:367−71.

[166] Liu Y, Chen S, Zhong L, Wu GZ. Preparation of high-stable silver nanoparticle dispersion by using sodium alginate as a stabilizer under gamma radiation. Radiat Phys Chem 2008;78:251−5.

[167] Anh NT, Phu DV, Duy NN, Du BD, Hien NQ. Synthesis of alginate stabilized gold nanoparticles by c-irradiation with controllable size using different Au3+ concentration and seed particles enlargement. Radiat Phys Chem 2010;79:405−8.

[168] Corot C, Robert P, Idee JM, Port M. Recent advances in iron oxide nanocrystal technology for medical imaging. Adv Drug Deliv Rev 2006;58:1471−504.

[169] Gupta AK, Gupta M. Synthesis and surface engineering of iron oxide nanoparticles for biomedical applications. Biomaterials 2005;26:3995−4021.

[170] Llanes F, Ryan DH, Marchessault RH. Magnetic nanostructured composites using alginates of different M/G ratios as polymeric matrix. Int J Biol Macromol 2000;27:35−40.

[171] Morales MA, Finotelli PV, Coaquira JAH, Rocha LMHM, Diaz AC, Baggio SEM, Rossi AM. In situ synthesis and magnetic studies of iron oxide nanoparticles in calcium-alginate matrix for biomedical applications. Mater Sci Eng C 2008;28:253−7.

[172] Ma HL, Xu YF, Qi XR, Maitani Y, Nagai T. Preparation and characterization of super paramagnetic iron oxide nanoparticles stabilized by alginate. Int J Pharm 2007;333:177−86.

[173] Ma HL, Xu YF, Qi XR, Maitani Y, Nagai T. Super paramagnetic iron oxide nanoparticles stabilized by alginate: pharmacokinetics, tissue distribution, and applications in detecting liver cancers. Int J Pharm 2008;354:217−26.

[174] Hall KK, Gattás-Asfura KM, Stabler CL. Microencapsulation of islets within alginate/poly (ethylene glycol) gels cross-linked via Staudinger ligation. Acta Biomater 2011;7:614−24.

[175] Polyak B, Geresh S, Marks RS. Synthesis and characterization of a biotin-alginate conjugate and its application in a biosensor construction. Biomacromolecules 2004;5:389−96.

[176] Rabeah AK, Marks RS. Impedance study of the hybrid molecule alginate-pyrrole: demonstration as host matrix for the construction of a highly sensitive amperometric glucose biosensor. Sens Actuators B Chem 2009;135:516−22.

[177] Eltzov E, Pavluchkov V, Burstin M, Marks RS. Creation of a fiber optic based biosensor for air toxicity monitoring. Sens Actuators B 2011;155:859−67.

[178] Ichijo H, Hirasa O, Kishi R, Oowada M, Sahara K, Kokufuta E, Kohno S. Thermoresponsive gels. Radiat Phys Chem 1995;46:185−90.

[179] Yu J, Gu Y, Du KT, Mihardja S, Sievers RE, Lee RJ. The effect of injected RGD modified alginate on angiogenesis and left ventricular function in a chronic rat infarct model. Biomaterials 2009;30:751−6.

[180] Cullen MSD, Ramaswamy S, Clarke LI, Gorga RG. Nanofibrous composites for tissue engineering applications. Nanobiotechnology 2009;1:369−90.

[181] Abidian MR, Martin DC. Multifunctional nano biomaterials for neural interfaces. Adv Funct Mater 2009;19:573−85.

[182] Chou AI, Akintoye SO, Nicoll SB. Photo-crosslinked alginate hydrogels support enhanced matrix accumulation by nucleus pulpous cells in vivo. Osteoarthr Cartil 2009;17:1377−84.

[183] Boerckel JD, Kolambkar YM, Dupont KM, Uhrig BA, Phelps EA, Stevens HY, García AJ, Guldberg RE. Effects of protein dose and delivery system on BMP-mediated bone regeneration. Biomaterials 2011;32:5241−51.

[184] Kulkarni AR, Soppimath KS, Aminabhavi TM. Controlled release of diclofenac sodium from sodium alginate beads crosslinked with glutaraldehyde. Pharm Acta Helv 1999;74:29−36.

[185] Kulkarni AR, Soppimath KS, Aminabhavi TM, Rudzinski WE. In vitro release kinetics of cefadroxil-loaded sodium alginate interpenetrating networks beads. Eur J Pharm Biopharm 2001;51:127−33.

[186] Karewicz A, Zasada K, Szczubiałka K, Zapotoczny S, Lach R, Nowakowska M. Smart alginate-hydroxypropylcellulose microbeads for controlled release of heparin. Int J Pharm 2010;385:163−9.

[187] Chan LW, Heng PWS, Wan LSC. Effect of cellulose derivatives on alginate microspheres prepared by emulsification. J Microencapsul 1997;14:545−55.

[188] Dong Z, Wang Q, Du Y. Alginate/gelatin blend films and their properties for drug controlled release. J Membr Sci 2006;280:37−44.

[189] Ciofani G, Raffa V, Pizzorusso T, Menciassi A, Dario P. Characterization of an alginate-based drug delivery system for neurological applications. Med Eng Phys 2008;30:848−55.

[190] Xu Y, Zhan C, Fan L, Wang L, Zheng H. Preparation of dual cross linked alginate chitosan blend gel beads and in vitro controlled release in oral site-specific drug delivery systems. J Pharm 2007;336:329−37.

[191] De S, Robinson D. Polymer relationships during preparation of chitosan-alginate and poly(L-lysine)-alginate nanospheres. J Control Release 2003;89:101−12.

[192] Razem D, Katusin-Razem B. The effects of irradiation on controlled drug delivery/controlled drug release systems. Radiat Phys Chem 2008;77:288−344.

[193] Colinet I, Dulong V, Hamaide T, Le Cerf D, Picton L. New amphiphilic modified polysaccharides with original solution behaviour in salt media. Carbohydr Polym 2009;75:454−62.

[194] Li Q, Liu CG, Huang ZH, Xue FF. Preparation and characterization of nanoparticles based on hydrophobic alginate derivative as carriers for sustained release of vitamin D3. J Agric Food Chem 2011;59:1962−7.

[195] Joshi GV, Pawar RR, Kevadiya BD, Bajaj HC. Mesoporous synthetic hectorites: a versatile layered host with drug delivery application. Microporous Mesoporous Mater 2011;142:542−8.

[196] Schmidt JJ, Rowley J, Kong HJ. Hydrogels used for cell-based drug delivery. J Biomed Mater Res Part A 2008;87:1113−22.

[197] Kong HJ, Kaigler D, Kim K, Mooney DJ. Controlling rigidity and degradation of alginate hydrogels via molecular weight distribution. Biomacromolecules 2004;5:1720−7.

[198] Hori Y, Winans AM, Irvine DJ. Modular injectable matrices based on alginate solution/microsphere mixtures that gel in situ and co-deliver immunomodulatory factors. Acta Biomater 2009;5:969–82.

[199] Gong JP, Katsuyama Y, Kurokawa T, Osada Y. Double network hydrogels with extremely high mechanical strength. Adv Mater 2003;15:1155–8.

[200] Castro DM, Orive G, Hernandez RM, Gascon AR, Pedraz JL. Comparative study of microcapsules elaborated with three polycations (PLL, PDL, PLO) for cell immobilization. J Micro Encapsul 2005;22:303–15.

[201] Sakai S, Ono T, Ijima H, Kawakami K. Synthesis and transport characterization of alginate/aminopropyl-silicate/alginate microcapsule: application to bioartificial páncreas. Biomaterials 2001;22:2827–34.

[202] Orive G, Hernández RM, Gascón AR, Igartua M, Pedraz JL. Encapsulated cell technology: from research to market. Trends Biotechnol 2002;20:382–7.

[203] Bunger CM, Tiefenbach B, Jahnke A, Gerlach C, Freier TH, Schmitz KP. Deletion of the tissue response against alginate-PLL capsules by temporary release of coencapsulated steroids. Biomaterials 2005;26:2353–60.

[204] Sugiura S, Oda T, Izumida Y, Aoyagi Y, Satake M, Ochiai A, Ohkohchi N, Nakajima M. Size control of calcium alginate beads containing living cells using micronozzle array. Biomaterials 2005;26:3327–31.

[205] Ribeiro CC, Barrias CC, Barbosa MA. Calcium phosphate-alginate microspheres as enzyme delivery matrices. Biomaterials 2004;25:4363–73.

[206] Akkaya A, Uslan AH. Sequential immobilization of urease to glycidyl methacrylate grafted sodium alginate. J Mol Catal B Enzym 2010;67:195–201.

[207] Oussalah M, Caillet S, Salmieri S, Saucier L, Lacroix M. Antimicrobial effects of alginate-based film containing essential oils for the preservation of whole beef muscle. J Food Prot 2006;69:2364–9.

[208] Salmieri S, Lacroix M. Physicochemical properties of alginate/poly-caprolactone-based films containing essential oils. J Agric Food Chem 2006;54:10205–14.

[209] Conte A, Scrocco C, Sinigaglia M, Del Nobile MA. Innovative active packaging systems to prolong the shelf life of Mozzarella cheese. J Dairy Sci 2007;90:2126–31.

[210] Shin JW, Choi SH, Kim DE, Kim HS, Lee JH, Lee IS, Lee EY. Heterologous expression of an alginate lyase from *Streptomyces* sp. ALG-5 in *Escherichia coli* and its use for preparation of the magnetic nanoparticle-immobilized enzymes. Bioprocess Biosyst Eng 2011;201(34):113–9.

[211] Takeda H, Yoneyama F, Kawai S, Hashimoto W, Murata K. Bioethanol production from marine biomass alginate by metabolically engineered bacteria. Energy Environ Sci 2011;4:2575–81.

[212] Athanasekou CP, Romanos GE, Kordatos K, Kasselouri RV, Kakizisa NK, Sapalidisa AA. Grafting of alginates on UF/NF ceramic membranes for wastewater treatment. J Hazard Mater 2010;182:611–23.

[213] Moura MR, Guilherme MR, Campese GM, Radovanovic E, Rubira AF, Muniz EC. Porous alginate – Ca^{2+} hydrogels interpenetrated with PNIPAAm networks: interrelationship between compressive stress and pore morphology. Eur Polym J 2005;41: 2845–52.

FURTHER READING

[1] Paul DR, Newman S, editors. Polymer blendsvols. 1 and 2. New York: Academic Press; 1978.

CHAPTER

Algae-Based Polyurethane Blends and Composites 11

Fatima Zia[1], Mehdi Barikani[2], Farukh Jabeen[3], Mohammad Zuber[1], Khalid M. Zia[1], Muhammad K. Khosa[1], Awais Khatri[4]

Government College University Faisalabad, Faisalabad, Pakistan[1]; Iran Polymer & Petrochemical Institute, Tehran, Iran[2]; North Dakota State University, Fargo, ND, United States[3]; Mehran University of Engineering and Technology, Jamshoro, Pakistan[4]

11.1 INTRODUCTION

Past few years have witnessed the field of renewable polymers attracting wide interest across the scientific community due to their native biological properties, biodegradability, biorenewability, and biocompatibility. Despite the excellent durability, stability, and superior tailor-made properties of polymers derived from nonrenewable resources (petrochemicals), they are confronting serious rejection through their ubiquitous persistence in the environment, posing a threat to natural ecosystems. Hence there is a high demand for producing biobased polymers. In fact, biopolymers help to reduce the global dependence on nonrenewable polymers, enhance limited properties of synthetic polymers, and reduce the accumulation of waste. Introducing the community to commercially viable new biomimic materials will definitely find consumer acceptance in real-time applications.

Alginates (Alg) are a native anionic polymer usually extracted from brown algae. Their structural constituents are D-glucuronic acid (G) blocks, D-mannuronic acid (M) blocks, and alternate G and M (GM)-blocks. Being a gel-forming, emulsifying, and stabilizing agent, Alg is largely used in food, textile, biomedical, cosmetic, and paper industries, after its first description by Stanford (a British chemist) in 1881 [1]. Alg is readily processable into microcapsules, microspheres, sponges, hydrogels, fibers, and foams, and can be simply modified through chemical and physical reactions. In addition, it is a sustainable and abundant resource. Above all, the biocompatibility, biodegradability, and nontoxicity renders Alg hydrogels particularly useful for tissue-engineering, wound-healing, and drug-delivery applications. As these gels have structural resemblance to the ECM in tissues therefore, works to play numerous critical roles [2]. Thus incorporation of Alg in natural and synthetic polymers provides innovative properties for critical demands.

On the other hand, a special type of polymer which differs from most of the other petroleum-derived polymers is polyurethane (PU), characterized by urethane groups as repeating units. PU can be synthesized from a variety of starting monomers and offers versatile applications that are categorized into sealants, coatings, adhesives,

Algae Based Polymers, Blends, and Composites. http://dx.doi.org/10.1016/B978-0-12-812360-7.00011-2

binders, rigid thermosets, flexible foams, thermoplastic, waterborne emulsions, elastomers, insulators, elastic fiber, integral skins, etc. A huge amount of work is done on the synthesis methods, characterization, properties, and applications of PUs. In past decade both organic and inorganic materials are introduced as fillers in PU, blended with PU/grafted on PU main chain, to provide different structures and properties for high performance applications.

This chapter discusses about Alg-based polyurethane (Alg-PU) materials including those with a third component (chitosan, bentonite clay, PET, and jute fiber). Brief information of Alg and PU can be found in the coming sections as the peculiar properties of both Alg and PU are important for the fabrication of their blends, composites, grafts, etc. Both these polymers, despite belonging to different types of origin, share the similar biomedical-grade properties. Alg-PU combination can result in perfect materials for wound dressings, drug-delivery systems, and tissue-engineering applications. Though Alg-PU materials have so much to offer with their structure—property relationship, so literature reported so far on these materials is presented for future investigations.

11.2 ALGAE TO ALGINATE

There are three principal groups of marine algae classified by their colors: brown, red, and green seaweeds [3]. Brown algae contain alginic acid, fucoidan (sulfated fucose), laminarin (β-1,3 glucan), and sargassan; green algae have sulfuric acid polysaccharides (sulfated galactans, xylans); and red algae comprise agars, carrageenans, xylans, floridean starch (amylopectin such as glucan), and water-soluble sulfated galactan, in addition to porphyran (mucopolysaccharides) located in the intercellular spaces [4]. Each of them possesses the predominance of a typical polysaccharide [3]. The most prominent polyanionic polysaccharide Alg is located in the cell wall (Fig. 11.1) and in the matrix of brown algae (marine kelp), amounting up to 40% of the dry stuff [5,6], and functions to cement the cells together and giving some mechanical properties to the algae as a structural forming unit [7].

11.2.1 SOURCES AND STRUCTURAL CHEMISTRY OF ALGINATE

Though there are different species of brown algae that contain Alg, these all are not sufficiently abundant and suitable for commercial production. The most commercially exploited species of brown seaweed are: *Macrocystis pyrifera*, *Laminaria hyperborea*, and *Ascophyllum nodosum* found in Scotland, Ireland, Japan, South Africa, and South Australia [8—10]. Although seaweeds are the main source of Alg, acetylated Alg (on C-2 and/or C-3 of M units) are excreted by some telluric bacteria such as several species of the genus *Pseudomonas* and *Azotobacter*, e.g., *A. crococcum* [11,12].

Alg occur as a mixture of calcium, potassium, and sodium salts of alginic acid in the cell walls of algae. They constitute a family of linear copolymers of (1 → 4)-β-D-mannuronic acid (M) and (1 → 4)-α-L-guluronic acid (G) units [14—17]. Intrinsically, Alg do not retain ordered repeating units. Alg present two types of

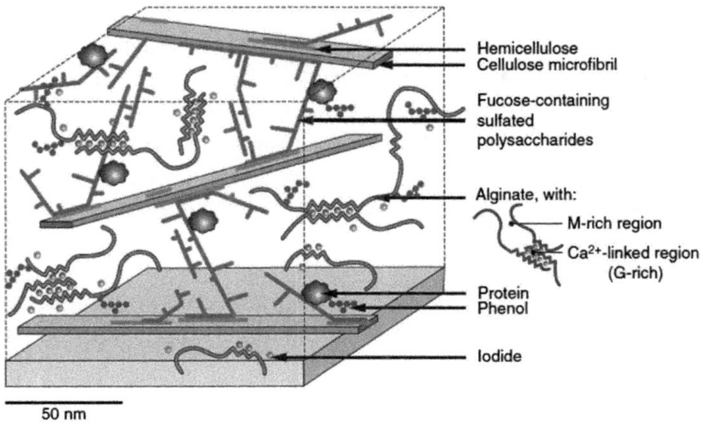

FIGURE 11.1

Cell wall model for brown algae from the order Fucales [13].

homopolymeric sequences: D-mannuronic acid (MM) blocks and L-guluronic acid (GG) blocks, and additionally, heteropolymeric sequences of M and G (MG) blocks depending on the biological source, growth, and stationary conditions [18–22] as shown in Fig. 11.2.

The X-ray diffraction analysis of different Alg (rich in M-block or G-block) suggested that saccharide units constitute the Alg chain taking up different conformations, i.e., $^4C_1$ in the case of mannuronic acid and $^1C_4$ in guluronic acid. A flat, ribbon-like chain conformation in M-blocks follows diequatorial (ee) linkages

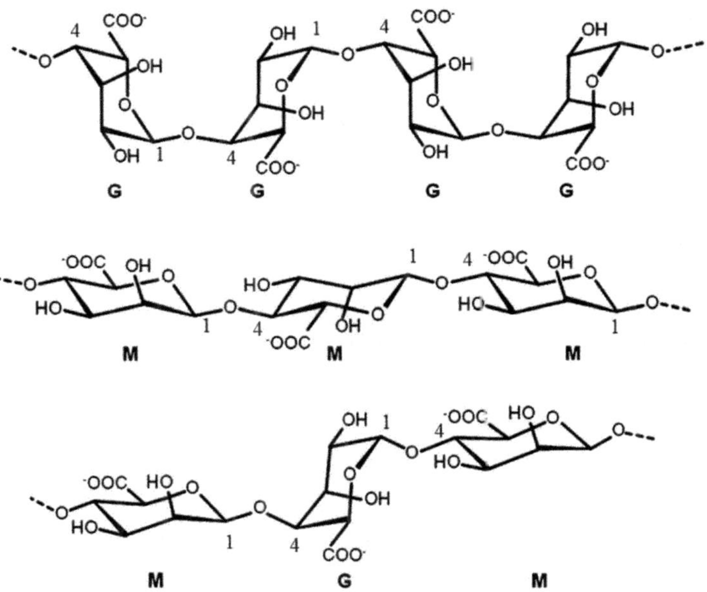

FIGURE 11.2

Types of structural block in Alg MM, GG, MG, and GM [2].

that join mannuronic acid unit, providing maximum separation between consecutive glycosidic oxygen atoms, while buckled and more rigid structure for the G-blocks arises due to diaxially (aa) linked guluronic acid units which actually result due to large hindered rotation around glycosidic bonds. MG blocks are illustrated by irregular tacticity of the bonds, i.e., axial–equatorial (ae) and equatorial–axial (ea) glycosidic bonds linking the G and M residues that account for the greater flexibility of the MG block compared to other arrangements. A decreasing trend of different sequences according to rigidity of the chain blocks is reported to be GG > MM > MG [23–26].

The functional group characterization of Alg of various molecular mass is done in different research studies. Fig. 11.3A presents FTIR spectrum of sodium Alg

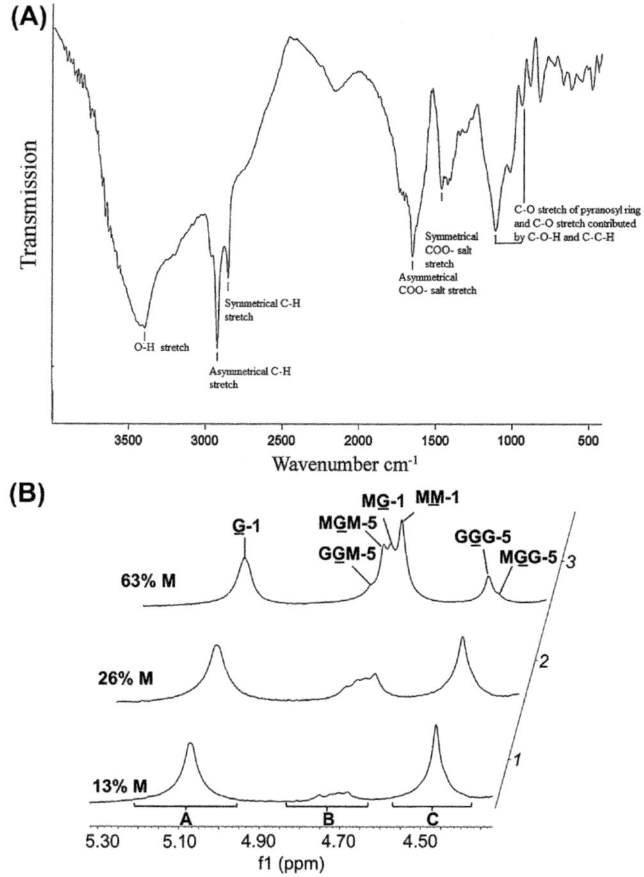

FIGURE 11.3

(A) FTIR spectra of Na-Alg [33] and (B) anomeric region in the ^1H NMR spectra of Alg containing varying M contents [34].

(Na-Alg) with M_n 12000–40000 (Aldrich). Stretching vibrations of –OH bonds of Alg appeared in the range of 3000–3600 cm^{-1}. Aliphatic asymmetric and symmetric stretching vibrations of C–H were observed at 2920 and 2851 cm^{-1}, respectively. Bands in 1649 and 1460 cm^{-1} were ascribed to respective asymmetric and symmetric stretching vibrations of COO^- salt ions. The C–O stretching vibration due to pyranosyl ring was observed at 1107 $cm^{-1}$and peak at 935 cm^{-1} were assigned to the C–O stretching with contributions from C–C–H and C–O–H deformation [27,28]. Sone et al. [29] reported the FTIR spectrum of Na-Alg where they assigned the peaks at 877 cm^{-1} and 814 cm^{-1} to β-mannuronic acid (M) and α-guluronic acid (G) units, respectively. The M- or G-block contents are calculated using peak ratios obtained from proton NMR (^{1}H-NMR) [30]. The ^{1}H-NMR spectrum showed four different dyads related to MM, GG, MG, and GM fractions [31]. Fig. 11.3B shows anomeric region in ^{1}H-NMR spectrum with varying M fraction [32].

11.2.2 PROPERTIES OF ALGINATES

Alkali metals (e.g., Na) Alg and ammonia Alg are soluble in water whereas divalent metal (Ca, Ba, Sr)-Alg, except magnesium, are water insoluble. Solubility limit of any of these Alg is not clearly defined, and the mixtures (Alg) with water convert from viscous liquids to pastes/plastic solids as the concentration of Alg increased depending on the DP. The pH of the medium plays an important part to solubilize Alg. For solubility the pH must be above a limiting value, where carboxylic acid group becomes deprotonated. Alkali metal-Alg solutions are normally polyelectrolytes and by changing the ionic strength, their properties such as conformation, viscosity, chain extension, and solubility can be modified. The ion-binding property of Alg is the basis of its gelling character. Characteristic ion-binding affinity of Alg to multivalent cations (e.g., Pb, Cu, Cd, Ba, Sr, and Ca), as shown in Fig. 11.4, depends on their composition through internal setting or diffusion methods. The pKa value of carboxylic group ranges to 3.4–4.4; therefore crosslinking is maintained in acidic conditions but the cross-linked cage (egg box model) breaks down under alkaline conditions owing to pKa values of carboxylic acid group as shown in the equation below:

$$\mathrm{COOH} + \mathrm{Ca}^{2+} + \mathrm{COOH} \xrightarrow{\mathrm{pH}\ 3-4} \mathrm{COO} - \mathrm{Ca} - \mathrm{COO} + \mathrm{H}_2$$

When pH of the Alg is less than pKa of uronic acids, the acid gels are produced. The stability of these gels is associated with intermolecular hydrogen bonds network. Addition of GDL (hydrolyzing lactone) in Na-Alg or the proton exchange in Ca-Alg gels are the two common methods to prepare acid gels. Polysaccharides often undergo acid hydrolysis under acidic conditions. Alginic acid degrades more rapidly than its salt due to C-5 carboxylic group intramolecular catalysis. Enzymatic degradation by lyase, followed by β-elimination (abstraction of proton at C-5 position) mechanism, is also possible in Alg. Another possibility is chain degradation of Alg due to phenolic groups in the presence of acids and bases, or any reducing agent.

This chemical property of Alg is important in modification step of Alg [34–41]. Although the in vitro/in vivo biocompatibility of Alg is studied extensively, discussions regarding the effect of different levels of purity and Alg composition are still needed. For instance, a study found that M-blocks were about 10 times more immunogenic, bringing cytokine production more potently than high G-block Alg [42], but others reported little or no immune response around Alg implants. Similarly, no major inflammatory behavior of Alg gel was observed when injected in mice [43,44].

11.2.3 EXTRACTION AND BIOSYNTHESIS OF ALGINATE

Due to the abundance of Alg in water bodies, plenty of it is available in nature. Annual industrial production of Alg is approximately 30,000 metric tons and is estimated to comprise less than 10% of biosynthesized Alg material. It is reported that by 2019 the market for Alg and its derivatives is projected to reach a value of $409.2 million, growing at a CAGR of 3.8% from 2014 to 2019 [15,45].

Biosynthetic pathway to get Alg (Fig. 11.4) includes oxidation of a carbon source to acetyl-CoA, its conversion to fructose-6-phosphate via gluconeogenesis by entering into TCA cycle. A series of biosynthetic transformation converted fructose-6-phosphate into GDP-mannuronic acid, which appears as a precursor to Alg synthesis. Overall, the biosynthetic process can be divided into four phases:

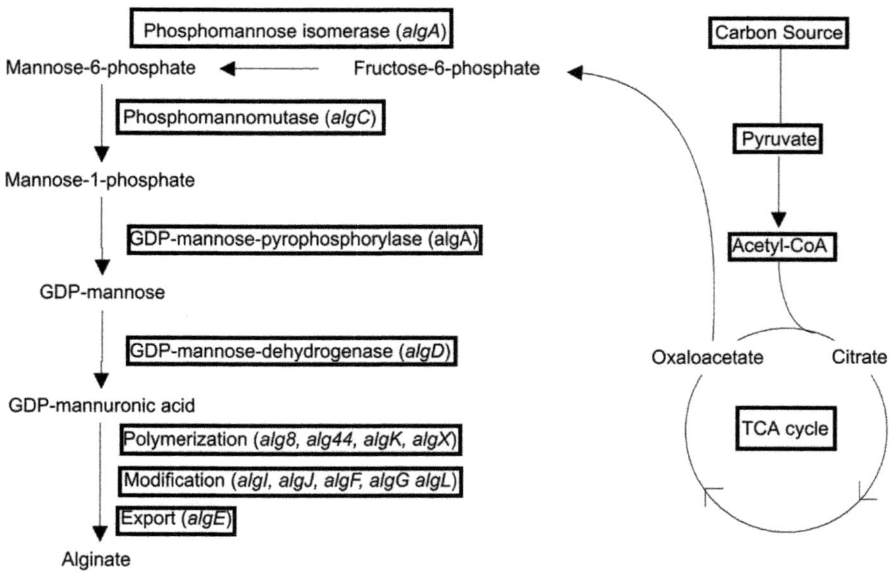

FIGURE 11.4

Bacterial alginate biosynthesis [12].

1. Preparation of GDP-mannuronic acid precursor
2. Cytoplasmic membrane transfer and polymerization to polymannuronic acid
3. Periplasmic transfer and post-polymerization modification
4. Export through the outer membrane

In the third stage, polymannuronic acid at the O-2 and/or O-3 positions are acetylated by numerous transacetylases under post-polymerization modification [12,46–48]. A family of epimerase enzymes performs epimerization to convert some non-acetylated M residues to G residues [47,49–51]. Lastly, Alg is released from the cell through transmembrane porins. Fig. 11.4 presents bacterial biosynthesis process of Alg production.

The extraction process of Alg is usually defined by separation and purification, facilitated by the capacity of Alg to vary between aqueous phase and gelation through cation-exchange reactions. The collected algae conventionally pass through the processes such as washing, drying, and milling. However, the wet algae processes are preferred due to the cost of drying. Washed algae are stabilized, followed by treatment with formalin (formaldehyde), mineral acids (HCl), and preservatives. The insoluble Ca/Mg-Alg present in cell wall converted into soluble Na-Alg ultimately permit extraction from seaweed residues. Thus the formalin-pretreated algae are converted into insoluble alginic acid by an acid pre-extraction treatment and then into Na-Alg. This way is more effective than the direct conversion of the original Ca-Alg. Acid treatment usually removes soluble algae constituents, e.g., amino acids, fucoidan, laminarin, and oligo-/monosaccharides leaving compact seaweed residues containing Alg, cellulose, pigments, and proteins. Hot alkaline solution treatment (sodium carbonate) results in the conversion of Alg into soluble Na-Alg. The Alg and pigments residue in liquid phase can be separated by filtration, centrifugation, or flotation steps from proteins and cellulose [52–54]. Precipitation of Na-Alg with ethanol, mineral acid, or $CaCl_2$, etc., converts highly diluted Na-Alg into insoluble alginic acid (mineral acid) and Ca-Alg ($CaCl_2$), and recovers Alg in a considerably smaller volume [34,55]. The product then undergoes drying and milling. Expensive equipment and long processing time makes this method unsuitable for large-scale production. Chemical method using Ba-Alg gels (as Ba has high affinity for Alg than Ca ions) involves disintegration under alkaline pH, followed by chemical treatment, to elute contaminants, and ethanol extraction results in pure Alg beads [56]. Fig. 11.5 shows the $CaCl_2$, mineral acid, and alcohol–derived extraction steps of Alg from seaweed.

11.2.4 ADVANCEMENT IN APPLICATIONS OF ALGINATE-BASED MATERIALS

The ability of Alg to emulsify and stabilize, to hold water, and form cross-linked gels led to their miscellaneous applications. However, it is observed that physicochemical and mechanical characteristics of Ca-Alg gels are influenced by the composition and the fabrication routes. The water-holding property of gels is employed to maintain smooth and soft texture in bakery and frozen food items, it firms the body of

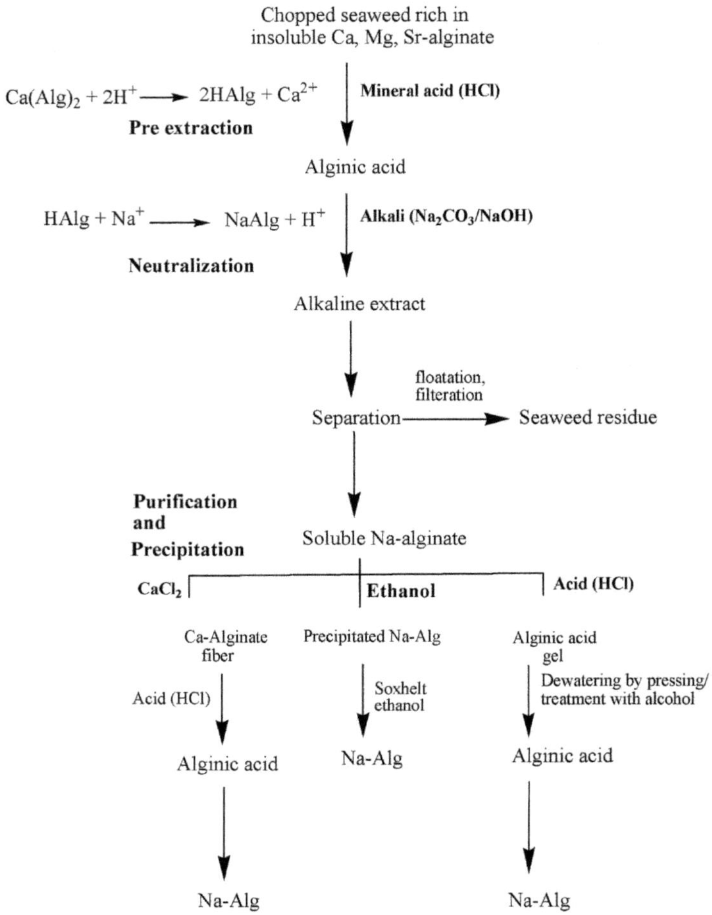

FIGURE 11.5

Scheme of extraction steps from algae to alginate [6,57].

puddings and pie fillings due to their gelling properties; used as emulsifier and stabilizer in salad dressing and sources, topping, and juices; in latex creaming and thickening of rubber products, surface sizing, coating material for paper industry, textile spinning, weaving, and printing, wound dressings such as adhesive strip, bandages, surgical sponges, theater curtains, pads of various kinds tampons, etc.; pharmaceutical and cosmetic uses include dental impression materials, tablet binder, tablet disintegrating agent, active compound in antibody formation, toothpaste, hand cream/lotions, shaving cream, liquid shampoos, etc. [12,58–62]. To control the poor mechanical stability in terms of strength, stability, and degradation due to water absorption and less cell adhesion/bioactivity, modification in Alg is significantly controlled by blending it with different organic and inorganic materials. The potential Alg modification in the last 3 years involves blending, grafting, or compositing it with a variety of organic/inorganic polymers or metal particles, reflecting the wide acceptance of this natural polymer for different industrial sectors, e.g., biomedicine, tissue engineering, food, membrane technology, and wastewater treatment. The Table 11.1 summarizes the applications of Alg-based materials (except Alg-based polyurethanes).

Table 11.1 List of Applications of Some Recent Alginate-Based Materials in Major Industrial Fields

Sr. No	Alginate-Based Materials	Field of Application	References
1	Alginate (Alg)-polyvinyl alcohol (PVA)	Wound-dressing material	[63]
2	Alg-chitosan	Potentially used for oral insulin delivery	[64]
3	Alg-chitosan	Drug-delivery applications	[65]
4	Alg-chitosan	Novel fiber for wound care applications	[66]
5	Alg-chitosan/titanium	Useful for tissue-engineering applications	[67]
6	Alg-collagen I	For wound healing	[68]
7	Na-Alg/polyacrylamide	For drug-delivery systems	[69]
8	Alg-polyacrylamide	Used as biomaterial for cartilage tissue	[70]
9	Alg-graft-poly(*N*,*N*′-dimethylacrylamide)	Biocompatible material useful for biomedical applications	[71]
10	Alg-polycaprolactone	Biomedical applications	[72]
11	Alg-lignin	Tissue-engineering scaffolds	[73]
12	Alg-halloysite	For bioprocessing and tissue engineering	[74]
13	Alg-halloysite	Scaffolds for tissue engineering	[75]
14	Methacrylated Alg-PEG	Bioadhesives for clinical purposes	[76]
15	Alg-sulfate/PVA	For enzyme immobilization	[77]
16	Alg-PEG	For encapsulation/delivery of bioactive agents	[78]
17	Mg-Al layered double hydroxide-Alg/PVA	Used for water remediation	[79]
18	Alg-PVA	For removal of phosphate	[80]
19	Na-Alg/PVA	Desalination/nanofiltration	[81]
20	Polyhydroxybutyrate—Ca-Alg/carbon nanotubes	Nanofibrous membrane for filtration	[82]
21	Alg-maghemite PVA	Radioactive wastewater treatment	[83]

Continued

Table 11.1 List of Applications of Some Recent Alginate-Based Materials in Major Industrial Fields—cont'd

Sr. No	Alginate-Based Materials	Field of Application	References
22	Alg-metal coordinated alginic acid	Defluoridation process application	[84]
23	Alg/zirconium	Useful for defluoridation of water	[85]
24	Alg-chitosan-cellulose nanocrystals	Emulsion stabilizer for polymer production	[86]
25	Alg-chitosan	Pickering emulsion	[87]
26	Alg/Alg-resistant starch	Controlled release carrier for food grade nisin	[88]
27	Alg-silver nanoparticles	Antimicrobial textile coatings	[89]
28	Alg-superadsorbent polymer	For textile effluents recycling	[90]
29	Cellulose-Alg	Food packaging applications	[91]
30	Agar-Alg-collagen	Bioactive food packaging films	[92]
31	Bentonite-Alg	For microbial fertilizer encapsulation	[93]

11.3 POLYURETHANES—A COMMERCIAL GRADE CLASS OF POLYMER

The last 79 years witnessed the progress and achievements of polyurethanes after its first discovery in 1937 by Otto Bayer. Polyurethane (PU) products were introduced in late 1940 and promptly established as quality-grade foam and elastomeric material [94]. The worldwide PU demand reached to 17 billion pounds in 2010 and is about 5% of total plastic consumption [95]. The global PU market was estimated to reach \$45 billion in the year 2014 and will be growing with CAGR of 6.3% up to 2021 [96]. PU is a significantly important class of specialty polymers as it possesses tunable physicochemical and thermomechanical properties with a wide array of commercial applications. A lot of research work has been done and reported so far to explore this polymer further. Waterborne PU [97], nonisocyanate PU [98], cross-linked PU [99], biopolymer (oils, polysaccharides, proteins, etc.)-based PU [100–102], antimicrobial PU [103] are the important achievements in PU chemistry. PUs have revolutionized the biomedical and tissue-engineering fields with their smart biologically active properties for the last one decade [104,105]. In this section we are presenting a short view of chemistry, properties, and applications of PUs.

11.3.1 SYNTHESIS AND STRUCTURAL CHEMISTRY OF POLYURETHANE

PUs include the polymers having multiple urethane moieties in the polymer backbone, irrespective of the rest of the chemical composition of chain. We usually find aliphatic/aromatic hydrocarbon, amides, esters, ethers, urea, biurets, isocyanate, and allophanates groups in addition to characteristic urethane groups [95,106]. The structural chemistry involves the reaction of isocyanates with polyols and diol/diamine resulting in the formation of respective urethane and urea linkages (Fig. 11.6). The carbamic acid which is an unstable product of reaction between polyurea and water decomposes to release free amine which further reacts with free-isocyanates to form various cross-linked products (biurets). The urethane moieties are linked to the “Soft” (S) and “Hard” (H) segments via covalent or hydrogen bonds. Hydrogen bonding results in self-assembled arrays of incalculable stacks, arranged both in parallel and antiparallel way [95,107].

The usual controlled route to synthesize polyurethane is step-growth polymerization (Fig. 11.7), where excess diisocyanate reacts with polyols (2/1) to form a prepolymer with isocyanate group at terminal points. In the second step, low molecular weight chain extenders (diol, diamine) react with the prepolymer to give higher molecular weight PUs. Using diamines/ureas as chain extenders instead of diols respective poly(urethane-urea)/poly(urethane-biuret) are obtained [108–113].

PU, as segmented polymer with their characteristic S- and H-blocks (Fig. 11.6), actually imparts flexibility and strength to the desired product. Soft segments of PU are usually made up of diols of long chain polyethers, polyesters, polycarbonates, polysiloxane (referred as macrodiol) which are responsible to provide flexibility, while hard segments arise due to isocyanate and chain extender (cross-linker). Usually higher contents of soft segments enhance the flexibility, while the hydrophilicity is associated with the nature of macrodiol [112,114–117]. By just varying the choice and amount of starting material an immense range of properties, e.g., hardness, softness, chemomechanical, hydrophilicity/phobicity, can be achieved [94,118,119]. PUs can thus acquire the chemical structure of elastomers, foams, thermosets, and thermoplastics [94,95].

Phase separation occurs in most of the PUs due to incompatibility between S and H segments, that differ in polarity and chemical nature. The degree of phase separation depends upon the nature of isocyanate, polyol, and chain extender in addition to molecular weight of macrodiol, reaction conditions, and extent of hydrogen bonding

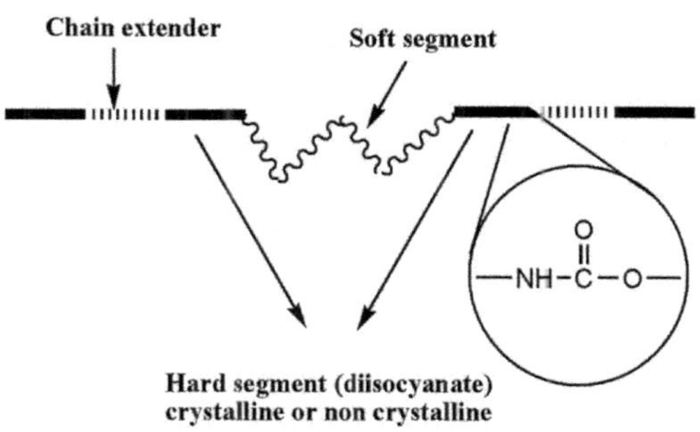

FIGURE 11.6

Structure of typical segmented polyurethane [107].

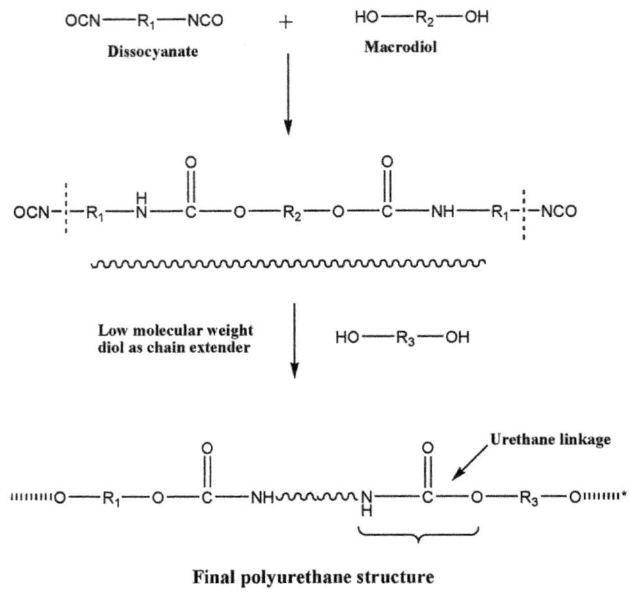

FIGURE 11.7

Typical step-growth polymerization route for PU synthesis.

between urethane groups. Generally, H segments can be organized in semicrystalline/amorphous domains, whereas S segment makes an elastic/amorphous matrix for dispersion of H segments [120,121]. Numerous characterization techniques, e.g., FTIR, SEM, SAXS, IR-dichrosim, neutron scattering, DMA, and DSC are employed to understand this phase separation behavior. Micro- and nanoscale investigation employed TEM, AFM, whereas micro-Raman spectroscopy investigated the complex phase segregated structure of segmented PU [122–131]. Understanding of crystallization, nano/micromorphology, and phase separation behavior provides a correlation between physics, chemistry, and engineering aspects of materials.

11.3.2 MISCELLANEOUS PROPERTIES OF POLYURETHANE

PU properties are majorly discussed with secondary bonding, for example, hydrogen/van der Waals interactions (Fig. 11.8). The partial positive charge of the hydrogen atom in the urethane group can attract the oxygen atom of various groups. Thus three types of hydrogen bonding can take place as free oxygen electron pair can come from ether groups (ether–urethane interaction), ester groups (ester–urethane interaction), or urethane groups (urethane–urethane interaction).

Contents of H segments significantly affect the properties of segmented PU. Thermoplastic PU behaves as brittle and hard high module TPU when hard segment contents increased from 30%–50% to 60%–70% [120,132–135]. Depending upon H segment ratio, type of synthesis technology and chemistry, the Young's modulus of PU varies in the range between 2 and 2000 MPa while the density lies between 0.2

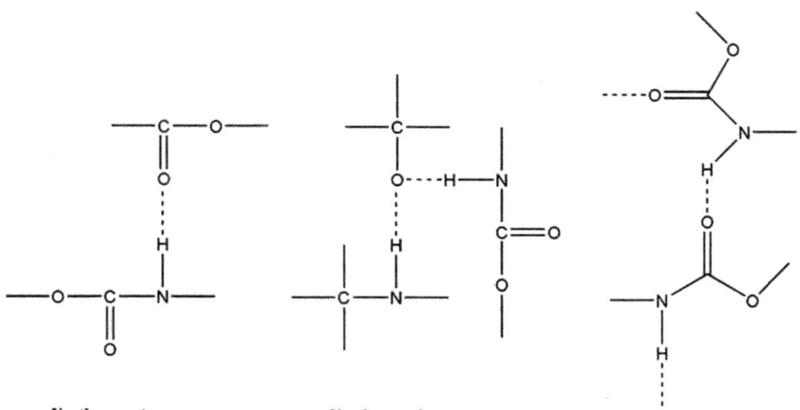

FIGURE 11.8

Hydrogen bonding interaction in polyurethanes [107].

and 1.2 g/cm^3. Tensile strength is very high 25–70 MPa and sensitive to molecular weight of polyol, degree of cross-linking and their symmetry.

The mechanical properties vary with the type of isocyanate, as aromatic rings and methylene units in isocyanate remarkably improve the tensile strength. TDI-PU when compared with MDI-PU showed more reactivity toward isocyanate and hence has high strength and lower strain values. The aliphatic isocyanate has lower reactivity [106,136,137]. Up to 50% diol (chain extender) contents increase the tensile strength [138]. Thermal stability of PU is determined by the type of isocyanate and hydrogen bonding [106]. Cross-linking often leads to high-temperature stability and microbial resistance. The highest suggested temperature for using Thermoplastic PU (TPU) is about 104°C (220°F), making it unsuitable for most of the high-temperature uses. PU is resistant to abrasion and crack propagation. Microbial degradation resistance of PU varies with their types, and primarily observed on free-isocyanate, biurettes, and urethane moieties. Higher rate of hydrolysis in polyester PU makes it more vulnerable to degrade than polyether PU [106,139,140]. Though TPU is affected by fuels, steam, esters, ketones, and strong acids/bases, but it is resilient to aliphatic hydrocarbons and dilute acids/bases [95].

11.3.3 APPLICATIONS AND ADVANCEMENTS OF POLYURETHANES

PUs are versatile and famous for their immensely varying mechanical and biological characteristics. The properties such as toughness, processability, durability, flexibility, surface functionality, biostability, and biocompatibility make them good candidates for medical grade usages. The incredible applications of PU are divided into different categories for convenience. The commercial array of PU applications include flexible slab (bedding, automotive, carpet underlayment, carrier media, furniture), rigid foams (appliances, insulation boards, automotive), flexible molded foams (automotive seating, bedding) reaction injection molding (mechanical, automotive), carpet backing (unitary, attached cushion), solid elastomers (adhesives, elastomers, medical), and two-component formulations (coatings, sealants, encapsulation) [95,141–148], but over the past 40 years PUs have preferably been used in biomedical devices (grafts, catheters, artificial arteries, heart valves, sutures, bandages, etc.) because of their abovementioned properties and segmented structure [149–153]. The most recent advances in PU chemistry involve the substitution of fossil-derived isocyanates and polyols with biobased molecules, because with the passage of time concerns over the use of environmentally benign feedstock have deepened. Another approach regarding this is the introduction of waterborne polyurethane (WPU), and nonisocyanate polyurethane (NI-PU). WPU are fully reacted water dispersible urethane polymers, which display numerous exceptional qualities than usual organic solvent–based PU such as nontoxicity, nonflammablility, and biodegradability and are thus employed to leather finishing, floor coverings, textile finishing, pressure sensitive adhesives, and adhesives [154–158]. NI-PU is a new arrival in PU materials made by reaction of cyclocarbonates and amines with no use of toxic isocyanates, and provide much improvement in water absorption, porosity, and thermochemical resistance compared to conventional ones. Potential applications of NI-PU consist of chemical-resistant coating, foam, sealants, etc. However, NI-PU has not attained commercial application so far. Methods for manufacturing of polymers or oligomers comprise cyclocarbonate groups in addition to reviewing their uses in thermoplastic and thermosetting systems is a subject of research nowadays [159,160]. Yet, it is worth mentioning that biobased and petroleum-based PUs do not ensure vital demands for high-temperature engineering appliances and in most commercial uses they need to be modified [161,162].

The more economical and ecofriendly way to modify PU materials is their mixing/blending/filling with biomass (biopolymers especially) which ensures to tackle the issues such as dominant consumption, limited petrochemical sources, high cost, as well as the concerns over climate change. Natural polymer–based PU materials provide improved properties, biodegradation, and lower the toxic waste accumulation in environment. There is a good miscibility or dispersion between PU and biopolymers in their blends/composites, owing to hydrogen bonding interactions between urethane groups and hydroxyl, carboxylic, and amine groups of various biopolymers [163–167].

11.4 ALGINATE-BASED POLYURETHANE MATERIALS

Alg polysaccharide is readily processable as hydrogels, microcapsules, microspheres, sponges, foams, and fibers, and can be easily modified via chemical and physical reactions to obtain different structures, properties, functions, and applications. Alg, therefore, can be prospectively blended/composited with natural and synthetic polymers for a wide array of improved functional properties and applications, e.g., for tissue engineering and enzyme entrapment/immobilization, as delivery vehicles for cells and drugs, as wound-dressing material in biomedicine, as metal chelators for wastewater treatment/purification, and as a food additive [168–170] (see Table 11.1).

Recently published papers have presented an overview of the advantageous impact of natural polymers, e.g., Alg [170], heparin [171], glucomannan [172], collagen [173], starch [174], and chitin/chitosan [175] on thermomechanical properties, biocompatibility, and biodegradation of polyurethanes. This section mainly discusses the synthesis routes, and structural and functional (in terms of applications) features of Alg-based polyurethane materials that will basically provide a way to develop novel strategies to further understand the behavior of the described polymers.

11.4.1 SYNTHESIS STRATEGY FOR ALG-PU BLENDS/COMPOSITES MATERIALS

The most favorable technique to synthesize Alg-based blends is the cross-linking of Alg with desired material. PU and Alg are found to be two incompatible polymers with different Tg, thus new methods to fabricate compatible compositions of these two polymers can be a fascinating attempt. Up-to-date reported literature demonstrates that aqueous polyurethane/waterborne polyurethane dispersions and foams can be hybridized [176], blended/composited [177–179], freeze-dried [180], and form semi-IPNs [29] with Alg followed by reaction with specific cross-linkers (e.g., $CaCl_2$) into final scaffolds, films, beads, porous adsorbent or elastomeric composite materials. However, in few publications, Alg is also used as a polyol for polyurethane synthesis to modify its properties [181,182], as in Fig. 11.9.

11.4.2 ALG-PU-BASED BLEND OR HYBRID FILMS/COATING MATERIALS

Native Alg has more than two −OH function groups in their main chain and is used as polyol for synthesis of Alg-PU hybrid materials. Particularly, if Na-Alg is used as polyol, the biocompatibility and hydrophilicity is considerably improved [183–187]. Kwon et al. [181] used Na-Alg as polyol for synthesis of polyurethane foam (PuF) with the aim of improving flexibility and hydrophilicity of foams. The PuF was prepared using fixed weight of polyols, i.e., PEG, glycerin, BDO, and Na-Alg and aliphatic diisocyanate (HDI) by following the previously reported procedure [188–190]. The possible reactions in the process of PuF synthesis are shown in Fig. 11.10. The presence of four −OH function groups on Alg provide the reaction sites for isocyanates (Fig. 11.11).

The urethane and urea peaks in the FTIR spectrum of Na-Alg/PuF confirmed that Na-Alg could be used as polyol in PU foams preparation. Kim et al. [182] studied the thermal and kinetic behavior of Alg-PU hybrid material using varying mole ratios i.e., 1/1, 1/0.50 of alginic acid (as polyol) and aromatic diisocyanate (TDI). The characteristic peaks in FTIR and solid-state NMR curves were used to confirm

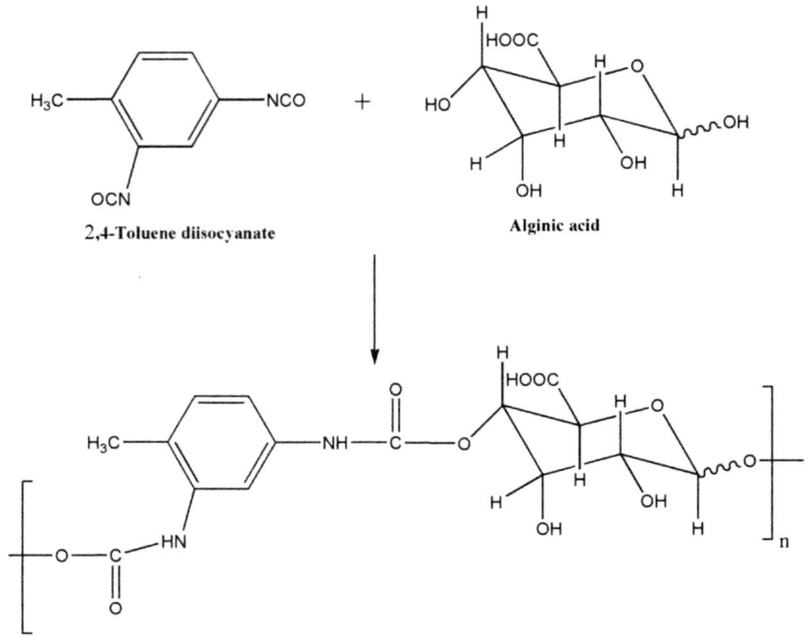

FIGURE 11.9

Reaction scheme for the synthesis of Alg-modified polyurethane [182].

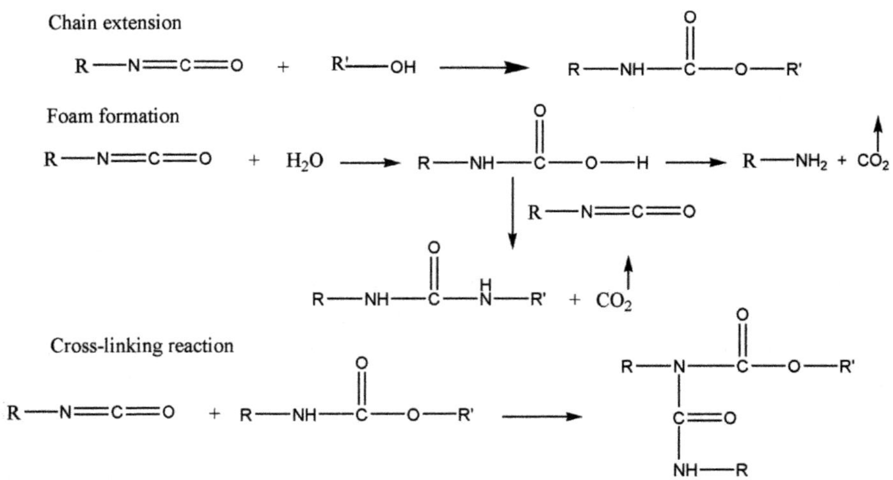

FIGURE 11.10

Possible reactions in PU foam formation [181].

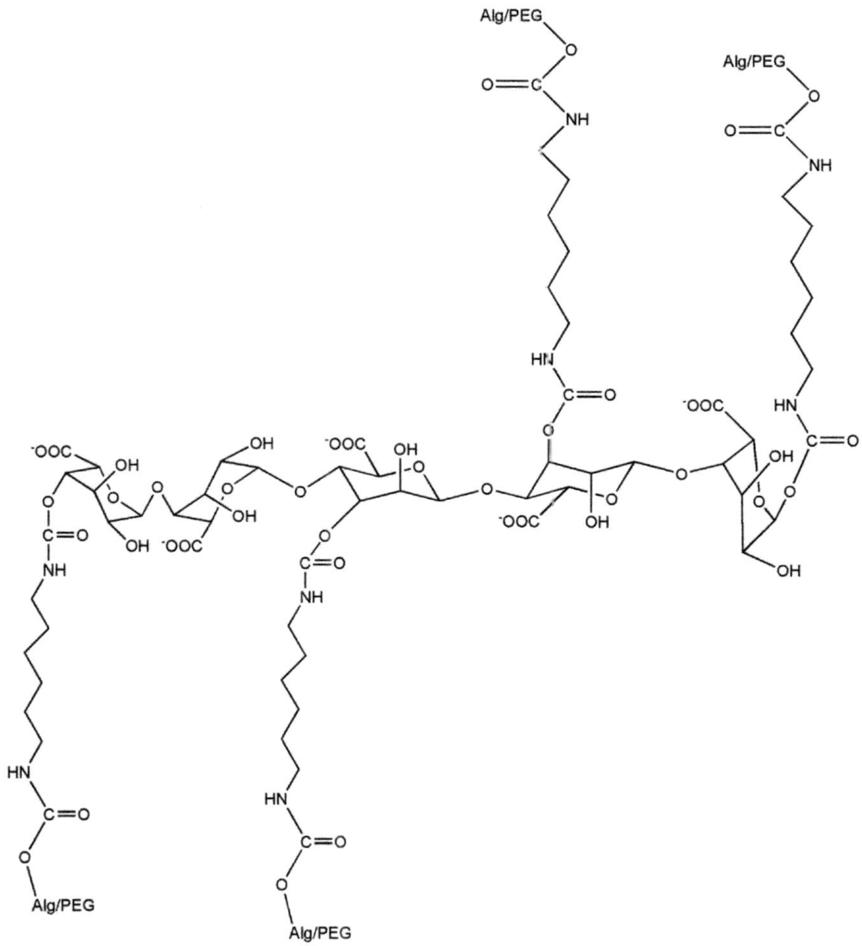

FIGURE 11.11

Possible reaction between isocyanate and alginate (Alg) [176].

the formation of Alg-PU hybrid materials. The SEM graphs showed a smooth surface morphology of the hybrids. The higher value of exothermic peak in DSC curve for Alg-PU hybrid as compared to neat alginic acid speculated the higher thermal stability. It was also noted that higher contents of Alg improved the thermal stability, as Alg-PU hybrid with 1 mol of Alg has higher value of exothermic peak temperature. On the other hand thermogravimetric analysis (TGA) curves showed that higher amount of PU to Alg increased the thermal stability of hybrids because PU is thermally more stable than Alg. Kinetic investigation was done that fit well with TGA data, where the whole degradation activity had been measured as three successive first-order reactions. In Alg-PU sample with higher PU contents, the highest activation energy value, i.e., 69.25 kJ/mol, was examined. Thus by adjusting the mole ratio one can manipulate the thermal stability, consequently developing the tunable Alg-PU hybrid materials.

In 2002 Fan and coworkers [177] successfully prepared Na-Alg and waterborne polyurethane (Na-Alg/WPU) blend films. Techniques such as FTIR, XRD, SEM, UV-spectra, TGA, and DTA were used to elucidate the structure and properties of blend films. A strong interaction and good miscibility or compatibility was observed between Na-Alg and WPU molecules. The breaking elongation and water resistivity of blend films showed higher values as compared to those of the neat Na-Alg film. However, the proper application of these materials was observed in 2007 when Yun and colleagues [178] used Na-Alg/WPU-Urea dispersion blends to improve the water vapor permeability (WVP) of coating materials using prepolymer-mixing process. It was found that stable Na-Alg/WPU-Urea blend film can be obtained below 30 wt% of Alg. The WVP quality of coated nylon fabric was improved due to an increase in Na-Alg amount that actually enhances the number and density of total micropores, i.e., tunnel-like/isolated micropores, formed after dissolution in water. These results obviously proved Na-Alg/WPU-Urea blends as coating materials.

11.4.3 ALG-PU-BASED CROSS-LINKED HYDROGEL MATERIALS

Hydrogels are produced by physicochemical cross-linking of certain polymers resulting in 3D hydrophilic expandable structure matrices. Alg forms hydrogels through chelating/selective binding/cross-linking between multivalent ions, e.g., Cu, Pb, Cd, Ca, Ba, Sr, Co, Ni, Zn, Mn, and carboxylic (−COOH) groups, absorbing huge amount of water/biological fluid [191−196]. Gel fabrication or formulation is often associated with G-block residue in Alg as it participates in the formation of tight junctions (egg box model) with cations via binding or cross-linking methods (diffusion and internal setting) as in Fig. 11.12. Thus Alg with high G-block contents resulted in stronger and brittle gels whereas higher contents of M-block yield softer and more elastic gels [35,193]. These hydrogels found tremendous applications in food, drug-delivery systems, and tissue engineering due to their biocompatibility and bioavailability, where at high pH (neutral or basic) the cross-linking between Alg and cation is broken due to pKa value of carboxylic functional group that leads to bursting of the hydrogel followed by drug release [40,197,198]. However, less stability and swelling are the two main issues often associated with Alg hydrogels. Functional groups grafting copolymerization and blending are the two modification ways to overcome the above issues currently, but the loss of chemical stability arises with the change in temperature, pH, and swelling degree owing to absence of covalent bonding with modifying agents [198−203].

Native and manmade (synthetic) hydrophilic polymers can be physically or chemically cross-linked with the aim of producing hydrogels [205]. Polyurethane on account of its crystallinity has great strength and anti-swelling properties. Grafting of PU on Alg improves the strength in terms of stability and swelling properties by increasing the cross-linking points [206]. Previous studies also confirmed that

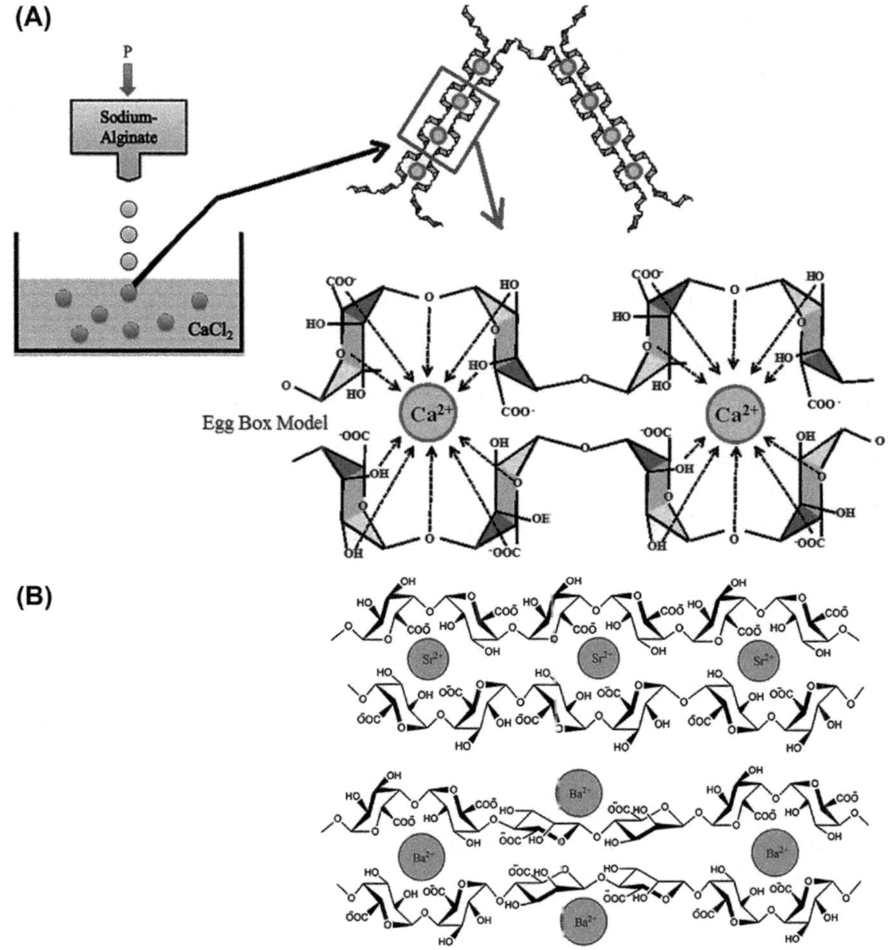

FIGURE 11.12

(A) Gelation process of alginate using $CaCl_2$ solution to fcrm a nonhomogenous microcapsule in an egg formation with Ca^{2+}, (B) binding of Sr^{2+} to G—G blocks solely, and Ba^{2+} to G—G and M—M blocks [204].

Alg acts as a thickener for aqueous polymer, e.g., anionic PU-based system [207]. This effect is linked to the formation of colloid structural hydrophilic covers that govern the viscous flow of dispersions reported by Travinskaya and Savelyev in 2006 [208]. They prepared anionic polyurethane water dispersion (APU-WD) gel compositions based on Na-Alg by cross-linking with Ca^{+2} ions using $CaCl_2$. First the APU-WD dispersions were prepared using oligooxytetramethylene glycol and HMDI in 1:2 ratios with dianhydride of pyromellitic acid and dihydrazide of dicarbonic acid in acetone solution and then carboxylic groups were transferred to a salt form followed by dispersion in water. Mixing of APU and Alg solutions was done with different composition ratios of APU-WD:Na-Alg (wt%), i.e., 95:5, 90:10, 85:15, 80:20, 50:50. Afterward, ionic cross-linking between Alg and PU chains was developed by treatment with $CaCl_2$ solution, providing the formation of a 3D-network arrangement with joined molecules in the network sections [209,210]. APU-WD were attached between cross-linked Na-Alg chains. Among the samples, APU-WD:Na-Alg with a 50:50 ratio, showed continuous increase in swelling behavior of gel with increase in pH due to electrostatic repulsion created by ionization of $-COOH$ groups into $-COO^-$ ions (Fig. 11.13).

By optimizing the degree of cross-linking, varying the composition ratio, and Ca^{2+} amount, systems with controlled thermal and pH-sensitivity, swelling ratio, and strength indexes can be obtained. An increase in gel cross-linking is observed with higher contents of Alg to anionic PU dispersions. Importantly, Alg contents increased the tensile strength of the material films due to increased network density and GG-blocks residue [211,212], on the other hand, thermal stability of the dispersions increased by the incorporation of APU.

Mixtures of APU-WD and Na-Alg formed structural non-Newtonian stable systems with higher viscosity in comparison with initial components in the absence of divalent cations. These gel compositions attained states of equilibrium swelling and de-swelling in 60 min and are prospectively useful for food and biomedical industries [208].

Oh and coworkers [212] combined the PU foam with Alg-bentonite hydrogel for sustainable drug release. Bentonite, a natural inorganic polymer, consisted of three layers of clay minerals. It is an inexpensive raw material that fulfills the key requirements of polymer-supported hydrogels. Bentonite generally has adsorption properties and become electrically charged when immersed in water. This effective adsorption property of bentonite reduced the protein drug release rate [213,214].

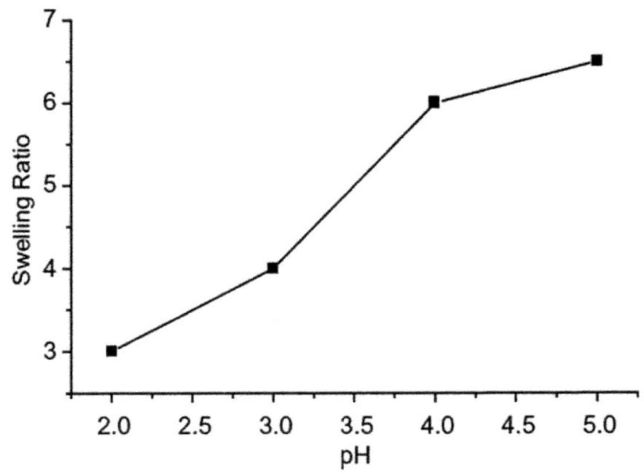

FIGURE 11.13

pH-dependent swelling behavior of APU-WD:Na-Alg (50:50) at 20°C [208].

HDI was added in a blend of polyols, i.e., glycerin, PEG, BDO, and Na-Alg to get hydrophilic PU foam (Fig. 11.14). Alg and bentonite blend hydrogel compositions were prepared with weighted percentage of bovine serum albumin followed by a hydrogel solution coating on PU foam under vacuum to obtain combined PU and Alg-bentonite hydrogel dressings. The insertion of hydrogel into the foam was done by lyophilization, bead mixing, and vacuum methods. However, vacuum was the most successful method among others for the preparation of hydrogel containing foam systems.

The characterization of the structure and properties of hydrogel containing PU foam was done via FTIR, SEM, mechanical testing, protein release, and cytotoxicity

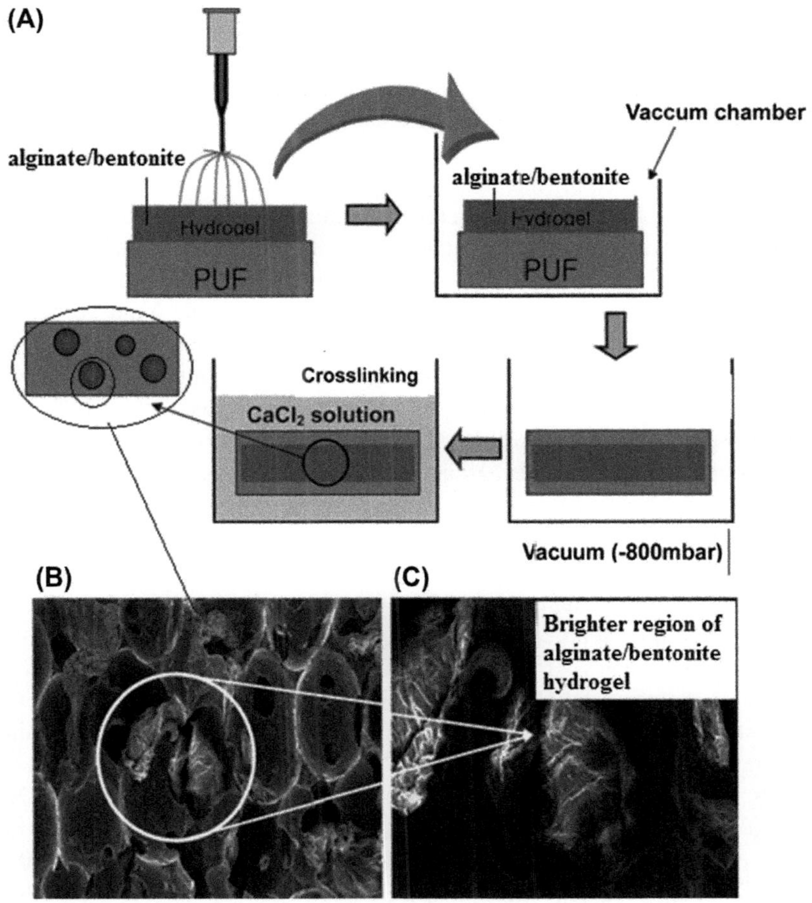

FIGURE 11.14

Preparation method of PU foam combined with Alg/bentonite hydrogel. Scanning electron micrographs of cross-sectional images. (A) PU foam combined with Alg/bentonite hydrogel (35×), and (B) PU foam combined with Alg/bentonite hydrogel (100×) [212].

tests. The mechanical properties of hydrogel containing PU foams were decreased because the soft chains become weak when PU foam absorbed the water from the hydrogel (a polymer comprising about 80%–90% water). That is why the mechanical properties of PU foam combined with hydrogel were decreased; however, this decrease in properties is not considered practically important due to good tearing resistance of material. It is also noted that increasing concentration of bentonite lowers the strength of Alg hydrogel, but does not affect the associated PU foam. The SEM results revealed that an even penetration of gel in the foam increases the wound-healing ability of dressings. Alg hydrogels are pH-sensitive, which allows it to be an effective drug-delivery carrier, while bentonite with negatively charged surface attracts the positively charged lysine present in BSA protein molecules and thus the protein drugs get absorbed over bentonite which slows down the release rate. The cytotoxicity tests using fibroblast (NIH3T3) cells confirmed the material as nontoxic and biocompatible [212].

Oh and coworkers [176] described $CaCl_2$ as the most appropriate cross-linking agent for Alg hydrogels with maximum 1 wt% concentration. Morphology of jute fiber-Alg hydrogel showed good interfacial adhesion due to their hydrophilic nature that ultimately improved the mechanical properties of hydrogel. The breaking stress and breaking strain of PU foams increased continuously with Alg contents. This result illustrates that higher Alg contents in PU foam enhanced the stress, strain, and fracture properties of foam by twofold. Kwon et al. [215] reported that higher Alg contents also increased the morphological stability of PU foam because pore size and shape changed from a deformed and elongated one to a spherical one [215]. Alg acts as a cross-linker among long PU chains owing to the side/end OH groups; however, when the Alg contents reached over 30 wt%, the PU foam was not fabricated and resultantly collapsed. The water and cell culture medium absorbance tests confirmed the higher absorbance of PU foams along with even penetration in PU foam surface (reported from SEM images) which is an important requirement for wound dressing. Thus PU foam hybrid with Alg-jute fiber hydrogel as a smart wound dressing with superior absorption and mechanical properties can be considered as a potential candidate for wound dressings [176].

Wang et al. [206] prepared calcium Alg (Ca-Alg) gel microspheres followed by grafting of polyurethane via free-radical graft polymerization technique (Fig. 11.15). PU was prepared using HEMA and IPDI while Ca-Alg hydrogel beads were prepared by dropping Alg in $CaCl_2$ cross-linking agent. The graft polymerization of the two reactants in the presence of AIBN initiator results in PU-grafted Ca-Alg hydrogels. Structural and morphological studies were done using FTIR, ^{13}CNMR, SEM, and XRD. TGA was done to study the thermal stability of gels. Grafting of PU in Ca-Alg formed highly ordered crystal region due to increase in intermolecular hydrogen bonding. The smooth and dense morphology of the PU-g-Ca-Alg was credited to PU side chains that formed crystalline area and increased the intermolecular attractions.

The better thermal stability of PU segments [216] increased the thermal decomposition rate of PU-g-Ca-Alg in Ca-Alg gels. Actually the amino and carbonyl

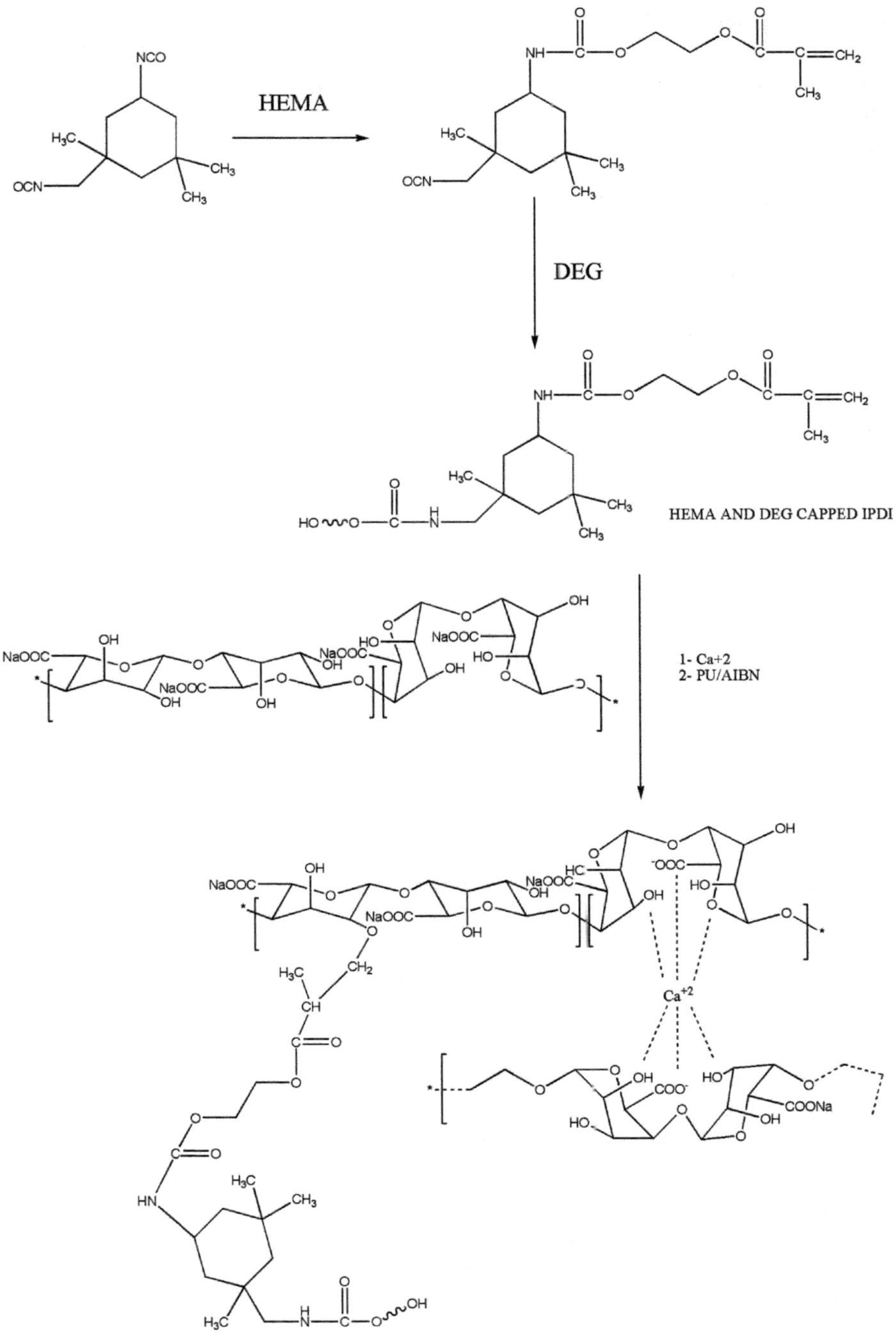

FIGURE 11.15

Procedure for synthesis of PU-g-Ca-Alg [206].

functional groups of PU intensified the intermolecular hydrogen bonding. The swelling property of hydrogel was determined as a function of temperature. The hydrophobic character inhibits water from inward diffusion, facilitated by crystallinity and intermolecular interactions. The minimum swelling degree of PU-grafted hydrogel was found to be decreased as low as 90%(w/w) compared to Ca-Alg-grafted polyvinyl acetate [203] and hydride with calcium phosphate [217] with swelling ratios as high as 300% and 900%, respectively.

Li et al. [218] reported smart PU-g-Ca-Alg hydrogel for protein imprinting, produced by radical-grafting polymerization. Hydrogel facilitates to form imprinted polymers in aqueous system, and maintains bioactivity and protein conformation which are otherwise a limiting factor in molecularly imprinted polymers (MIPs) [219]. Alg hydrogels are considered a perfect model for protein imprinting, but modified Alg polymer materials were developed either as blends, grafts or IPNs, in view of the shortcomings such as Alg strength and precise sites recognition [203,217,220]. Hence they [218] prepared PU-g-Ca-Alg using IPDI, HEMA, and DMPA instead of DEG used by Wang et al. [206]. A solution of BSA is mixed with PU-g-Ca-Alg followed by gelation with $CaCl_2$. Protein rebinding and recognition ability of PU-g-Ca-Alg of 20% grafting ratio were tested. Protein recognition ability is enhanced by incorporated side chains of PU which was confirmed by improved imprinting efficacy and selectivity factors achieved at high grafting ratio. Actually PU side chains created imprinting pockets and maintained recognition sites precision in rebinding procedure. In comparison to neat Ca-Alg hydrogels, the PU-g-Ca-Alg exhibited better rebinding selectivity, protein (BSA) recognition, and separation ability, and thus find application in bioseparation and chemical sensing [218].

Na-Alg/PU blend matrix was prepared for controlled oral delivery of proteins [221]. One important feature of these blends was the utilization of depolymerized product (bishydroxyethylene terephthalate) of waste PET material in synthesis of PU along with PEG and HMDI. PET utilization as a chain extender in PU synthesis and provides a valuable way to reduce waste load on environment. The prepared blend was turned into semi-IPN beads using $CaCl_2$ as cross-linker. A calculated percentage (0.1%) of BSA protein was encapsulated to the above solution of cross-linked Na-Alg/PU to prepare BSA/Na-Alg/PU beads. It was observed that increasing content of Alg (from 1% to 3%) with PU content at 9%−7% formed a more dense structure, which ultimately improved protein encapsulation efficiency from 61% to 94% while release rate of BSA was decreased. The diffusion of cation in Alg chain and its bonding with polymer increased with the higher contents of Alg, and increased number of binding sites, providing dense cross-linked packing of gel. In vitro swelling value was revealed to be higher at the pH 7.4 after 6 h. The pH-sensitive Alg/PU beads maintained better encapsulation, swelling, and sustained drug release than other polymeric systems such as gelatin [222], chitosan [223], and hydroxyapatite [224].

Studies were also carried out to estimate the effect of PEG and bishydroxyethylene contents in PU synthesis [225]. Researchers fixed the mole ratio of PU to alginate at 7/3 to study the concentration effect of PEG on the improvement of encapsulation. A decrease in swelling ratio was observed at acidic pH with increasing PEG ratio, but it was high in alkaline pH (7.4) due to porous morphology of the beads rich with PEG. The hydrogen bonding between −OH of PU and Alg might cause less swelling, while in alkaline medium the unreacted −OH groups of PU and Alg get deprotonated and ester bonds starts alkaline hydrolysis, therefore the increase in degree of swelling was probably due to electrostatic repulsions and disruption in cross-linking [221,226,227]. The blend beads with higher PEG contents in PU showed increase in protein encapsulation which decreases with lower PEG contents, i.e., from 98.7% to 88.9%. PU with higher PEG contents showed better protein release rate which actually depends on the permeation of dissolution fluid into blend beads, swelling and closure of the PU-Alg blend matrix, and the dissolution of drug following to leaching out of the swollen matrix [225,228]. Presence of phosphate buffer promotes the continuous erosion of BSA from PU-Alg blends for its greater affinity to Ca ions than Alg [227,229].

11.4.4 ALG-PU ELASTOMERIC COMPOSITES/NANOCOMPOSITES

Alg is a favorable polymer for the generation of nanoparticles, especially for the purpose of drug delivery and immobilization of enzymes because of its low cost, easy availability, and innocuous, biodegradable, and bioactive [230−234] character. Alg nanoparticles can be made simply by causing gelation with divalent cations such as Ca ions (ionic gelation method) [235−237]. Recently, nano-Alg materials have attracted significant attention in the area of heavy metal adsorption [238−240]. Addition of synthetic polymer usually enhances the mechanical strength of Alg composites. Daemi et al. [33] developed compatible composites based on anionic PU dispersions with Na-Alg. Reaction of polytetramethylene glycol (PTMG) and IPDI, extended with DMPA, produced aqueous PU dispersions, and solution blending of this prepared dispersion with Na-Alg resulted in Na-Alg/PU blend composite formation. Three different block ratios of PU compositions were prepared, out of which PU with block ratio 1:3:2 showed better properties. Thus Na-Alg/PU blend composites were based on this block ratio. Structural characterization of the films was performed through FTIR and NMR techniques that confirmed the proper miscibility of PU and Na-Alg as a result of Coulombic interactions and hydrogen bonding between hydrophilic and ionic groups in both PU and Na-Alg polymers. In contrast, it was obvious from storage modulus results that increase in Alg contents in the blend decreases the miscibility. Reduction of hard segment contents of PU in the blends actually lessens the hydrogen bond formation between urethane group of hard segments and Na-Alg. A DMPA study showed that melting temperature of soft segments is same in all blend compositions that also confirmed the compatibility between PU and Na-Alg. Surface morphology of samples using SEM presents an interspersed agglomeration of Na-Alg in anionic PU matrix. Agglomeration was developed due to similar negative charge of carboxylate (COO^-) ions in both Alg and PU polymers and their repulsive interactions. The Na peak in EDX curve ensured the presence of Na-Alg in PU matrix. The Na-Alg/PU composites exhibited greater hydrophilicity as compared to neat PU, credited to the presence of hydrophilic functional −OH and −COOH groups of Alg [33]. Daemi et al. [179] prepared stretchable Na-Alg/PU nanocomposites from PTMG, IPDI, MDEA, and Na-Alg. Cationic PU dispersions (Fig. 11.16) were blended with Na-Alg solution with PU to Alg mole ratios of 99/1%, 98/2%, and 96/4%. Presence of uronic vibrational bands of Alg in the anomeric region of FTIR spectrum of blends confirmed the existence of Alg in PU matrix. Two different shapes, i.e., rod and bead, of Alg nanoparticles were observed during SEM analysis correlating to the molecular weight. Excellent electrostatic interactions between quaternary-ammonium groups of PU (cationic) and carboxylate group of higher molecular weight Alg (anionic) resulted in quaternary-ammonium-carboxylate salt which appeared as nanorod-like particles on the bulk PU surface. Ring structure and uronic acid of Alg residue provide structural and thermal stability to the PU blends cleared from lower decomposition temperatures of nanocomposites. The ideal miscibility of nanocomposites arises due

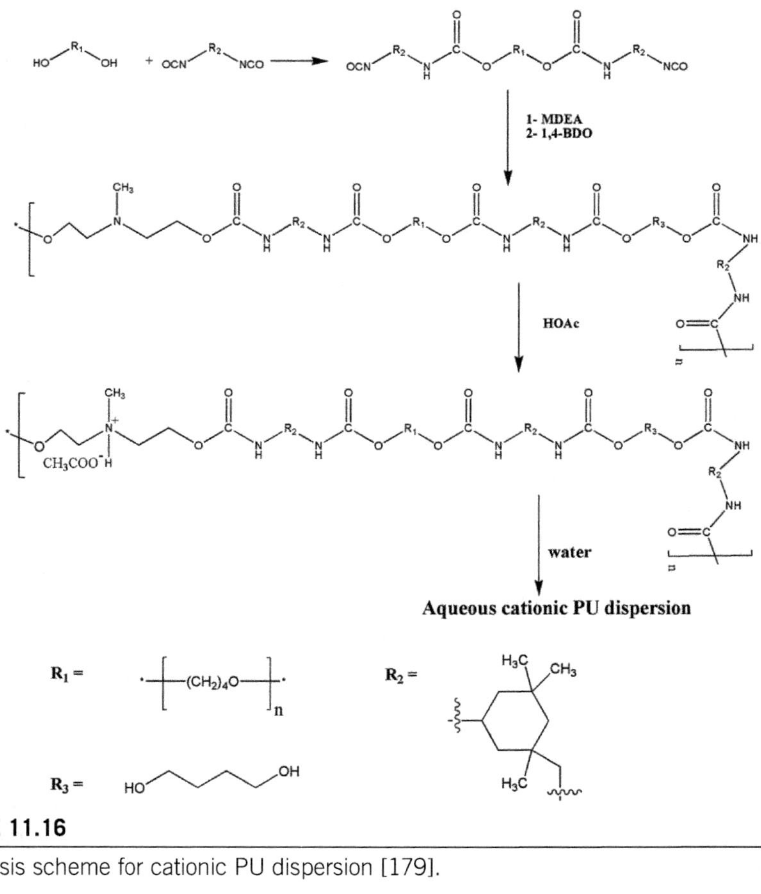

FIGURE 11.16

Synthesis scheme for cationic PU dispersion [179].

to opposite charge interaction between functional groups of two polymers and hydrogen bonding. Higher mechanical strength (26.66 MPa) of nanocomposites with high Alg contents (4%) was observed with lower elongation (1100%) that could be due to the decrease in phase separation of PU segments in the presence of Alg chains and the brittle nature of uronic residue. Other than this hydrophilicity of the PU increased by addition of Alg, i.e., the contact angle decreased from 75 degrees to 60 degrees from neat to Na-Alg composited PU [179].

Daemi et al. [241] prepared novel soluble Na-Alg/PU dispersions in aprotic solvents. Herein the Na-Alg is treated with tributylamine under controlled conditions to get soluble form of tributylammonium Alg (TBA-Alg), followed by TBA-Alg extended films. It is reported that the solubility of Alg in organic media required the formation of TBA salt of Alg [34]. The addition of chemically bonded nano-TBA-Alg into the PU backbone was made by the reaction of PU-prepolymer and an organic solution of TBA-Alg. The synthesized TBA-Alg-extended PUs presented unique FTIR spectra compared to common Alg with two different asymmetric stretching vibrations of uronic acid $-COO^-$ group residues of Alg. Moreover, ^{1}H-NMR peaks showed the existence of TBA-Alg into the backbone of PU. TG analysis results displayed that the thermal properties of sample films significantly depended upon the nature of the ionic functional groups of PU backbone and polyol, as $-COO^-$ group of PU ionomers have strong electrostatic interactions with ammonium groups of modified Alg. A continuous morphology of TBA-Alg-extended PU ionomers even at nanoscale creates a difference from nonionic samples represented by aggregate-like structures of Alg into the matrix of PU because of nonpolar character of soft segment.

In 2014 Daemi and coworkers [242] studied the role of emulsifiers and their nature on the ultimate properties of Na-Alg/PU ionomer composites. The two renowned emulsifiers MDEA and DMPA added hydrophilicity to PU backbone. The PU ionomers (cationic, anionic) were formed by the reaction of IPDI and PTMG, were extended with BDO, and formulated into Na-Alg/PU composites by solution blending of the PU dispersions and Na-Alg. The prepared PU dispersions and Na-Alg/PU composites were characterized via FTIR, SEM, DSC, DMTA, and TGA techniques. The effect of the type of emulsifier on the ultimate properties of the prepared elastomeric compositions was analyzed. The cationic PU and Na-Alg nanocomposite elastomers showed homogeneous dispersions and outstanding miscibility due to electrostatic interactions of oppositely charged functional groups of both polymers, while the anionic PU ones existed as incompatible composition (Alg agglomerates) owing to repulsive interactions between negatively charged groups, also reported in previous studies [179], as shown in SEM analysis of this work (Fig. 11.17).

It was also noted that using less amount of DMPA in this study as compared to previously reported one [179] raised the particle size. The FTIR and EDX studies verified the Na-Alg presence in the macro-/microstructure of the PU bulk, respectively. In addition, the thermal permanence of Na-Alg/PUs increased with increasing Alg content for both composites but it was greater in anionic PU-based composites.

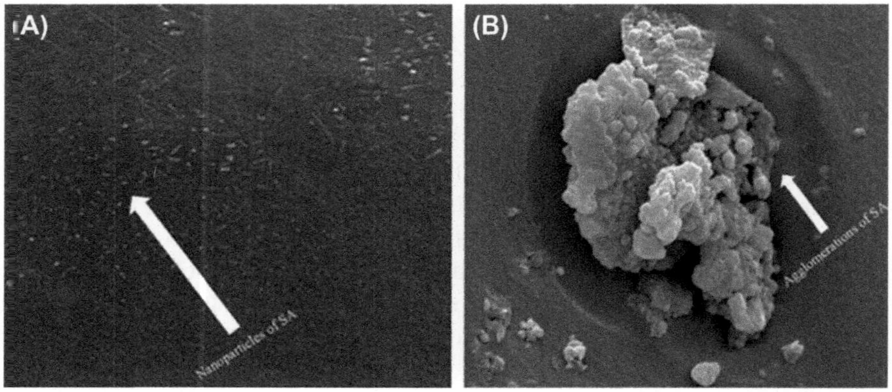

FIGURE 11.17

Morphology of alginate particles in PU ionomer matrix (A) cationic PU, and (B) anionic PU [242].

PU composites exhibited excellent mechanical properties with high elongation at break, whereas the elongation was considerably dropped for anionic PU/NA-Alg samples on account of their immiscibility [242].

Alg and PU foams were used as an immobilization matrix for *Klebsiella oxytoca* C3202 microbe for effective removal or degradation of benzoate [243]. Previous studies also reported that *Pseudomonas* sp. strain NGK1 immobilized Alg, agar, polyurethane foam, while polyacrylamide enhanced the degradation of naphthalene [244,245]. *K. oxytoca* C3202 immobilized matrices of Alg and PU extended the viability and degradability of 2 mM of benzoate in broth cultivation [243]. In the same way *Bacillus* sp. immobilized Alg and PU foam matrices used for the removal of *o*-phthalate. Higher operational stability and longevity of immobilized cells were observed in these matrices than free cells, and these immobilized matrices could be used repeatedly for 12 and 24 cycles [246]. However, the PU foam retained the immobilized cells for longer time than Alg due to its mechanical strength and porous structural morphology [245,247]. Sona et al. [29] reported flexible Alg/PU foam composite absorbents for effective removal of lead contamination from wastewater. The carboxylic groups of Alg acted as functional binding sites (ion-exchanger) for metal ions, especially for Pb^{+2} [248]. Poly(oxy C2-4 alkylene) diol and TDI (three isocyanate functionalities)-based prepolymer was employed to create immobilized Na-Alg composite foams. The FTIR and SEM-EDX results (Fig. 11.18A–C) presented specific vibration peaks, indicating grafting of Alg on PU backbone. The pH of sample solution participated in binding of lead, as lower pH convert carboxylate ions into carboxylic acid; therefore binding sites for lead ions decrease which ultimately reduced adsorption capacity of composite foams. The molar ratio of the available carboxyl group to lead ions is 2:1 compared to 6:1 of neat Ca-Alg beads [249]. SEM images established that like the common flexible polyurethane foams, the Alg/PU foams consist of three-dimensional (3D)

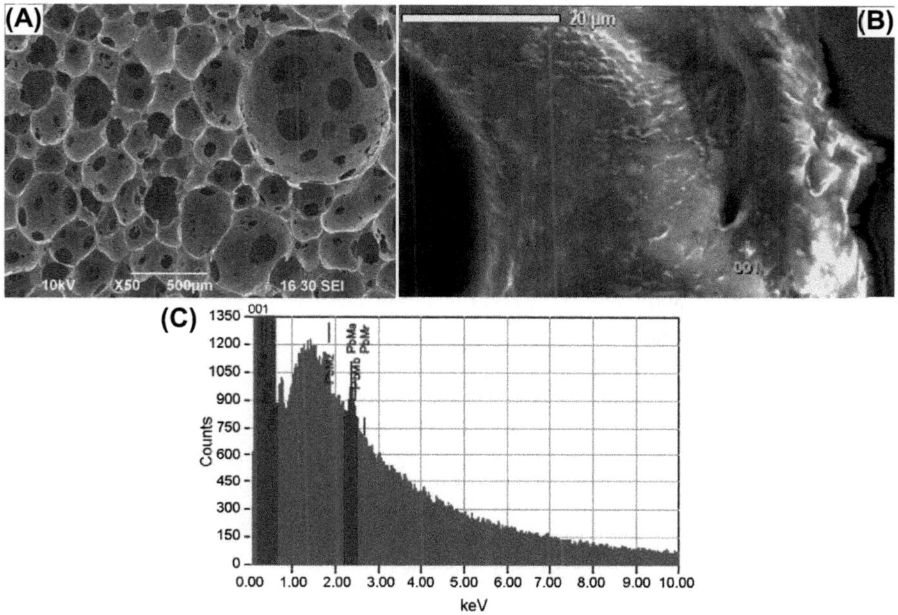

FIGURE 11.18

(A) SEM image of the flexible Na-Alg/PU composite foam at 50× magnification, (B), SEM image of the surface of Pb^{2+}-Na-Alg/PU sample, and (C) SEM-EDX analysis [29].

agglomerations of gas bubbles, separated from each other by thin sections of the host medium. The thin structures separating the bubble spaces are copolymers of Alg and the PU prepolymer (NB-9000B), made by means of different reactions. The ALG/PUCF is highly stable, flexible, and easy to use. Alg/PU composite foam can be regenerated with ethylenediamine-*N,N,N′,N′*-tetraacetic acid disodium salt (EDTA-2Na) making it reusable for wastewater treatment [29].

In 2015 Alg-PU is used as filler for chitosan matrix to fabricate ternary scaffolds. FTIR results suggested chemical interaction between PU, chitosan, and Alg by a slightly higher wave number shift in the range of 1700 cm^{-1} and H-bonded 1620 cm^{-1} than neat chitosan scaffold [180]. It is considered that the chemical interaction would take place through the establishment of NH—CO—NH groups between $-NH_2$, —NCO, and $-COO^-$ groups of chitosan, PU, and Alg, respectively [250]. The appearance of single melting peak in TG and DSC curves confirmed their blending with high thermal stability. The highly porous structure with even pore size facilitated cells' growth, as supported by previous study [251,252], as well. Increase in water uptake with decreasing retention-ability increased the in vitro cell proliferation and degradation rate, making them superior scaffolds for tissue-engineering use [180].

Table 11.2 Summary of Alginate-Based Polyurethane Materials

Sr. No	Name of Monomers	Form of Material	Method of Fabrication	Application	References
1	Alg/Anionic PU	Hydrogel films	Solution blending/ mixing	Promising material for biomedical/ food applications	[208]
2	Alg/PU	Composite foams	Prepolymer blending/ mixing	Adsorbent for Pb(II) ions from wastewater	[29]
3	Alg/PU	Hydrogel microspheres	Grafting	For protein imprinting applications	[218]
4	Alg/WPU	Blend dispersions	Prepolymer blending	Water permeable coating for fabric	[178]
5	Alg/PU-PET	Semi-IPN beads	Solution blending	Protein delivery system	[221]
6	Chitosan-Alg/PU	Ternary scaffolds	Freeze drying	For tissue engineering	[180]
7	Alg-bentonite/PU	Hydrogel containing foam	Vacuum method	For wound-dressing applications	[212]
8	Alg-jute fiber/PU	Hydrogel hybridized foam	Vacuum method	For wound-dressing applications	[176]

The Table 11.2 summarized fabrication methods, forms, and applications of some prominent Alg based PU materials.

11.5 SUMMARY

Integration of biopolymers into the synthetic ones not only improves the overall properties but also turns them into multifunctional environmentally benign biomaterial. The brief overview of the key features of Alg depicts this anionic native polymer for diverse range of properties due to presence of the functional moieties in its backbone. On the basis of this property, Alg is currently used to functionalize many natural and synthetic polymers for various industrial purposes. Polyurethanes, being a synthetic class of polymer, are awarded with superior tunable mechanical properties and stability, but are facing toxicity, biodegradability, and biocompatibility issues. Although waterborne PU/NI-PU provides a way for toxicity and degradation problems, the biocompatiblity of polyurethane still needs attention. Blending of polyurethane with natural polymers amended all these issues while imparting tremendous properties. Synthesis and properties of various Alg-derived polyurethane materials are reviewed in this chapter. Usually Alg is used as a reinforcing material for PU but is also used as a polyol in PU synthesis. Presence of ionic interactions and hydrogen bonding between −COOH/−OH groups of Alg and urethane linkage of PU/PU dispersions showed compact structure with good miscibility, dispersion-smooth morphology in blends and composite materials. However, nanorods, nanobeads, and agglomerates of Alg particles are observed in PU matrix depending on attractive and repulsive forces and molecular weight of Alg. The tensile strength with lower elongation at break was attributed to brittle nature of uronic residues and decrease in the phase separation of PU. Thermal stability of Alg/PU-based materials increased owing to the thermally stable ring structure of Alg. Currently three components, i.e., bentonite clay, chitosan, and jute fiber, are employed with Alg-PU-based materials to impart sophisticated properties such as absorption ability in drug-delivery systems, mechanical and thermal stability. The summary of the perspective applications of Alg/PU materials listed in Table 11.2 indicated this biomaterial to be used for different fields of application.

REFERENCES

[1] Draget KI, Østgaard K, Smidsrød O. Alginate-based solid media for plant tissue culture. Appl Microbiol Biotechnol 1989;31(1):79−83.

[2] Lee KY, Mooney DJ. Alginate: properties and biomedical applications. Prog Polym Sci 2012;37(1):106−26.

[3] Percival E, McDowell RH. In: Dey PM, editor. Methods in plant biochemistry. New York: Academic Press; 1990.

[4] Kraan S. Algal polysaccharides, novel applications and outlook. INTECH Open Access Publisher; 2012.

[5] Rinaudo M. Seaweed polysaccharides. In: Kamerling JP, editor. Comprehensive glycoscience, 2.2: polysaccharide functional properties. Oxford: Elsevier; 2007. p. 691−773.

[6] Skjak-Bræk G, Moe ST, Smidsrød O, Ingar Draget K. Alginates. In: Stephen AM, Phillips GO, Williams PA, editors. Food polysaccharides and their applications. CRC Press; 2006. p. 289−334.

[7] Larsen B, Salem DM, Sallam MA, Mishrikey MM, Beltagy AI. Characterization of the alginates from algae harvested at the Egyptian Red Sea coast. Carbohydr Res 2003; 338(22):2325−36.

[8] Rinaudo M. Main properties and current applications of some polysaccharides as biomaterials. Polym Int 2008;57(3):397−430.

[9] McHugh DJ. A guide to the seaweed industry. Rome, Italy: Food and Agriculture Organization of the United Nations; 2003.

[10] McHugh DJ. Prospects for seaweed production in developing countries. In: Fisheries circular. C968. Rome, Italy: FAO; 2002.

[11] Sabra W, Zeng AP, Deckwer WD. Bacterial alginate: physiology, product quality and process aspects. Appl Microbiol Biotechnol 2001;56(3−4):315−25.

[12] Remminghorst U, Rehm BHA. Bacterial alginates: from biosynthesis to applications. Biotechnol Lett 2006;28:1701−12.

[13] Deniaud-Bouët E, Kervarec N, Michel G, Tonon T, Kloareg B, Hervé C. Chemical and enzymatic fractionation of cell walls from Fucales: insights into the structure of the extracellular matrix of brown algae. Ann Bot 2014:mcu096.

[14] Hay ID, Rehman ZU, Moradali MF, Wang Y, Rehm BH. Microbial alginate production, modification and its applications. Microb Biotechnol 2013;6(6):637−50.

[15] Draget KI. Alginates. In: Phillips GO, Williams PA, editors. Handbook of hydrocolloids; 2009. p. 379−95.

[16] Sime WJ. In: Harris P, editor. Alginates, food gels. England: Elsevier Science Publishers Ltd; 1990.

[17] Haug A, Larsen B. Quantitative determination of the uronic acid composition of alginates. Acta Chem Scand 1962;16:1908−18.

[18] Gacesa P. Alginates. Carbohydr Polym 1988;8(3):161−82.

[19] Atkins ED, Mackie W, Parker KD, Smolko EE. Crystalline structures of poly-D-mannuronic and poly-L-guluronic acids. J Polym Sci Part B 1971;9(4):311−6.

[20] FAO. Sodium alginate, calcium alginate, alginic acid. In: 49th JECFA; 1997.

[21] Haug A, Larsen B, Smidsrød O. A study on the constitution of alginic acid by partial hydrolysis. Acta Chem Scand 1966;20:183−90.

[22] Haug A, Larsen B, Smidsrod O. Studies on the sequence of uronic acid residues in alginic acid. Acta Chem Scand 1967;21:691−704.

[23] Atkins ED, Mackie W, Smolko EE. Crystalline structures of alginic acids. Nature 1970;225:626−8.

[24] Smidsrød O, Whittington SG. Monte Carlo investigation of chemical inhomogeneity in polymers. Macromolecules 1969;2(1):42−4.

[25] Stokke BT, Smidsrød O, Brant DA. Predicted influence of monomer sequence distribution and acetylation on the extension of naturally occurring alginates. Carbohydr Polym 1993;22(1):57−66.

[26] Mander L, Liu HW. Comprehensive natural products II: chemistry and biology: 10 Volume Set. 2010. Newnes.

[27] Campos-Vallette MM, Chandía NP, Clavijo E, Leal D, Matsuhiro B, Osorio-Román IO, Torres S. Characterization of sodium alginate and its block fractions by surface-enhanced Raman spectroscopy. J Raman Spectrosc 2010;41(7):758−63.

[28] Leal D, Matsuhiro B, Rossi M, Caruso F. FT-IR spectra of alginic acid block fractions in three species of brown seaweeds. Carbohydr Res 2008;343(2):308−16.

[29] Sone H, Fugetsu B, Tanaka S. Selective elimination of lead (II) ions by alginate/polyurethane composite foams. J Hazard Mater 2009;162(1):423−9.

[30] Penman A, Sanderson GR. A method for the determination of uronic acid sequence in alginates. Carbohydr Res 1972;25:273−82.

[31] Grasdalen H, Larsen B, Smidsrød O. PMR study of the composition and sequence of uronate residues in alginates. Carbohydr Res 1979;68:23−31.

[32] Grasdalen H. High-Field, H-1-Nmr Spectroscopy of alginate − sequential structure and linkage conformations. Carbohydr Res 1983;118:255−60.

[33] Daemi H, Barikani M, Barmar M. Compatible compositions based on aqueous polyurethane dispersions and sodium alginate. Carbohydr Polym 2013;92(1): 490−6.

[34] Pawar SN, Edgar KJ. Alginate derivatization: a review of chemistry, properties and applications. Biomaterials 2012;33(11):3279–305.

[35] Draget KI, Taylor C. Chemical, physical and biological properties of alginates and their biomedical implications. Food Hydrocoll 2011;25(2):251–6.

[36] Draget KI, Skjåk-Bræk G, Smidsrød O. Alginic acid gels: the effect of alginate chemical composition and molecular weight. Carbohydr Polym 1994;25:31–8.

[37] Timell TE. The acid hydrolysis of glycosides: I. General conditions and the effect of the nature of the aglycone. Can J Chem 1964;42:1456–72.

[38] Smidsrød O, Haug A, Larsen B. The influence of pH on the rate of hydrolysis of acidic polysaccharides. Acta Chem Scand 1966;20:1026–34.

[39] Tsujino I, Saito T. A new unsaturated uronide isolated from alginase hydrolysate. Nature 1961;192:970–1.

[40] Fernandez-Perez M, González-Pradas E, Villafranca-Sanchez M, Flores-Cespedes F. Mobility of isoproturon from an alginate–bentonite controlled release formulation in layered soil. Chemosphere 2000;41(9):1495–501.

[41] Chen SC, Wu YC, Mi FL, Lin YH, Yu LC, Sung HW. A novel pH-sensitive hydrogel composed of N, O-carboxymethyl chitosan and alginate cross-linked by genipin for protein drug delivery. J Control Release 2004;96(2):285–300.

[42] Otterlei M, Ostgaard K, Skjakbraek G, Smidsrod O, Soonshiong P, Espevik T. Induction of cytokine production from human monocytes stimulated with alginate. J Immunother 1991;10:286–91.

[43] Zimmermann U, Klock G, Federlin K, Haning K, Kowaslski M, Bretzel RG, Horcher A, Entenmann H, Siebers U, Zekorn T. Production of mitogen contamination free alginates with variable ratios of mannuronic to guluronic acid by free flow electrophoresis. Electrophoresis 1992;13:269–74.

[44] Lee J, Lee KY. Local and sustained vascular endothelial growth factor delivery for angiogenesis using an injectable system. Pharm Res 2009;26:1739–44.

[45] http://www.prnewswire.com/news-releases/alginates–derivatives-market-worth-4092 million-by-2019–291648601.html.

[46] Franklin MJ, Ohman DE. Mutant analysis and cellular localization of the AlgI, AlgJ, and AlgF proteins required for O acetylation of alginate in *Pseudomonas aeruginosa*. J Bacteriol 2002;184:3000–7.

[47] Franklin MJ, Ohman DE. Identification of algF in the alginate biosynthetic gene cluster of *Pseudomonas aeruginosa* which is required for alginate acetylation. J Bacteriol 1993;175:5057–65.

[48] Franklin MJ, Douthit SA, McClure MA. Evidence that the algI/algJ gene cassette, required for O acetylation of *Pseudomonas aeruginosa* alginate, evolved by lateral gene transfer. J Bacteriol 2004;186:4759–73.

[49] Gimmestad M, Sletta H, Ertesvag H, Bakkevig K, Jain S, Suh SJ, et al. The *Pseudomonas fluorescens* AlgG protein, but not its mannuronan C-5- epimerase activity, is needed for alginate polymer formation. J Bacteriol 2003;185:3515–23.

[50] Franklin MJ, Chitnis CE, Gacesa P, Sonesson A, White DC, Ohman DE. *Pseudomonas aeruginosa* AlgG is a polymer level alginate C5-mannuronan epimerase. J Bacteriol 1994;176:1821–30.

[51] Chitnis CE, Ohman DE. Cloning of *Pseudomonas aeruginosa* algG, which controls alginate structure. J Bacteriol 1990;172:2894–900.

[52] Ons ø yen E. Alginates. In: Imeson AP, editor. Thickening and gelling agents for food. 2nd ed. New York: Springer; 1997. p. 22–44.

[53] Holtkamp AD, Kelly S, Ulber R, Lang S. Fucoidans and fucoidanases-focus on techniques for molecular structure elucidation and modification of marine polysaccharides. Appl Microbiol Biotechnol 2009;82(1):1−11.

[54] Hoek C, Mann D, Jahns HM. Algae: an introduction to phycology. Cambridge University Press; 1995.

[55] Cumashi A, Ushakova NA, Preobrazhenskaya ME, D'Incecco A, Piccoli A, Totani L, Tinari N, Morozevich GE, Berman AE, Bilan MI, Usov AI. A comparative study of the anti-inflammatory, anticoagulant, antiangiogenic, and antiadhesive activities of nine different fucoidans from brown seaweeds. Glycobiology 2007;17(5):541−52.

[56] Klöck G, Frank H, Houben R, Zekorn T, Horcher A, Siebers U, et al. Production of purified alginates suitable for use in immunoisolated transplantation. Appl Microbiol Biotechnol 1994;40:638−43.

[57] Lencina MS, Andreucetti NA, Gómez CG, Villar MA. Recent studies on alginates based blends, composites, and nanocomposites. In: Advances in natural polymers. Springer Berlin Heidelberg; 2013. p. 193−254.

[58] Reis RL, Neves NM, Mano JF, Gomes ME, Marques AP, Azevedo HS. Natural-based polymers for biomedical applications. Elsevier; 2008. p. 138.

[59] Sandford PA, Baird J. In: Aspinall G, editor. The polysaccharides. New York: Academic Press; 1983. p. 411.

[60] Nandini VV, Venkatesh KV, Nair KC. Alginate impressions: a practical perspective. J Conserv Dent 2008;11(1):37.

[61] Nallamuthu NA, Braden M, Patel MP. Some aspects of the formulation of alginate dental impression materials-setting characteristics and mechanical properties. Dent Mater 2012;28(7):756−62.

[62] Fabra MJ, Talens P, Chiralt A. Influence of calcium on tensile, optical and water vapour permeability properties of sodium caseinate edible films. J Food Eng 2010; 96(3):356−64.

[63] Kamoun EA, Chen X, Eldin MS, Kenawy ER. Crosslinked poly (vinyl alcohol) hydrogels for wound dressing applications: a review of remarkably blended polymers. Arab J Chem 2015;8(1):1−4.

[64] Mukhopadhyay P, Chakraborty S, Bhattacharya S, Mishra R, Kundu PP. pH-sensitive chitosan/alginate core-shell nanoparticles for efficient and safe oral insulin delivery. Int J Biol Macromol 2015;72:640−8.

[65] Ahmed TA, Khalid M. Development of alginate-reinforced chitosan nanoparticles utilizing W/O nanoemulsification/internal crosslinking technique for transdermal delivery of rabeprazole. Life Sci 2014;110(1):35−43.

[66] Sweeney IR, Miraftab M, Collyer G. Absorbent alginate fibres modified with hydrolysed chitosan for wound care dressings−II. Pilot scale development. Carbohydr Polym 2014;102:920−7.

[67] Wang Z, Zhang X, Gu J, Yang H, Nie J, Ma G. Electrodeposition of alginate/chitosan layer-by-layer composite coatings on titanium substrates. Carbohydr Polym 2014;103: 38−45.

[68] da Cunha CB, Klumpers DD, Li WA, Koshy ST, Weaver JC, Chaudhuri O, Granja PL, Mooney DJ. Influence of the stiffness of three-dimensional alginate/collagen-I interpenetrating networks on fibroblast biology. Biomaterials 2014;35(32):8927−36.

[69] Samanta HS, Ray SK. Synthesis, characterization, swelling and drug release behavior of semi-interpenetrating network hydrogels of sodium alginate and polyacrylamide. Carbohydr Polym 2014;99:666−78.

[70] Guo P, Yuan Y, Chi F. Biomimetic alginate/polyacrylamide porous scaffold supports human mesenchymal stem cell proliferation and chondrogenesis. Mater Sci Eng C 2014;42:622–8.

[71] Akın A, Işıklan N. Microwave assisted synthesis and characterization of sodium alginate-graft-poly (*N*, *N*′-dimethylacrylamide). Int J Biol Macromol 2016;82:530–40.

[72] Kim MS, Kim G. Three-dimensional electrospun polycaprolactone (PCL)/alginate hybrid composite scaffolds. Carbohydr Polym 2014;114:213–21.

[73] Quraishi S, Martins M, Barros AA, Gurikov P, Raman SP, Smirnova I, Duarte AR, Reis RL. Novel non-cytotoxic alginate–lignin hybrid aerogels as scaffolds for tissue engineering. J Supercrit Fluids 2015;105:1–8.

[74] Chiew CS, Poh PE, Pasbakhsh P, Tey BT, Yeoh HK, Chan ES. Physicochemical characterization of halloysite/alginate bionanocomposite hydrogel. Appl Clay Sci 2014; 101:444–54.

[75] Liu M, Dai L, Shi H, Xiong S, Zhou C. In vitro evaluation of alginate/halloysite nanotube composite scaffolds for tissue engineering. Mater Sci Eng C 2015;49:700–12.

[76] Jeon O, Samorezov JE, Alsberg E. Single and dual crosslinked oxidized methacrylated alginate/PEG hydrogels for bioadhesive applications. Acta Biomater 2014;10(1):47–55.

[77] Seker DC, Zain NA. Response surface optimization of glucose production from liquid pineapple waste using immobilized invertase in PVA–alginate–sulfate beads. Sep Purif Technol 2014;133:48–54.

[78] Miao T, Rao KS, Spees JL, Oldinski RA. Osteogenic differentiation of human mesenchymal stem cells through alginate-graft-poly (ethylene glycol) microsphere-mediated intracellular growth factor delivery. J Control Release 2014;192:57–66.

[79] Phuong NT. Entrapment of Mg–Al layered double hydroxide into alginate/polyvinyl alcohol beads for water remediation. J Environ Chem Eng 2014;2(2):1082–7.

[80] Hui B, Zhang Y, Ye L. Preparation of PVA hydrogel beads and adsorption mechanism for advanced phosphate removal. Chem Eng J 2014;235:207–14.

[81] Bano S, Mahmood A, Kim SJ, Lee KH. Chlorine resistant binary complexed NaAlg/PVA composite membrane for nanofiltration. Sep Purif Technol 2014;137:21–7.

[82] Guo J, Zhang Q, Cai Z, Zhao K. Preparation and dye filtration property of electrospun polyhydroxybutyrate–calcium alginate/carbon nanotubes composite nanofibrous filtration membrane. Sep Purif Technol 2016;161:69–79.

[83] Majidnia Z, Idris A. Evaluation of cesium removal from radioactive waste water using maghemite PVA–alginate beads. Chem Eng J 2015;262:372–82.

[84] Pandi K, Viswanathan N. A novel metal coordination enabled in carboxylated alginic acid for effective fluoride removal. Carbohydr Polym 2015;118:242–9.

[85] Prabhu SM, Meenakshi S. Novel one-pot synthesis of dicarboxylic acids mediated alginate–zirconium biopolymeric complex for defluoridation of water. Carbohydr Polym 2015;120:60–8.

[86] Rescignano N, Fortunati E, Armentano I, Hernandez R, Mijangos C, Pasquino R, Kenny JM. Use of alginate, chitosan and cellulose nanocrystals as emulsion stabilizers in the synthesis of biodegradable polymeric nanoparticles. J Colloid Interface Sci 2015;445:31–9.

[87] Nan F, Wu J, Qi F, Liu Y, Ngai T, Ma G. Uniform chitosan-coated alginate particles as emulsifiers for preparation of stable Pickering emulsions with stimulus dependence. Colloids Surf A 2014;456:246–52.

[88] Hosseini SM, Hosseini H, Mohammadifar MA, German JB, Mortazavian AM, Mohammadi A, Shojaee-Aliabadi S, Khaksar R. Preparation and characterization of

alginate and alginate-resistant starch microparticles containing nisin. Carbohydr Polym 2014;103:573–80.

[89] Zahran MK, Ahmed HB, El-Rafie MH. Surface modification of cotton fabrics for antibacterial application by coating with AgNPs–alginate composite. Carbohydr Polym 2014;108:145–52.

[90] Dhanapal V, Subramanian K. Recycling of textile dye using double network polymer from sodium alginate and superabsorbent polymer. Carbohydr Polym 2014;108:65–74.

[91] Sirviö JA, Kolehmainen A, Liimatainen H, Niinimäki J, Hormi OE. Biocomposite cellulose-alginate films: promising packaging materials. Food Chem 2014;151:343–51.

[92] Wang LF, Rhim JW. Preparation and application of agar/alginate/collagen ternary blend functional food packaging films. Int J Biol Macromol 2015;80:460–8.

[93] He Y, Wu Z, Tu L, Han Y, Zhang G, Li C. Encapsulation and characterization of slow-release microbial fertilizer from the composites of bentonite and alginate. Appl Clay Sci 2015;109:68–75.

[94] Krol P. Synthesis methods, chemical structures and phase structures of linear polyurethanes. Properties and applications of linear polyurethanes in polyurethane elastomers, copolymers and ionomers. Prog Mater Sci 2007;52(6):915–1015.

[95] Szycher M. In: Szycher's handbook of polyurethanes. CRC press; 2012.

[96] beforeitsnews.com/.../2016/.../global-polyurethane-markets.

[97] Du Y, Zhang J, Zhou C. Synthesis and properties of waterborne polyurethane-based PTMG and PDMS as soft segment. Polym Bull 2016;73(1):293–308.

[98] Rokicki G, Parzuchowski PG, Mazurek M. Non-isocyanate polyurethanes: synthesis, properties, and applications. Polym Adv Technol 2015;26(7):707–61.

[99] Oprea S. Structure and properties of cross-linked polyurethane copolymers. Adv Polym Technol 2009;28(3):165–72.

[100] Zia KM, Anjum S, Zuber M, Mujahid M, Jamil T. Synthesis and molecular characterization of chitosan based polyurethane elastomers using aromatic diisocyanate. Int J Biol Macromol 2014;66:26–32.

[101] Santayanon R, Wootthikanokkhan J. Modification of cassava starch by using propionic anhydride and properties of the starch-blended polyester polyurethane. Carbohydr Polym 2003;51(1):17–24.

[102] Chen R, Huang C, Ke Q, He C, Wang H, Mo X. Preparation and characterization of coaxial electrospun thermoplastic polyurethane/collagen compound nanofibers for tissue engineering applications. Colloids Surf B 2010;79(2):315–25.

[103] Jeong EH, Yang J, Youk JH. Preparation of polyurethane cationomer nanofiber mats for use in antimicrobial nanofilter applications. Mater Lett 2007;61(18):3991–4.

[104] Zhou L, Yu L, Ding M, Li J, Tan H, Wang Z, Fu Q. Synthesis and characterization of pH-sensitive biodegradable polyurethane for potential drug delivery applications. Macromolecules 2011;44(4):857–64.

[105] Shirota T, He H, Yasui H, Matsuda T. Human endothelial progenitor cell-seeded hybrid graft: proliferative and antithrombogenic potentials in vitro and fabrication processing. Tissue Eng 2003;9(1):127–36.

[106] Dodiuk H, Goodman SH, editors. Handbook of thermoset plastics. William Andrew; 2013.

[107] Prisacariu C. Polyurethane elastomers: from morphology to mechanical aspects. Springer Science & Business Media; 2011.

[108] Cohn D, Stern T, González MF, Epstein J. Biodegradable poly (ethylene oxide)/poly (ε-caprolactone) multiblock copolymers. J Biomed Mater Res 2002;59(2):273–81.

[109] Shelke NB, Aminabhavi TM. Synthesis and characterization of methoxypolyethyleneglycol and lauric acid grafted novel polyurethanes for controlled release of nifedipine. J Appl Polym Sci 2007;105(4):2155–63.

[110] Sathiskumar PS, Madras G. Synthesis, characterization, degradation of biodegradable castor oil based polyesters. Polym Degrad Stab 2011;96(9):1695–704.

[111] Pissis P, Kanapitsas A, Savelyev YV, Akhranovich ER, Privalko EG, Privalko VP. Influence of chain extenders and chain end groups on properties of segmented polyurethanes. II. Dielectric study. Polymer 1998;39(15):3431–5.

[112] Woo GL, Mittelman MW, Santerre JP. Synthesis and characterization of a novel biodegradable antimicrobial polymer. Biomaterials 2000;21(12):1235–46.

[113] Adhikari R, Gunatillake PA, Mccarthy SJ, Meijs GF. Effect of chain extender structure on the properties and morphology of polyurethanes based on H12MDI and mixed macrodiols (PDMS–PHMO). J Appl Polym Sci 1999;74(12):2979–89.

[114] Martin DJ, Warren LA, Gunatillake PA, McCarthy SJ, Meijs GF, Schindhelm K. Polydimethylsiloxane/polyether-mixed macrodiol-based polyurethane elastomers: biostability. Biomaterials 2000;21(10):1021–9.

[115] Ma Z, Hong Y, Nelson DM, Pichamuthu JE, Leeson CE, Wagner WR. Biodegradable polyurethane ureas with variable polyester or polycarbonate soft segments: effects of crystallinity, molecular weight, and composition on mechanical properties. Biomacromolecules 2011;12(9):3265–74.

[116] Skarja GA, Woodhouse KA. Synthesis and characterization of degradable polyurethane elastomers containing an amino acid-based chain extender. J Biomater Sci Polym Ed 1998;9(3):271–95.

[117] Guan J, Sacks MS, Beckman EJ, Wagner WR. Biodegradable poly (ether ester urethane) urea elastomers based on poly (ether ester) triblock copolymers and putrescine: synthesis, characterization and cytocompatibility. Biomaterials 2004;25(1):85–96.

[118] Park JH, Park KD, Bae YH. PDMS-based polyurethanes with MPEG grafts: synthesis, characterization and platelet adhesion study. Biomaterials 1999;20(10):943–53.

[119] Lin YH, Chou NK, Chen KF, Ho GH, Chang CH, Wang SS, Chu SH, Hsieh KH. Effect of soft segment length on properties of hydrophilic/hydrophobic polyurethanes. Polym Int 2007;56(11):1415–22.

[120] Oertel G, Abele L. Polyurethane handbook: chemistry, raw materials, processing, application, properties. USA: Hanser Publishers Macmillan; 1985.

[121] Petrović ZS, Ferguson J. Polyurethane elastomers. Prog Polym Sci 1991;16(5): 695–836.

[122] Clough SB, Schneider NS, King AO. Small-angle X-ray scattering from polyurethane elastomers. J Macromol Sci Part B 1968;2(4):641–8.

[123] Nakayama K, Ino T, Matsubara I. Infrared spectra and structure of polyurethane elastomers from polytetrahydrofuran, diphenylmethane-4, 4′-diisocyanate, and ethylenediamine. J Macromol Sci Chem 1969;3(5):1005–20.

[124] Visser SA, Pruckmayr G, Cooper SL. Small-angle neutron scattering analysis of model polyurethane ionomers. Macromolecules 1991;24(25):6769–75.

[125] Paik Sung CS, Hu CB, Wu CS. Properties of segmented poly (urethaneureas) based on 2, 4-toluene diisocyanate. 1. Thermal transitions, X-ray studies, and comparison with segmented poly (urethanes). Macromolecules 1980;13(1):111–6.

[126] Koenig JL. Fourier transform infrared spectroscopy of polymers. In: Spectroscopy: NMR, fluorescence, FT-I. Berlin Heidelberg: Springer; 1984. p. 87–154.

[127] Bajsic EG, Rek V, Sendijarevic A, Sendijarevic V, Frisch KC. DSC study of morphological changes in segmented polyurethane elastomers. J Elastom Plast 2000;32(2): 162–82.

[128] Blackwell J, Ross M. X-ray studies of the structure of polyurethane hard segments. J Polym Sci 1979;17(7):447–51.

[129] Chan-Chan LH, Solis-Correa R, Vargas-Coronado RF, Cervantes-Uc JM, Cauich-Rodríguez JV, Quintana P, Bartolo-Pérez P. Degradation studies on segmented polyurethanes prepared with HMDI, PCL and different chain extenders. Acta Biomater 2010;6(6):2035–44.

[130] Tocha E, Janik H, Debowski M, Vancso GJ. Morphology of polyurethanes revisited by complementary AFM and TEM. J Macromol Sci Part B 2002;41(4–6): 1291–304.

[131] Janik H, Pałys B, Petrovic ZS. Multiphase-separated polyurethanes studied by micro-raman spectroscopy. Macromol Rapid Commun 2003;24(3):265–8.

[132] Randall D, Lee S, editors. Huntsman international LLC. John Wiley & Sons; 2002.

[133] Wirpsza Z. Poliuretany: chemia, technologia, zastosowanie. Wydawnictwa Naukowo-Techniczne; 1991.

[134] Trifan DS, Terenzi JF. Extents of hydrogen bonding in polyamides and polyurethanes. J Polym Sci 1958;28(117):443–5.

[135] Seymour RW, Estes GM, Cooper SL. Infrared studies of segmented polyurethane elastomers. I. Hydrogen bonding. Macromolecules 1970;3(5):579–83.

[136] Zdrahala RJ, Gerkin RM, Hager SL, Critchfield FE. Polyether-based thermoplastic polyurethanes. I. Effect of the hard-segment content. J Appl Polym Sci 1979;24(9): 2041–50.

[137] Abouzahr S, Wilkes GL, Ophir Z. Structure-property behaviour of segmented polyether-MDI-butanediol based urethanes: effect of composition ratio. Polymer 1982;23(7):1077–86.

[138] Desai S, Thakore IM, Sarawade BD, Devi S. Structure-property relationship in polyurethane elastomers containing starch as a crosslinker. Polym Eng Sci 2000; 40(5):1200–10.

[139] Krasowska K, Janik H, Gradys A, Rutkowska M. Degradation of polyurethanes in compost under natural conditions. J Appl Polym Sci 2012;125(6):4252–60.

[140] Rutkowska M, Krasowska K, Heimowska A, Janik H, Balas A. Proc. Gordon Conference, California. 1999.

[141] Meier-Westhues U. Polyurethanes: coatings, adhesives and sealants. Vincentz Network GmbH & Co. KG; 2007.

[142] Li S, Vatanparast R, Lemmetyinen H. Cross-linking kinetics and swelling behaviour of aliphatic polyurethane. Polymer 2000;41(15):5571–6.

[143] Chattopadhyay DK, Raju KV. Structural engineering of polyurethane coatings for high performance applications. Prog Polym Sci 2007;32(3):352–418.

[144] Liaw D, Huang C, Liaw B. Synthesis and properties of polyurethanes based on bisphenol-S derivatives. Polymer 1998;39(15):3529–35.

[145] Wang W, Wang C. Polyurethane for biomedical applications: a review of recent developments. In: Davim JP, editor. The design and manufacture of medical devices. Woodhead Publishing Ltd.; 2012. p. 115–51.

[146] Zhou X, Zhang T, Guo D, Gu N. A facile preparation of poly (ethylene oxide)-modified medical polyurethane to improve hemocompatibility. Colloids Surf A 2014;441:34–42.

[147] https://polyurethane.americanchemistry.com/Introduction-to-Polyurethanes/Applications.

[148] http://urethaneproducts.com/.

[149] Lamba NM, Woodhouse KA, Cooper SL. Polyurethanes in biomedical applications. CRC press; 1997.

[150] Zdrahala RJ, Zdrahala IJ. Biomedical applications of polyurethanes: a review of past promises, present realities, and a vibrant future. J Biomater Appl 1999;14(1):67–90.

[151] Guelcher SA, Srinivasan A, Dumas JE, Didier JE, McBride S, Hollinger JO. Synthesis, mechanical properties, biocompatibility, and biodegradation of polyurethane networks from lysine polyisocyanates. Biomaterials 2008;29(12):1762–75.

[152] Zia KM, Barikani M, Khalid AM, Honarkar H. Surface characteristics of UV-irradiated chitin-based polyurethane elastomers. Carbohydr Polym 2009;77(3):621–7.

[153] Ding M, Li J, Tan H, Fu Q. Self-assembly of biodegradable polyurethanes for controlled delivery applications. Soft Matter 2012;8(20):5414–28.

[154] Wicks ZW, Wicks DA, Rosthauser JW. Two package waterborne urethane systems. Prog Org Coat 2002;44(2):161–83.

[155] Naghash HJ, Abili B. Synthesis of a silicone containing allylic monomer and its uses in the waterborne polyurethane/vinyl acetate–acrylic hybrid emulsion copolymers. Prog Org Coat 2010;69(4):486–94.

[156] Lee SJ, Kim BK. Covalent incorporation of starch derivative into waterborne polyurethane for biodegradability. Carbohydr Polym 2012;87(2):1803–9.

[157] Brinkman E, Vandevoorde P. Waterborne two-pack isocyanate-free systems for industrial coatings. Prog Org Coat 1998;34(1):21–5.

[158] Lin Y, Hsieh F. Water-blown flexible polyurethane foam extended with biomass materials. J Appl Polym Sci 1997;65(4):695–703.

[159] Guan J, Song Y, Lin Y, Yin X, Zuo M, Zhao Y, Tao X, Zheng Q. Progress in study of non-isocyanate polyurethane. Ind Eng Chem Res 2011;50(11):6517–27.

[160] Kathalewar M, Sabnis A, Waghoo G. Effect of incorporation of surface treated zinc oxide on non-isocyanate polyurethane based nano-composite coatings. Prog Org Coat 2013;76(9):1215–29.

[161] Chattopadhyay DK, Webster DC. Thermal stability and flame retardancy of polyurethanes. Prog Polym Sci 2009;34(10):1068–133.

[162] Belgacem MN, Gandini A, editors. Monomers, polymers and composites from renewable resources. Elsevier; 2011.

[163] Wang Y, Lue A, Zhang L. Rheological behavior of waterborne polyurethane/starch aqueous dispersions during cure. Polymer 2009;50(23):5474–81.

[164] Cao X, Zhang L. Effects of molecular weight on the miscibility and properties of polyurethane/benzyl starch semi-interpenetrating polymer networks. Biomacromolecules 2005;6(2):671–7.

[165] Zeng M, Zhang L, Zhou Y. Effects of solid substrate on structure and properties of casting waterborne polyurethane/carboxymethylchitin films. Polymer 2004;45(10):3535–45.

[166] Zia KM, Mahmood K, Zuber M, Jamil T, Shafiq M. Chitin based polyurethanes using hydroxyl terminated polybutadiene. Part I: molecular engineering. Int J Biol Macromol 2013;59:320–7.

[167] Sussman G. Management of the wound environment with dressings and topical agents. Wound care: a collaborative practice manual for health professionals. Philadelphia, PA: Lippincott Williams & Wilkins; 2007. p. 250–67.

[168] Yamasaki M, Ogura K, Hashimoto W, Mikami B, Murata K. A structural basis for depolymerization of alginate by polysaccharide lyase family-7. J Mol Biol 2005;352(1):11–21.

[169] Wang F, Lu X, Li XY. Selective removals of heavy metals (Pb^{2+}, Cu^{2+}, and Cd^{2+}) from wastewater by gelation with alginate for effective metal recovery. J Hazard Mater 2016;308:75–83.
[170] Zia KM, Zia F, Zuber M, Rehman S, Ahmad MN. Alginate based polyurethanes: a review of recent advances and perspective. Int J Biol Macromol 2015;79:377–87.
[171] Zia F, Zia KM, Zuber M, Tabasum S, Rehman S. Heparin based polyurethanes: a state-of-the-art review. Int J Biol Macromol 2016;84:101–11.
[172] Zia F, Zia KM, Zuber M, Ahmad HB, Muneer M. Glucomannan based polyurethanes: a critical short review of recent advances and future perspectives. Int J Biol Macromol 2016;87:229–36.
[173] Zuber M, Zia F, Zia KM, Tabasum S, Salman M, Sultan N. Collagen based polyurethanes-A review of recent advances and perspective. Int J Biol Macromol 2015;80:366–74.
[174] Zia F, Zia KM, Zuber M, Kamal S, Aslam N. Starch based polyurethanes: a critical review updating recent literature. Carbohydr Polym 2015;134:784–98.
[175] Usman A, Zia KM, Zuber M, Tabasum S, Rehman S, Zia F. Chitin and chitosan based polyurethanes: a review of recent advances and prospective biomedical applications. Int J Biol Macromol 2016;86:630–45.
[176] Oh ST, Kim SH, Jeong HY, Lee JM, Cho JW, Park JS. The mechanical properties of polyurethane foam wound dressing hybridized with alginate hydrogel and jute fiber. Fibers Polym 2013;14(2):173–81.
[177] Fan LH, Du YM, Tang RP, Huang RH, Zheng H. Structure and properties of sodium alginate/waterborne polyurethane blend films. J Anal Sci 2002;18(6):441–4.
[178] Yun JK, Yoo HJ, Kim HD. Preparation and properties of waterborne polyurethane-urea/sodium alginate blends for high water vapor permeable coating materials. J Appl Polym Sci 2007;105(3):1168–76.
[179] Daemi H, Barikani M, Barmar M. Highly stretchable nanoalginate based polyurethane elastomers. Carbohydr Polym 2013;95(2):630–6.
[180] Yuvarani I, Senthilkumar S, Venkatesan J, Kim SK, Al-Kheraif AA, Anil S, Sudha PN. Chitosan modified alginate-polyurethane scaffold for skeletal muscle tissue engineering. J Biomater Tissue Eng 2015;5(8):665–72.
[181] Kwon OJ, Oh ST, Lee SD, Lee NR, Shin CH, Park JS. Hydrophilic and flexible polyurethane foams using sodium alginate as polyol: effects of PEG molecular weight and cross-linking agent content on water absorbency. Fibers Polym 2007;8(4):347–55.
[182] Kim JS, Pathak TS, Yun JH, Kim KP, Park TJ, Kim Y, Paeng KJ. Thermal degradation and kinetics of alginate polyurethane hybrid material prepared from alginic acid as a polyol. J Polym Environ 2013;21(1):224–32.
[183] Ge J, Zhong W, Guo Z, Li W, Sakai K. Biodegradable polyurethane materials from bark and starch. I. Highly resilient foams. J Appl Polym Sci 2000;77(12):2575–80.
[184] Garcon R, Clerk C, Gesson JP, Bordado J, Nunes T, Caroco S, Gomes PT, da Piedade MM, Rauter AP. Synthesis of novel polyurethanes from sugars and 1, 6-hexamethylene diisocyanate. Carbohydr Polym 2001;45(2):123–7.
[185] Ryabov S, Kotelnikova N, Kercha Y, Laptiy S, Gaiduk R, Kosenko L, Yakovenko A. Preparation of polymer composites on the base of polyurethanes and natural polysaccharides. In: Macromolecular symposia. WILEY-VCH Verlag GmbH; 2001. p. 421–8.
[186] Kendaganna Swamy BK. Structure–property relationship of starch-filled chain-extended polyurethanes. J Appl Polym Sci 2003;90(11):2945–54.

[187] Kwon OJ, Yang SR, Kim DH, Park JS. Characterization of polyurethane foam prepared by using starch as polyol. J Appl Polym Sci 2007;103(3):1544–53.

[188] Beneš H, Rösner J, Holler P, Synková H, Kotek J, Horák Z. Glycolysis of flexible polyurethane foam in recycling of car seats. Polym Adv Technol 2007;18(2):149–56.

[189] Mason GL, Green JG, Hall KG, Durst BP. Rigid polyurethane foam testing for mine neutralization and vehicle traction performance. J Terramech 2000;37(2):87–98.

[190] Farrés IF, Norton IT. Formation kinetics and rheology of alginate fluid gels produced by in-situ calcium release. Food Hydrocoll 2014;40:76–84.

[191] Grant GT, Morris ER, Rees DA, Smith PJ, Thom D. Biological interactions between polysaccharides and divalent cations: the egg-box model. FEBS Lett 1973;32(1):195–8.

[192] Sikorski P, Mo F, Skjåk-Bræk G, Stokke BT. Evidence for egg-box-compatible interactions in calcium-alginate gels from fiber X-ray diffraction. Biomacromolecules 2007;8:2098–103.

[193] Donati I, Holtan S, Mørch YA, Borgogna M, Dentini M, Skjåk-Bræk G. New hypothesis on the role of alternating sequences in calcium-alginate gels. Biomacromolecules 2005;6:1031–40.

[194] Mørch ÝA, Donati I, Strand BL, Skjåk-Bræk G. Effect of Ca2+, Ba2+, and Sr2+ on alginate microbeads. Biomacromolecules 2006;7(5):1471–80.

[195] Skjåk-Bræk G, Grasdalen H, Smidsrød O. Inhomogeneous polysaccharide ionic gels. Carbohydr Polym 1989;10:31–54.

[196] Peppas NA, Bures P, Leobandung W, Ichikawa H. Hydrogels in pharmaceutical formulations. Eur J Pharm Biopharm 2000;50(1):27–46.

[197] Gray CJ, Dowsett J. Retention of insulin in alginate gel beads. Biotechnol Bioeng 1988;31(6):607–12.

[198] Iemma F, Spizzirri UG, Puoci F, Muzzalupo R, Trombino S, Cassano R, Leta S, Picci N. pH-Sensitive hydrogels based on bovine serum albumin for oral drug delivery. Int J Pharm 2006;312(1):151–7.

[199] Işıklan N, Kurşun F. Synthesis and characterization of graft copolymer of sodium alginate and poly (itaconic acid) by the redox system. Polym Bull 2013;70(3):1065–84.

[200] Meena R, Chhatbar M, Prasad K, Siddhanta AK. Development of a robust hydrogel system based on agar and sodium alginate blend. Polym Int 2008;57(2):329–36.

[201] Meng XW, Ha W, Cheng C, Dong ZQ, Ding LS, Li BJ, Zhang S. Hollow nanospheres based on the self-assembly of alginate-graft-poly (ethylene glycol) and α-cyclodextrin. Langmuir 2011;27(23):14401–7.

[202] Yang L, Zhang B, Wen L, Liang Q, Zhang LM. Amphiphilic cholesteryl grafted sodium alginate derivative: synthesis and self-assembly in aqueous solution. Carbohydr Polym 2007;68(2):218–25.

[203] Ying X, Qi L, Li X, Zhang W, Cheng G. Stimuli-responsive recognition of BSA-imprinted poly vinyl acetate grafted calcium alginate core-shell hydrogel microspheres. J Appl Polym Sci 2013;127(5):3898–909.

[204] Juárez GP, Spasojevic M, Faas MM, de Vos P. Immunological and technical considerations in application of alginate-based microencapsulation systems. Front Bioeng Biotechnol 2014;2(26):1–5.

[205] Caló E, Khutoryanskiy VV. Biomedical applications of hydrogels: a review of patents and commercial products. Eur Polym J 2015;65:252–67.

[206] Wang J, Ying X, Li X, Zhang W. Preparation, characterization and swelling behaviors of polyurethane-grafted calcium alginate hydrogels. Mater Lett 2014;126:263–6.

[207] Sukhorukova SA, Travinskaya TV, Chumak LA. Influence of water soluble polymers on rheological properties of anionic polyurethane latexes. Plasticheskie Massy 1988: 10 [in Russian].

[208] Travinskaya TV, Savelyev YV. Aqueous polyurethane-alginate compositions: peculiarities of behavior and performance. Eur Polym J 2006;42(2):388–94.

[209] Larson RG. The structure and rheology of complex fluids. New York: Oxford University Press; 1999.

[210] Ferry JD. Viscoelastic properties of polymers. New York: Wiley; 1980.

[211] Gombotz WR, Wee SF. Protein release from alginate matrices. Adv Drug Deliv Rev 1998;31:267–85.

[212] Oh ST, Kim WR, Kim SH, Chung YC, Park JS. The preparation of polyurethane foam combined with pH-sensitive alginate/bentonite hydrogel for wound dressings. Fibers Polym 2011;12(2):159–65.

[213] Oh ST, Kwon OJ, Chun BC, Cho JW, Park JS. The effect of bentonite concentration on the drug delivery efficacy of a pH-sensitive alginate/bentonite hydrogel. Fibers Polym 2009;10(1):21–6.

[214] Tien YM, Wu PL, Huang WH, Kuo MF, Chu CA. Wall friction measurement and compaction characteristics of bentonite powders. Powder Technol 2007;173(2):140–51.

[215] Kwon OJ. Water-absorptive polyurethane foams for wound dressings using sodium-alginate as a polyol. Seoul National University; 2009.

[216] Wu XY, Yang XH, Liu YH, Wang XL. J Shanghai Jiaotong Univ (Sci) 2005;10:193–6.

[217] Zhao K, Cheng G, Huang J, Ying X. Rebinding and recognition properties of protein-macromolecularly imprinted calcium phosphate/alginate hybrid polymer microspheres. React Funct Polym 2008;68(3):732–41.

[218] Li L, Ying X, Liu J, Li X, Zhang W. Molecularly imprinted polyurethane grafted calcium alginate hydrogel with specific recognition for proteins. Mater Lett 2015; 143:248–51.

[219] Hillberg AL, Tabrizian M. Biomolecule imprinting: developments in mimicking dynamic natural recognition systems. IRBM 2008;29(2):89–104.

[220] Zhang F, Cheng G, Ying X. Emulsion and macromolecules templated alginate based polymer microspheres. React Funct Polym 2006;66(7):712–9.

[221] Bhattacharyya A, Mukhopadhyay P, Kundu PP. Synthesis of a novel pH-sensitive polyurethane-alginate blend with poly (ethylene terephthalate) waste for the oral delivery of protein. J Appl Polym Sci 2014;131(16).

[222] Mladenovskat K, Janevic E, Dodov MG. BSA—loaded gelatin microspheres: comparative studies on biodegradation and drug release in presence of collagenase and trypsin. Maced Pharm Bull 2002;48(1–2):9–14.

[223] Dai YN, Li P, Zhang JP, Wang AQ, Wei Q. Swelling characteristics and drug delivery properties of nifedipine-loaded pH sensitive alginate-chitosan hydrogel beads. J Biomed Mater Res Part B Appl Biomater 2008;86(2):493–500.

[224] Boonsongrit Y, Abe H, Sato K, Naito M, Yoshimura M, Ichikawa H, Fukumori Y. Controlled release of bovine serum albumin from hydroxyapatite microspheres for protein delivery system. Mater Sci Eng B 2008;148(1):162–5.

[225] Bhattacharyya A, Mukhopadhyay P, Pramanik N, Kundu PP. Effect of Polyethylene glycol on bis (2-hydroxyethyl) terephthalate-based Polyurethane/alginate pH-sensitive blend for oral protein delivery. Adv PolymTechnol 2015;(1):35.

[226] Bueno VB, Bentini R, Catalani LH, Petri DF. Synthesis and swelling behavior of xanthan-based hydrogels. Carbohydr Polym 2013;92(2):1091–9.

[227] Bennour S, Louzri F. Study of swelling properties and thermal behavior of poly (N, N-dimethylacrylamide-co-maleic acid) based hydrogels. Adv Chem 2014:2014.

[228] Mandal S, Kumar SS, Krishnamoorthy B, Basu SK. Development and evaluation of calcium alginate beads prepared by sequential and simultaneous methods. Braz J Pharm Sci 2010;46(4):785–93.

[229] El-Kamel AH, Al-Gohary OM, Hosny EA. Alginate-diltiazem hydrochloride beads: optimization of formulation factors, in vitro and in vivo availability. J Microencapsul 2003;20(2):211–25.

[230] Ahmad Z, Pandey R, Sharma S, Khuller GK. Alginate nanoparticles as antituberculosis drug carriers: formulation development, pharmacokinetics and therapeutic potential. Indian J Chest Dis Allied Sci 2006;48(3):171.

[231] Finotelli PV, Silva DD, Sola-Penna M, Rossi AM, Farina M, et al. Microcapsules of alginate/chitosan containing magnetic nanoparticles for controlled release of insulin. Colloids Surf B 2010;81:206–11.

[232] Simpson NE, Grant SC, Gustavsson L, Peltonen VM, Blackband SJ, Constantinidis I. Biochemical consequences of alginate encapsulation: a NMR study of insulin-secreting cells. Biomaterials 2006;27(12):2577–86.

[233] Simpson NE, Stabler CL, Simpson CP, Sambanis A, Constantinidis I. The role of the $CaCl_2$-guluronic acid interaction on alginate encapsulated βTC3 cells. Biomaterials 2004;25(13):2603–10.

[234] Raj NKK, Sharma CP. Oral insulin-a perspective. J Biomater Appl 2003;17(3):183–96.

[235] Li X, Kong X, Shi S, Zheng X, Guo G, Wei Y, et al. Preparation of alginate coated chitosan microparticles for vaccine delivery. BMC Biotechnol 2008;8:89–99.

[236] Machado AH, Lundberg D, Ribeiro AJ, Veiga FJ, Lindman B, Miguel MG, et al. Preparation of calcium alginate nanoparticles using water-in-oil (W/O) nanoemulsions. Langmuir 2012;28:4131–41.

[237] Rajaonarivony M, Vauthier C, Couarraze G, Puisieux F, Couvreur P. Development of a new drug carrier made from alginate. J Pharm Sci 1993;82:912–7.

[238] Googerdchian F, Moheb A, Emadi R. Lead sorption properties of nanohydroxyapatite-alginate composite adsorbents. Chem Eng J 2012;200:471–9.

[239] Badruddoza AZM, Tay ASH, Tan PY, Hidajat K, Uddin MS. Carboxymethyl-beta-cyclodextrin conjugated magnetic nanoparticles as nano-adsorbents for removal of copper ions: synthesis and adsorption studies. J Hazard Mater 2011;185(2–3):1177–86.

[240] Soltani RDC, Khorramabadi GS, Khataee AR, Jorfi S. Silica nanopowders/alginate composite for adsorption of lead (II) ions in aqueous solutions. J Taiwan Inst Chem Eng 2014;45(3):973–80.

[241] Daemi H, Barikani M. Molecular engineering of manipulated alginate-based polyurethanes. Carbohydr Polym 2014;112:638–47.

[242] Daemi H, Barikani M, Barmar M. A simple approach for morphology tailoring of alginate particles by manipulation ionic nature of polyurethanes. Int J Biol Macromol 2014;66:212–20.

[243] Kim JH, Jeong WH, Karegoudar TB, Kim CK. Stable degradation of benzoate by *Klebsiella oxytoca* C302 immobilized in alginate and polyurethane. Biotechnol Bioprocess Eng 2002;7(6):347–51.

[244] Oh YS, Maeng J, Kim SJ. Use of microorganism-immobilized polyurethane foams to absorb and degrade oil on water surface. Appl Microbiol Biotechnol 2000;54(3):418–23.

[245] Manohar S, Kim C, Karegoudar T. Enhanced degradation of naphthalene by immobilization of *Pseudomonas* sp. strain NGK1 in polyurethane foam. Appl Microbiol Biotechnol 2001;55(3):311–6.

[246] Patil NK, Veeranagouda Y, Vijaykumar MH, Nayak SA, Karegoudar TB. Enhanced and potential degradation of o-phthalate by *Bacillus* sp. immobilized cells in alginate and polyurethane. Int Biodeterior Biodegrad 2006;57(2):82–7.

[247] Patil NK, Sharanagouda U, Niazi JH, Karegoudar TB. Stable degradation of catechol by *Pseudomonas* sp. strain NGK 1 encapsulated in alginate and polyurethane foam. Indian J Biotechnol 2004;3(4):568–72.

[248] Papageorgiou SK, Katsaros FK, Kouvelos EP, Nolan JW, Deit HL, Kanellopoulos NK. Heavy metal sorption by calcium alginate beads from *Laminaria digitata*. J Hazard Mater B 2006;137:1765–72.

[249] PIbanez J, Umetsu Y. Potential of protonated alginate beads for heavy metal uptake. Hydrometallurgy 2002;64:89–99.

[250] Silva SS, Menezes SM, Garcia RB. Synthesis and characterization of polyurethane-g-chitosan. Eur Polym J 2003;39(7):1515–9.

[251] Yang S, Leong KF, Du Z, Chua CK. The design of scaffolds for use in tissue engineering. Part I. Traditional factors. Tissue Eng 2001;7(6):679–89.

[252] Zhang Y, Zhang M. Three-dimensional macroporous calcium phosphate bioceramics with nested chitosan sponges for load-bearing bone implants. J Biomed Mater Res 2002;61(1):1–8.

CHAPTER

Algae-Derived Polyester Blends and Composites

12

Aqdas Noreen[1], Mudassir Jabeen[1], Shazia Tabasum[1], Anbreen Anjum[1], Rehana Naseer[2], Tomás J. Madera-Santana[3], Tayyaba Sultana[1], Salma Sultana[1]

Government College University Faisalabad, Faisalabad, Pakistan[1]; Government College Women University, Faisalabad, Pakistan[2]; Centro de Investigación en Alimentación y Desarrollo A.C. Hermosillo, México[3]

12.1 INTRODUCTION

There is an increasing demand for energy and chemicals because of the rapid increase in the world's population, which leads to a decline in fossil fuel reserves and an increase in the cost of crude oil [1–5]. Therefore, it is crucial to explore sustainable resources for the production of fuels, chemicals, polymers, heat, and electricity [6–10]. An additional motivation for reliance on renewable resources is to reduce the emission of dangerous and greenhouse gases produced from fossil fuels [11]. Biomass (trees, crops, aquatic plants, agricultural food, feed crop residue, wood, wood residue, animal wastes, and other waste materials) is a chief renewable reserve to generate energy and value-added products [12]. Approximately 75% of the biomass is made up of carbohydrates (saccharose, starch, and cellulose) [13], 20% contains lignin, and the remaining 5% is composed of natural compounds such as oils, proteins, and other substances. Currently, synthesis of polymers from biobased monomers offers a novel way to generate green polymers from sustainable resources and the properties of these biobased polymers can be improved by blending and composite formation [14].

The polyester industry is one of the main industries that have massive influence on the healthy and sustainable progress of society. Polyesters can be synthesized either by lactone ring opening polymerization or by polycondensation of diacid, hydroxyacid, or anhydride with secondary alcohol [15,16]. Polyesters are generally employed in surface coatings, fibers, rubbers, plasticizers, laminating resins, films, molding compositions, composites, etc. [17,18]. Biomass-based polyesters are used in biomedical field, such as polyhydroxyalkanoate (PHA) [19,20] and polylactic acid (PLA) [21,22]. Biodegradable polyesters such as PLA, poly(glycolic acid) (PGA), poly(ε-caprolactone) (PCL), and their copolymers are widely used in drug release and tissue engineering (TE) [23]. In addition, biomass-based diols and carboxylic acids are used to synthesize thermoplastic polyesters, for instance, poly(ethyleneterephthalate) [24–26], poly(butylene terephthalate) [27], and poly(trimethyleneterephthalate) [28]. Biobased polyesters are divided

Algae Based Polymers, Blends, and Composites. http://dx.doi.org/10.1016/B978-0-12-812360-7.00012-4

into two groups: (1) biopolyesters, i.e., totally synthesized from biomaterial (100% biobased) and (2) those prepared from biobased monomers obtained from renewable materials [29].

PHAs are bioplastics that are manufactured fully by using microorganisms. Over 30% of soil-inhabiting bacteria are used in the preparation of PHAs [30]. PHAs are used in green plastics for packaging material, in biocompatible and biodegradable implants, and in controlled drug delivery systems [31]. PLA is an aliphatic polyester (100% biobased) synthesized from lactic acid and employed in biomedical applications such as sutures, stents, dental implants, bone screws, vascular grafts, and pins [32]. It can also be used as a vector for long-lasting delivery of antimicrobial drugs and as a scaffold for cell and tissue growth [33].

Biomass-derived monomers are used for manufacturing biobased polyester, thereby substituting the petrochemical-based polyesters [34]. Few biobased polyesters synthesized from sustainable biobased monomers are summarized in Table 12.1 [5,33,35–38].

12.2 ALGAE AS ALTERNATIVE BIOMATERIAL FOR BIOBASED POLYESTERS

Monomers of biobased polyesters are generally produced from food and sugar compounds, which will lead to food supply demand stress by disturbing the supply–demand relationship. Thus, it is important to explore nonfood biomass-based monomers to synthesize polymers [17].

In this view, algal biomass is a low-cost, abundant, and unexploited renewable resource that can be explored to generate biobased polyesters and other polymers [39]. Several salient features [40–42] that make alga an excellent candidate for sustainable source are as follows:

- High growth rate
- High photosynthetic efficiency
- Great potential for carbon dioxide (CO_2) fixation
- Low amount of lignin
- High percentage of carbohydrates

Algae are a renowned viable resource of high-value chemical compounds such as carbohydrates, proteins, cellulose, starch, lipids, polyunsaturated fatty acids (PUFAs), antioxidants, pharmaceuticals, and pigments [43–45]. In addition, algae contain unique metabolites such as agar, alginate, carrageenan, fucoidan, ulvan, and laminarin [46–51]; therefore, algal biomass is believed to be a novel raw material that is converted into numerous products.

Table 12.1 Biomass-Derived Polyesters

S. No.	Biobased Monomer	Source of Monomer	Other Monomers Involved in Polyester Synthesis	Structure of Biobased Polyester	References
1.	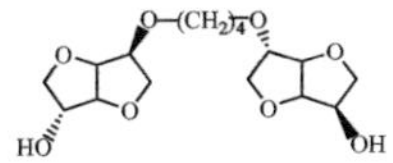 Isosorbide-based diol-ether	D-glucose	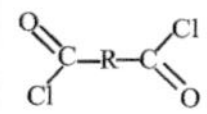	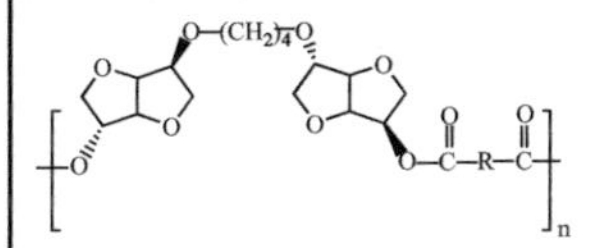 Poly(ether–ester)	[35]
2.	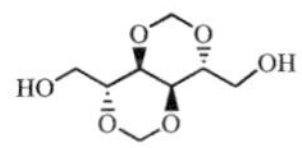 2,4:3,5-Di-O-methylene-D-mannitol (Manx)	Mannitol	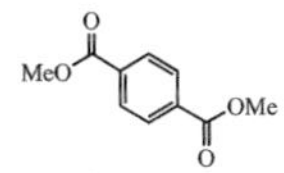 and 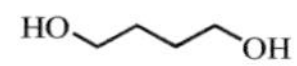	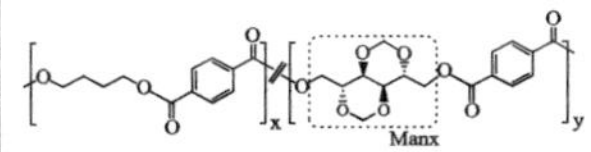 PB_xManx_yT	[5]
3.	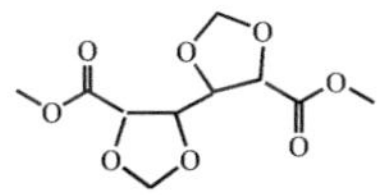 Bicyclic diacetalyzed galactaric acid (Galx)	Galactaric acid	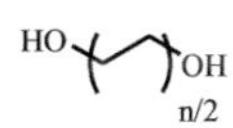 n = 6,8,10,12	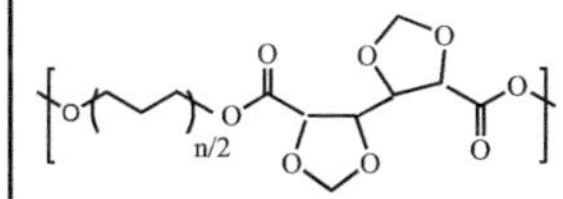 PEnGalx	[33]
4.	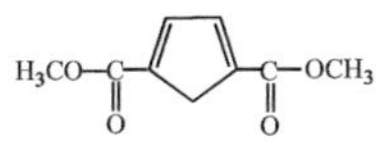 Dimethyly-2,5-furandicarboxylate	Furan	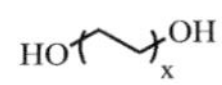 x = 2,3,4	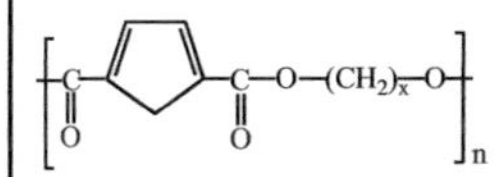 2,5-Furandicarboxylate polyesters	[36]

Continued

Table 12.1 Biomass-Derived Polyesters—cont'd

S. No.	Biobased Monomer	Source of Monomer	Other Monomers Involved in Polyester Synthesis	Structure of Biobased Polyester	References
5.	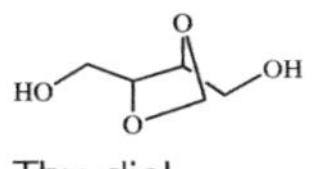 Thx-diol	Tartaric acid	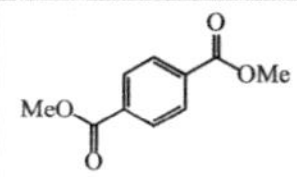 and 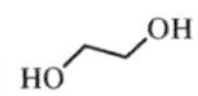	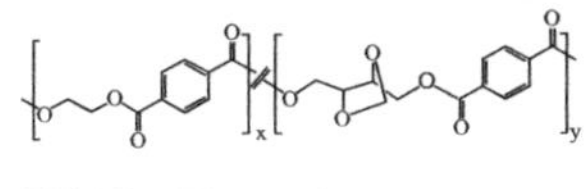 PE_xThx_yT copolyesters	[37]
6.	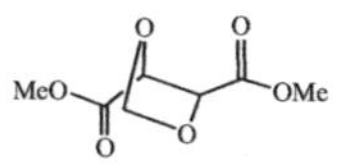 Thx-diester	Tartaric acid	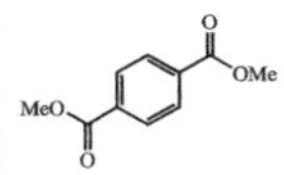 and 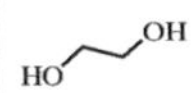	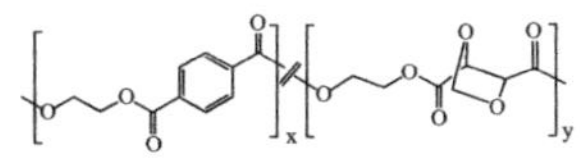 PET_xThx_y copolyesters	[37]
7.	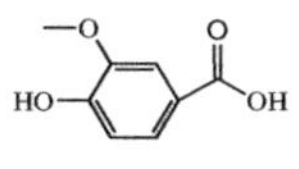 Vanillic acid	Lignin		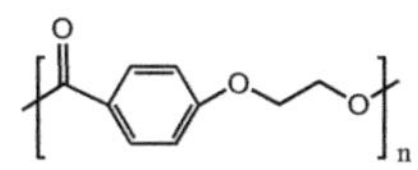 Polyethylene vanillate	[38]

12.3 POLY-β-HYDROXYBUTYRATE PRODUCTION FROM ALGAE

Approximately, 140 million tons of plastic is consumed annually all over the world, which uses up about 150 million tons of petrochemicals. Such a massive plastic production produces a huge quantity of waste materials that can require thousands of years for total biodegradation [52]. Therefore, greener and biodegradable plastic that originates from renewable materials is a good alternative of petrochemical-based plastic.

Poly(3-hydroxyalkanoate) (PHA) is considered a green substitute for conventional nonbiodegradable plastic that comes from sustainable resources. Of the PHAs, poly(3-hydroxybutyrate) (PHB) has a melting point close to that of polypropylene (PP), mechanical properties similar to those of PP, and oxygen barrier property better than that of PP.

The property that discriminates PHB from conventional petroleum-based plastic is its complete biodegradability [53–56]. Biodegradation of PHB in an aerobic environment produces CO_2 and H_2O, whereas under anaerobic conditions, the degradation results in CO_2 and CH_4. PHB is degraded in a wide variety of temperatures, still at a maximum of about 60°C with 55% of moisture levels. Biodegradation depends on various factors, i.e., moisture, pH, temperature, exposed surface area, and microbial activity of the environment. PHB is synthesized from acetyl coenzyme A by three enzymatic reactions (Fig. 12.1) [57].

Bacteria can generate up to 80% PHB of cellular dry weight in optimal conditions, but the price tag for PHB synthesis by bacterial fermentation is very high, which placed an emphasis on using plants as a photosynthetic low-priced production system [58–61]. Bacterial enzymes were targeted to different compartments or uttered in the cytosol of the plant cell, which produced PHB up to 40% of dry weight in the plastids of *Arabidopsis thaliana* [62,63]. However, owing to small growth and infertility, these plants were not appropriate for large-scale PHB production. At present, the maximum level of PHB synthesis in plants with fertile offspring is achieved in the plastid of *Nicotiana tabacum* (up to 18% PHB of cellular dry weight) [64]. Plant-based production systems compete with crops for agricultural land and the propagation of transgenic plants is hard to control. The main shortcoming for a plant-based expression system is the reasonably high growth rate resulting in a low profitability.

Microalgae have all the advantages of plant-driven systems along with high growth rates, ease of handling, and less requirement of light and water for cultivation [65]. Being a low-cost expression system, microalga has huge potential particularly for the biosynthesis of recombinant proteins important in several industrial, diagnostic, or therapeutic applications [66–69]. Hempel et al. [70] obtained 10.6% PHB of algal dry weight by establishing the bacterial PHB pathway in the cytosolic compartment of the diatom *Phaeodactylum tricornutum*.

The blue-green algae (cyanobacteria) show various structural features in common with bacteria. Conversely, they are classified with algae because they have

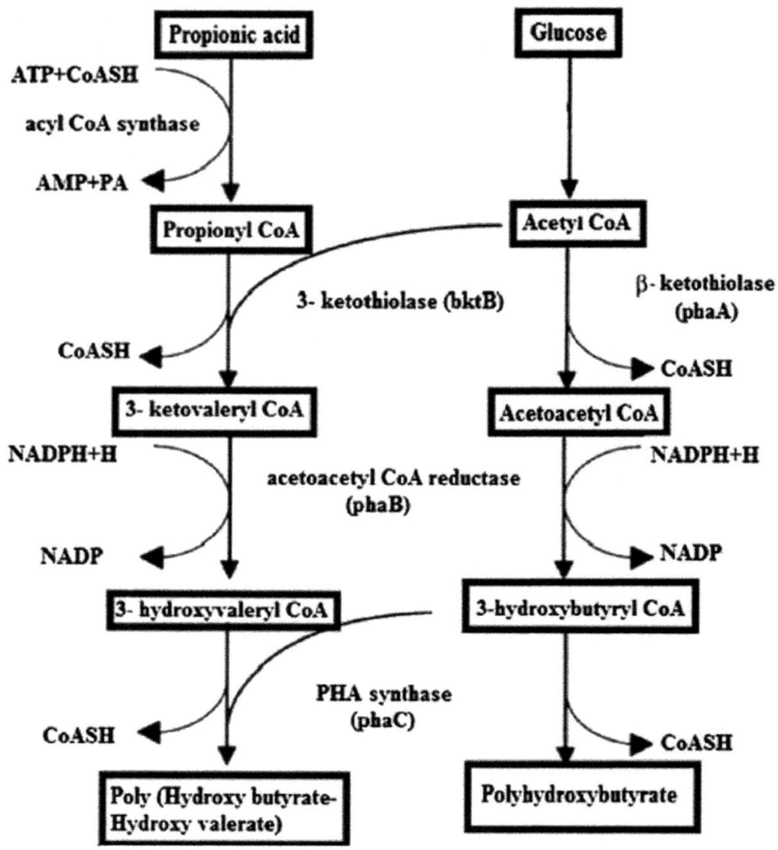

FIGURE 12.1

Poly-β-hydroxybutyrate production pathway [57].

chlorophyll a and related compounds [71]. These organisms are photoautotrophic and require the lowest amount of nutrients, thereby offering an alternative host system for PHB production. Cyanobacteria, a group of oxygen-evolving photosynthetic bacteria with a short generation time, require few simple inorganic nutrients (nitrate, phosphate, calcium, magnesium, ferrous, manganese, zinc, cobalt, and copper) for their growth and multiplication [72,73]. In 1966 the presence of PHB was reported for the first time in *Chlorogloea fritschii* (cyanobacterial species). So far, the existence of PHB has been confirmed for numerous cyanobacteria, such as *Aphanothece* spp., *Gloeothece* spp., *Spirulina* spp., and *Synechococcus* spp. [74–78].

Wastewater coming from rubber industries, fish farms, farmyards, and sewage treatment plants is a rich source of phosphorous and nitrogen [79,80]. Cyanobacteria can efficiently be cultivated in wastewater owing to their capability to use inorganic phosphorous and nitrogen. Different species of cyanobacteria accumulate significant quantities of PHB. Industrial utilization of cyanobacteria as PHB producers has the advantage of converting waste carbon dioxide, a greenhouse gas, to environmental friendly plastics using the energy of sunlight. Non-PHB-accumulating cyanobacteria have been genetically engineered by transformation with genes involved in the PHB pathway [81].

Genetically engineered cyanobacteria that accumulate PHB are transformed with the genes encoding PHB synthesis, viz., 3-ketothiolase, acetoacetyl-CoA reductase, and PHB synthase [82–84]. Metabolic engineering is enormously investigated to initiate novel pathways to enhance PHB synthesis and to increase the exploitable substrate range. Wanga et al. [85] assembled and optimized the 3HB biosynthesis pathways in the cyanobacterium *Synechocystis* 6803 and produced (R)- and (S)-3HB from CO_2 in the engineered cyanobacteria.

12.4 SYNTHESIS OF ALGAE-DERIVED POLYESTERS

Algae are believed to be novel raw materials to assemble a biorefinery because of their capacity to produce multiple inimitable metabolites [86–88]. Biorefining is conversion of renewable biomass to biofuels, bioenergy, and value-added products [89]. Biorefining is similar to the petroleum refineries in which crude oil is used as a starting material for production of chemicals and fuels.

12.4.1 ALGAL CARBOHYDRATE–BASED POLYESTERS

Biological or chemical transformation of lignocellulosic biomass is more difficult than that of carbohydrate-containing algal biomass [90]. Algal biomass has a high carbohydrate content, no lignin, and simplicity of extraction of carbohydrate from the raw feedstock. For example, approximately 75%–80% of the total dried mass is carbohydrates in *Gelidium amansii* (red algae).

12.4.1.1 Agar

Agar is an algal polysaccharide derived commonly from macroalgae, which is converted into 5-hydroxymethylfurfural (HMF) (Fig. 12.2A) [91]. Algal polysaccharides depolymerize to C_5 and C_6 sugars that can readily produce value-added monomers for polymer synthesis. The key constituent of agar is agarose, which has high reactivity because of the existence of a highly strained extra cyclic unit linked to 3,6-anhydrogalactose. In addition, agarose has a unique linkage pattern ($C_1 \rightarrow C_3$ linkage) between monomers, whereas in amylose, a component of starch, monomers are linked by $C_1 \rightarrow C_4$. Agarose cleaves in acidic media to produce a carbocation at the C_3 position, which in turn gives 3-deoxy-2-keto galactose. Finally, 5-membered ring closure generates a furanose-type intermediate as a precursor of HMF [93].

Simultaneous or successive oxidation of the CH_2OH and CHO functional groups of HMF produces 2,5-furandicarboxylic acid (FDCA) [94,95]. The structure of FDCA is related to that of terephthalic acid, and it is believed to be one of the

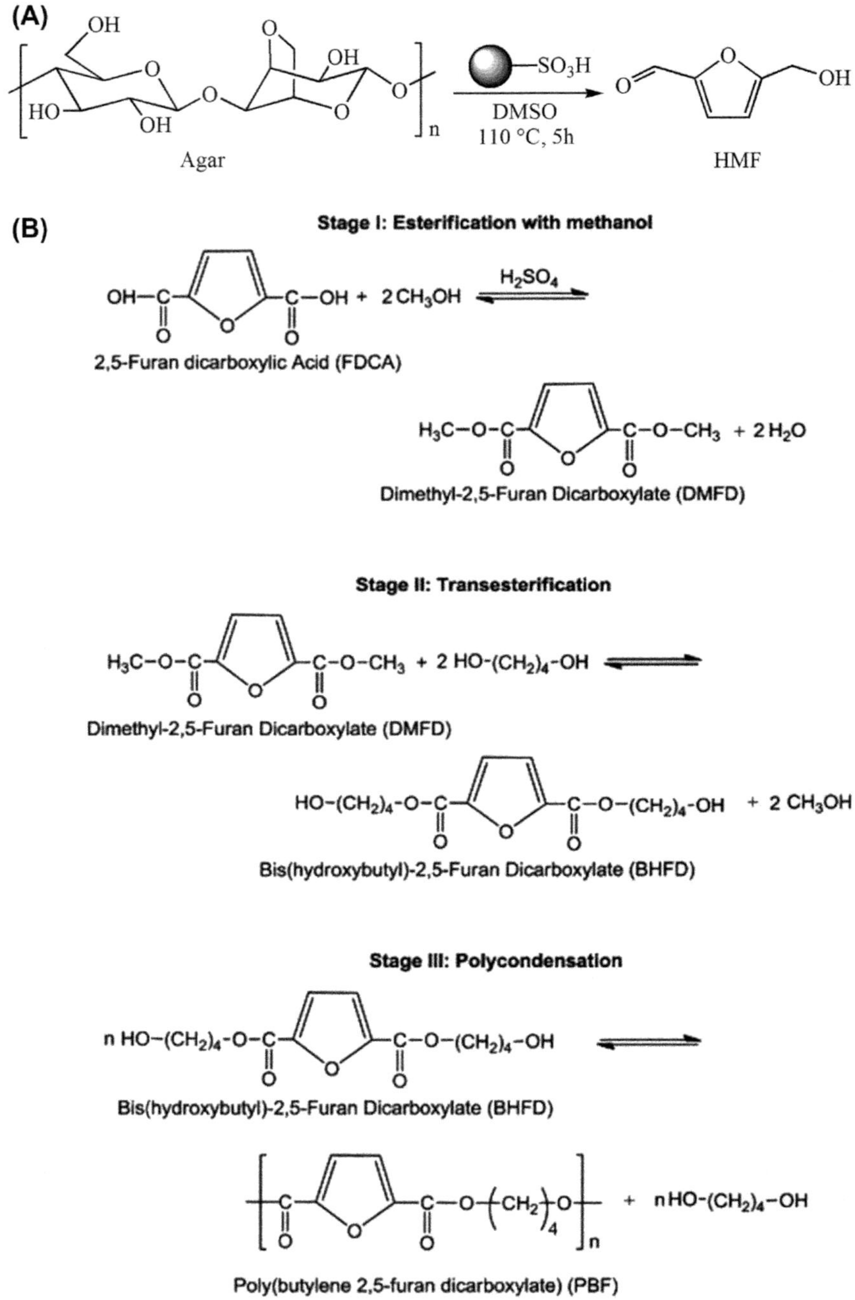

FIGURE 12.2

(A) Single-step conversion of macroalgae-derived agar into hydroxymethylfurfural (HMF) in the presence of solid Brönsted acid [91]. (B) Synthetic route for production of 2,5-furandicarboxylate polyesters [92]. *DMSO*, dimethyl sulfoxide.

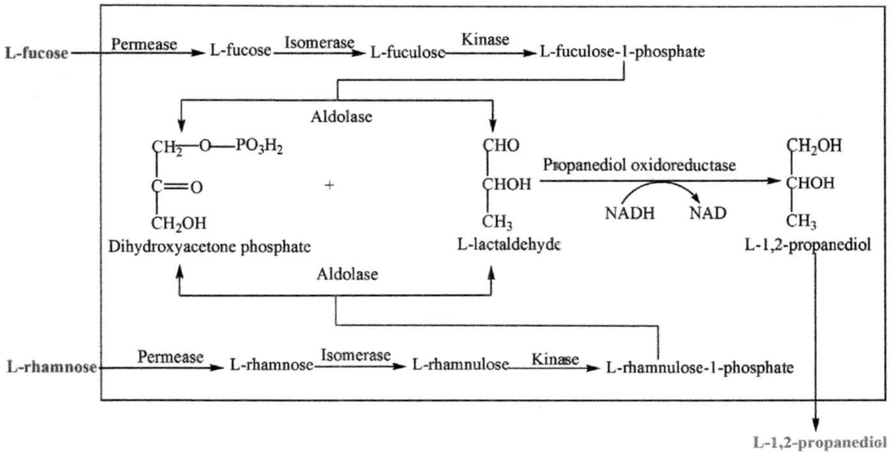

FIGURE 12.3

Schematic representation of metabolic pathways for the production of 1,2-propane diol from fucose/rhamnose [100].

most appropriate monomers for the synthesis of biobased polyesters [96]. Poly(ethylene 2,5-furandicarboxylate), poly(propylene 2,5-furandicarboxylate) (PPF), and poly(butylenes 2,5-furandicarboxylate) are biobased polyesters prepared from FDCA (Fig. 12.2B) [9,92,97–99].

12.4.1.2 Fucose and Rhamnose

Biological conversions of algae-derived unusual C_5 and C_6 monomers (fucose and rhamnose) give a higher yield of 1,2-propanediol (Fig. 12.3) compared with biological transformation of glucose [100]. 1,2-Propanediol is widely employed as a building block for polyester synthesis.

12.4.2 ALGAE OIL–DERIVED POLYESTERS

Higher animals and plants are deficient in enzymes necessary for the synthesis of PUFAs of more than 18 carbon atoms [101,102]. Fish oils and algae oils are the common sources of long-chain PUFAs [103]. Table 12.2 presents the microalgal PUFAs

Table 12.2 Particularly Interesting Microalgal Polyunsaturated Fatty Acids (PUFAs) [104]

PUFA	Structure	Microorganism Producer
γ-Linolenic acid	18:3ω6, 9, 12	*Arthrospira*
Arachidonic acid	20:4ω6, 9, 12, 15	*Porphyridium*
Eicosapentaenoic acid	20:5ω3, 6, 9, 12, 15	*Nannochloropsis, Phaeodactylum, Nitzschia*
Docosahexaenoic acid	20:6ω3, 6, 9, 12, 15, 18	*Crypthecodinium, Schizochytrium*

of particular interest [104]. Algae produce lipids that differ from those obtained from higher animals and plants [105]. Fatty acids of traditional plant oils are in the form of triacylglycerides [106–108], whereas in algal oil, fatty acids are mainly in the form of diacylglycerides in which the third hydroxy group of glycerol is substituted with a polar moiety, such as galactosyl or phosphate (Fig. 12.4A) [109]. Algae can also produce triacylglycerides, particularly when they are in the steady growth phase for a

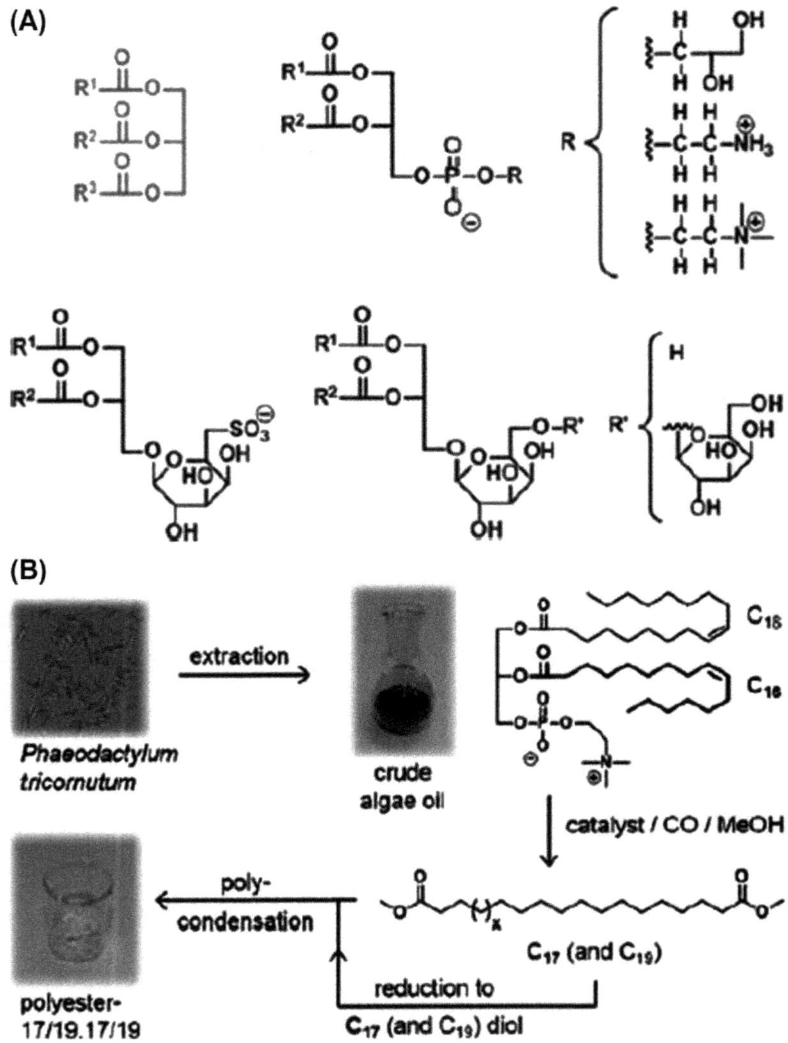

FIGURE 12.4

(A) Fatty acids as they appear in plants (gray, top left) and algae (black) [109]. (B) Approach to α,ω-dicarboxylic acid intermediates and polyester materials from algae oil [110].

longer period. Algae oil is converted into difunctional novel monomers for the synthesis of biobased polyesters (Fig. 12.4B) [110].

12.5 ADVANCEMENTS IN ALGAE-BASED POLYESTERS AND THEIR NOVEL APPLICATIONS

Hydrogels have excellent biocompatibility, good nutrient/oxygen transport ability, and resemblance to biological tissues because of their water swelling characteristics [111,112]. Consequently, they have been extensively used as therapeutic biomaterials for transplantation, cell immobilization, drug delivery systems, and wound healing [113,114]. However, hydrogels have poor mechanical properties; therefore, composites have been established to improve mechanical stiffness and durability [115,116]. Reinforcement shares the external load with the surrounding matrices [117]. Therefore, composites are believed to be excellent substitutes of hydrogels with reasonable strength and structural biocompatibility.

Nanofibers are gaining huge attention as efficient mechanical reinforcements in hydrogels because of their outstanding mechanical behavior and large surface area to volume ratios [118–121]. Therefore, keeping in view the mechanical strength and structural biocompatibility of composites and nanocomposites, algal-polymer/polyester blends and composites with tunable biodegradability and biocompatibility have been developed. These blends and composites are used widely for novel applications, which are discussed in the following sections.

12.5.1 ALGAE-DERIVED POLYESTERS IN TISSUE ENGINEERING

TE attempts to recover or replace injured organs or tissues by using a combination of specific cell types, scaffold materials, and growth factors [122,123]. Natural tissues generally hold a cellular structural part that is packed by a much softer matrix having the required quantity of cells [124]. Composites are fabricated using prearranged scaffolds and a suitable number of cells to mimic the native tissue design [125]. A general approach to use porous scaffolds for TE is demonstrated in Fig. 12.5. To accomplish the goal of tissue restoration, scaffolds have to serve the following functions [126]:

- Allow cell attachment and movement
- Transport and hold cells and biochemical factors
- Allow delivery of cell nutrients and expressed products
- Put forth certain biological and mechanical influences

Fiber-based techniques such as electrospinning, weaving, knitting, braiding, and direct writing are used for making three-dimensional (3D) tissue constructs that can meet the abovementioned challenges. An overview of the instruments and techniques used for making fibrous structures and scaffolds and representative microstructures are shown in Fig. 12.6 [127].

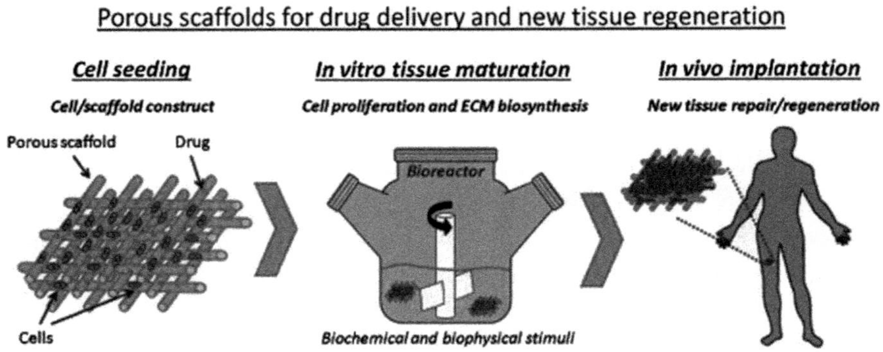

FIGURE 12.5

Diagram of the use of porous scaffolds for the repair or regeneration of tissues and organs in tissue engineering [126]. *ECM*, extracellular matrix.

A number of uses of hybrid scaffold are recognized, varying from musculoskeletal tissues to blood vessels [128,129].

12.5.1.1 Bone Tissue Engineering

Cell-laden scaffolds have been fabricated with multilayer structures using various computer-aided bottom-up processes [130–135]. Kim et al. [136] described a new hybrid cell-printing method to manage the internal microporous structure and to mechanically reinforce the 3D hydrogel scaffolds. First, single layers composed of micron-scale perpendicular PCL/alginate struts were fabricated by using a melt-plotting method to get a structure that can retain the mechanical strength of the scaffold [137]. Second, a bioink made up of a mixture of cell-laden hydrogels was printed onto the PCL/alginate struts. These two steps can be repeated several times in a layer-by-layer manner to give a 3D microporous cell-laden structure. The laden cells showed a significantly more developed cytoskeleton compared with those on a control scaffold consisting of the same material composition. The printed cells displayed a more homogeneous distribution within the scaffold compared with the control. Moreover, the cell-laden scaffold showed considerably better osteogenic activity compared with the control scaffold.

PLA and poly(lactic-co-glycolic acid) (PLGA) polyesters are the most commonly used biopolymers for bone TE [138–140]. PLA and PLA-calcium phosphate scaffolds can be prepared by using a gas foaming system and implanted in vivo for bone regeneration [141].

Fucoidan has anticoagulant, antiviral, and immunomodulatory behavior and performs several biological functions. A new biocomposite, prepared by the incorporation of fucoidan in elctrospun PCL, can be used for bone tissue regeneration [142,143]. Ulvan/poly(D,L-lactic acid) (PDLLA) and ulvan/PCL scaffolds are used for bone TE, wound dressings, and drug delivery systems [144,145].

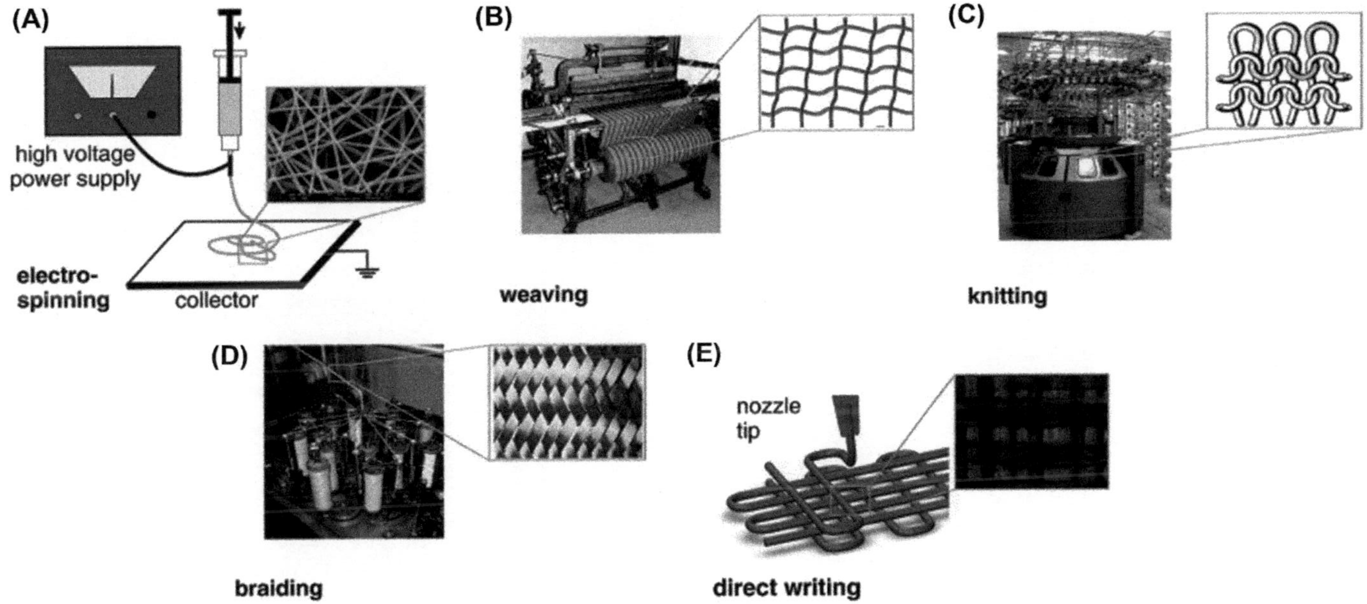

FIGURE 12.6

(A) Electrospinning of polymeric fibers and their deposition on a surface to compose a fibrous scaffold; (B) weaving of scaffolds by interlacing warp and weft fibers in vertical directions; (C) knitting of yarns in a sequence of connected loops; (D) braiding of fibers accomplished by intertwining three or more fiber strands; (E) direct writing of fibers by extruding fibers through a nozzle [127].

The purpose of cholesteatoma surgery is to get a safe, dry, and self-cleaning ear. Unluckily, growth of a problematic mastoid cavity is a hurdle of this therapeutic treatment [146,147]. The aim of mastoid obliteration is to replace the open mastoid cavity with a material that is compatible and free of infection and cholesteatoma. A variety of materials, both biological and alloplastic, have been employed for mastoid obliteration [148–154]. Osteoinductive potential in mastoid obliteration of a novel scaffold composed of PCL, alginate, and recombinant human bone morphogenetic protein-2 (rhBMP-2) was reasonably higher compared with control material. Microcomputerized tomography exhibited osteogenesis in the pores of the 3D-PCL scaffolds in the experimental group, compared with the control group (Fig. 12.7) [155].

Human umbilical cord serum (hUCS) has many growth factors. hUCS/PCL/alginate has great osteogenetic ability in mastoid obliteration than the control group [156]. Wang et al. [157] employed alginate and chitosan as surface modifiers to compose oppositely charged PLGA nanoparticles. Cohesive colloidal gels can be prepared by simply mixing the oppositely charged PLGA nanoparticles at different ratios, which exhibit attractive rheological behavior for facile injection as tissue scaffolds (Fig. 12.8). These gels are highly compatible with human umbilical cord mesenchymal stem cells [157].

12.5.1.2 Cartilage Tissue Engineering

Human adipose tissue–derived stem cells (hASCs) are transduced with a replication-deficient adenovirus carrying hTGF b2(Ad5-hTGF b2), and then the transduced cells are seeded and cultured in PLGA/alginate compounds [158]. The hASCs have a chondrogenic potential, hence making them a promising source of seeding cells for cartilage TE.

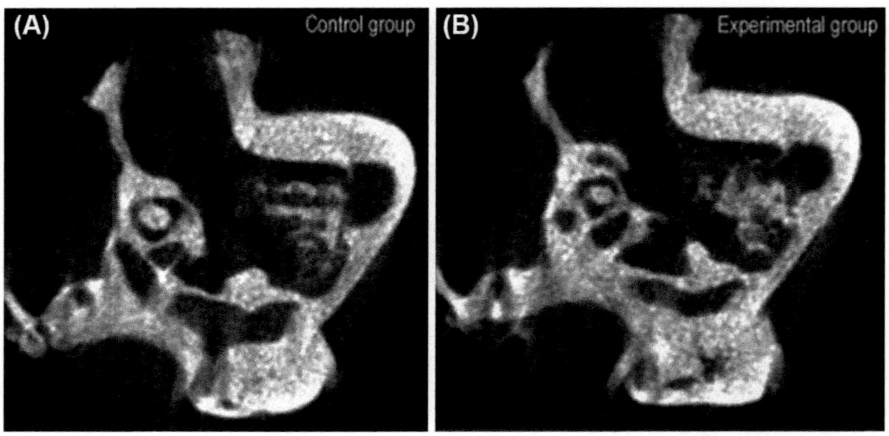

FIGURE 12.7

Ex vivo microcomputerized tomography shows osteogenesis in the porous scaffolds. Some pores of the poly(ε-caprolactone)/alginate/BMP-2 scaffold are filled with bony tissue, whereas in the control group, most pores are unoccupied [155].

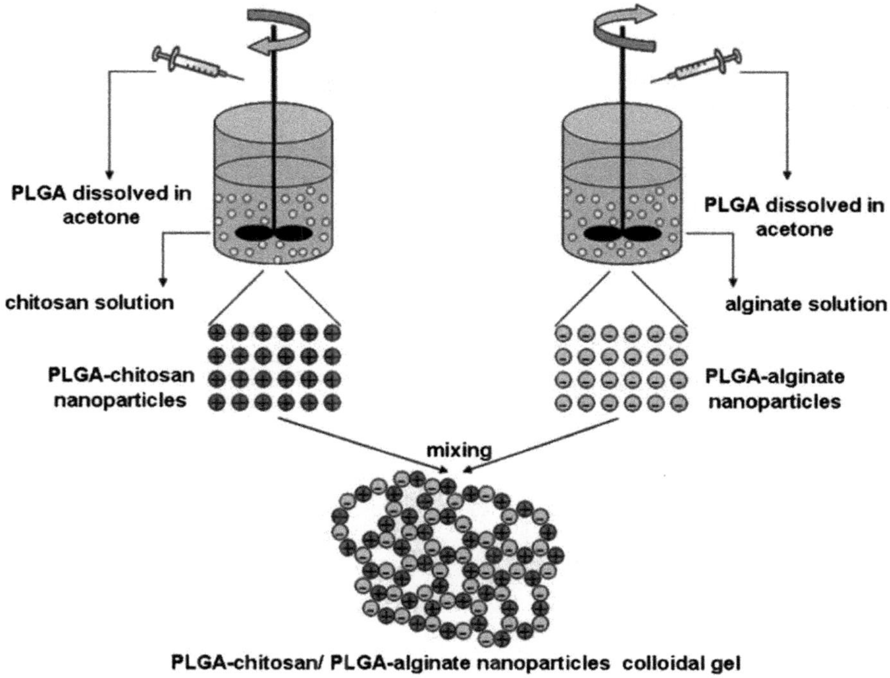

FIGURE 12.8

Schematic presentation of the process for fabrication of oppositely charged nanoparticles and formation of colloidal gel. *PLGA*, poly(lactic-co-glycolic acid).

12.5.1.3 Cardiac Tissue Engineering

The development of scaffolds for cardiac TE is a big challenge. Poly(mannitol fumarate-co-sebacate) (pFMSA), an unsaturated polyester, is copolymerized with alginate hydrogels. Three bimodal hydrogel scaffolds pFMSA-AA, pFMSA-MA, and pFMSA-NMBA can be prepared by cross-linking the alginate segments with calcium ions (Ca^{2+}) and the unsaturation of the esters with different vinyl monomers viz., acrylic acid (AA), methacrylic acid (MA), and N,N0-methylene bisacrylamide (NMBA), respectively. pFMSA-AA exhibits better physiochemical properties compared with the other two hydrogels. The physiochemical and biological favorability of pFMSA-AA hydrogels indicates its aptness for cardiac TE [159] (Fig. 12.9).

A graft comacromer of alginate and polypropylene fumarate (PPF) is used to prepare monomodal network hydrogel (HPAS-NO) and bimodal network hydrogel (HPAS-AA) (Fig. 12.10) [160]. The HPAS-AA hydrogel is more stable than HPAS-NO in the culture medium, with slow degradation in vitro in phosphate buffered saline. The HPAS-AA has a nanoporous structure and gets structured water along with free water. The HPAS-AA hydrogel is cytocompatible and supports 3D cell growth of L929 fibroblast even after 18 days and H9C2 cardiomyoblasts. The enhanced and long-term cellular growth of HPAS-AA hydrogel is due to the cell-responsive feature of structured water. Therefore, HPAS-AA hydrogel has a great potential for cardiac TE.

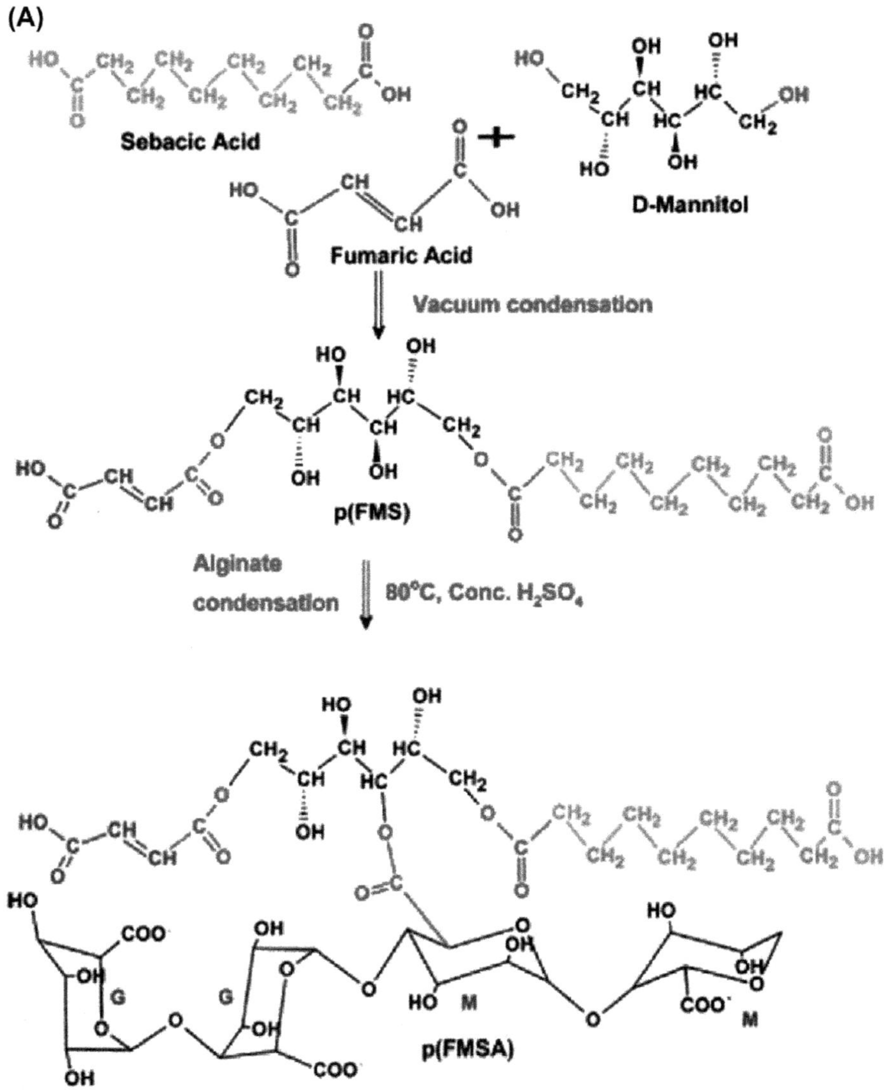

FIGURE 12.9

Synthesis of poly(mannitol fumarate-co-sebacate) (pFMSA)-AA, pFMSA-MA, and pFMSA-NMBA hydrogel scaffolds [159].

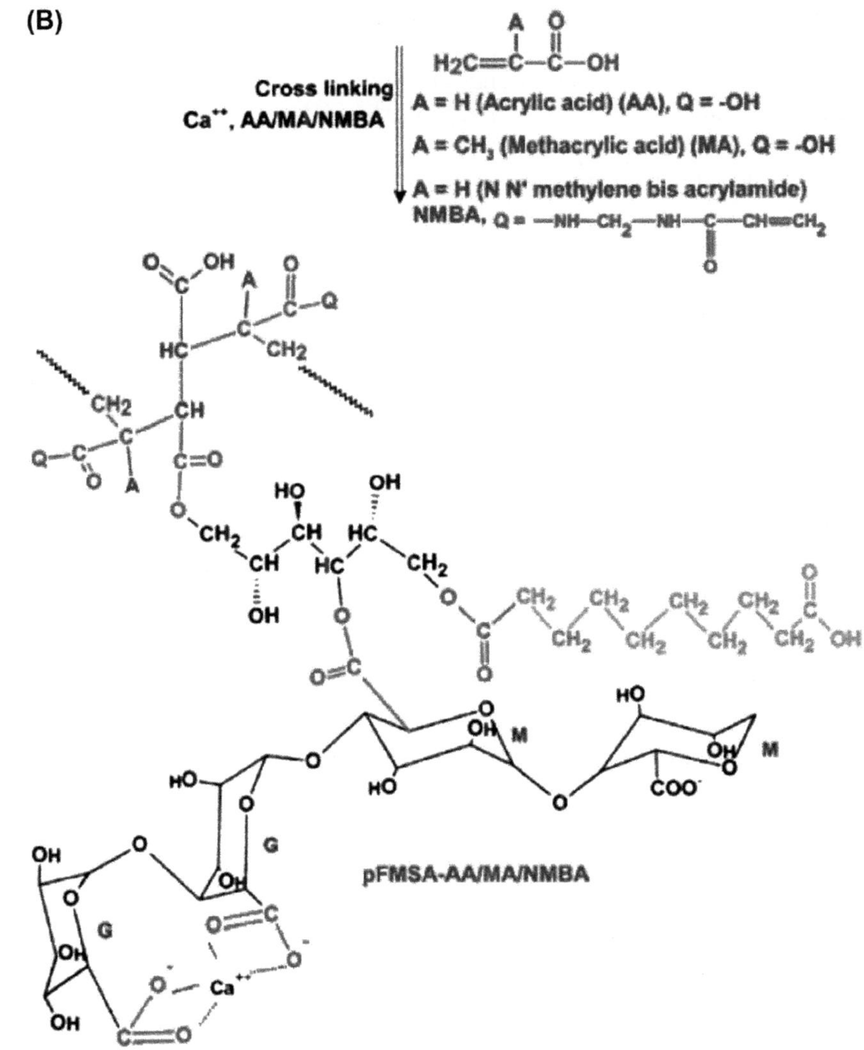

FIGURE 12.9 cont'd.

12.5.1.4 Adipose Tissue Engineering

Adipocytes derived from the multipotential population of adult human bone marrow stromal cells (HBMSCs) are an important cell source for adipose TE [161–164]. Morgan et al. [165] assessed the role of poly(DL-lactide-co-glycolide) ($P_{DL}LGA$) hollow fibers in the development of HBMSC-derived adipocytes for progress of a connected adipocyte layer. The large surface area of the $P_{DL}LGA$ fibers assisted the fast development of extensive adipocyte aggregates from an undifferentiated HBMSC monolayer.

12.5.1.5 Regeneration of Damaged Nerve

Restoration of damaged nerves is a normal medical problem. Direct reconnection of proximal and distal nerve ends and transplantation of autologous nerve grafts are used for injured peripheral nerve repair. Although the former method is used just for small defects, the later process requires the second surgical step for removal of the donor nerve. Other drawbacks [166,167] for this method are as follows:

- Permanent failure of the donor nerve function
- Inadequate supply of grafts
- Difference between defect nerve and graft nerve dimension

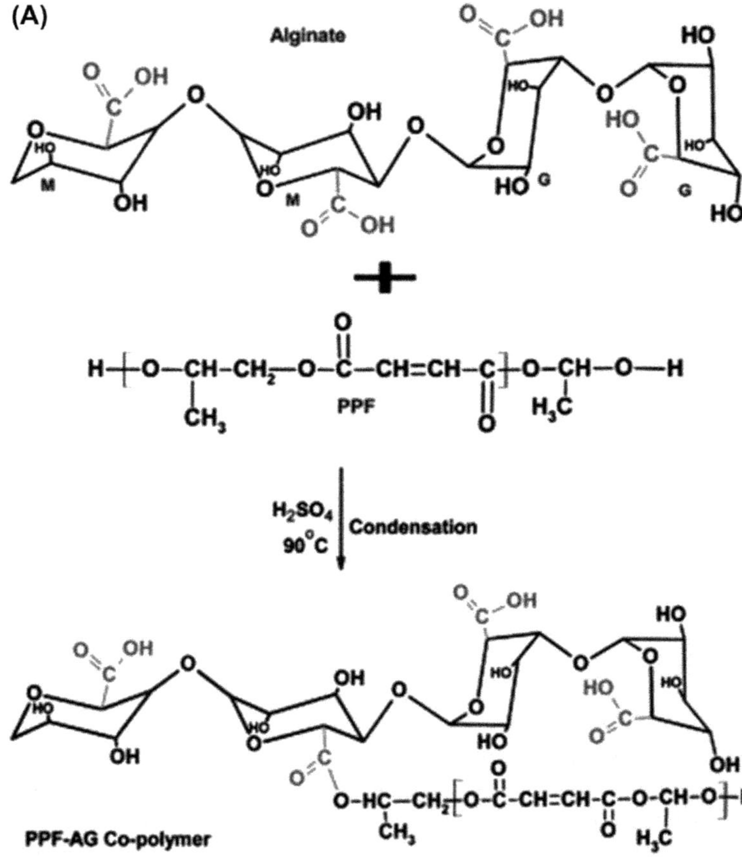

FIGURE 12.10

(A) Synthesis of poly(propylene 2,5-furandicarboxylate) (PPF)-alginate (ALG) graft comacromer and formation of monomodal network hydrogel (HPAS-NO) and (B) bimodal network hydrogel (HPAS-AA) [160].

(B)

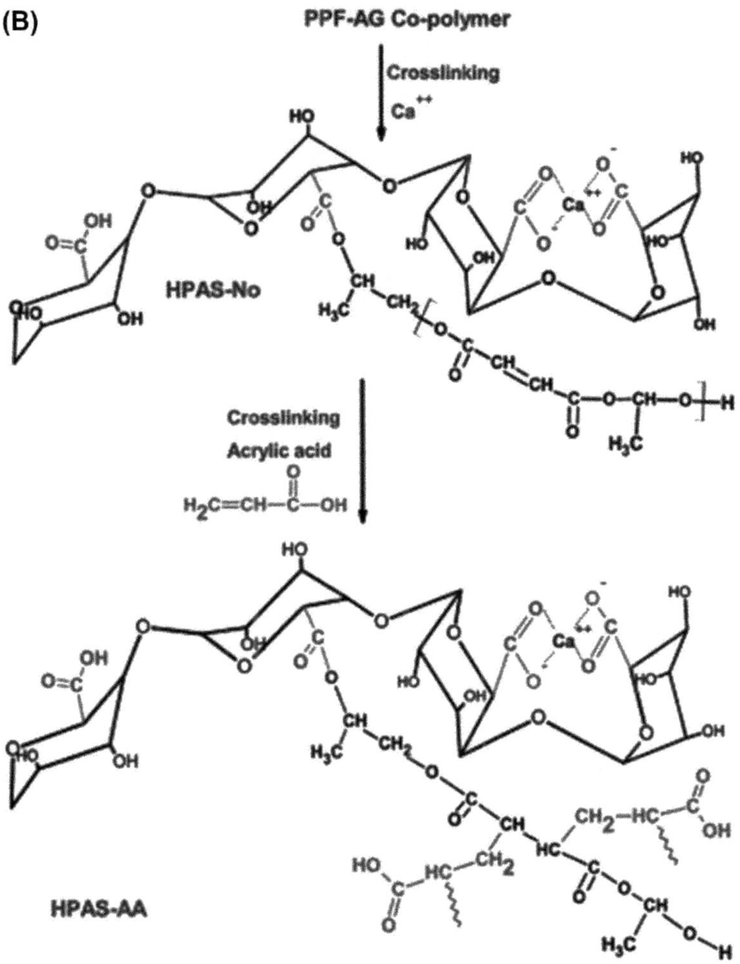

FIGURE 12.10 cont'd.

Consequently, for nerve regeneration, artificial nerve guide conduit (NGC) is extensively used to connect the gap between the detached peripheral nerve ends [168,169]. Biodegradable PLA [170], PGA [171], poly(lactic-co-glycolic acid) (PLGA) [172,173], and poly(phosphoester) are commonly used for NGC fabrications. An immersion precipitation method is developed to fabricate the porous NGC with selective permeability and hydrophilicity to avoid fibrous scar tissue penetration [174]. In this method, porous PLGA tubes such as NGC are prepared by immersing a water-saturated rod-shaped alginate hydrogel into a PLGA/Pluronic F127 mixture solution (Fig. 12.11). The hydrophilized PLGA/F127 tubes have efficient infiltration of nutrients and good mechanical strength and therefore maintained a stable support structure for nerve regeneration.

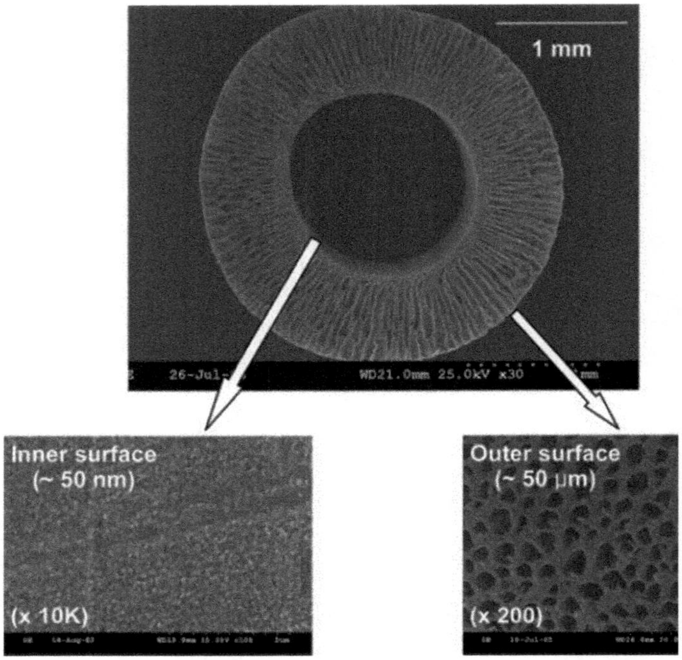

FIGURE 12.11

Scanning electron micrograph showing the morphologies of cross-sectional, inner, and outer surfaces of PLGA/F127 (3 wt%) tubes such as a nerve guide conduit (immersion time of alginate hydrogel rod in PLGA/F127 mixture solution, 5 min) [174].

12.5.2 ALGAE-DERIVED POLYESTERS IN WOUND HEALING

Since the early 1980s, a lot of wound dressings have been developed to promote wound healing [175]. The ideal dressing requires ensuring that the wound remains moist with exudates, but not macerated and free of infection, while accomplishing prerequisites concerning structure and biocompatibility [176]. In addition, it should be noncytotoxic and nonantigenetic, retain cell viability and phenotype, assure homogeneous cell distribution, and encourage movement and propagation of epithelial cells, fibroblasts, and endothelial cells, as well as the synthesis of extracellular matrix components required for wound repair [177]. Besides, wound dressings should display ease of application and removal and proper adherence [178].

Fucoidans are recognized to attach heparin-binding growth factors, such as fibroblast growth factor (FGF)-1 and FGF-2. Low-molecular-weight fucoidans increase FGF-2 activity and FGF-2-induced tube formation in endothelial cells [179,180]. Fucoidans also change the effects of transforming growth factor b1 on wound repair [181]. A composite hydrogel sheet composed of alginate, chitin/chitosan, and fucoidan (ACF-HS) produced from blended powder possesses all the advantages as mentioned earlier [182] (Figs. 12.12 and 12.13).

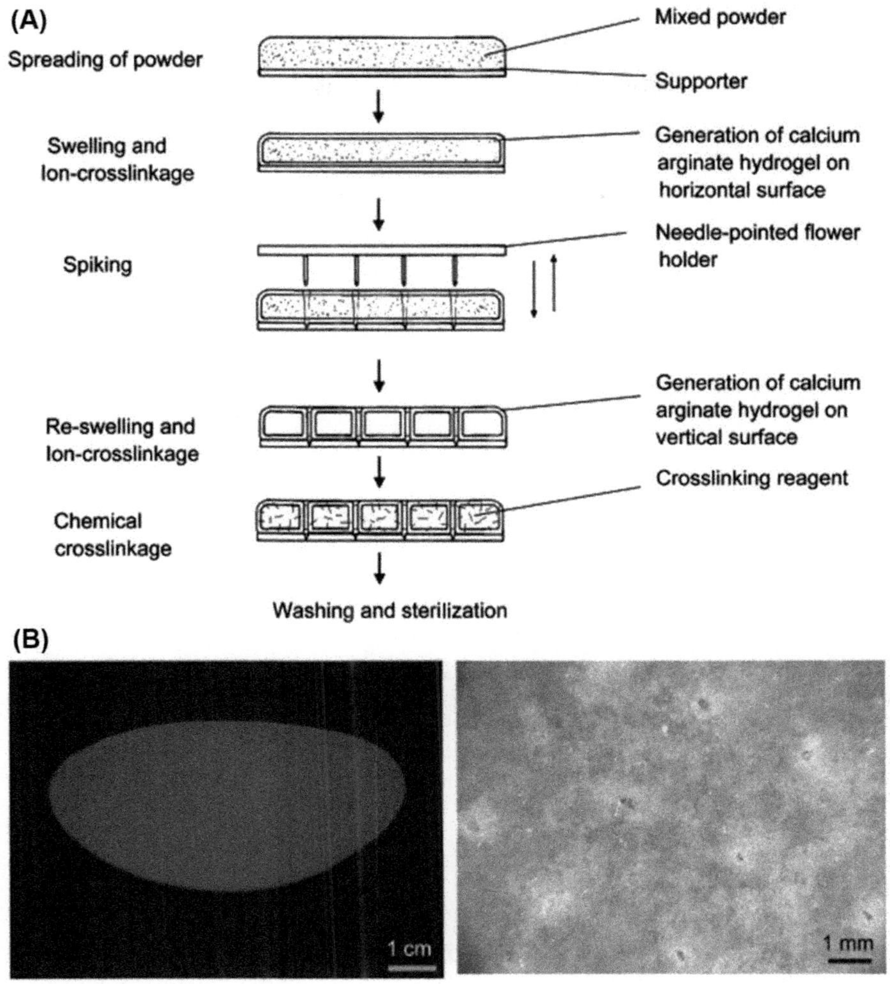

FIGURE 12.12

(A) Preparative procedure for hydrogel sheet composed of alginate, chitin/chitosan, and fucoidan (ACF-HS) and (B) photographic appearance of ACF-HS.

12.5.3 ALGAE-DERIVED POLYESTERS IN DRUG DELIVERY SYSTEM

Polymer-based drug delivery systems have been used for increasing the efficiency and specific targeting of new and already existing drugs. In this regard, the biodegradable polymers are of main concern because of their major advantage of elimination of polymer metabolites out of the body by innate metabolic processes. Microcapsules and microspheres of biodegradable polymers are prepared for controlled release of various pharmaceutical agents and macromolecules, which permits a constant release of drug and produces both humoral and cellular immunity

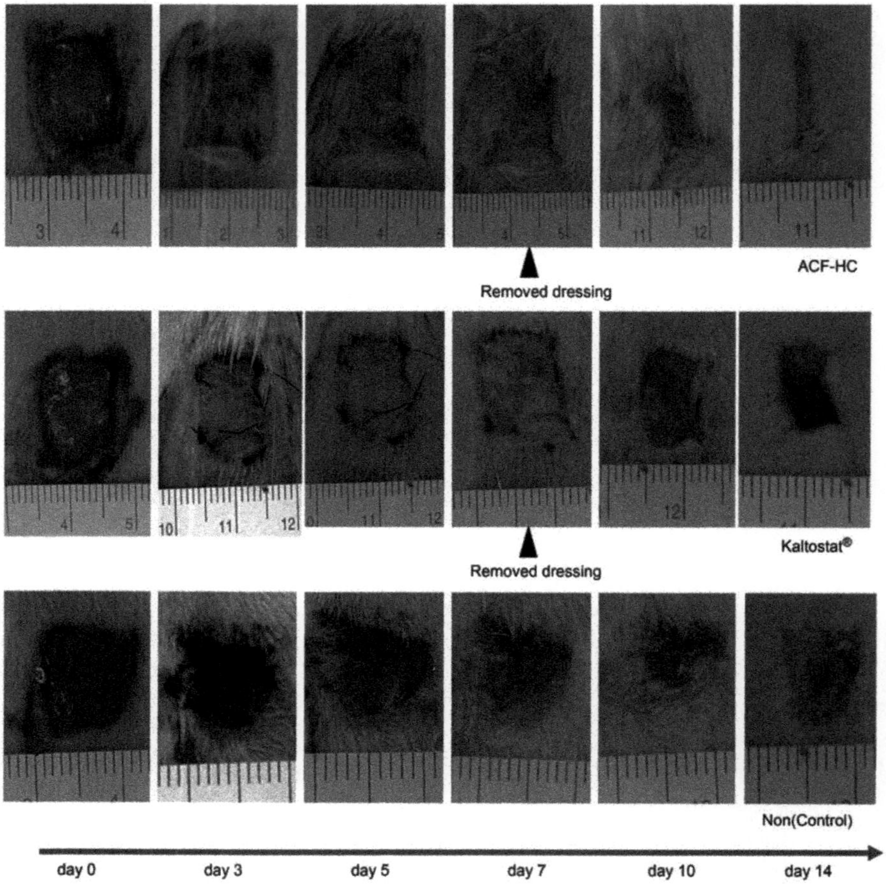

FIGURE 12.13

Photographic findings of wound covered with hydrogel sheet composed of alginate, chitin/chitosan, and fucoidan (ACF-HS) or Kaltostat or control (not covered). Each wound on the indicated day is representative of eight wounds (four rats) covered with ACF-HS or Kaltostat or control.

with efficient intracellular delivery [183]. Biodegradable polymeric microparticles act as an efficient delivery vehicle in a number of lower and higher vertebrates [184–187].

Controlled drug release offers several advantages over conventional drugs, such as better efficiency, reduced side effects, and enhanced patient compliance [188–190]. Table 12.3 illustrates the algae-based polyesters loaded with various drugs and their promising applications [145,191–199].

Silymarin is a combination of three flavonolignans, i.e., silybin, silychristin, and silydianin, and is extracted from the milk thistle. Silymarin has antioxidant activity,

Table 12.3 Algae-Based Polyesters Loaded With Various Drugs and Their Promising Applications

S. No.	Drug	Composition of Drug Carrier	Promising Applications	References
1.	Dexamethasone	PDLLA–Ulvan composite	Suitable for localized drug delivery	[145]
2.	Metronidazole	PCL/ALG composite	Rings for dental implants with controlled localized delivery of antibacterial agents	[191]
3.	Dexamethasone	PLGA NPs embedded in ALG hydrogel	Therapeutic agents for neural electrodes	[192]
4.	Tobramycin	ALG/PLGA nanoparticles	For pulmonary delivery of antibiotics	[193]
5.	Rifampicin	PLGA microspheres/ Na-ALG gel	Useful for improving antituberculosis treatment	[194]
6.	Resveratrol	ALG/CS-coated PLGA nanoparticles	Controlled release of resveratrol, which is used for (nano) chemoprevention/ chemotherapy	[195]
7.	Ibuprofen	PLGA/ALG-PLL microcapsules	For continuous release of antiinflammatory drugs	[196]
8.	Streptomycin	PLGA/ALG nanoparticles	Used for the design of new drug delivery systems	[197]
9.	Docetaxel	Mannitol-PLGA/D-α-tocopheryl polyethylene glycol nanoparticles	Applied in developing novel nanoformulations for breast cancer therapy	[198]
10.	Paclitaxel	PLGA microspheres entrapped in an alginate gel matrix	Local glioma chemotherapy	[199]
11.	Hepatitis B surface antigen	Alginate–chitosan–PLGA composite microspheres	Single-shot hepatitis B vaccination	[200]

ALG, *alginate;* CS, *chitosan;* NP, *nanoparticle;* PCL, *poly(ε-caprolactone);* PDLLA, *poly(D,L-lactic acid);* PLGA, *poly(lactic-co-glycolic acid);* PLL, *poly (L-lactide).*

supports cell regeneration, promotes protein synthesis, and inhibits carcinogenesis [201–204]. Silymarin-loaded PLGA nanoparticles encapsulated into alginate-based pH-responsive hydrogels permit a continuous release of silymarin mainly in the intestinal region [205] (Fig. 12.14).

Many efforts had been made to enhance protein stability and release kinetics of the PLGA by altering the physicochemical properties of the polymer [206]. A composite microsphere is prepared from PLGA, alginate, and chitosan for antigen delivery. This microcapsule not only provides antigen/protein a proper microenvironment by dual hydrophilic protection but also makes the initial release very slow. Both

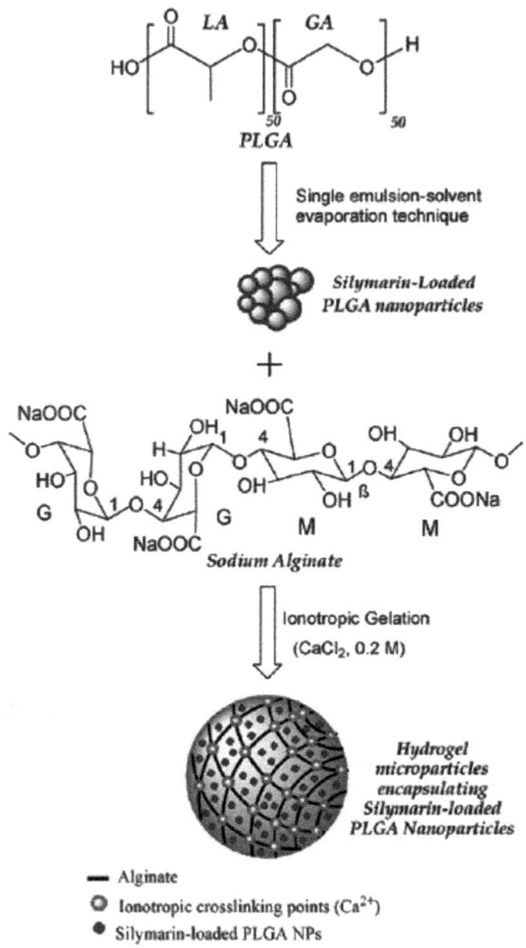

FIGURE 12.14

A schematic illustration for the preparation of the hydrogel microparticles encapsulating silymarin-loaded poly(lactic-co-glycolic acid) nanoparticles (PLGA NPs) [205].

alginate and chitosan are compatible with hydrophilic proteins, show mucoadhesiveness and stability in proteolytic and acid conditions, and are biodegradable, less toxic, abundant, and comparatively inexpensive biodegradable materials [207,208]. These two polymers are also immunostimulant in several higher vertebrates, including fish [209,210]. PLGA/alginate/chitosan composite microspheres show innate and adaptive immune response in fish *Labeo rohita* H. against *Aeromonas hydrophila* infections. A novel composite microsphere of PLGA/alginate/chitosan encapsulating OMP antigen increases the low entrapment efficiency, increases the initial burst release of the conventional PLGA microspheres, stimulates a long-lasting innate and adaptive immunity, and gives protection from *A. hydrophila* in fish compared with conventional PLGA microparticles [211].

Bovine serum albumin–loaded alginate/chitosan/PLGA composite microspheres have a higher protein encapsulation efficiency and reduce its burst release, which is used as a promising delivery system for water-soluble proteins and peptides [212].

Shi et al. [213] prepared hydrophobically modified sodium alginate–grafted PLGA (ALG-g-PLGA) by direct one-step polymerization of sodium alginate, glycolic acid, and lactic acid. ALG-g-PLGA hydrogel microspheres are obtained by cross-linking ALG-g-PLGA nanoparticles in the solution of calcium chloride, which can be used as the drug delivery vehicle for hydrophobic ibuprofen. ALG-g-PLGA as a drug carrier has a high drug-loading rate and encapsulation efficiency and prolongs drug release [213] (Fig. 12.15).

Dental implants support dental crowns and bridges by acting as abutment for the replacement of missing teeth. Systemic antibiotics and localized delivery of

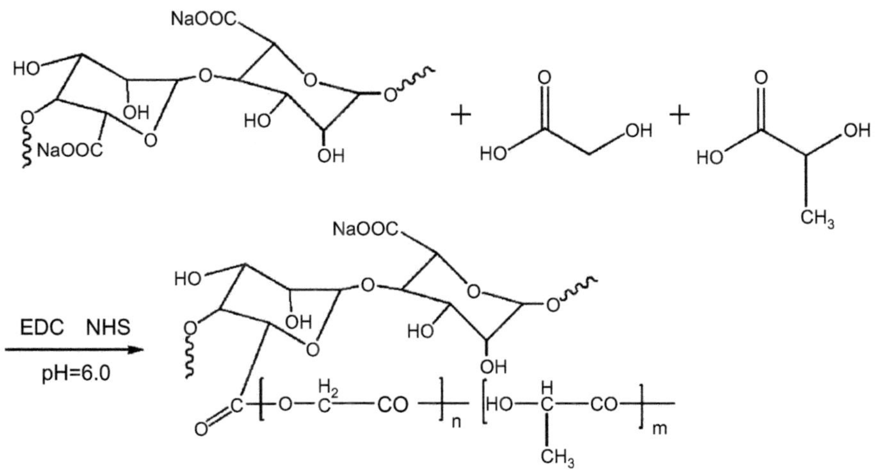

FIGURE 12.15

Synthesis of sodium alginate–grafted PLGA (ALG-g-PLGA) through direct polycondensation. *EDC*, 1-(3-dimethylaminopropyl)-3-ethyl-carbodiimide hydrochloride; *NHS*, N-hydroxysuccinimide.

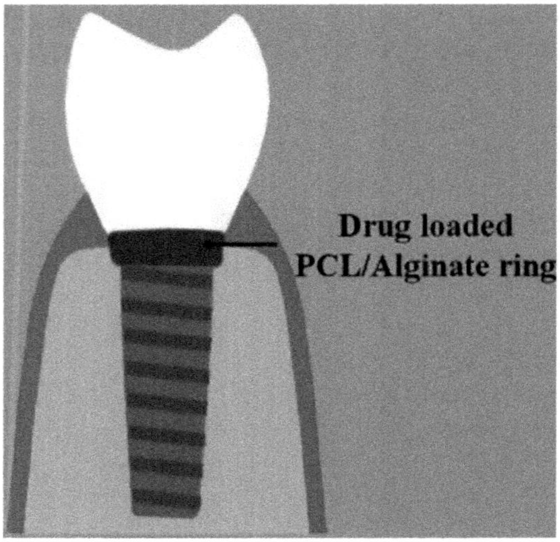

FIGURE 12.16

Schematic showing placement of metronidazole-loaded poly(ε-caprolactone) (PCL)/alginate ring on implant [191].

bactericidal agents are commonly used to prevent bacterial growth at the site of implantation. A ring-like implantable device containing an antibacterial agent is prepared from a composite of metronidazole-loaded PCL and calcium alginate hydrogel [191]. This guarantees high absorption of the drug around the implant and minimizes the side effects together with a systemic drug application. A viable site of the annular ring would be around the body of a root form dental implant adjacent to the bone crest as shown in Fig. 12.16.

One of the main reasons of morbidity and mortality in the world is heart failure, principally myocardial infarction [214]. Surgery is one of the established curing methods for myocardial infarction, but it has some limitations [215], such as follows:

- Thrombosis
- Partial durability
- Reoperation

Various biomedical approaches, including drug delivery and TE, have been used to cure myocardial infarction. TAT-HSP27-loaded PLGA microsphere and alginate hydrogel—based carrier are used to control release of TAT-HSP27 [216]. Heat shock protein 27 (HSP27) is fused with transcriptional activator (TAT) from human immunodeficiency virus (HIV) to increase HSP27 diffusion across the cell membranes. In fact, the fusion protein (TAT-HSP27) maintains its bioactivity and recovers the increase of cardiomyoblasts cultured in hypoxic conditions, which is useful for treating myocardial infarction in a minimally invasive manner.

The nasal cavity possesses numerous advantages as a site for drug delivery, i.e., ease of administration, a large surface area for absorption, and applicability for long-term treatments. One main limiting feature for nasal drug delivery is the limited time available for absorption within the nasal cavity because of mucociliary clearance. Several drug delivery systems have been used to entrap the drug and to increase the residence time in the nasal cavity. Alginate-PLGA microspheres have the highest mucoadhesion properties and are suitable nasal delivery systems. The limiting step for the nasal clearance of nasally administered particulate systems is their dislocation from the initial site of deposition, and the following interactions with the mucous layer in the rest of the nasal passage do not significantly affect the clearance time [217].

12.6 CONCLUSION AND FUTURE PERSPECTIVES

In polyester industry, the conversion of biomass components into green polyesters is one of the promising and economical techniques to overcome fossil fuel crisis. Among various biomaterials, algal biomass is gaining great attention because of its unique chemical composition, versatility, and low cost. Cyanobacteria offer a superb feedstock for production of plastic as a result of their high yield and the aptitude to develop in a range of environments. The consumption of cyanobacterial plastics kicks off the possibility of using carbon source and reduces the emission of greenhouse gases from factories and power plants. Agar and algae oil are transformed to novel monomers for polyester synthesis. Algae-derived polyester scaffolds have great potential for cartilage and bone TE, wound dressing, and drug delivery system. Therefore, algal-polymer-polyester blends and composites will be exploited in TE and controlled and continuous drug delivery for healthy development of the world.

REFERENCES

[1] Fernando S, Adhikari S, Chandrapal C, Murali N. Biorefineries: current status, challenges, and future direction. Energy Fuels 2006;20(4):1727−37.

[2] Nielsen JRR. Making fuels from biomass. Science 2005;308:1421−2.

[3] Alonso DM, Bond JQ, Dumesic JA. Catalytic conversion of biomass to biofuels. Green Chem 2010;12(9):1493−513.

[4] Pillai C. Challenges for natural monomers and polymers: novel design strategies and engineering to develop advanced polymers. Des Monomers Polym 2010;13: 87−121.

[5] Chen GQ, Patel MK. Plastics derived from biological sources: present and future: a technical and environmental review. Chem Rev 2012;112(4):2082−99.

[6] Amarasekara AS. Renewable polymers: synthesis processing, and technology. 1st ed. Hobkon (NJ): John Wiley & Sons, Inc.; 2011. p. 381−422.

[7] Gubbels E, Walc LJ, Noordover BAJ, Koning CE. Linear and branched polyester resins based on dimethyl-2,5-furandicarboxylate for coating applications. Eur Polym J 2013;49(9):3188−98.

[8] Wu L, Mincheva R, Xu Y, Raquez JM, Dubois P. High molecular weight poly(butylene succinate-co-butylene furandicarboxylate) copolyesters: from catalyzed polycondensation reaction to thermomechanical properties. Biomacromolecules 2012;13(9):2973−81.

[9] Ma J, Pang Y, Wang M, Xu J, Ma H, Nie X. The copolymerization reactivity of diols with 2,5-furandicarboxylic acid for furan-based copolyester materials. J Mater Chem 2012;22(8):3457−61.

[10] Gandini A, Coelho D, Gomes M, Reis B, Silvestr A. Materials from renewable resources based on furan monomers and furan chemistry: work in progress. J Mater Chem 2009;19(45):8656−64.

[11] Ragauskas AJ, Williams CK, Davison BH, Britovsek G, Cairney J, Eckert CA. The path forward for biofuels and biomaterials. Science 2006;311(5760):484−9.

[12] Ohara H. Biorefinery. Appl Microbiol Biotechnol 2003;62(5):474−7.

[13] Corma A, Iborra S, Velty A. Chemical routes for the transformation of biomass into chemicals. Chem Rev 2007;107(6):2411−502.

[14] Yu L, Deana K, Li L. Polymer blends and composites from renewable resources. Prog Polym Sci 2006;31(6):576−602.

[15] Doulabi ASH, Mirzadeh H, Imani M, Sharifi S, Atai M, Ataei SM. Synthesis and preparation of biodegradable and visible light crosslinkable unsaturated fumarate-based networks for biomedical applications. Polym Adv Technol 2008;19(9):1199.

[16] Guner FS, Yagci Y, Erciyes AT. Polymers from triglyceride oils. Prog Polym Sci 2006;31(7):633−70.

[17] Zhanga J, Li J, Tang Y, Lin L, Long M. Advances in catalytic production of bio-based polyester monomer 2,5-furandicarboxylic acid derived from lignocellulosic biomass. Carbohydr Polym 2015;130:420−8.

[18] Bakare IO, Pavithran C, Okieimen FE. Pillai CKS Polyesters from renewable resources: preparation and characterization. J Appl Polym Sci 2006;100(5):3748−55.

[19] Hiroe A, Ushimaru K, Tsuge T. Characterization of polyhydroxyalkanoate (PHA) synthase derived from *Delftia acidovorans* DS-17 and the influence of PHA production in *Escherichia coli*. J Biosci Bioeng 2013;115(6):633−8.

[20] Yang C, Zhang W, Liu R, Zhang C, Gong T, Li Q. Analysis of polyhydroxyalkanoate (PHA) synthase gene and PHA-producing bacteria in activated sludge that produces PHA containing 3-hydroxydodecanoate. FEMS Microbiol Lett 2013;346(1):56−64.

[21] Rhim JW. Preparation and characterization of vacuum sputter silver coated PLA film. LWT Food Sci Technol 2013;54(2):477−84.

[22] Zhang N, Zeng C, Wang L, Ren J. Preparation and properties of biodegradable poly(lactic acid)/poly(butylene adipate-co-terephthalate) blend with epoxy-functional styrene acrylic copolymer as reactive agent. J Polym Environ 2013;21(1):286−92.

[23] Chen Q, Zhu C, Thouas GA. Progress and challenges in biomaterials used for bone tissue engineering: bioactive glasses and elastomeric composites. Prog Biomater 2012;1:1−22.

[24] Ding W, Li L, Zhang L, Ju D, Peng S, Chai W. An XPS study on the chemical bond structure at the interface between SiO(x)N(y) and N doped polyethylene terephthalate. J Chem Phys 2013;138:104706.

[25] Foolmaun RK, Ramjeeawon T. Comparative life cycle assessment and social life cycle assessment of used polyethylene terephthalate (PET) bottles in Mauritius. Int J Life Cycle Assess 2013;18:155−71.

[26] Ke H, Li D, Zhang H, Wang X, Cai Y, Huang F. Electrospun form-stable phase change composite nanofibers consisting of capric acid-based binary fatty acid eutectics and polyethylene terephthalate. Fibers Polym 2013;14(1):89–99.

[27] Zhang DY, Liu P, Ouyoung CF, Gao Q, Zheng KS, Hu HF. Influence of PNA012 on crystallization and mechanical properties of polybutylene terephthalate. Adv Mater Res 2013;624:264–8.

[28] Engelmann G, Bonatz E, Ganster J, Pinnow M, Bohn A. Preparation of PTT/clay nanocomposites with solid-state polymerized polytrimethylene terephthalate. J Appl Polym Sci 2013;127(3):1585–94.

[29] Noreen A, Zia KM, Zuber M, Ali M, Mujahid M. Int J Biol Macromol 2016. http://dx.doi.org/10.1016/j.ijbiomac.2016.01.067.

[30] Chen GQ, Patel MK. Recent developments and future prospects on bio-based polyesters derived from renewable resources: a review. Chem Rev 2012;112:2082–99.

[31] Wu Q, Sun SQ, Yu PHF, Chen AXZ, Chen GQ. Acta Polym Sin 2000;6:751–6.

[32] Albertsson AC, Varma IK. Recent developments in ring opening polymerization of lactones for biomedical applications. Biomacromolecules 2003;4(6):1466–86.

[33] Williams CK. Synthesis of functionalized biodegradable polyesters. Chem Soc Rev 2007;36(10):1573–80.

[34] Zia KM, Noreen A, Zuber M, Tabasum S, Mujahid M. Recent developments and future prospects on bio-based polyesters derived from renewable resources: a review. Int J Biol Macromol 2016;82:1028–40.

[35] Gartnera H, Li Y, Almenar E. Improved wettability and adhesion of polylactic acid/chitosan coating for bio-based multilayer film development. Appl Surf Sci 2015;332:488–93.

[36] Fenouillot F, Rousseau A, Colomines G, Loup RS. Polymers from renewable 1,4:3,6-dianhydrohexitols (isosorbide, isomannide and isoidide): a review. Prog Polym Sci 2010;35(5):578–622.

[37] Chatti S, Bortolussi M, Bogdal D, Blais JC, Loupy A. Synthesis and properties of new poly(ether–ester)s containing aliphatic diol based on isosorbide. Effects of the microwave-assisted polycondensation. Eur Polym J 2006;42(2):410–24.

[38] Gandini A, Silvestre AJD. In: Belgacem MN, Gandini A, editors. Terpenes: major sources, properties and applications. Amsterdam: Elsevier Ltd; 2008. p. 17–38.

[39] Lavilla C, Alla A, Ilarduya AM, Benito E, Martin MGG, Galbis JA, Guerra SM. Carbohydrate-based polyesters made from bicyclic acetalized galactaric acid. Biomacromolecules 2011;12:2642–52.

[40] Nigam PS, Singh A. Production of liquid biofuels from renewable resources. Prog Energy Combust Sci 2010;37(1):52–68.

[41] Subhadra B, Edwards M. An integrated renewable energy park approach for algal biofuel production in United States. Energy Policy 2010;38(9):4897–902.

[42] Matsumoto M, Yokouchi H, Suzuki N, Ohata H, Matsunaga T. Saccharification of marine microalgae using marine bacteria for ethanol production. Appl Biochem Biotechnol 2003;105:247–54.

[43] Yen HW, Hu IC, Chen CY, Ho SH, Le DJ, Chang JS. Microalgae based biorefinery—from biofuels to natural products. Bioresour Technol 2013;135:66–174.

[44] Griffith MJ, Hille RPV, Harrison STL. Lipid productivity, settling potential and fatty acid profile of 11 microalgal species grown under nitrogen replete and limited conditions. J Appl Phycol 2011;24:989–1001.

[45] Mata TM, Martinsa AA, Caetano NS. Microalgae for biodiesel production and other applications: a review. Renew Sustain Energy Rev 2010;14:217–32.

[46] Chen YH, Tu CJ, Wu HT. Growth-inhibitory effects of the red alga *Gelidium amansii* on cultured cells. Biol Pharm Bull 2004;27:180–4.

[47] Andrade LR, Salgado LT, Farina M, Pereira MS, Mourao PAS, Filho GMA. Ultrastructure of acidic polysaccharides from the cell walls of brown algae. J Struct Biol 2004; 145:216–25.

[48] Goncalves AG, Ducatti DR, Duarte ME, Noseada MD. Sulfated and pyruvylated disaccharide alditols obtained from a red seaweed galactan: ESIMS and NMR approaches. Carbohydr Res 2002;337:2443–53.

[49] Berteau O, Mulloy B. Sulfated fucans, fresh perspectives: structures, functions, and biological properties of sulfated fucans and an overview of enzymes active toward this class of polysaccharide. Glycobiology 2003;13(6):29–40.

[50] Deville C, Gharbi M, Dandrifosse G, Peulen O. Study on the effects of laminarin, a polysaccharide from seaweed, on gut characteristics. J Sci Food Agric 2007;87(9): 1717–25.

[51] Lahaye M, Ray B. Cell-wall polysaccharides from the marine green alga Ulva "rigida" (Ulvales, Chlorophyta) — NMR analysis of ulvan oligosaccharides. Carbohydr Res 1996;283:161–73.

[52] Suriyamongkol P, Weselake R, Narine S, Moloney M, Shah S. Biotechnological approaches for the production of polyhydroxyalkanoates in microorganisms and plants - a review. Biotechnol Adv 2007;25(2):148–75.

[53] Steinbuchel A, Valentin HE. Diversity of bacterial polyhydroxyalkanoic acids. FEMS Microbiol Lett 1995;128(3):219–28.

[54] Bear MM, Renard E, Randriamahef S, Langlois V. Preparation of a bacterial polyester with carboxy groups in side chains. CR Acad Sci 2001;4(4):289–93.

[55] Diard S, Carlier JP, Ageron E, Langlois V. Accumulation of Poly(3-hydroxybutyrate) from octanoate in different *Pseudomonas* belonging to the rRNA homology group I. Syst Appl Microbiol 2002;25(2):183–8.

[56] Hofer P, Choi YJ, Osborne MJ, Miguel CB, Vermette P, Groleau D. Production of functionalized polyhydroxyalkanoates by genetically modified *Methylobacterium extorquens* strains. Microb Cell Fact 2010;9:70–5.

[57] Balaji S, Gopi K, Muthuvelan B. A review on production of poly β hydroxybutyrates from cyanobacteria for the production of bio plastics. Algal Res 2013;2:278–85.

[58] Poirier Y, Dennis DE, Klomparens K, Somerville C. Polyhydroxybutyrate, a biodegradable thermoplastic, produced in transgenic plants. Science 1992;256(5056):520–3.

[59] Mittendorf V, Robertson EJ, Leech RM, Kruger N, Steinbuchel A, Poirier Y. Synthesis of medium-chain-length polyhydroxyalkanoates in *Arabidopsis thaliana* using intermediates of peroxisomal fatty acid beta-oxidation. Proc Natl Acad Sci USA 1998; 95(23):13397–402.

[60] Snell KD, Peoples OP. Polyhydroxyalkanoate polymers and their production in transgenic plants. Metab Eng 2002;4(1):29–40.

[61] Van Beilen JB, Poirier Y. Production of renewable polymers from crop plants. Plant J 2008;54(4):684–701.

[62] Bohmert K, Balbo I, Kopka J, Mittendorf V, Nawrath C, Poirier Y, Tischendorf G, Trethewey RN, Willmitzer L. Transgenic *Arabidopsis* plants can accumulate polyhydroxybutyrate to up to 4% of their fresh weight. Planta 2000;211(6):841–5.

[63] Nawrath C, Poirier Y, Somerville C. Targeting of the polyhydroxybutyrate biosynthetic pathway to the plastids of *Arabidopsis thaliana* results in high levels of polymer accumulation. Proc Natl Acad Sci USA 1994;91(26):12760–4.

[64] Bohmert-Tatarev K, McAvoy S, Daughtry S, Peoples OP, Snell KD. High levels of bioplastic are produced in fertile transplastomic tobacco plants engineered with a synthetic operon for the production of polyhydroxybutyrate. Plant Physiol 2011; 155(4):1690–708.

[65] Bozarth A, Maier UG, Zauner S. Diatoms in biotechnology: modern tools and applications. Appl Microbiol Biotechnol 2009;82(2):195–201.

[66] Mayfield SP, Manuell AL, Chen S, Wu J, Tran M, Siefker D, Muto M, Marin-Navarro J. *Chlamydomonas reinhardtii* chloroplasts as protein factories. Curr Opin Biotechnol 2007;18(2):126–33.

[67] Potvin G, Zhang Z. Strategies for high-level recombinant protein expression in transgenic microalgae: a review. Biotechnol Adv 2010;28(6):910–8.

[68] Rasala BA, Muto M, Lee PA, Jager M, Cardoso RM, Behnke CA, Kirk P, Hokanson CA, Crea R, Mendez M. Production of therapeutic proteins in algae, analysis of expression of seven human proteins in the chloroplast of *Chlamydomonas reinhardtii*. Plant Biotechnol J 2010;8(6):719–33.

[69] Walker TL, Purton S, Becker DK, Collet C. Microalgae as bioreactors. Plant Cell Rep 2005;24(11):629–41.

[70] Hempel F, Bozarth AS, Lindenkamp N, Klingl A, Zauner S, Linne U, Steinbuchel A, Maier UG. Microalgae as bioreactors for bioplastic production. Microb Cell Fact 2011; 10:81.

[71] Gamal AAE. Biological importance of marine algae. Saudi Pharm J 2010;18:1–25.

[72] Dias JML, Oehmen A, Serafim LS, Lemos PC, Reis MAM, Oliveira R. Metabolic modelling of polyhydroxyalkanoate copolymers production by mixed microbial cultures. BMC Syst Biol 2008;2(59):1–21.

[73] Jyotsana D, Priyangshusarma M, Meeta L, AjoyMandal K, Banwari L. Evaluation of bacterial strains isolated from oil-contaminated soil for production of polyhydroxyalkanoic acids (PHA). Pedobiologia 2010;54:25–30.

[74] Vincenzini M, Sili C, Philippis R, Ena A, Materassi R. Occurrence of poly-β-hydroxybutyrate in *Spirulina* species. J Bacteriol 1990;172:2791–2.

[75] Miyake M, Erata M, Asada Y. A thermophilic cyanobacterium, *Synechococcus* sp. MA19, capable of accumulating poly-β-hydroxybutyrate. J Ferment Bioeng 1996; 82:512–4.

[76] Wu GF, Wu QY, Shen ZY. Accumulation of poly-β-hydroxybutyrate in cyanobacterium *Synechocystis* sp PCC6803. Bioresour Technol 2001;76:85–90.

[77] Nava YF, Maranon E, Soons J, Castrillon L. Denitrification of wastewater containing high nitrate and calcium concentrations. Bioresour Technol 2008;99:7976–81.

[78] Quillaguaman J, Guzman H, Thuoc DV, Kaul RH. Synthesis and production of polyhydroxyalkanoates by halophiles: current potential and future prospects. Appl Microbiol Biotechnol 2010;85:1687–96.

[79] Sudesh K, Abe H, Doi Y. Synthesis, structure and properties of polyhydroxyalkanoates: biological polyesters. Prog Polym Sci 2000;25:1503–55.

[80] Andreessen B, Lange AB, Robenek H, Steinbuchel A. Conversion of glycerol to poly (3-hydroxypropionate) in recombinant *Escherichia coli*. Appl Environ Microbiol 2010;76:622–6.

[81] Sharma L, Mallick N. Enhancement of poly-β-hydroxybutyrate accumulation in *Nostoc muscorum* under mixotrophy, chemoheterotrophy and limitations of gas exchange. Biotechnol Lett 2005;27:59−62.

[82] Bhattacharya D, Sarma PM, Krishnan S, Mishra S, Lal B. Evaluation of genetic diversity among *Pseudomonas citronellolis* strains isolated from oily sludge-contaminated sites. Appl Environ Microbiol 2003;69(7):1435−41.

[83] Chen GQ, Wu Q. The application of polyhydroxyalkanoates as tissue engineering materials. Biomaterials 2005;26(33):6565−78.

[84] Tao Y, He Y, Wu Y, Liu F, Li X, Zong W, Zhou Z. Characteristics of a new photosynthetic bacterial strain for hydrogen production and its application in wastewater treatment. Int J Hydrog Energy 2008;33(3):963−73.

[85] Wanga B, Pugh S, Nielsen DR, Zhang W, Meldrum DR. Engineering cyanobacteria for production of PHB directly from CO_2. Metab Eng 2013;16:68−77.

[86] Demirbas A. Biorefineries: current activities and future developments. Energy Convers Manag 2009;50:2782−801.

[87] Cherubini F. The biorefinery concept: using biomass instead of oil for producing energy and chemicals. Energy Convers Manag 2010;51(7):1412−21.

[88] Trivedi J, Aila M, Bangwal DP, Kaul S, Garg MO. Algae based biorefinery—How to make sense? Renew Sustain Energy Rev 2015;47:295−307.

[89] Dario GDA, Viatcheslav K. CT F-Cienc Tecnol Futuro 2011:5−22.

[90] McMillan JD. ACS symposium series, vol. 566; 1994. p. 292−324. Washington, DC.

[91] Kim B, Jeong J, Shin S, Lee D, Kim S, Yoon HJ. Facile single-step conversion of macroalgal polymeric carbohydrates into biofuels. Chem Sus Chem 2010;3(11):1273−5.

[92] Papageorgiou GZ, Tsanaktsis V, Papageorgiou DG, Exarhopoulos S, Papageorgiou M, Bikiaris DN. Evaluation of polyesters from renewable resources as alternatives to the current fossil-based polymers. Phase transitions of poly(butylenes 2,5-furan-dicarboxylate). Polymer 2014;55:3846−58.

[93] Dutta S, De S, Saha B. Advances in biomass transformation to 5-hydroxymethylfurfural and mechanistic aspects. Biomass Bioenergy 2013;55:355−69.

[94] Papageorgiou GZ, Tsanaktsis V, Papageorgiou DG, Chrissafis K, Exarhopoulos S, Bikiaris DN, Papageorgiou GZ, Tsanaktsis V, Papageorgiou DG, Chrissafis K, Exarhopoulos S, Bikiaris DN. Eur Polym J 2015;67:383−96.

[95] Dutta S, De S, Saha B. A brief summary of the synthesis of polyester building-block chemicals and biofuels from 5-hydroxymethylfurfural. Chem Plus Chem 2012;77(4):1−15.

[96] Gomes M, Gandini A, Silvestre AJ, Reis B. Synthesis and characterization of poly(2,5-furan dicarboxylate)s based on a variety of diols. J Polym Sci Part A Polym Chem 2011;49:3759−68.

[97] Jiang M, Zhang QQL, Ye C, Zhou G. A series of furan-aromatic polyesters synthesized via direct esterification method based on renewable resources. J Polym Sci Part A Polym Chem 2012;50:1026.

[98] Zeng C, Seino H, Hatanaka JKR, Yoshie N. Bio-based furan polymers with self-healing ability. Macromolecules 2013;46(5):1794−802.

[99] Tsanaktsis V, Vouvoudi E, Papageorgiou GZ, Papageorgiou DG, Chrissafis K, Bikiaris DN. Thermal degradation kinetics and decomposition mechanism of polyesters based on 2,5-furandicarboxylic acid and low molecular weight aliphatic diols. J Anal Appl Pyrolysis 2015;112:369−78.

[100] Saxena RK, Anand P, Saran S, Isar J, Agarwal L. Microbial production and applications of 1, 2-propanediol. Indian J Microbiol 2010;5:2−11.

[101] Certik M, Shimizu S. Biosynthesis and regulation of microbial polyunsaturated fatty acid production. J Biosci Bioeng 1999;87:1−14.

[102] Gill I, Valivety R. Polyunsaturated fatty acids, part 1: occurrence, biological activities and applications. Trends Biotechnol 1997;15:401−9.

[103] Apt KE, Behrens PW. Commercial developments in microalgal biotechnology. J Phycol 1999;35:215−26.

[104] Spolaore P, Cassan CJ, Duran E, Isambert A. Commercial applications of microalgae. J Biosci Bioeng 2006;101(2):87−96.

[105] Sargent JR, Bell MV, Bell JG, Henderson RJ, Tocher DR. Origins and functions of n-3 polyunsaturated fatty acids in marine organisms. In: Cevc G, Paltauf F, editors. Phospholipids: characterization, metabolism and novel biological applications. Champaign (Illinois): AOCS; 1995. p. 248.

[106] Anneken DJ, Both S, Christoph R, Fieg G, Steinberner U, Westfechtel A. Ullmanns encyclopedia of industrial chemistry. Weinheim: Wiley-VCH; 2006. p. 73−116.

[107] Biermann U, Bornscheuer U, Meier MAR, Metzger JO, Schfer HJ. Oils and fats as renewable raw materials in chemistry. Angew Chem 2011;123:3938−56.

[108] Chikkali S, Mecking S. Refining of plant oils to chemicals by olefin metathesis. Angew Chem 2012;124:5902−9.

[109] Goss R, Wilhelm C. Lipids in photosynthesis. Dordrecht: Springer; 2010. p. 117−37.

[110] Roesle P, Stempfle F, Hess SK, Zimmerer J, Bartulos CR, Lepetit B, Eckert A, Kroth PG, Mecking S. Synthetic polyester from algae oil. Angew Chem Int 2014; 53(26):6800−4.

[111] Hoffman AS. Hydrogels for biomedical applications. Adv Drug Deliv Rev 2002;54(1): 3−12.

[112] Peppas NA, Hilt JZ, Khademhosseini A, Langer R. Hydrogels in biology and medicine: from molecular principles to bionanotechnology. Adv Mater 2006;18(11): 1345−60.

[113] Jen AC, Wake MC, Mikos AG. Review: hydrogels for cell immobilization. Biotechnol Bioeng 1996;50(4):357−64.

[114] Jeon O, Bouhadir KH, Mansour JM, Alsberg E. Photocrosslinked alginate hydrogels with tunable biodegradation rates and mechanical properties. Biomaterials 2009; 30(14):2724−34.

[115] Butler DL, Goldstein SA, Guilak F. Functional tissue engineering: the role of biomechanics. J Biomech Eng 2000;122:570.

[116] Ramakrishna S, Mayer J, Wintermantel E, Leong KW. Biomedical applications of polymer-composite materials: a review. Compos Sci Technol 2001;61(9):1189−224.

[117] Summerscales J. Microstructural characterisation of fibre-reinforced composites. CRC; 1998.

[118] Agarwal S, Greiner A, Wendorff JH. Electrospinning of manmade and biopolymer nanofibers-progress in techniques, materials, and applications. Adv Funct Mater 2009;19(18):2863−79.

[119] Dzenis Y. Spinning continuous fibers for nanotechnology. Science 2004;304(5679): 1917.

[120] Zucchelli A, Focarete ML, Gualandi C, Ramakrishna S. Electrospun nanofibers for enhancing structural performance of composite materials. Polym Adv Technol 2011;22(3):339−49.

[121] Huang ZM, Zhang YZ, Kotaki M, Ramakrishna S. A review on polymer nanofibers by electrospinning and their applications in nanocomposites. Compos Sci Technol 2003; 63(15):2223−53.

[122] Stupp SI. Biomaterials for regenerative medicine. MRS Bull 2005;30:546−53.

[123] Salerno A, Netti PA. Optimal design and manufacture of biomedical foam pore structure for tissue engineering applications. In: Biomedical foams for tissue engineering applications. Woodhead Publishing; 2014. p. 71.

[124] Langer R, Vacanti JP. Tissue engineering. Science 1993;260:920−6.

[125] Hutmacher DW, Biomater J. Scaffold design and fabrication technologies for engineering tissues - state of the art and future perspectives. Sci Polym Ed 2001;12(1): 107−24.

[126] Salerno A, Pascual CD. Bio-based polymers, supercritical fluids and tissue engineering. Process Biochem 2015;50:826−38.

[127] Tamayol A, Akbari M, Annabi N, Paul A, Khademhosseini A, Juncker D. Fiber-based tissue engineering: progress, challenges, and opportunities. Biotechnol Adv 2013;31: 669−87.

[128] Ma PX, Elisseeff JF. Scaffolding in tissue engineering. Boca Raton (FL, USA): CRC Press; 2006.

[129] Hollister SJ. Porous scaffold design for tissue engineering. Nat Mater 2005;4:518−24.

[130] Cohen DL, Lo W, Tsavaris A, Peng D, Lipson H, Bonassar LJ. Increased mixing improves hydrogel homogeneity and quality of three-dimensional printed constructs. Part C Tissue Eng Methods 2010;17:239.

[131] Murphy SV, Atala A. 3D bioprinting of tissues and organs. Nat Biotechnol 2014;32: 773.

[132] Ahn SH, Lee HJ, Bonassar LJ, Kim GH. Cells (MC3T3-E1)-laden alginate scaffolds fabricated by a modified solid-freeform fabrication process supplemented with an aerosol spraying. Biomacromolecules 2012;13:2997.

[133] Lee HJ, Ahn SH, Bonassar LJ, Kim GH. Cell (MC3T3-E1)-printed poly (ε-caprolactone)/alginate hybrid scaffolds for tissue regeneration. Macromol Rapid Commun 2013;34:142.

[134] Yeo MG, Kim GH. Cell-printed hierarchical scaffolds consisting of micro-sized polycaprolactone (PCL) and electrospun PCL nanofibers/cell-laden alginate struts for tissue regeneration. J Mater Chem B 2014;2:314−24.

[135] Lee HJ, Kim YB, Ahn SH, Lee JS, Jang CH, Yoon H, Chun W, Kim GH. New approach for fabricating collagen/ECM-based bioinks using preosteoblasts and human adipose stem cells. Adv Healthc Mater 2015;4(9):1359−68.

[136] Kim YB, Lee H, Yang GH, Choi CH, Lee D, Hwang H, Jung WK, Yoon H, Kim GH. Mechanically reinforced cell-laden scaffolds formed using alginate based bioink printed onto the surface of a PCL/alginate mesh structure for regeneration of hard tissue. J Colloid Interface Sci 2016;461:359−68.

[137] Kim YB, Kim GH. PCL/alginate composite scaffolds for hard tissue engineering: fabrication, characterization, and cellular activities. ACS Comb Sci 2015;17(2):87−99.

[138] Salerno A, Gutiérrez MF, Barrio JSR, Domingo C. Macroporous and nanometre scale fibrous PLA and PLA−HA composite scaffolds fabricated by a bio safe strategy. RSC Adv 2014;4(106):61491−502.

[139] Salerno A, Gutiérrez MF, Barrio JSR, Domingo C. Bio-safe fabrication of PLA scaffolds for bone tissue engineering by combining phase separation, porogen leaching and $scCO_2$ drying. J Supercrit Fluids 2015;97:238−46.

[140] Jabbarzadeh E, Deng M, Lv Q, Jiang T, Khan YM, Nair LS. VEGF-incorporated biomimetic poly (lactide-co-glycolide) sintered microsphere scaffolds for bone tissue engineering. J Biomed Mater Res B Appl Biomater 2012;100:2187–96.

[141] Montjovent M, Mathieu L, Hinz B, Applegate LL, Bourban P, Zambelli P. Biocompatibility of bioresorbable poly(L-lactic acid) composite scaffolds obtained by supercritical gas foaming with human fetal bone cells. Tissue Eng 2005;11(11):1640–9.

[142] Lee JS, Jin GH, Yeo MG, Jang CH, Lee H, Kim GH. Fabrication of electrospun biocomposites comprising polycaprolactone/fucoidan for tissue regeneration. Carbohydr Polym 2012;90(1):181–8.

[143] Jin GH, Kim GH. Rapid-prototyped PCL/fucoidan composite scaffolds for bone tissue regeneration: design, fabrication, and physical/biological properties. J Mater Chem 2011;21(44):17710–8.

[144] Kikionis S, Ioannou E, Toskas G, Roussis V. Electrospun biocomposite nanofibers of ulvan/PCL and ulvan/PEO. J Appl Polym Sci 2015;132:421–53.

[145] Alves A, Duarte ARC, Mano JF, Sousa RA, Reis RL. PDLLA enriched with ulvan particles as a novel 3D porous scaffold targeted for bone engineering. J Supercrit Fluids 2012;65:32–8.

[146] Palva T. Reconstruction of ear canal in surgery for chronic ear. Arch Otolaryngol 1962; 75(4):329–34.

[147] Sade J, Weinberg J, Berco E, Brown M, Halevy A. The marsupialized (radical) mastoid. J Laryngol Otol 1982;96(10):869–75.

[148] Mahendran S, Yung MW. Mastoid obliteration with hydroxyapatite cement: the Ipswich experience. Otol Neurotol 2004;25:19–21.

[149] Yung M, Smith P. Mid-temporal pericranial and inferiorly based periosteal flaps in mastoid obliteration. Otolaryngol Head Neck Surg 2007;137:906–12.

[150] Matsuno T, Nakamura T, Kuremoto K, Notazawa S, Nakahara T, Hashimoto Y. Development of b-tricalcium phosphate/collagen sponge composite for bone tissue regeneration. Dent Mater J 2006;25:138–44.

[151] Jang CH, Park H, Cho YB, Song CH. Mastoid obliteration using a hyaluronic acid gel to deliver a mesenchymal stem cells-loaded demineralized bone matrix: an experimental study. Int J Pediatr Otorhinolaryngol 2008;72:1627–32.

[152] Jang CH, Cho YB, Yang HC, Kim JS, Choi CH, Jang SJ. Effect of piperacillin-tazobactam coated b-tricalcium phosphate for mastoid obliteration in otitis media. Int J Pediatr Otorhinolaryngol 2011;75:631–4.

[153] Jang CH, Cho YB, Bae CS. Evaluation of bioactive glass for mastoid obliteration: a guinea pig model. In Vivo (Brooklyn) 2007;21:651–5.

[154] Yanagihara N, Komori M, Hinohira Y. Total mastoid obliteration in staged canal up tympanoplasty for cholesteatoma facilitates tympanic aeration. Otol Neurotol 2009; 30:766–70.

[155] Janga CH, Kimb MS, Cho YB, Jang YS, Kim GH. Mastoid obliteration using 3D PCL scaffold in combination with alginate and rhBMP-2. Int J Biol Macromol 2013;62: 614–22.

[156] Jang CH, Cho YB, Choi CH, Jang YS, Jung W, Lee H, Kim GH. Effect of umbilical cord serum coated 3D PCL/alginate scaffold for mastoid obliteration. Int J Pediatr Otorhinolaryngol 2014;78:1061–5.

[157] Wang Q, Jamal S, Detamore MS, Berkland C. PLGA-chitosan/PLGA-alginate nanoparticle blends as biodegradable colloidal gels for seeding human umbilical cord mesenchymal stem cells. J Biomed Mater Res A 2011;96A(3):520–7.

[158] Jin XB, Sun YS, Zhang K, Wang J, Shi TP, Ju XD. Ectopic neocartilage formation from predifferentiated human adipose derived stem cells induced by adenoviral-mediated transfer of hTGF beta2. Biomaterials 2007;28(9):2994–3003.

[159] Thankam FG, Muthu J. Alginate–polyester comacromer based hydrogels as physiochemically and biologically favorable entities for cardiac tissue engineering. J Colloid Interface Sci 2015;457:52–61.

[160] Finosh GT, Jayabalan M, Vandana S, Raghu KG. Hybrid alginate-polyester bimodal network hydrogel for tissue engineering – influence of structured water on long-term cellular growth. Colloids Surf B Biointerfaces 2015;135:855–64.

[161] Tare RS, Babister JC, Kanczler J, Oreffo RO. Skeletal stem cells: phenotype, biology and environmental niches informing tissue regeneration. Mol Cell Endocrinol 2008; 288(1–2):11–21.

[162] Bianco P, Kuznetsov SA, Riminucci M, Robey GP. Postnatal skeletal stem cells. Methods Enzym 2006;419:117–48.

[163] Alhadlaq A, Tang M, Mao JJ. Engineered adipose tissue from human mesenchymal stem cells maintains predefined shape and dimension: implications in soft tissue augmentation and reconstruction. Tissue Eng 2005;11(3–4):556–66.

[164] Stosich MS, Mao JJ. Adipose tissue engineering from human adult stem cells: clinical implications in plastic and reconstructive surgery. Plast Reconstr Surg 2007;119(1): 71–83.

[165] Morgan SM, Ainsworth BJ, Kanczler JM, Babister JC, Chaudhuri JB, Oreffo ROC. Formation of a human-derived fat tissue layer in PDLLGA hollow fibre scaffolds for adipocyte tissue engineering. Biomaterials 2009;30:1910–7.

[166] Wang S, Cai Q, Hou J, Bei J, Zhang T, Yang J. Acceleration effect of basic fibroblast growth factor on the regeneration of peripheral nerve through a 15-mm gap. Biomed Mater Res A 2003;66A(3):522–31.

[167] Widmer MS, Gupta PK, Lu L, Meszlenyi RK, Evans GRD, Brandt K. Manufacture of porous biodegradable polymer conduits by an extrusion process for guided tissue regeneration. Biomaterials 1998;19(21):1945–55.

[168] Maquet V, Martin D, Malgrange B, Franzen R, Schoenen J, Moonen G. Peripheral nerve regeneration using bioresorbable macroporous polylactide scaffolds. J Biomed Mater Res 2000;52(4):639–51.

[169] Heath CA, Rutkowski GE. The development of bioartificial nerve grafts for peripheral-nerve regeneration. Trends Biotechnol 1998;16(4):163–8.

[170] Evans GR, Brandt K, Widmer MS, Lu L, Meszlenyi RK, Gupta PK. In vivo evaluation of poly(l-lactic acid) porous conduits for peripheral nerve regeneration. J Biomater 1999;20(12):1109–15.

[171] Mackinnon SE, Dellon AL. Clinical nerve reconstruction with a bioabsorbable polyglycolic acid tube. Plast Reconstr Surg 1990;85(3):419–24.

[172] Hadlock T, Elisseeff J, Langer R, Vacanti J, Cheney MA. Tissue-engineered conduit for peripheral nerve repair. Arch Otolaryngol Head Neck Surg 1998;124(10):1081–6.

[173] Bini TB, Gao S, Xu X, Wang S, Ramakrishna S, Leong KW. Peripheral nerve regeneration by microbraided poly(L-lactide-*co*-glycolide) biodegradable polymer fibers. J Biomed Mater Res Part A 2004;68A(2):286–95.

[174] Oh SH, Lee JH. Fabrication and characterization of hydrophilized porous PLGA nerve guide conduits by a modified immersion precipitation method. J Biomed Mater Res Part A 2007;80A(3):530–8.

[175] Paddle-Ledinek JE, Nasa Z, Cleland HJ. Effect of different wound dressings on cell viability and proliferation. Plast Reconstr Surg 2006;117(Suppl. 7):110–8.

[176] Purna SK, Babu M. Collagen based dressings–a review. Burns 2000;26:54–62.

[177] Balakrishnan B, Mohanty M, Umashankar PR, Jayakrishnan A. Evaluation of an in situ forming hydrogel wound dressing based on oxidized alginate and gelatin. Biomaterials 2005;26:6335–42.

[178] Quinn KJ, Courtney JM, Evans JH, Gaylor JDS, Reid WH. Principles of burn dressings. Biomaterials 1985;6:369–77.

[179] Nakamura S, Nambu M, Kishimoto S, Ishizuka T, Hattori H, Kanatani Y, Takase B, Kishimoto S, Amano Y, Aoki H, Kiyosawa T, Ishihara M, Maehara T. Effect of controlled release of fibroblast growth factor-2 from chitosan/fucoidan micro complex hydrogel on in vitro and in vivo vascularization. J Biomed Mater Res A 2008;85A: 619–27.

[180] Belford DA, Hendry IA, Parish CR. Investigation of the ability of several naturally occurring and synthetic polyanions to bind to and potentiate the biological activity of acidic fibroblast growth factor. J Cell Physiol 1993;157:184–9.

[181] Leary RO, Rerek M, Wood EJ. Fucoidan modulates the effect of transforming growth factor (TGF)-b1 on fibroblast proliferation and wound repopulation in in vitro models of dermal wound repair. Biol Pharm Bull 2004;27:266–70.

[182] Murakami K, Aoki H, Nakamura S, Nakamura SI, Takikawa M, Hanzawa M, Kishimoto S, Hattori H, Tanaka Y, Kiyosawa T, Sato Y, Ishihara M. Hydrogel blends of chitin/chitosan, fucoidan and alginate as healing-impaired wound dressings. Biomaterials 2010;31:83–90.

[183] Men Y, Thomasin C, Merkle HP, Gander B, Corradin G. A single administration of tetanus toxoid in biodegradable microspheres elicits T cell and antibody responses similar or superior to those obtained with aluminum hydroxide. Vaccine 1995;13(7): 683–9.

[184] Eldridge JH, Hammond CJ, Meulbroek JA, Staas JK, Gilley RM, Tice TR. Controlled vaccine release in the gut-associated lymphoid tissues. I. Orally administered biodegradable microspheres target the peyer's patches. J Controll Release 1990;11(1–3): 205–14.

[185] Eldridge JH, Staas JK, Meulbroek JA, McGhee JR, Tice TR, Gilley RM. Biodegradable microspheres as a vaccine delivery system. Mol Immunol 1991; 28(3):287–94.

[186] O'Hagan DT, Jeffery H, Roberts MJJ, McGee JP, Davis SS. Controlled release microparticles for vaccine development. Vaccine 1991;9(10):768–71.

[187] Behera T, Nanda PK, Mohanty C, Mohapatra D, Swain P, Das BK. Parenteral immunization of fish, *Labeo rohita* with Poly D, L-lactide-co-glycolic acid (PLGA) encapsulated antigen microparticles promotes innate and adaptive immune responses. Fish Shellfish Immunol 2010;28(2):320–5.

[188] Uhrich K, Cannizzaro S, Langer R, Shakesheff K. Polymeric systems for controlled drug release. Chem Rev 1999;99(11):3181–98.

[189] Malam Y, Loizidou M, Seifalian AM. Liposomes and nanoparticles: nanosized vehicles for drug delivery in cancer. Trends Pharmacol Sci 2009;30(11):592–9.

[190] Zhang J, Misra R. Magnetic drug-targeting carrier encapsulated with thermosensitive smart polymer: core–shell nanoparticle carrier and drug release response. Acta Biomater 2007;3(6):838–50.

[191] Lana SF, Kehindec T, Zhanga X, Khajotia S, Schmidtke DW, Starly B. Controlled release of metronidazole from composite poly-ε-caprolactone/alginate (PCL/alginate) rings for dental implants. Dent Mater 2013;29(6):656–65.

[192] Kim DH, Martin DC. Sustained release of dexamethasone from hydrophilic matrices using PLGA nanoparticles for neural drug delivery. Biomaterials 2006;27(15): 3031–7.

[193] Ungaro F, Angelo I, Coletta C, Bianca REV, Sorrentino R, Perfetto B. Dry powders based on PLGA nanoparticles for pulmonary delivery of antibiotics: modulation of encapsulation efficiency, release rate and lung deposition pattern by hydrophilic polymers. J Control Release 2012;157(1):149–59.

[194] Hu C, Feng H, Zhu C. Preparation and characterization of rifampicin-PLGA microspheres/sodium alginate in situ gel combination delivery system. Colloids Surf B Biointerfaces 2012;95:162–9.

[195] Sanna V, Roggio AM, Siliani S, Piccinini M, Marceddu S, Mariani A. Development of novel cationic chitosan-anionic alginate coated poly(D,L-lactide-co-glycolide) nanoparticles for controlled release and light protection of resveratrol. Int J Nanomed 2012;7:5501–16.

[196] Baruch L, Benny O, Gilert A, Ukobnik M, Itzhak OB, Machluf M. Alginate-PLL cell encapsulation system co-entrapping PLGA-microspheres for the continuous release of anti-inflammatory drugs. Biomed Microdevices 2009;11:1103–13.

[197] Asadi A. Streptomycin-loaded PLGA-alginate nanoparticles: preparation, characterization, and assessment. Appl Nanosci 2014;4(4):455–60.

[198] Tao W, Zeng X, Liu T, Wang Z, Xiong Q, Ouyang C. Docetaxel-loaded nanoparticles based on star-shaped mannitol-core PLGA-TPGS diblock copolymer for breast cancer therapy. Acta Biomater 2013;9(11):8910–20.

[199] Ranganath SH, Kee I, Krantz WB, Chow PKH, Wang CH. Hydrogel matrix entrapping PLGA-paclitaxel microspheres: drug delivery with near zero-order release and implantability advantages for malignant brain tumour chemotherapy. Pharm Res 2009;26(9):2101–14.

[200] Zheng X, Huang Y, Zheng C, Dong S, Liang W. Alginate–chitosan–PLGA composite microspheres enabling single-shot hepatitis B vaccination. AAPS J 2010;12(4): 519–24.

[201] Knopp SB, Rietbrock S, Weyhenmeyer R, Bocker RH, Beckurts KT, Lang W. Inhibitory effects of silibinin on cytochrome P-450 enzymes in human liver microsomes. Pharmacol Toxicol 2000;86(6):250–6.

[202] Sonnenbichler J, Scalera F, Sonnenbichler I, Weyhenmeyer R. Stimulatory effects of silibinin and silicristin from the milk thistle *Silybum marianum* on kidney cells. J Pharmacol Exp Ther 1999;290(3):1375–83.

[203] Flora K, Hahn M, Rosen H, Benner K. Milk thistle (*Silybum marianum*) for the therapy of liver disease. Am J Gastroenterol 1998;93:139–43.

[204] Zi X, Feyes DK, Agarwal R. Anticarcinogenic effect of a flavonoid antioxidant, silymarin, in human breast cancer cells MDA-MB 468: induction of G1 arrest through an increase in Cip1/p21 concomitant with a decrease in kinase activity of cyclin-dependent kinases and associated cyclins. Clin Cancer Res 1998;4:1055–64.

[205] El-Sherbiny IM, Abdel-Mogib M, Dawidar AM, Elsayed A, Smyth HDC. Biodegradable pH-responsive alginate-poly (lactic-co-glycolic acid) nano/micro hydrogel matrices for oral delivery of silymarin. Carbohydr Polym 2011;83(3):1345–54.

[206] Sahoo SK, Panyam J, Prabha S, Labhasetwar V. Residual polyvinyl alcohol associated with poly (D, L-lactide-co-glycolide) nanoparticles affects their physical properties and cellular uptake. J Control Release 2002;18:105–14.

[207] Tian JY, Sun XQ, Chen XG. Formation and oral administration of alginate microspheres loaded with pDNA coding for lymphocystis disease virus (LCDV) to Japanese flounder. Fish Shellfish Immunol 2008;24:592–9.

[208] Martins S, Sarmento B, Souto EB, Ferreira DC. Insulin-loaded alginate microspheres for oral delivery effect of polysaccharide reinforcement on physicochemical properties and release profile. Carbohyr Polym 2007;69:725–31.

[209] Fujiki K, Matsuyamas H, Yano T. Protective effect of sodium alginates against bacterial infection in common carp, *Cyprinus carpio* L. J Fish Dis 1994;17:349–55.

[210] Zaharoff DA, Rogers CJ, Hance KW, Schlom J, Greiner JW. Chitosan solution enhances both humoral and cell-mediated immune responses to subcutaneous vaccination. Vaccine 2007;25:2085–94.

[211] Behera T, Swain P. Alginate chitosan PLGA composite microspheres induce both innate and adaptive immune response through parenteral immunization in fish. Fish Shellfish Immunol 2013;35:785–91.

[212] Zhenga CH, Gaoa JQ, Zhang YP, Liang WQ. Protein delivery system: biodegradable alginate–chitosan–poly(lactic-co-glycolic acid) composite microspheres. Biochem Biophys Res Commun 2004;323:1321–7.

[213] Shi G, Ding Y, Zhang X, Wu L, He F, Ni C. Drug release behavior of poly (lactic-glycolic acid) grafting from sodium alginate (ALG-g-PLGA) prepared by direct polycondensation. J Biomater Sci 2015;26:1152–62.

[214] Kleiner DH, Landmesser U, Drexler H. Molecular mechanism in heart failure. J Am Coll Cardiol 2006;48:56–66.

[215] Lange R, Weipert J, Homann M, Mendler N, Paek SU, Holper K. Performance of allografts and xenografts for right ventricular outflow tract reconstruction. Ann Thorac Surg 2001;71(5):365–7.

[216] Lee J, Tan CY, Lee SK, Kim YH, Lee KY. Controlled delivery of heat shock protein using an injectable microsphere/hydrogel combination system for the treatment of myocardial infarction. J Control Release 2009;137(3):196–202.

[217] Tafaghodi M, Tabassi SAS, Jaafari MR, Zakavi SR, Momen-Nejad M. Evaluation of the clearance characteristics of various microspheres in the human nose by gamma-scintigraphy. Int J Pharm 2004;280:125–35.

CHAPTER

Algae-Based Polyolefins 13

Mohammad Zuber[1], Khalid M. Zia[1], Aqdas Noreen[1], Shazia A. Bukhari[1], Nosheen Aslam[1], Neelam Sultan[1], Mudassir Jabeen[1], Bo Shi[2]

Government College University Faisalabad, Faisalabad, Pakistan[1]; Kimberly-Clark Corporation, Neenah, WI, United States[2]

13.1 INTRODUCTION

Polyolefins are synthetic polymers comprising olefinic monomers. Several million metric tons of polyolefins are produced and consumed annually all over the world. Owing to wide range of properties, many potential applications, and comparatively low cost, they have gained much attention in recent years. They are recyclable and their properties can be improved significantly via blending and composite technologies [1]. They can be categorized on the basis of their monomeric unit and chain structure as ethylene-based polyolefins, propylene-based polyolefins, higher polyolefins, and polyolefin elastomers. Linear chains of ethylene-based polyolefins are mostly obtained at low pressure using transitions metals as catalysts, whereas branched chains of different crystallinity levels and densities are produced at high pressure with oxygen or peroxide initiators [2,3]. Propylene-based polyolefins and higher polyolefins are produced using transition metal catalysts resulting in stereospecific and linear chain structures [4]. By combining ethylene and propylene, polyolefin elastomers [5] are produced with metal or single-site catalysts in the presence or absence of dienes (for cross-linking) and they are usually amorphous with high molecular weight and heterogeneous in phase structures. Polyolefins have diverse applications and are used to manufacture various products such as containers, home appliances, automotive parts, toys, engineering plastics, grocery bags, adhesives, and prosthetic implants. They behave as thermoplastic elastomers, thermosets, or thermoplastics and may be amorphous or highly crystalline [6–8]. They can be extruded as films, pipes/profiles, and fibers; molded into different shapes, and coated onto other materials [9].

Depending on the number of olefinic monomers, polyolefins may be classified into homopolymers, copolymers [10], or terpolymers [11], and on the basis of chain conformation, configuration, and processing conditions, they may be categorized as crystalline or amorphous. Homopolymers, copolymers, and terpolymers are the base materials for polyolefinic blends and composites, which are obtained by radical or

Algae Based Polymers, Blends, and Composites. http://dx.doi.org/10.1016/B978-0-12-812360-7.00013-6

Table 13.1 Structures of Commercial Polyolefin Homopolymers [1]

S. No.	Name (Other Name)	Chemical Structure (Repeat Unit)
1.	Polyethylene (polyethene, polymethylene)	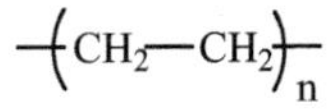
2.	Polypropylene (polypropene)	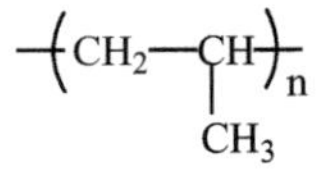
3.	Polybutylene (polybutene-1)	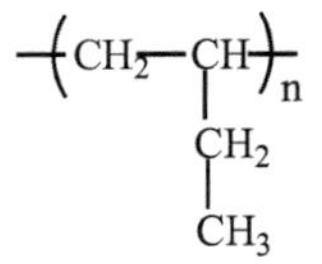
4.	Polyisobutylene	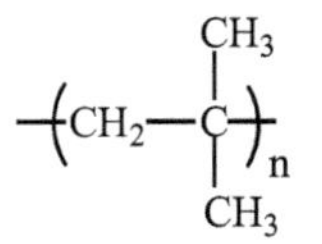
5.	Polybutadiene	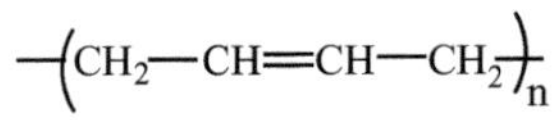
6.	Poly-4-methylpentene-1	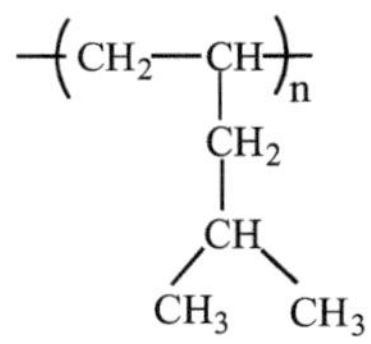
7.	Polyisoprene	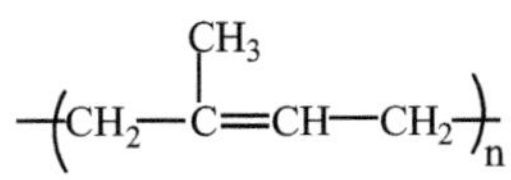

ionic chain growth polymerization of olefins using free radicals and organometallic complexing (Ziegler–Natta and metallocenes) catalysts. Polyethylene (PE), polypropylene (PP), polybutene-1, polymethylpentene-1, and higher polyolefins are homopolymers and their structures are given in Table 13.1 [1]. The global polyolefin companies produce PE and PP in large amounts compared with other homopolyolefins [2,6]. PE may be present in several forms that differ in density levels, chain structures, and crystallinity, such as low-density PE (LDPE), linear low-density PE (LLDPE), high-density PE (HDPE), ultrahigh-molecular-weight PE, and ultralow-density PE. PP is present in three stereospecific forms (isotactic, syndiotactic, and atactic) of varying densities [12–14]. Polyolefin copolymers are random or block copolymers of the same or different monomers and may be single phase or heterophasic based on the quantity of comonomer, the polymerization catalyst, and the processes, e.g., ethylene–propylene rubber [15–17]. A renowned example of terpolymer is ethylene–propylene–diene monomer [18–20].

13.2 SYNTHESIS OF BIOBASED OLEFINIC MONOMERS

The products that come from biobased feedstock represent only a small fraction of the chemical industry's yield. Efforts are carried out to shift the dependence of chemical industry from nonrenewable to renewable raw materials for chemical production owing to the rapid increase in the mineral oil price and a rising concern about the dwindling of these resources in the near future. The production of plant biomass all over the world is vast and is more than enough to meet the demand for the coproduction of materials, fuels, and chemicals along with the demand for feeds and foods [21]. Optimal exploitation of agricultural and forest residue, waste biomass, and nonused land permit the further development of the biobased industry provided the economic and technological limitations are overcome [22,23]. Much attention is being paid to the preparation and application of biobased polymers (derived from living organisms such as plants, trees, and algae or any polymer applicable for biomaterials) [24–30], which are classified into three classes [31] as given below:

1. Polymers directly extracted from biomass (with or without modification), e.g., polymers derived from cellulose and starch.
2. Polymers directly synthesized by microorganisms in their natural or genetically modified form, such as polyhydroxyalkanoates.
3. Polymers derived from intermediates obtained from renewable feedstock, such as bionylons through diacids come from biomass; biopolyurethanes by incorporating vegetable oil-based polyols, and biopolyethylene from bioethylene produced from bioethanol.

Polyolefins, including PE, PP, polyisobutylene, polyisoprene, and polybutadiene, have been synthesized from biobased monomers instead from petroleum-derived monomers to solve the environmental issues caused by petrochemical-derived plastics [31–33].

13.2.1 BIOETHYLENE

In the early 20th century in Western Europe, biobased ethylene was produced by dehydration of ethanol and till the 1980s it was used only in few countries for industrial applications [34,35]. A part of this ethylene was used to prepare PE, whereas the remaining was used to produce polyvinyl chloride and styrene. The polymerization of biobased ethylene is similar to the method for polymerizing petrochemical ethylene, and the first company that offered biobased PE at a commercial scale is a Brazilian company, Braskem. Bioethanol-based PE has similar properties as petroleum-based PE [36].

Nowadays, microbial ethanol is being produced on large scale by fermentation of sugar (from sugar cane), starch (from corn), and lignocellulosic feedstocks (from corn stalks, wood, and other plant wastes) [37]. Bioethanol is further dehydrated to ethylene by using different catalysts such as activated alumina, activated clay,

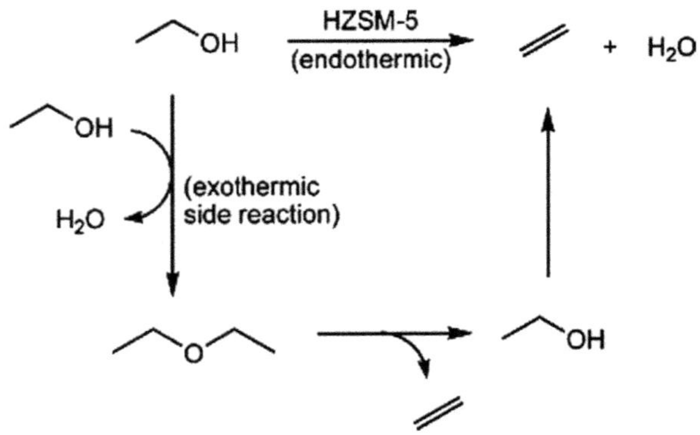

FIGURE 13.1

Synthesis of bioethylene via the dehydration of bioethanol obtained from fermenting biomass [40–42].

transition metal oxides, heteropolyacid zeolites, sulfuric acid, and phosphoric acid [38]. Ethanol dehydration to ethylene is an endothermic reaction [39]. Fig. 13.1 shows the synthesis of bioethylene from bioethanol by using zeolites (HZSM-5) or mesoporous silica (Ni-MCM-41) as catalysts [40–42].

Natural and synthetic plastics are degradable by enzymes and microorganisms such as fungi and bacteria [43–45]. Under different soil conditions the biodegradation proceeds differently, depending on their different properties. The most important process for initiating environmental degradation of synthetic polymers such as polylactic acid, PE, poly-(trimethylene terephthalate), and their copolymers is abiotic hydrolysis [46,47]. Two mechanisms, such as oxobiodegradation and hydrobiodegradation, are used for biodegradation of PE [48].

13.2.2 BIOPROPYLENE

The production of biopropylene is technologically more challenging than that of bioethylene. The process for the preparation of biobased propylene consists of the following steps: (1) biobased synthesis of isobutanol [49], (2) dehydration to butylene, (3) isomerization to 2-butylene [50], and (4) metathesis with ethylene to propylene [51]. In another method, 1,2-propanediol or acetone is first produced by fermentation, then converted to 2-propanol, and finally dehydrated to propylene [32,52]. To produce propylene through sustainable and renewable paths, several strategies have been proposed (fermentation, gasification, and cracking of biomass), which are shown in Fig. 13.2 [42,53–62].

13.2.3 BIOISOBUTYLENE

Biological and chemical catalyses play important roles in the synthesis of isobutylene. Biobased isobutanol (Fig. 13.3) [63] is dehydrated by using solid acids or

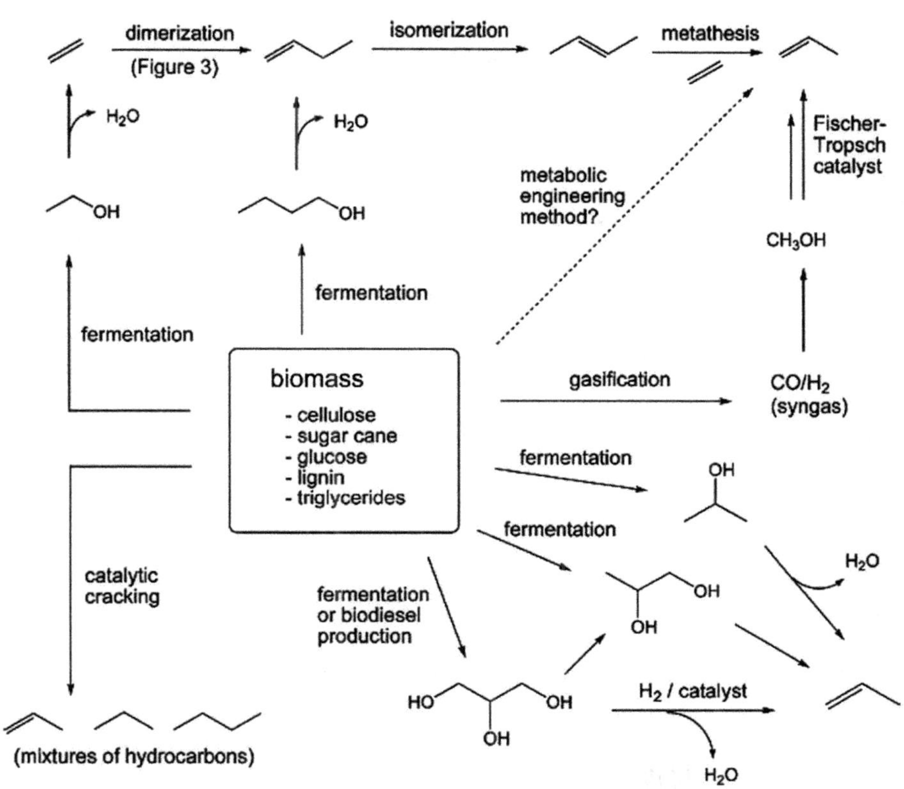

FIGURE 13.2

Possibilities for synthesis of propylene from biomass using fermentation, gasification, or cracking strategies [42,53–62].

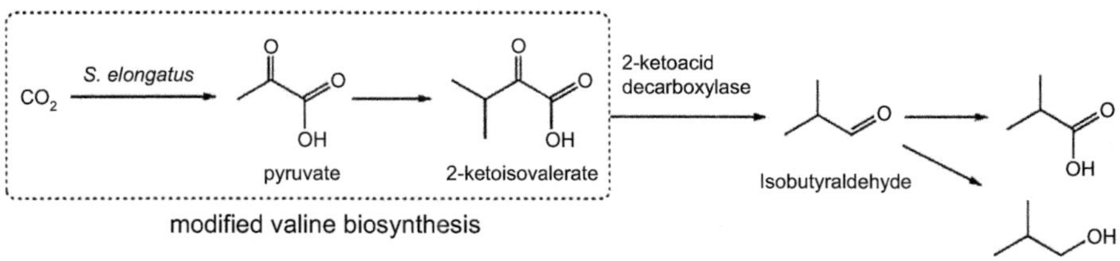

FIGURE 13.3

Synthesis of isobutanol for bioisobutylene production [63].

heterogeneous catalyst to isobutylene. Another nonfermentation method for the production of isobutanol is metabolizing glucose and 2-keto-acid decarboxylase by engineered *Escherichia coli* [49]. An isomerization-metathesis of β-pinene is another renewable method for bioisobutylene production (Fig. 13.4) [64,65].

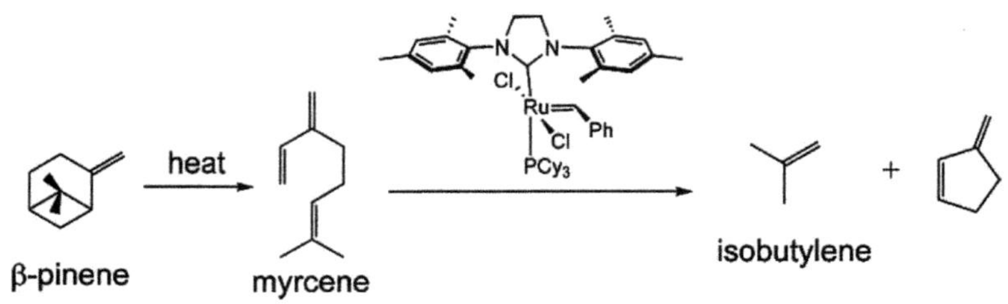

FIGURE 13.4

Thermal isomerization of β-pinene and subsequent intramolecular metathesis for the synthesis of isobutylene and an isoprene substitute [64,65].

13.2.4 BIOISOPRENE

Methods for obtaining isoprene (C_5) from renewable materials have been used in synthetic rubber manufacturing, e.g., pyrolysis of natural rubber (1860), turpentine (1884), and DL-limonene (1911) to get isoprene in low yields [66]. In 1911, Staudinger had optimized the transformation of DL-limonene to isoprene by exploiting an electrically inflamed coil of wire recognized as an isoprene lamp to attain isoprene in 67% yield. Currently, Goodyear and Genencor developed a fermentation process for isoprene production. In the bioisoprene approach, fermentation of sugars, for example glucose, or other biomass takes place with engineered bacteria [67]. This approach gives gaseous isoprene, which is purified and recovered for polymerization [68]. The manufacture of bioisoprene includes the metabolism of glucose with recombinant *E. coli* via two feasible routes (Fig. 13.5). Both mevalonate (MVA) and 5-methyl erythritol phosphate (MEP) pathways transfer the isopentenyl pyrophosphate (IPP) to 1,1-dimethylallyl pyrophosphate (DMAPP). During the production of bioisoprene, the alkenes isomerization of IPP to DMAPP is parallel to the natural rubber (NR) biosynthesis. Unlike NR biosynthesis, which involves propagation of additional IPP units, plant-based isoprene synthase converts the DMAPP to isoprene [69].

The assessment of natural rubber trees and bioisoprene procedures recommended that this fermentation method gives a significant amount of isoprene. For example, the annual production of natural rubber from clones of *Hevea* trees might be as high as 2500 kg/ha [70,71]. The polymerization of bioisoprene in hexane through heterogeneous $TiCl_4$/(iso-butyl)$_3$Al gave a high *cis*-1,4 content (98.5%) [67]. Glass transition temperatures and viscosity values [72] of biobased polyisoprene are very similar to those of petroleum-based polymers.

13.2.5 BIOBUTADIENE

Traditionally, butadiene has been obtained from bioethanol by using different heterogeneous catalysts [73], for example, catalysts based on tantala/silica [74], alumina/ZnO [75], and MgO/silica [76]. Currently, trimetallic (Cu:Zr:Zn) and bimetallic (Zr:Zn) catalysts based on silica generate up to 67% 1,3-butadiene and also give acetone and ethylene as residues [77]. In the multistep process, first oxidation

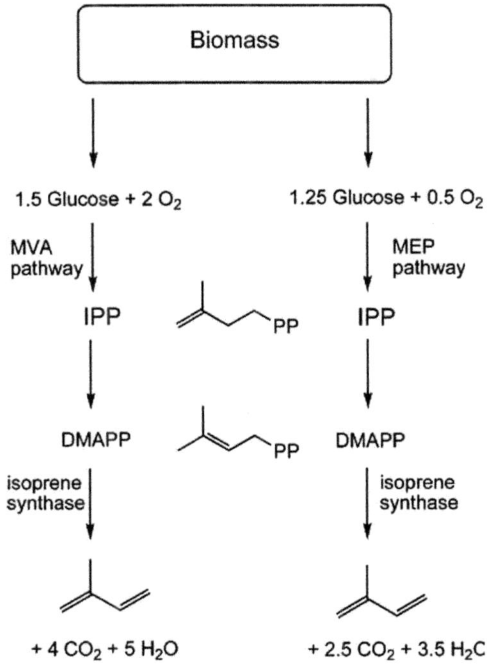

FIGURE 13.5

Synthesis of bioisoprene from glucose using the mevalonate (MVA) or 5-methyl erythritol phosphate (MEP) pathways [67]. *DMAPP*, 1,1-dimethylallyl pyrophosphate; *IPP*, isopentenyl pyrophosphate; *PP*, polypropylene.

of ethanol to acetaldehyde takes place followed by an aldol reaction, which gives acetaldol. After that, a four-carbon molecule undergoes dehydration reactions, giving 1,3-butadiene. Because fermentation methods produce butadiene and bioethanol, these can be regarded as renewable monomers.

13.3 FATTY ACIDS–BASED POLYOLEFINS

Renewable feedstocks are exploited commonly because of the increasing awareness about the greenhouse effect [78], which is normally caused by the use of fossil fuels. Utilization of biomass for fuels, chemicals, and energy is basically carbon neutral. The amount of CO_2 released during the alteration of biomass by pyrolysis, gasification, anaerobic digestion, combustion, and fermentation represents the fixed amount in it. Mostly, olefins are synthesized by the steam cracking process. They may also be synthesized from natural or biogas through methanol [79–81] and oxidative coupling process [82,83]. Coal- [84–86] and biomass-based processes [87–92] have also been described. Traditional methods used the lowest energy, whereas coal and biomass-based processes used about 60%–150% higher energy and methane-based processes used 30% higher energy [93]. Other procedures for manufacturing biobased olefins,

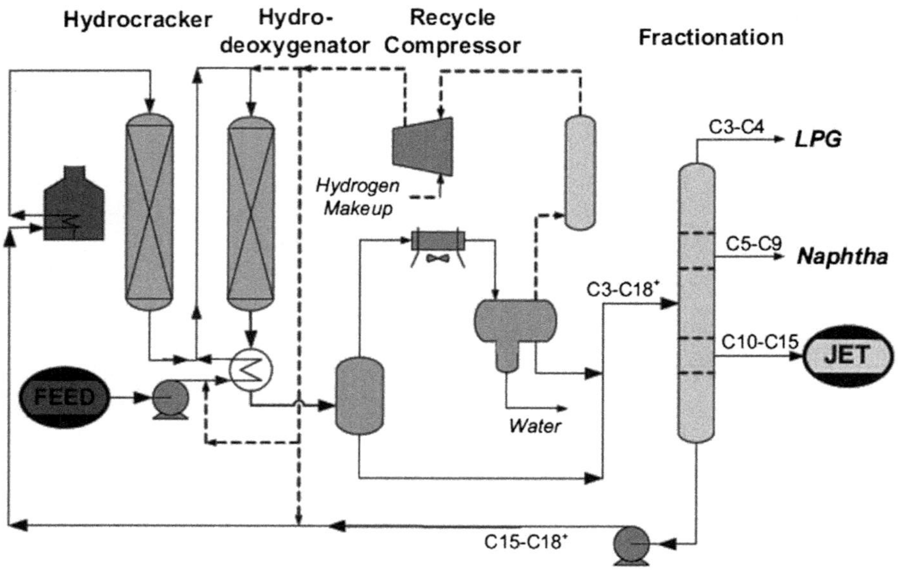

FIGURE 13.6

Schematic representation of the Bio-Synfining process [98].

such as fluidized catalytic cracking of vegetable oils, have been reported in the literature [94]. The use of biobased hydrocarbons to supplement petroleum-based feeds to existing steam crackers thus seems the most viable near-term approach to the transition of the commodity organic chemicals industry from petroleum toward renewable feeds, with little or no capital costs to olefin producers. The bio-synfining method provides possibilities for synthesizing hydrocarbon feeds for traditional steam crackers. The patent-pending method [95–97] is illustrated schematically in Fig. 13.6. In this method, fatty acids, algae, or vegetable oils are converted into paraffinic hydrocarbon fractions having no residual oxygenates [98].

The first reaction is the most important reaction during biooil hydrotreating known as hydrodeoxygenation. Initially the fatty acid chains are hydrogenated and deoxygenated into straight-chain paraffins. Propane and n-paraffins are generated as fatty acids connected to a glycerol backbone. In the next step, the extended straight-chain paraffins are converted into smaller branched and straight isoparaffins in the presence of water, and the product consists of a wide distribution of mostly C_3–C_{18} hydrocarbons, the C_4–C_{10} hydrocarbons being renewable naphtha. The latter can be impressively converted into the preferred light olefins in a steam cracker.

13.4 VEGETABLE OIL–BASED POLYOLEFINS

Industrial and academic research have paid attention on vegetable oils as a platform for polymeric materials owing to their intrinsic biodegradability, low price, and universal availability [99–112]. Polyolefins, for example, PP and PE, have been utilized in surgery since the 1950s because of good biocompatibility, high chemical stability, and favorable mechanical properties [113]. Vegetable oil–based polyolefins are synthesized by different synthetic means owing to the existence of double bonds in the fatty acid chains. In the presence of boron trifluoride diethyl etherate, soybean oil is cationically polymerized in a supercritical dioxide medium and has potential applications as medical lubricants [114]. Some less-poisonous metal compounds such as Ni(carboxylate)$_2$ are highly essential for the polymerization of vegetable oils to fabricate biobased resins appropriate for printing inks as a binder [115]. Acyclic diene metathesis polymerization of double bonds present in unsaturated fatty acid derivatives is a well-organized process to produce high-molecular-weight polymers from vegetable oils (Fig. 13.7A) [116]. It has been shown that ethenolysis of the major fatty acid esters from meadow foam oil produced a number of polymerizable compounds (Fig. 13.7B) [117]. Likewise, thermosets can be produced from norbornene-functionalized castor oil or linseed oil by using the distinctive ring-opening metathesis polymerization method [118]. Heavy metal catalysts such as tungsten and ruthenium are used in both these metathesis polymerization methods and provide highly specific new materials that would otherwise be complicated to produce by using common radical polymerization. The common process to integrate the vegetable oils into a polymeric backbone is the copolymerization of vegetable

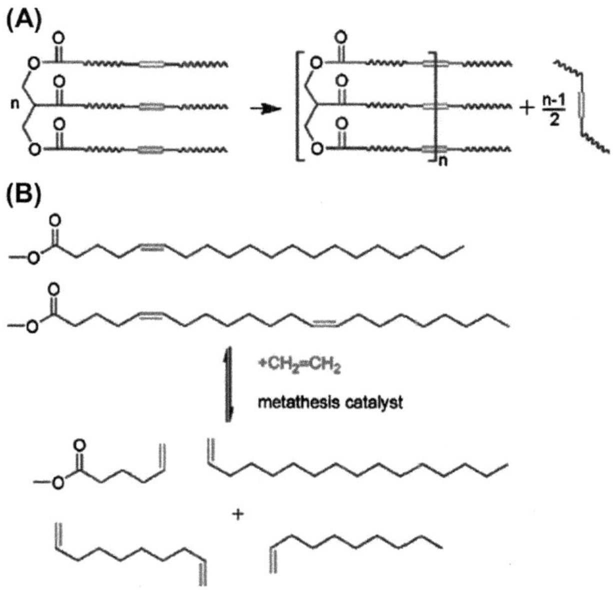

FIGURE 13.7

(A) Acyclic diene metathesis (ADMET) polymerization of double bonds present in unsaturated fatty acid derivatives [116] and (B) metathetical ethenolysis of fatty acid esters from meadowfoam oil [117].

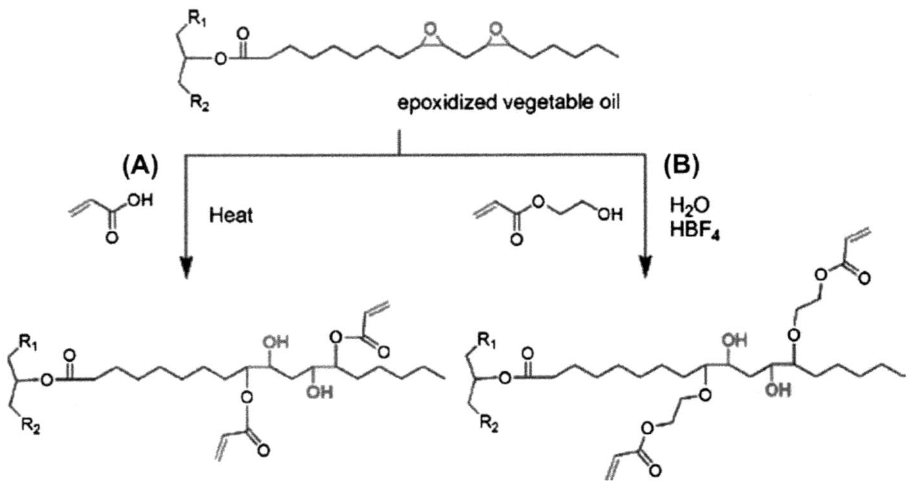

FIGURE 13.8

Synthesis of acrylated epoxidized soybean oil by combination of epoxidized vegetable oil with (A) acrylic acid and (B) 2-hydroxyethyl acrylate [127].

oils with highly reactive monomers, such as methyl acrylic acid and styrene [119–121]. In vegetable oils, functional groups are used to increase the polymerizable reactivity [122,123]. Highly reactive double bonds are commonly used for the combination of acrylic acid and epoxidized vegetable oils [124–126]. Through the epoxy ring-opening process, the carboxyl group of acrylic acid can react with the epoxy group of epoxidized vegetable oils (Fig. 13.8A) and generate a hydroxyl group instead of an ester group. Moreover, self-polymerization of epoxy groups also takes place during this epoxy ring-opening process [125]. Glass transition temperatures of vegetable oil–based polymers are comparatively low; thus copolymerization with styrene can be utilized to increase the transition temperature up to 90°C [124].

The glass transition temperature can also be increased by increasing the double bond functionality and, so increase the cross-linking density. For instance, reactive double bonds might be added by reacting the hydroxyl group produced in the epoxy ring-opening method with maleic anhydride (MA). Polymers synthesized by this process have high glass transition temperatures and good mechanical properties [125]. In 2013, a new class of acrylate-functional monomers/oligomers was obtained from epoxidized soy methyl ester and epoxidized soybean oil by exploiting 2-hydroxyethyl acrylate as an acrylating agent (Fig. 13.8B) [127]. It was found that the acrylic functional products synthesized through this method have high acrylic contents and low viscosity. Olefin metathesis offers a well-established method to synthesize terminal olefins. Transition metals are usually used as catalysts. Ethylene generally reacted with unsaturated fatty acid esters to form terminal olefins and to cleave their double bonds.

Oleic acid methyl ester can react with ethylene via metathesis and generate 1-decene and 9-decenoic acid methyl ester. Terminal olefins of different molecular weights can be produced by using different kinds of fatty acids having double bonds in diverse locations on their alkane chains. These terminal olefins might be copolymerized by other polymerizable olefins to enhance the properties and directly polymerized to produce brushlike polyolefin [128].

Addition of highly reactive double bonds by using the carboxyl is a simple method to produce fatty acid–based polymerizable monomers. Oleyloxyethyl methacrylate can be synthesized by reacting 2-hydroxyethyl methacrylate with oleoyl chloride and has potential applications in bone cement formulations (Fig. 13.9A) [129]. Similarly, structural monomers can be produced by reacting fatty acids with glycidyl methacrylate [131]. Chains of the fatty acids greatly affect the polymer properties. Increasing the chain length can increase the polymer viscosity, whereas double bonds can decrease the viscosity.

These monomers can also be produced by the lipase-catalyzed green method. The 2-hydroxyethyl methacrylate reacts with methyl oleate to generate oleyloxyethyl methacrylate in the presence of Novozym 435 lipase as catalyst [132]. This enzymatic reaction is carried out at about 50°C, and the catalyst simply removed by filtration. Moreover, oleic acid–based epoxy monomers can be produced by reacting 2-hydroxylethyl acrylate and epoxy stearic acid (Fig. 13.9B) [130]. The double bonds of the fatty acid are converted into an epoxy, and the polymer becomes more oxidatively stable and therefore further cross-linked by epoxy groups [128].

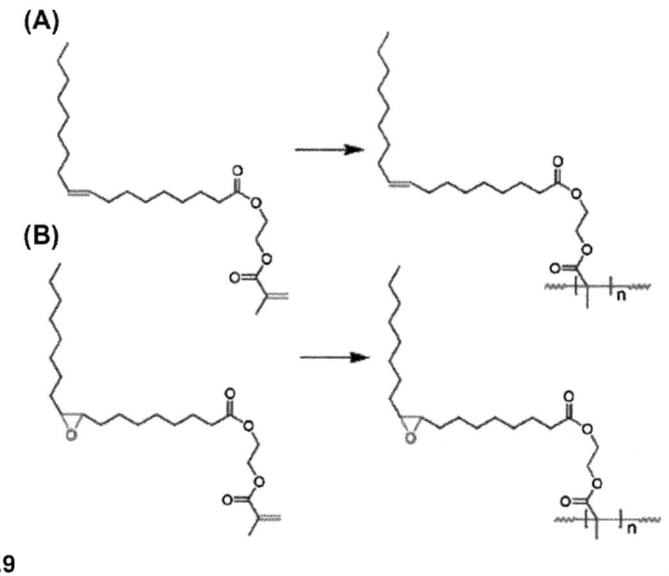

FIGURE 13.9

Oleic acid–based vinyl monomers (oleyloxyethyl methacrylate) obtained by reacting 2-hydroxyethyl methacrylate with (A) oleoyl chloride and (B) methyl oleate [129,130].

13.5 ALGAL POLYSACCHARIDE/POLYOLEFIN BIOCOMPOSITES

The improvements of the polymer composites industry have been very encouraging in the past. The manufacturing price of the composite products can be reduced by using natural fibers as fillers; therefore, polymer biocomposites have received significant attention [133,134]. Furthermore, the consumption of synthetic fibers and polymers can also be minimized, and this indirectly can preserve the environment [135,136]. In addition, biocomposites may be outstanding alternatives to petroleum-based polymer composites with conventional reinforcing materials like carbon fiber and glass [137–139].

13.5.1 AGAR/HIGH DENSITY POLYETHYLENE BIOCOMPOSITES

Production of biocomposites using PE as a matrix may have a positive significance because this polymer usually holds favorable properties such as chemical resistance, nontoxicity, and biocompatibility [140]. HDPE has high tensile stress in comparison with LDPE, is easily processed, and is recyclable and inexpensive [141]. HDPE uses low energy for the processing of biocomposites and avoids thermal breakdown of the natural fibers because of the low melting point compared with PP.

Agar is a polysaccharide that consists of agaropectin and agarose. It is abundantly present in nature, inexpensive, renewable, and nonhazardous and also has gelling property; therefore, it is used as fillers for many polymer composite systems [142,143]. HDPE/agar biocomposites can also be utilized to produce biodegradable and biocompatible products. However, HDPE/agar biocomposites have very low compatibility, which reduces the performance, especially the mechanical properties such as tensile extension, tensile strain, and impact strength of the finishing composite. HDPE has low compatibility because it is a nonpolar, thermoplastic, and hydrophobic polyolefin with little free energy surface [144,145]. On the other hand, agar is a polar, hydrophilic biopolymer that has high free energy surface [146,147]; hence, the significance of compatibilization of the biocomposites cannot be ignored.

Therefore, different chemicals such as acids, alkalis, compatibilizers, and coupling agents were used to improve the physicomechanical properties of biocomposites [148–151]. However, some of these chemicals are harmful and needed more energy and time; therefore, they show some limitations. Ionic liquids were successfully utilized in polymeric composites as compatibilizers, especially for hydrophilic filler and hydrophobic polymer [152].

A combination of ionic liquids and urea can be utilized as a coupling agent for agarose/talc composite films to enhance interfacial adhesion between talc and agarose particles [153]. Furthermore, the uses of ionic liquids are very attractive because they can dissolve many organic and inorganic (liquids/solids) compounds such as polymers and biopolymers [154–157]. Ionic liquids are thermally stable, chemically inert, biocompatible, nonvolatilizable, and noncombustible; can be modified according to the needs; and are also a very promising replacement for the traditional toxic volatile

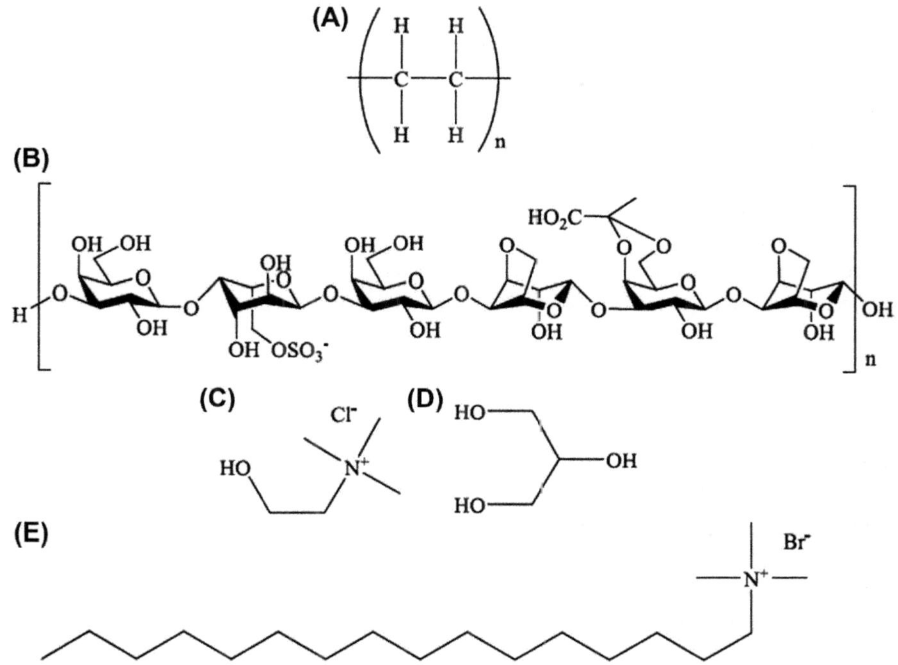

FIGURE 13.10

Chemical structures of (A) high-density polyethylene, (B) agar, (C) choline chloride, (D) Gly, and (E) HTAB [163].

organic solvent [158–161]. Choline chloride/glycerol (ChCl/Gly) is a eutectic-based ionic liquid, which is biocompatible, environmental friendly, and inexpensive and also has a very low freezing point (−40°C) [162]. Shamsuri et al. [163] combined ChCl/Gly and a cationic surfactant (hexadecyltrimethylammonium bromide) to increase the compatibility between agar and HDPE (Fig. 13.10).

Surfactants are amphiphilic compounds that have both nonpolar and polar parts (hydrophobic and hydrophilic alkyl groups, HTAB) [164,165] and stimulate the polymer composite components to cooperate with each other [166–168]. They also improved the compatibility of the biocomposites. In addition, ionic liquid is also used to solvate the surfactant [169] and to help the HTAB to react with both agar and HDPE in the biocomposites.

13.5.2 CALCIUM ALGINATE/LINEAR LDPE BIOCOMPOSITES

Environmental polices of developed countries are increasing the pressure on manufacturers to consider the environmental impact of their products. Therefore, natural fibers are used as a reinforcement of polymer-based composites because of their renewable sources. They are also used as reinforcement in automotive industry.

Natural fibers have many advantages, such as low density, low cost, biodegradability, and high specific properties, compared with conventional ones [170]. Natural fiber composites have high durability and stability against insects and fungi compared with wood. Furthermore, they have high thermal and mechanical properties compared with synthetic fibers. A combination of natural fibers and polymers noticeably increased the stiffness and strength of the resulting materials even at low fiber content. Nevertheless, owing to the poor fiber/matrix compatibility, the interfacial adhesion is not adequate, which decreased the mechanical properties. Therefore, a compatibilizing agent is needed. Many compatibilizing techniques are used, especially in the case of MA-grafted polymers for reactive blending with the fibers and polymers.

The development of the interfacial adhesion clearly enhanced the mechanical properties of the composites [171–176]. The combination of natural fibers and biodegradable polymers give degradable and environment-friendly composites. Natural fiber thermoplastic composite is comparatively a new group of composite materials in which natural fibers/fillers, for example, jute fiber, kenaf fiber, wood fiber, and sisal, are combined with a thermoplastic, such as PP, PE, polyvinyl alcohol, and polyvinyl chloride, to construct the composite. Owing to the improvement in the processing equipment and new production techniques, natural fiber thermoplastic composites are becoming very attractive and used in residential, building, and automotive applications.

Alginate is a polymeric acid obtained from brown seaweeds. It consists of linear copolymers α-(1-4)-linked L-guluronic acid (G) and β-(1-4)-linked D-mannuronic acid (M) [177] and has the properties of both G and M units. The contents of MM, GG, and GM segments are considerably different for alginate obtained from different types of seaweeds. Alginate gives salts with monovalent metal ions, for example, potassium and sodium give water-soluble salts [178–183]. Most divalent ions, for example, zinc and calcium, give water-insoluble alginate salts. Alginate fibers might be synthesized by squeezing out the solutions of sodium alginate into a bath of calcium ions. Compositing is a valuable process to improve the physicochemical performance of polymeric materials.

Thus polymer composites are extensively used in both industrial and scientific applications. LLDPE was utilized as a matrix and calcium alginate fiber was utilized as the reinforcing material to synthesize a reinforced composite. LLDPE has shorter chain branching and low molecular weight distribution; therefore, it is less shear sensitive. During a shearing method, such as extrusion, LLDPE remains viscous and is thus difficult to process. During extrusion, the lesser shear sensitivity of LLDPE gives quicker stress relaxation of the polymer chains and thus the physical properties changed the blow-up ratios. During melt extension, LLDPE has lower viscosity at all strain rates. Owing to the short chain branching, LLDPE permits the chains to “slide by” one another upon elongation without becoming entangled.

This property is essential for film applications because LLDPE films might be downgauged smoothly while retaining high toughness and strength. The rheological properties of LLDPE can be summed up as "stiff in shear" and "soft in extension." They can be recycled into other things such as landscaping ties, trash can liners, lumber, compost bins, shipping envelopes, and floor tiles. The quality of the interface is important for fiber-reinforced polymer composites, as they manage the mechanical performance of the resulting composites [181,182]. The single fiber fragmentation test (SFFT) is used to measure the interfacial properties of fiber-reinforced composites. Convenient monitoring of failure processes directly, replication of the stress transfer properties in real composites, and simple specimen handling are the advantages of the SFFT process. It is an important process to determine the interfacial properties of the fiber-reinforced polymer composites [184–187]. Degradation tests of composites showed that they lose their mechanical properties about 30% and, therefore, show biodegradability.

13.5.3 CALCIUM ALGINATE/POLYACRYLAMIDE–POLYPROPYLENE/FIBRONECTIN BIOCOMPOSITES

Biomaterials have obtained significant consideration in the fields of bioengineering and bioscience to distinguish definite molecules. Biomolecules such as enzymes and protein antibodies are frequently utilized to develop biological substitutes for the therapeutic replacement and tremendous need for organs and tissues [188]. But these biomolecules are unstable and expensive, and their affinity decreases significantly under artificial conditions [189,190]. In the molecular imprinting method, synthetic recognition sites are able to memorize the chemical properties and shape of the target molecules. Molecularly imprinted polymers (MIP) have distinctive properties, for example, easy preparation, low cost, and high stability, in comparison with the affinity matrices synthesized by using enzymes or antibodies [191]. The MIPs have been utilized in a variety of applications, such as separations [192], catalysis, solid-phase extraction [193], biomimetic sensor [194], and drug delivery [195]. Pan et al. [196] employed a molecular imprinting method to establish the cell-adhesive peptide RGDS onto a thermoresponsive cell culture substrate, and used to produce cell sheets. Fukazawa et al. [197] recommended a novel molecular imprinting method by using the cell-adhesive protein fibronectin (FN) as the template for cell capturing. The interaction between FN and fetal calf serum was attained on the FN imprinting substrate, which increased the cell adhesion. The imprinting of low-molecular-weight compounds has been well recognized. In the imprinting of proteins, there are many conflicts owing to the structural complexity, large size, and elastic conformation of the biomacromolecules [198]. Polyacrylamide (PAM) hydrogels have been used for protein imprinting [199]. But because of the poor regeneration and poor mechanical properties of the surface, their applications are limited. Bovine serum albumin (BSA)-imprinted PP fiber-grafted PAM hydrogel was developed, in which acrylamide was used as a functional monomer, BSA as a template molecule, and nonwoven PP fiber as a matrix [200]. BSA-imprinted calcium alginate/PAM hydrogel film was developed in which BSA act as a template and sodium alginate and acrylamide act as functional monomers to increase the strength of the PAM hydrogel. Acetic acid solution and sodium dodecyl sulfate (SDS) were employed to eliminate BSA. But it is difficult to totally eliminate the SDS, and the remaining SDS can cause cell death [201].

FN-imprinted PP nonwoven holding calcium alginate/polyacrylamide hydrogel films (PP-s-CA/PAM MIP) were produced by using nonwoven PP fiber as matrix through ultraviolet radiation-reduced polymerization. The PP-s-CA/PAM MIP has a high adsorption capability for FN compared with nonimprinted polymer (NIP) and is utilized for the culture of mouse fibroblast cells (L929). Results suggested that PP-s-CA/PAM MIP has improved cell adherence property in comparison with NIP and is widely used in tissue engineering and cell culture [202] (Fig. 13.11).

13.6 SEAWEED RESIDUE/POLYOLEFIN BIOCOMPOSITES

Seaweed (SW) might be used for the production of biomass and also provide the productive land for other uses [203]. SW is a pure natural material that has many properties such as excellent heat capacity, heat insulation, strong carbon dioxide fixation, and biodegradability. Owing to fibrils, it also has energy-absorbing properties, which increased the sound-absorption and insulation properties. In addition, it also functions as an excellent flame retardant [204]. In spite of these properties, SW obtained from chemical reactions has many dumping problems. Therefore, SW was used as reinforcing material for biocomposites because it is a favorable material. It is divided into three main classes of red algae, brown algae, and green algae. Lipids, proteins, carbohydrates, ash, nitrogen, and fibers are the most important

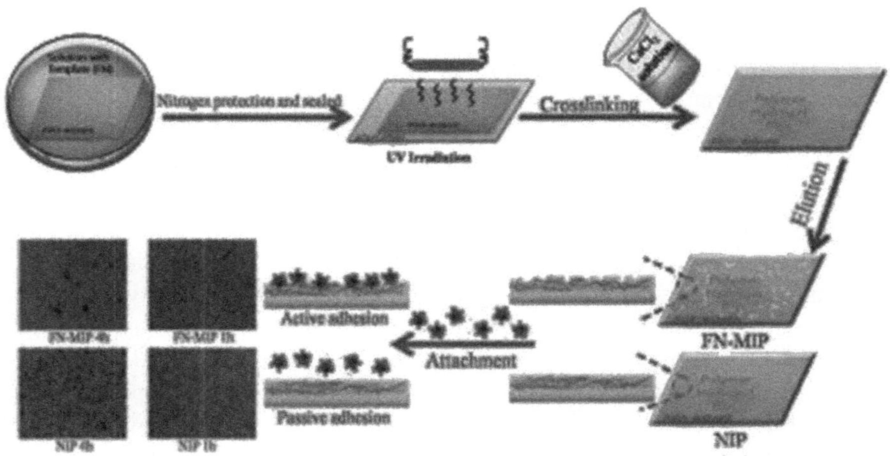

FIGURE 13.11

Schematic representation of the fabrication procedure of imprinted PP-s-CA/PAM and cell adhesion [202].

chemical components of SW. For the structural constituent of cell walls, SW also consists of crystalline cellulose [205,206]. SW has been successfully studied as reinforcement in biocomposites, for example, red algae fiber/PBS [207,208], SW/PP [209–211], and SW residues/HDPE composites [212]. The outcomes illustrated that it is a brilliant reinforcement for biocomposites and might be utilized in a variety of applications such as in the construction and automotive industries.

13.6.1 HIGH DENSITY POLYETHYLENE/SEAWEED RESIDUES BIOCOMPOSITES

Exploitation of marine species in bone implanting has become effectively essential in the biomedical field. In contrast, biomedicine and bioengineering application using HDPE filled with marine residues particularly seaweed residues (SR) is limited. Nevertheless, research on the direct applications of hydroxyapatite and corals and implantation in persons and animals has been developed [213–216]. Guillemin et al. [217,218] studied various types of coralline structures as bone-embedded substitutes. Bone-embedded substitutes were absorbed and gradually replaced by growing bone, showing that corals are osteoconductive materials that act as a straight platform for osteoblast.

Mora and Ouhayoun [219] carried out histological studies of hydroxyapatite and coral implants in bone damage and found outstanding outcome when implants were studied after 12 months. Papacharalambous and Anastasoff [220] also studied four different coral genders, such as *Porites*, *Acropora*, *Pocillopora*, and *Fungia*, and exploited them to enhance the facial factions in three patients (younger than 30 years). They suggested that coral skeleton has excellent biocompatibility and can be shaped easily and their shape remains for a long time.

Albano et al. [212] studied the mechanical, thermal, rheological, and toxicological behavior of HDPE composite filled with SR using Lica 12 as the coupling agent, which is appropriate for bone replacement material. The addition of SR in polymeric matrix shows that it is biocompatible with animal tissues with no unfavorable reactions and therefore has potential applications in biomedicine. Catano et al. [221] examined the effect of gamma rays on thermal, mechanical, and rheological properties of HDPE filled with SR and recommended that high-filler-content composite is the most excellent selection for industrial and biomedical applications.

13.6.2 SEAWEED/POLYPROPYLENE BIOCOMPOSITES

Jang et al. [222] synthesized SW-reinforced PP biocomposites and studied the SW groups and their pretreatment effects on the properties of biocomposites. The SW was obtained from the residue during bioenergy production processing. *Laminaria japonica* present in brown algae (BA) form and *Enteromorpha crinite* present in green algae (GA) form were characterized by pretreatments with sulfuric acid, acetone, and ethanol. SW/PP biocomposites were prepared using different amounts of SW. GA/PP biocomposites have superior impact strength and thermal stability compared with BA/PP biocomposites. The pretreatment of GA with sulfuric acid results in excellent thermal stability, whereas sulfuric acid–treated GA/PP biocomposites have better mechanical and thermal properties.

13.7 *CHLORELLA*–POLYETHYLENE BIOCOMPOSITE

Biological fixation of CO_2 by using photosynthetic microalga such as *Chlorella* is an excellent process to decrease CO_2 in the atmosphere due to the high consumption rate (10%–20%) of light energy with *Chlorella* than common plants (0.1%–0.5%). For such fixation process, efficient utilization of the microalgae is important after fixation of CO_2. So, *Chlorella*–PE composite is used in building materials. Owing to the poor affinity of *Chlorella* for PE, it is necessary to modify *Chlorella* with PE. This modified PE can effectively interact with *Chlorella* by chemical bonds. It has been reported that MA is a valuable modifier for PP and PE [223]. Otsuki et al. [224] synthesized a new composite that consists of PE and *Chlorella* via chemical alteration of PE with MA (Fig. 13.12). *Chlorella*-MA-modified PE composites (Ch–MPE) have tensile strength about two times greater than that of composites resulting from unmodified PE. Fig. 13.13 illustrates a mechanism for

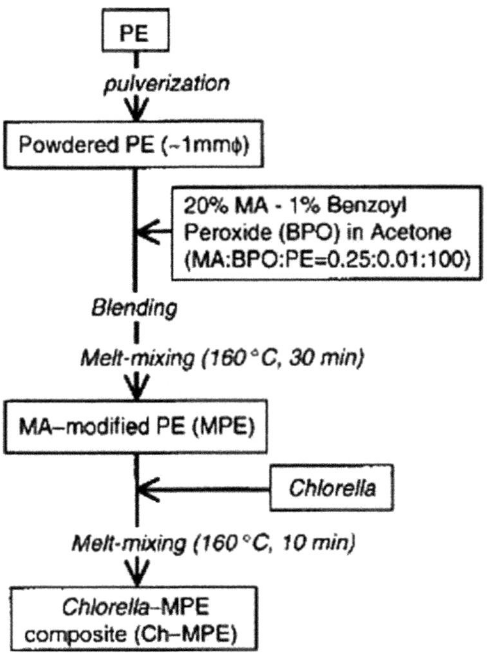

FIGURE 13.12

Synthesis of *Chlorella*-polyethylene composite through solid-phase reactions [224].

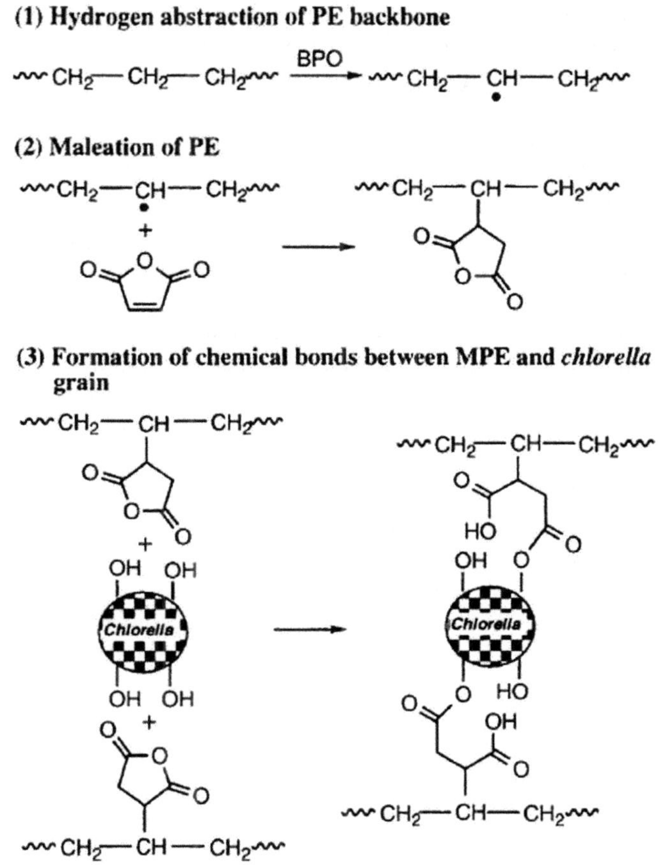

FIGURE 13.13

A proposed reaction mechanism for the formation of chemical bonds between chlorella grains and an MPE matrix [225].

the formation of chemical bonds between MPE and chlorella grains that increased the tensile strength [225]. Owing to the suitable thermal flexibility of Ch–MPE, it might be simply shaped into dish and plates through a heat-pressurizing process.

13.8 CONCLUSION AND FUTURE PERSPECTIVES

Biobased polyolefins derived from intermediates obtained from renewable resources are gaining much attention. Biobased ethylene can be produced by endothermic dehydration of bioethanol by using different catalysts. Several strategies are involved to produce biopropylene through sustainable and renewable paths such as fermentation, gasification, and cracking of biomass. Isobutanol produced by fermentative or nonfermentative method is dehydrated by using solid acids or heterogeneous catalyst to bioisobutylene. Isomerization-metathesis of β-pinene is another method for bioisobutylene production. Bioisoprene can be prepared from glucose using the MVA or MEP routes. Bioderived butadiene is synthesized from bioethanol by using different heterogeneous catalysts and trimetallic and bimetallic catalysts supported on silica. Properties of biobased polyolefins are very similar to petroleum-based polyolefins. In the bio-synfining method, fatty acids, algae, or vegetable oils are converted into paraffinic hydrocarbon fractions having no residual oxygenate. Vegetable oil–based polyolefins are synthesized by different synthetic means owing to the existence of double bonds in the fatty acid chains. The fabrication cost of the composites can be reduced by using natural fibers as fillers; therefore, polymer biocomposites have received great attention. Natural fibers have many advantages such as low density, low cost, biodegradability, and high specific properties compared with conventional ones. Natural fiber composites have high durability and stability against insects and fungi compared with wood. Furthermore, they have better thermal and mechanical properties compared with synthetic fibers. Biocomposites produced using PE as a matrix may have great importance because this polymer shows chemical resistance, nontoxicity, and biocompatibility. HDPE/agar biocomposites are utilized to produce biodegradable and biocompatible products. ChCl/Gly and HTAB can be used to increase the compatibility between agar and HDPE. Calcium alginate/linear LDPE biocomposites lose their mechanical properties about 30% and show biodegradable behavior. FN-imprinted PP nonwoven holding calcium alginate/PAM hydrogel films are widely used in tissue engineering and cell culture. HDPE/SR composite is appropriate for bone replacement material. *Chlorella*-MA-modified PE composites have tensile strength about two times greater than that of composites resulting from unmodified PE.

REFERENCES

[1] Domasius N. Polyolefin composites. Hoboken (New Jersey): John Wiley & Sons, Inc.; 2008. p. 1–5.

[2] Gahleitner M. Melt rheology of polyolefins. Prog Polym Sci 2001;26:895–944.

[3] Bersted BH, Slee JD, Richter CA. J Appl Polym Sci 1999;26:1001–14.

[4] Moore EP, editor. Polypropylene handbook. Munich: Hanser; 1996.

[5] Aenderkerk J, Michels D, editors. Proceedings of metallocenes 96 conference, Schotland, Dusseldorf, vol. 4; 1996. p. 01–12.

[6] Jim P. Introduction to polyolefins. In: Soares JBP, McKenna TFL, editors. Polyolefin reaction engineering. 1st ed. Wiley-VCH Verlag GmbH & Co. KGaA; 2012.

[7] Sanchis R, Fenollar O, Garcia D, Sanchez L, Balart R. Improved adhesion of LDPE films to polyolefin foams for automotive industry using low-pressure plasma. Int J Adhes Adhes 2008;28:445–51.

[8] Samuel COU. Polyolefin fibres: industrial and medical applications. Abington Hall, Granta Park, Great Abington Cambridge CB21, 6AH, England: Elsevier, Technology & Engineering, Woodhead Publishing Limited; 2009. p. 428.

[9] Cornelia V. Handbook of polyolefins. 2nd ed. Abington Hall, Granta Park, Great Abington Cambridge CB21, 6AH, England: CRC Press, Technology & Engineering, Woodhead Publishing Limited; 2000. p. 1032.
[10] Chung TC. Synthesis of functional polyolefin copolymers with graft and block structures. Prog Polym Sci 2002;27(1):39–85.
[11] Yongsok S, Taejin K, Soon MH, Hyoung JC. Nonisothermal crystallization behaviors of a polyolefin terpolymer and its foam. Polymer 2007;48:3844–9.
[12] Eckstein A, Suhm J, Friedrich C, Maier RD, Sassmannshausen J, Bochmann M, Mülhaupt R. Improved adhesion of LDPE films to polyolefin foams for automotive industry using low-pressure plasma. Macromolecules 1998;31(4):1335–40.
[13] Weiqing W, Weiguo H, Dekmezian AH, Charles JR. Long chain branched isotactic polypropylene. Macromolecules 2002;35(10):3838–43.
[14] Abbas R. Syndiotactic/atactic block polyolefins, catalysts and processes for producing the same. US 6448349 B1. September 10, 2002. p. 1–9.
[15] Wang B, Luebkert MR. Hot melt adhesive based on ethylene-propylene rubber (EPR) and semi crystalline olefinic polymers. US 6143818 A. November 7, 2000.
[16] Naoki H, Hirotaka O, Arimitsu U. Preparation and properties of ethylene propylene rubber (EPR)–clay nanocomposites based on maleic anhydride-modified EPR and organophilic clay. J Appl Polym Sci 2004;93(2):758–64.
[17] Marianna K, Yiqun L, Jeremy RA, Scott Parent J. The dynamics of montmorillonite clay dispersion and morphology development in immiscible ethylene–propylene rubber/polypropylene blends. Polymer 2007;48:4520–8.
[18] Ismail H, Pooria P, Ahmad FMN, Abu B. Morphological, thermal and tensile properties of halloysite nanotubes filled ethylene propylene diene monomer (EPDM) nanocomposites. Polym Test 2008;27:841–50.
[19] Konstantinos GG, Ralf T, József KK. Characteristics of ethylene propylene diene monomer rubber/organoclay nanocomposites resulting from different processing conditions and formulations. Polym Int 2004;53(8):1191–7.
[20] Ismail H, Shaari SM. Curing characteristics, tensile properties and morphology of palm ash/halloysite nanotubes/ethylene-propylene-diene monomer (EPDM) hybrid composites. Polym Test 2010;29:872–8.
[21] Dale BE. 'Greening' the chemical industry: research and development priorities for biobased industrial products. J Chem Technol Biotechnol 2003.
[22] Hill K. Fats and oils as oleochemical raw materials. Pure Appl Chem 2000;72: 1255–64.
[23] Rajni HK, Ulrika T, Linda G, Pal B. Industrial biotechnology for the production of biobased chemicals – a cradle-to-grave perspective. Trends Biotechnol 2007;25(3): 119–24.
[24] Gandini A. Polymers from renewable resources: a challenge for the future of macromolecular materials. Macromolecules 2008;41(24):9491–504.
[25] Raquez JM, Deleglise M, Lacrampe MF, Krawczak P. Thermosetting (bio)materials derived from renewable resources: a critical review. Prog Polym Sci 2010;35(4): 487–509.
[26] Mohanty A, Misra M, Drzal L. Sustainable bio-composites from renewable resources: opportunities and challenges in the green materials world. J Polym Environ 2002; 10(1):19–26.
[27] Mohanty AK, Misra M, Drzal LT, Selke SE, Harte BR, Hinrichsen G. Natural fibers, biopolymers, and biocomposites: an introduction. London: Taylor & Francis; 2005. p. 1–36.

[28] Van BJ, Poirier Y. Prospects for biopolymer production in plants. In: Green gene technology. Berlin: Springer Verlag; 2007. p. 133−51.

[29] Wool RP, Sun XS. Bio-based polymers and composites. London: Academic Press; 2005.

[30] Ragauskas AJ, Williams CK, Davison BH, Britovsek G, Cairney J, Eckert CA. The path forward for biofuels and biomaterials. Science 2006;311(5760):484−9.

[31] Queiroz AUB, Collares-Queiroz FP. Innovation and industrial trends in bioplastics. J Macromol Sci Part C Polym Rev 2009;49(2):65−78.

[32] Chen GQ, Patel MK. Plastics derived from biological sources: present and future: a technical and environmental review. Chem Rev 2012;112:2082−99.

[33] Yoshiyasu N, Shunji Y, Masaaki S. Liquid scintillation counting of solid-state plastic pellets to distinguish bio-based polyethylene. Polym Test 2014;33:13−5.

[34] Morschbacker AJ. Macromol Sci Part C Polym Rev 2009;49:79.

[35] Shen L, Patel MK. Lenzinger Ber 2010;88:1.

[36] Chen GQ. Plastics from bacteria, natural functions and applications. Heidelberg: Springer; 2010. p. 1−16.

[37] Huang H. Natural functions and applications. Heidelberg: Springer; 2010. p. 389−404.

[38] Huang YM, Li H, Huang XJ, Hu YC, Hu YC. J Bioprocess Eng 2008;6:1.

[39] Tsao U, Zasloff HB. Production of ethylene from ethanol. US Patent 4134926. 1979.

[40] Iwamoto M, Kasai K, Haishi T. Conversion of ethanol into polyolefin building blocks: reaction pathways on nickel ion-loaded mesoporous silica. ChemSusChem 2011;4: 1055−8.

[41] Hu Y, Zhan N, Dou C, Huang H, Han Y, Yu D, Hu Y. Biotechnol J 2010;5:1186−91.

[42] Robert TM. How well can renewable resources mimic commodity monomers and polymers? J Polym Sci Part A Polym Chem 2012;50:1−15.

[43] Göpferich A. Handbook of biodegradable polymers. Amsterdam: Harwood Academic; 1997. p. 451−71.

[44] Heidary S, Gordon B. Hydrolyzable poly(ethylene terephthalate). J Environ Polym Degrad 1994;2:19.

[45] Hiltunen K, Seppälä JV, Itävaara M, Härkönen M. The biodegradation of lactic acid−based poly(ester-urethanes). J Environ Polym Degrad 1997;5:167−73.

[46] Albertsson AC, Barenstedt C, Karlsson S. Abiotic degradation products from enhanced environmentally degradable polyethylene. Acta Polym 1994;45:97.

[47] Cruz Pinto JJC, Carvalho MES, Ferreira JFAA. The kinetics and mechanism of polyethylene photo-oxidation. Angew Makromol Chem 1994;216:113.

[48] Luinstra GA. Poly(propylene carbonate), old copolymers of propylene oxide and carbon dioxide with new interests: catalysis and material properties. Polym Rev 2008;48: 192.

[49] Atsumi S, Hanai T, Liao JC. Non-fermentative pathways for synthesis of branched-chain higher alcohols as biofuels. Nature 2008;451:86−9.

[50] Luo XL, Tang DY, Li MJ. Quantum investigation on the mechanism of isomerization of 1-butylene catalyzed by Rh-complex. J Mol Struct Theo Chem 2005;731:139.

[51] Li X, Guan J, Zheng AM, Zhou DH, Han XW, Zhang WP, Bao XH. DFT studies on the reaction mechanism of cross-metathesis of ethylene and 2-butylene to propylene over heterogeneous Mo/HBeta catalyst. J Mol Catal A Chem 2010;330:99.

[52] van Haveren J, Scott E, Sanders J. Bulk chemicals from biomass. Biofuels Bioprod Biorefin 2008;2:41.

[53] Smith C, editor. European plastics news: Braskem commits to producing bio-based polypropylene. 2010. Available at: http://plasticsnews.com/headlines2.html?id¼420130.
[54] Jagger A. Dow chemical explores methods for producing its key feedstocks from renewable resources. 2011. Available at: http://www.icis.com/Articles/2011/02/28/9438198/dow-studiesbio-based-propylene-routes.html.
[55] (a) Fadigas JCS, Gambetta R, Mota CJA, Goncalves VLC. WO2010012060 A2;56.(b) Petroquimica SAQ, Brazil, 2010.
[56] (a) Hulteberg C, Brandin J. In: WO2010052208.(b) Swed AB. Biofuel-Solution, Sweden, 2010.
[57] Wang Z, Zhuge J, Fang H, Prior BA. Glycerol production by microbial fermentation: a review. Biotech Adv 2001;19:201–23.
[58] Zhou CH, Beltramini JN, Fan YX, Lu GQ. Chemoselective catalytic conversion of glycerol as a biorenewable source to valuable commodity chemicals. Chem Soc Rev 2008;37:527–49.
[59] Ramey D, Yang ST. Production of butyric acid and butanol from biomass. Morgantown (WV): U.S. Department of Energy; 2004.
[60] Ooi YS, Zakaria R, Mohamed AR, Bhatia S. Catalytic conversion of palm oil-based fatty acid mixture to liquid fuel. Biomass Bioenergy 2004;27:477–84.
[61] Onwudili JA, Williams PT. Hydrothermal reforming of bio-diesel plant waste: products distribution and characterization. Fuel 2010;89:501–9.
[62] Yin S, Dolan R, Harris M, Tan Z. Subcritical hydrothermal liquefaction of cattle manure to bio-oil. Bioresour Technol 2010;101:3657–64.
[63] Benson EE, Kubiak CP, Sathrum AJ, Smieja JM. Electrocatalytic and homogeneous approaches to conversion of CO_2 to liquid fuels. Chem Soc Rev 2009;38:89–99.
[64] Delancey JM, Cavazza MD, Rendos MG, Ulisse CJ, Palumbo SG, Mathers RT. Sustainable polymers: a joint collection from green chemistry and polymer chemistry. J Polym Sci Part A Polym Chem 2017;5:3117–8 (in press).
[65] Kobayashi S, Lu C, Hoye TR, Hillmyer MA. J Am Chem Soc 2009;131:7960–1.
[66] Whitby GS, editor. Synthetic rubber. New York: Wiley; 1954.
[67] Whited GM, Feher FJ, Benko DA, Cervin MA, Chotani GK, McAuliffe JC, LaDuca RJ, Ben Shoshan EA, Sanford KJ. Development of a gas phase bio process for isoprene-monomer production using metabolic pathway engineering. Ind Biotechnol 2010;6:152–63.
[68] Cervin M, Whited GM, Chotani GK, Valle F, Fioresi C, Sanford KJ, McAuliffe JC, Ferher FJ, Puhala AS, Miasnikov A, Aldor IS. Compositions and methods for producing isoprene. WO2009/076676 A2. 2009.
[69] Miller B, Oschinski C, Zimmer W. First isolation of an isoprene synthase gene from poplar and successful expression of the gene in *Eschericha coli*. Planta 2001;213:483–7.
[70] Hayashi Y. Production of natural rubber from para rubber tree. Plant Biotechnol 2009;26:67–70.
[71] Crement DA, Priyadarshan PM, Hoa TTT, Venkatachalam P, editors. Hevea rubber breeding and genetics. Hoboken: Wiley; 2007.
[72] Asadullah M, Ito S, Kunimori K, Yamada M, Tomishige K. Energy efficient production of hydrogen and syngas from biomass: development of low-temperature catalytic process for cellulose gasification. Environ Sci Technol 2002;36:4476–81.
[73] Sun HN, Wristlers JP. Butadiene. New York: Wiley; 2007.

[74] Stahly EE, Jones HE, Corson BB. Butadiene from ethanol. Ind Eng Chem 1948;40: 2301–3.

[75] Bhattacharyya SK, Avasthi BN. One-step catalytic conversion of ethanol to butadiene in a fluidized bed. Ind Eng Chem Process Des Dev 1963;2:45–51.

[76] Kvisle S, Aguero A, Sneeden RPA. Transformation of ethanol into 1,3-butadiene over magnesium oxide/silica catalysts. Appl Catal 1988;42:117–31.

[77] Jones MD, Keir CG, Iulio CD, Robertson RAM. Catal Sci Technol 2011;1:267–72.

[78] Houghton JT, Ding Y, Griggs DJ, Noguer M, Van der linden PJ, Xiaosu D. Climate change 2001: the scientific basis: contributions of working group I to the third assessment report of the intergovernmental panel on climate change. Cambridge University Press; 2001.

[79] Fung SC, Kuechler KH, Smith JS, Coute NP, Vaughn SN, Cao C. In: U.S.P. Office, editor. Catalytic production of olefins at high methanol partial pressures. USA: ExxonMobil Chemical Patents Inc.; 2003.

[80] Chen JQ, Bozzano A, Glover B, Fuglerud T, Kvisle S. Recent advancements in ethylene and propylene production using the UOP/Hydro MTO process. Catal Today 2005;106:103–7.

[81] Roberts M. UOP hydro breakthrough on MTO process. Chem Week 1995;157: 15–115.

[82] Hoebink J, Venderbosch HM, Van GPC, Den OPF, Marin GB. Economics of the oxidative coupling of methane as an add-on unit for naphtha cracking. Chem Eng Technol 1995;18:12–6.

[83] Swanenberg G. Cogeneration of ethylene and electricity with oxidative methane coupling: a study of technical feasibility and economic merit conducted for DSM. Eindhoven (The Netherlands): Sparqle International and SEP, Eindhoven University of Technology; 1998.

[84] Bechtel. Gasification plant cost and performance optimization task 2 topical report coke/coal gasification with liquids coproduction. Washington (DC): US Department of Energy; 2003.

[85] Gray D. Producing liquid fuels from coal. In: National research council board on energy and environmental systems workshop on trends in oil supply and demand. Washington, DC, USA; 2005.

[86] Zhang Y. Shenhua coal conversion technology and industry development. In: Global climate and energy project international workshop on exploring the opportunities for research to integrate advanced coal technologies with CO_2 capture and storage in China; 2005. Palo Alto (USA).

[87] Hamelinck C. Outlook for advanced biofuels. Utrecht: Utrecht University; 2004.

[88] Patel M, Crank M, Dornburg V, Hermann B, Roes L, Husing B. Medium and long-term opportunities and risks of the biotechnological production of bulk chemicals from renewable resources. Utrecht: Universiteit Utrecht; 2005.

[89] BR Inc.. Gasification–fermentation pilot facility in Arkansas. Fayetteville (Arkansas): BRO Energy LLC and Bioengineering Resources, Inc.; 2005.

[90] Consultants TSS. Gridley ethanol demonstration project utilizing biomass gasification technology: pilot plant gasifier and syngas conversion testing. National Renewable Energy Laboratory; 2005.

[91] Coskata. Energy positive process. Warrenville (Illinois): Coskata Inc.; 2008.

[92] Range Fuels. Our first plant. Broomfield (USA). 2008.

[93] Ren T, Daniels B, Patel MK, Blok K. Petrochemicals from oil natural gas coal and biomass: production costs in 2030–2050. Resour Conserv Recycl 2009;53:653–63.

[94] Marker TL. Production of olefins from biorenewable feedstocks. Des Plaines (IL, USA): UOP LLC; 2007.

[95] Abhari R, Havlik PZ. Hydrodeoxygenation process. USA. 2009.

[96] Abhari R, Tomlinson HL, Havlik PZ, Jannasch NL. Process for co-producing jet fuel and LPG from renewable sources. USA. 2008.

[97] Abhari R, Tomlinson HL, Roth EG. Biorenewable naphtha. USA. 2009.

[98] Steven PP, Carl MS, Marie FR, Ramin A, Guy BM, Kevin MVG. Biomass to olefins: cracking of renewable naphtha. Chem Eng J 2011;176–177:178–87.

[99] Lligadas G, Ronda JC, Galia M, Cadiz V. Poly(ether urethane) networks from renewable resources as candidate biomaterials: synthesis and characterization. Biomacromolecules 2007;8(2):686–92.

[100] Miao S, Sun L, Wang P, Liu R, Su Z, Zhang S. Soybean oil based polyurethane networks as candidate biomaterials: synthesis and biocompatibility. Eur J Lipid Sci Tech 2012;114:1165–74.

[101] Guo B, Chen Y, Lei Y, Zhang L, Zhou WY, Rabie ABM, et al. Biobased poly(propylene sebacate) as shape memory polymer with tunable switching temperature for potential biomedical applications. Biomacromolecules 2011;12(4):1312–21.

[102] Meier MA, Metzger JO, Schubert US. Plant oil renewable resources as green alternatives in polymer science. Chem Soc Rev 2007;36(11):1788–802.

[103] Xia Y, Larock RC. Vegetable oil-based polymeric materials: synthesis, properties, and applications. Green Chem 2010;12(11):1893–909.

[104] Biermann U, Friedt W, Lang S, Luhs W, Machmuller G, Metzger JO, et al. New syntheses with oils and fats as renewable raw materials for the chemical industry. Angew Chem Int Ed 2000;39(13):2206–24.

[105] Sharma V, Kundu P. Condensation polymers from natural oils. Prog Polym Sci 2008; 33(12):1199–215.

[106] Ronda JC, Lligadas G, Galia M, Cadiz V. Vegetable oils as platform chemicals for polymer synthesis. Eur J Lipid Sci Tech 2011;113(1):46–58.

[107] Petrović ZS. Polyurethanes from vegetable oils. Polym Rev 2008;48(1):109–55.

[108] Montero de Espinosa L, Meier MA. Plant oils: the perfect renewable resource for polymer science. Eur Polym J 2011;47(5):837–52.

[109] Galia M, de Espinosa LM, Ronda JC, Lligadas G, Cadiz V. Vegetable oil-based thermosetting polymers. Eur J Lipid Sci Tech 2010;112(1):87–96.

[110] Desroches M, Escouvois M, Auvergne R, Caillol S, Boutevin B. From vegetable oils to polyurethanes: synthetic routes to polyols and main industrial products. Polym Rev 2012;52(1):38–79.

[111] Lligadas G. Renewable polyols for polyurethane synthesis via thiol-ene/yne couplings of plant oils. Macromol Chem Phys 2013;214:415–22.

[112] Noreen A, Zia KM, Zuber M, Tabasum S, Zahoor AF. Bio-based polyurethane: an efficient and environment friendly coating systems: a review. Prog Org Coat 2016;91: 25–32.

[113] Zaikov G, Gumargalieva K, Polishchuk AY, Adamyan A, Vinokurova T. Polyolefin biodegradation. Int J Polym Mater 1999;44(1–2):107–33.

[114] Liu Z, Sharma BK, Erhan SZ. From oligomers to molecular giants of soybean oil in supercritical carbon dioxide medium: 1. Preparation of polymers with lower molecular weight from soybean oil. Biomacromolecules 2007;8(1):233–9.

[115] Mello VM, Oliveira GV, Mandarino JM, Carrao-Panizzi MC, Suarez PA. New metal catalysts active for thermal polymerization of vegetable oils. Ind Crop Prod 2013; 43:56–60.

[116] Tian Q, Larock RC. Model studies and the ADMET polymerization of soybean oil. J Am Oil Chem Soc 2002;79(5):479–88.
[117] Warwel S, Bruse F, Demes C, Kunz M. Polymers and polymer building blocks from meadow foam oil. Ind Crop Prod 2004;20(3):301–9.
[118] Lu Y, Larock RC. Novel polymeric materials from vegetable oils and vinyl monomers: preparation, properties, and applications. ChemSusChem 2009;2(2):136–47.
[119] Andjelkovic DD, Valverde M, Henna P, Li F, Larock RC. Novel thermosets prepared by cationic copolymerization of various vegetable oils-synthesis and their structure–property relationships. Polymer 2005;46(23):9674–85.
[120] Li F, Hanson MV, Larock RC. Soybean oil-divinylbenzene thermosetting polymers: synthesis, structure, properties and their relationships. Polymer 2001;42(4):1567–79.
[121] Kundu PP, Larock RC. Novel conjugated linseed oil-styrene-divinylbenzene copolymers prepared by thermal polymerization. 1. Effect of monomer concentration on the structure and properties. Biomacromolecules 2005;6(2):797–806.
[122] Mauldin T, Haman K, Sheng X, Henna P, Larock R, Kessler M. Ring-opening metathesis polymerization of a modified linseed oil with varying levels of crosslinking. J Polym Sci Pol Chem 2008;46(20):6851–60.
[123] Esen H, Kusefoğlu SH. Photolytic and free-radical polymerization of cinnamate esters of epoxidized plant oil triglycerides. J Appl Polym Sci 2003;89(14):3882–8.
[124] La Scala J, Wool RP. Property analysis of triglyceride-based thermosets. Polymer 2005;46(1):61–9.
[125] Luo Q, Liu M, Xu Y, Ionescu M, Petrović ZS. Thermosetting allyl resins derived from soybean oil. Macromolecules 2011;44(18):7149–57.
[126] Bunker SP, Wool RP. Synthesis and characterization of monomers and polymers for adhesives from methyl oleate. J Polym Sci Polym Chem 2002;40(4):451–8.
[127] Rengasamy S, Mannari V. Development of soy-based UV-curable acrylate oligomers and study of their film properties. Prog Org Coat 2013;76:78–85.
[128] Miao S, Wang P, Su Z, Zhang S. Vegetable-oil-based polymers as future polymeric biomaterials. Acta Biomater 2014;10:1692–704.
[129] Vazquez B, Deb S, Bonfield W, Roman JS. Characterization of new acrylic bone cements prepared with oleic acid derivatives. J Biomed Mater Res 2002;63(2):88–97.
[130] Miao S, Callow N, Zhang S, Su Z, Wang P, Liu Y. Enzymatic synthesis of oleic acid-based epoxy monomer for the production of value added polymers. Biotechnol Lett 2013. http://dx.doi.org/10.1007/s10529-013-1162-1.
[131] Scala JJL, Sands JM, Orlicki JA, Robinette EJ, Palmese GR. Fatty acid-based monomers as styrene replacements for liquid molding resins. Polymer 2004;45(22):7729–37.
[132] Cho HG, Park SY, Jegal J, Song BK, Kim HJ. Preparation and characterization of acrylic polymers based on a novel acrylic monomer produced from vegetable oil. J Appl Polym Sci 2010;116(2):736–42.
[133] Santos AS, Farina MZ, Pezzin PT, et al. The application of peach palm fibers as an alternative to fiber reinforced polyester composites. J Reinf Plast Compos 2008;27:1805–16.
[134] Singha AS, Rana RK. Preparation and properties of agave fiber-reinforced polystyrene composites. J Thermoplast Compos Mater 2011;26:513–26.
[135] Pan M, Mei C, Song Y. A novel fire retardant affects fire performance and mechanical properties of wood flour-high density polyethylene composites. BioResources 2012;7:1760–70.

[136] Bodirlau R, Spiridon I, Teaca CA. Influence of components ratio upon mechanical properties of wood/thermoplastic polymer composites. Cell Chem Technol 2009;43: 205–9.

[137] Jawaid M, Abdul Khalil HPS, Alattas OS. Woven hybrid biocomposites: dynamic mechanical and thermal properties. Compos Part A 2012;43:288–93.

[138] Han SO, Karevan M, Bhuiyan MA, Park JH, Kalaitzidou K. Effect of exfoliated graphite nanoplates on the mechanical and viscoelastic properties of poly(lactic acid) biocomposites reinforced with kenaf fibers. J Mater Sci 2012;27(8): 3535–43.

[139] Han YH, Han SO, Cho DH, Kim HI. Kenaf/polypropylene biocomposites: effects of electron beam irradiation and alkali treatment on kenaf natural fibers. Compos Interfaces 2007;14(5–6):559–78.

[140] Rocha M, Mansur A, Mansur H. Characterization and accelerated ageing of UHMWPE used in orthopedic prosthesis by peroxide. Materials 2009;2:562–76.

[141] Chaharmahali M, Mirbagheri J, Tajvidi M, et al. Mechanical and physical properties of wood-plastic composite panels. J Reinf Plast Compos 2010;29:310–9.

[142] Madera-Santana TJ, Robledo D, Azamar JA, et al. Preparation and characterization of low density polyethylene-agar biocomposites: torque-rheological, mechanical, thermal and morphological properties. Polym Eng Sci 2010;50:585–91.

[143] Oprea S. Preparation and characterization of the agar/polyurethane composites. J Compos Mater 2011;45:2039–45.

[144] Sharif A, Aalaie J, Shariatpanahi H, et al. Study on the structure and properties of nanocomposites based on high-density polyethylene/starch blends. J Polym Res 2011;18:1955–69.

[145] Najafi SK, Sharifnia H, Tajvidi M. Effects of water absorption on creep behavior of wood–plastic composites. J Compos Mater 2008;42:993–1002.

[146] Shamsuri AA, Abdullah DK, Daik R. Fabrication of agar/biopolymer blend aerogels in ionic liquid and co-solvent mixture. Cell Chem Technol 2012;46:45–52.

[147] Taha I, Ziegmann G. A comparison of mechanical properties of natural fiber filled biodegradable and polyolefin polymers. J Compos Mater 2006;40:1933–46.

[148] El-Shekeil YA, Sapuan SM, Khalina A, et al. Effect of alkali treatment on mechanical and thermal properties of Kenaf fiber-reinforced thermoplastic polyurethane composite. J Therm Anal Calorim 2012;109:1435–43.

[149] Li X, Panigrahi S, Tabil LG. A study on flax fiber reinforced polyethylene biocomposites. Appl Eng Agric 2009;25:525–31.

[150] Niu P, Liu B, Wei X, et al. Study on mechanical properties and thermal stability of polypropylene/hemp fiber composites. J Reinf Plast Compos 2011;30:36–44.

[151] Zabihzadeh SM, Ebrahimi G, Enayati AA. Effect of compatibilizer on mechanical, morphological, and thermal properties of chemimechanical pulp-reinforced PP composites. J Thermoplast Compos Mater 2011;24:221–31.

[152] Zhao L, Li Y, Cao X, et al. Multifunctional role of an ionic liquid in melt-blended poly(methyl methacrylate)/multi-walled carbon nanotube nanocomposites. Nanotechnology 2012;23:1–8.

[153] Shamsuri AA, Daik R. Utilization of ionic liquid/urea mixture as a physical coupling agent for agarose/talc composite films. Materials 2013;6:682–98.

[154] Adsul MG, Terwadkar AP, Varma AJ, et al. Cellulases from *Penicillium janthinellum* mutants: solid-state production and their stability in ionic liquids. BioResources 2009; 4:1670–81.

[155] Shamsuri AA, Abdullah DK. A preliminary study of oxidation of lignin from rubber wood to vanillin in ionic liquid medium. Oxid Commun 2012;35:767–75.

[156] Pezoa R, Cortinez V, Hyvärinen S, et al. The use of ionic liquids in the pretreatment of forest and agricultural residues for the production of bioethanol. Cell Chem Technol 2010;44:165–72.

[157] Shamsuri AA, Abdullah DK. Protonation and complexation approaches for production of protic eutectic ionic liquids. J Phys Sci 2010;21:15–28.

[158] Li Z, Khimyak YZ, Taubert A. Lessons from a "failed" experiment: zinc silicates with complex morphology by reaction of zinc acetate, the ionic liquid precursor (ILP) tetrabutylammonium hydroxide (TBAH), and glass. Materials 2008;1:3–24.

[159] Shamsuri AA, Abdullah DK. Isolation and characterization of lignin from rubber wood in ionic liquid medium. Mod Appl Sci 2010;4:19–27.

[160] Gupta M, Paul S, Gupta R. $FeCl_3$-catalysed selective oxidation of alkyl/benzyl alcohols in the presence of molecular oxygen using recyclable room-temperature ionic liquid. Oxid Commun 2011;34:922–31.

[161] Shamsuri AA, Dzulkefly KA. Synthesizing of ionic liquids from different chemical reactions. Singap J Sci Res 2011;1:246–52.

[162] Abbott AP, Harris RC, Ryder KS, et al. Glycerol eutectics as sustainable solvent systems. Green Chem 2011;13:82–90.

[163] Shamsuri AA, Daik R, Zainudin ES, Tahir PM. Compatibilization of HDPE/agar biocomposites with eutectic-based ionic liquid containing surfactant. J Reinf Plast Compos 2014;33(5):440–53.

[164] Menger FM. Amphiphiles I have known. C R Chim 2009;12:54–60.

[165] Dzulkefly K, Khoh HF, Ahmad FBH, et al. Solvent-free esterification process for the synthesis of glucose bolaform surfactants. Orient J Chem 2010;26:747–52.

[166] Islam MN, Islam MS. Mechanical properties of chemically treated sawdust-reinforced recycled polyethylene composites. Ind Eng Chem Res 2011;50:11124–9.

[167] Silviya EK, Unnikrishnan G, Varghese S, et al. Surfactant effects on poly (ethylene-*co*-vinyl acetate)/cellulose composites. Compos Part B Eng 2013;47:137–44.

[168] Shamsuri AA, Daik R, Ahmad I, et al. Nylon-6/liquid natural rubber blends prepared via emulsion dispersion. J Polym Res 2009;16:381–7.

[169] Kimizuka N, Nakashima T. Molecular self-assembly in ionic liquids. In: Ohno H, editor. Electrochemical aspects of ionic liquids. 2nd ed. New Jersey: John Wiley & Sons; 2011. p. 169–82.

[170] Kamol D, Ruhul AK, Sarwaruddin Chowdhury AM. Study on the degradation and interfacial properties of calcium alginate fiber reinforced polyethylene oxide composites. J Thermo Compos Materials 2012;25(7):807–19. http://dx.doi.org/10.1177/0892705711413543.

[171] Khan RA, Khan MA, Sultana S, Khan MN, Shubhra QTH, Farhana G, Noor FG. Mechanical, degradation and interfacial properties of synthetic degradable fiber reinforced polypropylene composites. J Reinf Plast Compos February 23, 2009.

[172] Idicula M, Joseph K, Thomas S. Mechanical performance of short banana/sisal hybrid fiber reinforced polyester composites. J Reinf Plast Compos April 24, 2009.

[173] Khan MA, Khan RA, Zaman HU, Alam MN, Hoque MA. Effect of surface modification of jute with acrylic monomers on the performance of polypropylene composites. J Reinf Plast Compos May 21 , 2009;29(8):1195–205.

[174] Khan MA, Ghoshal S, Khan RA, Pervin SA, Mustafa AI. Preparation and characterization of jute fiber reinforced shellac biocomposites: effect of additive. J Chem Chem Technol 2008;2(3):231–4.

[175] Moran J, Alvarez V, Prtrucci R, Kenny J. Mechanical properties of polypropylene composites based on natural fibers subjected to multiple extrusion cycles. J Appl Polym Sci 2007;103:228–37.

[176] Garcia M, Garmendia I, Garcia J. Influence of natural fiber type in eco-composites. J Appl Polym Sci 2008;107:2994–3004.

[177] Aqdas N, Khalid MZ, Mohammad Z, Muhammad A, Mohammad M. A critical review of algal biomass: a versatile platform of bio-based polyesters from renewable resources. Int J Biol Macromol 2016;86:937–49.

[178] Qin Y. Gel swelling properties of alginate fibers. J Appl Polym Sci 2004;91:1641–5.

[179] Mishra S, Mohanty AK, Drzal LT, Misra M, Parija S, Nayak SK, Tripathy SS. Studies on mechanical performance of biofibre/glass reinforced polyester hybrid composites. Compos Sci Technol 2003;63:1377–85.

[180] Fan L, Du Y, Huang R, Wang Q, Zhang L. Preparation and characterization of alginate/gelatin blend fibers. J Appl Polym Sci 2005;96:1625–9.

[181] Fan L, Zhu H, Zheng H, Xu Y, Zhang C. Structure and properties of blend fibers prepared from alginate and konjac glucomannan. J Appl Polym Sci 2007;106:3903–7.

[182] Lee YJ, Shin DS, Kwon OW, Park WH, Choi HG, Lee YR, Han SS, Noh SK, Lyoo WS. Preparation of atactic poly(vinyl alcohol)/sodium alginate blend nanowebs by electrospinning. J Appl Polym Sci 2007;106:1337–42.

[183] Qin Y. The characterization of alginate wound dressings with different fiber and textile structures. J Appl Polym Sci 2006;100:2516–20.

[184] Feresenbet E, Raghavan D, Holmes GA. The influence of silane coupling agent composition on the surface characterization of fiber and on fiber-matrix interfacial shear strength. J Adhes 2003;79:643–65.

[185] Greenfield MJ, Pedicini A, Penn LS. Development of a single fiber fragmentation test for high strain rates. Inter J Adhes Adhes 2000;20:403–7.

[186] Carbal FS, Paiva MC, Nunes JP, Bernardo CAA. Novel technique for the interfacial characterization of glass fiber-polypropylene systems. Polym Test 2003;22:907–13.

[187] Ruhul AK, Mubarak AK, Anjan KD, Kajal KD, Kamol D, Avik K, Suvasree S, Tanzina H, Nazia N, Bapi S, Saha M. Thermo-mechanical and interfacial properties of calcium alginate fiber-reinforced linear low-density polyethylene composite. Polym Plast Technol Eng 2010;49:602–8.

[188] Li HY, Li H, Wang BJ, et al. Synthesis and properties of poly(3-hydroxybutyrate-*co*-3-hydroxyvalerate)/chitin nanocrystals composite scaffolds for tissue engineering. Chin Chem Lett 2014;25:1635–8.

[189] Wei Q, Becherer T, et al. Protein interactions with polymer coatings and biomaterials. Angew Chem Int Ed 2014;53:8004–31.

[190] Ballotta V, Smits AIPM, Driessen Mol A, Bouten CVC, Baaijens FPT. Synergistic protein secretion by mesenchymal stromal cells seeded in 3D scaffolds and circulating leukocytes in physiological flow. Biomaterials 2014;35:9100–13.

[191] Qin L, He XW, Zhang W, Li WY, Zhang YK. Macroporous thermosensitive imprinted hydrogel for recognition of protein by metal coordinate interaction. Anal Chem 2009; 81:7206–16.

[192] Zhao KY, Cheng GX, Huang JJ, Ying XG. Rebinding and recognition properties of protein-macromolecularly imprinted calcium phosphate/alginate hybrid polymer microspheres. React Funct Polym 2008;68:732–41.

[193] Zhao KY, Feng LZ, Lin HQ, et al. Adsorption and photocatalytic degradation of methyl orange imprinted composite membranes using TiO_2/calcium alginate hydrogel as matrix. Catal Today 2014;236:127–34.

[194] Lee MH, Thomas JL, Tseng HY, et al. Sensing of digestive proteins in saliva with a molecularly imprinted poly (ethylene-*co*-vinyl alcohol) thin film coated quartz crystal microbalance sensor. ACS Appl Mater Interfaces 2011;3:3064–71.

[195] Yang YQ, He XW, Wang YZ, Li WY, Zhang YK. Epitope imprinted polymer coating CdTe quantum dots for specific recognition and direct fluorescent quantification of the target protein bovine serum albumin. Biosens Bioelectron 2014;54:266–72.

[196] Pan GQ, Guo QP, Ma Y, Yang HL, Li B. Thermo-responsive hydrogel layers imprinted with RGDS peptide: a system for harvesting cell sheets. Angew Chem Int Ed 2013;52: 6907–11.

[197] Fukazawa K, Ishihara K. Fabrication of a cell-adhesive protein imprinting surface with an artificial cell membrane structure for cell capturing. Biosens Bioelectron 2009;25: 609–14.

[198] Deng QL, Li YL, Zhang LH, Zhang YK. Molecularly imprinted macroporous monolithic materials for protein recognition. Chin Chem Lett 2011;22:1351–4.

[199] Qin L, He XW, Zhang W, Li WY, Zhang YK. Surface-modified polystyrene beads as photografting imprinted polymer matrix for chromatographic separation of proteins. J Chromatogr A 2009;1216:807–14.

[200] Zhao KY, Lin BB, Cui WK, et al. Preparation and adsorption of bovine serum albumin-imprinted polyacrylamide hydrogel membrane grafted on non-woven polypropylene. Talanta 2014;121:256–62.

[201] Zhao KY, Chen T, Lin BB, et al. Adsorption and recognition of protein molecular imprinted calcium alginate/polyacrylamide hydrogel film with good regeneration performance and high toughness. React Funct Polym 2015;87:7–14.

[202] Dun WZ, Zhuo C, Kong YZ, Bo HK, Lan XL, Xia D, Hai W, Chao Z, Xi GL, Lin HZ. Polypropylene non-woven supported fibronectin molecular imprinted calcium alginate/polyacrylamide hydrogel film for cell adhesion. Chin Chem Lett 2015;26: 807–10.

[203] Park JI, Woo HC, Lee JH. Production of bio-energy from marine algae: status and perspectives. Chem Eng Res (Korean) 2008;46(5):833–44.

[204] Hassan MM, Mueller M, Wagners MH. Exploratory study on seaweed as novel filler in polypropylene composite. J Appl Polym Sci 2008;109:1242–7.

[205] Gupta VK, Rastogi A. Biosorption of hexavalent chromium by raw and acid-treated green alga *Oedogonium hatei* from aqueous solutions. J Hazard Mater 2009;163: 396–402.

[206] Parameswari E, Lakshmanan A, Thilagavathi T. Effect of pretreatment of blue green algal biomass on bioadsorption of chromium and nickel. J Algal Biomass Utln 2009; 1(1):9–17.

[207] Lee MO, Han SO, Seo YB. Red algae fibre/poly(butylene succinate) biocomposites: the effect of fibre content on their mechanical and thermal properties. Compos Sci Technol 2008;68:1266–72.

[208] Sim KJ, Han SO, Seo YB. Dynamic mechanical and thermal properties of red algae fiber reinforced poly(lactic acid) biocomposites. Macromol Res 2010;18(5):489–95.

[209] Hassan MM, Mueller M, Tartakowska DJ, Wagner MH. Mechanical performance of hybrid rice straw/sea weed polypropylene composites. J Appl Polym Sci 2011;120: 1843–9.

[210] Luan L, Wu W, Wagner MH, Mueller M. Seaweed as novel biofiller in polypropylene composites. J Appl Polym Sci 2010;118:997–1005.

[211] Luan L, Wu W, Wagner MH. Rheological behavior of lubricating systems in polypropylene/seaweed composites. J Appl Polym Sci 2011;121:2143–8.

[212] Albano C, Karam A, Domínguez N, Sánchez Y, González J, Aguirre O, Cataño L. Thermal, mechanical, morphological, thermogravimetric, rheological and toxicological behavior of HDPE/seaweed residues composites. Compos Struct 2005;71:282–8.

[213] Bajaria S, Bellare A. Deformation morphology and wear behavior of polyethylene used in orthopedic implants. J Appl Med Polym 1999;3(1):1–5.

[214] Cerrai P, Guerra G, Tricoli M. Periodontal membranes from composites of hydroxyapatite and bioresorbable block copolymers. J Mater Sci 1999;10(12):677–82.

[215] Dalby MJ, Di Silvio L, Harper EJ, Bonfield W. In vitro evaluation of a new polymethylmethacrylate cement reinforced with hydroxyapatite. J Mater Sci 1999;10:793–6.

[216] Harper EJ, Braden M, Bonfield W. Mechanical properties of hydroxyapatite reinforced poly(ethylmethacrylate) bone cement after immersion in physiological solution: influence of a silane coupling agent. J Mater Sci 2000;11:491–7.

[217] Guillemin G, Meunier A, Dallant P, Christel P, Pouliquen JC, Sedel L. Comparison of coral resorption and bone apposition with two natural corals of different porosities. J Biomed Mater Res 1989;23(7):765–9.

[218] Guillemin G, Patat JL, Fournie J, Chetail M. The use of coral as a bone graft substitute. J Biomed Mater Res 1987;21(5):557–67.

[219] Mora F, Ouhayoun JP. Clinical evaluation of natural coral and porous hydroxyapatite implants in periodontal bone lesions: results of a 1-year follow-up. J Clin Periodontol 1995;22(11):877–84.

[220] Papacharalambous SK, Anastasoff KI. Natural coral skeleton used as onlay graft contour augmentation of the face. Int J Oral Maxillofac Surg 1993;22:260–4.

[221] Catano L, Albano C, Karam A, Dominguez N, Sanchez Y, Gonzalez J. Effect of gamma irradiation on mechanical, thermal and rheological behavior of HDPE filled with seaweed residues. Nucl Instrum Methods Phys Res B 2005;236:348–53.

[222] Jang YH, Han SO, Sim IN, Kim H. Pretreatment effects of seaweed on the thermal and mechanical properties of seaweed/polypropylene biocomposites. Compos Part A 2013;47:83–90.

[223] (a) Karnani R, Krishnan M, Narayan R. Polym Eng Sci 1997;37:476.
(b) Gaylord NG, Mishra MK. J Polym Sci Polym Lett Ed 1983;21:23.

[224] Toshi O, Farao Z, Hiroshi K, Takahiro H. Synthesis and tensile properties of a novel composite of chlorella and polyethylene. J Appl Polym Sci 2004;92:812–6.

[225] Farao Z, Hiroshi K, Ryouichi K, Takahiro H, Masatada Y, Toshi O. Preparation and characterization of a novel polyethylene-*chlorella* composite. Chem Mater 1999;1: 1952–6.

FURTHER READING

[1] Nakayama A, Kawasaki N, Arvanitoyannis I, Aiba S, Yamamoto N. Synthesis and biodegradation of poly(γ-butyrolactone-*co*-L-lactide). J Environ Polym Degrad 1996;4: 205–11.

CHAPTER

Chlorella-Based Composites

14

Takahiro Hirotsu[1], Toshi Otsuki[2]

National Institute of Advanced Industrial Science and Technology (AIST), Takamatsu, Japan[1]; IHI Corporation, Yokohama, Japan[2]

14.1 INTRODUCTION

Reducing the discharge of carbon dioxide (CO_2) into the atmosphere is one of the crucial goals to be achieved in this century, since the increased CO_2 concentration is causing global warming. Consequently upon a number of studies intensively carried out on atmospheric fixation of CO_2, particularly since the 1990s, biological fixation of CO_2 with microalgae attracted a lot of attention as the most sustainable method [1,2]. Recently several life-cycle assessment (LCA) studies were explored to verify the feasibility of the production of biodiesel from CO_2 fixed in microalgae [3,4].

Starting in 1990, a 10-year intensive research project on biological CO_2 fixation and utilization was carried out in Japan by Research Institute of Innovative Technology for the Earth (RITE) , with the aim of developing new technology for fixation and utilization of CO_2 in industrial exhaust gases by aquatic microorganisms, including microalgae and photosynthetic bacteria [5–7]. These microorganisms exhibit some distinctive characters: (1) higher efficiency of CO_2 fixation using solar energy compared to that of terrestrial plants, (2) photosynthesis of useful substances, including carbohydrates, lipids, and proteins, and (3) feasible incorporation of the CO_2 fixation system into industrial plants. In this project, extensive screening of microalgae collected from hot springs, lakes, and oceans was performed in view of the capability of fixing CO_2, photosynthesis, and storage of useful substances such as hydrocarbons, polysaccharides, and lipids, and the tolerance to CO_2 concentration and temperature of flue gas, and several promising microalgae were found, including *Chlorella* sp. UK001 [7]. Furthermore, photo-bioreactor systems with microalgae were also developed, demonstrating a CO_2 fixation rate of 50 g CO_2/m^2 d (10-h irradiation per day).

Such biological CO_2 fixation systems will produce a large amount of microalgae instead of discharging CO_2 into the atmosphere. Consequently resultant microalgae require their effective utilization as fodder, building materials, as well as in other applications. Microalgae will otherwise putrefy and decompose to liberate CO_2 again into the atmosphere if we do not utilize them effectively. We have, therefore, developed several novel composites of chlorella with versatile plastics, including polyvinylchloride (PVC), polyethylene (PE), and poly(propene) (PP), particularly

Algae Based Polymers, Blends, and Composites. http://dx.doi.org/10.1016/B978-0-12-812360-7.00014-8

aiming at their application to building materials. It is noteworthy that these plastics are hydrophobic, an opposing character against hydrophilic chlorella. The development of composites with the purpose of extensive utilization of chlorella can only be realized if the compatibility issue between chlorella and plastics is addressed.

This chapter first describes the feasibility of fixation of CO_2 from flue gas on the basis of an LCA study of a hypothetical system, incorporated in a 1000-MW liquefied natural gas power station, indicating that such CO_2 fixation system could be practical with an efficient utilization of resultant microalgae as fodder. Then we describe the preparation of a novel composite of chlorella and PVC with slight polarity for its utilization as building material. Based on some findings about PVC–chlorella composite, we challenge the compounding of chlorella with hydrophobic PE and demonstrate that it is realized through slight graft modification of PE with maleic anhydride (MA). Finally, it is described that MA-grafted PP can be successfully applied as a compatibilizer for compounding of chlorella and another hydrophobic PP.

14.2 FEASIBILITY OF A BIOLOGICAL CO_2 FIXATION AND UTILIZATION SYSTEM

Feasibility of biological CO_2 fixation and utilization is explored on the basis of an LCA study of a model system under several requirements described in Table 14.1 [8]. In this analysis, the source of CO_2 is assumed to be the flue gas discharged from a 1000-MW liquefied natural gas (LNG) power station, the volume of which is 8.87×10^9 Nm^3/y and corresponds to 1.74×10^6 t CO_2/y due to the CO_2

Table 14.1 Assumptions for Evaluation of the System

Items	Value	Remarks
CO_2 fixation rate	50 g $CO_2/m^2/d$	
Setup area of the system	3068 m^2	2.3 m × 2.3 m × 580 units
Total culture liquid volume	497 m^3	
CO_2 emission per power generated	0.51 kg CO_2/kWh	Liquefied natural gas power station
Irradiation time (operation period)	10 h/d	
Indirect effect of CO_2 fixation	10.1 kg CO_2/kg feed	5.5 C effect [10]
Space velocity of the gas	0.01 min^{-1}	5.0 m^3/min
Carbon content of microalgae	50 wt%	
Culture liquid temperature	35°C	
Formation enthalpy of glucose	3.0 kWh/kg CO_2	112 kcal/mol CO_2
Operation rate of the system	100%	365 days/year (continuous)

Reprinted from Otsuki T. A study for the biological CO_2 fixation and utilization system. Sci Total Environ 2001;277:21–25.

concentration of 10 v/v%. The flue gas from LNG power stations has great advantages including its quality as well as the abovedescribed capacities [9]. The cultured microalgae are separated and dried for utilization as fodder. The energy necessary for working the system is supplied from the power station.

From the assumptions noted above, and those listed in Table 14.1, we can evaluate the CO_2 balance and the energy balance of the system. Thus the fixation rate of CO_2 by the system is (a) 5.60×10^4 kg CO_2/y, corresponding to 3.2×10^{-3}% of the total amount of CO_2 in the flue gas, while the production rate of feed (dried microalgae with a water content = 5.0 wt%) is (b) 3.05×10^4 kg feed/y. Considering the emission of CO_2 by operation of the system, the emission rate is calculated to be (c) 2.33×10^4 kg CO_2/y, with a value of 125.4 kWh/d (50% for culture, 21% for supply of culture medium, 6% for separation of resulting microalgae, and 23% for drying separated microalgae) as the power for operation of the system [9]. The result (c) < (a) indicates that the CO_2 balance of the system is positive, i.e., the system decreases the concentration of CO_2 in the atmosphere.

Another source of CO_2 emission is related to the system itself; construction of the system requires materials as well as electric power, both responsible for CO_2 emission into the atmosphere. The initial CO_2 emission due to construction of the system is calculated to be (d) 1.92×10^6 kg CO_2 [8], using a sophisticated method [9]. Based on the production rate of feed, the initial emission is evaluated to be 63.0 kg CO_2 y/kg feed, calculated as (d)/(b). Similarly the relevant fixation and emission of CO_2 are (e) 1.83 and (f) 0.76 kg CO_2/kg feed as (a)/(b) and (c)/(b). This shows the net fixation of CO_2 by the system is (g) 1.07 kg CO_2/kg feed, equal to the difference (e)−(f), much smaller than the value of the initial emission.

Utilization of microalgae as an alternative fodder for crops causes the reduction in emission of greenhouse gases into the atmosphere, since crop production causes emission of greenhouse gases such as methane and nitrous oxide [10]. From the molar conversion factors of the greenhouse gases CO_2:CH_4:N_2O = 1:21:270, the total emission of the greenhouse gases, normalized by the value of CO_2, can be estimated as equivalent to 5.91 ton C of CO_2 for a production of 1.0 ton C of crops [10]. Assuming that the reduction effect in emission of CO_2 is 5.5 ton C of CO_2 per ton C of microalgae as a fodder instead of crops, the resulting effect in this study is evaluated as follows:

$$5.5 \times (\mathrm{e}) = 10.1\ \mathrm{kg\ CO_2/kg\ feed\ (h)}. \qquad (14.1)$$

Considering not only the direct (g) but also the indirect effect (h), the following mass balance is satisfied:

$$63.0 - (\mathrm{g}) \times t = (\mathrm{h}) \times t, \qquad (14.2)$$

where t is the time of operating the system. This relation is given in Fig. 14.1, where the solid and broken lines are corresponding to the left- and right-hand sides of Eq. (14.2). Fig. 14.1 clearly shows that the indirect effect is much greater than the direct one in reduction of the greenhouse effect. The time at the intersection point of the two lines is named as payback period, c.5.6 years; accordingly, beyond the payback time the CO_2 balance becomes positive, even if the contribution of CO_2 emission

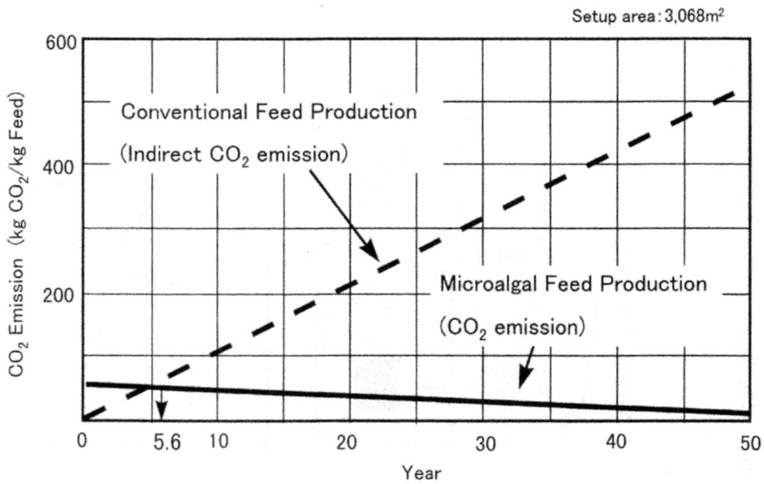

FIGURE 14.1

Payback period of the system.

Reprinted from Otsuki T. A study for the biological CO_2 fixation and utilization system. Sci Total Environ 2001; 277:21–25.

from the construction of the system is taken into account. This demonstrates that substitution of widely used fodders exhibiting huge greenhouse effects with microalgae is significant from a practical point of view.

Briefly referring to the energy balance, the energy consumption of the system is calculated to be 0.82 kWh/kg CO_2, which is much smaller than the formation enthalpy (3.0 kWh/kg CO_2) of glucose from $6CO_2$ and $6H_2O$, because of effective use of solar energy in the system.

14.3 CHARACTERISTICS OF CHLORELLA AGGREGATES

Before describing compounding of chlorella with plastic resins, it is essential to understand principal properties of chlorella that will affect the characteristics of the resultant composites [11]. Chlorella used in our study, purchased from Yaeyama Shokusan Co., Ltd. (Japan), contains mainly protein (53–60 wt%), lipid (6–15 wt%), and carbohydrates (10–20 wt%). The cell wall of chlorella weighs c.13.6 wt% of the whole cell, and mainly consists of α-cellulose (c.15.4 wt%), hemicellulose (c.31 wt%), protein (c.27 wt%), and fatty acids (c.9.2 wt%). Chlorella exhibits a unique behavior of aggregation. Micrographs of spray-dried chlorella are shown in Fig. 14.2. Chlorella grains are nearly spherical with different particle sizes (Fig. 14.2A). A grain is an aggregate of a large number of chlorella cells tightly bound to each other (Fig. 14.2B). Fig. 14.2C reveals that a grain contains a hollow sphere with a wall thickness several times greater than the dimension of one cell

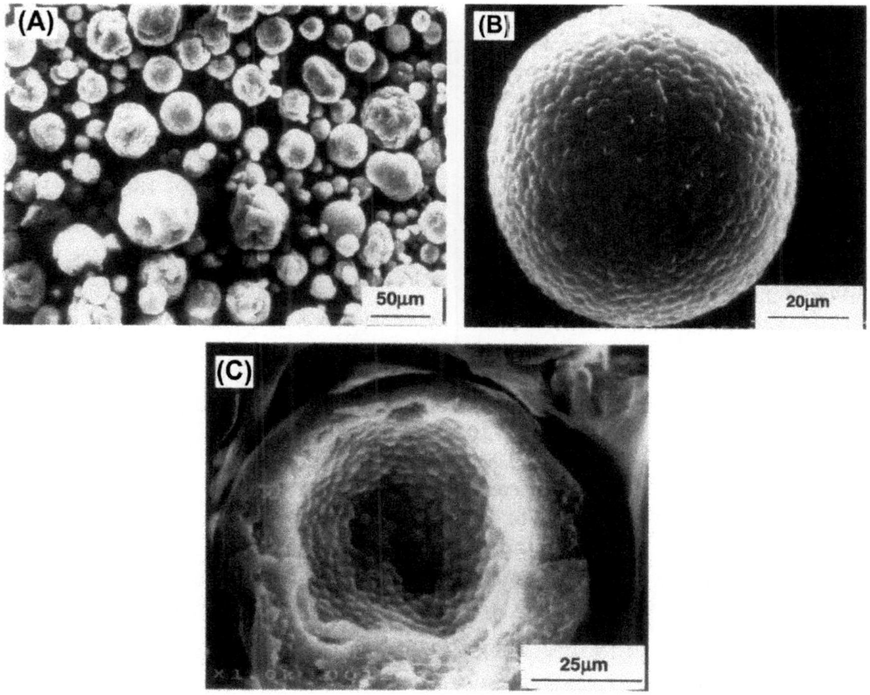

FIGURE 14.2

Scanning electron micrographs of chlorella: overall appearance (A), a typical aggregate grain (B), a profile of a chlorella grain displaying its hollow spherical shape and wall, consisting of chlorella cells of about 3–5 μm (C).

Reprinted from Zhang F, Kabeya H, Kitagawa R, Hirotsu T, Yamashita M, Otsuki T. A novel polyethylene–chlorella composite. I. Characterization of chlorella biologically fixing CO_2. J Appl Polym Sci 2000;77: 2278–2284.

(3–5 μm). Fig. 14.3 shows particle-size distributions of chlorella grains on standing under ultrasonic dispersion in water and ethanol for 0–120 min. There are two peaks around particle sizes of 4 and 100 μm corresponding to chlorella cells and chlorella aggregates, respectively. After ultrasonic treatment in water (Fig. 14.3A), the peak due to chlorella aggregates decreases more in intensity, shifting to a smaller size, with time, while the peak due to chlorella cells increases in intensity with no shift. This indicates that the chlorella grains break down into chlorella cells in water, consistent with the hydrophilic property of chlorella.

The particle-size distribution in ethanol, on the other hand, shows hardly any change, with a slight change even after 120-min treatment (Fig. 14.3B). Accordingly, the average particle size of chlorella changes only slightly from 63.8 to 53 μm in ethanol, compared to a marked decrease from 63.8 to 5.7 μm in water. This strongly suggests that the binding between chlorella cells is primarily due to

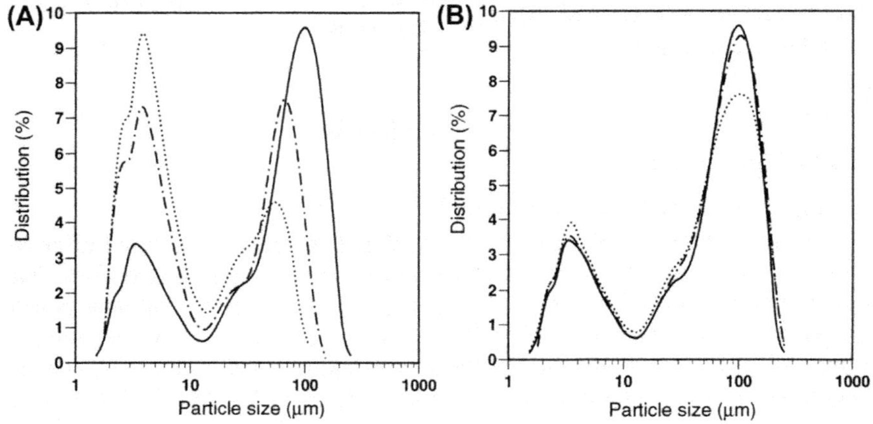

FIGURE 14.3

Changes in particle-size distribution of chlorella with ultrasonic-dispersion time in water (A) and ethanol (B). The dispersion times are 0 min (———), 30 min (–·–··), and 120 min (-----).

Reprinted from Zhang F, Kabeya H, Kitagawa R, Hirotsu T, Yamashita M, Otsuki T. A novel polyethylene–chlorella composite. I. Characterization of chlorella biologically fixing CO_2. J Appl Polym Sci 2000;77: 2278–2284.

hydrogen bonds or ionic interactions, considering the constitution of the cell wall described above.

It is noted that chlorella cells, derived through ultrasonic dispersion of chlorella grains in water, aggregate again to form hollow spheres on drying (Fig. 14.4). This unique character has a significant impact on thermal and structural stabilities of chlorella, both determining the primary properties of chlorella composites with plastic resins.

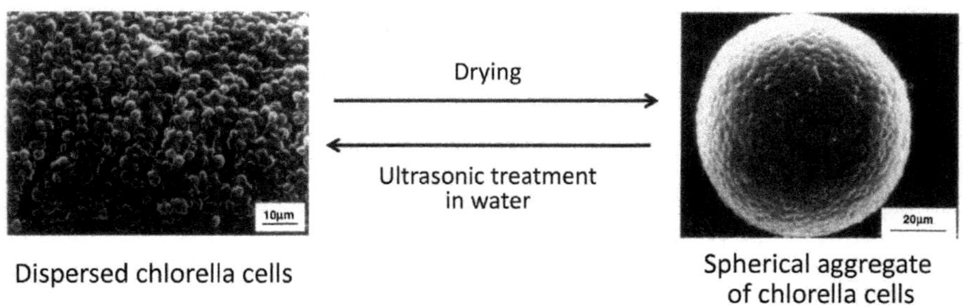

FIGURE 14.4

A reversible change between dispersion and aggregation states of chlorella cells.

Reproduced from Zhang F, Kabeya H, Kitagawa R, Hirotsu T, Yamashita M, Otsuki T. A novel polyethylene–chlorella composite. I. Characterization of chlorella biologically fixing CO_2. J Appl Polym Sci 2000;77: 2278–2284.

Thermogravimetric (TG) curves of up to 550°C vacuum-dried chlorella in air and nitrogen environments are shown in Fig. 14.5, compared with those of leucine and cellulose. The initial weight loss (c.4 wt%) is due to the remaining water. The slow weight loss of chlorella reflects the diversity of substances in chlorella. Greater amount of residues of chlorella compared to cellulose and leucine at 350°C indicates that chlorella is more stable than leucine and cellulose, which decompose drastically at c.300°C and 350°C, respectively. SEM observations reveal that chlorella grains retain the spherical shape, almost equivalent to the original, at 250°C in air as well as in nitrogen gas. Furthermore, the shape of chlorella grains does not exhibit any change under a load of 75 MPa, but shows a slight deformation at 150 MPa.

The high resistance to cracking and thermal stability of chlorella grains makes us expect that chlorella will resist deformation and thermal decomposition in the process of compounding with thermal plastic resins, which is probably due to the spherical and unique assembly of chlorella cells. IR spectroscopy confirms the existence of cellulosic substances and proteins by the appearance of bands at 3440 and 1072 cm^{-1} due to $-OH$ groups as well as at 3312, 1654, 1543, and 1242 cm^{-1} assigned to aliphatic secondary amide groups [11]. These main functional groups exhibit hydrophilic interactions between chlorella cells causing the unique behavior of aggregation, as well as thermal and mechanical stabilities of the resulting aggregates.

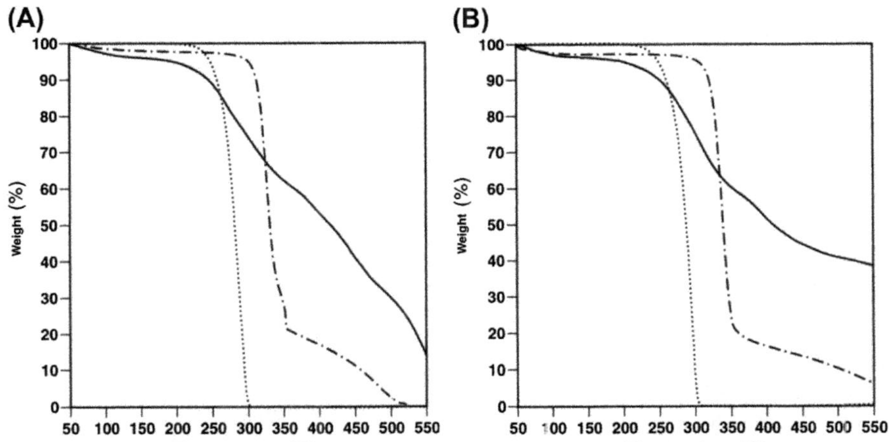

FIGURE 14.5

Thermogravimetric analysis of chlorella (———), cellulose (—·—··), and leucine (-----) in air (A) and nitrogen (B).

Reprinted from Zhang F, Kabeya H, Kitagawa R, Hirotsu T, Yamashita M, Otsuki T. A novel polyethylene–chlorella composite. I. Characterization of chlorella biologically fixing CO_2. J Appl Polym Sci 2000;77: 2278–2284.

14.4 CHLORELLA COMPOSITE WITH POLYVINYLCHLORIDE

A biological CO_2 fixation system described in Section 14.2 will produce a large amount of chlorella instead of discharging CO_2 into the atmosphere. Resultant chlorella has to be further utilized as materials otherwise it will putrefy and decompose to again release CO_2 into the atmosphere. One of the solutions is to develop composites of chlorella with plastic resins so that the fixed CO_2 can be retained firmly for a long period. This section describes the compounding of chlorella with poly(1-chloroethene) (PVC) aiming at its application as ecofriendly building materials [12]. PVC is expected to be more favorable for compounding with chlorella because of its slight polarity.

14.4.1 OPTIMUM MOLDING CONDITIONS

Chlorella, purchased from Yaeyama Shokusan Co., Ltd., is spray-dried after being strained by centrifugal filtration. A chlorella sample was vacuum-dried at 60°C for 6 days before being molded with PVC. The average particle size is c.62.7 μm. Poly(1-chloroethene) used here is commercially available, with a polymerization degree of 1100 and a grain size of 50–150 μm. Poly(1-chloroethene) chlorella composites (PVC–CH composites) were prepared by a heat-pressurizing method. Eight grams of a mixture composed of 80 wt% PVC and 20 wt% chlorella (80PVC–20CH) was thoroughly mixed, put into a rectangular mold made of aluminum (length × width × thickness: $120 \times 20 \times 7\ mm^3$), and then heat-pressed under selected molding conditions. Test samples for measuring tensile strength were arranged according to the JIS [13].

The optimum conditions for molding were determined by considering not only the tensile strength but also color change of resultant PVC–CH composite. The tensile strength of an 80PVC–20CH composite increases with molding temperature to exhibit a maximal at c.200°C. The color of chlorella begins to change from deep green to brown, when the molding temperature is raised to 190°C, consistent with a critical change in a TG curve of chlorella at around 190°C, probably due to volatilization and/or decomposition of volatile substances. The molding temperature should be restricted below 190°C. On the basis of these findings, the optimum molding pressure and time were determined by comparing their effects on the tensile strength of the 80PVC–20CH composite at 180 and 190°C. The optimum conditions for preparing the 80PVC–20CH composite are as follows: temperature = 180°C, pressure = 4.4 MPa, and time = 5 min. These conditions yield the PVC–CH composite with high tensile strength and no color change of chlorella.

14.4.2 INFLUENCE OF PHYSICAL PROPERTIES OF CHLORELLA

We consider the influences of the water content and the size distribution of chlorella on the tensile strength of PVC–CH composite, as both properties are closely related with the formation of the interface of the composite. Chlorella samples with different contents of water were arranged by adding relevant amounts of water to chlorella vacuum-dried at 60°C for 6 days, which is considered to have no water. The tensile strength of 80PVC−20CH composite gets a maximum value at a water content of 2 wt%, beyond which it decreases, and sharply above 13 wt%. The existence of a little water (2 wt%) in chlorella possibly prevents the evaporation of volatile substances in chlorella, which might otherwise be trapped in the interfaces. SEM observations clearly show the existence of large cracks in an 80PVC−20CH composite with a water content of 16.7 wt% in contrast to no crack in an 80PVC−20CH composite with 5.66 wt% water.

Chlorella samples with different average particle sizes were prepared by sieving of original chlorella. Fig. 14.6 demonstrates that the optimum range of average particle size occurs at 30−80 μm, beyond or below which the tensile strength of 80PVC−20CH composite is reduced. Furthermore, the distribution of particle sizes also has a great influence on the tensile strength of 80PVC−20CH composite. As an example, sieved chlorella, with an average size of 59.8 μm and a narrower size distribution of 52−75 μm, has a tensile strength even greater than the original unsieved chlorella having an average size of 62.7 μm and a wider size distribution of 3−150 μm (marked ★ in Fig. 14.6).

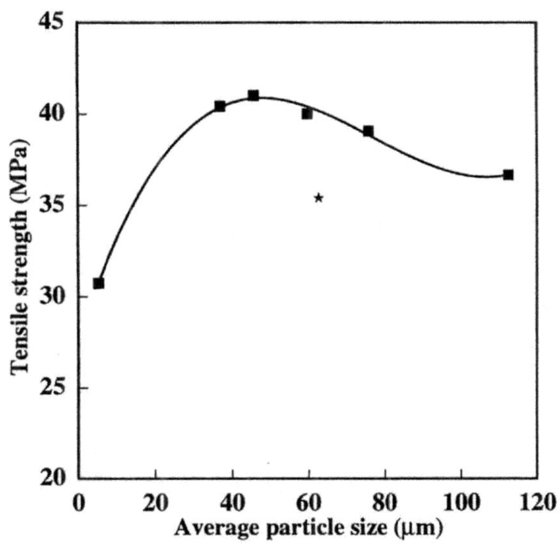

FIGURE 14.6

Tensile strength of 80PVC−20CH composite as a function of the average particle size of chlorella at molding temperature 190°C, pressure 2.2 MPa, and time 5 min; chlorella sieved (■), and original chlorella before sieving with an average particle size of 62.7 μm (★). *CH*, chlorella; *PVC*, polyvinyl chloride.

Reprinted from Zhang F, Kabeya H, Kitagawa R, Hirotsu T, Yamashita M, Otsuki T. An exploratory research of PVC−chlorella composite material (PCCM) as effective utilization of chlorella biologically fixing CO_2. J Mater Sci 2000;35:2603−2609.

14.4.3 MICROSTRUCTURE OF PVC–CH COMPOSITE

SEM micrographs of the surface and cross section of an 80PVC–20CH composite are given in Fig. 14.7. The chlorella particles are inlaid in the matrix of PVC; and the chlorella grains retain their original shapes without being crushed or deformed during molding. Gas voids are observed around the chlorella grains, and furthermore, unmelted primary particles of PVC (white parts in the photo) can be seen around the chlorella grains. The gas voids existing around chlorella grains are possibly due to air remaining in the material and/or gas from evaporation of water or volatile substances in chlorella while molding. The primary particles of PVC probably remain around the chlorella grains in the PVC–CH composite because of the poor heat conductivity of gas voids around chlorella. Furthermore, the distribution of molding stress is not homogeneous around chlorella grains as the shape of chlorella is not regular. Accordingly, melting of primary particles of PVC around the chlorella grains to form a domain or micro-domain matrix is prevented significantly by smaller molding stress.

In comparison with PVC, the 80PVC–20CH composite exhibits much smaller plastic elongation. It is only 1.86%, quite different from PVC having an elongation of about 180%. The contrast is probably due to the unmelted primary particles remaining around the chlorella grains in the 80PVC–20CH composite. The specific volume of the 80PVC–20CH composite is 22% greater than that of PVC, indicating that chlorella is a good filler for increasing the material volume.

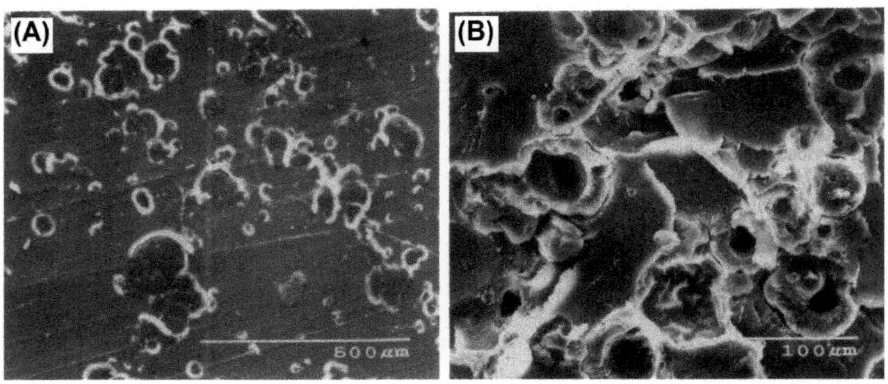

FIGURE 14.7

SEM micrographs of an 80PVC–20CH composite: surface (A) and fracture (B). *CH*, chlorella; *PVC*, polyvinyl chloride; *SEM*, scanning electron microscopy.

Reprinted from Zhang F, Kabeya H, Kitagawa R, Hirotsu T, Yamashita M, Otsuki T. An exploratory research of PVC–chlorella composite material (PCCM) as effective utilization of chlorella biologically fixing CO_2. J Mater Sci 2000;35:2603–2609.

14.4.4 EFFECT OF CHLORELLA CONTENT ON TENSILE STRENGTH

Let us theoretically consider the tensile strength of a polymer–CH composite based on several assumptions. We assume that the total section area (A) of the composite is equal to the summation of the section areas of the polymer matrix (A_p) and chlorella part (A_c). The tensile strength of the composite can be given by Eq. (14.3):

$$\sigma_{(p-c)} = F/A, \tag{14.3}$$

where F is the tension load added on the composite. Provided that the tensile strength of the composite is given only by the contribution from the polymer matrix, its tension load F_0 can be calculated by Eq. (14.4):

$$F_0 = \sigma_p A_p, \tag{14.4}$$

where σ_p is the tensile strength of the polymer. Introducing Eq. (14.4) into Eq. (14.3) yields Eq. (14.5):

$$\sigma_{0(p-c)} = \sigma_p A_p/A. \tag{14.5}$$

If the gas holes in the composite are ignored, the volume of the composite (V) is equal to the sum of the volumes of the polymer matrix (V_p) and chlorella phase (V_c). V_p and V_c can be defined as

$$V_p = W_p/D_p \tag{14.6}$$

and

$$V_c = W_c/D_c, \tag{14.7}$$

where W_p and D_p are the weight and density of polymer, respectively; and W_c and D_c are the weight and density of chlorella, respectively Assuming that all the cross sections of the composite have the same values of A_p and A_c,

$$A_p/A = V_p/V. \tag{14.8}$$

Using $V = V_p + V_c$ and Eqs. (14.6) and (14.7), Eq. (14.8) can be rewritten as

$$A_p/A = 1/(1 + D_p W_c/W_p D_c). \tag{14.9}$$

Substituting this relation into Eq. (14.5) yields the following equation:

$$\sigma_{0(p-c)} = \sigma_p/(1 + D_p W_c/W_p D_c). \tag{14.10}$$

This is a theoretical relation that expresses the tensile strength of the polymer–CH composite with the tight interface but no interactions between the chlorella grains and the polymer matrix.

For the PVC–CH composite, using the weight content of chlorella in the composite (wt%), $C\%$, $W_c/W_p = C\%/(100-C\%)$. With the experimental D_c and D_p values of 0.7 and 1.4, respectively, $D_p/D_c = 2.0$. Thus for the PVC–CH composite with these relations, Eq. (14.11) is finally obtained:

$$\sigma_{0(pvc-c)} = \sigma_{pvc}[1/\{1 + 2.0 \times C\%/(100 - C\%)\}]. \tag{14.11}$$

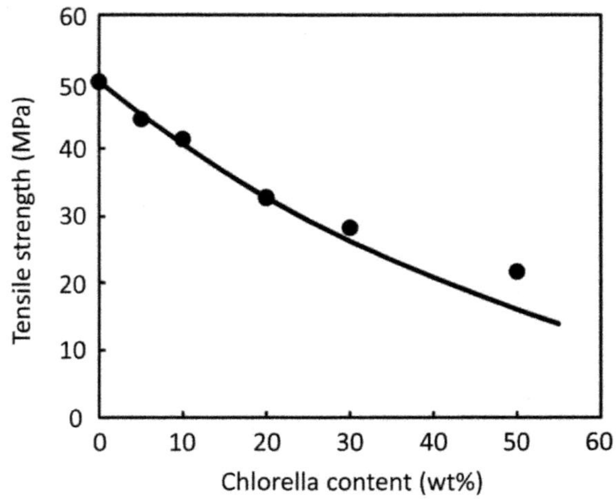

FIGURE 14.8

Relation between tensile strength and chlorella content in PVC–CH composite (●) at molding temperature 180°C, pressure 4.4 MPa, and time 5 min, using KM-55 with a 0.5 wt% content. The solid curve represents the theoretical result according to Eq. (14.11). *CH*, chlorella; *PVC*, polyvinyl chloride.

Reproduced from Zhang F, Kabeya H, Kitagawa R, Hirotsu T, Yamashita M, Otsuki T. An exploratory research of PVC–chlorella composite material (PCCM) as effective utilization of chlorella biologically fixing CO_2. J Mater Sci 2000;35:2603–2609.

Fig. 14.8 shows experimental results of the tensile strength of the PVC–CH composite against the chlorella content, with a solid curve expressing the theoretical relationship described in Eq. (14.11). The experimental values are well put on the theoretical curve calculated with the content of chlorella ($C\%$) smaller than 20 wt %. In the $C\%$ values greater than 20 wt%, however, the experimental values become greater than theoretical expectations, with the difference being increased gradually with increasing the $C\%$. As described in Section 14.4.3, the gas voids appear around chlorella grains in the PVC–CH composite, the content increasing with the content of chlorella. Since such voids in the composite cause the section area of the PVC matrix A_{pvc} to be smaller than the theoretical value to be obtained by Eq. (14.9), we expect that the tensile strength of the composite will be smaller than the theoretical value obtained by Eq. (14.11), with the difference increasing with the chlorella content. The expectation, however, is opposite to the results shown in Fig. 14.8. This can be caused by some interactions of hydrophilic chlorella with slightly polar PVC matrix. The contribution of such interactions to the tensile strength of the chlorella composite will be fully dealt with in the following sections.

Consequently, the PVC–CH composite has a tensile strength greater than 30 MPa, which satisfies the requirement of rigid PVC products [14], if the chlorella content is smaller than 20 wt%. With the chlorella content smaller than 50 wt%, the tensile strength is greater than 15 MPa, which meets the requirement of plasticized PVC [15].

14.5 CHLORELLA COMPOSITE WITH POLYETHYLENE

On the basis of findings in the preceding section, we describe composites of chlorella with polyethylene (PE), one of the extensively used plastics, since such compounding has great impacts not only on stabilizing a large amount of CO_2-fixed chlorella in the PE matrix for a long term but also on reusing wasted PE or replacing virgin PE. Hydrophilic chlorella, however, hardly exhibits compatibility with hydrophobic polymers, including PE and poly(propene) (PP). How to make these opposing materials compatible is the principal subject here. Chemical modification of cellulosic substances or polymer resins, for both components to exhibit higher affinity with each other, is a prominent strategy for enhancing the compatibility between them [16]. MA is reported to be an effective modifier for PE and PP [17,18]. In this section we propose a method for improving the compatibility of chlorella and PE through chemical modification of PE with MA [19]. The consequent chlorella composite with MA-modified PE (MPE) exhibits a marked increase in tensile strength, compared to unmodified PE [19].

14.5.1 MODIFICATION OF PE WITH MA BY A MELT-MIXING METHOD

Polyethylene (high density), of a commercially available grade (M_w: c.40,000, density: 0.95 g/cm^3; Mitsui Sekiyu Kagaku, Japan), was mechanically sieved into spherical powder with a size of c.1 mm before modification. Chemical modification of PE with MA was performed in a roller mixer with a chamber volume of 60 cm^3 (R60, Toyo Seiki Seisakusho, Japan). MA (Wako Chemicals, Japan) and benzoyl peroxide (BPO) (Nacalai Tesque, Japan) as an initiator were dissolved in acetone in turn, and then the acetone solution was sprayed onto the PE powders until the weight ratio of MA:BPO:PE became 0.25:0.02:100 (the molar ratio = 0.072:0.0023:100, where the last number is a molar number of ethylene units). After it was well mixed, the mixture was fed into the roller mixer and treated at 160°C for 30 min to modify the PE. It is worth mentioning here that to avoid cross-linking reactions between PE macroradicals as well as the homopolymerizations of MA monomers during the grafting of PE, extremely low molar quantities of MA and BPO were employed, compared to that of monomer units of PE [17,20].

Fig. 14.9 shows IR spectra of an MPE sample, together with those of a reference PE-graft-MA (rMPE) from Aldrich (0.85 wt% MA, catalog no. 43720-4) for comparison. Absorption bands at 2918, 2850, 1473, and 719 cm^{-1} in both spectra are due to the same backbone of PE. The magnified spectrum of the sample exhibits clear absorption bands at 1792 and 1715 cm^{-1} due to single-succinic anhydride grafts and their hydrolysis products, respectively [20,21], with no bands due to poly(MA) and free MA which would appear at 1784 and 1780 cm^{-1} [22]. Little reduction in the weight of the sample, after its dissolution in boiling 1,2-dichlorobenzene followed by precipitation in acetone, is also consistent with nonexistence of acetone-soluble poly(MA) and free MA. We can, therefore, conclude that a majority of MAs added are grafted as the single-succinic anhydride type onto PE chains

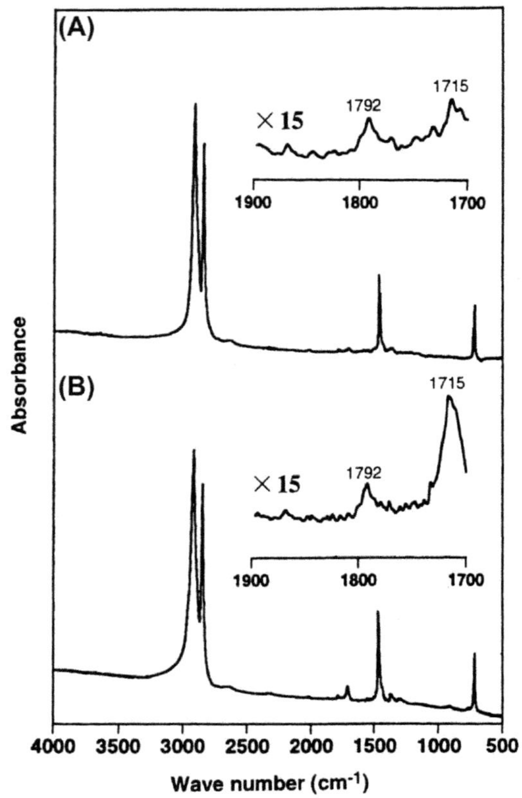

FIGURE 14.9

Fourier transform infrared spectroscopy spectra of MPE prepared in this study (A) and rMPE from Aldrich (0.85 wt% MA, catalog number 43720-4) (B).

Reprinted from Zhang F, Kabeya H, Kitagawa R, Hirotsu T, Yamashita M, Otsuki T. Preparation and characterization of a novel polyethylene-chlorella composite. Chem Mater 1999;11:1952–1956.

(hereafter designated as MPE), a part of which readily undergoes hydrolysis (see a band at 1715 cm^{-1} in Fig. 14.9). The amount of MA grafts (c.0.23 wt%) corresponds to the utilization of 92% of added MA, the residual MA being volatilized during the melt-mixing.

14.5.2 COMPOUNDING OF CHLORELLA WITH MPE BY A MELT-MIXING METHOD

Following the modification of PE as described above, a composite of chlorella with an MPE, i.e., MPE–CH composite, was prepared by feeding chlorella into the MPE in a chamber and mixing them at 160°C for 10 min; the product was finally shaped into pellets with a size of less than 3.36 mm. For the synthesis of a composite of

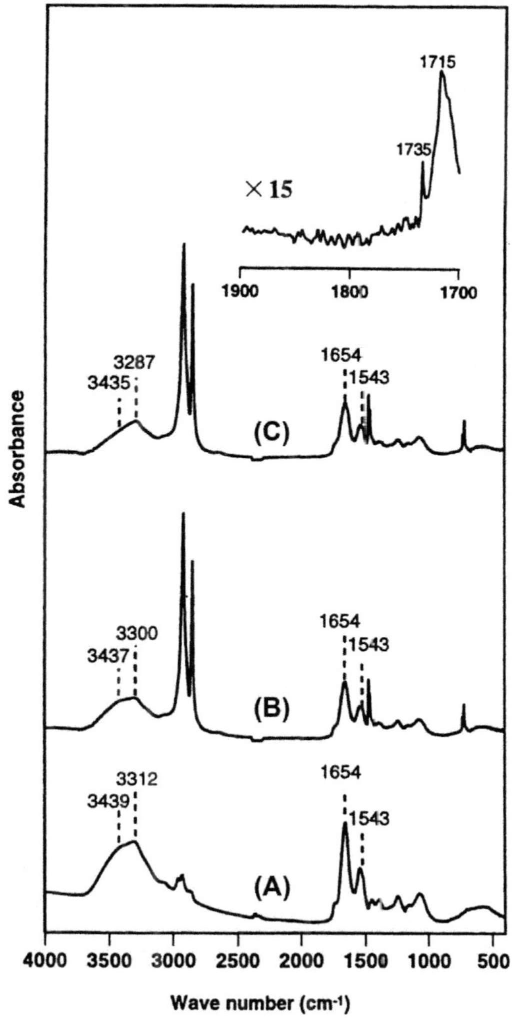

FIGURE 14.10

Fourier transform infrared spectroscopy spectra of chlorella (A), 70UPE–30CH composite (B), and 70MPE–30CH composite (C). The top-right spectrum represents the subtraction spectrum between the 70MPE–30CH composite (Fig. 14.10C) and chlorella (Fig. 14.10A).

Reprinted from Zhang F, Kabeya H, Kitagawa R, Hirotsu T, Yamashita M, Otsuki T. Preparation and characterization of a novel polyethylene-chlorella composite. Chem Mater 1999;11:1952–1956.

chlorella with unmodified PE, i.e., UPE–CH composite, PE and chlorella are directly fed into the mixer and melt-mixed at 160°C for 10 min.

IR spectra given in Fig. 14.10 reveal that a 70MPE–30CH composite exhibits a smaller intensity of absorption at 3440 cm^{-1}, a band assigned to –OH groups as already mentioned in Section 14.3, compared to the original chlorella. We evaluate the intensity ratio I_{3440}/I_{1654}, where I_{3440}, and I_{1654} denote the absorbances at 3440 and 1654 cm^{-1}, respectively; the latter due to second amide groups of chlorella. Compared with the value of 0.810 for chlorella, the I_{3440}/I_{1654} value for the 70MPE−30CH composite shows a significant drop to 0.787, while remains almost unchanged (0.812) for a 70UPE−30CH composite. This suggests that a part of the −OH groups of chlorella in the 70MPE−30CH composite takes part in a reaction with MA groups grafted onto PE.

A difference spectrum between the spectra of the 70MPE−30CH composite and chlorella with respect to 1543 cm^{-1} is given in the top right of Fig. 14.10. From this difference spectrum, it is certain that the signal at 1792 cm^{-1} almost disappears and the signal at 1715 cm^{-1} becomes a little stronger, although a peak at 1735 cm^{-1} may be accidental because of its narrow width. IR observations of the formation of ester bonds in cellulosic composites prepared by melt-mixing are certainly difficult, even when XPS observation can verify such formation [23]. Later a weak shoulder has been observed at c.1735 cm^{-1}, as reported elsewhere [24]. From these results, we can deduce that MA grafts in MPE primarily cause ester bonds with −OH groups of cellulose or hemicellulose components on the surface of chlorella grains in the MPE−CH composite, generating the equimolar carboxylic groups (see Fig. 14.11).

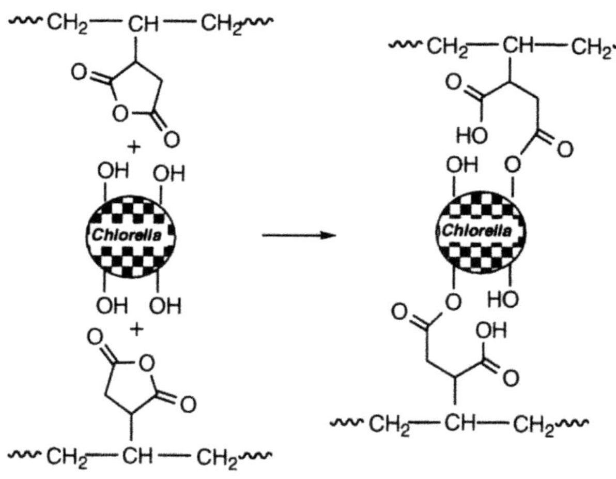

FIGURE 14.11

A proposed reaction mechanism for the formation of chemical bonds between chlorella grains and MPE matrix.

Reprinted from Zhang F, Kabeya H, Kitagawa R, Hirotsu T, Yamashita M, Otsuki T. Preparation and characterization of a novel polyethylene-chlorella composite. Chem Mater 1999;11:1952–1956.

14.5.3 TENSILE PROPERTIES OF MPE–CH COMPOSITE

Tensile strengths of MPE–CH composites are plotted against their chlorella contents in Fig. 14.12A, and compared with those of UPE–CH composites. It is emphasized that the MPE–CH composites possess much greater tensile strengths than do the UPE–CH composites with the chlorella contents smaller than 50 wt%. Another finding is that the tensile strengths of the MPE–CH composites are greater than the theoretical expectations from Eq. (14.10) with $\sigma_p = 24.4$ MPa and $D_p/D_c = 1.5$, while the opposite relation occurs for the UPE–CH composites. This suggests strong interactions between the chlorella grains and the MPE matrix. Accordingly,

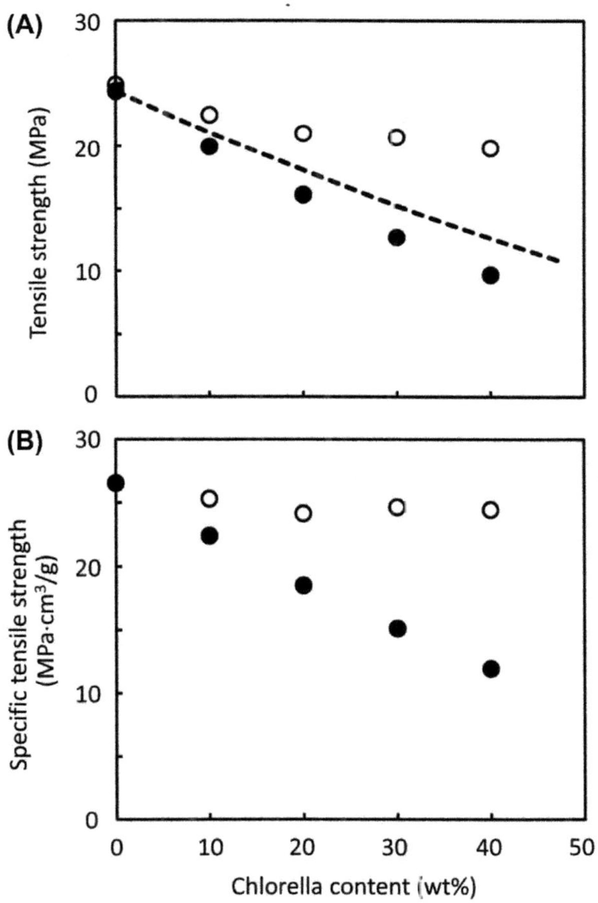

FIGURE 14.12

Tensile properties of MPE–CH (O) and UPE–CH composites (●). The broken curve is based on Eq. (14.10) with $\sigma_p = 24.4$ MPa and $D_p/D_c = 1.5$.

Reproduced from Zhang F, Kabeya H, Kitagawa R, Hirotsu T, Yamashita M, Otsuki T. Preparation and characterization of a novel polyethylene-chlorella composite. Chem Mater 1999;11:1952–1956.

the marked increases in tensile strength of the MPE–CH composites with increasing the chlorella content, compared with the UPE–CH composites, are probably due to stronger interactions of the chlorella grains with the MPE matrix. Thus the specific tensile strength is almost constant for the MPE–CH composite, even with increasing the chlorella content, while decreasing rapidly for the UPE–CH composite (Fig. 14.12B).

Both the decrease in elongation and the increase in Young's modulus of the MPE–CH composite, compared with the UPE–CH composite (Fig. 14.13), are consistent with the strong interactions between the chlorella grains and the MPE matrix.

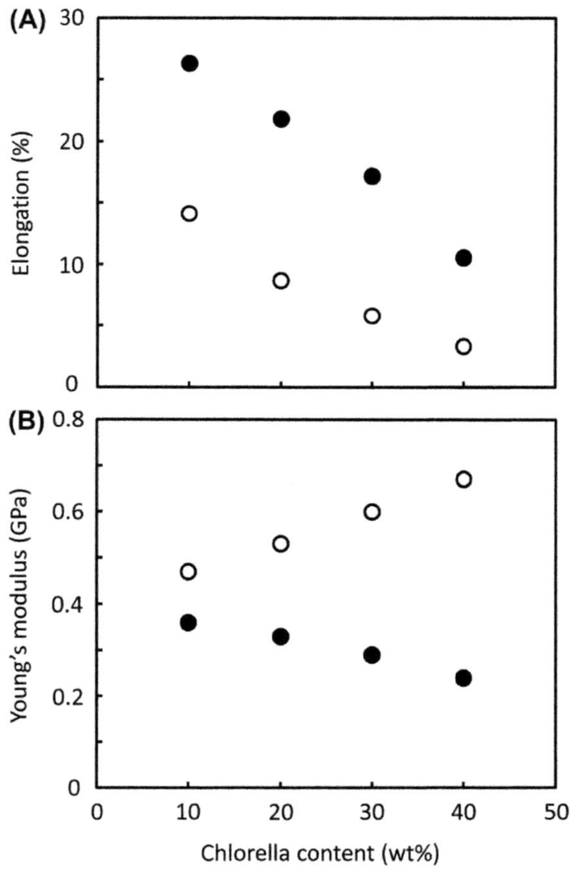

FIGURE 14.13

Elongations (A) and Young's modulus (B) of MPE–CH (○) and UPE–CH composites (●).

Reproduced from Zhang F, Kabeya H, Kitagawa R, Hirotsu T, Yamashita M, Otsuki T. Preparation and characterization of a novel polyethylene-chlorella composite. Chem Mater 1999;11:1952–1956.

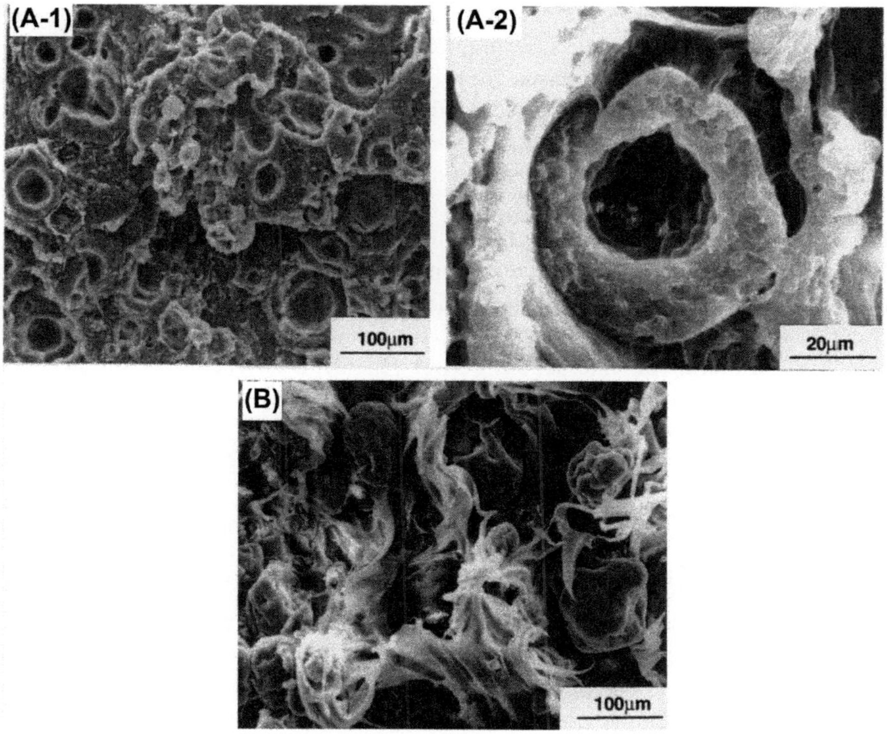

FIGURE 14.14

Scanning electron microscopy micrographs of fractures of 70PE–30CH (A) and 70UPE–30CH composites (B). A chlorella grain tightly bound with the MPE matrix (A-2).

Reprinted from Zhang F, Kabeya H, Kitagawa R, Hirotsu T, Yamashita M, Otsuki T. Preparation and characterization of a novel polyethylene-chlorella composite. Chem Mater 1999;11:1952–1956.

SEM micrographs of composite fractures reveal that chlorella grains in a 70MPE–30CH composite are well inlaid in the MPE matrix, with no intervening air gaps between the chlorella grains and the MPE matrix (Fig. 14.14A-1). This is in contrast to those in a 70UPE–30CH composite, which shows no adhesion with the UPE matrix leaving air gaps in the rough interface (Fig. 14.14B). Fig. 14.14A-2 shows the binding between a chlorella grain and its surrounding MPE matrix in the 70MPE–30CH composite, yielding the tight interface between them. The finding is convincing to show that upon the fracture of the composite, the each chlorella grain breaks into two pieces when an MPE–CH composite is strained to failure. This means that the interactions between the MPE matrix and chlorella grains, $\sigma_{(p-c)}$, are stronger than those between chlorella cells, σ_c, originated primarily from hydrogen bonds. The stronger $\sigma_{(p-c)}$ interactions are consistent with the formation of chemical bonds between chlorella grains and the MPE matrix. The formation of such chemical bonds is also supported by the phenomenon that the chlorella grains on the surface or upon fracture of an MPE–CH composite do not fall away from the MPE matrix, even after immersion in water for a week or after repeated flexing, in contrast to the chlorella grains that readily break off from a UPE–CH composite under the same conditions.

The theoretical formula of tensile strength $\sigma_{0(\text{p}-\text{c})}$ given by Eq. (14.10) is based on the assumption that the tensile strength is due only to the contribution from the polymer matrix with tight interfaces with chlorella grains ignoring the interactions. From the SEM observations, the tight interfaces between chlorella grains and the MPE matrix result from strong interactions there. Such interactions, therefore, are naturally being added to the contribution from the matrix. Consequently, the tensile strength of the MPE–CH composite, $\sigma_{(\text{mpe}-\text{c})}$, is increased with increasing chlorella content. Considering the interactions between the MPE matrix and chlorella grains $\sigma_{(\text{p}-\text{c})}$ stronger than those between chlorella cells themselves σ_{c}, this relation is roughly expressed by Eq. (14.12):

$$\sigma_{(\text{mpe}-\text{c})} = \sigma_{\text{mpe}}A_{\text{mpe}}/A + \sigma_{\text{c}}A_{\text{c}}/A, \tag{14.12}$$

where $A = A_{\text{mpe}} + A_{\text{c}}$. The first term of the right-hand side corresponds to the $\sigma_{0(\text{mpe}-\text{c})}$ expressed by Eq. (14.10). As the second term increases with A_c/A, Eq. (14.12) well explains the nature of the tensile strength of the MPE–CH composite shown in Fig. 14.12A, where the discrepancy between the experimental result and the theoretical expectation appears more significant with increase in the chlorella content.

On the contrary, the UPE–CH composites possess smaller tensile strength than the theoretical expectations, the difference being increased with the chlorella content as shown in Fig. 14.12A. It is recalled here that air gaps appear extensively at the boundaries between chlorella grains and the UPE matrix (Fig. 14.14B). This is principally due to no interaction between the two phases, and causes decrease in the section area contributed from the matrix. Accordingly, the real section area from the UPE matrix, A_{pe}', becomes smaller than its ideal section area, i.e., A_{pe}, since the difference between them corresponds to the contribution from the air gaps. Thus, the tensile strength of the UPE–CH composite, $\sigma_{(\text{pe}-\text{c})}$, can be expressed by Eq. (14.13)

$$\sigma_{(\text{pe}-\text{c})} = \sigma_{\text{pe}}A'_{\text{pe}}\Big/A. \tag{14.13}$$

We can expect the following relation:

$$\sigma_{(\text{pe}-\text{c})} < \sigma_{0(\text{pe}-\text{c})}, \tag{14.14}$$

since $A_{\text{pe}}' < A_{\text{pe}}$. The increasing difference between $\sigma_{(\text{pe}-\text{c})}$ and $\sigma_{0(\text{pe}-\text{c})}$, with increasing the chlorella content with concomitant contribution from air gaps, is well consistent with the experimental results in Fig. 14.12A.

We have demonstrated the preparation of a novel composite consisting of highly hydrophilic chlorella grains and highly hydrophobic PE. The adhesion between chlorella grains and PE matrix is realized through the formation of ester bonds between chlorella grains and PE matrix, slightly modified with MA. Such a strong

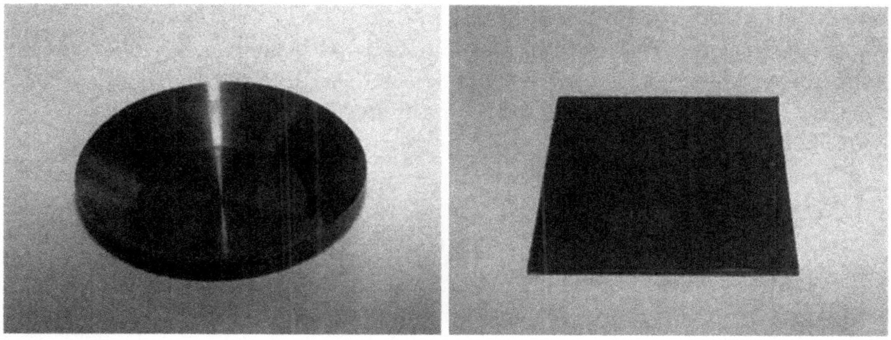

FIGURE 14.15

Moldings of a 70MPE–30CH composite.

Reprinted from Otsuki T, Zhang F, Kabeya H, Hirotsu T. Synthesis and tensile properties of a novel composite of chlorella and polyethylene. J Appl Polym Sci 2004;92:812–816.

adhesion causes significant elevation of the tensile strength, much higher than the theoretical expectation that is based on the assumptions of tight and smooth interface but considering no interactions between the two phases.

An MPE–CH composite has been demonstrated to be readily shaped into a plate-type [size: 15 cm × 15 cm × 0.3–0.5 cm (width × length × thickness)] and a dish-type molding [size: 10 cm × 6 cm × 1.5 cm (upper diameter × bottom diameter × height)] by heat-pressuring method because of its thermal plasticity [24]. Examples are given in Fig. 14.15. Another characteristic of this composite is its interesting structure with hollow grains incorporated in the PE matrix. These characters of the MPE–CH composite will induce its extensive utilization in functional materials including building materials.

14.6 CHLORELLA COMPOSITE WITH POLY(PROPENE)

Compounding of chlorella with poly(propene) PP, another major plastic, is very important for stabilizing a large amount of CO_2-biofixed chlorella for a long period in the resultant composite that has a high potential of extensive utilization for building materials. Poly(propene) is hydrophobic, similar to PE described in the preceding section, and effective compounding of hydrophilic chlorella with PP requires some compatibilization as described in Section 14.5. Furthermore PP is semicrystalline, and such compounding probably causes some changes in crystallinity of the resultant matrix. In this section, we deal with preparation of composites of chlorella with PP by a melt-mixing method using MA-grafted PP as a compatibilizer, and discuss in detail how the chemical bonds between chlorella and the compatibilized matrix, as well as the crystallinity of the matrix, affect the tensile strength of the resultant composite [25].

Chlorella samples (from Yaeyama Shokusan Co., Ltd. Japan) are originally hollow, spherical aggregate grains with their walls 5–20 μm thick, and an average size of c.50 μm, consisting of chlorella cells. The chlorella samples (OCH) were ground to yield ground samples (GCH) as fragments of the grain walls with an average particle size of about 4 μm. Poly(propene), purchased from Aldrich Chemical Co. Inc., is isotactic PP with M_w = c.250,000 and a density of 0.900 g/cm^3. MA-grafted PP (MPP), also purchased from Aldrich Chemical Co. Inc., is characterized with an MA content of c.0.6 wt% and a density of 0.950 g/cm^3. Composites of original chlorella with compatibilized PP (CPP–OCH) were synthesized by melt-mixing chlorella and PP, with MPP as a compatibilizer, fixing the weight ratio of MPP/chlorella at 0.5. Uncompatibilized PP–chlorella composites (UPP–OCH) were prepared without MPP. All the composites were obtained using a roller mixer described in Section 14.5, at a speed of 80 rpm at 180°C for 20 min, and were further hot-pressurized (180°C, 3 MPa, 2 min) into sample sheets for the standard tensile test. With a ground chlorella sample, MPP–, CPP–, and UPP–GCH composites were synthesized by the same methods for investigating fundamental properties of the composites.

14.6.1 SOLID-PHASE ESTERIFICATION BETWEEN CHLORELLA AND MPP

IR spectral comparison of MPP and PP is shown in Fig. 14.16. A major difference between them appears at 1800–1770 cm^{-1}. A broad band at 1800–1770 cm^{-1}, constituted of three fine bands, is probably assigned to MA grafts of MPP [26]. Graft reactions of MA onto PP are complicated and still controversial [17], because of β-scission of tertiary carbon radicals in PP chains. Heinen et al. determined the structures of three types of grafts on PP based on the information obtained from ^{13}C NMR spectroscopy (Fig. 14.17) [20]. One type is the single-succinic anhydride grafts (**1**) on tertiary carbons in PP chains, and the other two are itaconic anhydride terminals (**2**) [17,20]. Accordingly, the broad band at 1800–1770 cm^{-1} can be assigned to the grafts **1** (1792 cm^{-1}) and **2** (1780–1770 cm^{-1}). A band at 1716 cm^{-1} is due to hydrolysis products of the grafts.

Fig. 14.18 shows IR spectra of 80MPP–20GCH and 80UPP–20GCH composites (consisting of 80 wt% MPP and 80 wt% PP, respectively, with 20 wt% GCH), together with a spectrum of chlorella. Compared with chlorella (Fig. 14.18A), the 80MPP–20GCH composite exhibits a lower intensity of absorption at 3440 cm^{-1} (Fig. 14.18C), in accordance with the result of an MPE–CH composite described in Section 14.5. In contrast, the 80UPP–20GCH composite causes almost no change in absorption intensity at the same wave number (Fig. 14.18B). A difference spectrum between the 80MPP–20GCH composite and chlorella, with respect to the band at 1543 cm^{-1} originated from chlorella, is shown in Fig. 14.18E. The most significant differences in the spectrum of MPP in Fig. 14.16B exist in complete disappearance of the band at c.1790 cm^{-1}, followed by a decrease in intensity at 1780–1750 cm^{-1}, and appearance of a new band around 1730 cm^{-1}, probably

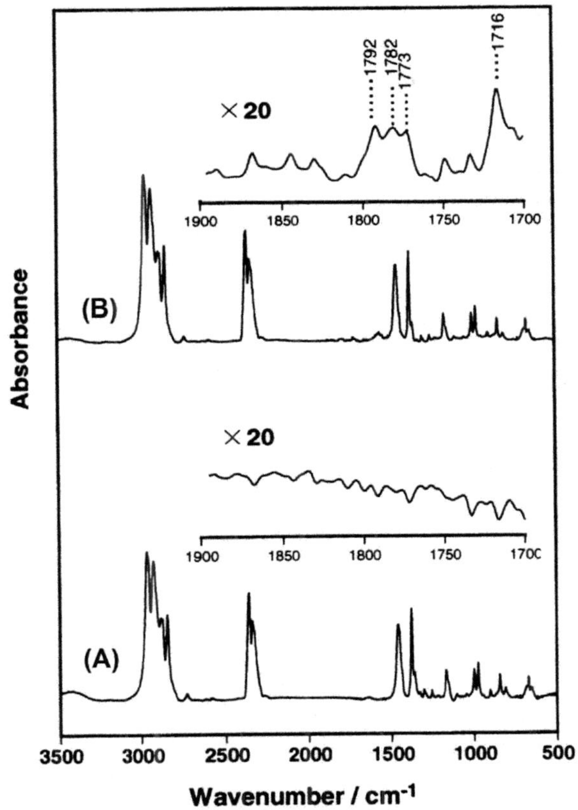

FIGURE 14.16

Fourier transform infrared spectroscopy spectra of polypropene (A) and maleic anhydride–grafted polypropene (B).

Reprinted from Zhang F, Endo T, Kitagawa R, Kabeya H, Hirotsu T. Synthesis and characterization of a novel blend of polypropylene with chlorella. J Mater Chem 2000;10:2666–2672.

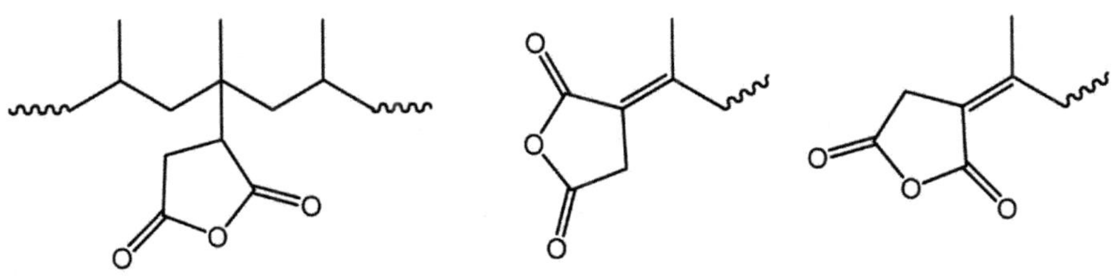

1 2

FIGURE 14.17

Graft structures in polypropene, determined by ^{13}C nuclear magnetic resonance spectroscopy.

Reproduced from Heinen W, Rosenmöller CH, Wenzel CB, de Groot HJM, Lugtenburg J, van Dunin M. ^{13}C NMR study of the grafting of maleic anhydride onto polyethene, polypropene, and ethene-propene copolymers. Macromolecules 1996;29:1151–1157.

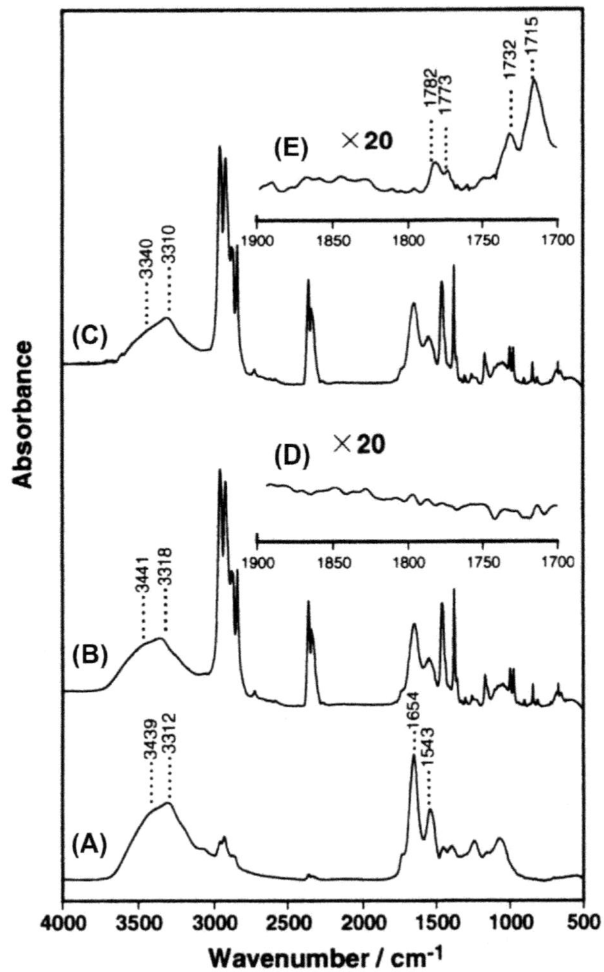

FIGURE 14.18

FTIR spectra of chlorella (A), 80UPP–20GCH composite (B), and 80MPP–20GCH composite (C). The magnified spectra represent the difference spectra between 80UPP–20GCH composite and chlorella (D), and between 80MPP–20GCH composite and chlorella (E).

Reprinted from Zhang F, Endo T, Kitagawa R, Kabeya H, Hirotsu T. Synthesis and characterization of a novel blend of polypropylene with chlorella. J Mater Chem 2000;10:2666–2672.

due to ester groups. The latter was not distinctly observed in the MPE–PE composite as described in Section 14.5.

We have found that the single-succinic anhydride grafts, observed at c.1790 cm^{-1}, are more reactive than the other grafts in MPP series prepared by a melt-mixing method [26], supporting the almost complete disappearance of the band at c.1790 cm^{-1} (Fig. 14.18E). Furthermore, ball-milling of cellulose with MPE or MPP intensively induces esterification between −OH groups of cellulose and single-succinic anhydride grafts of MPE or MPP, primarily due to the activation of −OH groups of cellulose, so that −OH groups strongly bound to each other through hydrogen bonding are liberated mechanically to be free from each other [18,23,27,28]. It is emphasized that ball-milling of crystalline cellulose generates a great number of activated −OH groups on the resulting microparticles of amorphous cellulose that can be subjected to extensive esterification even through melt-mixing with MPP [28]. As noted above, the GCH chlorella sample consists of fragments of the grain walls as a result of grinding the original chlorella OCH; hence the GCH sample probably possesses a greater number of activated −OH groups on the surfaces of the fragments so that they can cause extensive reaction with MA grafts on MPP to yield more ester bonds between them, in contrast to MPE−CH composites which are prepared with the original chlorella OCH sample. The clear appearance of the signal at 1732 cm^{-1} in Fig. 14.18E is probably due to the use of the ground chlorella GCH with a larger number of activated −OH groups.

14.6.2 CHANGES IN CRYSTALLINITY OF THE PP MATRICES

Investigating the effect of the chemical bonds between MPP and chlorella on the crystallinity of the PP matrix is very interesting because PP is a semicrystalline polymer. This polymer possesses not only ordered but also disordered phases, and its crystallinity depends on the content of the ordered phase, which is defined as the molar fraction of chain atoms having a specified steric configuration related to their immediate neighbors along the chain. The crystallinity of PP, based on the IR spectroscopy of PP, can be estimated from the absorbance ratio of the band at 998 cm^{-1} to the band at 974 cm^{-1}, A_{998}/A_{974} [29]:

$$\text{Crystallinity (wt\%)} = 109 \times (A_{998} - A_{920})/(A_{974} - A_{920}) - 31.4, \qquad (14.15)$$

where the absorbance at 920 cm^{-1}, A_{920}, is for the background correction for best reproducibility.

Table 14.2 shows corrected absorbance ratios and crystallinities of MPP, UPP, and their composites with chlorella. It is obvious that compounding with GCH results in a decrease in crystallinity of MPP, while that with UPP almost no change. The marked decrease in crystallinity of MPP is probably due to the difficulty in steric ordering of MPP chains covalently bonded to rigid chlorella.

14.6.3 PHASE TRANSITIONS OF THE PP MATRICES

DSC results of 80UPP−20GCH and 80MPP−20GCH composites are listed, compared with those of PP and MPP, in Table 14.3. Since chlorella does not undergo any significant change below 200°C [11], the melting and solidification phase transitions of the composites are entirely due to the PP and MPP matrices. The melting and solidification enthalpies of 80UPP−20GCH composites are consistent with 80%

Table 14.2 Crystallinity Index of Polypropene Calculated From Fourier Transform Infrared Spectroscopy Results

	$(A_{998}-A_{920})/(A_{974}-A_{920})$	Crystallinity (wt%)
PP	0.805	56.3
80UPP–20GCH composite	0.798	55.6
MPP	0.792	54.9
80MPP–20GCH composite	0.732	48.8

Reprinted from Zhang F, Endo T, Kitagawa R, Kabeya H, Hirotsu T. Synthesis and characterization of a novel blend of polypropylene with chlorella. J Mater Chem 2000;10:2666–2672.

Table 14.3 Melting and Solidification Data of Polypropene Obtained From Differential Scanning Calorimetry Analysis

Polymers and Their Composites	Melting		Solidification	
	Onset (°C)	Enthalpy (J/g)	Onset (°C)	Enthalpy (J/g)
PP	155.8	90.5	120.7	−94.2
80UPP–20GCH	156.2	72.8	121.5	−76.5
MPP	155.1	93.8	126.9	−94.0
80MPP–20GCH	153.2	69.6	126.0	−70.9

Reprinted from Zhang F, Endo T, Kitagawa R, Kabeya H, Hirotsu T. Synthesis and characterization of a novel blend of polypropylene with chlorella. J Mater Chem 2000;10:2666–2672.

of those of neat PP, with almost the same onset temperatures of the corresponding phase transitions. This indicates that the PP matrix in the 80UPP–20GCH composite is almost identical to neat PP; thus the PP matrix is independent from the GCH phase in the composite.

The 80MPP–20GCH composite, on the other hand, reveals smaller enthalpies of melting as well as solidification than the values predicted for 80% of neat MPP (Table 14.3), with a broader DSC signal in melting at a little lower onset temperature. The enthalpy of a phase transition of a semicrystalline polymer depends primarily on the crystallinity of the polymer. Accordingly, the decreases in melting and solidification enthalpies of the 80MPP–20GCH composite are attributed to the increase in content of the amorphous phase of MPP after the compounding with chlorella as described in Table 14.2.

The glass transition temperature, T_g, of a composite is also useful for understanding the interaction modes between the components. Such temperatures of the 80MPP– and the 80UPP–20GCH composites can be determined by dynamical mechanical analysis (DMA). The T_g value of PP is c.1°C, and is almost equal to that of MPP. The T_g values of the 80UPP– and the 80MPP–20GCH composites decrease by about 3°C and 8.5°C from those of neat PP and MPP, respectively. Considering that chlorella itself is a kind of rigid matter with no viscoelasticity, these phenomena may result from some characteristics of chlorella and/or the formation of chemical bonds between the MPP and chlorella phases. Generally the glass transition of a polymer is related to the movement of segments of polymer chains in free spaces. Although it does not undergo any significant change below 200°C, chlorella still contains a small amount of volatile and/or decomposable substances at 180°C [11], the melt-mixing temperature. During the melt mixing with PP or MPP, some volatile and/or decomposable substances may penetrate into the amorphous part of PP or MPP, causing swelling of some parts of the PP molecular chains. This may induce movement of the segments, leading to the depression in T_g of PP in the 80UPP–20GCH composite. For the 80MPP–20GCH composite, characterized by the formation of chemical bonds between the MPP and GCH phases described earlier, the chemical bonds significantly restrict the stable arrangement of MPP chains bound onto GCH particles and cause their coarse entanglement with other MPP and PP chains. Thus the resultant volume expansion near the surfaces of the GCH particles will induce more active motion of the segments of MPP and PP chains over there, leading to a further decrease in T_g of the 80MPP–20GCH composite than that observed in the 80UPP–20GCH composite lacking the chemical bonds between UPP and chlorella.

14.6.4 TENSILE PROPERTIES OF THE COMPOSITES

To understand the effects of MPP as a compatibilizer in compounding of PP with chlorella, a comparison of the tensile properties of CPP–OCH and UPP–OCH composites with varying chlorella contents is shown in Fig. 14.19. With a fixed MPP/chlorella weight ratio of 0.5, the tensile strength of the CPP–OCH composite decreases with an increase in chlorella content. The effect is more gradual than that observed in the UPP–OCH composite without MPP (Fig. 14.19A). Moreover, the Young's modulus of the CPP–OCH composite increases with an increase in chlorella content, while that of the UPP–OCH composite decreases (Fig. 14.19B). These results are very similar to those of the MPE– and UPE–chlorella composites described in Section 14.5.

Micrographs of fractures of 70PP–30CH composites are presented in Fig. 14.20. Chlorella grains in a 70CPP–30OCH composite (55PP–15MPP–30OCH) are well inlaid in the CPP matrix, with no air gaps between the chlorella grains and CPP matrix (Fig. 14.20B). In contrast, the grains in a 70UPP–30OCH composite show no adhesion with the PP matrix leaving air gaps in the interface (Fig. 14.20A). It is emphasized in Fig. 14.20B that interactions between the chlorella grains and the CPP matrix in the 70CPP–30OCH composite $\sigma_{(p-c)}$ are stronger than hydrogen bond interactions among chlorella cells σ_c, since each chlorella grain is split into two pieces on the fracture inlaid in each matrix retaining their original shapes. This explains the formation of chemical bonds well described earlier. These features of the fractures of the PP–CH composites are quite similar to those of the PE–CH composites described in Section 14.5. GCH particles are inlaid so completely that the interfaces between the GCH particles and CPP matrix can hardly be distinguished (Fig. 14.20C).

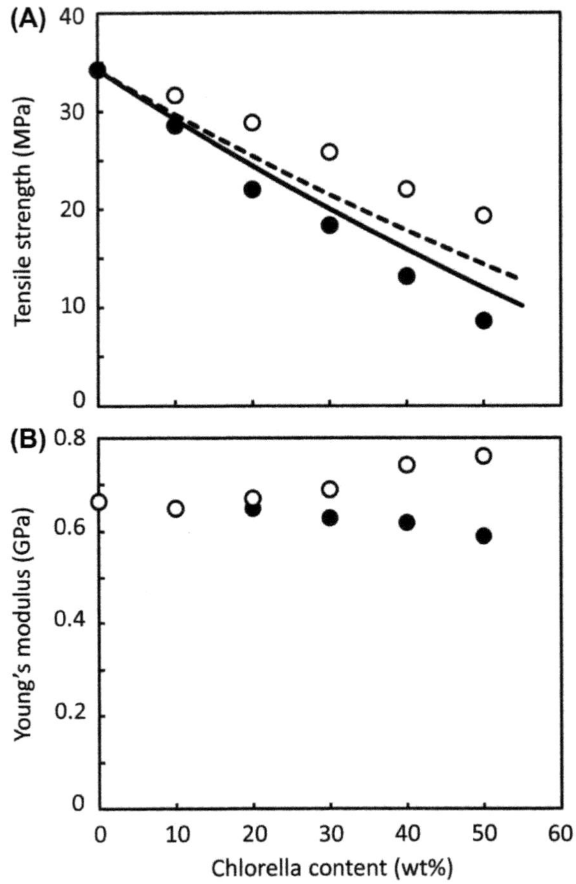

FIGURE 14.19

Tensile properties of CPP–CH (O) and UPP–CH composites (●). The solid and broken curves are based on Eq. (14.16) with $\sigma_{pp} = 34.3$ MPa and $\sigma_{mpp} = 23.8$ MPa and Eq. (14.10) with $\sigma_p = 34.3$ MPa and $D_p/D_c = 1.4$.

Reproduced from Zhang F, Endo T, Kitagawa R, Kabeya H, Hirotsu T. Synthesis and characterization of a novel blend of polypropylene with chlorella. J Mater Chem 2000;10:2666–2672.

14.6.5 EFFECTS OF THE ESTERIFICATION AND THE CRYSTALLINITY ON THE TENSILE PROPERTIES

Let us consider the tensile strengths of the UPP– and the CPP–OCH composites, based on the same assumptions as in Section 14.4.4. All the notations used here have the same meanings as in Section 14.4.4, unless specified otherwise. For a chlorella composite, with its polymer matrix composed of two components p1

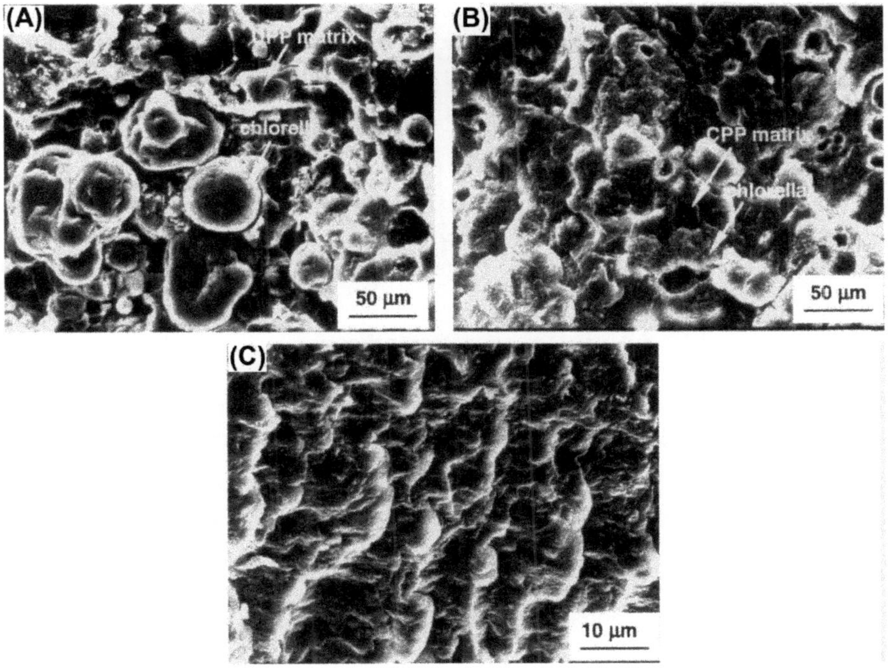

FIGURE 14.20

Micrographs of fractures of a 70UPP–30OCH composite (A), a 70CPP–30OCH composite (B), and a 70CPP–30GCH composite (C).

Reprinted from Zhang F, Endo T, Kitagawa R, Kabeya H, Hirotsu T. Synthesis and characterization of a novel blend of polypropylene with chlorella. J Mater Chem 2000;10:2666–2672.

and p2 having no interactions with the chlorella phase, the tensile strength is expressed by the following equation instead of Eq. (14.5):

$$\sigma_{0(\mathrm{p1-p2-c})} = \sigma_{\mathrm{p1}} A_{\mathrm{p1}}/A + \sigma_{\mathrm{p2}} A_{\mathrm{p2}}/A, \tag{14.16}$$

where $A = A_{\mathrm{p1}} + A_{\mathrm{p2}} + A_{\mathrm{c}}$. Consistent with the idea in Section 14.4.4, the volume of each component (i) and total volume are defined as

$$V_{\mathrm{i}} = W_{\mathrm{i}}/D_{\mathrm{i}} \tag{14.17}$$

and

$$V = \Sigma_{\mathrm{i}} V_{\mathrm{i}}, \tag{14.18}$$

where $W = \Sigma_i W_i$, and the subscript i denote p1, p2, and c. Since all the cross sections of the composite have the same values of A_i, Eq. (14.19) is satisfied:

$$A_{\mathrm{i}}/A = V_{\mathrm{i}}/V. \tag{14.19}$$

Substituting Eqs. (14.17) and (14.18) into Eq. (14.19),

$$A_{p1}/A = 1/[1 + (D_{p1}/D_{p2}) \times (W_{p2}/W_{p1}) + (D_{p1}/D_c) \times (W_c/W_{p1})] \quad (14.20)$$

and

$$A_{p2}/A = 1/\left[\left(D_{p2}/D_{p1}\right) \times \left(W_{p1}/W_{p2}\right) + 1 + \left(D_{p2}/D_c\right) \times \left(W_c/W_{p2}\right)\right] \quad (14.21)$$

Making the polymers p1 and p2 correspond to PP and MPP, respectively, the weight ratios between PP, MPP, and chlorella required in Eqs. (14.20) and (14.21) are given as functions of the chlorella content $C\%$ (wt%), with the experimental condition $W_{mpp}/W_c = W_{p2}/W_c = 0.5$:

$$W_{p1}/W_{p2} = W_{pp}/W_{mpp} = 2.0 \times (100 - 1.5 \times C\%)/C\% \quad (14.22)$$

and

$$W_{p1}/W_c = W_{pp}/W_c = (100 - 1.5 \times C\%)/C\%. \quad (14.23)$$

The tensile strength of PP (σ_{pp}) is 34.3 MPa [25], while the tensile strength of MPP (σ_{mpp}) is 23.8 MPa [30]. First calculating Eqs. (14.20) and (14.21) with the values of Eqs. (14.22) and (14.23) as well as considering $D_c = 0.65$ g/cm^3, $D_{pp} = 0.900$ g/cm^3, and $D_{mpp} = 0.950$ g/cm^3, and introducing those values into Eq. (14.16) yields the theoretical values of the tensile strength $\sigma_{0(pp-mpp-c)}$. The result is given by a solid curve in Fig. 14.19A.

Another equation equivalent to Eq. (14.16) is derived, provided that the tensile strength for a mixed polymer (p12) of polymers p1 and p2 is given by the volume-weighted mean of the tensile strengths of p1 and p2:

$$\sigma_{p12} = \sigma_{p1} \times (V_{p1}/V_{p12}) + \sigma_{p2} \times (V_{p2}/V_{p12}), \quad (14.24)$$

where the weight and volume of the mixed polymer are assumed as follows:

$$W_{p12} = W_{p1} + W_{p2} \quad (14.25)$$

and

$$V_{p12} = V_{p1} + V_{p2}, \quad (14.26)$$

and the density of the mixed polymer, D_{p12}, is defined by Eq. (14.27):

$$D_{p12} = W_{p12}/V_{p12}. \quad (14.27)$$

The tensile strength of the chlorella composite with the mixed polymer p12 is expressed by the following equation:

$$\sigma_{0(p12-c)} = \sigma_{p12}/(1 + D_{p12}W_c/W_{p12}D_c). \quad (14.28)$$

This equation is equivalent to Eq. (14.16). It is noted that $\sigma_{0(p12-c)}$ has the same form as $\sigma_{0(p-c)}$ in Eq. (14.10), the former being derived just by introducing σ_{p12}, D_{p12}, and W_{p12} into the latter instead of σ_p, D_p, and W_p. The tensile strength of the mixed polymer (σ_{p12}) will be very important to qualitatively understand the tensile property of the CPP–CH composite below.

As the polymer matrix consists of one component, Eq. (14.10) is satisfied for the UPP–OCH composite and the result with $D_p/D_c = 0.90/0.65$ yields the broken curve in Fig. 14.19A. It is to be noted again that these theoretical equations $\sigma_{0(p-c)}$ and $\sigma_{0(p1-p2-c)}$ or $\sigma_{0(p12-c)}$ are applied for chlorella composites with no interaction between chlorella and the matrices, and no air gaps in relevant interfaces. Furthermore, it is assumed for the CPP–OCH composite that the compatibilized PP (CPP) causes no change in crystallinity of PP and MPP themselves. In Fig. 14.19A, the theoretical tensile strength $\sigma_{0(cpp-c)}$ (a solid curve) is smaller than that of $\sigma_{0(pp-c)}$ (a broken curve). The difference between them is greater with more chlorella content. This is due to the fact that the tensile strength of MPP is smaller than that of PP.

It is obvious from Fig. 14.19A that the tensile strength of the UPP–OCH composite is smaller than that of the theoretical expectation $\sigma_{0(pp-c)}$, similar to the result of the PE–CH composite. The UPP–OCH composite has a lot of air gaps between chlorella grains and PP matrix, as demonstrated by the microphotograph of a fracture of a 70UPP–30OCH composite (Fig. 14.20A), because of no chemical affinity between the chlorella and PP phases. Accordingly, the real section area of the PP matrix (A_{pp}') becomes smaller than the ideal one (A_{pp}), with the difference increasing with an increase in the chlorella content. The difference between the experimental and theoretical results for the UPP–OCH composite is probably ascribed to the decrease in real section area of the PP matrix because of $\sigma_{pp} A_{pp}'/A < \sigma_{pp} A_{pp}/A$ in Eq. (14.5).

The tensile strength of the CPP–OCH composite, on the other hand, is a little complicated in comparison to the UPP–OCH composite, since the CPP–OCH composite forces us to consider a mixed effect of crystallinity and esterification, both of which characterize the CPP–OCH composite. As described earlier, the formation of ester bonds in the CPP–OCH composite causes a decrease in crystallinity of the CPP matrix. Such decrease in crystallinity will give rise to the smaller tensile strength of the CPP matrix σ_{cpp}' than that of the original matrix σ_{cpp}. According to Eq. (14.28), the tensile strength of the CPP–OCH composite $\sigma_{0(cpp-c)}$ is determined by the tensile strength of the CPP matrix σ_{cpp}, as σ_{cpp} is related to σ_{p12} in Eq. (14.28). Then we expect that the decrease in crystallinity of the CPP matrix reduces the tensile strength of the corresponding CPP–OCH composite: $\sigma_{0(cpp'-c)} < \sigma_{0(cpp-c)}$, where the former is for the CPP matrix with a reduced crystallinity. The experimental results, however, are greater than the theoretical expectations $\sigma_{0(cpp-c)}$, as is evident in Fig. 14.19A. This suggests that the tensile strength of the CPP–OCH composite is governed more by the interfacial interactions between chlorella grains and the CPP matrix than by the reduction in crystallinity of the CPP matrix. Thus, the tensile strength of the CPP–OCH composite depends not only on that of the CPP matrix but also on the contribution by the chlorella grains σ_c, because of the stronger interactions between MPP and chlorella $\sigma_{(p-c)}$ than those among chlorella cells alone σ_c, i.e., $\sigma_{(p-c)} > \sigma_c$. The contribution from the esterification between chlorella grains and CPP matrix substantially increases the tensile strength leading to the following relation:

$$\sigma_{0(cpp-c)} < \sigma_{0(cpp'-c)} + \sigma_c A_c / A. \tag{14.29}$$

The contribution of the second term on the right-hand side in Eq. (14.29) is so pronounced that the CPP—OCH composite can exhibit a tensile strength greater than the theoretical expectations, not only for the CPP—OCH composite $\sigma_{0(cpp-c)}$ but also for the UPP—OCH composite $\sigma_{0(pp-c)}$ as clearly shown in Fig. 14.19.

14.7 CONCLUDING REMARKS

An LCA study has demonstrated that a practical system of biological CO_2 fixation with microalgae necessitates effective use of the resultant microalgae. Compounding of chlorella with versatile plastics is a new strategy for retaining a large amount of the biologically fixed CO_2 in resultant composites for long period with a potential application in building materials. Heat-pressuring chlorella with slightly polar PVC produces a novel composite that satisfies the requirement of rigid or plasticized PVC, depending on the chlorella content. Compounding chlorella with hydrophobic PE or PP, however, requires slight grafting of PE or PP with MA for acquisition of compatibility between the components. Melt-mixing PE with 0.25 wt% of MA, followed by further melt-mixing the resultant MPE with chlorella, yields a novel MPE—chlorella composite with tight interfaces between chlorella grains and MPE matrix, because of the formation of ester bonds between chlorella and MPE. On the basis of this finding, compounding chlorella with another hydrophobic PP is successfully attained through melt-mixing them with MPP as a compatibilizer, to yield a novel compatibilized PP—chlorella composite. These chlorella composites are readily shaped into required moldings. Therefore, the compounding of chlorella with these versatile plastics will be an essential base for realizing a practical system of biological CO_2 fixation.

ACKNOWLEDGMENTS

This project was funded by New Energy and Industrial Technology Development Organization (NEDO) in Japan. A major part of the results described in this chapter was attained by a collaboration research between the former Shikoku National Industrial Research Institute (SNIRI) and Research Institute of Innovative Technology for the Earth (RITE) during 1995—2000. The authors express their sincere thanks to all members who concentrated their interests and efforts on this collaboration research, including Dr. F. Zhang, Dr. H. Kabeya, Mr. R. Kitagawa, and Mr. M. Yamashita. T. O. expresses his thanks to the engineers of the Total System Working Group, RITE, who contributed actively to the solution of LCA issues. T. H. greatly appreciates Dr. J. Hosokawa, a former Chief of the Research Planning Section, SNIRI, for his encouragement and considerate support.

REFERENCES

[1] Pires JCM, Alvim-Ferraz MCM, Martins FG, Simões M. Carbon dioxide capture from flue gases using microalgae: engineering aspects and biorefinery concept. Renew Sustain Energy Rev 2012;16:3043—53.

[2] Zhao B, Su Y. Process effect of microalgal-carbon dioxide fixation and biomass production: a review. Renew Sustain Energy Rev 2014;31:121–32.

[3] Sander K, Murthy GS. Life cycle analysis of algae biodiesel. Int J Life Cycle Assess 2010;15:704–14.

[4] Stephenson AL, Kazamia E, Dennis JS, Howe CJ, Scott SA, Smith AG. Life-cycle assessment of potential algal biodiesel production in the United Kingdom: a comparison of raceways and air-lift tubular bioreactors. Energy Fuels 2010;24:4062–77.

[5] Michiki H. Biological CO_2 fixation and utilization project. Energy Convers Manage 1995;36(6–9):701–5.

[6] Usui N, Ikenouchi M. The biological CO_2 fixation and utilization project by RITE(1) – highly-effective photobioreactor system. Energy Convers Manage 1997;38:S487–92.

[7] Murakami M, Ikenouch M. The biological CO_2 fixation and utilization project by RITE(2) – screening and breeding of microalgae with high capability in fixing CO_2. Energy Convers Manage 1997;38:S493–7.

[8] Otsuki T. A study for the biological CO_2 fixation and utilization system. Sci Total Environ 2001;277:21–5.

[9] Otsuki T. The doctor thesis titled a study on novel environmentally conscious materials. The University of Tokushima; 2004.

[10] Watanabe Y, Ohmura N, Saiki H. Isolation and determination of cultural characteristics of microalgae which functions under CO_2 enriched atmosphere. Energy Convers Manage 1992;33:545–52.

[11] Zhang F, Kabeya H, Kitagawa R, Hirotsu T, Yamashita M, Otsuki T. A novel polyethylene – chlorella composite. I. Characterization of chlorella biologically fixing CO_2. J Appl Polym Sci 2000;77:2278–84.

[12] Zhang F, Kabeya H, Kitagawa R, Hirotsu T, Yamashita M, Otsuki T. An exploratory research of PVC – chlorella composite material (PCCM) as effective utilization of chlorella biologically fixing CO_2. J Mater Sci 2000;35:2603–9.

[13] Japan Standard Society. Testing method for tensile properties of plastics. JIS K7113–1995. 1995. p. 1.

[14] Japan Standard Society. Designation of rigid polyvinyl chloride compounds. JIS K6740–1976. 1976. p. 1.

[15] Japan Standard Society. Plasticized polyvinyl chloride compounds. JIS K6723–1995. 1995. p. 2.

[16] Al-Malaika S, editor. Reactive modifiers for polymers. London: Chapman & Hall; 1997.

[17] Moad G. The synthesis of polyolefin graft copolymers by reactive extrusion. Prog Polym Sci 1999;24:81–142.

[18] Hirotsu T, Endo T, Umemura M. Ball-milling promoted chemical bonding between cellulose and plastics. In: Ranu B, Stolle A, editors. Ball milling towards green synthesis: applications, projects, challenges. London: RSC; 2015 [chapter 9].

[19] Zhang F, Kabeya H, Kitagawa R, Hirotsu T, Yamashita M, Otsuki T. Preparation and characterization of a novel polyethylene-chlorella composite. Chem Mater 1999;11: 1952–6.

[20] Heinen W, Rosenmöller CH, Wenzel CB, de Groot HJM, Lugtenburg J, van Dunin M. ^{13}C NMR study of the grafting of maleic anhydride onto polyethene, polypropene, and ethene-propene copolymers. Macromolecules 1996;29:1151–7.

[21] Yang L, Zhang F, Endo T, Hirotsu T. Microstructure of maleic anhydride grafted polyethylene by high-resolution solution-state NMR and FTIR spectroscopy. Macromolecules 2003;36:4709–18.

[22] De Roover B, Sclavons M, Carlier V, Devaux J, Legras R, Momtaz A. Molecular characterization of maleic anhydride-functionalized polypropylene. J Polym Sci Polym Chem 1995;33:829–42.

[23] Qiu W, Endo T, Hirotsu T. Interfacial interactions of a novel mechanochemical composite of cellulose with maleated polypropylene. J Appl Polym Sci 2004;94:1326–35.

[24] Otsuki T, Zhang F, Kabeya H, Hirotsu T. Synthesis and tensile properties of a novel composite of chlorella and polyethylene. J Appl Polym Sci 2004;92:812–6.

[25] Zhang F, Endo T, Kitagawa R, Kabeya H, Hirotsu T. Synthesis and characterization of a novel blend of polypropylene with chlorella. J Mater Chem 2000;10:2666–72.

[26] Qiu W, Endo T, Hirotsu T. Interfacial interaction, morphology, and tensile properties of a composite of highly crystalline cellulose and maleated polypropylene. J Appl Polym Sci 2006;102:3830–41.

[27] Zhang F, Qiu W, Yang L, Endo T, Hirotsu T. Mechanochemical preparation and properties of a cellulose-polyethylene composite. J Mater Chem 2002;12:24–6.

[28] Qiu W, Zhang F, Endo T, Hirotsu T. Milling-induced esterification between cellulose and maleated polypropylene. J Appl Polym Sci 2004;91:1703–9.

[29] Nishikida K, Iwamoto R. Fundamentals and applications of infra-red spectroscopy on materials analysis. Tokyo: Kodansha Scientific; 1986. p. 214–5.

[30] Qiu W, Zhang F, Endo T, Hirotsu T. Preparation and characterization of composites of high-crystalline cellulose with polypropylene: effects of maleated polypropylene and cellulose content. J Appl Polym Sci 2003;87:337–45.

CHAPTER

15 Alginate Blends of Poly(vinyl alcohol) and Poly(*N*-vinyl-2-pyrrolidone) for Higher Physicomechanical Properties: Rationale of Making Heteropolymers

Tuncer Caykara[1], Yasar Saleem[2]

Gazi University, Beşevler, Turkey[1]; Biotechnology Research Center, PCSIR, Laboratories Complex, Lahore, Pakistan[2]

15.1 ALGINATE BLENDS WITH POLY(VINYL ALCOHOL)

Polymer blends are widely used in biomedical applications. However, blend materials from either synthetic or natural polymers alone are not always able to meet all the complex demands of biomaterials. The success of synthetic polymers as biomaterials relies on their wide range of mechanical properties and transformation processes that allow a variety of different shapes to be easily obtained in addition to their low production cost. Biological polymers present good biocompatibility, but their mechanical properties are often poor. The necessity of preserving biological properties complicates their processability, and their production or recovery costs are very high [1,2]. Therefore, biologically important polymeric materials based on blends of synthetic and natural polymers are prepared, such as poly(*N*-vinyl-2-pyrrolidone) (PVP)/kappa-carrageenan, PVP/iota-carrageenan [3], poly(ethylene oxide)/hydroxypropyl methylcellulose [4], and poly(vinyl alcohol) (PVA)/chitosan [5].

Alginate, a naturally occurring polysaccharide obtained from marine brown algae, comprises linear chains of (1,4)-β-D-mannuronic acid and (1,3)-α-L-guluronic acid [6]. Sodium alginate (NaAlg), a polyelectrolyte having a rigid molecular chain [7], and good film-forming ability, has been extensively exploited and studied in detail on biomedical applications as a drug carrier [8,9]. However, NaAlg has some rigidity and brittleness. For this reason, it is expected that mechanical properties of NaAlg can be improved through blending with a synthetic polymer such as PVA and PVP. PVA is highly elastic in nature and is also known as a biocompatible material. Accordingly, the blend of NaAlg and PVA is good candidate for new

Algae Based Polymers, Blends, and Composites. http://dx.doi.org/10.1016/B978-0-12-812360-7.00015-X

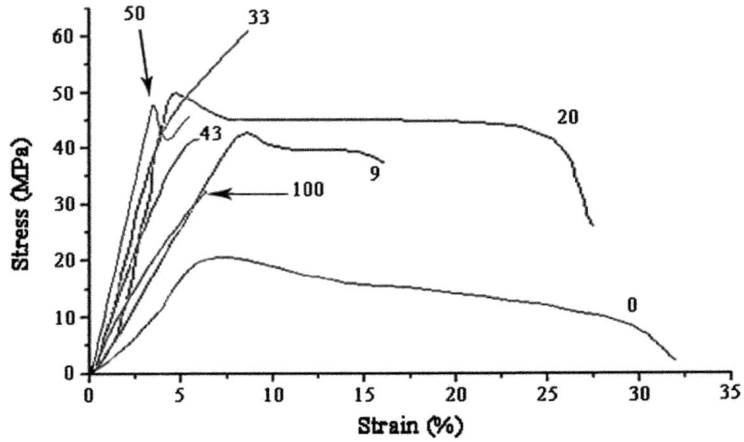

FIGURE 15.1

Stress–strain curves of the PVA/NaAlg blend films. The values given in the figure are wt% NaAlg [10].

biomaterial with enhanced mechanical properties [10]. NaAlg also forms optically clear homogeneous blends with PVA, and the Young modulus and the tensile stress of the blends are greater than that of the homopolymers. The results of the stress–strain measurements of PVA, NaAlg, and PVA/NaAlg blend films are shown in Fig. 15.1. NaAlg exhibits a tensile strength of 33 MPa and elongation at break of 6.5%. Upon mixing with PVA, the tensile strength and elongation at break increase with the increase of the NaAlg content. This result suggests the existence of specific interactions between NaAlg and PVA chains, such as those reported for the blend films of silk fibroin/NaAlg [11] and poly(acrylamide)/NaAlg [12].

Fourier transform infrared (FT-IR) spectroscopy is one of the most powerful techniques for investigation of a multicomponent system, because it provides information on the blend composition and on the polymer–polymer interaction. Our discussion will mainly concentrate on the hydroxyl and carbonyl stretching vibration bands in Fig. 15.2, as expected to be affected by the hydrogen bonding interactions. Fig. 15.2 shows the FT-IR spectra for pure PVA and NaAlg samples and their blends. The bands of NaAlg appear at 3500 cm^{-1} for the hydroxyl groups and at 1613 and 1415 cm^{-1} for the asymmetric and symmetric carboxylate stretching vibrations, respectively. The spectrum of the PVA/NaAlg blends is characterized by the presence of the absorption bands of the pure components. Moreover, the intensity of these bands is roughly proportional with the blending ratio. The characteristic band of NaAlg appeared at 1613 cm^{-1} shifts to a lower frequency range, indicating a stronger hydrogen bonding formation between the hydroxyl groups of PVA and the carboxylate groups of NaAlg in the blend (Fig. 15.3). In addition, it is noticed that the hydroxyl stretching bands become much broader with the increase of NaAlg content. This also strongly supports the hydrogen bonding formation.

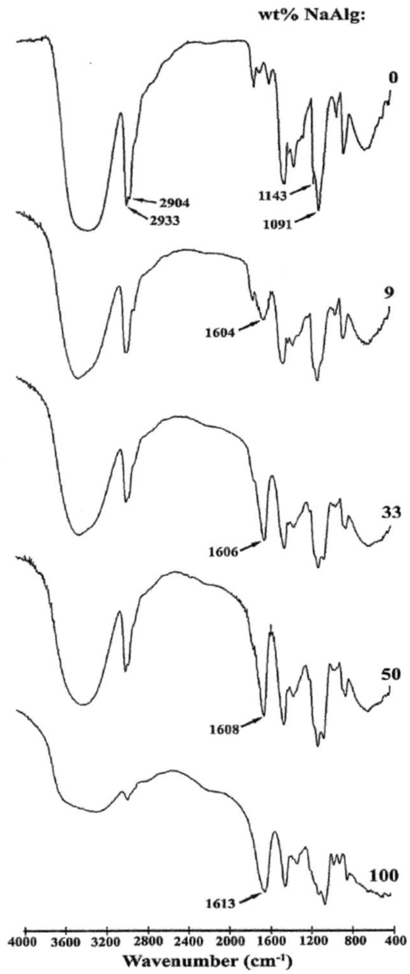

FIGURE 15.2

Fourier transform infrared spectra of the poly(vinyl alcohol)/NaAlg blends [10].

Typical weight loss (TG) and derivative of weight loss (DTG) curves of PVA, NaAlg, and PVA/NaAlg blends are presented in Fig. 15.4. As shown this figure, the mass loss of pure PVA begins at 230°C and reaches a maximum at 254°C. However, the mass loss of pure NaAlg starts at 229°C and reaches a maximum at 243°C. In this case, PVA and NaAlg have almost the same thermal stability. The TG curves of both NaAlg and PVA also indicate one reaction stage, which is reflected as a single peak in the DTG curves (Fig. 15.4). Similarly, PVA/NaAlg blends degrade in one step because the thermal degradation temperatures of the components are rather close to each other. This is evidenced by the appearance of only one degradation reaction peak in DTG thermograms. However, the initial thermal degradation temperature of the blends is greater than that

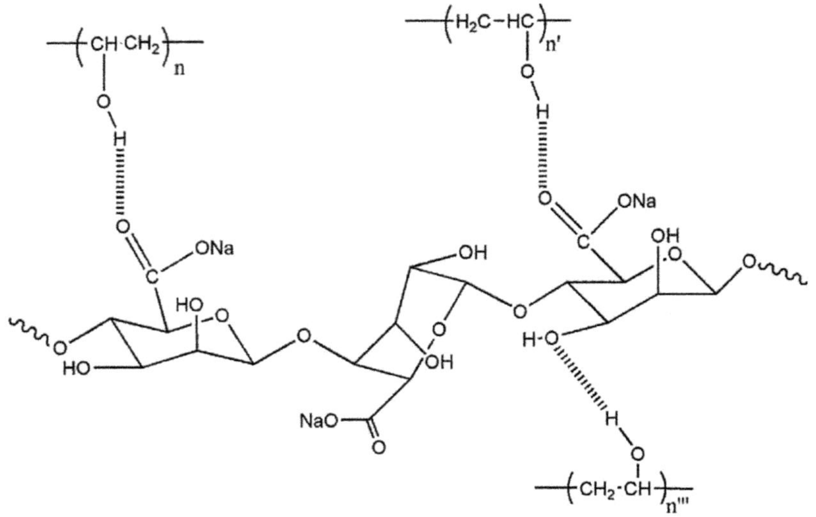

FIGURE 15.3

Possible interactions in the poly(vinyl alcohol)/NaAlg blends [10].

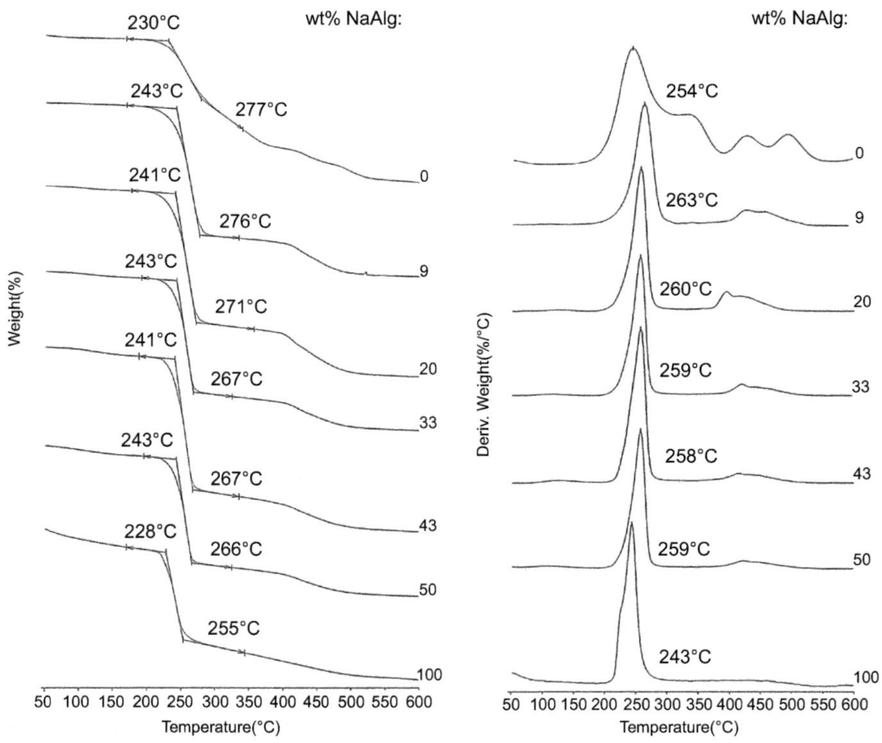

FIGURE 15.4

Typical weight loss (left) and derivative of weight loss (right) curves of the poly(vinyl alcohol)/NaAlg blends [10].

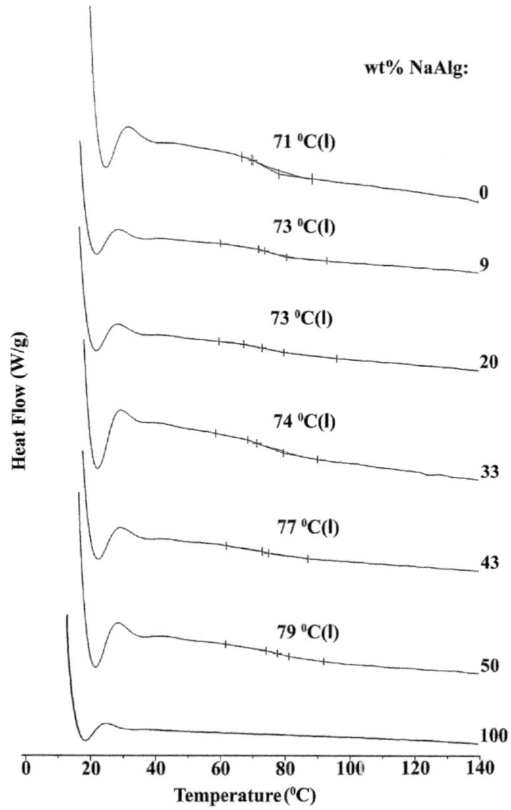

FIGURE 15.5

Differential scanning calorimetry curves of the poly(vinyl alcohol)/NaAlg blends [10].

of the homopolymers, which indicates that there is a specific interaction between NaAlg and PVA in the PVA/NaAlg blends. Differential scanning calorimetry (DSC) curves of PVA, NaAlg, and PVA/NaAlg blends are shown in Fig. 15.5. The observed glass transition temperatures (T_g) are depicted by arrows in this figure. It is well known that the miscible blends show a single T_g intermediate to those of the pure components. For pure PVA and all PVA/NaAlg blends, a single T_g value is clearly observed, and each observed T_g depends on the blend composition (Fig. 15.5).

This suggests that the blends are miscible for all compositions. On the other hand, in the same figure, it is difficult to detect the T_g of pure NaAlg. However, it can be seen that the T_g is low for the PVA homopolymer ($T_g = 71°C$) compared with that of the PVA/NaAlg blends. However, the T_g for these blends increases from 72 to 79°C with the increase of the NaAlg content in the blend. This behavior is also attributed to the hydrogen bond formation between PVA and NaAlg units of these blends. The mobility of the chains decreases because of the presence of hydrogen bonds that acts as cross-links.

15.2 ALGINATE BLENDS WITH POLY(*N*-VINYL-2-PYRROLIDONE)

PVP has excellent transparency and biocompatibility. It has been used as a main component of temporary skin covers and wound dressings. PVP is also a very good choice for making hydrogels, but the hydrogel of PVP itself is of limited applicability because of its inferior mechanical properties. Therefore, PVP blending with other polymers has played a significant role in biomedical materials [13]. PVP is also a water-soluble polymer and the hydrogen bonding may form via the interaction of the carbonyl groups of PVP with the hydroxyl groups of NaAlg. Because of these properties, PVP is a suitable candidate to form a blend with NaAlg. By casting from aqueous solutions of PVP and NaAlg, transparent, colorless, and flexible films are generated [14]. These properties are a rough combination of those of the pure components. The results of the stress—strain measurements of PVP, NaAlg, and PVP/NaAlg blend films are shown in Fig. 15.6. When the mechanical properties of the pure components are compared with those of their blends, the Young modulus and the stress of the blends decrease and the elongation at break increases with the increase in the NaAlg content. This may be due to the formation of the hydrogen bonding between NaAlg and PVP components.

Fig. 15.7 shows the FT-IR spectra of the pure and blend films in the wavelength range of 1000—2000 cm^{-1}. The characteristic bands in the spectrum of the blends are observed at 1656 and 1025 cm^{-1} for PVP and NaAlg components, respectively. As also shown in Fig. 15.7, there are a 15 cm^{-1} shift to a lower wavenumber for the carbonyl stretching band of PVP and a shoulder at 1604 cm^{-1} for the asymmetric carboxylate stretching vibration band for NaAlg of the blends compared with the pure components.

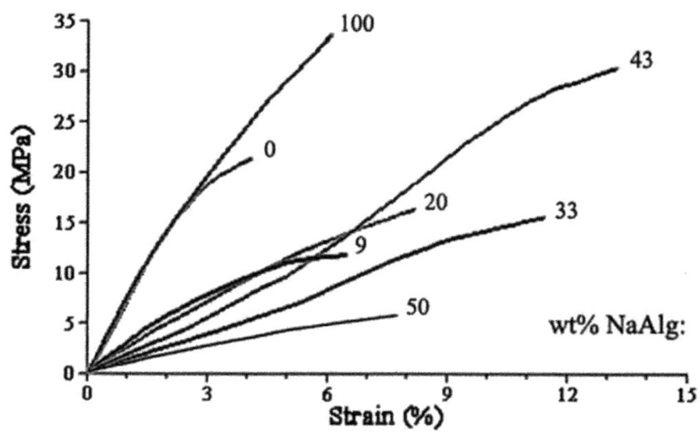

FIGURE 15.6

Stress—strain curves of the poly(*N*-vinyl-2-pyrrolidone)/NaAlg blend films [14].

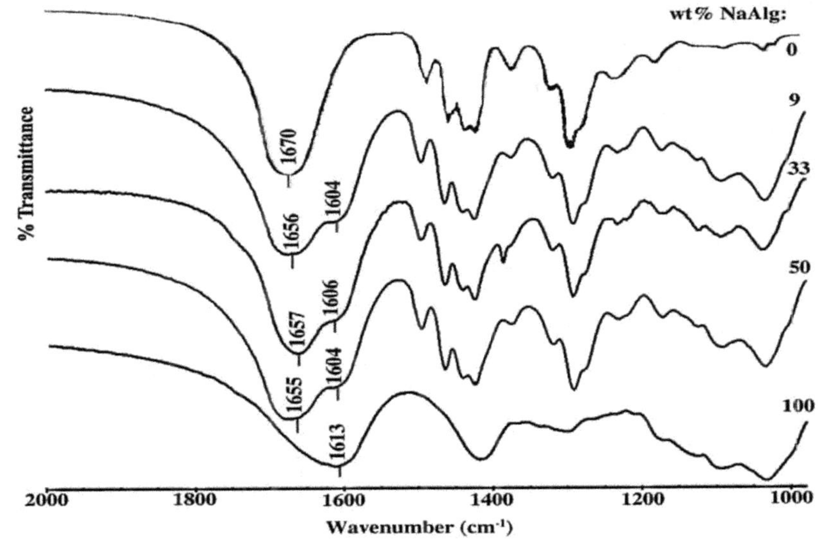

FIGURE 15.7

Fourier transform infrared spectra of poly(*N*-vinyl-2-pyrrolidone)/NaAlg blend films (1000–2000 cm^{-1}) [14].

This strongly supports the idea that a hydrogen bond can form between the carbonyl groups of PVP and the hydroxyl groups of NaAlg (Fig. 15.8).

TG and DTG curves of PVP, NaAlg, and PVP/NaAlg blends are shown in Fig. 15.9. The mass loss of pure PVP begins at 411°C and reaches a maximum at

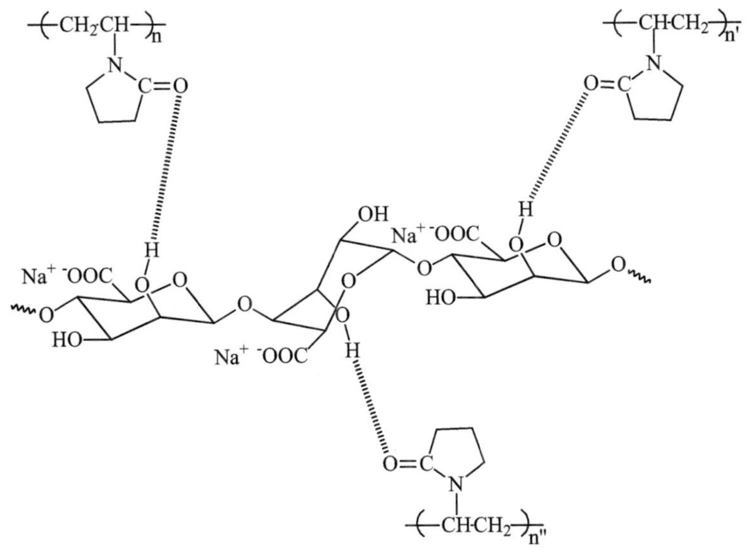

FIGURE 15.8

Possible interactions in the poly(*N*-vinyl-2-pyrrolidone)/NaAlg blends [14].

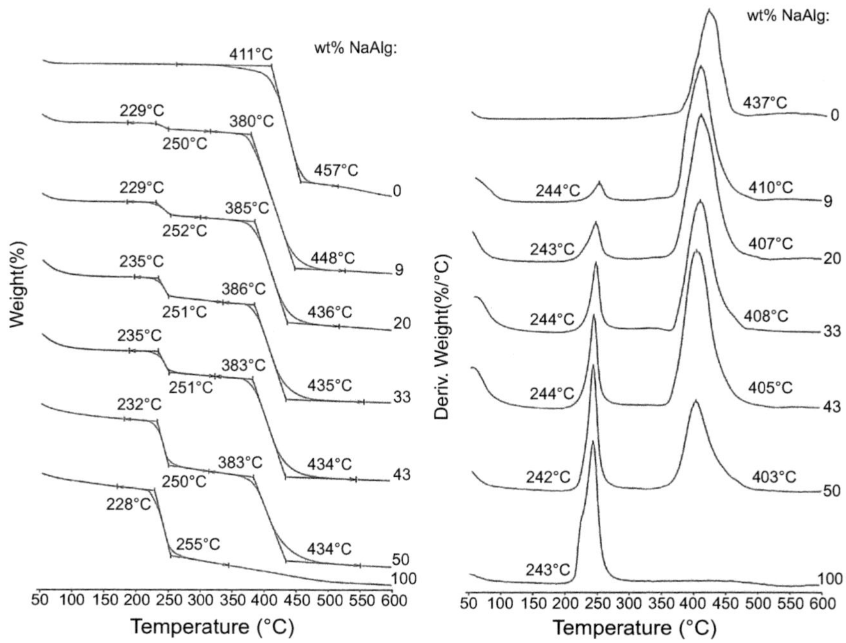

FIGURE 15.9

Typical weight loss (left) and derivative of weight loss (right) curves of the poly(*N*-vinyl-2-pyrrolidone)/NaAlg blends [14].

435°C. However, the mass loss of pure NaAlg starts at 229°C and reaches a maximum at 243°C. The PVP shows better thermal stability than NaAlg. The TG curves of both NaAlg and PVP also indicate one reaction stage, which is reflected as a single peak in the DTG curves. However, PVP/NaAlg blends degrade in two steps. This is evidenced by the appearance of distinct peaks in DTG thermograms. Two distinct reaction peaks at around 243 and 400°C are identified in the DTG thermograms of PVP/NaAlg blends. These peaks are attributed to the thermal degradation of NaAlg and PVP, respectively. This behavior shows that the thermal degradation reaction mechanism of PVP/NaAlg blends is the same as that of pure PVP and NaAlg homopolymers.

DSC curves of PVP, NaAlg, and PVP/NaAlg blends are shown in Fig. 15.10. The observed values of T_g are depicted by arrows in this figure. In the literature, several conflicting values of T_g of PVP were found, ranging from 54 to 175°C [15–20]. These may be attributed to the large influence of sorbed moisture due to the hygroscopic nature of the material, as will be shown by present measurements. Hence for hydrophilic polymers such as PVP, careful drying at the highest possible temperature, preferably under a nitrogen blanket, is necessary to obtain correct T_g values. To completely remove the moisture, PVP and PVP/NaAlg blends are maintained at 120°C for 3 h under nitrogen flux before DSC measurements. The miscible blends show a single T_g

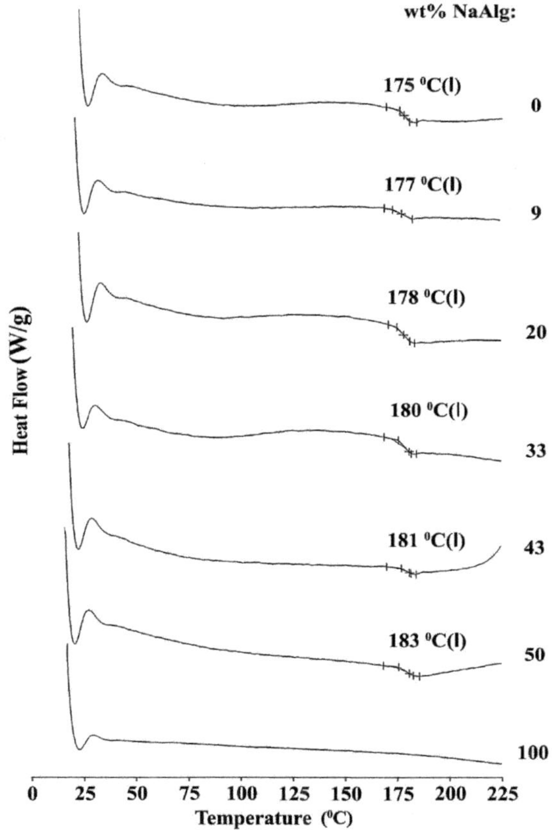

FIGURE 15.10

Differential scanning calorimetry curves of the poly(vinyl alcohol)/NaAlg blends [14].

intermediate to those of the pure components. For pure PVP and all PVP/NaAlg blends, a single T_g value is clearly observed, and each observed value of T_g depends on the blend composition (Fig. 15.10). This suggests that the blends are miscible for all compositions. On the other hand, in the same figure it is difficult to detect the T_g of pure NaAlg. However, it can be seen that the T_g is low for the PVP homopolymer ($T_g = 172°C$) compared with that of the PVP/NaAlg blends. However, the T_g for these blends were slightly increased from 176 to 181°C with the increase of NaAlg content in the blend. This behavior is also attributed to the hydrogen bond formation between PVP and NaAlg units of these blends.

15.3 SURFACE PROPERTIES OF ALGINATE BLENDS WITH PVA AND PVP

The surface characterization of the PVP/NaAlg and PVA/NaAlg blend films gives an idea of the thermodynamic work of adhesion between the natural and synthetic polymer chains. Also, it helps with the selection of the appropriate pairs. As known from acid–base adhesion theory, when the surface acidity of one component is high and the surface basicity of the other component is also high, strong adhesion between the components and better compatibility occur [21]. However, in actual practice, thermodynamic-based adhesion calculations are not sufficient for choosing a suitable pair of natural and synthetic polymer components. An experimental investigation of the adhesion and compatibility is also required because of the different chain diffusion and viscoelastic behaviors of the natural and synthetic polymer components. Therefore, the surface characterization of the polymeric component is another important guide for this pair selection process [21,22].

Contact angles are characteristic constants of liquid–solid systems and provide valuable information on the surface energy of solids. The free energy components of the surface energy of solids in contact with a liquid can be represented by the following complete Young equation comprising both apolar and polar interactions [23,24]:

$$\gamma_L(1+\cos\theta) = 2\left(\sqrt{\gamma_s^{LW}\gamma_L^{LW}} + \sqrt{\gamma_s^{-}\gamma_L^{+}} + \sqrt{\gamma_s^{+}\gamma_L^{-}}\right) \tag{15.1}$$

In this equation, the superscript LW represents the apolar Lifshitz–van der Waals forces. Subscripts L and s refer to liquid and solid, respectively. γ_L is the total surface free energy of liquid, θ is the contact angle of a liquid drop on solid surface, and γ_L^{LW} and γ_s^{LW} are apolar Lifshitz–van der Waals components of liquid and solid, respectively. γ_s^{+} and γ_L^{+} are electron-acceptor surface free energy components of solid and liquid, respectively. γ_s^{-} and γ_L^{-} are electron-donor surface free energy components of solid and liquid, respectively. Here, γ_L^{LW} can be determined first with an apolar liquid (paraffin). For an apolar liquid, $\gamma_L^{+} = \gamma_L^{-} = 0$ and $\gamma_L^{LW} = \gamma_L$; therefore, the last two terms of the right-hand side of Eq. (15.1) become 0, so this equation can be written in the following form:

$$\gamma_L(1+\cos\theta) = 2\sqrt{\gamma_s^{LW}\gamma_L^{LW}} \tag{15.2}$$

Consequently, the γ_s^{LW} value can be determined directly from the apolar liquid (paraffin) contact angle. When two polar liquids are used, two equations of the form of Eq. (15.1) constitute a set of two simultaneous equations that can be solved for the two unknown properties of the solid, γ_s^{-} and γ_s^{+} followed by averaging the γ_s^{-} and γ_s^{+} results to a single value. Water-drop contact angle values are always used in these sets to avoid large discrepancies between simultaneous equation solutions [25].

The contact angle values (θ) of paraffin oil, water, glycerol, ethylene glycol, and formamide drops on the PVP/NaAlg and PVA/NaAlg blend films are listed in Table 15.1 [22]. These data were evaluated to determine the surface free energy components of PVP/NaAlg and PVA/NaAlg blend films with the Van Oss–Good method through Eqs. (15.1) and (15.2). The surface free energy component values of the liquids used were taken from Ref. [25] and are given in Table 15.2 [22].

Table 15.1 Contact Angle (θ) Values for Poly(*N*-vinyl-2-pyrrolidone) (PVP)/NaAlg and Poly(vinyl alcohol) (PVA)/NaAlg Blend Films [22]

Blend	NaAlg (wt%)	Paraffin	Water	Glycerol	Ethylene Glycol	Formamide
PVP	0	11.8 ± 0.4	50.0 ± 0.4	54.6 ± 0.8	68.0 ± 0.4	65.9 ± 0.4
PVA	0	15.9 ± 0.6	53.9 ± 0.4	54.0 ± 0.6	47.1 ± 0.7	35.0 ± 0.5
NaAlg	100	46.1 ± 0.4	78.0 ± 0.5	68.0 ± 0.8	47.2 ± 0.4	25.0 ± 0.6
PVP/NaAlg	9	15.8 ± 0.4	55.0 ± 0.0	50.7 ± 0.5	64.1 ± 0.4	58.0 ± 0.6
PVP/NaAlg	20	21.0 ± 0.6	61.8 ± 0.4	54.6 ± 0.5	58.2 ± 0.4	45.0 ± 0.0
PVP/NaAlg	33	36.3 ± 0.5	65.7 ± 0.5	58.1 ± 0.3	54.0 ± 0.6	37.0 ± 0.3
PVP/NaAlg	43	40.0 ± 0.0	69.0 ± 0.6	60.9 ± 0.4	50.0 ± 0.0	34.1 ± 0.4
PVP/NaAlg	50	43.5 ± 0.5	71.1 ± 0.3	65.0 ± 0.0	48.3 ± 0.5	30.0 ± 0.0
PVA/NaAlg	9	20.9 ± 0.4	62.7 ± 0.6	57.6 ± 0.5	47.2 ± 0.4	35.0 ± 0.0
PVA/NaAlg	20	28.0 ± 0.6	69.2 ± 0.7	61.8 ± 0.4	47.1 ± 0.3	35.0 ± 0.0
PVA/NaAlg	33	31.0 ± 0.7	71.0 ± 0.5	63.8 ± 0.4	47.2 ± 0.4	35.1 ± 0.3
PVA/NaAlg	43	35.0 ± 0.0	73.9 ± 0.4	65.0 ± 0.0	47.2 ± 0.5	35.0 ± 0.0
PVA/NaAlg	50	41.8 ± 0.7	75.7 ± 0.5	66.8 ± 0.4	47.2 ± 0.4	35.0 ± 0.0

Table 15.2 Surface Free Energy Component Values of the Liquids (mJ/m^2)

Liquid	γ_s^{LW}	γ_s^+	γ_s^-	γ_s^{TOT}
Paraffin	28.9	0	0	28.9
Water	21.8	25.5	25.5	72.8
Glycerol	34.0	3.92	57.4	64.0
Ethylene glycol	29.0	1.92	47.0	48.0
Formamide	39.0	2.28	39.6	58.0

The data were taken from Van Oss CJ. Interfacial forces in aqueous Media. New York: Marcel Dekker; 1994.

The LW surface free energy component of a solid material shows the contribution of apolar interactions to the surface free energy of the solid. This component forms the greater part of the total surface free energy of a material. The LW components of the surface free energies of the PVP/NaAlg and PVA/NaAlg blend films are calculated with Eq. (15.2) and the paraffin-oil-drop contact angle values are given in Table 15.3. The γ_s^{LW} values of PVP, PVA, and NaAlg are 28.29, 27.80, and 20.72 mJ/m^2, respectively. As expected, the γ_s^{LW} values of both PVP/NaAlg and PVA/NaAlg blend films decreased with an increase of the weight percentage of NaAlg in the blend. This can be explained by the fact that the alginate units of the PVP/NaAlg and PVA/NaAlg blends have a higher dispersion force than both the PVP and PVA units. It is well known that there is an inverse relationship between the intermolecular distance and the dispersion forces present. The γ_s^+ and γ_s^- values

Table 15.3 Surface Free Energy Components of Poly(*N*-vinyl-2-pyrrolidone) (PVP)/NaAlg and Poly(vinyl alcohol) (PVA)/NaAlg Blend Films (mJ/m^2) [22]

Blend	NaAlg (wt%)	γ_s^{LW}	γ_s^+	γ_s^-	γ_s^{TOT}
PVP	0	28.29	1.15	51.00	43.61
PVA	0	27.80	2.08	28.35	43.16
NaAlg	100	20.72	7.71	5.40	33.62
PVP/NaAlg	9	27.82	1.36	36.65	41.94
PVP/NaAlg	20	27.01	1.87	22.23	39.90
PVP/NaAlg	33	23.56	3.74	16.20	39.13
PVP/NaAlg	43	22.53	4.78	12.23	37.82
PVP/NaAlg	50	21.51	5.72	10.58	37.07
PVA/NaAlg	9	27.03	2.84	17.81	41.25
PVA/NaAlg	20	25.62	3.67	11.64	38.69
PVA/NaAlg	33	24.92	3.99	10.30	37.74
PVA/NaAlg	43	23.91	4.64	7.99	36.09
PVA/NaAlg	50	22.01	5.50	7.00	34.42

of these blends were calculated with the general contact angle equation (Eq. 15.1), into which previously found γ_s^{LW} values were inserted. Water/glycerol, water/ethylene glycol, and water/formamide sets were simultaneously solved, and the results are also given in Table 15.3 [22].

As shown in this table, the most important difference was found to be in the surface basicity, γ_s^-, or the electron donor coefficient of the surface free energies of the PVP and PVA homopolymers. The PVP and PVA homopolymers have a basic character ($\gamma_s^- = 51.00\ \text{mJ/m}^2$ for PVP and $\gamma_s^- = 28.35\ \text{mJ/m}^2$ for PVA), whereas the surface of NaAlg has a slightly acidic character ($\gamma_s^+ = 7.71\ \text{mJ/m}^2 > \gamma_s^- = 5.40\ \text{mJ/m}^2$). When the γ_s^- value of the PVP homopolymer is compared with that of PVA, PVP is seen to have a higher γ_s^- value. This could be explained by the fact that PVA has the highest tendency of hydrogen bond formation because of the interaction between hydroxyl groups. On the other hand, the γ_s^- values of both the PVP/NaAlg and PVA/NaAlg blend films decrease with the increase of the weight percentage of NaAlg in the blends. The total surface energies (γ_s^{TOT}) of the PVP/NaAlg and PVA/NaAlg blend films are calculated with Eq. (15.3) and are given in Table 15.3.

$$\gamma_s^{TOT} = \gamma_s^{LW} + 2\sqrt{\gamma_s^+ \gamma_s^-} \tag{15.3}$$

As shown in this table, the decrease of the γ_s^- values is also reflected in the decrease of the γ_s^{TOT} of the PVP/NaAlg and PVA/NaAlg blend films with the increase of the NaAlg content. In addition, both PVP and PVA have a higher γ_s^{TOT} value than that of NaAlg (Table 15.3). This is why NaAlg is enhanced on the surfaces of all blend systems. It is well known that some surface enrichment of the low surface free energy component takes place in blend systems [26]. The total surface free energy (γ_s^{TOT}) of the blends is calculated via the following equation, assuming an additivity rule [22]:

$$\gamma_s^{TOT}(\text{Blend}) = \varphi_1 \gamma_s^{TOT}(1) + \varphi_2 \gamma_2^{TOT}(2) \tag{15.4}$$

where φ_1 and φ_2 denote the weight percentages of PVP or PVA and NaAlg in the blend and $\gamma_s^{TOT}(1)$ and $\gamma_s^{TOT}(2)$ are the total surface energies of pure PVP or PVA and NaAlg, respectively. The ratios of the experimental total surface energy values to the calculated total surface energy values $\gamma_{Exp}/\gamma_{Theo}$ first increased and then remained almost constant with the increase of NaAlg in the blend and are plotted in Fig. 15.11. As shown in this figure, the $\gamma_{Exp}/\gamma_{Theo}$ values of these blends are always higher than 1. This is mainly due to an increase in the component with the low surface free energy of the blends at the surface. Fig. 15.11 shows also that the enrichment of NaAlg is higher at the PVA/NaAlg surface in comparison with the PVP/NaAlg blends. This behavior is attributed to the presence of strong acid–base interactions between the PVP and NaAlg chains ($\gamma_s^- = 51.00\ \text{mJ/m}^2$ for PVP and $\gamma_s^+ = 7.71\ \text{mJ/m}^2$ for NaAlg) in comparison with the PVA/NaAlg pair.

Atomic force microscopy (AFM) is used in taping mode to determine the surface morphologies of polymer films. The AFM images of PVP, PVA, PVP/NaAlg, and PVA/NaAlg films are given in Fig. 15.12. The surface morphologies of the PVP

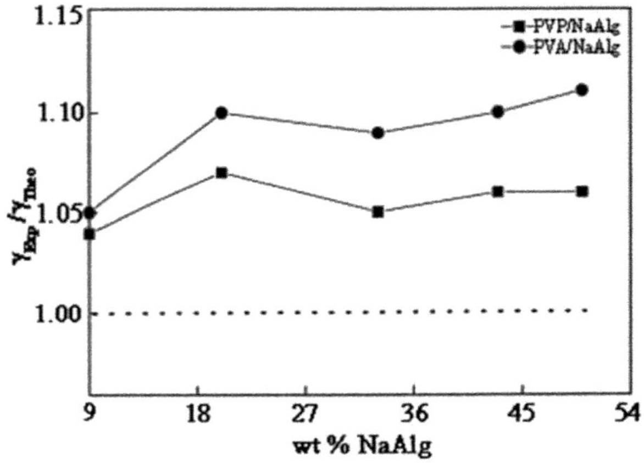

FIGURE 15.11

Changes in the ratios of the blends with the weight percentage of NaAlg [22]. *PVA*, poly(vinyl alcohol); *PVP*, poly(*N*-vinyl-2-pyrrolidone).

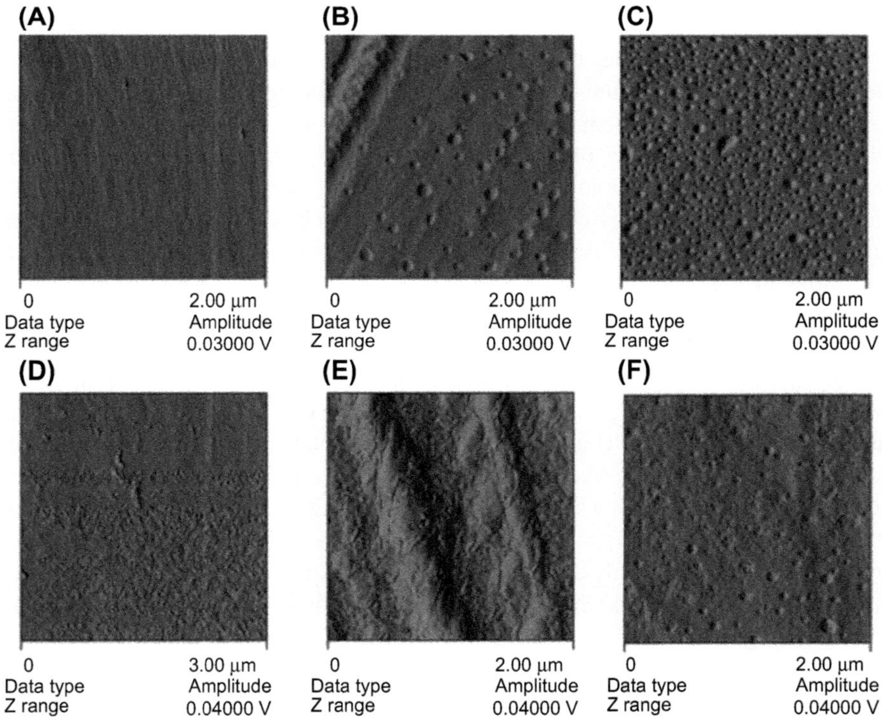

FIGURE 15.12

Atomic force microscopy images of (A) pure PVP, (B) PVP/NaAlg (9 wt% NaAlg), (C) PVP/NaAlg (50 wt% NaAlg), (D) pure PVA, (E) PVA/NaAlg (9 wt% NaAlg), and (F) PVA/NaAlg (50 wt% NaAlg) [22].

and PVA films and their blends with NaAlg are different. The PVP and PVA films have relatively smooth and flat surfaces, whereas their blends show two separated phases. This observation agrees very well with the contact angle observations.

REFERENCES

[1] Xiao C, Lu Y, Liu H, Zhang L. J Macromol Sci Pure Appl Chem 2000;37:1663.
[2] Maria GC. Polym Int 1997;43:55.
[3] Relleve L, Yoshii F, Rosa DA, Kume T. Angew Makromol Chem 1999;273:63.
[4] Fuller CS, MacRae RJ, Walther M, Cameron RE. Polymer 2001;42:9583.
[5] Miya M, Iwamoto R. J Polym Sci Part B Polym Phys 1984;22:1149.
[6] Moe ST, Skjak-Brek G, Ichijo H. J Appl Polym Sci 1994;51:1773.
[7] Hirano S, Mizutani C, Yamaguchi K, Miura O. Biopolymers 1978;17:805.
[8] Yuk SH, Cho SH, Lee HB. J Control Release 1995;7:69.
[9] Hari PR, Chandy T, Sharma CP. J Appl Polym Sci 1996;59:1795.
[10] Caykara T, Demirci S. J Macromol Sci Pure Appl Chem 2006;43:1121.
[11] Liang CX, Hirabayashi K. J Appl Polym Sci 1992;45:1937.
[12] Xiao C, Lu Y, Liu H, Zhang L. J Macromol Sci Pure Appl Chem 2000;A37(12):16631.
[13] Vijayasekaran S, Chirila TV, Hong Y, Tahija SG, Dalton PD, Constable IJ, McAllister IL. J Biomater Sci Polym Ed 1996;7:685.
[14] Caykara T, Demirci S. Polym Plast Technol Eng 2007;46:737.
[15] Brandrup J, Immergut EH. Polymer handbook. New York: Wiley; 1975. pp. III−155.
[16] Kinsinger JB, Panchak JR, Kelso RL, Bartlett JS, Graham RK. J Appl Polym Sci 1965; 9:429.
[17] Griffin-Lewis O. Physical constants of linear homopolymers. Berlin: Springer Verlag; 1965. p. 128.
[18] Van Krevelen DV, Hoftijzer PJ. Properties of polymers, correlations with chemical structure. Amsterdam: Elsevier; 1972. p. 99.
[19] Sidelkovskaya FP. The chemistry of *N*-vinylpyrrolidone and its polymers. Moscow. 1970. p. 99.
[20] Nielsen LE. Mechanical properties of polymers. New York: Reinhold; 1962. p. 21.
[21] Doğan M, Eroğlu MS, Erbil HY. J Appl Polym Sci 1999;74:2848.
[22] Caykara T, Demirci S, Eroğlu MS, Güven O. J Polym Sci Part B Polym Phys 2006;44: 426.
[23] Van Oss CJ, Good RJ. Langmuir 1992;8:2877.
[24] Erbil HY. Langmuir 1994;10:276.
[25] Van Oss CJ. Interfacial forces in aqueous media. New York: Marcel Dekker; 1994.
[26] Caykara T, Eroğlu MS, Güven O. J Appl Polym Sci 1998;69:1551.

CHAPTER

Alginate-Poly(Ethylene) Glycol and Poly(Ethylene) Oxide Blend Materials

16

Fatima Zia[1], Muhammad N. Anjum[1], Muhammad J. Saif[1], Tahir Jamil[2], Kausar Malik[2], Shazia Anjum[3], Ismat BiBi[3], Muhammad A. Zia[4]

Government College University Faisalabad, Faisalabad, Pakistan[1]; University of the Punjab, Lahore, Pakistan[2]; Islamia University Bahawalpur, Bahawalpur, Pakistan[3]; University of Agriculture Faisalabad, Faisalabad, Pakistan[4]

16.1 INTRODUCTION

The academic and applied/industrial areas of research signify the importance of polymer blend technology. The development in polymer and material science leads to effective combination of several polymeric materials using various methods for a wide range of applications. These blend materials offer possibilities to produce innovative materials with tunable properties, low cost, and improved processing conditions that can enhance the properties of the individual polymers in the blends [1–6]. However, morphological and functional properties of blends depend on the miscibility and compatibility among the reacting polymers, which control the dispersion and interfacial interaction between them [7,8]. The primary interactions found in polymer blends are secondary forces with no covalent interactions. The most common secondary interaction existing among blends are hydrogen bonding, pi-electrons, ionic and dipole, and charge transfer complex. Therefore, the selection of reacting polymers is a matter of great concern and several methods have been used to evaluate the interaction between blending polymers and then their miscibility [9–15].

Concern has been raised regarding the negative environmental impacts of synthetic polymers. Strategies have been developed by researchers through the modification of materials via blending with natural materials, and compositing or designing of environmentally benign synthetic polymers. The introduction of water-soluble synthetic polymers has overshadowed the outbreaks regarding contamination caused by synthetic polymers because of their nontoxic and environmental friendly character. However, the natural or synthetic polymers used in blends are not at all times able to fulfill all the demands of the multifaceted material. Ultimately, researchers blended the two categorized polymers to take advantage of the desired properties of each group of polymer. These blends are structurally and functionally modified materials, having biodegradability and biocompatibility. For example, natural polymers such as kappa carrageenan, iota carrageenan,

Algae Based Polymers, Blends, and Composites. http://dx.doi.org/10.1016/B978-0-12-812360-7.00016-1

hydroxypropyl methylcellulose, chitosan, alginate, keratin, silk fibroin, collagen, and elastin are blended with synthetic polymers such as poly(*N*-vinyl-2-pyrrolidone), poly(*N*-vinyl-2-pyrrolidone), polycaprolactone, poly(ethylene oxide), poly(vinyl alcohol) (PVA), polyurethanes, polyethylene terethalate, and poly lactic acid [16–38].

Water-soluble synthetic polymers, such as poly(ethylene) oxide (PEO), poly(ethylene) glycol (PEG), and PVA, are nonionic, nontoxic, and biocompatible and are successfully used in blends. Water is sometimes proved to be a good solvent for native polymers but not always; therefore, water-soluble polymers are used in blends with native polymers and alginate is one of such natural polymer [39–48]. PEG and PEO are well studied by physical, morphological, micelle formation, solution dynamics, and rheological methods [49–54]. Similarly, alginate (a polyelectrolyte having a rigid molecular chain of guluronnic and mannuronic blocks) is also established as a cheap, safe (nontoxic), biodegradable/compatible, and hydrophilic biopolymer for tissue engineering applications [55–59]. The colloidal properties of alginate makes it a thickening, film forming, suspension and gel producing, emulsion stabilizing agent, and therefore it is considered for various applications [60,61] in food industry. The chapter focuses on the alginate-based PEO and PEG blend membrane, nanofiber, composite, microsphere/beads, etc. and their uses particularly for biomedical or drug delivery applications.

16.2 POLY(ETHYLENE) OXIDE AND POLY(ETHYLENE) GLYCOL

Polyethylene oxides (PEOs) and polyethylene glycols (PEGs) are polymers with the subunit C–O–C as shown in Fig. 16.1. These synthetic polyethers are easily accessible and available in a range of molecular weights. These polymers are soluble in

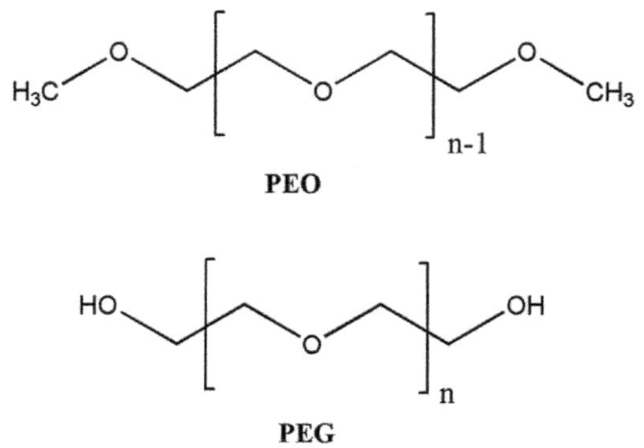

FIGURE 16.1

Structure of poly(ethylene) oxide (PEO) and poly(ethylene) glycol (PEG).

both water and organic solvents (for instance, ethanol, methylene chloride, acetone, toluene, and chloroform) and are amphiphilic in nature [62,63].

PEOs are water soluble, nonionic, hydrophilic, electrically and thermally stable, inert to many organic reagents, and nontoxic polymers synthesized commercially through catalytic polymerization of ethylene oxide using various catalytic systems. The average molecular weight of PEOs ranges from as low as 200 to $5 \times 10^6/7 \times 10^6$, represented by the formula: $(OCH_2CH_2)_n$, where n denotes the average number of oxyethylene groups. However, PEO products exist as viscous liquids or waxy solids with molecular weights below 25,000 and are categorized as PEGs.

PEOs have characteristic features such as thickening, flocculent, dispersing, lubrication, binding, sustained release, and water-holding properties. Chemically, PEO resins are found as hydrogels, thermoplastic compounds, hydrodynamic drug reduction systems, and association complexes. Morphologically, PEOs have a sharp crystalline melting point (M_p) at about 65°C and maintain a high level of crystallinity far above the melting point. PEOs are biocompatible, non immunogenic, and have established biodegradable (aerobically and anaerobically) polymers. Extensive testing on animals showed that the PEOs in the dry or aqueous form have a very low and manageable toxicity. The most favorable applications of PEOs are in textile impregnation, ink jet printing media, laundry detergents, superabsorbents, flocculation and coagulation, food additive, and in biomedical field. Chemical modification of PEO resins by grafting, block polymerization, branching, and reactive extrusion enhances their processability in the melt. Modified PEOs and water-insoluble polymers/native polymer blends have been proposed and synthesized successfully with desirable properties [64–72].

PEG has an $O(CH_2)_2$ monomer unit, characterized by both a polar oxygen atom and a nonpolar $(CH_2)_2$ group, and is usually prepared by anionic chain transfer polymerization. PEG is thus soluble in a wide variety of polar and nonpolar solvents. PEGs are viscous and colorless liquids, whereas higher molecular-weight PEGs are waxy, white solids with melting points proportional to their molecular weights to an upper limit of about 67°C. The terminal −OH groups of PEG present covalent attachment sites for molecules. Per molecule of PEG (M_w 3350) has the capacity to create hydrogen bonds with 100 molecules of water. PEG (molecular weight from few hundreds to 20,000) used enormously in the biomaterials for drug delivery applications and also used extensively as a hydrophobic drug carrier to improve the aqueous solubility or dissolution features. It is currently used as plasticizer in the polymeric materials as well [73–81].

16.3 ALGINATE-BASED POLY(ETHYLENE) GLYCOL MATERIALS

Immobilized alginate system proves to be suitable for mammalian cell preservation under mild processing conditions. Alginate meets the principal requirements (easy modification, sterilization, and storage) to be used as matrix material; however, the high-density network in the alginate matrix limits cell growth. PEG, acrylic acid, and cystein are conjugated to alginate backbone so as to improve alginate characteristics. Poly-L-lysine and PEG-based alginate entrapment systems with adequate pore size have been reported to enhance the density of growing cells [82–89]. Alginate gels work as a primary material in protein delivery systems, whereas PEG-coated membranes show minimal toxicity, protein resistance and conservation of biological properties, which assist to increase the biocompatibility of polymers [90–92]. PEG is used as cross-linker, because of its easy functionalization, established cell compatibility, and demonstrated use in islet encapsulation [93–96]. Taking the advantageous interaction between alginate and PEG, a network gel entrapment system was developed within alginate beads to offer an environment for high-density growth of mammalian cells [89]. The PEG–alginate system, in contrast to the usual alginate system, allowed cell growth in pores; consequently the alginate matrix worked as a structural matrix for cells. The PEG contents were varied between 10 and 20 wt%, which controlled the pore size, providing radially well-defined areas for cell growth. This concentration variation of PEG provided radial diffusion paths for products in the neighboring alginate matrix without the need for critical changes in the process. The illustrated monoclonal antibody production rate increased 1.6-fold and 3-fold for the slower growing batch-immobilized culture and immobilized perfusion culture, respectively, compared with the basic suspension batch culture.

Alginate–graft–PEG copolymer was prepared for gel entrapment designs and microencapsulation techniques [82]. The covalent grafting of functionalized PEG (obtained on reacting PEG with succinic anhydride using pyridine catalyst) on amine-functionalized 2,3-dioctylamine alginate was done via a well-defined series of steps at neutral pH (Fig. 16.2). The grafting degree was tested by selective solvent extraction and the most reliable estimation of unreacted PEG gave a value of 20 wt% when the PEG and alginate were coupled following the carbodiimide chemistry. The PEG-grafted copolymer showed higher water retention than alginate on cross-linking with divalent cations.

A dually cross-linked encapsulation system consisting of a mixture of alginate and PEG–acrylate was developed for the encapsulation of islets of Langerhans [97]. The dually cross-linked gels provided advantageous chemical stability of the subsequent gels owing to the covalent bonds that retained the integrity of the gels than the conventional alginate-based encapsulation system with reversible ionic linkages mode. Dextrans of known molecular weights were diffused through these gels to understand porosity and permeability. The gels showed in vitro viability, biocompatibility, and insulin secretory function.

Diacrylate/dimethacrylate group functionalization of PEG followed by cross-linking has been established to make nondegradable hydrogels by UV photoinitiation. PEG having acrylate end groups were employed for the ability of acrylate end group to interact with the mucin glycoproteins [98–102]. Alginate-based acrylated-PEG using cystein spacer molecule was also reported as a mucoadhesive polymer [103]. This polymer reflects the ease, strength, and gelation ability of alginate, whereas the mucoadhesive character comes from the characteristics and acrylate functionality of PEG. Alginate-based acrylated-PEG was prepared using thiol

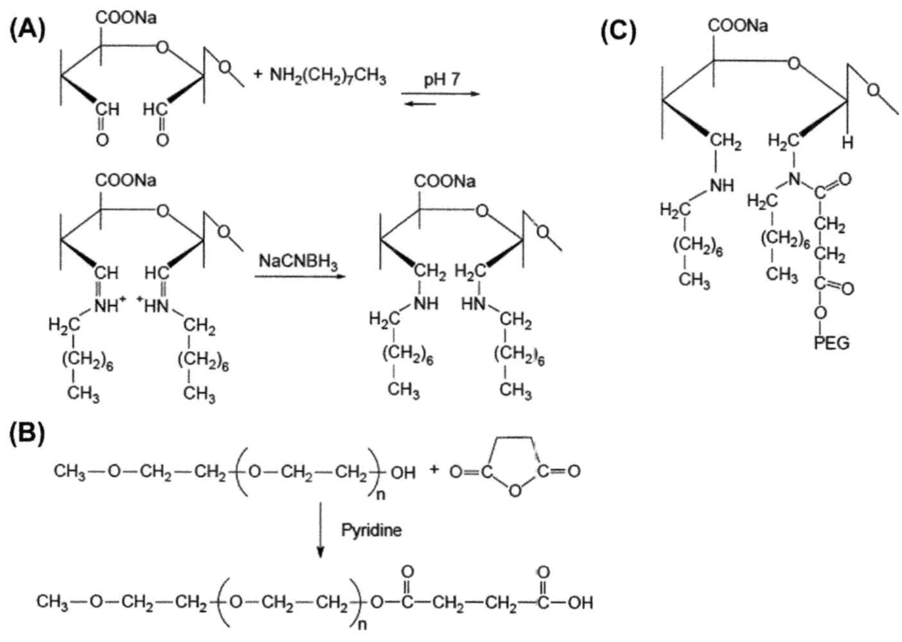

FIGURE 16.2

Scheme (A) synthesis of 2,3-dioctylamine alginate; (B) functionalization of poly(ethylene) glycol (PEG); (C) alginate-graft-PEG copolymer [82].

alginate of two different types [high G-block content (LF200S) and medium G-block content (HF120)]. Thiol-alginate reactions involve 1-ethyl-3-(3-dimethylamino-propyl)-carbodiimide hydrochloride (EDAC) to activate −COOH groups on alginate, followed by replacement of amine group by amide group using L-cystein. Finally thiol alginate is dissolved in tris(2-carboxyethyl)phosphine hydrochloride (TCEP) solution to prevent disulfide interactions and then conjugated with PEG-DA. An increase in adhesion of thiolated alginate-HF120 was noted compared with alginate LF200S sample, which was due to a significantly higher disulfide proportion interacting with mucin glycoproteins. The basic chemistry of Michael-type addition reaction, involving nucleophilic addition of the thiols on the thiolated-alginate to the vinyl group on the acrylate−PEG, shown in Fig. 16.3. Alginate-based acrylated-PEG did not show any cytotoxic effect when cultured in human foreskin fibroblasts for 24 h by means of fluorescent calcein and ethidium homodimer labeling assays followed by fluorescent microscopy imaging [103].

Using the Staudinger ligation, alginate−PEG hydrogel was reported using both ionic and covalent cross-linking phenomena for the cellular encapsulation [104]. This ligation scheme proved to be a core way for the covalent linkage of polymers (e.g., alginate and PEG), to avoid unfavorable effects on cellular functioning and viability when compared with the former illustrated chemoligation schemes, i.e., photopolymerization (which suffered from cell cytotoxicity and the production of free radicals) [105−108].

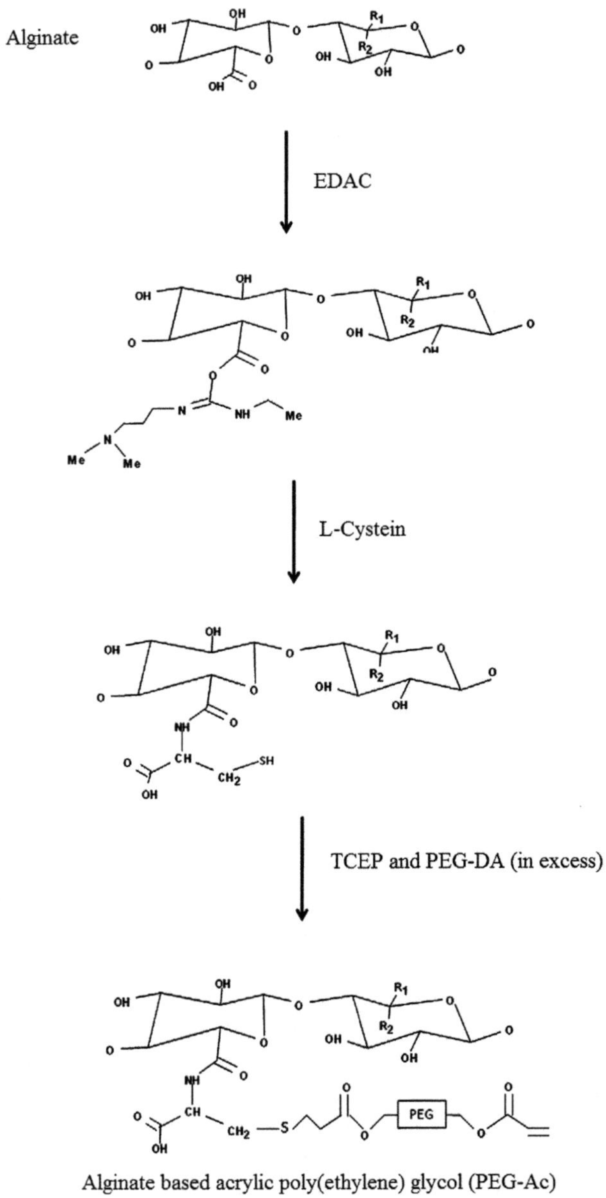

FIGURE 16.3

Schematic illustration of the synthesis of alginate–PEGAc [103].

Covalently cross-linked alginate-PEG (XAlgPEG) hydrogel microbeads involved chemoselective cross-linking between phosphine-terminated PEG and azide-functionalized alginate followed by ionic cross-linking using barium ions (Fig. 16.4). The microbeads displayed higher resistance to osmotic swelling (five-fold decrease in swelling), higher resistance to degradation by chelation solution, and comparable porosity and diffusion than conventional Ba-cross-linked alginate beads. The microbeads of alginate—PEG hydrogels had high cell compatibility and metabolic activity in insulinoma cell lines and rat and human pancreatic islet cells. The viability and function of cells in alginate—PEG gels are comparable with that in Ba-alginate controls. The covalent cross-linking remarkably enhanced gel stability, in addition to maintaining gel integrity and cellular compatibility, thus making the gel useful for a wide range of tissue engineering applications [104].

Ca-alginate—PEG hybrid microspheres were prepared via a one-step extrusion process under physiological conditions [111]. A mixture of Na-alginate and multi-arm vinyl sulfone—terminated PEG was extruded and delivered to Ca^{+2} ions and a thiol cross-linker bath. Vinyl sulfone—terminated PEG undergoes covalent cross-linking in the swiftly gelled spherical Ca—alginate matrix. It was noted that the cross-linked PEG maintained spherical morphology after liquefaction of the Ca-alginate. The stability of PEG gels depend on the stoichiometric ratio thiol/vinyl sulfone, whereas the hydrogel permeability could be adjusted by suitable choice of the arm length of PEG—vinyl sulfone. The concentration of PEG—vinyl sulfone, quality of the storage solvent, and presence/absence of Ca—alginate matrix influenced the swelling property of the gel. However, no significant change in the mechanical resistance was noted after the dissolution of Ca-alginate.

In continuation of this basic theme line, alginate—PEG hybrid microspheres were fabricated for cell microencapsulation of primary human foreskin fibroblasts [112]. The microsphere involved ionotropic gelation of Na-alginate with Ca ions and covalent cross-linking of vinyl sulfone—terminated multiarm-PEG. Microencapsulation of prescribed cells in gel microsphere sustained their viability and proliferation.

Alginate, alginate-graft-PEG, and alginate graft-S-S-arginine-glycine-aspartic acid microspheres were fabricated for the encapsulation and intracellular delivery of a bioactive vascular endothelial growth factor A (VEGFA) into human mesenchymal stem cells (MSCs) [113]. All the microspheres were of moderate size and smooth shape with adjustable surface modification, allowing internalization into human MSCs [114,115]. The process involved mild and fast gelation to synthesize these microspheres, so drug and protein protection is better in the encapsulated microspheres. Covalent linkages developing from the oxidation of two sulfhydryl (SH) groups and the SH terminal group on PEG (SH-PEG-NH_2) result in the chemical conjugation of S-S-arginine-glycine-aspartic acid onto PEG. Disulfide bonds provide reversibility and relative stability in plasma, and turn out to be attractive for drug delivery systems. The three prepared microspheres showed a significant increase in osteogenic differentiation than control. Treatment involving neat microspheres or nonencapsulated VEGFA, regulating osteoprogenitor cell differentiation and bone tissue homeostasis [113].

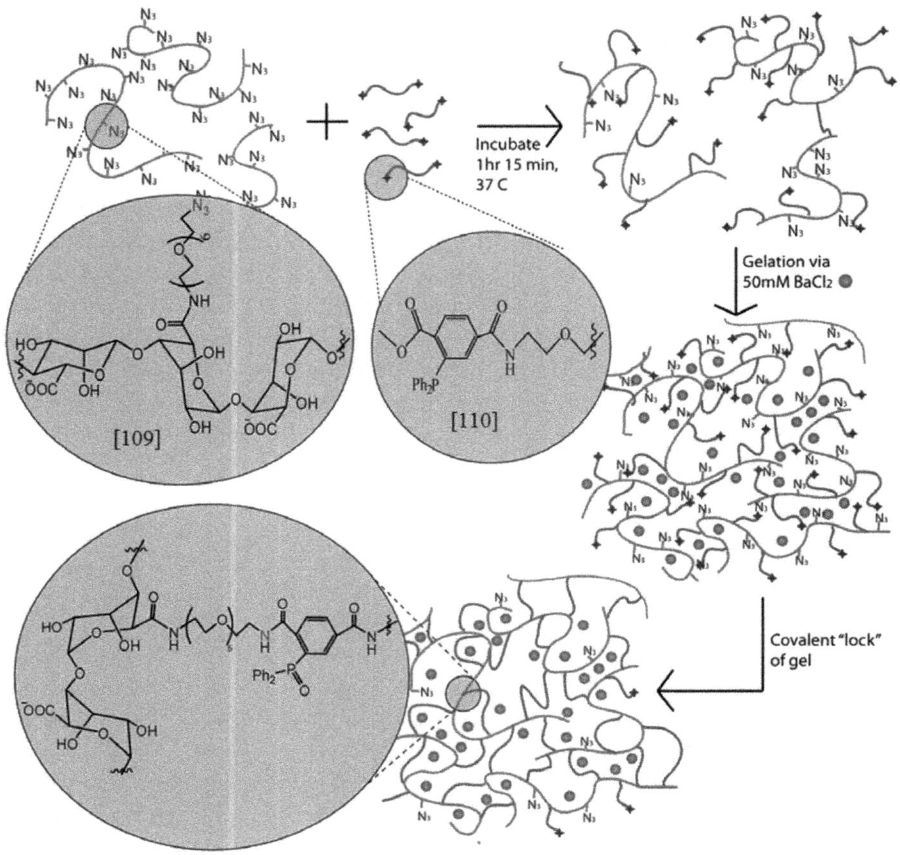

FIGURE 16.4

Schematic representation of XAlgPEG microbead fabrication. Azide-functionalized alginate polymer [109] is premixed with 1-methyl-2-diphenylphosphino-terephthalate (MDT)—poly(ethylene) glycol polymer [110] for 1 h 15 min before cellular incorporation and instantaneous gel formation via subsequent exposure to divalent cations (Ba^{2+}). Additional cross-linking of gel is achieved via Staudinger ligation.

Alginate and PEG fibers were obtained by spinning their solution through spinneret into a coagulating bath containing aqueous $CaCl_2$ and ethanol [116]. The fibers were used for the delivery of different concentrations of salicylic acid drug. With 5 wt% of PEG, the tensile strength of 13.41 cN/tex and breaking elongation at 23.13% of blend fibers were achieved, whereas the water swelling ratio of alginate—PEG blend fibers increased with increasing PEG concentration. The controlled release tests of salicylic acid showed that the increasing content of PEG enhanced the amount of released salicylic acid. However, the drug release rate decreased with an increasing amount of the drug loaded in the fiber, but collectively the release amount is increased. The drug is released at a faster rate at pH 7.4 than at pH 1.0, being concurrently speeded up by a higher ionic strength; thus all the abovementioned properties made alginate/PEG fiber potentially a useful material in drug delivery systems.

16.4 ALGINATE-BASED POLY(ETHYLENE) OXIDE MATERIALS

PEO is found to be suitable for Na-alginate blends, as the hydroxyl (−OH) groups on Na-alginate can interact through the hydrogen bond with the ether oxygen in PEO just as the primary −OH group on cellulose/methylcelluloses forms hydrogen bond with ether oxygen in PEO. Moreover, both polymers are water soluble, so chances of allergic reactions are minimized [17,117].

PEO and Na-alginate blend films were prepared by the solution casting method to investigate the miscibility between two polymers [32]. Fourier transform infrared spectroscopy (FTIR) revealed that the broadening of −OH stretching bands with increasing Na-alginate content was due to hydrogen bond interaction between ether oxygen atoms of PEO and −OH groups of Na-alginate (Fig. 16.5A). Young modulus, stress, and the elongation at break of PEO/Na-alginate were also increased because of this interaction with the increasing Na-alginate content. Surface free energy values decreased with increasing Na-alginate contents. Atomic-force microscopy indicated microscopic phase separation and islands on the surfaces of the PEO/Na-alginate blends (Fig. 16.5B−D) because of the aggregation of Na-alginate, which enhanced with an increase of the Na-alginate (50 wt%) content.

To improve the practical applicability of Na-alginate-based PEO blend films synthesized by the casting method, methyl acrylate monomer and gamma radiation were used [118]. Modification of the films was done through glycerol and mustard oil. The Na-alginate-based films were effectively prepared using 12 kGy of gamma radiation, PEO (10%), mustard oil (20%), glycerol (15%), and methyl acrylate (7%). The thermal, mechanical, and structural properties of the films were evaluated. An outstanding improvement in the tensile strength (30%) and tear strength (67%) is reported; on the other hand, elongation at break was reduced to 40% of the Na-alginate-based films using methyl acrylate and gamma irradiation. These films were found to have considerable decrease in the moisture content, water vapor permeability and also showed the anticipated stability.

Electrospinning is another most viable method used to make electrospun mats comprising ultrafine fibers, characterized by a high specific surface area, aspect ratio, and porosity, which are therefore used favorably for wound dressings, drug delivery carriers, tissue engineering scaffolds, as sensor, filter, and for protective clothing. In 2009, it was reported that almost 100 polymers (natural/synthetic) have been electrospun for fiber formation [9,16,119,120]. Na-alginate is a polyelectrolyte with high conductivity and offers a wide range of viscosity in solution. However, the electrospinning process of this biopolymer is hindered by the repulsive force among the polyanions. Blending of Na-alginate with biocompatible, nontoxic, and nonionic synthetic polymers (PEO/PVA) reduced the repulsive force

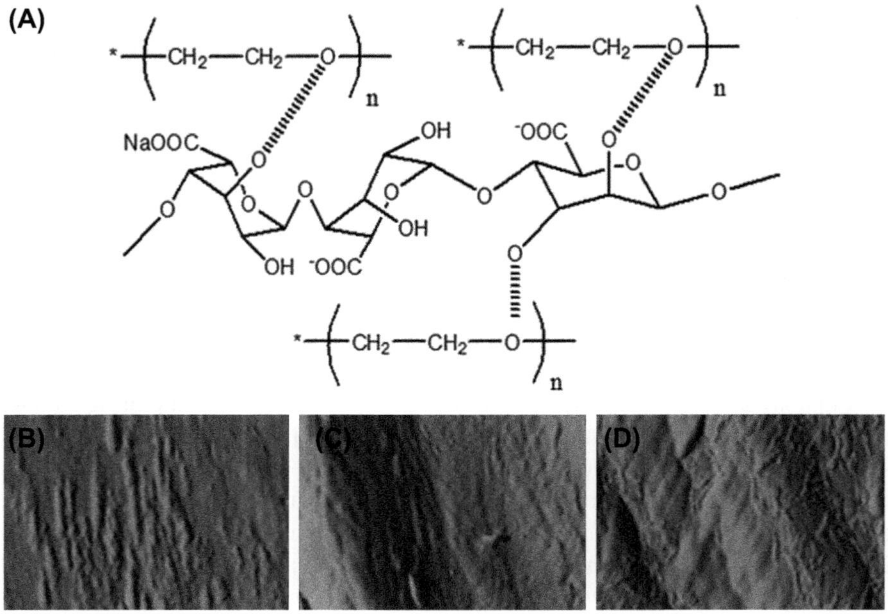

FIGURE 16.5

(A) Hydrogen bond interaction between alginate and PEO; atomic-force microscopy images of (B) pure PEO, (C) PEO/NaAlg (9 wt% NaAlg), and (D) PEO/NaAlg (50 wt% NaAlg) [32]. *PEO*, poly(ethylene)oxide.

among Na-alginate molecules, and consequently permits a favorable electrospinning of Na-alginate blends. Both synthetic polymers can coordinate with metal cations through the −OH group (PVA) and oxygen of the ether group (PEO). With higher molecular-weight PEOs (pseudocrown ethers) the coordination becomes stronger than the −OH of PVA [121−124]. It was found that the conductivity of a 2% solution of Na-alginate is decreased by 16% after the addition of PEO, owing to the interaction between PEO and alginate that decrease the degree of ionization of carboxylate salt [48]. Morphologically, smooth fibers of diameter 228 nm provided the highest tensile strength (4.5 MPa) with Na-alginate/PEO in a 1:1 ratio, whereas the highest elongation at break (8.5%) was shown by spindle-like beads of diameter 114 nm with Na-alginate/PEO in a 3:1 ratio. Water resistance of the blend fibers was improved using hexamethylene diisocyanate and aqueous $CaCl_2$ cross-linker. Electrospinning of Na-alginate aqueous solution is also hindered because of inadequate chain entanglement between Na-alginate chains [48].

Electrospun Na-alginate nanofibers were obtained by blending with PEO using lecithin surfactant [125]. Lecithin, as a natural surfactant, assisted in the fabrication of the uniform nanofiber and was found to reduce the size of alginate beads by decreasing the interfacial tension between the surfactant and external alginate phase [126,127]. It could be noted that inappropriate or less amount of PEO resulted in beaded fiber. Therefore, the best composition was set on Na-alginate/PEO 1:2 and 2: 2 solutions, and 0.3 wt% lecithin. The Na-alginate/PEO blend nanofibers exhibited good water absorption and structural morphology. The fibrous morphology of blend fibers became clearer with increasing alginate contents. The Na-alginate/PEO blend nanofibers showed biocompatible behavior proved by culturing human dermal fibroblasts for 24 h and have potential of wound dressing for exuding wounds.

This chapter investigated the role of PEO and PVA as a carrier for making alginate fiber from its PEO blends. It was already found that the low molecular-weight PEO2 (20 kDa) did not improve the electrospinning of alginate fibers owing to insufficient chain entanglement sites [46]. The higher molecular-weight PEO100 (1000 kDa) facilitated entanglement sites to alginate chains and resulted in a physical network that actually acts as a cage to enclose alginate chains (Fig. 16.6). However, low molecular-weight PEO imparted flexibility to alginate chains. Keeping the alginate/higher molecular-weight PEO at 80/20 (3.5 w/v%) effective chain entanglement for electrospinning is attained.

On the other hand, PEO has the capability of coordinating by its ether group with the Na cation of Na-alginate; so PEO as a carrier polymer exhibited molecular entanglement, triggering electrospinning of alginate fibers. A polymer, as carrier, should deliver the required entanglement i.e., it must be electrospinnable, entangled, and of high molecular weight (M_w) (PEO > 600 kDa) [128]. On the basis of these requirements, we stipulate that the PEO−PEO interaction of the high-molecular-weight entangled PEO is fitted well to the abovementioned requirements and key to facilitate alginate fiber formation during electrospinning of solution. The addition of 1 wt % of Triton X-100 (surfactant) did not affect the solution dynamics of alginate−PEO blends prominently. Fibers (nonbeaded) can be obtained only with 0.5 wt% of surfactant, keeping the alginate to PEO ratio at 80/20. Thus a nonionic surfactant

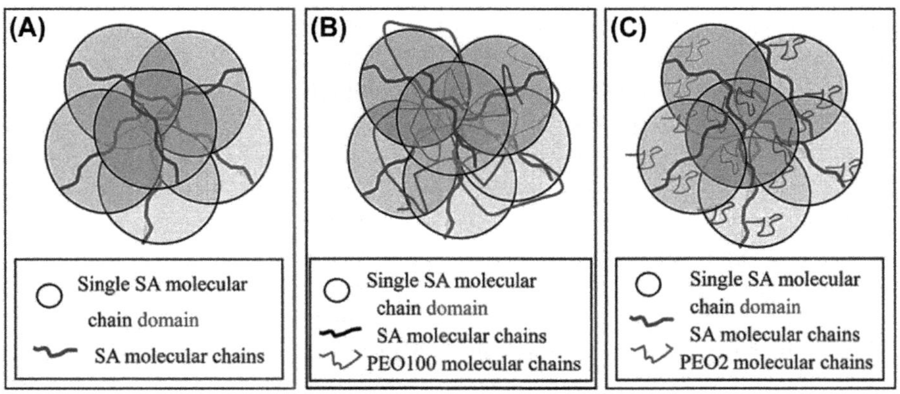

FIGURE 16.6

Schematic drawing of the chain conformation in alginate (SA) and SA/poly(ethylene)oxide (PEO) systems: (A) pure SA solution; (B) SA/PEO100 blended solution; (C) SA/PEO2 blended solution [46].

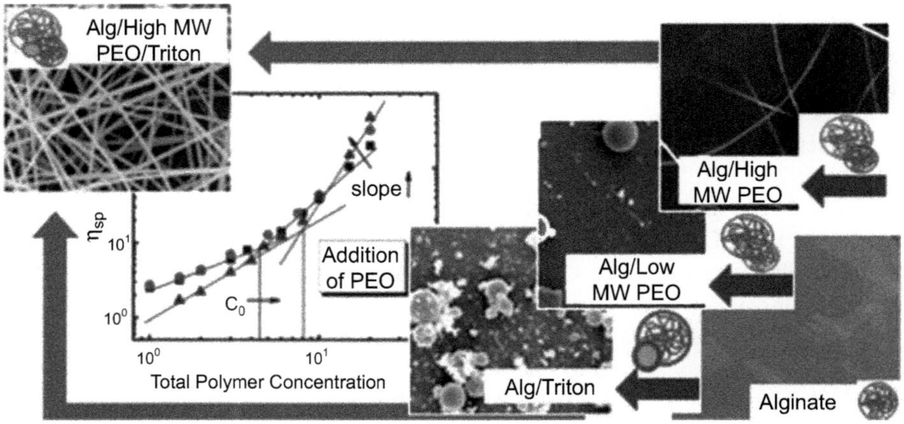

FIGURE 16.7

Effect of carrier polymer and nonionic surfactant on alginate-poly(ethylene)oxide (PEO) blend fiber formation [128].

contributes to increase the concentration of alginate. Overall, the alginate concentration was increased up to 70 wt% by employing a higher molecular-weight PEO (2000 kDa), and after the addition of Triton X-100 surfactant (Fig. 16.7), the concentration reached up to 85 wt% alginate, higher than previously reported.

By comparing the system (Na-alginate/PEO) using PVA (205 kDa) as the carrier polymer, alginate—PVA (ratio of 30:70) produced beaded fibers that inferred the independence of electrospin fiber from coordination of carrier polymer and metal cation [128].

In 2012 the electrospinning method was used to synthesize the nanofiber membranes of a core—shell structure from a solution of PEO and sodium alginate (Na-alginate/PEO) using the $CaCl_2$ cross-linker [129]. Fig. 16.8 showed the schematic setup for the preparation of Na-alginate (SA)/PEO core shell nanofibers. It was noted in FTIR analysis that the asymmetrical peak of COO^- ions and —OH peak of alginate shifted, predicting the hydrogen bond interaction between PEO and alginate. The transmission electron microscopy, X-ray photoelectron spectroscopy, and scanning electron microscopy results suggested that SA was properly wrapped by the PEO, with uniform diameter and rough surface morphology after cross-linking with $CaCl_2$. Evaluation of indirect cytotoxicity of Na-alginate/PEO nanofiber membranes has proved them nontoxic to fibroblast cells, whereas cell culture proposed that Na-alginate/PEO nanofibers have a tendency to support fibroblast cells proliferation and attachment and therefore have potential to be used as scaffold material.

Academic and industrial research showed a tremendous progress with the introduction of fiber-reinforced composites, and because of their superior strength and bending features these composites found huge utility in industrial, civil, and military sectors. Natural fiber—reinforced composites/nanocomposites have attracted much attention because of their low density, CO neutrality, biodegradability, low cost, and recyclability and therefore offers a valuable replacement of synthetic fiber—reinforced composite materials [130,131].

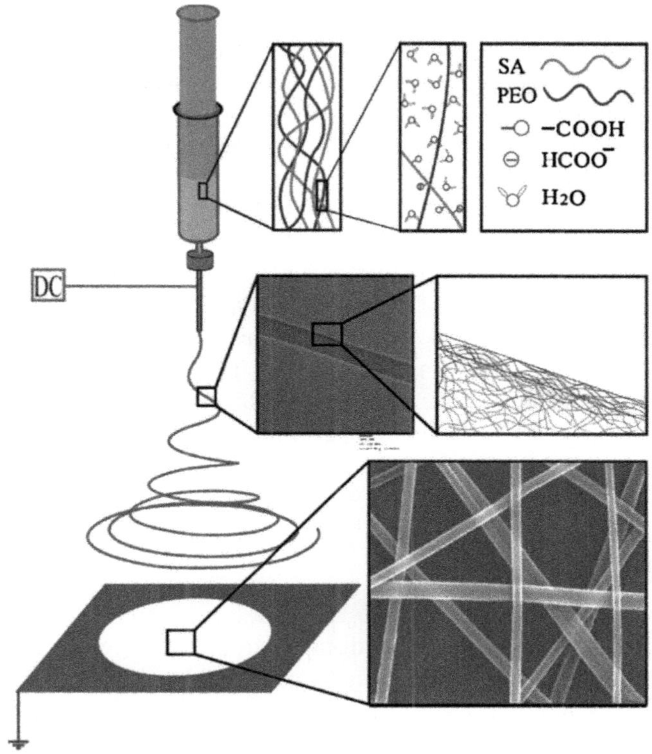

FIGURE 16.8

Schematic electrospinning setup of the core–shell nanofibers. *PEO*, poly(ethylene)oxide; *SA*, alginate.

Very few studies have been conducted to investigate the fiber as a reinforcing agent. Ca-alginate fiber–reinforced PEO composites were prepared via compression molding [132]. The Ca-alginate fibers were used at 10 wt% as a filler material in these unidirectional composites. Tensile, bending, and impact strengths of the composites were increased compared with the PEO matrix material. Tensile and bending strengths of the PEO/Ca-alginate fiber composite were found to be 11 and 18 MPa, respectively, which were 57% and 80% higher than those of the PEO matrix material. The interfacial shear strength of the composite fiber composites was 0.47 MPa, calculated by the single fiber fragmentation test, indicating good fiber–matrix adhesion.

16.5 SOME ALGINATE-PEO/PEG–BASED MIXED MATERIALS

Na-alginate-pectin-PEO electrospun fibers were produced to study in vitro folic acid release. The effect of releasing media was also studied [133]. The release rate of folic acid from electrospun fiber is maximum at a higher pH (7.8); however, the fibers were degraded under highly acidic conditions (pH 1.2). The swelling of the fiber caused the highest release at pH 7.8, as confirmed by SEM. The release of folic acid from Na-alginate–pectin electrospun fibers was not affected by gastric juice and acidified water when compared with simulated intestinal fluid at the same pH. In conclusion, because almost half of the folic acid is released from electrospun fibers in intestinal fluid, therefore the fibers would be a good carrier of folic acid for selective release in the small intestine.

Alginate (SA) is among one of the most studied anionic polymers (polyelectrolyte) to form complex with chitosan (CHI), because the final product not only preserves its biodegradable and biocompatible properties but also mechanically strong [134]. The coaxial nanofibers SA/CHI/PEO of the core–shell type have been effectively obtained using two blends: SA/PEO 1:1 (w/w) in an aqueous solution of 5% C_2H_5OH as the core and CHI/PEO, 70:30 (w/w) as the shell [38]. The thermal stability of the CHI/PEO membrane is better than the SA/PEO and SA/CHI/PEO membranes. After 4, 10, 24 h of SA/CHI/PEO incubation in water, the degree of weight loss was same, ascribed only to the presence of 40 wt% of PEO that was leached by this solvent.

16.6 CONCLUSION

Modification of traditional materials to attain desirable multifunctional properties for industrial end uses especially those designed from natural resources are of great interest. Alginate/PEO electrospun fibers found huge biomedical and drug delivery applications. PEO improved the inherit (native) gelling process and provided flexibility to alginate, while retaining biocompatibility. Solution casting blends of alginate/PEO showed phase separation with 50 wt% of alginate thus providing alginate-enriched blend surface. Similarly, alginate has been modified with PEG for the fabrication of microcapsule coatings, self-assembling, nanospheres, mucoadhesive polymers, and hybrid microspheres. Different surfactants are used in the PEG/PEO–alginate material to enhance interfacial interaction. Some other biopolymers such as chitosan and pectin improve the properties of these blend materials further. More recent work showed the use of cinnamoyl Pluronic F127 surfactant for cinnamoyl alginate/cinnamoyl poly(ethylene glycol) nanogels. The nanogels exhibited in vivo residence time greater than 96 h, along with the sustained release of human growth hormone [135].

However, the PEO/PEG–alginate blend technology is still not well established, as much focus is paid on certain process technology such as electrospinning, which itself needs further studies to become controllable and reproducible. Therefore, new innovative process technology and adjustable conditions are required to make these materials practically more useful.

REFERENCES

[1] Aroguz AZ, Engin HH, Baysal BM. The assessment of miscibility and morphology of poly (ε-caprolactone) and poly (para-chlorostyrene) blends. Eur Polym J 2007;43(2):403–9.

[2] Panapitiya NP, Wijenayake SN, Huang Y, Bushdiecker D, Nguyen D, Ratanawanate C, Kalaw GJ, Gilpin CJ, Musselman IH, Balkus KJ, Ferraris JP. Stabilization of immiscible polymer blends using structure directing metal organic frameworks (MOFs). Polymer 2014;55(8):2028–34.

[3] Masegosa RM, Nava D, Prolongo MG, Salom C. Linear unsaturated polyester+poly (ε-caprolactone) blends: calorimetric behaviour and morphology. Thermochim Acta 2006;440(1):93–101.

[4] Khalf AI, Nashar DE, Maziad NA. Effect of grafting cellulose acetate and methylmethacrylate as compatibilizer onto NBR/SBR blends. Mater Des 2010;31(5):2592–8.

[5] Mourad AH. Thermo-mechanical characteristics of thermally aged polyethylene/polypropylene blends. Mater Des 2010;31(2):918–29.

[6] Rastin H, Jafari SH, Saeb MR, Khonakdar HA, Wagenknecht U, Heinrich G. On the reliability of existing theoretical models in anticipating type of morphology and domain size in HDPE/PA-6/EVOH ternary blends. Eur Polym J 2014;53:1–2.

[7] Woo EM, Barlow JW, Paul DR. Phase behavior of blends of aliphatic polyesters with a vinylidene chloride/vinyl chloride copolymer. J Appl Polym Sci 1986;32(3):3889–97.

[8] Thomas S, Shanks R, Chandrasekharakurup S, editors. Nanostructured polymer blends. William Andrew; November 28, 2013.

[9] Krause S. Polymer blends. New York: Academic Press; 1979.

[10] Varnell DF, Runt JP, Coleman MM. FT IR and thermal analysis studies of blends of poly (ε-caprolactone) with homo-and copolymers of poly (vinylidene chloride). Polymer 1983;24(1):37–42.

[11] Varnell DF, Coleman MM. FT IR studies of polymer blends: V. Further observations on polyester-poly (vinyl chloride) blends. Polymer 1981;22(10):1324–8.

[12] Sionkowska A, Wisniewski M, Skopinska J, Kennedy CJ, Wess TJ. Molecular interactions in collagen and chitosan blends. Biomaterials 2004;25:795–801.

[13] Sionkowska A, Skopinska J, Wisniewski M. Collagen synthetic polymer interactions in solution and thin films. J Mol Liq 2009;145:135–8.

[14] Cascone MG. Dynamic–mechanical properties of bioartificial polymeric materials. Polym Int 1997;43:55–69.

[15] Pingping Z. A new criterion of polymer–polymer miscibility detected by viscometry. Eur Polym J 1997;33:411–4.

[16] Relleve L, Yoshii F, Kume T. Radiation-modified hydrogel based on poly (*N*-vinyl-2-pyrrolidone) and carrageenan. Die Angew Makromol Chem 1999;273(1):63–8.

[17] Kondo T, Sawatari C. Intermolecular hydrogen bonding in cellulose/poly (ethylene oxide) blends: thermodynamic examination using 2, 3-di-*O*-and 6-*O*-methylcelluloses as cellulose model compounds. Polymer 1994;35(20):4423–8.

[18] Miya M, Iwamoto R, Mima S. FT-IR study of intermolecular interactions in polymer blends. J Polym Sci Polym Phys 1984;22(6):1149–51.

[19] Sajeev US, Anoop AK, Menon D, Nair S. Control of nanostructures in PVA, PVA/chitosan blends and PCL through electrospinning. Bull Mater Sci 2008;31:343–51.

[20] Zhang X, Hua H, Shen X, Yang Q. In vitro degradation and biocompatibility of poly(*l*-lactic acid)/chitosan fiber composites. Polymer 2007;48:1005–11.

[21] Lee KH, Baek DH, Ki CS, Park YH. Preparation and characterization of wet spun silk fibroin/poly(vinyl alcohol) blend filaments. Int J Biol Macromol 2007;41:168–72.

[22] Li M, Lu S, Wu Z, Tan K, Minoura N, Kuga S. Structure and properties of silk fibroin–poly(vinyl alcohol) gel. Int J Biol Macromol 2002;30:89–94.

[23] Aluigi A, Vineis C, Varesano A, Tonin C, Ferrero F, Canetti M, Mazzuchetti G, Ferrero F, Tonin C. Structure and properties of keratin/PEO blend nanofibres. Eur Polym J 2008;44:2465–75.

[24] Tonin C, Aluigi A, Vineis C, Varesano A, Montarsolo A, Ferrero F. Thermal and structural characterization of poly(ethyleneoxide)/keratin blend films. J Therm Anal Calorim 2007;89:601–8.

[25] Chandy T, Mooradian DL, Rao GH. Chitosan/polyethylene glycol-alginate microcapsules for oral delivery of hirudin. J Appl Polym Sci 1998;70(11):2143–53.

[26] Alexy P, Bakoš D, Hanzelová S, Kukolíková L, Kupec J. Poly(vinyl alcohol)-collagen hydrolysate thermoplastic blends. I. Experimental design optimisation and biodegradation behaviour. Polym Test 2003;22:801–9.

[27] Yilmaz E, Ozalp D, Yilmaz O. Miscibility study of chitosan/poly(vinyl pyrrolidone) blends in dilute solution. Int J Polym Anal Charact 2005;10:329–39.

[28] Sionkowska A, Kaczmarek H, Kowalonek J, Wisniewski M, Skopinska J. Surface state of UV irradiated collagen/PVP blends. Surf Sci 2004;566–568:608–12.

[29] Marsano E, Vicini S, Skopinska J, Sionkowska A. Chitosan and poly(vinyl pyrrolidone): compatibility and miscibility of blends. Macromol Sym 2004;218:251–60.

[30] Al Meslmani B, Mahmoud G, Strehlow B, Mohr E, Leichtweiß T, Bakowsky U. Development of thrombus-resistant and cell compatible crimped polyethylene terephthalate cardiovascular grafts using surface co-immobilized heparin and collagen. Mater Sci Eng 2014;43:538–46.

[31] Zeng M, Fang Z, Xu C. Effect of compatibility on the structure of the microporous membrane prepared by selective dissolution of chitosan/synthetic polymer blend membrane. J Membr Sci 2004;230:175–81.

[32] Caykara T, Demirci S, Eroğlu MS, Güven O. Poly (ethylene oxide) and its blends with sodium alginate. Polymer 2005;46(24):10750–7.

[33] Yas A. Preparation and characterization of in situ cross–linking alginate/polyethylene glycol 6000 blend microparticles for controlling salbutamol sulphate pulmonary delivery. Int J Pharm Pharm Sci November 2013;5(4):728–33.

[34] Li L, Ying X, Liu J, Li X, Zhang W. Molecularly imprinted polyurethane grafted calcium alginate hydrogel with specific recognition for proteins. Mater Lett 2015;143:248–51.

[35] Kamoun EA, Kenawy ER, Tamer TM, El-Meligy MA, Eldin MS. Poly (vinyl alcohol)-alginate physically crosslinked hydrogel membranes for wound dressing applications: characterization and bio-evaluation. Arab J Chem 2015;8(1):38–47.

[36] Villa C, Manzoli V, Abreu MM, Verheyen CA, Seskin M, Najjar M, Molano RD, Torrente Y, Ricordi C, Tomei AA. Effects of composition of alginate-polyethylene glycol microcapsules and transplant site on encapsulated islet graft outcomes in mice. Transplantation September 14, 2016.

[37] Wise SG, Byrom MJ, Waterhouse A, Bannon PG, Ng MK, Weiss AS. A multilayered synthetic human elastin/polycaprolactone hybrid vascular graft with tailored mechanical properties. Acta Biomater 2011;7(1):295–303.

[38] Nista SV, Bettini J, Mei LH. Coaxial nanofibers of chitosan–alginate–PEO polycomplex obtained by electrospinning. Carbohydr Polym August 20, 2015;127:222–8.

[39] Lee YJ, Shin DS, Kwon OW, Park WH, Choi HG, Lee YR, Han SS, Noh SK, Lyoo WS. Preparation of atactic poly(vinyl alcohol)/sodium alginate blend nanowebs by electrospinning. J Appl Polym Sci 2007;106:1337–42.

[40] Huang L, Nagapudi K, Apkarian RP, Chaikof EL. Engineered collagen–PEO nanofibers and fabrics. J Biomater Sci Polym Ed 2001;12(9):979–93.

[41] Lee CM, Jin HJ, Botsaris GD, Kaplan DL. Silk apatite composites from electrospun fibers. J Mater Res 2005;20:3374.

[42] Duan B, Dong C, Yuan X, Yao K. Electrospinning of chitosan solutions in acetic acid with poly(ethylene oxide). J Biomater Sci Polym Ed 2004;15(6):797–811.

[43] Yi F, Guo ZX, Hu P, Fang ZX, Yu J, Li Q. Mimetics of eggshell membrane protein fibers by electrospinning. Macromol Rapid Commun 2004;25(10):1038–43.

[44] Bhattarai N, Li Z, Edmondson D, Zhang M. Alginate-based nanofibrous scaffolds: structural, mechanical, and biological properties. Adv Mater 2006;18(11):1463–7.

[45] Jeong SI, Krebs MD, Bonino CA, Khan SA, Alsberg E. Electrospun alginate nanofibers with controlled cell adhesion for tissue engineering. Macromol Biosci 2010;10(8): 934–43.

[46] Nie H, He A, Wu W, Zheng J, Xu S, Li J, Han CC. Effect of poly(ethylene oxide) with different molecular weights on the electrospinnability of sodium alginate. Polymer 2009;50(20):4926–34.

[47] Safi S, Morshed M, Hosseini Ravandi SA, Ghiaci M. Study of electrospinning of sodium alginate, blended solutions of sodium alginate/poly(vinyl alcohol) and sodium alginate/poly(ethylene oxide). J Appl Polym Sci 2007;104(5):3245–55.

[48] Lu JW, Zhu YL, Guo ZX, Hu P, Yu J. Electrospinning of sodium alginate with poly(-ethylene)oxide. Polymer 2007;47:8026–31.

[49] Cheng SZD, Chen J. Nonintegral and integral folding crystal-growth in low molecular mass poly(ethylene oxide) fractions. J Polym Sci B Polym Phys 1991;29:311–27.

[50] Balijepalli S, Schultz J, Lin J. Phase behavior and morphology of poly(ethylene oxide) blends. Macromolecules 1996;29:6601–66.

[51] Chen J, Cheng SZD, Wu SS, Lotz B, Wittmann J-C. Polymer decoration study in chain folding behavior of solution-grown poly(ethlene oxide) crystals. J Polym Sci Pol Phys 1995;33:1851–5.

[52] Merdas A, Gindre M, Ober R, Nicot C, Urbach W, Waks M. Nonionic surfactant reverse of $C_{12}E_4$ in dodecane: temperature dependence of size and shape. J Phys Chem 1996;100:15180–6.

[53] Bieze TW, Vandermaarel JR, Eisenbach CD, Leyte JC. Polymer dynamics in aquarious poly(ethylene oxide) solutions an NMR study. Macromolecules 1994;27:1355–66.

[54] Briscoe B, Luck P, Zhu S. Rheological study of poly(ethylene oxide) in aqueous salt solutions at high temperature and pressure. Macromolecules 1996;29:6208–11.

[55] Sennerby L, Röstiund T, Albrektsson B, Albrektsson T. Acute tissue reactions to potassium alginate with and without colour/flavour additives. Biomaterials 1987;8(1): 49–52.

[56] Hashimoto T, Suzuki Y, Yamamoto E, Tanihara M, Kakimaru Y, Suzuki K. Biomaterials 2004;25:1407–14.

[57] Mierisch CM, Cohen SB, Jordan LC, Robertson PG, Balian G, Diduch DR. J Arthrosc Relat Surg 2002;18:892–900.

[58] Dar A, Shachar M, Leor J, Cohen S. Optimization of cardiac cell seeding and distribution in 3D porous alginate scaffolds. Biotechnol Bioeng 2002;80(3):305–12.

[59] Dausse Y, Grossin L, Miralles G, Pelletier S, Mainard D, Hubert P, Baptiste D, Gillet P, Dellacherie E, Netter P, Payan E. Cartilage repair using new polysaccharidic biomaterials: macroscopic, histological and biochemical approaches in a rat model of cartilage defect. Osteoarthr Cartil 2003;11(1):16–28.

[60] Zactiti EM, Kieckbusch TG. Potassium sorbate permeability in biodegradable alginate films: effect of the antimicrobial agent concentration and crosslinking degree. J Food Eng 2006;77(3):462–7.

[61] Davis TA, Volesky B, Mucci A. A review of the biochemistry of heavy metal biosorption by brown algae. Water Res 2003;37(18):4311–30.

[62] Bailey FE, Koleske JV. Poly(ehtylene oxide). New York: Academic Press; 1976.

[63] Harris JM, Chess RB. Effect of pegylation on pharmaceuticals. Nat Rev Drug Discov 2003;2(3):214–21.

[64] Subbu C, Mathew CM, Kesavan K, Rajendran S. Electrochemical, structural and optical studies on poly(vinylidene chloride-co-acrylonitrile) based polymer blend membranes. Int J Electrochem Sci 2014;9:4944–58.

[65] Akhtar MS, Cheralathan KK, Chun JM, Yang OB. Composite electrolyte of heteropolyacid (HPA) and polyethylene oxide (PEO) for solid-state dye-sensitized solar cell. Electrochim Acta 2008;53(22):6623–8.

[66] Rolland J, Brassinne J, Bourgeois JP, Poggi E, Vlad A, Gohy JF. Chemically anchored liquid-PEO based block copolymer electrolytes for solid-state lithium-ion batteries. J Mater Chem A 2014;2(30):11839–46.

[67] Son WK, Youk JH, Lee TS, Park WH. The effects of solution properties and polyelectrolyte on electrospinning of ultrafine poly (ethylene oxide) fibers. Polymer 2004; 45(9):2959–66.

[68] Idris NH, Majid SR, Khiar AS, Hassan MF, Arof AK. Conductivity studies on chitosan/PEO blends with LiTFSI salt. Ionics 2005;11(5–6):375–7.

[69] NPCS Board of Consultants & Engineers. The complete book on water soluble polymers. Asia Pacific Business Press Inc.; 2009.

[70] Kuila T, Acharya H, Srivastava SK, Samantaray BK, Kureti S. Enhancing the ionic conductivity of PEO based plasticized composite polymer electrolyte by $LaMnO_3$ nanofiller. Mater Sci Eng B 2007;137(1):217–24.

[71] Fink JK. Handbook of engineering and specialty thermoplastics, water soluble polymers. John Wiley & Sons; 2011.

[72] Zhang HH, Maitra P, Wunder SL. Preparation and characterization of composite electrolytes based on PEO (375)-grafted fumed silica. Solid State Ion 2008;178(39): 1975–83.

[73] Harris JM, editor. Poly (ethylene glycol) chemistry: biotechnical and biomedical applications. Springer Science & Business Media; 2013.

[74] Dinç CÖ, Kibarer G, Güner A. Solubility profiles of poly (ethylene glycol)/solvent systems. II. Comparison of thermodynamic parameters from viscosity measurements. J Appl Polym Sci 2010;117(2):1100–19.

[75] French AC, Thompson AL, Davis BG. High-purity discrete PEG-oligomer crystals allow structural insight. Angew Chem Int Ed 2009;48(7):1248–52.

[76] Chiou WL, Riegelman S. Pharmaceutical applications of solid dispersion systems. J Pharm Sci 1971;60(9):1281–302.

[77] Zhang M, Li XH, Gong YD, Zhao NM, Zhang XF. Properties and biocompatibility of chitosan films modified by blending with PEG. Biomaterials 2002;23(13):2641–8.

[78] Serajuddin A. Solid dispersion of poorly water-soluble drugs: early promises, subsequent problems, and recent breakthroughs. J Pharm Sci 1999;88(10):1058–66.

[79] Bernier JJ, Donazzolo Y. Effect of low-dose polyethylene glycol 4000 on fecal consistency and dilution water in healthy subjects. Gastroenterol Clin Biol 1997; 21:7–11.

[80] Zalipsky S. Functionalized poly(ethylene glycols) for preparation of biologically relevant conjugates. Bioconj Chem 1995;6(2):150–65.
[81] Leuner C, Dressman J. Improving drug solubility for oral delivery using solid dispersions. Eur J Pharm Biopharm 2000;50(1):47–60.
[82] Laurienzo P, Malinconico M, Motta A, Vicinanza A. Synthesis and characterization of a novel alginate–poly (ethylene glycol) graft copolymer. Carbohydr Polym 2005; 62(3):274–82.
[83] d'Ayala GG, Malinconico M, Laurienzo P. Marine derived polysaccharides for biomedical applications: chemical modification approaches. Molecules 2008;13(9): 2069–106.
[84] Sungsanit K, Kao N, Bhattacharya SN. Properties of linear poly (lactic acid)/polyethylene glycol blends. Polym Eng Sci 2012;52(1):108–16.
[85] Laurienzo P, Malinconico M, Mattia G, Russo R, La Rotonda MI, Quaglia F, Capitani D, Mannina L. Novel alginate–acrylic polymers as a platform for drug delivery. J Biomed Mater Res Part A 2006;78(3):523–31.
[86] Bernkop-Schnurch A, Kast CE, Richter MF. Improvement in the mucoadhesive properties of alginate by the covalent attachment of cysteine. J Control Release 2001;71: 277–85.
[87] Hua S, Wan A. Synthesis, characterization and swelling behaviors of sodium alginate-g-poly(acrylic acid)/sodium humate superabsorbent. Carbohydr Polym 2009;75:79–84.
[88] Rowley JA, Madlambayan G, Mooney DJ. Alginate hydrogels as synthetic extracellular matrix materials. Biomaterials 1999;20(1):45–53.
[89] Seifert DB, Phillips JA. Porous alginate–poly(ethylene glycol) entrapment system for the cultivation of mammalian cells. Biotechnol Prog 1997;13(5):569–76.
[90] Maguire T, Novik E, Schloss R, Yarmush M. Alginate-PLL microencapsulation: effect on the differentiation of embryonic stem cells into hepatocytes. Biotechnol Bioeng 2006;93(3):581–91.
[91] Tu R, Lu CL, Thyagarajan K, Wang E, Nguyen H, Shen S, Hata C, Quijano RC. Kinetic study of collagen fixation with polyepoxy fixatives. J Biomed Mater Res 1993; 27(1):3–9.
[92] Merril EW, Salzman EW. Polyethylene oxide as biomaterial. ASAIO 1983;6:60–6.
[93] Han DK, Jeong SY, Kim YH. Evaluation of blood compatibility of PEO grafted and heparin immobilized polyurethanes. J Biomed Mater Res 1989;23(A2 Suppl.): 211–28.
[94] Weber LM, He J, Bradley B, Haskins K, Anseth KS. PEG-based hydrogels as an in vitro encapsulation platform for testing controlled beta-cell microenvironments. Acta Biomater 2006;2:1–8.
[95] Cruise GM, Hegre OD, Lamberti FV, Hager SR, Hill R, Scharp DS, et al. In vitro and in vivo performance of porcine islets encapsulated in interfacially photopolymerized poly(ethylene glycol) diacrylate membranes. Cell Transpl 1999;8:293–306.
[96] Kizilel S, Scavone A, Liu X, Nothias JM, Ostrega D, Witkowski P, Millis M. Encapsulation of pancreatic islets within nano-thin functional polyethylene glycol coatings for enhanced insulin secretion. Tissue Eng Part A 2010;16(7):2217–28.
[97] Desai NP, Sojomihardjo A, Yao Z, Ron N, Soon-Shiong P. Interpenetrating polymer networks of alginate and polyethylene glycol for encapsulation of islets of Langerhans. J Microencapsul 2000;17(6):677–90.
[98] Bryant SJ, Anseth KS. The effects of scaffold thickness on tissue engineered cartilage in photocrosslinked poly(ethylene oxide) hydrogels. Biomaterials 2001;22:619–26.

[99] Lin-Gibson S, Bencherif S, Cooper JA, Wetzel SJ, Antonucci JM, Vogel BM, Horkay F, Washburn NR. Synthesis and characterization of PEG dimethacrylates and their hydrogels. Biomacromolecules 2004;5:1280–7.

[100] Almany L, Seliktar D. Biosynthetic hydrogel scaffolds made from fibrinogen and polyethylene glycol for 3D cell cultures. Biomaterials 2005;26(15):2467–77.

[101] West JL, Hubbell JA. Polymeric biomaterials with degradation sites for proteases involved in cell migration. Macromolecules 1999;32:241–4.

[102] Davidovich-Pinhas M, Bianco-Peled H. Novel mucoadhesive system based on sulfhydryl–acrylate interactions. J Mater Sci Mater Med 2010;21:2027–34.

[103] Davidovich-Pinhas M, Bianco-Peled H. Alginate–PEGAc: a new mucoadhesive polymer. Acta Biomater 2011;7(2):625–33.

[104] Hall KK, Gattás-Asfura KM, Stabler CL. Microencapsulation of islets within alginate/ poly (ethylene glycol) gels cross-linked via Staudinger ligation. Acta Biomater 2011; 7(2):614–24.

[105] Cruise GM, Hegre OD, Scharp DS, Hubbell JA. A sensitivity study of the key parameters in the interfacial photopolymerization of poly(ethylene glycol) diacrylate upon porcine islets. Biotechnol Bioeng 1998;57:655–65.

[106] Williams CG, Malik AN, Kim TK, Manson PN, Elisseeff JH. Variable cytocompatibility of six cell lines with photoinitiators used for polymerizing hydrogels and cell encapsulation. Biomaterials 2005;26:1211–8.

[107] Wilson JT, Chaikof EL. Challenges and emerging technologies in the immunoisolation of cells and tissues. Adv Drug Deliv Rev 2008;60:124–45.

[108] Hu X, Gao C. Photoinitiating polymerization to prepare biocompatible chitosan hydrogels. J Appl Polym Sci 2008;110:1059–67.

[109] Shoichet MS, Winn SR. Cell delivery to the central nervous system. Adv Drug Deliv Rev 2000;42:81.

[110] Hasse C, Bohrer T, Barth P, Stinner B, Cohen R, Cramer H, et al. Parathyroid xenotransplantation without immunosuppression in experimental hypoparathyroidism: long-term in vivo function following microencapsulation with a clinically suitable alginate. World J Surg 2000;24:1361–6.

[111] Mahou R, Wandrey C. Alginate–poly (ethylene glycol) hybrid microspheres with adjustable physical properties. Macromolecules 2010;43(3):1371–8.

[112] Mahou R, Meier RP, Bühler LH, Wandrey C. Alginate-poly (ethylene glycol) hybrid microspheres for primary cell microencapsulation. Materials 2014;7(1):275–86.

[113] Miao T, Rao KS, Spees JL, Oldinski RA. Osteogenic differentiation of human mesenchymal stem cells through alginate-graft-poly (ethylene glycol) microsphere-mediated intracellular growth factor delivery. J Control Release 2014;192:57–66.

[114] Huang SY, Pooyan S, Wang J, Choudhury I, Leibowitz MJ, Stein S. A polyethylene glycol copolymer for carrying and releasing multiple copies of cysteine-containing peptides. Bioconj Chem 1998;9(5):612–7.

[115] Mosmann T. Rapid colorimetric assay for cellular growth and survival: application to proliferation and cytotoxicity assays. J Immunol Methods 1983;65(1–2):55–63.

[116] Wang Q, Zhang N, Hu X, Yang J, Du Y. Alginate/polyethylene glycol blend fibers and their properties for drug controlled release. J Biomed Mater Res Part A 2007;82(1):122–8.

[117] Valizadeh M, Ravandi SA, Ramakrishna S. Recent advances in electrospinning of some selected biopolymers. J Text Polym 2013;1(2):70–84.

[118] Sumi SA, Rahman W, Alam J, Dafader NC, Manir S, Khan MR. Irradiated sodium–alginate/poly (ethylene oxide) blend films improved by methyl acrylate monomer. J Appl Polym Sci 2016;133(26):43562.

[119] Jayaraman K, Kotaki M, Zhang Y, Mo X, Ramakrishna S. Recent advances in polymer nanofibers. J Nanosci Nanotechnol 2004;4(1−2):52−65.

[120] Liang D, Hsiao BS, Chu B. Functional electrospun nanofibrous scaffolds for biomedical applications. Adv Drug Deliv Rev 2007;59(14):1392−412.

[121] Kim JH, Min BR, Won J, Kim CK, Kang YS. Structure and coordination properties of facilitated olefin transport membranes consisting of crosslinked poly (vinyl alcohol) and silver hexafluoroantimonate. J Polym Sci Part B 2004;42(4):621−8.

[122] Longenberger L, Mills G. Formation of metal particles in aqueous solutions by reactions of metal complexes with polymers. J Phy Chem 1995;99(2):475−8.

[123] Kim JH, Min BR, Lee KB, Won J, Kang YS. Coordination structure of various ligands in crosslinked PVA to silver ions for facilitated olefin transport. Chem Commun 2002; 22:2732−3.

[124] Saquing CD, Manasco JL, Khan SA. Electrospun nanoparticle−nanofiber composites via a one-step synthesis. Small 2009;5(8):944−51.

[125] Park SA, Park KE, Kim W. Preparation of sodium alginate/poly (ethylene oxide) blend nanofibers with lecithin. Macromol Res 2010;18(9):891−6.

[126] Esquisabel A, Hernandez RM, Igartua M, Gascon AR, Calvo B, Pedraz JL. Effect of lecithins on BCG-alginate-PLL microcapsule particle size and stability upon storage. J Microencapsul 2000;17(3):363−72.

[127] McKee MG, Layman JM, Cashion MP, Long TE. Phospholipid nonwoven electrospun membranes. Science 2006;311(5759):353−5.

[128] Saquing CD, Tang C, Monian B, Bonino CA, Manasco JL, Alsberg E, Khan SA. Alginate−polyethylene oxide blend nanofibers and the role of the carrier polymer in electrospinning. Ind Eng Chem Res 2013;52(26):8692−704.

[129] Ma G, Fang D, Liu Y, Zhu X, Nie J. Electrospun sodium alginate/poly (ethylene oxide) core−shell nanofibers scaffolds potential for tissue engineering applications. Carbohydr Polym 2012;87(1):737−43.

[130] Bledzki AK, Gasan J. Composites reinforced with cellulose based fibers. Prog Polym Sci 1999;24:221−74.

[131] Khan MA, Bhattacharia SK. Effect of novel coupling agent on the mechanical and thermal properties of unidirectional jute−vinyl ester composites. J Reinf Plast Compos 2007;26(6):617−27.

[132] Dey K, Khan RA, Chowdhury AS. Study on the mechanical, degradation, and interfacial properties of calcium alginate fiber-reinforced polyethylene oxide composites. J Thermoplas Compos Mater 2012;25(7):807−19.

[133] Alborzi S, Lim LT, Kakuda Y. Release of folic acid from sodium alginate-pectin-poly (ethylene oxide) electrospun fibers under in vitro conditions. LWT Food Sci Technol 2014;59(1):383−8.

[134] Hamman JH. Chitosan based polyelectrolyte complexes as potential carrier materials in drug delivery systems. Mar Drugs 2010;8:1305−22.

[135] Yoon DY, Kim JC. In vivo residence duration of human growth hormone loaded in nanogels comprising cinnamoyl alginate, cinnamoyl Pluronic F127 and cinnamoyl poly (ethylene glycol). Int J Pharm 2016;509(1):229−36.

CHAPTER

Alginate-Based Hybrid Nanocomposite Materials

17

Fatima Zia[1], Hadi Sobhani[2], Mohsen Mohammadi[3], Makshoof Athar[4], Muhammad Afzal[5], Tayyaba Sultana[1], Yury Shchipunov[6]

Government College University Faisalabad, Faisalabad, Pakistan[1]; Hamadan University of Technology, Hamadan, Iran[2]; Qom University of Technology (QUT), Old Tehran, Iran[3]; University of the Punjab, Lahore, Pakistan[4]; National Institute of Biotechnology and Genetic Engineering, Faisalabad, Pakistan[5]; Russian Academy of Sciences, Vladivostok, Russia[6]

17.1 INTRODUCTION

Biohybrid materials are in demand over the last decade and established as a prevalent strategy in both scientific and technical research fields with the aim to meet challenges in performance properties of the subsequent materials. The importance of these materials increased as the green technology became more vital in areas such as aerospace, biomedical, tissue engineering, automotive, sensors, food packaging, and electronics. This chapter deals with hybrid materials prepared from polysaccharide alginate with graphene−graphite and/or titania nanoparticles. This type of composite, in accordance with the classification suggested in Ref. [1], can be categorized as a bionanocomposite because of its formation from biopolymers and nano-sized inorganic particles. It is believed that combining nanoparticles with biopolymers, which ranked after synthetic polymers in most of the properties, improve them in significant manner and introduce new functionalities [1−11]. The main interest in natural polymeric materials stemmed from their biocompatibility, biodegradability, and their sustainability. Such biopolymers are synthesized as polysaccharides from photosynthesizing plants, animals, and microorganisms. Their application, compared to synthetic polymers produced from fossil-fuel and nonrenewable sources, further allows decreasing the ecological problem of carbon dioxide content in the atmosphere [4,6,7,12,13]. Inorganic fillers are widely used in polymer industry with the aim of reinforcing and improving the stiffness and toughness, enhancing the barrier properties, resistance to fire and ignition, or simply increasing the bulk and reducing the cost [14−16]. The term is usually applied to solid micro-sized additives added in large amounts (typically 20−50 wt%). Nano-sized particles provided a more notable effect in comparison with common micro-sized additives. It was systematically investigated by researchers at the Toyota research center in the early 1990s. Their findings about the improvement of mechanical properties along with dimensional stability, water and gas−barrier properties of nylon-6 by properly

Algae Based Polymers, Blends, and Composites. http://dx.doi.org/10.1016/B978-0-12-812360-7.00017-3

exfoliated nanoplatelets of clay montmorillonite initiated a great interest in nanofillers [17–19]. One important fact is that nanoparticle effect was observed at content of a few percentage points instead of tens. Composites made from synthetic polymers with nanofillers are separated into an individual group called nanocomposites [1].

Main conclusions of researchers from Toyota were borne out of lots of team results of which are summarized in numerous reviews (e.g., [20–35]). Readers will find exhaustive discussion of disparate nanoparticles of various origins and shapes such as clays, carbon, and allotrops. Comparison of functional properties of nanocomposites shows that improvement by graphene is superior to all these nanofillers. The same is valid for bionanocomposites that are considered in two recent special reviews [36,37]. Graphene's uniqueness consists in extraordinary mechanical, thermal, and electric properties, high chemical stability, and biocompatibility. These outstanding features acquired by (bio)nanocomposites after combining with (bio) polymers came to the widespread attention of numerous teams.

Titanium dioxide (TiO_2), titanium (IV) oxide, or simply titania, present another kind of nanoparticles. They attract tremendous interest because of potential application as very efficient photocatalysts, solar-energy converters, ductile ceramics, and sensors [38–41]. Furthermore, titania possesses high chemical stability and low toxicity that makes it suitable for pharmaceuticals, drug delivery, and cosmetics. Therefore titania is combined with (bio)polymers, with the aim to gain novel functionalities.

The chapter is devoted to bionanocomposites of carbon allotrops and titania only with alginate (alg). One can find very short information about this polysaccharide in recent reviews devoted to biopolymers with graphene [36,37]. As regards titania, so far as we know, there are original works of its composite materials with alginate but its bionanocomposites are reviewed only with cellulose [42–45]. Alg is one of the main structural components of marine algae [46–48]. This polysaccharide is unique in that it is a natural block copolymer. It consists of residues of 1,4-linked β-D-mannuronic and α-L-guluronic acids. Difference in their conformations results in distinctions in the block stiffness and cross-linking by multivalent metal cations through carboxylic groups [10,49–52]. These properties are of great importance for alginate bionanocomposite with nanoparticles of various origins. It is aimed to survey thoroughly in this chapter.

17.1.1 ALGINATE (ALG)

This polysaccharide is composed of residues of two uronic acids one of which is a 1,4-linked β-D-mannuronic (*M*) acid and another its C-5 epimer, α-L-guluronic acid (*L*) (Fig. 17.1). They are arranged in different proportions and sequences in carbohydrate macromolecule depending on the source and maturation of plant [47,48,53]. There are homopolymeric regions or blocks, *MMM* or *GGG*, and alternating sequences *GMGMGM* not in strictly regular manner. The presence of blocks makes alg unique among biopolymers. An important point is that they differ in

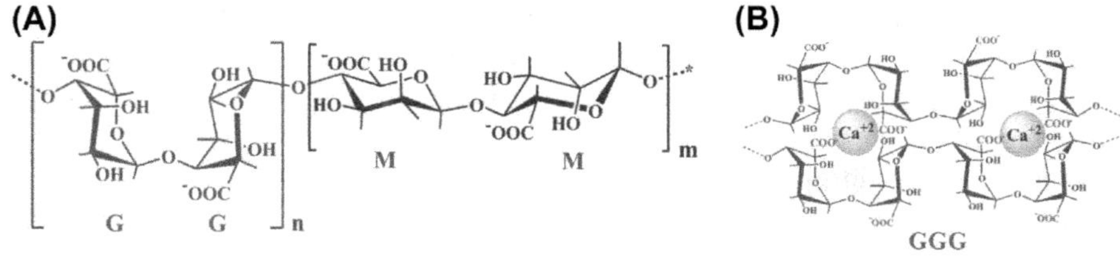

FIGURE 17.1

(A) Structure of alginate having (1 → 4) β-D-mannuronic acid (M) and (1 → 4) α-L-guluronic acid (G)-residues arranged in two blocks. (B) Complex with calcium cations of GGG-blocks in accordance with the "egg-box model."

conformations. *M*-residues in the *MMM*-blocks are covalently linked together through diequatorial links, while *G*-residues in the *GGG*-blocks through diaxial ones (Fig. 17.1A). The conformational difference leads to distinctions in block stiffness. As shown by various authors [47,49,50,52,54,55], *GGG* sequence is stiffer than *MMM* and alternating ones. Increased stiffness also makes *G*-blocks more extended.

The reason for the stiffness of *GGG*-blocks is its ribbonlike structure of pleated chain (Fig. 17.1A) [56,57]. This conformation is very suited for chelating of multivalent metal cations, as shown in Fig. 17.1B by an example of calcium cations. Complex formation results in cross-linking of two neighbor chains through *GGG*-blocks in accordance with the "egg-box model," whereas *M*-residues do not participate in the process [46,57,58]. It is an important property for regular alg use in preparation of hydrogels, films, capsules, scaffolds, etc. [46–48,59–62]. Novel applications include the stabilization of colloids and nanoparticles in solutions, controlling nanoparticle size, phase behavior and nanomaterial morphology, serving as template, reducing metal ions, and synthesizing nano-sized metal oxides and ceramics [10,63–66]. Here alg bionanocomposites with graphene-graphite and titania nanoparticles are considered in details.

17.1.2 TWO-DIMENSIONAL CARBON ALLOTROPES

The IUPAC defined graphene in 1995 as "a single carbon layer of the graphite structure, describing its nature by analogy to a polycyclic aromatic hydrocarbon of quasi infinite size" when no one had a chance of handling with it in the form of individual sheets [67]. The team led by Geim and Novoselov was the first to find a way to separate a single-atom-thick carbon layer from the parent matter in 2004 [68]. It was possible because graphite, which according to IUPAC definition is "an allotropic form of the element carbon consisting of layers of hexagonally arranged carbon atoms in a planar condensed ring system (graphene layers)," forms a lamellar structure in the parallel-stacked sheets attached through the van der Waals forces [69–74]. These forces are not very strong and that is why graphene layers can glide over each other as well as can be isolated from the parent graphite. Explosively growing number of publications, starting in 2004, in which various methods of preparation, treatment, and modification were used, resulted in a notable extension of the family of two-dimensional carbon allotrops. Initially, only two allotrops graphite and graphene were recognized, as recommended by the IUPAC in 1995, whereas exfoliated sheetlike or flakelike ultrathin multilayers, graphene oxide, and chemically modified graphenes, as well as derived materials prepared by stacking, folding, creasing, wrinkling, and pillaring were also discovered later on. In view of the absence of common names, resulting in confusion and inconsistency in the terminology observed in the literature, the international editorial team of the journal *Carbon* suggested an extended nomenclature for the two-dimensional carbon materials in 2013 [75]. Following it, the term "graphene" was used for the separated single-atom-thick carbon layer. It can be a freestanding film or sticking to a substrate. In case of nonisolated, graphite-like carbon structures, preference is given to the term "graphene layer." A monolayer derivative prepared by oxidation of graphene is called "graphene oxide," frequently abbreviated as *GO*. Reducing a *GO* yields "reduced graphene oxide" or *rGO*. The term "graphene nanosheet" is often seen in publications, but it is needless to use "nano" because of graphene's nanosized dimensions [75]. "Graphene microsheet" rather makes more sense to take account of the lateral dimension exceeding 100 nm.

It is practically impossible to prepare freestanding graphene in the form of a single-atom-thick carbon layer, and, furthermore, the monolayer tends to go into restacking immediately after separation [69—74]. It is common practice to deal with *GO* instead of graphene due to its easy dispersion in solutions [76—78] or its stacked monolayer structure. It is a rather widespread belief that a stack consisting of less than 10 single-atom layers can be considered as graphene [36,79], as this is a critical thickness. When becoming thinner, multilayer film becomes different in electronic structure from graphite [80]. It should be pointed out that Geim and Novoselov in the cited article give particular emphasis to graphene uniqueness. Its differences from the few-layer (3 to <10) counterparts are obvious from electronic spectra that are complicated for stacks owing to the appearance of various charge carriers and overlapping of the conduction and valence bands, whereas graphene and its bilayer in some approximation have only electrons and holes of one types [68,81]. Hence, the international editorial team of the *Carbon* suggested the term "multilayer graphene" for material consisting of 2 to about 10 of stacked graphene layers, or its subset as "few-layer graphene" in case of certainty of two and three layers [75].

There are rather numerous graphitic products consisting 10 and more stacked graphene layers. These are nano-sized graphite that differs from the common micro-sized fine powder of milled graphite. Therefore it is assigned to two-dimensional allotrops with less than 100 nm thickness. They are termed as "graphite nanoplates," "graphite nanosheets," or "graphite nanoflakes." "Exfoliated graphite" frequently used in the literature is a multilayer nano-sized product prepared by partial exfoliation.

Intense interest in graphene is caused by a number of unique properties. It combines unusual stiffness (Young modulus = 0.5—1 TPa) with elasticity (spring constants = 1—5 N/m), carrier mobility ($>$200,000 cm^2/V s) determining thermal (up to 5000 W/m K) and electrical conductivities, specific surface area

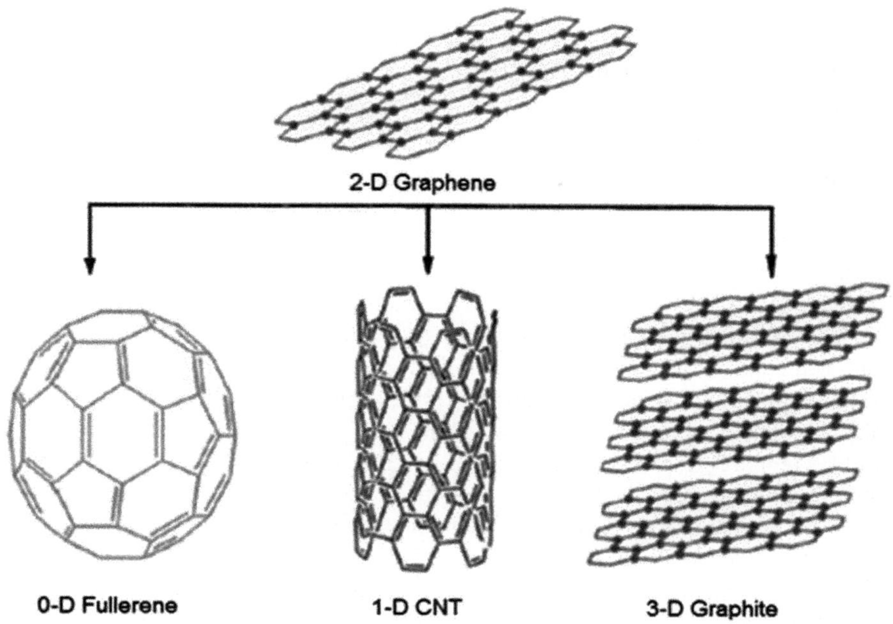

FIGURE 17.2

Structure of graphene and its derivatives [82].

ranging up to about 2500 m^2/g, optical transmittance, chemical stability that all are characterized by epithets such as very high, extremely, extraordinary. Abnormal elasticity of graphene allows it wrapping up into a zero-dimensional fullerene and rolling up into a zero-dimensional carbon nanotube shown schematically in Fig. 17.2. These extraordinary properties can be explained by a tight two-dimensional (2D) honeycomb lattice (Fig. 17.2) in which each carbon atom is bonded with three neighbors through sigma (σ) bonds in sp^2 hybridization. Rests of its electrons are in the p_z-orbitals delocalized with other π-electrons over the graphene layer, forming an easy movable two-dimensional electron gas [69,70,73,74,80,81]. Fig. 17.3 showed the SEM images of (A) stalked sheets of graphite and (B) exfoliated sheets of graphene oxide.

The major practical issue is the biocompatibility and toxicity of graphene and *GO*. It is of great importance for bionanocomposites because of their biomedical applications. As a carbon allotrop, toxicity of *GO* is mainly studied owing to its dispersibility in water and current applications as a drug-delivery system. This problem is given meticulous attention in few recent reviews [83–85]. Several studies dealt with various aspects of *GO*'s biological influence. They were performed both in vitro and in vivo with cells, cellular organelles, and experimental organisms (mice, rabbits, and zebrafish). *GO* was accumulated in lung, liver, spleen, and kidney transferred through the plasma membrane, and translocated within the endosome/lysosome systems. Chronic toxicities were also mentioned. *GO* could

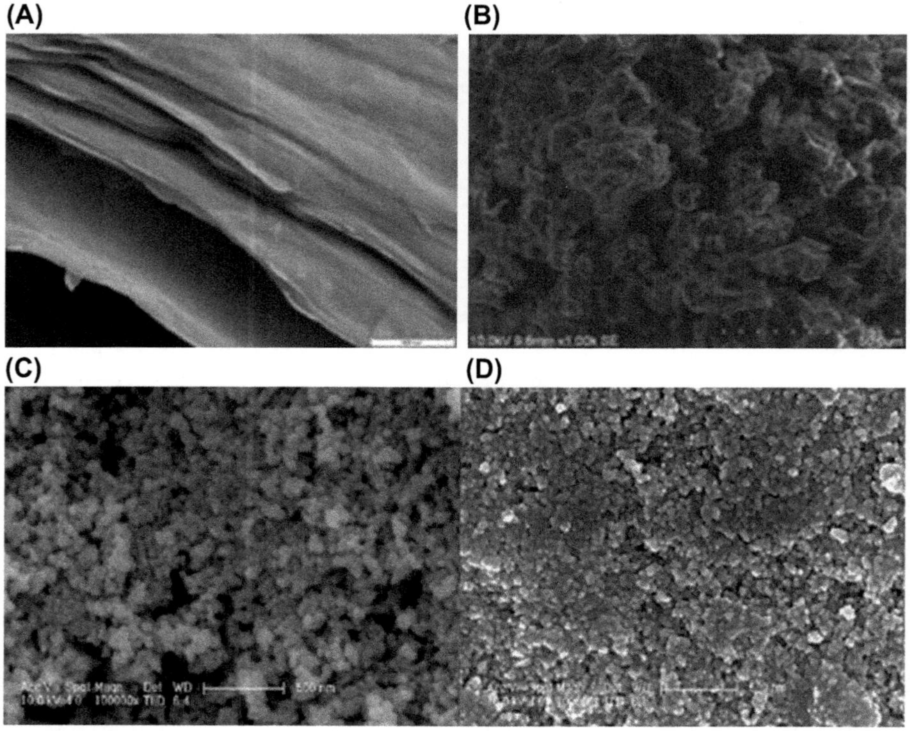

FIGURE 17.3

SEM images of (A) graphite and (B) graphene oxide produced by Hummer's method [182], (C) TiO_2 NPs synthesized by sol–gel method, and (D) TiO_2 NPs synthesized by hydrothermal method [183]. *NP*, nanoparticle; *SEM*, scanning electron microscopy.

bring about cell and macrophage apoptosis, lung granuloma formation, hemolysis, inflammation, immune system suppression, and mice death. On the other hand, it induced immunoactivation [85]; did not cause pathological changes in the liver, spleen and kidney [84]; and was selectively accumulated in tumor cells. The disagreement in data can be explained by differences in testing conditions and *GO* preparation, its dimension, oxidation degree, modification, and purity. There is a well-documented dose- and size-dependency of its biological influence [83–85]. Although there are numerous experimental results to date concerning the *GO* toxicity, its biocompatibility is still a crucial issue.

Proven biocompatibility and nonimmunogenicity of alginate and its resorbable biodegradation products are beyond question (e.g., Refs. [59,86,87]). Biopolymers mixed with *GO* usually reduce its toxicity [84,85]. One might expect a similar mitigation of possible toxic effects and improved biocompatibility of carbon allotropes in their bionanocomposites with alginate.

17.1.3 TITANIA NANOPARTICLES (TIO_2 NPS)

Titania (titanium dioxide, TiO_2) exists in three native forms, i.e., anatase, rutile, and brookite in the igneous rocks and sediments on earth as a less-expensive, stable, and nontoxic material. Titania gained popularity in 1972 when the first photocatalytic splitting process of water was reported. In due course of time nano form of titania (titania nanoparticles, TiO_2 NPs) was recognized as a well-known functional material for numerous applications such as pigments in paints and cosmetics, additives for food packaging and paper and pulp industry, as a photocatalyst, as electrode of photovoltaic cell, and as sensor material. TiO_2 is also employed for preparing self-decontamination textiles and worked well for UV protection and antibacterial action. Various conventional approaches used for the synthesis of TiO_2 NPs, including sol, sol−gel, solvo-thermal, and hydrothermal are currently being modified, accompanied with novel synthesis approaches of TiO_2 nanotubes, nanorods, nanowires, along with mesoporous and photonic structural configurations [39,88−95]. SEM images of TiO_2 prepared from most common sol gel and hydrothermal process are shown in Fig. 17.3C and D. Nanoscale titania maintains its photocatalytic ability, and surface reactivity, resulting in selective interactions/imprinting of biological molecules (phosphorylated proteins and peptides), in addition to some nonspecific binding with DNA. Anatase TiO_2 NPs (below 20 nm in size) is the predominant form with the most pronounced surface reactivity/interactions with cells in culture, phosphoproteins, and tissues in vivo [96−99]. Recent work reported the effective ROS generation mechanism of TiO_2 and zinc oxide NPs as food preservatives against different bacterial strains that proved their biocompatible nature [100]. According to the United States Food and Drug Administration (USFDA), the amount of TiO_2 as a food color additive is not to exceed 1% by body weight and is approved as food contact substance for packaging materials [101].

Despite the gradual increase in different applications, there are great concerns about the risks/critical impacts of TiO_2 NPs to human health and environment. Upon exposure to animals TiO_2 is found to prompt adverse pulmonary responses; however, the most probable dangerous biological effects of TiO_2 NPs are still not entirely understood compared to their nano-sized counterparts [102,103]. Conferring a strict estimate, anatase TiO_2 (crystals of eight-faced tetragonal dipyramids) at 5−10 μg/mL concentration stimulates some signs of cytotoxicity [99,104], and causes oxidative DNA damage, micronuclei formation, and lipid peroxidation. Moreover, TiO_2-NPs in aggregated form enter into cells when present in culture medium, mainly by endocytosis and result in cytotoxicity and genotoxicity in WISH cells culture. TiO_2 NPs can cause DNA damage, cytotoxicity, and apoptosis of A549 cells, and thus could induce cell injury and health hazards. Limited data are available in which carcinogenicity of TiO_2 NPs included the production of reactive oxygen species and signaled alterations of particular cancer-related genes [103,105,106].

In short, comprehensive toxicokinetics analyses of human body are required on distribution, metabolism, absorption, accumulation, and excretion of TiO_2 NPs via several exposures. Future research should concentrate on assessing the unclear molecular mechanisms by which TiO_2 NPs may cause cancer [105].

As a result, the modifications of TiO_2 NPs surface through compositing with polymers will be essential to prevent aggregation of nanoparticles, cytotoxic action, and oxidation. TiO_2 composites will probably behave as better agents than pure TiO_2 for abovementioned applications. In this regard, biopolymer matrix materials provide the best way to minimize the toxicological issues by tuning the structure, properties, and consequently functionality of these NPs as in the case of TiO_2-alginate bionanocomposites.

17.2 GRAPHENE/GRAPHITE—ALGINATE NANOCOMPOSITES

Graphene-based polymeric nanocomposites displayed a considerable improvement in mechanical properties (tensile strength, modulus, fracture toughness, fatigue, creep, wear resistance) along with superior thermoelectrical properties at a very low graphene wt% than other usual fillers, i.e., CNT, hallocyanite, clays. However, the enhancement in properties very much depends on the processing method, nature of polymer matrix, and types or forms of filler. It is found that pure graphene is usually incompatible with organic polymers resulting in nonhomogeneous composites. The *GO* sheets provide a large number of oxy functional groups that can change the van der Waals interaction, and increase the compatibility with organic polymers due to functionalization enhancement. Additionally, low cost, massive production, and good water dispersability also adds to the value of *GO* filler [107—111]. A large number of synthetic polymers, i.e., high density polyethylene, poly(methyl methacrylate), polystyrene, polycarbonate, polyvinyl alcohol, nylon, polypropylene, epoxy, silicon rubber, polyaniline, and polyimide are used as matrix material for graphene/*GO*-nanocomposites [112,113]. Similarly, *GO*-functionalized polysaccharides, such as alginate [114], chitosan [115,116], cellulose [117] are also documented. *GO*-based bionanocomposites offered improvement in mechanical properties along with biocompatibility and biodegradability. Increase in strength and toughness of chitosan by adding 1 wt% *GO* [115], and enhancement in elastic modulus of chitosan (over 200%) with the incorporation of just 0.1—0.3 wt% of graphene [116] are reported. *GO*—cellulose nanocomposite paper [117] showed that tensile strength increases by 1.2 and 2.3 folds using only 0.3 wt% of graphene, compared to neat cellulose and *GO* paper. Furthermore, *GO*-based nanocomposites were found noncytotoxic where *GO* acts as a biological activator in hydrophilic biopolymers, for example, gelatin/*GO* films were made with enhanced bioactivity and mechanical properties [118]. Using the wet-spinning process, sodium alginate (Na-alg)/*GO* fibers were synthesized through a $CaCl_2$ coagulating bath [119]. The presence of large number of hydrophilic groups in both *GO* nanofiller and Na-alg matrix such as carboxylate groups (COO^{-1}), ensured good compatibility that directs uniform dispersion of *GO* in Na-alg. The Na-alg/*GO* fibers showed an increase in tensile strength by 94%, while the modulus is 4.3 GPa with 4 wt% *GO*-loading and

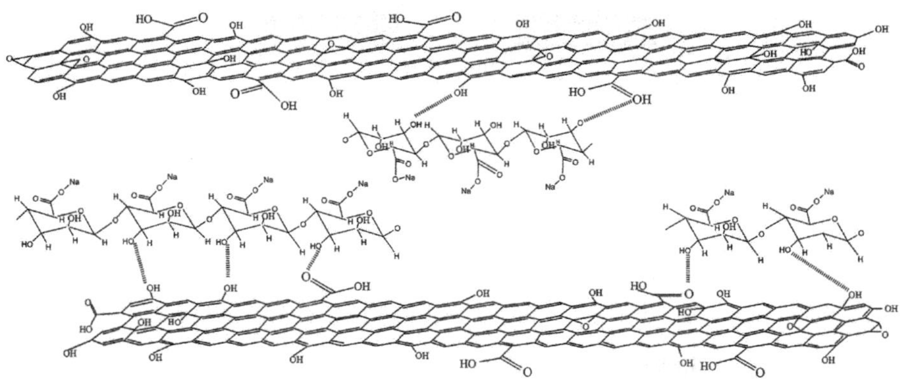

FIGURE 17.4

Schematic representation *GO*/Na-alg nanocomposite films structure [120]. *GO*, graphene oxide.

9.39 GPa with 2 wt% *GO* at 100% draw ratio. In 2013 alginate was composited to produce films by solution mixing with 0–6 wt% of *GO* [120]. Actually, the basal planes and the borders of *GO* sheet offer numerous oxygen functional groups which could permit homogeneous dispersion of *GO* sheets and produce interfacial interaction with hydroxyl groups of alginate. The schematic interaction between GO and Na-alg is shown in Fig. 17.4. Nanocomposite films of Na-alginate and GO (GO of 6 wt%) exhibited the highest value of tensile strength 71 to 113 MPa and Young's modulus from 0.85 to 4.18 GPa. In another study.

The fourier transform infra-red spectroscopy (FTIR) peaks of O–H stretching vibrations at 3347 and 3401 cm^{-1} broadened and shifted to smaller wavelengths of 3331 cm^{-1} which can be ascribed to the interaction of alginate and *GO* through intermolecular hydrogen bonds that ultimately induced good interfacial adhesion at the interface of *GO* and alginate, and resulted in an enhanced mechanical properties. X-ray diffraction (XRD) spectroscopy results also suggested that *GO* sheets, comprising numerous functional moieties on the surface, form linkage with the alginate matrix. The exceptional thermal stability of *GO*, which normally exhibits a decomposition temperature around 200°C, significantly improved the thermal stability of *GO*/Na-alg composite films compared to neat alginate film. In another study [114], bionanocomposite films of Na-alg (SA) were prepared with pure and modified *GO* by solution casting method. Tetraethylenepentamine-modified *GO* (*MGO*), due to addition of $-NH_2$ groups, improved its compatibility and interfacial interaction with alginic matrix. Fig. 17.5 showed preparation route and interaction between pure graphite, GO and modified GO with SA. It is already reported that the protonation of N-functional groups of modified *GO* converts them into multivalent nanoparticle and form electrostatic interactions with negatively charged carboxylate groups on Na-alg chains quickly [121]. In the FTIR study, the shifting of stretching vibration peaks of composite films from 3355, 1593 cm^{-1} to lower wave numbers

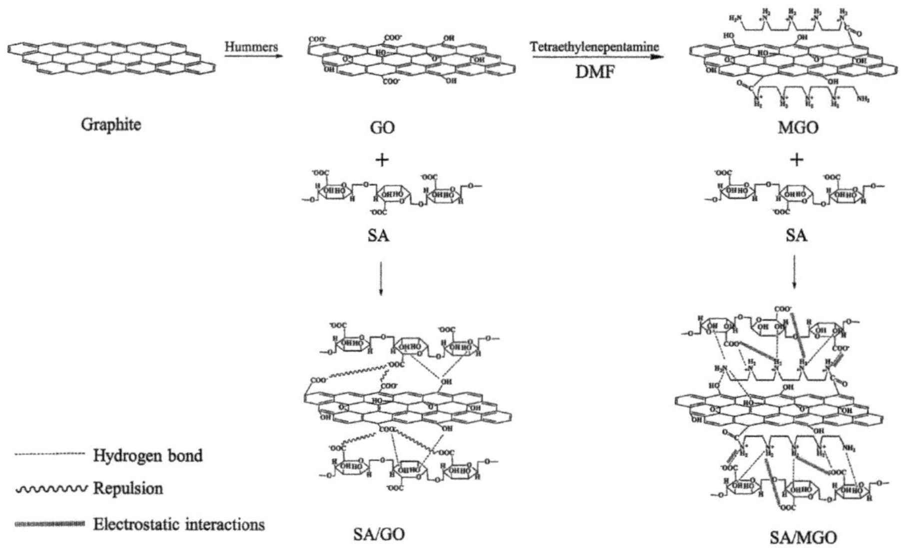

FIGURE 17.5

Preparation routes of MGO, SA/*GO*, and SA/MGO-n biocomposites, and schematic diagram of interaction between SA and fillers [114]. *GO*, graphene oxide; *MGO*, modified graphene oxide; *SA*, sodium–alginate.

ascribed that MGO retains $-NH_2$ and $-NH$ groups, directing stronger interaction (hydrogen bonds and electrostatic interactions) with alginate. The FTIR peaks of N–H stretching (amino groups), C–H, C=O bending and stretching (amide group vibrations), and C–N, C–OH stretching vibration also confirmed the modification of alginate [122]. The SEM images showed that the fillers disperse well at lower wt%, but at higher loading levels the aggregation of *GO* filler was more pronounced than MGO into Na-alg matrix. The tensile strength reached 59.28 and 69.32 MPa (improved by 44% and 68.4%) for unmodified and modified *GO*-based alginate films respectively, than neat alginate films i.e., 41.16 MPa, indicating a considerable reinforcing effect from *GO* and MGO nanoparticles. Modified *GO*/Na-alg films also have higher thermal stability and lower water uptake [114].

Besides improving the mechanical properties of polymeric material, *GO* with multivalent ions can also produce the association at edges by means of intense coordination with –COOH while a weak attachment is also observed on the *GO* planes with the carbonyl and epoxy groups [123,124]. On the other hand alginate forms cross-linked networks via ionic linkage with multivalent cations e.g., Ca^{2+}/Ba^{2+}/Fe^{3+} (Fig. 17.6). Alginate on cross-linking with polyvalent cations assists to increase their water barrier properties/resistance, mechanical resistance, rigidity, and cohesiveness [125,126]. Multivalent ion cross-linked Na-alg/*GO* nanocomposite films were produced by a two-step synthesis approach which provides outstanding mechanical properties [127]. Na-alg and *GO* are self-assembled due to hydrogen bonding and

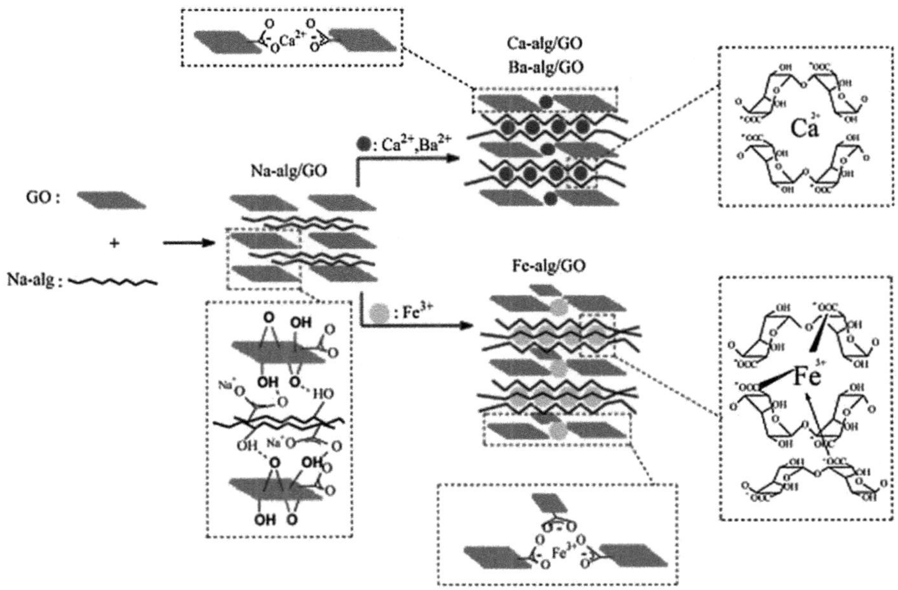

FIGURE 17.6

Schematic models of the alginate/*GO* composite films cross-linked by multivalent cations [127]. *GO*, graphene oxide.

interfacial adhesion, increasing the interlayer spacing of *GO* sheets from 0.83 to 1.08 nm which enhances after cross-linking of composite films with multivalent cations further. The resultant multivalent cation cross-linked Na-alg/*GO* films have higher thermal stability than Na-alg/*GO*. High tensile strength of the composited films is assigned to ionic cross-linking of alginate by appropriate ion-exchange strategy, hydrogen bonds between *GO* and Na-alg and interactions between multivalent cations and *GO*. The multivalent Fe-alg/*GO* composite films were found to be stronger than divalent Na-alg/*GO* films. There is no doubt that these novel nanocomposite films are expected to be used in packing and structural materials.

17.2.1 GRAPHENE–ALGINATE ADSORBENTS

Ionotropic metal alginate complexes, e.g., nickel alginate (Ni-alg), copper alginate (Cu-alg), and cobalt alginate (Co-alg) depicted the chelation ability of alginate with metal ions [128–132]. Ca-alg has been extensively utilized to immobilize various host materials such as TiO_2 nanoparticles [133], activated carbon [134], maghemite nanoparticles [135], kaolin [136], CNT [137], chitosan [138], to produce adsorbents for the removal of heavy metals and dyes from aqueous media. Usually, Ca-alg is characterized with two different binding sites relative to the different –COOH group present on different polymeric block chains. The carboxyl groups on G- and M-blocks of alginate participated in adsorption process via egg-box structure (G-block) or by direct adsorption process (M-block) [139]. Graphene/*GO* materials are also used for the adsorption of fluoride ions [140], Cu, Cd, Pb, etc. [141,142], heavy metals (As) [143], dyes (methylene blue, methyl violet, orange-G, rhodamine-B) [144] and display good adsorption ability.

So using the sol–gel process *GO*/Ca-alg fibers were prepared for the removal of methylene blue dye [145]. The adsorption capacity of *GO*/Ca-alg was found to be higher due to the presence of *GO* which actually provides larger surface area, and more exposed binding sites (functional groups) for surface complexation and so improved the adsorption capacity of the composites [137,146]. The adsorption procedure followed the Langmuir and the Freundlich isotherms, however, the former was more appropriate with maximum adsorption ability value of 181.8 mg/g. This model also suggested that adsorption of methylene blue occurs on the uniform monolayer surface of *GO*/Ca-alg. The kinetic study showed that the adsorption process follows pseudo-second-order equation while dye adsorption comes out of intraparticle diffusion mechanism. The composite adsorption ability increased with temperature and the process is reported to be exothermic and thermodynamically favorable. *GO*/Ca-alg has high adsorption and desorption (78.6% at 1.5 pH) capacity in the entire pH range. So it is advantageous to avoid the secondary pollution problem due to easy separation from aqueous solution [145].

Effect of the *GO* loading on Ca-alg and macroporous alginic beads structure was investigated for the removal of acridine orange dye [147]. Graphical presentation of the dye adsorption mechanism is illustrated in Fig. 17.7. The adsorption behavior followed the Langmuir isotherm model. The disappearance of characteristic peak of *GO* in XRD proved the well-homogeneous distribution of *GO* in alginate matrix. Encapsulated *GO* provides active sites for ion exchanging of dye molecules and a much improvement in removal efficacy in Ca-alg/*GO* adsorbents is observed i.e., from 62% to 87% but with slow adsorption rate. This was primarily due to closed structure of Ca-alg/*GO* that to some extent disadvantageous for ion diffusion compared to relative open pore structure of alginic beads as provided by SEM analysis. However, introduction of *GO* in Ca-alg beads doubled the adsorption capacity and enhanced by 15% in alginic beads that are competitive to previously reported adsorbents [148–152]. It was also concluded that *GO* adsorption relied on pH-dependent dissociation oxygen functional groups and also on $\pi-\pi$ stacking effect, etc. So on the whole this alginate/*GO* hybrid material would offer a definite adsorbent system for water [147].

Polymeric hydrogels incorporated with nanomaterials represent an interesting approach to tailor the mechanical characteristics of the hydrogels which at the same time responsive to various stimuli [155]. Using the sol–gel process, *GO* was used to synthesize Ca-alg hybrid gel material for heavy metal adsorption [153]. The kinetic study of adsorption of Cu ions on Ca-alg and *GO* encapsulated Ca-alg followed the pseudo-first-order kinetic equation and the adsorption isotherm fitted well into Langmuir isotherm. When *GO* was encapsulated in Ca-alg, the q_m (maximum adsorption capacity) increased to 60.2 mg/g. This q_m value is significantly higher than for *GO* alone which is 46.6 mg/g. The large surface area and

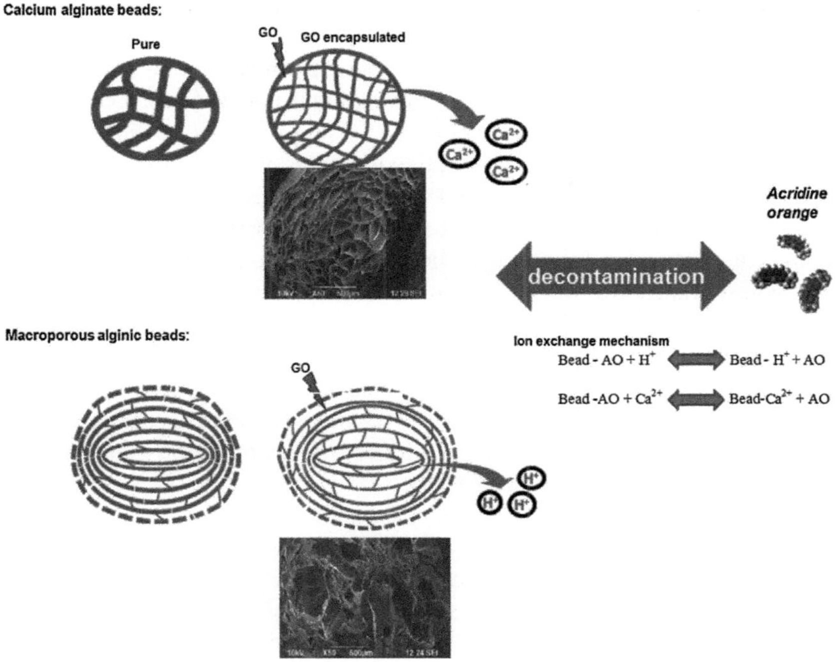

FIGURE 17.7

Graphical abstract of acridine orange adsorption on *GO*-encapsulated Ca-alginate and macroporous alginic beads [147]. *GO*, graphene oxide.

oxygen-containing functional groups on the surface and edges of *GO* facilitate the adsorption of Cu^{+2} [142], and thus become a reason of enhanced adsorption capacity of *GO* encapsulated Ca-alg beads. The adsorption kinetics showed a slower adsorption process for *GO* hybrid Ca-alg beads as the distribution of Cu^{+2} binding sites are hindered by the presence of *GO* nanofiller while a uniform distribution of Cu^{+2} binding sites is observed in Ca-alg beads [154]. The *GO*-encapsulated Cu-alg beads have more defined pore structure as observed in SEM, but are thermally less stable, owing to the increased thermal conductivity of *GO*, promoting bond cleavage [147].

The previous studies [145,153,156] reported *GO*-incorporated alginate composites characterized by single network. Gong et al. [157] fabricated the composited double network hydrogel for the first time that provides better strength and mechanical properties when compared to single networks. The double network gels are characterized with two interpenetrating polymer networks in which one is highly stretched and cross-linked, brittle, and stiff, while the other is less densely packed, soft, flexible, and stretchable. The two networks are formed via covalent linkages and thus provide gel with higher bond energy. This study is actually a continuation of previous research works [145,158,159] for comparison of single and double network gel. Novel nanocomposited graphene/alginate double network beads were also produced [160]. Batch and column adsorption isotherms and kinetics were studied to compare double and single network graphene/alginate nanocomposites. A more wrinkled surface of double network beads is examined through SEM which supports adsorption phenomenon by providing more contact sites. The surface area, average pore volume, and diameter of double network were higher when compared to single network owing to cross-linked embedded *GO* particles in 3D network with alginate. The thermal stability of double network is higher than single network nanocomposite gels. The double network graphene/alginate nanocomposites showed better maximum batch adsorption capacity, i.e., 2.3 g/g, for methylene blue dye than single network, i.e., 1.84 g/g, provided by Langmuir adsorption model which is quite superior to already reported alginate-based materials [135,149,161–163]. In column adsorption, the removal rate of dye is reduced but still carries considerable values even after 10 adsorption/desorption cycles. Kinetic studies of adsorption followed pseudo-second-order model. Both the network gels are stable up to pH 10 but at pH 12 the adsorption capacity of single network falls. Thus these graphene–alginate double network hybrid gels could be more useful for water-treatment process [160].

Recently, *GO* was reported for the quick removal of radioactive long-lived anthropogenic radionuclides from wastewater [164]. Moreover, it effectively controls the stability, prevents aggregation (agglomeration), and increases the usage (recovery) of magnetic NPs in water-treatment process. In situ synthesis of magnetic PB/*GO* nanocomposites was reported [165] because the direct use of magnetic NPs for broader water-treatment applications is impractical. Therefore anchoring Fe_3O_4 magnetic NPs onto large surface area of *GO* and in situ coating of Fe_3O_4 magnetic NPs with Prussian blue (PB) is a promising method to overcome radioactive water contamination [165]. Ca-alg microbeads were designed to encapsulate/immobilized magnetic PB/Fe_3O_4/*GO* nanocomposites for Cs^+ removal in nuclear wastewater and soil [166]. By fixing the magnetic PB on *GO*, effective reduction in the aggregation of PB was observed and the magnetism was sustained. The PB/Fe_3O_4/*GO* adsorbents displayed high selectivity to Cs^+ and could remove its trace amounts while these novel nanocomposites conferred 80% of Cs^+ removal in less than 2 h with maximum adsorption capacities of 43.5 mg/g. Langmuir isotherm fitted well with adsorption values and kinetics study directed the pseudo-second-order kinetic model. Higher temperature increased the adsorption ability of cesium; elaborating the adsorption process is endothermic and materials are thermally stable. The microbeads prevent the leaching of PB within the pH of 4–10 so can be effectively used for the treatment of large-scale cesium-contaminated water and soil.

Biopolymers also offered characteristic aspects of aerogels such as open porous structure with high specific pore volume and surface, therefore, huge interest is developed for biopolymer (cellulose, alginate, chitin/chitosan)-based aerogels [167–170]. Na-alg can also be formulated in many forms i.e., films, beads, fibers, and aerogels, attributed to egg-box structure followed by 3-D network assembly between different polymer chains [171–174]. Na-alg aerogel with its native properties is used in drug delivery, tissue engineering, oil absorptive materials, sewage treatment, and thermal insulation. But in addition to structural strength problems,

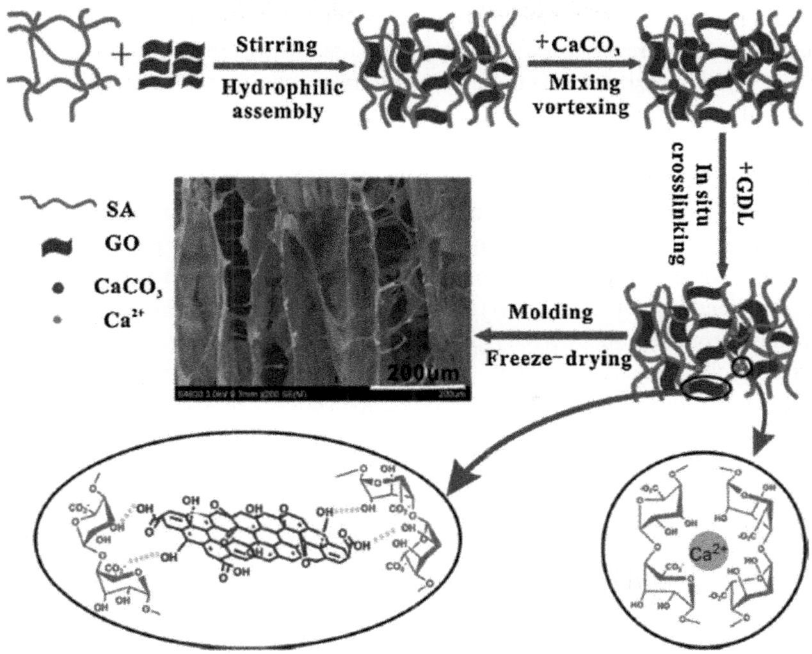

FIGURE 17.8

Schematic illustration of fabrication process and formation mechanism of Na-alginate/*GO* aerogel [184]. *GO*, graphene oxide; *SA*, sodium–alginate.

alginate gels also suffer from brittle and heterogeneous pore size due to quick gelation and fragile collapse [175–181]. So using the in situ cross-linker, GDL, 3D mold –shaped Na-alg-*GO* aerogels (Fig. 17.8) were synthesized to improve the mechanical and elastic performance of neat Na-alg aerogels [184]. The FTIR and X-ray photoelectron spectroscopy characterized the formation of strong intermolecular hydrogen bonds between alginate (–OH) and *GO* (–COOH). The pure Na-alg aerogel reported to be collapsed at 60% strain [185] but Na-alg/*GO* aerogels demonstrated a notable improvement in compressive strength (from 233 to 324 kPa) even after the fifth compressive cycle the strength was 245 kPa. This improvement in strength and toughness is also described by SEM analysis which reported the homogeneous dispersion, and miscibility of the components. Various "riband bridges" that acted as spring plates to oppose the deformation along with the load-transfer ability were developed due to interconnect porous architecture. From the Langmuir adsorption isotherms study, the q_m value of Na-alg/*GO* aerogels was found to be 98 mg/g and 267 mg/g for Cu^{2+} and Pb^{2+} respectively which indicates their potential for waste water-treatment process [184].

Graphene aerogels have also been extensively studied for highly flexible strain sensors. Yet, most of the graphene aerogels are found to be very weak and delicate to manage because of the presence of $\pi-\pi$ stacking and van der Waals interactions between adjoining graphene sheets that directs to a brittle structure. Therefore generation of highly sensitive and conductive graphene network skeleton with excellent mechanical properties is quiet a great challenge. Herein fabrication of low density bio-based graphene/Na-alg aerogel with high porosity of 99.61% is reported through freeze drying and chemical reduction methods [186]. Presence of effective H-bonding between graphene and Na-alg actually improved both strength and elasticity. The subsequent aerogels display exceptional durability i.e., greater than 6000 loading/unloading cycles, outstanding sensitivity to compression i.e., gauge factor is 1.01 and bending sensitivity is of 0.172 rad^{-1}. Thus the provided mechanical flexibility and electronic conductivity proposed these bio-based aerogels as powerful candidate in flexible strain sensors.

Wet-spinned biocomposite fibers of *GO*-doped Ca-alg was synthesized and used for the adsorption of Ciprofloxacin from waste stream [156]. The adsorption values fitted well with Langmuir isotherm and followed pseudo-second-order kinetic model. With 6% *GO*/Ca-alg showed high q_m value, i.e., 66.3 mg/g at a pH of 5.9. The ciprofloxacin removal rate increased with increasing percentage of *GO* and reached approximately 79% with 6% of *GO* contents. This significant increment in adsorption is related to micro surface channels and additional functional groups provided by *GO* reinforced Ca-alg composite fibers than neat Ca-alg fibers.

17.2.2 GRAPHENE–ALGINATE SCAFFOLDS

Appropriate pore size, mechanical strength, and stability are the requirements of tissue engineering scaffolds. *GO*/Na-alg composite films and *GO*/Na-alg hybrid gel beads without the interconnected porous structure are reported [120,187]. Fan et al. [188] formulated *GO*/Na-alg/polyacrylamide (PAM) nanocomposite hydrogel of less than 10 μm pore size with increased dye-adsorption capacity. However, toxic acrylamide, the degraded product of PAM, and its smaller pore size limited its application in tissue engineering. Therefore 3D porous *GO*/Na-alg composite scaffolds (Fig. 17.9) with superior mechanical strength and proper pore size are developed for tissue engineering applications [189]. The incorporation of *GO* did not affect the porosity and interconnention of Na-alg scaffolds but changed the pore size. The Young's modulus and compressive strength of *GO*/Na-alg with 5 wt% of *GO* increased by 91.6% (2.33 MPa), and 42.8% (0.140 MPa), respectively, than bare Na-alg (1.2 and 0.098 MPa). Interfacial adhesion due to hydrogen bonds among functional groups of *GO* and Na-alg (Fig. 17.9) is one of the principal reason of this improvement. Moreover, presence of large number of functional moieties on *GO* surface improved the hydrophilic character [190] of composite which is helpful in drugs and nutrients uptake, and hence made it feasible for tissue engineering.

17.2.3 GRAPHENE–ALGINATE MEMBRANES

Numerous efforts were done to fabricate various kinds of membranes, mainly for pharmaceutical, water purification, and electronics applications. Inorganic

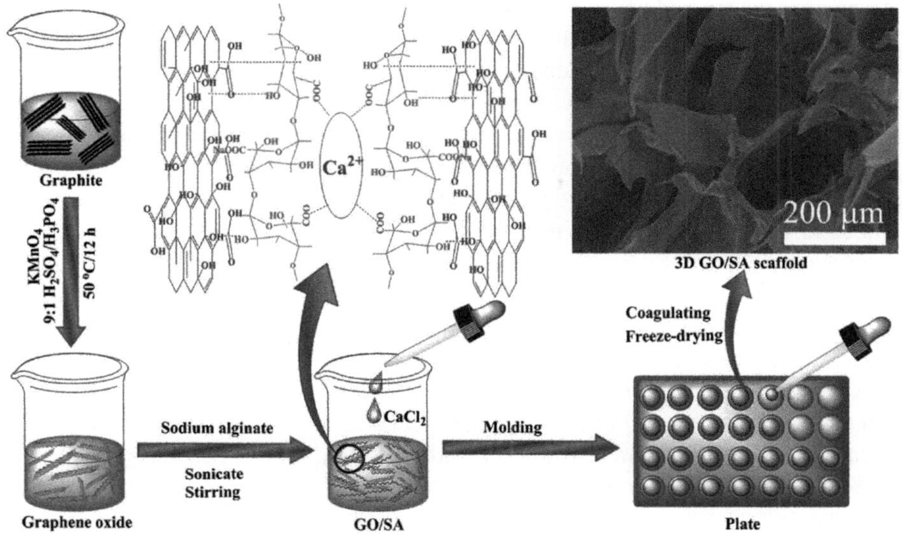

FIGURE 17.9

Fabrication process of 3D porous *GO*/Na-alg composite scaffolds and schematic illustration of hydrogen bonding and covalent linking between *GO* and Na-alg [189]. *GO*, graphene oxide.

membranes, e.g., ceramics, silica, and zeolites, have exceptional performance but limited commercial applications due to high costs involved in their processing. Instead, organic polymeric materials owing to their low-cost, good film-forming property, and easy physical and chemical modification proved to be the most appropriate and useful candidate for barrier membranes [191–195]. Thus the production of hybrid membranes of both polymer and inorganic constituent provides uniqueness of inorganic constituents with exceptional film-forming properties of polymer. Continuous attempts are made, including grafting, cross-linking, and blending, to fabricate new kinds of composite membranes through homogeneous distribution of fillers in polymer matrix [196–199].

Distillation is a conventional method of separation using these membranes. However, some economic and purity issues of this method drew the attention of researchers toward pervaporation (PV) method. In this perspective, *GO* filler at very low content ratio is recognized to generate a remarkable improvement in the mechanical, morphological, and barrier properties of the resulting membrane [110,133,200–270].

Numerous types of *GO* ultrathin membranes are made and immensely used in water purification, gas separation, and PV [201–203]. Among organic polymers, Na-alg-modified membranes were also used extensively in PV [204]. These properties of *GO* encourage to develop nanocomposite membranes using *GO* as a filler material in the Na-alg matrix and are used in the pervaporation dehydration of organic solvents. The selectivity and permeability of water channels in hybrid membrane interface, especially if we talk about *GO*-Na-alg interface, was attributed to the transport of water molecules toward the low-friction carbon surface (nonoxide region of *GO*). Water molecules slide along nanosheet to next interlayer rather its movement through edges and defects, in contrast to ethanol molecules as it move around the same interlayer, and the proper size of the structural defects and edge to edge slits acted as molecular sieve to block organic solvent (ethanol) molecules [202,205,206]. Highly oxygenated surface functionalized graphene sheets (FGSs) were incorporated in Na-alg to develop water-selective intercalation structure membranes for the PV dehydration of isopropanol alcohol [207]. Increase in FGS contents leads to more confined crystalline structure and hydrogen bond interaction that improves the selectivity and permeation of neat Na-alg membranes. But above 2 wt% FGS, the selectivity is lost due to agglomeration of filler particles. A four-times increase in water selectivity is observed by kinetic and thermodynamic factors. Additionally, both hydrophilicity and thermal stability are improved by the homogeneous dispersion of highly functionalized FGS fillers while a possible water-selective interface is formed within the Na-alg matrix.

Similarly, hybrid membranes using Na-alg as matrix material were prepared by blending with pristine graphene oxide (pGO) and reduced graphene oxide (rGO) nanosheets [208]. Both pGO and rGO showed different physicochemical structures as rGO is characterized by smaller nanosheet size (0.9 μm), less negative charges, more structural defects, more non-oxide regions, and less oxygen-carrying groups. All these factors along with increased water channels of rGO-filled membranes (1.6 wt%) displayed higher permeation flux [1699 g/(m^2 h)] and separation factor (1566).

The well-aligned brick-and-mortar structural design of rGO and pGO was observed through field emission scanning electron microscopy. However, agglomeration is pronounced at higher contents of 2 wt% rGO than 2.4 wt% of pGO leading to stronger van der Waals interaction in rGO. Moreover, the membranes loaded with rGO have high crystallinity than pGO membranes. This fact is mainly due to decreased hydrogen bond effect together with the elimination of oxygen-carrying groups from basal planes of nanosheet. Na-alg having 2 wt% of pGO showed a respective increase of 45% (98.8−143 MPa) and 98% (3.35−6.64 GPa) in tensile strength and tensile modulus. While a slightly lower mechanical strength was seen in Na-alg with 2 wt% of rGO further strengthening the result that agglomeration of rGO nanosheets occur at high content [208]. This increment was ascribed to the brick-and-mortar morphology along with interfacial interaction [209,210] that could provide effective load-transfer property. Thus both pristine and reduced *GO*/Na-alg hybrid membranes exhibited higher pervaporation performance for dehydration of ethanol−water system than the previously reported Na-alg membranes [204].

Highly aligned *GO* or graphene sheets incorporated in Na-alg and Ca-alg matrix, respectively, with different compositions were studied [211] as shown in Fig. 17.10. The order parameter $\langle P_2 \rangle$ values from affine deformation model for *GO*/Na-alg composites showed slight dependence on the graphene sheets concentration (above ~10 wt%), with a maximum $\langle P_2 \rangle$ value of 0.8 at 25 wt% *GO* inside the Na-alg matrix. However, the lower alignment of graphene sheets inside the Ca-alg matrix was

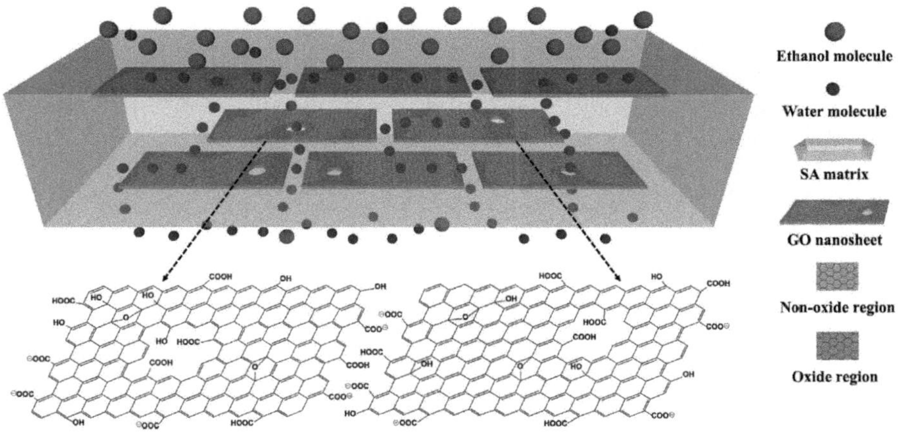

FIGURE 17.10

Schematic representation of mechanism for selective water permeation through hybrid membrane with well-aligned *GO* nanosheets [208]. *GO*, graphene oxide; *SA*, sodium–alginate.

observed, finding the average $\langle P_2 \rangle$ value of 0.7. Orientation of graphene sheets within the Ba-alg matrix is not well defined. The formation of a highly aligned *GO*/Na-alg composite structure was described through the affine deformation model, while the developed yield stress causes sheets to align in-plane with the polymer matrix upon drying. The impaired orientation of graphene sheets inside the Ca-alg matrix and absence of orientation in the Ba-alg matrix was determined through metal ion cross-linking, developing the structure in the polymer matrix itself.

Sodium alginate (Na-alg) membrane, carrying Zwitterionic-*GO* nanosheets as filler, were produced for effective water-permeation and water–alcohol separation process [212]. Presence of high-density zwitterionic groups on *GO* provides electrostatic interaction sites in combination with water molecules, directing to high water affinity and ethanol repellency, while the 2D *GO* confers constant path offering the membrane with efficient water permeation (Fig. 17.11). These nanohybrid membranes displayed supreme separation using the model of dehydration of water–alcohol mixture system. When compared to Na-alg and the unmodified-*GO*- nanohybrid membrane, the Zwitterionic-*GO*-nanosheets-filled Na-alg hybrid membranes showed optimum permeation flux (2140 g/m^2 h) and separation factor (1370). Thus the potential of zwitterionic graphene materials in membrane technology (especially for dense water-permeation membranes) delivers a facile method to make economical sorption and water-transport pathway in membrane for pronounced applications.

17.2.4 GRAPHENE–ALGINATE ENCAPSULATING SYSTEM

Graphene materials (scaffolds, 2D, 3D) showed to improve the proliferation, adhesion, and differentiation of mesenchymal stem cells, rate of maturation from murine

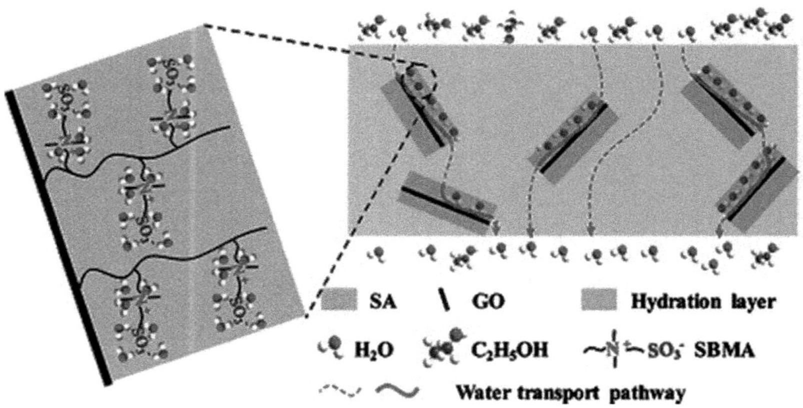

FIGURE 17.11

Schematic illustration of water-transport pathway through Zwitterionic-*GO*-nanosheets filled Na-alg (SA) [212]. *GO*, graphene oxide.

myoblasts into myotubes, and enhance gelatin methacrylate mechanical and electrical properties, while not affecting the viability of encapsulated fibroblast cells [213–216].

These properties of *GO* are combined with microencapsulation technology but for the long-term benefits of this system, the selection of appropriate polymer matrices is vital. In a previous study, *GO*-containing alginate beads of 50 mm of diameter were effectively used as copper-adsorbent material [153]. Recent studies reported the reinforcement of *GO* filler into alginate microcapsules (160 mm) to encapsulate C_2C_{12} myoblasts to secrete erythropoietin (EPO) [217]. It was observed that presence of *GO* inside the alginate core confers irregular surface appearance on the alginate microcapsule, more intensely with higher *GO* concentration. On addition of poly-L-lysine (PLL), i.e., auxiliary to *GO*/alg microcapsules, a convincing constancy of the surfaces was observed. Addition of *GO* NPs at a concentration of 10–50 mg/mL detected a minor number of C_2C_{12}-EPO myoblasts cells apoptosis, probably due to different mechanisms of interaction between *GO* NPs and cells in the alg matrix (3D) compared to a culture medium (2D). It was noted that the concentration of *GO* between 25 and 50 mg/mL enhanced the metabolic activity, viability, and membrane integrity of encapsulated C_2C_{12}-EPO myoblasts, and can be desirably used for clinical application of cell-microencapsulation.

17.2.5 GRAPHENE–ALGINATE FILMS FOR ELECTRONICS AND SENSORS

Synthetic and native polymers comprising carboxyl functional groups, such as (polyacrylic acid and carboxymethyl cellulose) are normally used as polymer binders for porous composites in energy-storage devices, for example, in electrodes (in primary and secondary batteries), supercapacitors, double-layer/electrochemical capacitors, battery hybrid devices. Alginate is a desirable material (binder) for energy-storage devices owing to the presence of high percentage of –COOH groups and easy dissolution in water. Carboxylic groups of alginate promptly participate to form rigid hydrogen bonds with oxidized surfaces, leading to improved stability and mechanical strength of device parts [218,219].

Biodegradable Na-alg and graphene oxide (ALG) nanocomposite films were fabricated by solution casting method and then stacked between an indium tin oxide (ITO)-coated poly(ethylene terephthalate) (PET) and a PET sheet for triboelectric generator [219]. A rough surface morphology of films was observed by SEM and physical interaction between the *GO* and Na-alg is dominant. The triboelectric generator produced with the highest *GO* (5 wt%) content results in short-circuit current density of 1×10^{-4} mA/cm^2, open-circuit voltage up to 1.3 V, corresponding to a peak power density of 1.33 mW/m^2. The results are not corresponding to triboelectric nanogenerators studied previously [220,221] but the methodology provides first step toward the application of self-powered bioelectronics devices.

Numerous graphene-based sensors and biosensors are established effectively in electrochemistry. Graphene–chitosan composite film modified electrode is produced for immobilizing the hemoglobin [222]. Reports are also available in which graphene-based electrodes are used for the detection of dopamine, NADH, glucose, ethanol, Cd^{2+}, Pb^{2+} [223–226], and for the study of proteins [227] as well. A new way to understand direct electrochemistry of proteins is the casting of protein-incorporated films, either made up of biopolymers, hydrogel polymers, insoluble surfactants, and inorganic/organic composite materials. Surface of the solid electrodes provides microenvironment to maintain the structure of redox proteins and stimulate the electron transfer between electro-active center of proteins and the electrode [228–232].

An attractive methodology described here is the introduction of *GO* sheets into the porous Ca-alg films on the surface of electrode, that is used to immobilize enzyme (myoglobin) [233]. These novel, porous *GO*/Ca-alg films may produce synergy on electro–catalytic activity and therefore improve the sensitivity of the biosensor. The porosity of alginate is already studied to be enhanced using PEG as porogenic and $CaCl_2$ as cross-linking agent [234]. The charge transfer resistance (R_{ct}) values of ordinary *GO*/Ca-alg film, when compared with R_{ct} of porous *GO*/Ca-alg, was found to be reduced over 700 Ω, indicating that the porous morphology of film supports the direct electron transfer between enzymes and electrodes. On the other hand, myoglobin *GO*/Ca-alg films indicated much larger value of R_{ct}, i.e., 4596.5 Ω. Good conductive and biocompatible character of *GO* resulted in higher catalytic efficiency of the porous myoglobin-immobilized-*GO*/Ca-alg films than neat porous myoglobin-Ca-alg and myoglobin-*GO*-Nafion [235,236] films toward reduction of O_2 and H_2O_2, and could sense a minor concentration of NO_2. Fig. 17.12 shows cyclic voltammograms for electrocatalytic reduction of NO_2^- (A); O_2 (B); and H_2O_2. This concludes that the films could be potentially a useful candidate for biosensors/reactors (Fig. 17.12).

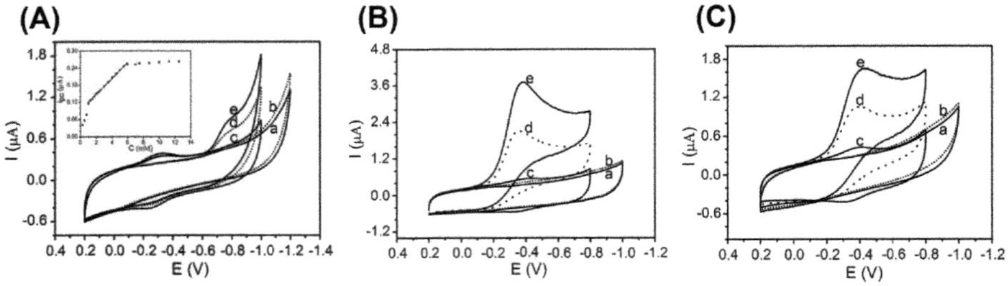

FIGURE 17.12

(A) Cyclic voltammograms at a scan rate of 0.1 V/s in 8 mL of pH 5.0 buffers for porous *GO*/Ca-alg films with (a) no $NaNO_2$, (b) 2.5 mM $NaNO_2$, and for porous Mb–*GO*–CA films with (c) no $NaNO_2$, (d) 2.5 mM $NaNO_2$, and (e) 5.0 mM $NaNO_2$. Inset is the calibration curve of reduction peak currents and concentrations of $NaNO_2$; and cyclic voltammograms at a scan rate of 0.1 V/s in 8 mL of pH 7.0 buffers for porous *GO*/Ca-alg films; (B) with (a) no O_2 present, (b) 20 mL of air was injected into the sealed cell, and for porous myoglobin-*GO*/Ca-alg films with (c) no O_2 present, (d) 20 mL, and (e) 40 mL of air was injected, respectively, (C) with (a) no/H_2O_2, (b) 18.75 μm H_2O_2, and for porous myoglobin–*GO*/Ca-alg films with (c) no H_2O_2, (d) 18.75 μM H_2O_2, and (e) 37.5 μM H_2O_2 [233].

17.2.6 GRAPHITE–ALGINATE CONDUCTIVE COMPOSITE

The conductive property of graphite material is usefully employed to make conductive composite hydrogel. Carbonaceous material actually improves the mechanical toughness of these hydrogels which otherwise is a noticeable problem with these materials. CNT and graphite were already reported to improve the strength of polyacrylamide and chitosan-conductive hydrogels [237,238], but the synthesis routes are complicated. Hence there is a need to develop a simple process involving mixing of these carbon filler into polymeric hydrogels to make them conductive. Therefore graphite is homogenously dispersed in sodium alginate matrix via in situ release of Ca^{+2} ions into conductive composite hydrogels [239]. As the graphite contents are increased between 10 and 30 wt%, the conductivity of composite hydrogel increases, though, it decrease above 30 wt% of graphite contents, which is higher when compared to others [240]. Actually graphite microparticles incline to form agglomerates at high content, resulting in irregular dispersion of graphite and reducing conductivity of the composite hydrogels. Swelling behavior of the composite hydrogels at various graphite contents were observed and found to be decreased at a higher graphite contents.

17.2.7 GRAPHITE–ALGINATE ADSORBENT

Nanographite-immobilized/impregenated alginate composite gel beads were obtained and used for the separation of Co(II) and Ni(II) from aqueous solution [241]. The sorption process of these ions is elaborated by kinetic results and followed pseudo-second-order kinetic model. The Co(II) adsorption is well explained by Freundlich isotherm whereas the Ni(II) adsorption is presented well by Langmuir isotherm. The q_m values for Co(II) and Ni(II) ions were 11.63 and 11.48 mg/g, respectively. The alginate matrix actually provides ion-exchange sites and removal of Co(II) and Ni(II) ions from solution was possibly due to ion-exchange reaction between Ca(II) ions present in the beads and Co(II) and Ni(II) ions. Hence the above cited literature fully explored the value of ALG based GO nanocomposites for diverse applications. Although appropriate wt.% of GO is crucial to attain the respective properties. Table 17.1 summarized the impact of filler percentage and synthesis method on the mechanical properties of GO/ALG hybrid nanocomposites.

17.3 TITANIA–ALGINATE HYBRID NANOCOMPOSITES

17.3.1 TITANIA–ALGINATE ADSORBENT

Sodium alginate and titania NPs (Na-alg/TiO_2) composites were synthesized and investigated for dye (DR80 and AG25) adsorption from textile wastewater [133]. The FTIR peaks indicating the stretching vibration of COOH groups in alg shifted to lower wave numbers, i.e., from 3442 to 3396 cm^{-1} and appeared broader and weaker, which showed the strong interaction between these groups and TiO_2 NPs. The decrease in the band showing primary –OH groups (2151 cm^{-1}) reflects immobilization of TiO_2 nanoparticle onto the Na-alg [242,243]. Dye adsorption increased with composite amount, attributing to availability and much improvement in adsorbent surface. At pH 2, Na-alg/TiO_2 displayed notably high dye adsorption due to electrostatic attraction between the ionized, positively charged surface functionalities of the adsorbent and negatively charged anionic dye. However, as the pH increased, the negatively charged sites became more pronounced causing the electrostatic repulsion and did not favor dye adsorption [244]. The dye adsorption and kinetic studies showed that obtained results followed Langmuir isotherm and the pseudo-second-order kinetic model.

Organic–inorganic nanocomposite hydrogels are studied as a result of their improved thermomechanical stability and swelling properties than the single component hydrogels [245–247]. TiO_2-incorporated Na-alg cross-linked polyacrylic (AA) hydrogels were fabricated (Fig. 17.13) by free-radical copolymerization and used for the removal of cationic methylene blue dye [248]. SEM micrographs confirmed the formation of porous 3-D internal structure of the chemically cross-linked nanocomposite hydrogel. The prepared hydrogels have high swelling ratio due to increasing concentration of AA onto Na-alg, which intensify the intermolecular spaces between Na-alg chains and promotes H-bonding with water molecules. At 1.2 M of AA, the maximum swelling capacity was approximately 413 g/g. Methylene blue adsorption proposed firstly electrostatic interaction between the anionic surface of absorbent (COO^-) and the positively charged dye, and secondly, hydrogen-bonding interaction between the reactive hydroxyl groups and TiO_2^- surface/OH groups of polymer chains and the imines groups of the dye [248]. On increasing the amount of TiO_2 (0.05–0.2), the dye removal rate is

Table 17.1 Preparation Methods, Mechanical Properties, and Applications of Some *GO*-Reinforced Na-Alginate Nanohybrid Systems

Sr. No	Wt% Filler/Matrix	Synthesis Method	Mechanical Properties			Applications	References
			Tensile Strength	Young/Compressive Modulus	Compressive Strength		
1	4 *GO*/Na-alginate	Wet spinning	0.62 GPa	4.30 GPa	–	Fibers for wound dressings	[119]
2	5 *GO*/Na-alginate	Solution mixing and freeze drying	–	2.33 MPa	0.14 MPa	Scaffolds for tissue engineering	[189]
3	4 *GO*/Na-alginate	In situ cross-linking and freeze-drying method	–	–	324 kPa	Aerogels for heavy metal separation from wastewater	[184]
4	2 *GO*/Na-alginate	Physical blending	143.1 MPa	6.64 GPa	–	Separation membranes for water/organic solvent mixture	[208]
5	1 Modified *GO*/Na-alginate	Solution casting	69.32 MPa	3.8 GPa	–	–	[114]
6	*GO*/Na-alginate	Solution casting	113 MPa	4.18 GPa	–	–	[120]

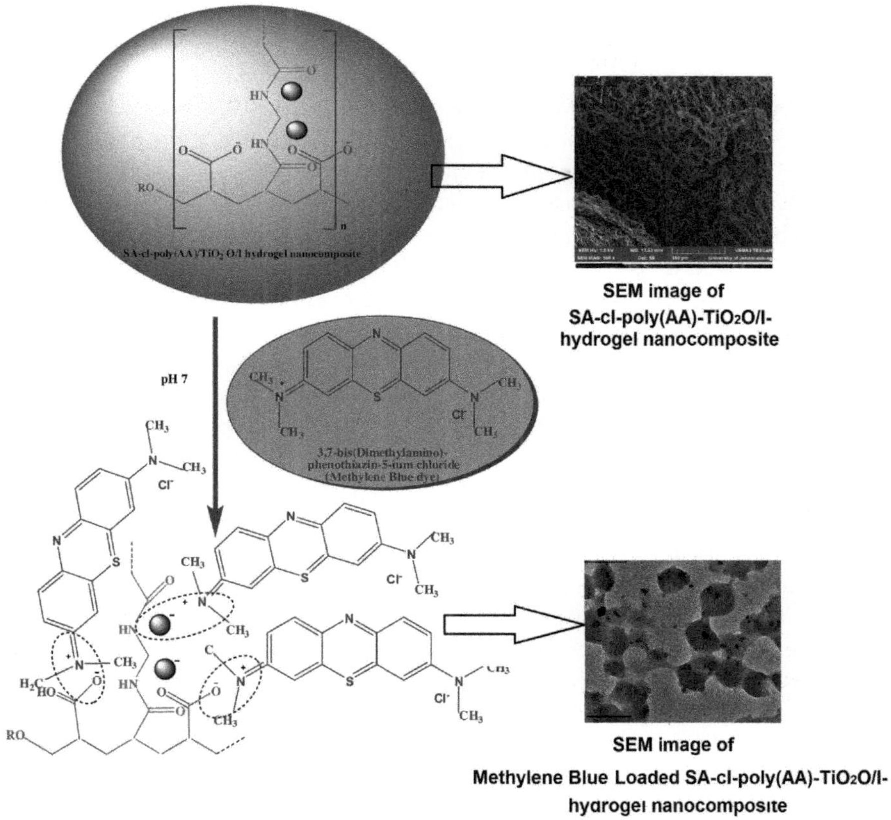

FIGURE 17.13

General mechanism for SA-cross-linked-poly (AA)-TiO_2O/I-hydrogel nanocomposite synthesis by KPS-initiated free-radical polymerization of AA onto SA in the presence of MBA [248]. *KPS*, potassium persulfate; *SA*, sodium–alginate.

enhanced owing to increase in ionic centers and intrinsic charge repulsion inside the hydrogel network, providing more surface area for methylene blue adsorption [249–251]. The percent dye removal by hydrogel (without TiO_2) is 80%, while it is 99.4% for nanocomposites containing 0.2 g of TiO_2 NPs. The adsorption process of dye agrees with pseudo-second-order kinetics and followed Langmuir isotherm, resulting in a very high adsorption capacity q_m 2257 (mg/g) while negative values of G° confirming that adsorption is spontaneous [248].

17.3.2 TITANIA–ALGINATE COATINGS

Alginate is limitedly used in combination with electrophoretic deposition to produce bioactive coatings. TiO_2 and alginate were chosen due to the expected advantages of their combined use and also to make a comparison with similar, electrophoretically deposited coatings using other substrates as reported elsewhere [252–257]. In this regard, electrophoretic alginate-based coatings were prepared by introducing TiO_2 NPs and bioactive glass microparticles as inorganic phases for stainless-steel substrates [258]. The resultant coatings have structural stability and oppose deformation with no microcracking. The optimal ceramic content was 6 g/L for both TiO_2 NPs-alg and TiO_2 NPs/bioactive glass-alg systems. TiO_2 NPs/bioactive glass-alg coatings exhibited better adhesion strength than TiO_2 NPs-alg coatings to the substrate. This could be described by the presence of large bioactive glass particles in TiO_2 NPs/bioactive glass-alg coatings as higher load is required to cause spallation of the material. TiO_2 NPs/bioactive glass-alg showed low E_{corr} value than TiO_2 NPs-alg coatings due to the reaction of highly reactive bioactive glass particles with the medium, producing their partial dissolution; while TiO_2 in fact did not show this dissolution. Actually, reduced particle size of TiO_2 supports the effective blocking of the liquid penetration, sealing the pores left by the bioactive glass particles. However, the coatings displayed high E_{corr} and low i_{corr} values, compared to the naked metal, directing the protective action of attained coatings. This work led to soft, bioactive coatings for bone replacing devices and tissue engineering applications.

Alginate and TiO_2 are appropriate materials for the EPD process on account of being electrically charged after placing in proper solvents. A well-adhering and protective nano-TiO_2/alg composite coating was synthesized for Mg-alloy (AZ91D) substrates in a single-step EPD process [259]. AZ91D was chosen as reference material, as it was studied broadly in bone replacement field of applications [260–264]. The coating is made by a matrix of alginate that acts as a binder for the TiO_2 (filler) particles, leading to the clusters formation of TiO_2 NPs/alg during EPD. It enhanced the effective size of the particles, as estimated by SEM. The coated samples displayed the initial corrosion protection that is 3–7 times higher than the bared alloy. The corrosion resistance was further increased when immersed in DMEM for various time intervals, directed to the corrosion products. These products insulated the Mg from the corrosion medium. Using the same alloy, some other works were also reported, e.g., calcium-phosphate coatings produced by electrodeposition [265].

17.3.3 TITANIA–ALGINATE ENCAPSULATION SYSTEM

Oxide-NPs are also applied to immobilize biomolecules. The semiconductor oxide-NPs are used as labels/tracers in electrochemical evaluation [266]. Alginate/alginate-TiO_2 nanocomposite materials were used as matrix to immobilize myelin basic protein (MBP) using *N*-hydroxysuccinimide/1-ethyl-3-(3-dimethylaminopropyl) carbodiimide (NHS/EDC) cross-linkers as mentioned in Fig. 17.14 [267].

Well dispersion and strong adherence of TiO_2 NPs were observed in alg matrix along with improvement in thermal properties of alg. The MBP immobilization is enhanced due to large surface area available after the incorporation of TiO_2 NPs to

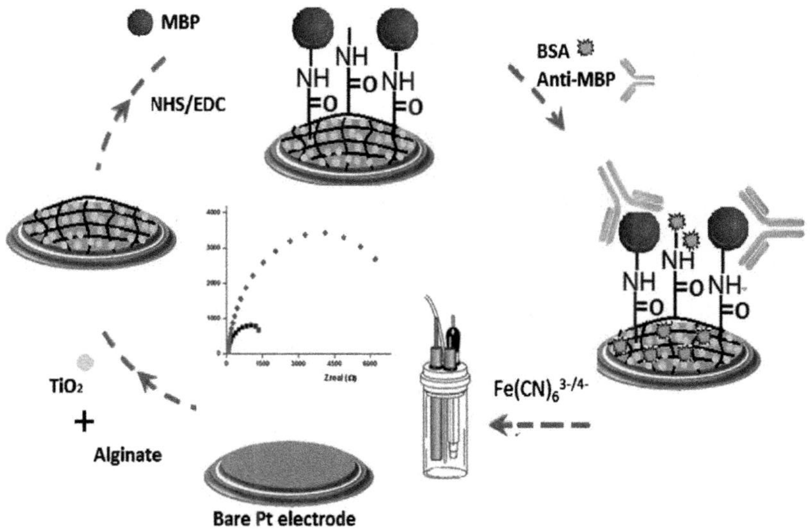

FIGURE 17.14

Fabrication process of the alginate-TiO_2-MBP immunosensor [267]. *MBP*, myelin basic protein.

the alg. The fabricated impedimetric immunosensors are able to determine the anti-MBP. The results were calculated in varied linear ranges and low detection limits, i.e., 0.25 ng m/L for alg-MBP and 0.18 ng/L for alg-TiO_2-MBP immunosensor. The designed immunosensors when compared to ELISA method are less time-consuming, i.e., 104 s for the alg-MBP and 79 s for the alg-TiO_2-MBP immunosensor.

17.3.4 TITANIA–ALGINATE PHOTOCATALYTIC SYSTEM

TiO_2 is immobilized under UV-radiation over Ca-alg for the abstraction of Hg(II) ions from aqueous solutions [268]. The photoreduction process of Hg(II) followed the pseudo-first-order kinetics. The presence of Fe(III) and citric acid considerably increased the photoreduction of Hg(II) ions owing to enhancement in charge separation of photo-induced hole/electron pairs; probably due to simultaneous redox reactions [4], by the initial citric acid concentration, and the fast reaction between Hg(II) and citric acid in the presence of Fe(III). The optimum conditions for mercury photoreduction were reported to be 2 g/L of the photocatalyst and 100 mg/L of initial Hg(II) concentration at pH 8 that maintained photoreduction activity up to 75% for at least four cycles without sacrificing their effectiveness.

TiO_2-impregnated glutaraldehyde-cross-linked alginate beads (TIGCAB) were prepared (Fig. 17.15A) and applied for the photoreduction of Cr(VI) from acid aqueous solution using a photocatalytic process [269]. The results showed that

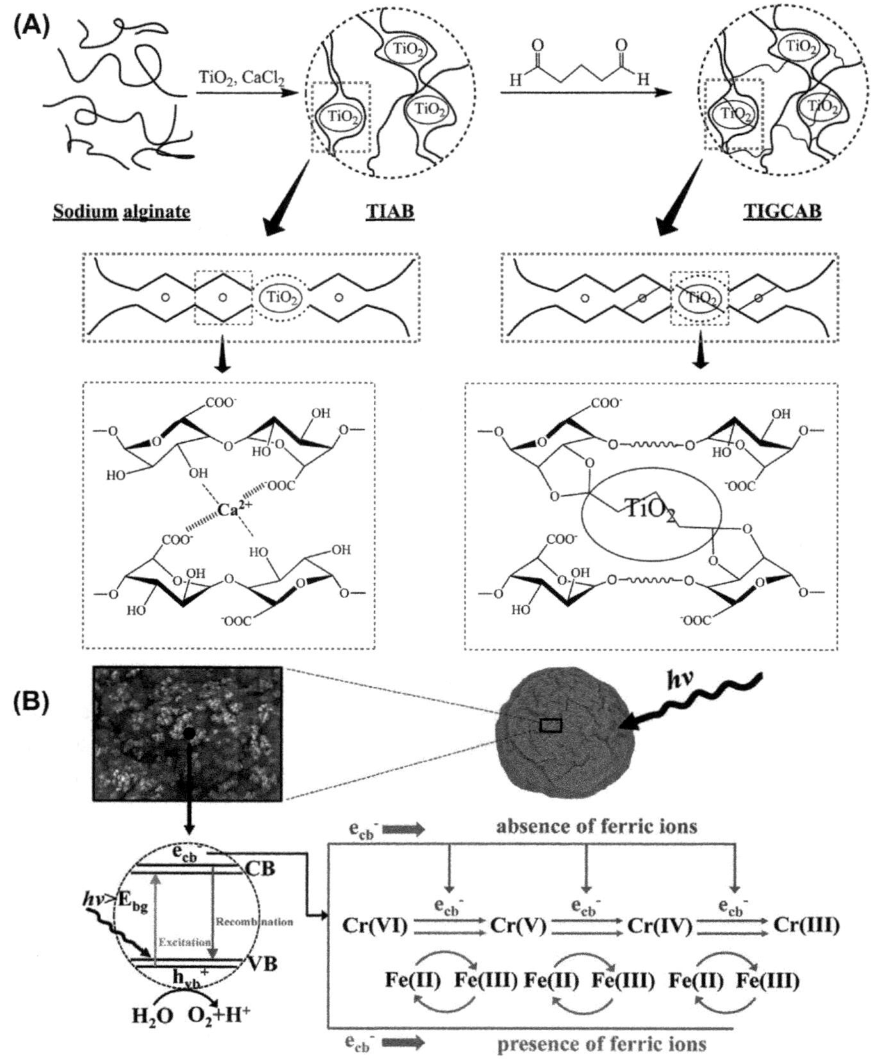

FIGURE 17.15

(A) Preparation of TIGCAB, (B) The proposed photoreduction mechanism of Cr(VI) on the TIGCAB [269]. *TIGCAB*, TiO_2^- impregnated glutaraldehyde–cross-linked alginate bead.

UV-light energy plays a crucial role in Cr(VI) reduction. It was proposed that photocatalytic activity for the reduction of Cr(VI) ions over TiO_2 of the TIGCAB may occur through direct reduction by the photo-generated electrons [269]. The increase in the photoreduction of Cr(VI) ions is because of prevention of electron–hole recombination and the reaction between Fe(II) and Cr(VI) in the presence of Fe(III) ions (Fig. 17.15B) which may function as an effective electron acceptor [270]. The photoreduction was completed in lower pH solution, where Cr(VI) anions adsorbed on the surface of the TiO_2 capture the photo-generated conduction band electrons and their reduction occurs. Water is oxidized by the photo-generated valence band holes, thus completing the redox cycle [271]. As pH increased, the interaction between chromium ions and TiO_2 becomes weaker and led to the decrease in photocatalytic reaction.

The photoreduction of Cr(VI) is completed in 12 h when the initial Cr(VI) concentration was 5 mg/L. However, only about 90% and 37% of photoreduction reactions were completed when the Cr(VI) concentrations were 10 and 20 mg/L, respectively. This may be due to the fact that for a fixed dosage of TIGCAB, the total existing electron–hole pairs were limited to get a higher reduction efficiency of Cr(VI) at higher concentration. Finally, TIGCAB can be reused without losing effectiveness for at least three cycles in batch experiments [269].

17.3.5 TITANIA–ALGINATE MEMBRANE

Polyaniline-coated titanium dioxide [PANI (TiO_2)] fillers were introduced into Na-alg matrix and cross-linked with glutaraldehyde to synthesize nanocomposite membranes for PV dehydration of 1,4-dioxane [272]. PANI-inorganic particles composite materials revealed better physico-chemical and mechanical properties in addition to improving dispensability of polymer and stability of nanofiller [273,274]. Na-alg [PANI (TiO_2)] membranes exhibited high values of water selectivity (maximum selectivity is 12,848) having 1 wt% of TiO_2 but have less flux than the other membranes. This increment in selectivity is due to reduction in intersegmental spacing after adding TiO_2 filler that acts as reinforcing bridge, which improves the selective permeation property by reducing the swelling effect of the membrane. However, permeation selectivity and swelling depends on the feeding liquid molecules [272].

17.4 GRAPHENE AND TITANIA-BASED ALGINATE MIXED MATRIX MATERIALS

This chapter mainly enlists the recent trends in graphene- and titania-based alginate nanocomposites but literature also shows the use of mixed *GO* fillers/hydroxyapatite nanosheets–reinforced alginate materials [275]. Similarly, *GO*-reinforced alginate/carboxymethyl cellulose (CMC) [276] and alginate/konjac glucomannan (KGM) [187] mixed matrix materials for biomedical and drug-delivery uses are also synthesized. The incorporation of *GO* improved strength and modulus (stiffness) of 1wt% *GO*/25%CMC/75%Alg by 40% and 1128% ,respectively, than bare 25%CMC/75% Alg films [276]. *GO* performed as good drug-binding effector for controlled release rate of drugs from KGM/SA/*GO* hydrogel carrier for the drug-delivery uses [187].

Just like graphene, TiO_2 is also incorporated in PVA/alginate [277,278], gum acacia/alginate-mixed matrix materials [279] for the photoreduction of lead, mercury, and iodine, and adsorbent for Reactive Blue H5 (dye). In addition to its use as matrix material, alginate also acts as binder for graphite aiding to enhance electrochemical performance [280], and as sacrificing template [281] and stabilizer [282] for nano TiO_2. However, the detailed discussion of these materials is not the part of this chapter.

17.5 CONCLUSION

Graphene/graphene oxide and titania nanofillers have produced a spectacular improvement in the properties of polymeric materials. Alginate is shining as a star material and effectively used matrix for various bionanocomposites. The above discussion clearly pointed out that *GO* is a more appropriate form of graphene derivatives and reinforcement of *GO* within 1–6 wt% contents modified the thermomechanical, electric, conductive, and biological properties of neat alginate dramatically in addition to prompting the emerging applications such as adsorbents (separation of metal ions, dyes, antibiotics etc.), membranes for separation, encapsulating systems for biomolecules, electronic sensors/biosensors, coatings. Reported discussion found that titania NPs–based alginate bionanocomposites did not get so much exposure as graphene-reinforced alginate nanocomposites did. But their usage is gradually increasing in high-performance miscellaneous applications (Table 17.2) with superior electric, optical, conductive, adsorptive, and catalytic properties. Electrostatic interactions, hydrogen bonding, van der Waals interactions, hydrophobic interactions, and coordination complex formations played a key role in the interactions of these fillers (graphene/*GO*, TiO_2) and alginate, and ultimately impart particular properties for a specific use. Currently, composites of TiO_2 nanotubes [283] are also explored and hopefully will be used with alginate as reinforcement material. Still the agglomeration of nanoparticles up to 5–6 wt% of *GO* needs to be further

Table 17.2 List of the TiO_2/Alginate Nanomaterials and Their Applications

Sr. No	Type of Material	Applications	References
1	Film coating	Immunosensor	[267]
2	Coatings	Coatings for biomedical grade alloys	[259]
3	Adsorbent solution	Adsorbent for elimination of acid/direct dyes from colored wastewater	[133]
4	Beads	Photoreduction of Cr(VI) in environmental pollution cleanup	[269]
5	Beads	Elimination of mercury from polluted water	[268]
6	Coatings	Applicable in bone-substituting devices and bone tissue engineering	[258]
7	Hydrogel	Methylene blue adsorption from wastewater	[248]
8	Membrane	PV dehydration of 1,4-dioxane	[272]

investigated. Although concerns over the cytotoxicity/toxicology levels of these nanofillers are satisfied using biopolymer matrix. Investigations to understand the relation between amount, size, and shape of nanofillers and toxicity level in human body will be required for anticipated biomedical applications.

REFERENCES

[1] Shchipunov Y. Bionanocomposites: green sustainable materials for the near future. Pure Appl Chem 2012;84(12):2579–607.

[2] Ramakrishna S, Huang ZM, Kumar GV, Batchelor AW, Mayer J. An introduction to biocomposites. London: Imperial College Press; 2004.

[3] Ray SS, Bousmina M. Biodegradable polymers and their layered silicate nanocomposites: in greening the 21st century materials world. Prog Mater Sci 2005;50(8): 962–1079.

[4] Ruiz-Hitzky E, Aranda P, Darder M. Bionanocomposites. In: Kirk-Othmer encyclopedia of chemical technology. Wiley; 2008. p. 1–28.

[5] Visakh PM, Thomas S, Pothan LA. Fully green bionanocomposites. In: A handbook of applied biopolymer technology: synthesis, degradation and applications. London: Royal Society of Chemistry; 2011. p. 102–28.

[6] Pilla S. Engineering applications of bioplastics and biocomposites – an overview. In: Pilla S, editor. Handbook of bioplastics and biocomposites engineering applications. Salem (Massachusetts): Scrivener; 2011. p. 1–15.

[7] Chen GQ, Patel MK. Plastics derived from biological sources: present and future: a technical and environmental review. Chem Rev 2012;112(4):2082–99.

[8] Koronis G, Silva A, Fontul M. Green composites: a review of adequate materials for automotive applications. Compos B Eng 2013;44(1):120–7.

[9] Rhim JW, Park HM, Ha CS. Bio-nanocomposites for food packaging applications. Prog Polym Sci 2013;38(10–11):1629–52.

[10] Schnepp Z. Biopolymers as a flexible resource for nanochemistry. Angew Chem Int Ed 2013;52(4):1096–108.

[11] Cheng Q, Lei J, Tang Z. Bioinspired layered materials with superior mechanical performance. Acc Chem Res 2014;47(4):1256–66.

[12] Darder M, Aranda P, Ruiz-Hitzky E. Bionanocomposites: a new concept of ecological, bioinspired, and functional hybrid materials. Adv Mater 2007;19(10):1309–19.

[13] Mohanty AK, Misra M, Hinrichsen G. Biofibres, biodegradable polymers and biocomposites: an overview. Macromol Mater Eng 2000;276–277(1):1–24.

[14] Handbook of fillers for plastics. New York: Van Nostrand Reinhold; 1987.

[15] Brydson JA. Plastics materials. Oxford: Butterworth-Heinemann; 1999.

[16] Wolf R, Kaul BL. Plastics, additives. In: Ullmann F, editor. Ullmann's encyclopedia of industrial chemistry. Wiley-VCH; 2007. p. 1–51.

[17] Kojima Y, Usuki A, Kawasumi M, Okada A, Kurauchi T, Kamigaito O. Sorption of water in nylon-6 clay hybrid. J Appl Polym Sci 1993;49(7):1259–64.

[18] Kojima Y, Usuki A, Kawasumi M, Okada A, Kurauchi T, Kamigaito O. Synthesis of nylon-6/clay hybrid by montmorillonite intercalated with ε-caprolactam. J Polym Sci A Polym Chem 1993;31(4):983–6.

[19] Okada A, Usuki A. The chemistry of polymer-clay hybrids. Mater Sci Eng C 1995; 3(2):109–15.

[20] Alexandre M, Dubois P. Polymer-layered silicate nanocomposites: preparation, properties and uses of a new class of materials. Mater Sci Eng R 2000;28(1–2):1–63.

[21] Caseri W. Nanocomposites of polymers and metals or semiconductors: historical background and optical properties. Macromol Rapid Commun 2000;21(11):705–22.

[22] Garces JM, Moll DJ, Bicerano J, Fibiger R, McLead DG. Polymeric nanocomposites for automotive applications. Adv Mater 2000;12(23):1835–9.

[23] Ke YC, Stroeve P. Polymer-layered and silica nanocomposites. Amsterdam: Elsevier; 2005.

[24] Leuteritz A, Kretzschmar B, Pospiech D, Costa FR, Wagenknecht U, Heinrich G. Industry-relevant preparation, characterization, and applications of polymer nanocomposites. In: Nalwa HS, editor. Polymeric nanostructures and their applications. North Lewis Way (CA): American Scientific Publishers; 2007. p. 99–151.

[25] Utracki LA. Clay-containing polymeric nanocomposites. Shawbury (UK): Rapra Technology; 2004.

[26] Xie XL, Mai YW, Zhou XP. Dispersion and alignment of carbon nanotubes in polymer matrix: a review. Mater Sci Eng R 2005;49(4):89–112.

[27] Chowdhury SR, Kar S, Ha CS. Polymer nanocomposites: a promising class of materials for a wide range of applications. In: Nalwa HS, editor. Polymeric nanostructures and their applications. (North Lewis Way (CA): American Scientific Publishers; 2008. p. 201–41.

[28] Rozenberg BA, Tenne R. Polymer-assisted fabrication of nanoparticles and nanocomposites. Prog Polym Sci 2008;33(1):40–112.

[29] Zou H, Wu SS, Shen J. Polymer/silica nanocomposites: preparation, characterization, properties, and applications. Chem Rev 2008;108(9):3893–957.

[30] Zhao YL, Stoddart JF. Noncovalent functionalization of single-walled carbon nanotubes. Acc Chem Res 2009;42(8):1161–71.

[31] Schexnailder P, Schmidt G. Nanocomposite polymer hydrogels. Colloid Polym Sci 2009;287(1):1–11.

[32] Schaefer HE. Nanoscience. The science of the small in physics, engineering, chemistry, biology and medicine. Berlin: Springer; 2010.

[33] Kim JK, Pal K, Sridhar V. Role of different nanoparticles in elastomeric nanocomposites. In: Mittal V, Kim JK, Pal K, editors. Recent advances in elastomeric nanocomposites. Advanced structured materials. Berlin: Springer; 2011. p. 3–55.

[34] Sahoo NG, Rana S, Cho JW, Li L, Chan SH. Polymer nanocomposites based on functionalized carbon nanotubes. Prog Polym Sci 2010;35(7):837–67.

[35] Hu K, Kulkarni DD, Choi I, Tsukruk VV. Graphene-polymer nanocomposites for structural and functional applications. Prog Polym Sci 2014;39(11):1934–72.

[36] Rouf TB, Kokini JL. Biodegradable biopolymer–graphene nanocomposites. J Mater Sci 2016;51(22):9915–45.

[37] Terzopoulou Z, Kyzas GZ, Bikiaris DN. Recent advances in nanocomposite materials of graphene derivatives with polysaccharides. Materials 2015;8(2):652–83.

[38] Carp O, Huisman CL, Reller A. Photoinduced reactivity of titanium dioxide. Prog Solid State Chem 2004;32(1–2):33–177.

[39] Chen X, Mao SS. Titanium dioxide nanomaterials: synthesis, properties, modifications, and applications. Chem Rev 2007;107(7):2891–959.

[40] Fujishima A, Zhang X, Tryk DA. TiO_2 photocatalysis and related surface phenomena. Surf Sci Rep 2008;63(12):515–82.

[41] Colombo P, Vakifahmetoglu C, Costacurta S. Fabrication of ceramic components with hierarchical porosity. J Mater Sci 2010;45(20):5425–55.

[42] Dastjerdi R, Montazer M. A review on the application of inorganic nano-structured materials in the modification of textiles: focus on anti-microbial properties. Colloid Surf B 2010;79(1):5–18.

[43] Tung WS, Daoud WA. Self-cleaning fibers via nanotechnology: a virtual reality. J Mater Chem 2011;21(22):7858–69.

[44] Radetic M. Functionalization of textile materials with TiO_2 nanoparticles. J Photochem Photobiol C Photchem Rev 2013;16:62–76.

[45] Liu H, Gao SW, Cai JS, He CL, Mao JJ, Zhu TX, Chen Z, Huang JY, Meng K, Zhang KQ, Al-Deyab SS, Lai YK. Recent progress in fabrication and applications of superhydrophobic coating on cellulose-based substrates. Materials 2016;9(3):124.

[46] Clare K, Algin. In: Whistler RL, BeMiller JN, editors. Industrial gums: polysaccharides and their derivatives. San Diego: Academic Press; 1993. p. 105–43.

[47] Moe ST, Draget KI, Skjak-Braek G, Smidsrod O. Alginates. In: Stephen AM, editor. Food polysaccharides and their applications. New York: Marcel Dekker; 1995. p. 245–66.

[48] Day DF. Alginates. In: Kaplan DL, editor. Biopolymers from renewable resources. Berlin: Springer; 1998. p. 119–43.

[49] Maurstad G, Danielsen S, Stokke BT. Analysis of compacted semiflexible polyanions visualized by atomic force microscopy: influence of chain stiffness on the morphologies of polyelectrolyte complexes. J Phys Chem B 2003;107(32):8172–80.

[50] Vold IMN, Kristiansen KA, Christensen BE. A study of the chain stiffness and extension of alginates, in vitro epimerized alginates, and periodate-oxidized alginates using size-exclusion chromatography combined with light scattering and viscosity detectors. Biomacromolecules 2006;7(7):2136–46.

[51] Russo R, Malinconico M, Santagata G. Effect of cross-linking with calcium ions on the physical properties of alginate films. Biomacromolecules 2007;8(10):3193–7.

[52] Shchipunov YA, Postnova IV. Water-soluble polyelectrolyte complexes of oppositely charged polysaccharides. Compos Interf 2009;16:251–79.

[53] Cottrell IW, Kovacs P. Alginates. In: Davidson RL, editor. Handbook of water-soluble gums and resins. New York: McGraw-Hill Book Company; 1980. pp. 2-1–2-43.

[54] Mancini M, Moresi M, Rancini R. Mechanical properties of alginate gels: empirical characterisation. J Food Eng 1999;39:369–78.

[55] Smidsrod O, Haug A. Properties of poly(1,4-hex urinates) in the gel stage. ii. comparison of gels of different chemical composition. Acta Chem Scand 1972;26(1):79–88.

[56] Rees DA. Polysaccharide shapes. London: Chapman and Hall; 1977.

[57] Rees DA, Welsh EJ. Secondary and tertiary structure of polysaccharides in solutions and gels. Angew Chem Int Ed 1977;16(4):214–24.

[58] Grant GT, Morris ER, Rees DA, Smith PJC, Thom D. Biological interactions between polysaccharides and divalent cations: the egg-box model. FEBS Lett 1973;32(1):195–8.

[59] Augst AD, Kong HJ, Mooney DJ. Alginate hydrogels as biomaterials. Macromol Biosci 2006;6(8):623–33.

[60] Smidsrod O, Skjak-Braek G. Alginate as immobilization matrix for cells. Trends Biotechnol 1990;8(3):71–8.

[61] Drury JL, Mooney DJ. Hydrogels for tissue engineering: scaffold design variables and applications. Biomaterials 2003;24(24):4337–51.

[62] Ma PX. Alginate for tissue engineering. In: Ma PX, Elisseeff JH, editors. Scaffolding in tissue engineering. New York: CRC Press; 2006. p. 13–25.
[63] Shimizu T, Takada A. Preparation of Bi-based superconducting. Fiber by metal biosorption of Na-alginate. Polym Gels Netw 1997;5(3):267–83.
[64] Darder M, Lopez-Blanco M, Aranda P, Leroux F, Ruiz-Hitzky E. Bionanocomposites based on layered double hydroxides. Chem Mater 2005;17(8):1969–77.
[65] Brayner R, Vaulay MJ, Fievet F, Coradin T. Alginate-mediated growth of Co, Ni, and CoNi nanoparticles: influence of the biopolymer structure. Chem Mater 2007;19(5): 1190–8.
[66] Burkova Y, Beleneva I, Shchipunov Y. Bactericidal sodium alginate films containing nanosized silver particles. Colloid J 2015;77(6):707–14.
[67] Fitzer E, Kochling KH, Boehm HP, Marsh H. Recommended terminology for the description of carbon as a solid. Pure Appl Chem 1995;67(3):473–506.
[68] Novoselov KS, Geim AK, Morozov SV, Jiang D, Zhang Y, Dubonos SV, Grigorieva IV, Firsov AA. Electric field effect in atomically thin carbon films. Science 2004; 306(5696):666–9.
[69] Krueger A. Carbon materials and nanotechnology. Weinheim: Wiley-VCH; 2010.
[70] Allen MJ, Tung VC, Kaner RB. Honeycomb carbon: a review of graphene. Chem Rev 2010;110(1):132–45.
[71] Hoheisel TN, Schrettl S, Szilluweit R, Frauenrath H. Nanostructured carbonaceous materials from molecular precursors. Angew Chem Int Ed 2010;49(37):6496–515.
[72] Delhaes P. Carbon-based solids and materials. London: ISTE-Wiley; 2011.
[73] Ivanovskii AL. Graphene-based and graphene-like materials. Russ Chem Rev 2012; 81(7):571–605.
[74] Alwarappan S, Kumar A. Graphene-based materials. Science and technology. Boca Raton: CRC Press; 2014.
[75] Bianco A, Cheng HM, Enoki T, Gogotsi Y, Hurt RH, Koratkar N, Kyotani T, Monthioux M, Park CR, Tascon JMD, Zhang J. All in the graphene family - a recommended nomenclature for two-dimensional carbon materials. Carbon 2013;65(1–6).
[76] Paredes JI, Villar-Rodil S, Martinez-Alonso A, Tascon JMD. Raphene oxide dispersions in organic solvents. Langmuir 2008;24(19):10560–4.
[77] Chen D, Feng HB, Li JH. Graphene oxide: preparation, functionalization, and electrochemical applications. Chem Rev 2012;112(11):6027–53.
[78] Perrozzi F, Prezioso S, Ottaviano L. Graphene oxide: from fundamentals to applications. J Phys Condens Matter 2015;27(1):1–21.
[79] Chung DDL. J Mater Sci 2016;51(1):554–68.
[80] Geim AK, Novoselov KS. The rise of graphene. Nat Mater 2007;6(3):183–91.
[81] Partoens B, Peeters FM. From graphene to graphite: electronic structure around the K point. Phys Rev B 2006;74(7):075404.
[82] Zhang B, Wang Y, Zhai G. Biomedical applications of the graphene-based materials. Mater Sci Eng C 2016;61:953–64.
[83] Wang K, Ruan J, Song H, Zhang JL, Wo Y, Guo SW, Cui DX. Biocompatibility of graphene oxide. Nanoscale Res Lett 2011;6(8).
[84] Kiew SF, Kiew LV, Lee HB, Imae T, Chung LY. Assessing biocompatibility of graphene oxide-based nanocarriers: a review. J Control Release 2016;226:217–28.
[85] Zhang B, Wei P, Zhou ZX, Wei TT. Adv Drug Deliv Rev 2016;105:145–62.
[86] Matthew HWT. Polymers for tissue engineering scaffolds. In: Dumitria S, editor. Polymeric biomaterials. New York: Marcel Dekker; 2002. p. 173–92.

[87] Rinaudo M. Main properties and current applications of some polysaccharides as biomaterials. Polym Int 2008;57(3):397–430.
[88] Naicker PK, Cummings PT, Zhang H, Banfield JF. Characterization of titanium dioxide nanoparticles using molecular dynamics simulations. J Phy Chem B 2005;109(32): 15243–9.
[89] Hashimoto K, Irie H, Fujishima A. TiO_2 photocatalysis: an historical overview and future prospects. Jpn J Appl Phys 2005;44(12R):8269–85.
[90] Fujishima A. Electrochemical photolysis of water at a semiconductor electrode. Nature 1972;238:37–8.
[91] Hoffmann MR, Martin ST, Choi W, Bahnemann DW. Environmental applications of semiconductor photocatalysis. Chem Rev 1995;95(1):69–96.
[92] Ko S, Pekarovic J, Fleming PD, Ari-Gur P. High performance nano-titania photocatalytic paper composite. Part I: experimental design study for TiO_2 composite sheet using a natural zeolite microparticle system and its photocatalytic property. Mater Sci Eng B 2010;166:127–31.
[93] Galkina OL, Vinogradov VV, Vinogradov AV, Agafonov AV. Development of the low-temperature sol-gel synthesis of TiO_2 to provide self-cleaning effect on the textile materials. Nanotechnol Russ 2012;7:604–14.
[94] Senić Ž, Bauk S, Vitorović-Todorović M, Pajić N, Samolov A, Rajić D. Application of TiO_2 nanoparticles for obtaining self- decontaminating smart textiles. Sci Tech Rev 2011;61:63–72.
[95] Montazer M, Seifollahzadeh S. Enhanced self-cleaning, anti-bacterial and UV protection properties of nano TiO_2 treated textile through enzymatic pretreatment. J Photochem Photobiol 2011;87:877–83.
[96] Liang SS, Makamba H, Huang SY, Chen SH. Nano-titanium dioxide composites for the enrichment of phosphopeptides. J Chromatogr A 2006;1116(1):38–45.
[97] Ichinose I, Kikuchi T, Lee SW, Kunitake T. Imprinting and selective binding of di-and tri-peptides in ultrathin TiO_2-gel films in aqueous solutions. Chem Lett 2002;(1): 104–5.
[98] Brown EM, Paunesku T, Wu A, Thurn KT, Haley B, Clark J, Priester T, Woloschak GE. Methods for assessing DNA hybridization of peptide nucleic acid–titanium dioxide nanoconjugates. Anal Biochem 2008;383(2):226–35.
[99] Arora H, Doty C, Yuan Y, Boyle J, Petras K, Rabatic B, Paunesku T, Woloschak G. Titanium dioxide nanocomposites. Nanotechnologies for the life sciences. 2007.
[100] Venkatasubbu GD, Baskar R, Anusuya T, Seshan CA, Chelliah R. Toxicity mechanism of titanium dioxide and zinc oxide nanoparticles against food pathogens. Colloids Surf B Biointerfaces 2016;148:600–6.
[101] FDA. Listing of color additives exempt from certification. Code of Federal Regulations Title 21-food and drugs. 21 CFR 73.2575. (Washington, DC): US Government Printing Office; 2002.
[102] Iavicoli I, Leso V, Bergamaschi A. Toxicological effects of titanium dioxide nanoparticles: a review of in vivo studies. J Nanomater 2012;2012:5.
[103] Wang Y, Cui H, Zhou J, Li F, Wang J, Chen M, Liu Q. Cytotoxicity, DNA damage, and apoptosis induced by titanium dioxide nanoparticles in human non-small cell lung cancer A549 cells. Environ Sci Pollut Res 2015;22(7):5519–30.
[104] Gurr JR, Wang AS, Chen CH, Jan KY. Ultrafine titanium dioxide particles in the absence of photoactivation can induce oxidative damage to human bronchial epithelial cells. Toxicology 2005;213(1):66–73.

[105] Shi H, Magaye R, Castranova V, Zhao J. Titanium dioxide nanoparticles: a review of current toxicological data. Part Fibre Toxicol 2013;10(1):1.

[106] Saquib Q, Al-Khedhairy AA, Siddiqui MA, Abou-Tarboush FM, Azam A, Musarrat J. Titanium dioxide nanoparticles induced cytotoxicity, oxidative stress and DNA damage in human amnion epithelial (WISH) cells. Toxicol In Vitro 2012;26(2):351−61.

[107] Vickery L, Patil AJ, Mann S. Fabrication of graphene-polymer nanocomposites with higher-order three-dimensional architectures. Adv Mater 2009;21:2180−4.

[108] Stankovich S, Dikin DA, Piner RD, Kohlhaas KA, Kleinhammes A, et al. Synthesis of graphene-based nanosheets via chemical reduction of exfoliated graphite oxide. Carbon 2007;45:1558−65.

[109] Becerril H, Mao J, Liu Z, Stoltenberg M, Bao Z, Chen Y. Evaluation of solution-processed reduced graphene oxide films as transparent conductors. ACS Nano 2008;2:463−70.

[110] Huang X, Qi X, Boey F, Zhang H. Graphene-based composites. Chem Soc Rev 2012;41(2):666−86.

[111] Xu Y, Sheng K, Li C, Shi G. Self-assembled graphene hydrogel via a one-step hydrothermal process. ACS Nano 2010;4(7):4324−30.

[112] Kuilla T, Bhadra S, Yao D, Kim NH, Bose S, Lee JH. Recent advances in graphene based polymer composites. Prog Polym Sci 2010;35(11):1350−75.

[113] Cai M. Nanoscale mechanical characterization of graphene/polymer nanocomposites using atomic force microscopy. 2013.

[114] Nie L, Liu C, Wang J, Shuai Y, Cui X, Liu L. Effects of surface functionalized graphene oxide on the behavior of sodium alginate. Carbohydr Polym 2015;117:616−23.

[115] Han D, Yan L, Chen W, Li W. Preparation of chitosan/graphene oxide composite film with enhanced mechanical strength in the wet state. Carbohydr Polym 2011;83(2):653−8.

[116] Fan H, Wang L, Zhao K, Li N, Shi Z, Ge Z, Jin Z. Fabrication, mechanical properties, and biocompatibility of graphene-reinforced chitosan composites. Biomacromolecules 2010;11(9):2345−51.

[117] Luong ND, Pahimanolis N, Hippi U, Korhonen JT, Ruokolainen J, Johansson LS, Nam JD, Seppälä J. Graphene/cellulose nanocomposite paper with high electrical and mechanical performances. J Mater Chem 2011;21(36):13991−8.

[118] Wan C, Frydrych M, Chen B. Strong and bioactive gelatin−graphene oxide nanocomposites. Soft Matter 2011;7(13):6159−66.

[119] He Y, Zhang N, Gong Q, Qiu H, Wang W, Liu Y, Gao J. Alginate/graphene oxide fibers with enhanced mechanical strength prepared by wet spinning. Carbohydr Polym 2012;88(3):1100−8.

[120] Ionita M, Pandele MA, Iovu H. Sodium alginate/graphene oxide composite films with enhanced thermal and mechanical properties. Carbohydr Polym 2013;94(1):339−44.

[121] Zhong D, Huang X, Yang H, Cheng R. New insights into viscosity abnormality of sodium alginate aqueous solution. Carbohydr Polym 2010;81(4):948−52.

[122] Sha J, Xie L, Ma Y, Dong S, Wang Y. Aminofunctional modification of graphene oxide sheets and its composite with epoxy resin. China Plast 2011;8:013.

[123] Cong HP, Wang P, Yu SH. Highly elastic and superstretchable graphene oxide/polyacrylamide hydrogels. Small 2014;10(3):448−53.

[124] Park S, Lee KS, Bozoklu G, Cai W, Nguyen ST, Ruoff RS. Graphene oxide papers modified by divalent ions—enhancing mechanical properties via chemical cross-linking. ACS Nano 2008;2(3):572−8.

[125] Kilan K, Warszyński P. Thickness and permeability of multilayers containing alginate cross-linked by calcium ions. Electrochim Acta 2014;144:254–62.

[126] Crossingham YJ, Kerr PG, Kennedy RA. Comparison of selected physico-chemical properties of calcium alginate films prepared by two different methods. Int J Pharm 2014;473(1):259–69.

[127] Zheng H, Yang J, Han S. The synthesis and characteristics of sodium alginate/graphene oxide composite films crosslinked with multivalent cations. J Appl Polym Sci 2016;133(27).

[128] Hassan RM. Alginate polyelectrolyte ionotropic gels. J Mater Sci 1991;26(21):5806–10.

[129] Soares JP, Santos JE, Chierice GO, Cavalheiro ET. Thermal behavior of alginic acid and its sodium salt. Eclética Quím 2004;29(2):57–64.

[130] Hassan RM, Summan AM, Hassan MK, El-Shatoury SA. Kinetics and mechanism of sol-gel transformation on polyelectrolytes of some transition metal ions, especially cobalt alginate ionotropic membranes. Eur Polym J 1989;25(12):1209–12.

[131] Iijima M, Hatakeyama T, Nakamura K, Hatakeyama H. Thermomechanical analysis of calcium alginate hydrogels in water. J Therm Anal Calorim 2002;70(3):807–14.

[132] Khairou KS, Al-Gethami WM, Hassan RM. Kinetics and mechanism of sol–gel transformation between sodium alginate polyelectrolyte and some heavy divalent metal ions with formation of capillary structure polymembranes ionotropic gels. J Membr Sci 2002;209(2):445–56.

[133] Mahmoodi NM, Hayati B, Arami M, Bahrami H. Preparation, characterization and dye adsorption properties of biocompatible composite (alginate/titania nanoparticle). Desalination 2011;275(1):93–101.

[134] Kim TY, Jin HJ, Park SS, Kim SJ, Cho SY. Adsorption equilibrium of copper ion and phenol by powdered activated carbon, alginate bead and alginate-activated carbon bead. J Ind Eng Chem 2008;14(6):714–9.

[135] Rocher V, Siaugue JM, Cabuil V, Bee A. Removal of organic dyes by magnetic alginate beads. Water Res 2008;42(4):1290–8.

[136] Li Y, Xia B, Zhao Q, Liu F, Zhang P, Du Q, Wang D, Li D, Wang Z, Xia Y. Removal of copper ions from aqueous solution by calcium alginate immobilized kaolin. J Environ Sci 2011;23(3):404–11.

[137] Li Y, Liu F, Xia B, Du Q, Zhang P, Wang D, Wang Z, Xia Y. Removal of copper from aqueous solution by carbon nanotube/calcium alginate composites. J Hazard Mater 2010;177(1):876–80.

[138] Gotoh T, Matsushima K, Kikuchi KI. Preparation of alginate–chitosan hybrid gel beads and adsorption of divalent metal ions. Chemosphere April 30, 2004;55(1):135–40.

[139] Papageorgiou SK, Katsaros FK, Kouvelos EP, Nolan JW, Le Deit H, Kanellopoulos NK. Heavy metal sorption by calcium alginate beads from *Laminaria digitata*. J Hazard Mater October 11, 2006;137(3):1765–72.

[140] Li Y, Zhang P, Du Q, Peng X, Liu T, Wang Z, Xia Y, Zhang W, Wang K, Zhu H, Wu D. Adsorption of fluoride from aqueous solution by graphene. J Colloid Interf Sci 2011;363(1):348–54.

[141] Deng X, Lü L, Li H, Luo F. The adsorption properties of Pb (II) and Cd (II) on functionalized graphene prepared by electrolysis method. J Hazard Mater 2010;183(1):923–30.

[142] Yang ST, Chang Y, Wang H, Liu G, Chen S, Wang Y, Liu Y, Cao A. Folding/aggregation of graphene oxide and its application in Cu^{2+} removal. J Colloid Interface Sci 2010;351(1):122–7.

[143] Mishra AK, Ramaprabhu S. Functionalized graphene sheets for arsenic removal and desalination of sea water. Desalination 2011;282:39–45.

[144] Ramesha GK, Kumara AV, Muralidhara HB, Sampath S. Graphene and graphene oxide as effective adsorbents toward anionic and cationic dyes. J Colloid Interface Sci 2011; 361(1):270–7.

[145] Li Y, Du Q, Liu T, Sun J, Wang Y, Wu S, Wang Z, Xia Y, Xia L. Methylene blue adsorption on graphene oxide/calcium alginate composites. Carbohydr Polym 2013; 95(1):501–7.

[146] He YQ, Zhang NN, Wang XD. Adsorption of graphene oxide/chitosan porous materials for metal ions. Chin Chem Lett 2011;22(7):859–62.

[147] Sun L, Fugetsu B. Graphene oxide captured for green use: influence on the structures of calcium alginate and macroporous alginic beads and their application to aqueous removal of acridine orange. Chem Eng J 2014;240:565–73.

[148] Lv G, Li Z, Jiang WT, Chang PH, Jean JS, Lin KH. Mechanism of acridine orange removal from water by low-charge swelling clays. Chem Eng J 2011;174(2):603–11.

[149] Rocher V, Bee A, Siaugue JM, Cabuil V. Dye removal from aqueous solution by magnetic alginate beads crosslinked with epichlorohydrin. J Hazard Mater 2010;178(1): 434–9.

[150] Lin YB, Fugetsu B, Terui N, Tanaka S. Removal of organic compounds by alginate gel beads with entrapped activated carbon. J Hazard Mater 2005;120(1):237–41.

[151] Zhang W, Zhou C, Zhou W, Lei A, Zhang Q, Wan Q, Zou B. Fast and considerable adsorption of methylene blue dye onto graphene oxide. Bull Environ Contam Toxicol 2011;87(1):86–90.

[152] Ai L, Zhang C, Chen Z. Removal of methylene blue from aqueous solution by a solvothermal-synthesized graphene/magnetite composite. J Hazard Mater 2011; 192(3):1515–24.

[153] Algothmi WM, Bandaru NM, Yu Y, Shapter JG, Ellis AV. Alginate–graphene oxide hybrid gel beads: an efficient copper adsorbent material. J Colloid Interface Sci 2013;397:32–8.

[154] Mata YN, Blázquez ML, Ballester A, González F, Munoz JA. Biosorption of cadmium, lead and copper with calcium alginate xerogels and immobilized *Fucus vesiculosus*. J Hazard Mater 2009;163(2):555–62.

[155] Biondi M, Borzacchiello A, Mayol L, Ambrosio L. Nanoparticle-integrated hydrogels as multifunctional composite materials for biomedical applications. Gels 2015;1(2): 162–78.

[156] Wu S, Zhao X, Li Y, Zhao C, Du Q, Sun J, Wang Y, Peng X, Xia Y, Wang Z, Xia L. Adsorption of ciprofloxacin onto biocomposite fibers of graphene oxide/calcium alginate. Chem Eng J 2013;230:389–95.

[157] Gong JP, Kurokawa T, Narita T, Kagata G, Osada Y, Nishimura G, Kinjo M. Synthesis of hydrogels with extremely low surface friction. J Am Chem Soc 2001;123(23): 5582–3.

[158] Xu R, Zhou G, Tang Y, Chu L, Liu C, Zeng Z, Luo S. New double network hydrogel adsorbent: highly efficient removal of Cd (II) and Mn (II) ions in aqueous solution. Chem Eng J 2015;275:179–88.

[159] Xu X, Lü S, Gao C, Wang X, Bai X, Gao N, Liu M. One-pot facile synthesis of silica reinforced double network hydrogels based on triple interactions. Chem Eng J 2014; 240:331–7.

[160] Zhuang Y, Yu F, Chen J, Ma J. Batch and column adsorption of methylene blue by graphene/alginate nanocomposite: comparison of single-network and double-network hydrogels. J Environ Chem Eng 2016;4(1):147–56.

[161] Ma T, Chang PR, Zheng P, Zhao F, Ma X. Fabrication of ultra-light graphene-based gels and their adsorption of methylene blue. Chem Eng J 2014;240:595–600.

[162] Hassan AF, Abdel-Mohsen AM, Fouda MM. Comparative study of calcium alginate, activated carbon, and their composite beads on methylene blue adsorption. Carbohydr Polym 2014;102:192–8.

[163] Wang Y, Wang W, Wang A. Efficient adsorption of methylene blue on an alginate-based nanocomposite hydrogel enhanced by organo-illite/smectite clay. Chem Eng J 2013;228:132–9.

[164] Romanchuk AY, Slesarev AS, Kalmykov SN, Kosynkin DV, Tour JM. Graphene oxide for effective radionuclide removal. Phys Chem Chem Phys 2013;15(7):2321–7.

[165] Yang HJ, Sun L, Zhai JL, Li HY, Zhao Y, Yu HW. In situ controllable synthesis of magnetic prussian blue/graphene oxide nanocomposites for removal of radioactive cesium in water. J Mater Chem A 2014;2(2):326–32. http://dx.doi.org/10.1039/C3TA13548A.

[166] Yang H, Li H, Zhai J, Sun L, Zhao Y, Yu H. Magnetic prussian blue/graphene oxide nanocomposites caged in calcium alginate microbeads for elimination of cesium ions from water and soil. Chem Eng J 2014;246:10–9.

[167] Gurikov P, Raman SP, Weinrich D, Fricke M, Smirnova I. A novel approach to alginate aerogels: carbon dioxide induced gelation. RSC Adv 2015;5(11):7812–8.

[168] Mallepally RR, Bernard I, Marin MA, Ward KR, McHugh MA. Superabsorbent alginate aerogels. J Supercrit Fluids 2013;79:202–8.

[169] Sescousse R, Gavillon R, Budtova T. Aerocellulose from cellulose–ionic liquid solutions: preparation, properties and comparison with cellulose–NaOH and cellulose–NMMO routes. Carbohydr Polym 2011;83(4):1766–74.

[170] Singh J, Dutta PK, Dutta J, Hunt AJ, Macquarrie DJ, Clark JH. Preparation and properties of highly soluble chitosan–L-glutamic acid aerogel derivative. Carbohydr Polym 2009;76(2):188–95.

[171] Watthanaphanit A, Supaphol P, Tamura H, Tokura S, Rujiravanit R. Wet-spun alginate/chitosan whiskers nanocomposite fibers: preparation, characterization and release characteristic of the whiskers. Carbohydr Polym 2010;79(3):738–46.

[172] Deze EG, Papageorgiou SK, Favvas EP, Katsaros FK. Porous alginate aerogel beads for effective and rapid heavy metal sorption from aqueous solutions: effect of porosity in Cu^{2+} and Cd^{2+} ion sorption. Chem Eng J 2012;209:537–46.

[173] Boguń M, Mikołajczyk T, Rabiej S. Effect of formation conditions on the structure and properties of nanocomposite alginate fibers. J Appl Polym Sci 2009;114(1):70–82.

[174] Alnaief M, Alzaitoun MA, García-González CA, Smirnova I. Preparation of biodegradable nanoporous microspherical aerogel based on alginate. Carbohydr Polym 2011;84(3):1011–8.

[175] Veronovski A, Knez Ž, Novak Z. Preparation of multi-membrane alginate aerogels used for drug delivery. J Supercrit Fluids 2013;79:209–15.

[176] Malafaya PB, Silva GA, Reis RL. Natural–origin polymers as carriers and scaffolds for biomolecules and cell delivery in tissue engineering applications. Adv Drug Deliv Rev 2007;59(4):207–33.

[177] García-González CA, Alnaief M, Smirnova I. Polysaccharide-based aerogels—promising biodegradable carriers for drug delivery systems. Carbohydr Polym 2011; 86(4):1425–38.

[178] Dashdorj U, Reyes MK, Unnithan AR, Tiwari AP, Tumurbaatar B, Park CH, Kim CS. Fabrication and characterization of electrospun zein/Ag nanocomposite mats for wound dressing applications. Inte J Biol Macromol 2015;80:1–7.

[179] Vijaya Y, Popuri SR, Boddu VM, Krishnaiah A. Modified chitosan and calcium alginate biopolymer sorbents for removal of nickel (II) through adsorption. Carbohydr Polym 2008;72(2):261–71.

[180] Jiang S, Zhang Z, Lai Y, Qu Y, Wang X, Li J. Selenium encapsulated into 3D interconnected hierarchical porous carbon aerogels for lithium–selenium batteries with high rate performance and cycling stability. J Power Sources 2014;267:394–404.

[181] Cheng Y, Lu L, Zhang W, Shi J, Cao Y. Reinforced low density alginate-based aerogels: preparation, hydrophobic modification and characterization. Carbohydr Polym 2012;88(3):1093–9.

[182] Paulchamy B, Arthi G, Lignesh BD. A simple approach to stepwise synthesis of graphene oxide nanomaterial. J Nanomed Nanotechnol 2015;6(1):253.

[183] Vijayalakshmi R, Rajendran V. Synthesis and characterization of nano-TiO_2 via different methods. Arch App Sci Res 2012;4(2):1183–90.

[184] Jiao C, Xiong J, Tao J, Xu S, Zhang D, Lin H, Chen Y. Sodium alginate/graphene oxide aerogel with enhanced strength–toughness and its heavy metal adsorption study. Int J Biol Macromol 2016;83:133–41.

[185] Gawryla MD, van den Berg O, Weder C, Schiraldi DA. Clay aerogel/cellulose whisker nanocomposites: a nanoscale wattle and daub. J Mater Chem 2009;19(15):2118–24.

[186] Yuan X, Wei Y, Chen S, Wang P, Liu L. Bio-based graphene/sodium alginate aerogels for strain sensors. RSC Adv 2016;6(68):64056–64.

[187] Wang J, Liu C, Shuai Y, Cui X, Nie L. Controlled release of anticancer drug using graphene oxide as a drug-binding effector in konjac glucomannan/sodium alginate hydrogels. Colloids Surf B 2014;113:223–9.

[188] Fan J, Shi Z, Lian M, Li H, Yin J. Mechanically strong graphene oxide/sodium alginate/polyacrylamide nanocomposite hydrogel with improved dye adsorption capacity. J Mater Chem A 2013;1(25):7433–43.

[189] Wan Y, Chen X, Xiong G, Guo R, Luo H. Synthesis and characterization of three-dimensional porous graphene oxide/sodium alginate scaffolds with enhanced mechanical properties. Mater Exp 2014;4(5):429–34.

[190] Ninan N, Muthiah M, Park IK, Elain A, Thomas S, Grohens Y. Pectin/carboxymethyl cellulose/microfibrillated cellulose composite scaffolds for tissue engineering. Carbohydr Polym 2013;98(1):877–85.

[191] Pabby AK, Rizvi SS, Requena AM, editors. Handbook of membrane separations: chemical, pharmaceutical, food, and biotechnological applications. CRC press; April 9, 2015.

[192] Wee SL, Tye CT, Bhatia S. Membrane separation process—pervaporation through zeolite membrane. Sep Purif Technol 2008;63(3):500–16.

[193] Li K. Ceramic membranes for separation and reaction. John Wiley & Sons; April 30, 2007.

[194] Shao P, Huang RY. Polymeric membrane pervaporation. J Membr Sci 2007;287(2):162−79.

[195] Mallada R, Menéndez M, editors. Inorganic membranes: synthesis, characterization and applications. Elsevier; May 1, 2008.

[196] Bhat SD, Aminabhavi TM. Novel sodium alginate−Na^+ MMT hybrid composite membranes for pervaporation dehydration of isopropanol, 1, 4-dioxane and tetrahydrofuran. Sep Purif Technol 2006;51(1):85−94.

[197] Magalad VT, Gokavi GS, Nadagouda MN, Aminabhavi TM. Pervaporation separation of water−ethanol mixtures using organic−inorganic nanocomposite membranes. J Phys Chem C 2011;115(30):14731−44.

[198] Magalad VT, Gokavi GS, Ranganathaiah C, Burshe MH, Han C, Dionysiou DD, Nadagouda MN, Aminabhavi TM. Polymeric blend nanocomposite membranes for ethanol dehydration—effect of morphology and membrane−solvent interactions. J Membr Sci 2013;430:321−9.

[199] Adoor SG, Prathab B, Manjeshwar LS, Aminabhavi TM. Mixed matrix membranes of sodium alginate and poly (vinyl alcohol) for pervaporation dehydration of isopropanol at different temperatures. Polymer 2007;48(18):5417−30.

[200] Zinadini S, Zinatizadeh AA, Rahimi M, Vatanpour V, Zangeneh H. Preparation of a novel antifouling mixed matrix PES membrane by embedding graphene oxide nanoplates. J Membr Sci 2014;453:292−301.

[201] Sun P, Zhu M, Wang K, Zhong M, Wei J, Wu D, Xu Z, Zhu H. Selective ion penetration of graphene oxide membranes. ACS Nano 2012;7(1):428−37.

[202] Kim HW, Yoon HW, Yoon SM, Yoo BM, Ahn BK, Cho YH, Shin HJ, Yang H, Paik U, Kwon S, Choi JY. Selective gas transport through few-layered graphene and graphene oxide membranes. Science 2013;342(6154):91−5.

[203] Yeh TM, Wang Z, Mahajan D, Hsiao BS, Chu B. High flux ethanol dehydration using nanofibrous membranes containing graphene oxide barrier layers. J Mater Chem A 2013;1(41):12998−3003.

[204] Bhat SD, Aminabhavi TM. Pervaporation separation using sodium alginate and its modified membranes—a review. Sep Purif Rev 2007;36(3):203−29.

[205] Nair RR, Wu HA, Jayaram PN, Grigorieva IV, Geim AK. Unimpeded permeation of water through helium-leak−tight graphene-based membranes. Science 2012;335(6067):442−4.

[206] Boukhvalov DW, Katsnelson MI, Son YW. Origin of anomalous water permeation through graphene oxide membrane. Nano Lett 2013;13(8):3930−5.

[207] Suhas DP, Raghu AV, Jeong HM, Aminabhavi TM. Graphene-loaded sodium alginate nanocomposite membranes with enhanced isopropanol dehydration performance via a pervaporation technique. RSC Adv 2013;3(38):17120−30.

[208] Cao K, Jiang Z, Zhao J, Zhao C, Gao C, Pan F, Wang B, Cao X, Yang J. Enhanced water permeation through sodium alginate membranes by incorporating graphene oxides. J Membr Sci 2014;469:272−83.

[209] Liang J, Huang Y, Zhang L, Wang Y, Ma Y, Guo T, Chen Y. Molecular-level dispersion of graphene into poly (vinyl alcohol) and effective reinforcement of their nanocomposites. Adv Funct Mater 2009;19(14):2297−302.

[210] Zhao Q, An QF, Liu T, Chen JT, Chen F, Lee KR, Gao CJ. Bio-inspired polyelectrolyte complex/graphene oxide nanocomposite membranes with enhanced tensile strength and ultra-low gas permeability. Polym Chem 2013;4(16):4298−302.

[211] Vilcinskas K, Norder B, Goubitz K, Mulder FM, Koper GJ, Picken SJ. Tunable order in alginate/graphene biopolymer nanocomposites. Macromolecules 2015;48(22):8323–30.

[212] Zhao J, Zhu Y, He G, Xing R, Pan F, Jiang Z, Zhang P, Cao X, Wang B. Incorporating zwitterionic graphene oxides into sodium alginate membrane for efficient water/alcohol separation. ACS Appl Mater Interface 2016;8(3):2097–103.

[213] Nayak TR, Andersen H, Makam VS, Khaw C, Bae S, Xu X, Ee PL, Ahn JH, Hong BH, Pastorin G, Ozyilmaz B. Graphene for controlled and accelerated osteogenic differentiation of human mesenchymal stem cells. ACS Nano 2011;5(6):4670–8.

[214] Lee WC, Lim CH, Shi H, Tang LA, Wang Y, Lim CT, Loh KP. Origin of enhanced stem cell growth and differentiation on graphene and graphene oxide. ACS Nano 2011;5(9):7334–41.

[215] Ku SH, Park CB. Myoblast differentiation on graphene oxide. Biomaterials 2013;34(8):2017–23.

[216] Shin SR, Aghaei-Ghareh-Bolagh B, Dang TT, Topkaya SN, Gao X, Yang SY, Jung SM, Oh JH, Dokmeci MR, Tang XS, Khademhosseini A. Cell-laden microengineered and mechanically tunable hybrid hydrogels of gelatin and graphene oxide. Adv Mater 2013;25(44):6385–91.

[217] Ciriza J, del Burgo LS, Virumbrales-Muñoz M, Ochoa I, Fernandez LJ, Orive G, Hernandez RM, Pedraz JL. Graphene oxide increases the viability of C_2C_{12} myoblasts microencapsulated in alginate. Int J Pharm 2015;493(1):260–70.

[218] Zhang Z, Bao W, Lu H, Jia M, Xie K, Lai Y, Li J. Water soluble polyacrylic acid as a binder for sulfur cathode in lithium-sulfur battery. ECS Electrochem Lett 2012;1:A34–7.

[219] Valentini L, Rescignano N, Puglia D, Cardinali M, Kenny J. Preparation of alginate/graphene oxide hybrid films and their integration in triboelectric generators. Eur J Inorg Chem 2015;2015(7):1192–7.

[220] Yang Y, Zhang H, Chen J, Jing Q, Zhou YS, Wen X, Wang ZL. Single-electrode-based sliding triboelectric nanogenerator for self-powered displacement vector sensor system. ACS Nano 2013;7(8):7342–51.

[221] Lin L, Wang S, Xie Y, Jing Q, Niu S, Hu Y, Wang ZL. Segmentally structured disk triboelectric nanogenerator for harvesting rotational mechanical energy. Nano Lett 2013;13(6):2916–23.

[222] Xu H, Dai H, Chen G. Direct electrochemistry and electrocatalysis of hemoglobin proteinentrapped in graphene and chitosan composite film. Talanta 2010;81(1):334–8.

[223] Wang Y, Li Y, Tang L, Lu J, Li J. Application of graphene-modified electrode for selective detection of dopamine. Electrochem Commun 2009;11(4):889–92.

[224] Shan C, Yang H, Han D, Zhang Q, Ivaska A, Niu L. Electrochemical determination of NADH and ethanol based on ionic liquid-functionalized graphene. Biosens Bioelectron 2010;25(6):1504–8.

[225] Li J, Guo S, Zhai Y, Wang E. High-sensitivity determination of lead and cadmium based on the Nafion-graphene composite film. Anal Chim Acta 2009;649(2):196–201.

[226] Shan C, Yang H, Han D, Zhang Q, Ivaska A, Niu L. Graphene/AuNPs/chitosan nanocomposites film for glucose biosensing. Biosens Bioelectron 2010;25(5):1070–4.

[227] Zuo X, He S, Li D, Peng C, Huang Q, Song S, Fan C. Graphene oxide-facilitated electron transfer of metalloproteins at electrode surfaces. Langmuir 2009;26(3):1936–9.

[228] Sun H, Hu N, Ma H. Direct electrochemistry of hemoglobin in polyacrylamide hydrogel films on pyrolytic graphite electrodes. Electroanalysis 2000;12(13):1064–70.

[229] Zhou Y, Yang H, Chen HY. Direct electrochemistry and reagentless biosensing of glucose oxidase immobilized on chitosan wrapped single-walled carbon nanotubes. Talanta 2008;76(2):419–23.

[230] Nassar AE, Bobbitt JM, Stuart JD, Rusling JF. Catalytic reduction of organohalide pollutants by myoglobin in a biomembrane-like surfactant film. J Am Chem Soc 1995; 117(44):10986–93.

[231] Zhang L, Zhang Q, Li J. Direct electrochemistry and electrocatalysis of hemoglobin immobilized in bimodal mesoporous silica and chitosan inorganic–organic hybrid film. Electrochem Commun 2007;9(7):1530–5.

[232] Zhan T, Xi M, Wang Y, Sun W, Hou W. Direct electrochemistry and electrocatalysis of myoglobin immobilized on Fe_2O_3 nanoparticle–sodium alginate–ionic liquid composite-modified electrode. J Colloid Interface Sci 2010;346(1):188–93.

[233] Ma Q, Sun H, Hou S. Application of graphene oxide sheets incorporated in the porous calcium alginate films on the glassy carbon electrode for biosensor construction based on myoglobin. J Appl Electrochem 2013;43(10):975–84.

[234] Wang M, Sheng Q, Zhang D, He Y, Zheng J. TiC nanoparticles-chitosan composite film for the direct electron transfer of myoglobin and its application in biosensing. Bioelectrochemistry 2012;86:46–53.

[235] Li Q, Sun H, Liu X, Zhao XS. Electroactive porous films of myoglobin within calcium alginate. J Solid State Electrochem 2012;16(4):1651–61.

[236] Guo C, Sun H, Zhao XS. Myoglobin within graphene oxide sheets and nafion composite films as highly sensitive biosensor. Sens Actuators B 2012;164(1):82–9.

[237] Tang Q, Lin J, Wu J. The preparation and electrical conductivity of polyacry- lamide/graphite conducting hydrogel. J Appl Polym Sci 2008;108:1490–5.

[238] Liu X, Huang Y, Sun X, Sheng G, Zhao F, Wang S, et al. Conductive carbon nanotube hydrogel as a bioanode for enhanced microbial electrocatalysis. ACS Appl Mater Interface 2014;6:8158–64.

[239] Qu B, Chen C, Qian L, Xiao H, He B. Facile preparation of conductive composite hydrogels based on sodium alginate and graphite. Mater Lett 2014;137:106–9.

[240] Boruah M, Phukon P, Jyoti Saikia B, Dolui SK. Electrical actuation of electro- responsive hydrogels based on poly(acrylamide–co-acrylic acid)/graphite suitable for biomedical applications. Polym Compos 2014;35:27–36.

[241] Jung W, Jeon BH, Cho DW, Roh HS, Cho Y, Kim SJ, Lee DS. Sorptive removal of heavy metals with nano-sized carbon immobilized alginate beads. J Ind Eng Chem 2015;26:364–9.

[242] Li Z, Du Y. Biomimic synthesis of CdS nanoparticles with enhanced luminescence. Mater Lett 2003;57(16):2480–4.

[243] Brugnerotto J, Lizardi J, Goycoolea FM, Argüelles-Monal W, Desbrieres J, Rinaudo M. An infrared investigation in relation with chitin and chitosan characterization. Polymer 2001;42(8):3569–80.

[244] Kumar MN. A review of chitin and chitosan applications. React Funct Polym 2000; 46(1):1–27.

[245] Pandey S, Goswami GK, Nanda KK. Nanocomposite based flexible ultrasensitive resistive gas sensor for chemical reactions studies. Sci Rep 2013;3:2082.

[246] Pandey S, Goswami GK, Nanda KK. Green synthesis of polysaccharide/gold nanoparticle nanocomposite: an efficient ammonia sensor. Carbohydr Polym 2013;94(1): 229–34.

[247] Kaur H, Banipal TS, Thakur S, Bakshi MS, Kaur G, Singh N. Novel biodegradable films with extraordinary tensile strength and flexibility provided by nanoparticles. ACS Sustain Chem Eng 2012;1(1):127–36.
[248] Thakur S, Pandey S, Arotiba OA. Development of a sodium alginate-based organic/inorganic superabsorbent composite hydrogel for adsorption of methylene blue. Carbohydr Polym 2016;153:34–46.
[249] Wang KH, Hsieh YH, Wu CH, Chang CY. The pH and anion effects on the heterogeneous photocatalytic degradation of o-methylbenzoic acid in TiO_2 aqueous suspension. Chemosphere 2000;40(4):389–94.
[250] Piscopo A, Robert D, Weber JV. Influence of pH and chloride anion on the photocatalytic degradation of organic compounds: Part I. Effect on the benzamide and para-hydroxybenzoic acid in TiO_2 aqueous solution. Appl Catal B 2001;35(2):117–24.
[251] Bouzaida I, Ferronato C, Chovelon JM, Rammah ME, Herrmann JM. Heterogeneous photocatalytic degradation of the anthraquinonic dye, Acid Blue 25 (AB25): a kinetic approach. J Photochem Photobiol A 2004;168(1):23–30.
[252] Chen Q, Cordero-Arias L, Roether JA, Cabanas-Polo S, Virtanen S, Boccaccini AR. Alginate/Bioglass® composite coatings on stainless steel deposited by direct current and alternating current electrophoretic deposition. Surf Coat Technol 2013;233:49–56.
[253] Cordero-Arias L, Cabanas-Polo S, Gao H, Gilabert J, Sanchez E, Roether JA, Schubert DW, Virtanen S, Boccaccini AR. Electrophoretic deposition of nanostructured-TiO_2/chitosan composite coatings on stainless steel. RSC Adv 2013;3(28):11247–54.
[254] Szklarska M, Dercz G, Simka W, Dudek K, Starczewska O, Łężniak M, Łosiewicz B. Alginate biopolymer coatings obtained by electrophoretic deposition on Ti15Mo alloy. Acta Phys Pol A 2014;4(125):919–23.
[255] Sangeetha K, Roy A, Singh S, Lee B, Kumta PN. Novel alginate based coatings on Mg alloys. Mater Sci Eng B 2011;176(20):1703–10.
[256] Cheong M, Zhitomirsky I. Electrodeposition of alginic acid and composite films. Colloids Surf A 2008;328(1):73–8.
[257] Zhitomirsky D, Roether JA, Boccaccini AR, Zhitomirsky I. Electrophoretic deposition of bioactive glass/polymer composite coatings with and without HA nanoparticle inclusions for biomedical applications. J Mater Process Technol 2009;209(4):1853–60.
[258] Cordero-Arias L, Cabanas-Polo S, Gilabert J, Goudouri OM, Sanchez E, Virtanen S, Boccaccini AR. Electrophoretic deposition of nanostructured TiO_2/alginate and TiO_2-bioactive glass/alginate composite coatings on stainless steel. Adv Appl Ceram 2014;113(1):42–9.
[259] Cordero-Arias L, Boccaccini AR, Virtanen S. Electrochemical behavior of nanostructured TiO_2/alginate composite coating on magnesium alloy AZ91D via electrophoretic deposition. Surf Coat Technol 2015;265:212–7.
[260] Hornberger H, Virtanen S, Boccaccini AR. Biomedical coatings on magnesium alloys—a review. Acta Biomater 2012;8(7):2442–55.
[261] Zhang J, Dai C, Wei J, Wen Z, Zhang S, Chen C. Degradable behavior and bioactivity of micro-arc oxidized AZ91D Mg alloy with calcium phosphate/chitosan composite coating in m-SBF. Colloids Surf B 2013;111:179–87.
[262] Zhang J, Dai CS, Wei J, Wen ZH. Study on the bonding strength between calcium phosphate/chitosan composite coatings and a Mg alloy substrate. Appl Surf Sci 2012;261:276–86.

[263] Abidin NI, Atrens AD, Martin D, Atrens A. Corrosion of high purity Mg, Mg2Zn0.2Mn, ZE41 and AZ91 in Hank's solution at 37°C. Corros Sci 2011;53(11):3542–56.

[264] Wu C, Wen Z, Dai C, Lu Y, Yang F. Fabrication of calcium phosphate/chitosan coatings on AZ91D magnesium alloy with a novel method. Surf Coat Technol 2010; 204(20):3336–47.

[265] Song YW, Shan DY, Han EH. Electrodeposition of hydroxyapatite coating on AZ91D magnesium alloy for biomaterial application. Mater Lett 2008;62(17):3276–9.

[266] Luo X, Morrin A, Killard AJ, Smyth MR. Application of nanoparticles in electrochemical sensors and biosensors. Electroanalysis 2006;18(4):319–26.

[267] Derkus B, Emregul E, Emregul KC, Yucesan C. Alginate and alginate-titanium dioxide nanocomposite as electrode materials for anti-myelin basic protein immunosensing. Sens Actuators B 2014;192:294–302.

[268] Baloyi J, Seadira T, Moutloali R, Raphulu M, Ochieng A. Kinetic investigation of Hg (II) photoreduction using TiO_2-immobilized on calcium alginate beads: effect of ferric ion and organic compound. Int'l conf. on chemical engineering & advanced computational technologies (ICCEACT'2014), Nov. 24-25, 2014, Pretoria (South Africa). http://iieng.org/siteadmin/upload/5662E1114006.pdf.

[269] Liu YG, Hu XJ, Wang H, Chen AW, Liu SM, Guo YM, He Y, Hu X, Li J, Liu SH, Wang YQ. Photoreduction of Cr (VI) from acidic aqueous solution using TiO_2-impregnated glutaraldehyde-crosslinked alginate beads and the effects of Fe (III) ions. Chem Eng J 2013;226:131–8.

[270] Chen D, Ray AK. Removal of toxic metal ions from wastewater by semiconductor photocatalysis. Chem Eng Sci 2001;56(4):1561–70.

[271] Papadam T, Xekoukoulotakis NP, Poulios I, Mantzavinos D. Photocatalytic transformation of acid orange 20 and Cr (VI) in aqueous TiO_2 suspensions. J Photochem Photobiol A 2007;186(2):308–15.

[272] Lokesh BG, Rao KK, Reddy KM, Rao KC, Rao PS. Novel nanocomposite membranes of sodium alginate filled with polyaniline-coated titanium dioxide for dehydration of 1, 4-dioxane/water mixtures. Desalination 2008;233(1):166–72.

[273] Xia H, Wang Q. Ultrasonic irradiation: a novel approach to prepare conductive polyaniline/nanocrystalline titanium oxide composites. Chem Mater 2002;14(5):2158–65.

[274] Chang KC, Hsu CH, Peng CW, Huang YY, Yeh JM, Wan HP, Hung WC. Preparation and comparative properties of membranes based on PANI and three inorganic fillers. Express Polym Lett 2014;8:207–18.

[275] Xiong G, Luo H, Zuo G, Ren K, Wan Y. Novel porous graphene oxide and hydroxyapatite nanosheets-reinforced sodium alginate hybrid nanocomposites for medical applications. Mater Charact 2015;107:419–25.

[276] Yadav M, Rhee KY, Park SJ. Synthesis and characterization of graphene oxide/carboxymethylcellulose/alginate composite blend films. Carbohydr Polym 2014; 110:18–25.

[277] Majidnia Z, Fulazzaky MA. Photoreduction of Pb (II) ions from aqueous solution by titania polyvinylalcohol–alginate beads. J Taiwan Inst Chem Eng 2016;66:88–96.

[278] Majidnia Z, Idris A. Photocatalytic reduction of iodine in radioactive waste water using maghemite and titania nanoparticles in PVA-alginate beads. J Taiwan Inst Chem Eng 2015;54:137–44.

[279] Singh V. Mesoporous titania spheres derived from sodium alginate-gum acacia composite beads: efficient adsorbent for "reactive blue H5G" dye. J Environ Chem Eng 2015;3(4):2727–37.

[280] Ni S, Ma J, Zhang J, Yang X, Zhang L. The electrochemical performance of commercial ferric oxide anode with natural graphite adding and sodium alginate binder. Electrochimica Acta 2015;153:546–51.

[281] Yu C, Li X, Liu Z, Yang X, Huang Y, Lin J, Zhang J, Tang C. Synthesis of hierarchically porous TiO_2 nanomaterials using alginate as soft templates. Mater Res Bull 2016; 83:609–14.

[282] Callegaro S, Minetto D, Pojana G, Bilanicová D, Libralato G, Ghirardini AV, Hassellöv M, Marcomini A. Effects of alginate on stability and ecotoxicity of nano-TiO_2 in artificial seawater. Ecotoxicol Environ Saf 2015;117:107–14.

[283] Manjula N, Balaji R, Ramya K, Dhathathreyan KS, Ramachandraiah A. Studies on development of titania nanotube (TNT) based ePTFE-Nafion®-composite membrane for electrochemical methanol reformation. Int J Hydrog Energy 2016;41(21): 8777–84.

CHAPTER

Characterization Techniques for Algae-Based Materials

18

Yolanda Freile-Pelegrín[1], Tomás J. Madera-Santana[2]

CINVESTAV-IPN, Unidad Mérida, Mérida, Mexico[1]; Centro de Investigación en Alimentación y Desarrollo A.C. Hermosillo, México[2]

18.1 SEAWEEDS (ALGAE) CLASSIFICATION

Marine algae, or seaweeds, represent a major component of the marine ecosystem and are crucial primary producers in the oceanic food web. They comprise a complex and heterogeneous group of organisms characterized by their photosynthetic nature and their simple reproductive structure. They are classified into three divisions according to their pigment composition: Rhodophyta (red algae), Heterokontophyta (brown algae), and Chlorophyta (green algae). Heterokontophyta is the largest group, ranging from giant kelps (20 m long) to smaller species (30–60 cm long). Red and green algae are usually smaller, ranging from a few centimeters to about a meter in length. Marine algae also contain large amounts of special polysaccharides needed to support the seaweed thallus in the water, as they are less rigid and strong compared to terrestrial plants. They have various neutral sugars and sugar acids as constituents that are found in terrestrial plants, but a few also contain acidic half ester–sulfate groups attached to hydroxyl groups of the sugars. These are not found in other plants but occur in animal polysaccharides.

18.2 SPECIAL POLYMERS FROM ALGAE AND THEIR APPLICATIONS

Phylogenetically, Rhodophyta are the oldest division of lower plants. They differ from all the other plants in the polysaccharide composition of their cell walls and the intercellular matrix, whose biological function is to give a flexible structure to the seaweeds, helping them withstand currents and wave motion [1]. Chlorophyta are a division closer to the higher plants and contain cell walls with remarkable structural and biochemical similarity to cell walls found in terrestrial plants (starch, cellulose, pectins, and hemicelluloses). The major polysaccharides, however, in all the species examined are polydisperse heteropolysaccharides in which at least some of the hydroxyl groups of the sugar residues are substituted by half ester–sulfate groups (glucuronoxylomannan) [2]. They all are extremely

Algae Based Polymers, Blends, and Composites. http://dx.doi.org/10.1016/B978-0-12-812360-7.00018-5

complex, highly branched molecules that do not appear to have a backbone or simple repeating unit. Quantitatively, the major polysaccharide of Heterokontophyta (brown algae) is alginic acid. It occurs in both the intercellular region and the cell walls, and its biological functions are considered to be principally of structural and ion-exchange type [3].

Rhodophyta contain sulfated galactans and two groups of these special polymers, known as phycocolloids agar and carrageenan, are manufactured industrially on a large scale. The use of phycocolloids in the food industry as functional food products is largely based on their ability to form gels and on the unique properties of these gels [4].

Agar is the strongest gelling hydrocolloid and is mainly isolated from *Gelidium*, *Pterocladia*, and *Gracilaria*, of which *Gracilaria* is typically more sulfated than the others with a pattern of sulfation dominated by the esterification of C-6 of the linked galactose L-unit. In this case, this L-galactose 6-sulfated residue is synthesized in *Gracilaria* as a biological precursor of the 3,6-anhydro-L-galactose and is enzymatically converted to the anhydrous form by sulfohydrolases [5]. However, this enzymatic activity seems to be lower than that occurs in *Gelidium* and *Pterocladia*. A higher number of unfinished 6-sulfated molecules are therefore found in agars from *Gracilaria*, producing nongelling or weak-gel-forming polysaccharides. Thus, the physical properties of the resulting agar (gel strength, gelling, and melting temperatures) are linked to their chemical structures. The gelling ability of agars from most *Gracilaria* species can be considerably improved by adopting an alkali treatment before extraction, which increases the proportion of 3,6-anhydrogalactose. This reaction occurs when 6-sulfated-L-galactose units are present; by heating the polysaccharide in strong alkaline media, the three OH groups can be ionized, producing an intramolecular nucleophilic displacement of the sulfate group in position 6. This reaction is highly specific and no other sulfate group is affected [5].

Carrageenan is a generic name for a family of gel-forming and viscosifying polysaccharides. It is a sulfated polygalactan with an ester-sulfate content of 15%–40% and an average relative molecular mass well above 100 kDa. The different degrees and patterns of sulfates impart the characteristic solubility properties. Some carrageenans are soluble in cold water with only viscosifying properties, whereas others are soluble only in hot water and have the ability to form a thermo-reversible gel with potassium or calcium ions. The carrageenans are grouped into three main families [6]: the kappa family, including kappa (κ) and iota (ι) carrageenans; the beta family, including beta (β) and omega (ω); and the lambda family, including lambda (λ), theta (θ), and pi (π). In each of these families, there are different related precursor structures. Natural carrageenans typically occur as mixtures of different hybrid types. Additionally, methyl or pyruvic acid acetal constituents and the presence of small amounts of other sugars can add to the structural complexity. Variations in these components influence the hydration, gel strength and texture, melting and setting temperatures, syneresis, and synergism. Kappa carrageenan forms stronger gels in the presence of potassium salts, followed by calcium salts. Potassium gives a rigid, elastic gel, while calcium produces a stiff, brittle gel. Kappa gives the strongest gels of all carrageenans, but they are also the ones most likely to bleed (subject to syneresis). If iota and lambda carrageenans are blended, the rigidity and brittleness of the gel can be reduced, but the gel strength may be lowered. Iota forms stronger gels with calcium salts, followed by potassium salts. Calcium gels are soft and resilient and are virtually free of bleeding. They can be frozen and thawed without destroying the gel [7].

Another biopolymer obtained from brown seaweeds is alginate. In their natural state, alginates are bonded with all of the salts present in the seawater, particularly Ca^{2+}, Na^{+}, Mg^{2+}, Sr^{2+}, and Ba^{2+} ions [8]. Alginate is a linear polysaccharide copolymer of (1-4)-linked β-D-mannuronic acid (M) and α-L-guluronic acid (G). The two residues are arranged randomly within the alginate molecule. Three types of segments or blocks can be distinguished within the alginate: two homopolymers (MM and GG), and one heterogeneous or alternating (MG). Different raw materials feature different M and G contents, and the alginate properties of particular seaweed are dependent on both the M/G ratio and the block distribution within the alginate molecule. Alginate has the ability to form heat-stable gels with some divalent metals. The physical properties of alginates depend on the proportion of the three types of blocks and are also related to the seaweed source. Alginates with high G have higher gelling properties, whereas those with high M have higher viscosity.

The uses of alginates are based on three main properties. First is their ability, when dissolved in water, to thicken the resulting solution (more technically described as their ability to increase the viscosity of aqueous solutions). Second is their ability to form gels, which occurs when a calcium salt is added to a solution of sodium alginate in water. The gel is formed by a chemical reaction when calcium displaces sodium from the alginate, holding the long alginate molecules together, resulting in a gel. No heat is required and the gels do not melt when heated. This is in contrast to the agar gels, where the water must be heated to about 80°C to dissolve the agar and the gel forms when cooled below about 40°C. Third property of alginates is the ability to form films of sodium or calcium alginates and fibers of calcium alginates [3].

Related to green algae, ulvan represents a class of sulfated heteropolysaccharide extracted from the cell wall of Chlorophyta belonging to *Ulva* species, which proliferate in eutrophicated areas [2]. This genus is commonly found on all seashores and can produce a significant amount of biomass in nutrient-enriched waters. Interest in the potential uses of this underexploited biomass focuses on its polysaccharides, which have unique functional properties and innovative structures. Ulvan represents about 8%—29% of the dry weight of the algae. The chemical heterogeneity of ulvan comprises rhamnose, glucuronic acid, iduronic acid, xylose, sulfate, and trace amounts of glucose, galactose, and mannose. This polyelectrolyte, referred to as ulvan, is able to gel in the presence of boron and other divalent cations [9,10]. The original physicochemical, rheological, and biological properties recently revealed for this complex, sulfated aldobiouronan open the way for novel potential applications. The most striking feature that distinguishes the chemical composition of ulvan from other polysaccharides of marine origin is, therefore, the presence of uncommon sugars such as iduronic and sulfate rhamnose, displaying a close similarity with mammalian glycosamino-glycans. This resemblance and the possibility of obtaining this material from cheap and renewable resources mean that a deeper investigation of the biological activity, and the feasibility of using this polysaccharide as new material for biomedical applications, is worthwhile. Materials suitable for biomedical applications, namely tissue engineering, regenerative medicine, and drug delivery, are required to be biocompatible and biodegradable, and their degradation products must be safe and easily cleared from the host organism.

18.3 POLYMERS: DEFINITIONS AND CLASSIFICATIONS

The polymers are macromolecules whose chain length can reach several orders of magnitude. They can be considered to belong to the normal area of solid-state physics. The polymer chain length is the number of monomeric units forming the macromolecule, and it is estimated that a macromolecule can have a length of the order of 1500–3000 Å, which is more than 1000 times greater than the length of a simple water molecule. Considering the aspects of chain length and equivalent diameter (molecular anisotropy), the relationship between these parameters is quite strong. These features are unique and fundamental for the polymer, and both have important implications for the properties and structures in the solid state. Polymers are mostly solid materials and have several applications such as structural materials, fibers, or thin films. This has opened up different applications in various industries (construction, transportation, textiles, household goods, packaging, etc.). However, for many possible applications of polymers, it is necessary to study their behavior under different conditions or by physical and chemical means to establish correlations between the microstructure and its properties. This involves various disciplines such as engineering, physics, chemistry, mathematics, and mechanics, among others.

Thermoplastic polymers can be melted or dissolved in solvents to form solutions. These polymers can be heated several times above the glass transition temperature (T_g) and around the melting temperature (T_m) to be shaped. Therefore, melting is the basis of most processing methods for polymeric materials [11]. However, the properties of thermoplastics are highly sensitive to heat. Some examples of thermoplastics are polyethylene (PE), polyvinyl chloride (PVC), polystyrene (PS), polycarbonate (PC), polymethyl methacrylate (PMMA), and polypropylene (PP). Among the synthetic biodegradable polymers are polycaprolactone (PCL), polyhydroxybutyrate (PHB), polylactic acid (PLA), polybutylene succinate (PBS), and copolymer of polybutylene adipate-co-terephthalate (PBAT) [12]. However, some polymers, called thermosetting polymers, have macromolecular chains that are chemically bonded together to form three-dimensional networks. This type of solid polymer can be molded at room temperature or above but when heated more strongly becomes hard and infusible [13]. Elastomers are rubbers or rubber-like materials. Whatever the type of bond between the polymer chains, it is known that the probabilistic nature of chemical reactions that give rise to polymers leads to two factors involved in their rigid body behavior when solidified: the chain configuration and the molecular weight distribution.

Temperature plays an important role in the practical uses of polymers, because it produces thermal breakdown of the polymer chains. Therefore, during polymer formulation, inhibitors, antioxidants, stabilizers, and other additives are incorporated. Nonetheless, the decomposition temperature (above 200−400°C) produces an uncatalyzed scission of the bonds in the main chain of the polymer, causing its degradation. To convert raw polymeric materials into finished products, the polymer must be processed. Polymer processing is defined as an "engineering activity concerned with operations carried out on polymeric materials or systems to increase their utility" [14]. Polymer processing is commonly analyzed in terms of the methods used, such as extrusion, injection molding, blow molding, calendaring, mixing and dispersion, rotational molding, and others [15].

18.4 PROCESSING OF COMPOSITE MATERIALS

Factors such as technology and cost of composite materials depend on the processability, for instance, how the components are combined to form the composite material. The processability is closely related to the ability of the components to join, thereby forming a cohesive material. The processing conditions usually involve high temperature, pressure, and time, where the polymeric matrix typically determines these parameters. The rigid bonding of the filler with the matrix at an elevated temperature has a disadvantage because bond weakening or debonding may occur during the subsequent cooling due to difference in the thermal contraction (related to the coefficient of thermal expansion) between filler and matrix. Fiber composites are usually fabricated by the impregnation of the matrix in molten or liquid state using short fibers or fiber preforms (woven fabric).

18.5 POLYMER-MATRIX COMPOSITES

Polymer-matrix composites (PMCs) can be classified based on whether the matrix is a thermoset or a thermoplastic polymer. Thermoset-matrix composites are traditionally far more common, but thermoplastic-matrix composites are currently undergoing rapid development. The advantages of thermoplastic-matrix composites compared to thermoset-matrix composites include lower manufacturing costs (no curing, unlimited shelf life, reprocessing, low moisture content, weldability, etc.). Some disadvantages of thermoplastic PMCs are high viscosities, high processing temperatures, and the need to perform fiber or particle-surface treatments.

PMCs are much easier to fabricate than metal-matrix, carbon-matrix, and ceramic-matrix composites, whether the polymer is a thermoset or a thermoplastic. This is because of the relatively low processing temperatures required to fabricate PMCs.

The molding methods are those conventionally used for thermoplastic polymers, and include injection molding (heating above the melting temperature of the thermoplastic and forcing the slurry into a closed die opening through the use of a screw mechanism). Other methods such as, extrusion (forcing the slurry through a die opening via a screw mechanism), calendering (pouring the slurry into a set of rollers with a small opening between adjacent rollers to form a thin sheet), and thermoforming (heating above the softening temperature of the thermoplastic and forming over a die using matching dies, a vacuum, or air pressure, or without a die using movable rollers) [16].

18.6 BIO-BASED COMPOSITE MATERIALS

The replacement of petroleum-derived plastics by novel bio-based materials from inexpensive, renewable natural resources greatly impacts the plastic industry (coatings, composites, etc.), where biodegradable matrices and bio-based by-products are used as fillers [17,18]. Bio-based products from renewable resources can exhibit similar and sometimes better properties than commercial petroleum-based products, and may have improved recycling and biodegradability capabilities [19]. A biodegradable polymeric matrix is PLA, which is a promising biodegradable aliphatic polyester produced by the condensation polymerization of lactic acid [20]. PLA is referred to as a "green" polymer because it is available from agricultural renewable resources; hence, its production requires less fossil resources compared to petroleum-based plastics. However, PLA has some drawbacks such as relatively high resin cost, poor impact strength, low distortion temperature, and brittleness. To overcome the existing limitations and improve the mechanical performance of PLA, some authors have studied the reinforcement of PLA with other renewable and biodegradable fillers (starch, sisal, hemp, and wood fibers) [21–23]. Apart from conventional fillers, considerable interest has recently been generated concerning the utilization of novel fillers derived from industrial waste.

The conventional industrial process for agar extraction includes seaweed pretreatment, extraction, filtration, concentration, and dehydration. The agar is extracted and dissolved in boiling water, often under pressure. The agar dissolved in water must be filtered to remove the residual seaweed, and the hot filtrate is cooled to form a gel containing approximately 1% agar [24]. Particular control of several parameters (i.e., pH, temperature, and time) is needed to obtain control of the molecular weight distribution of the polysaccharide.

In the current industrial practice of producing agar, large quantities of seaweed wastes (SWWs) are produced from the filtration step, where a "filter cake" is produced and eliminated. The filter cake represents approximately 70% of the raw material used. This filter cake or SWW consists mainly of residual seaweeds enriched with cellulose, hemicellulose, agar residues, and small amounts of floridean starch. The filter cake also contains diatomite earth, which is used as a filtration aid in the filtration step. Hence, SWW is a material enriched with different polysaccharides that can be used as inexpensive hybrid filler in biocomposites.

Some authors have reported the use of raw seaweed biomass as a reinforcement in biocomposites with PBS, PP, high density PE, and PLA [17,18,24–26]. These biocomposites were formulated using seaweeds from three divisions Chlorophyta, Phaeophyta, and Rhodophyta (green, brown, and red algae, respectively) as a biofiller in a polymeric matrix. According to the results reported by the previous authors, the seaweeds could produce composite materials using thermoplastic matrices. These composites would have potential applications in the automotive, construction, or packaging industries. The SWW obtained from agar extraction could be used as a biofiller in a polymeric matrix, and the formulations of PLA-SWW biocomposites are shown in Table 18.1.

In this chapter, the characterization techniques applied in polymer–algae-based biocomposites are reported, and the effect of filler addition on the structure and properties of the biocomposites is evaluated.

18.7 OPTICAL AND MORPHOLOGICAL PROPERTIES

18.7.1 COLOR MEASUREMENTS

Color is an important attribute for any commercial composite or biocomposite. Therefore, instruments can measure this physical attribute. The color can be read using the Commission International de L'Eclairage (CIE) Lab parameters (L^*, a^*, b^*) with a spectro-colorimeter. The scanner is calibrated with a standard tile. In this coordinate system, the L^* value is a measure of the lightness (brightness) ranging from 0 (black) to 100 (white), the a^* value is a measure of the redness ranging from −100 (green) to +100 (red), and the b^* value is a measure of the yellowness ranging from −100 (blue) to +100 (yellow). The color differences (ΔE) are also calculated by the following equation:

$$\Delta E = \sqrt{(\Delta L^*)^2 + (\Delta a^*)^2 + (\Delta b^*)^2} \tag{18.1}$$

Table 18.1 Processed Formulations of Biocomposites With Polylactic Acid (PLA) and Seaweed Wastes (SWWs)

Sample Code	PLA (% wt)	SWW (% wt)	Compatibilizer (% wt)
A	100	–	–
B	95	5	–
C	80	20	–
D	95	–	5
E	90	5	5
F	75	20	5

Compatibilizer, *Maleic anhydride + dycumyl peroxide; SWW, seaweed wastes.*

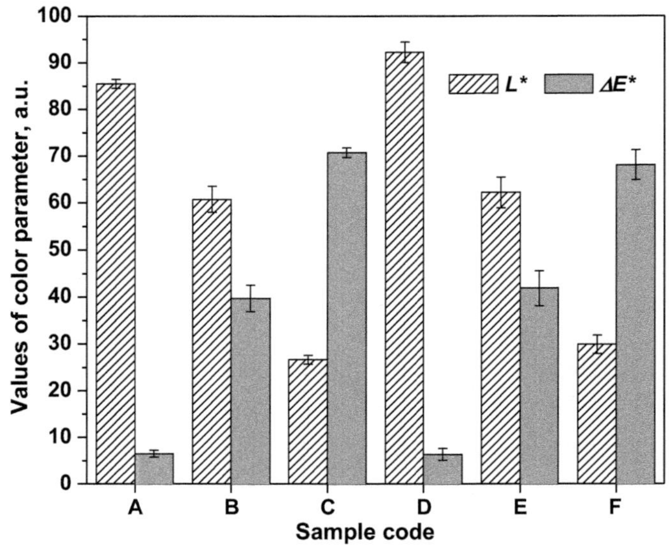

FIGURE 18.1

Color parameters measured in PLA-SWW biocomposites. *PLA*, polylactic acid; *SWW*, seaweed waste.

where $\Delta L^* = L^* - L_0^*$, $\Delta a^* = a^* - a_0^*$, and $\Delta b^* = b^* - b_0^*$; L_0^*, a_0^*, and b_0^* represent the color parameter values of the standard; and L^*, a^*, and b^* represent the color parameters of the sample. To determine the parameters in the samples, at least six measurements should be performed.

The color parameters (L^* and ΔE^*) of PLA-SWW biocomposites with and without compatibilizer are shown in Fig. 18.1. The color changes are due to the variation in SWW content in the PLA matrix. The luminosity values (L^*) showed a gradual decrease as the SWW content increased. To improve the adhesion of SWW particles in the PLA matrix, a compatibilization treatment with maleic anhydride–dicumyl peroxide (MAH-DCP) was performed in samples with codes D and F (Sample code in Table 18.1). The reduction of the L^* values is attributed to the close vicinity of the SWW filler in the PLA matrix, which produces low brightness in the biocomposites. However, the opposite behavior was observed in the color difference (ΔE). This result was somewhat expected because ΔE is closely related to the SWW content, and this parameter shows the combined effects of the three-color parameters (L, a, and b). The addition of SWW and compatibilizer (MAH-DCP) to the PLA matrix significantly modified ($P < .05$) the ΔE value obtained from the biocomposites.

18.7.2 OPTICAL MICROSCOPY

Several techniques are available for analysis of solid materials; the most simple is without doubt optical microscopy (Fig. 18.2). Two typical modes of optical microscopy are commonly used, based on the measurement of transmitted or reflected

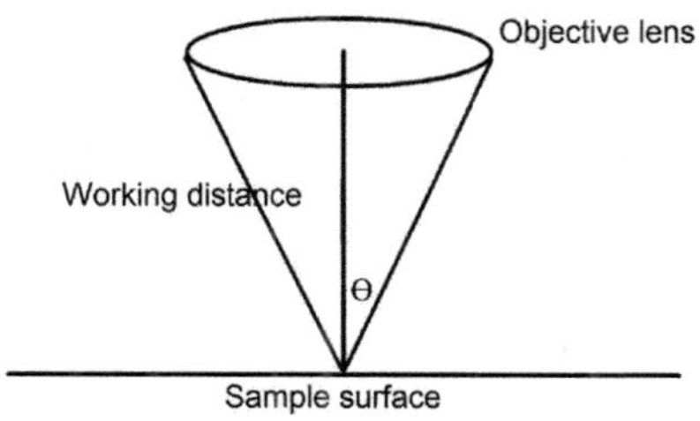

FIGURE 18.2

Schematic representation of optical microscope.

light from a transparent or opaque sample, respectively. Many solid-state materials are nontransparent in their as-grown/as-deposited states, and it is usually difficult to prepare thin cross sections for transmission microscopy. Therefore, some researchers use the reflection mode, also known as episcopic light differential interference contrast (DIC) microscopy. This technique is useful for imaging a variety of reflective samples including minerals, metals, semiconductors, glasses, polymers, and composites.

The degree of discernment of fine details within a magnified image is generally referred to as the resolution of a microscope. Light is used as a source of illumination in optical microscopy, and the resolution is expressed in the same units as the wavelength of light (nm). The theoretical resolution (R) for an optical system can be calculated using Abbe's Eq. (18.2):

$$R = \frac{0.61\lambda}{\eta(\sin\theta)} \tag{18.2}$$

The denominator of Eq. (18.2) represents the numerical aperture (NA) of the objective lens, which is related to its light gathering ability. Some primary factors that influence the resolution of the objective lens are the wavelength of light used, the index of refraction (η) of the environment surrounding the lens (e.g., 1.00 for air), and the angle of illumination (θ).

From the above equation, the resolution limit that can be reached using visible light is between 350 and 700 nm. Using high NAs (e.g., 1.3–1.4 for oil-immersion lenses where $\eta = 1.5$), the theoretical resolution using polychromatic visible light (c.500 nm) is of the order of 220 nm. Hence, any sample features that are less than 220 nm apart from one another will appear blurry. In practice, the observed resolution is often worse than the theoretical value, depending on the degree of inherent optical aberrations of the lens. It can be seen from Eq. (18.2) that higher spatial resolutions (i.e., smaller R values) are possible through the use of shorter wavelengths.

Fig. 18.3 shows the surface morphology of PLA-SWW biocomposites. The surface observed at a magnification of 10× using an American stereo microscope reveals the inclusions of SWW in the PLA matrix. Moreover, some details related to the adhesion characteristics between the filler and matrix cannot be observed by this technique, which is one of its limitations.

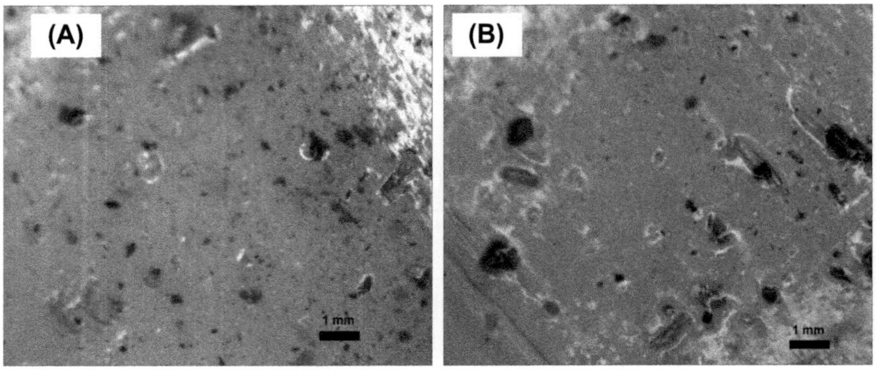

FIGURE 18.3

Surface morphology of biocomposites with an SWW content of 5 wt% without (A) and with (B) MAH-DCP compatibilizer. *MAH-DCP*, maleic anhydride–dicumyl peroxide; *SWW*, seaweed waste.

18.7.3 ELECTRON MICROSCOPY

Although optical microscopy can be extended into the nano-regime, other techniques must be used to clearly discern components below 100 nm. Indeed, the current "nanotechnology revolution" that we are experiencing would not have been possible if there were not suitable techniques for characterizing nanomaterials. As we saw in the previous section, to improve the resolution, we must use source radiation with as small a wavelength as possible.

The heart of any electron microscope is the electron source or gun. In fact, the electron gun within microscopes is identical to that operating within old televisions. These TV monitors contain a cathode ray tube (CRT) with a set of tungsten filaments. Electrons generated by these filaments are accelerated toward an anode that is coated with phosphor particles, which emit light of different colors when struck with electrons. The image observed on a monitor is a result of the electron guns being scanned sequentially from left to right in a fraction of a second; this process is known as rastering. In the same way, in the operation of a scanning electron microscope (SEM) or scanning transmission electron microscope (STEM), focused electron beam is allowed to scan across selected regions of a sample.

The SEM technique is very useful for observing the interfacial morphology. The morphological analysis of cryogenically fractured PLA-SWW makes it possible to investigate the SWW dispersion and interfacial adhesion of SWW particles in the PLA matrix. Fig. 18.4A and B show a comparison of two biocomposites of PLA-SWW with and without compatibilizer, respectively. Biocomposites that have undergone compatibilization treatment show an improvement in the dispersion and adhesion of SWW in the matrix [24].

18.8 THERMAL PROPERTIES

Thermal analysis techniques are used to characterize polymers (thermoplastics, thermosets, and elastomers). The four most important thermal analysis techniques are

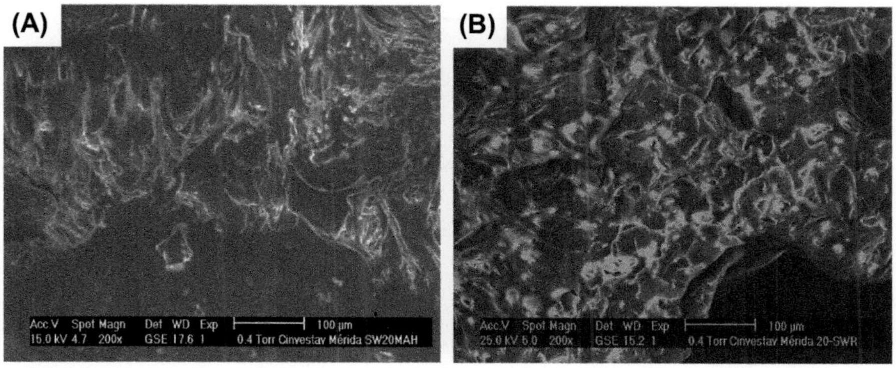

FIGURE 18.4

Morphology of cryogenic fracture surface of biocomposites with an SWW content of 20 wt%. *SWW*, seaweed waste.

differential scanning calorimetry (DSC), thermogravimetric analysis (TGA), thermomechanical analysis (TMA), and dynamic mechanical analysis (DMA). Sometimes a combination of two or three techniques provides complete information on the sample.

In DSC experiments, the heat flow to and from a sample is measured as a function of temperature as the sample is heated, cooled or held at constant temperature. The measurement signal registered by calorimeter is the energy absorbed or released by the sample. The calorimeter can detect endothermic or exothermic events, and transition and reaction enthalpies to measure the temperature of peaks or change of slope in second-order transition (glass transition) and determine the specific heat capacity. Usually, a nonisothermal DSC allows determination of these thermal events. The amorphous polymers show a glass transition temperature (T_g); however, semicrystalline polymers possess a crystallization temperature and melting temperature with their respective enthalpies. Nevertheless, these parameters can be altered by the polymerization process used or by the addition of additives, processing aids, fillers, compatibilizers, and so on [27].

Fig. 18.5 shows the DSC thermogram of PLA matrix. Different thermal parameters such as the glass transition (T_g), heat capacity (ΔC_p), crystallization temperature (T_c) and enthalpy (ΔH_c), and melting temperature (T_m) and enthalpy (ΔH_m) can be determined from this thermogram. Moreover, with the help of these thermal parameters, it is possible to calculate the phases of PLA-SWW biocomposites, such as the crystalline (χ_c), mobile amorphous (χ_{ma}), and rigid amorphous (χ_{ra}) phases, which can be done using the three-phase model reported by Ref. [28].

$$\chi_c(\%) = \frac{\Delta H_m - \Delta H_c}{\Delta H_m^{\circ}} \times \frac{100}{w} \tag{18.3}$$

$$\chi_{ma}\,(\%) = \frac{\Delta C_p}{\Delta C_p^{\circ}} \times 100 \tag{18.4}$$

$$\chi_{ra}\,(\%) = 100 - \chi_c - \chi_{ma} \tag{18.5}$$

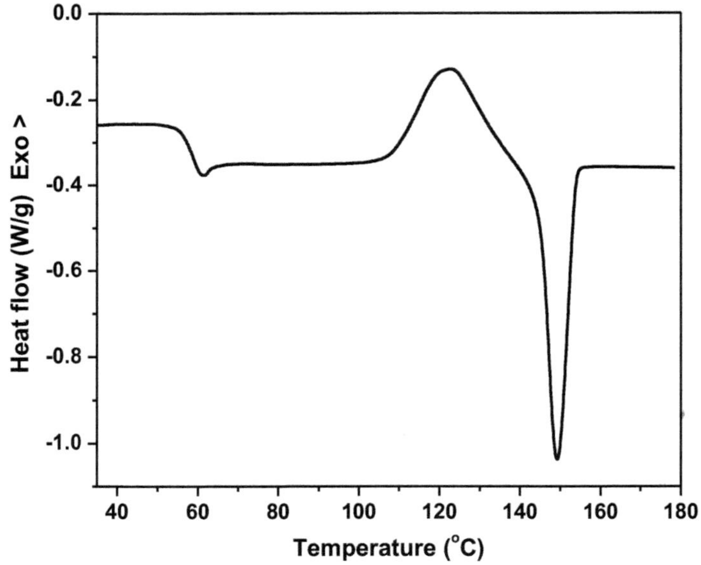

FIGURE 18.5

DSC thermogram of PLA matrix. *DSC*, differential scanning calorimetry; *PLA*, polylactic acid.

where w is the PLA weight fraction in the blend, and ΔH_m and ΔH_c are the experimental enthalpies of melting and crystallization just before melting, respectively. ΔH_m° is the enthalpy of melting for 100% crystalline PLA (93.6 J/g), ΔC_p is the specific heat change at T_g, and ΔC_p° is the specific heat change at T_g of fully amorphous PLA (0.628 J/gK) [28].

The thermal parameters of the neat PLA and PLA-SWW biocomposites are listed in Table 18.2. The DSC thermograms of the biocomposites show clear changes in the glass transition, an exothermal crystallization peak, and an endothermic

Table 18.2 Thermal Parameters of the Polylactic Acid—Seaweed Waste Biocomposites

Sample Code	T_g (°C)	ΔC_p (J/g°C)	T_c (°C)	ΔH_c (J/g)	T_m (°C)	ΔH_m (J/g)	χ_c (%)	χ_{ra} (%)	χ_{ma} (%)
A	58	0.57	105.6	28.4	144.7	31.4	3.2	6.0	90.8
B	57.3	0.51	98.8	27.6	151.3	32.3	5.3	13.4	81.3
C	56.8	0.13	93.6	22.6	150.9	23.1	0.7	78.6	20.7
D	57.8	0.48	102.3	27.7	151.8	31.5	4.0	19.6	76.4
E	57.5	0.37	79.1	23.9	151.5	29.4	6.5	33.5	60.0
F	57.2	0.28	89.5	21.8	150.9	22.7	1.2	53.9	44.9

melting peak during heating. The T_g values of the PLA-SWW biocomposites were affected by the presence of SWW, because a decrease from 58 to 56.6°C was observed. The crystallization behavior of the PLA-SWW biocomposites with and without compatibilizer was affected and showed lower crystallization temperature (T_c) and enthalpy (ΔH_c) in comparison with neat PLA. Moreover, neat PLA showed a narrow melting peak at 144.7°C, whereas the PLA-SWW biocomposites displayed a higher melting temperature (T_m) (150.9−152.1°C). The increase in T_m could be attributed to the presence of the SWW filler in the PLA matrix. However, the SWW filler restricts the mobility of the PLA chains; similar behavior was observed in seaweed fibers and PP composites by Luan et al. [18]. The increase in χ_c observed in samples with codes B and E could suggest that in these formulations the SWW served as nucleation sites for crystallization and enhanced the rate of crystallization. Furthermore, the rigid amorphous phase (χ_{ra}) in the PLA-SWW biocomposites increased from 13.4% to 78.6% compared with the neat PLA (6%). In contrast, the mobile amorphous phase (χ_{ma}) decreased from 90.8% (Sample code A) to 20.7% and 44.9% in samples with codes C and F, respectively. This result could indicate that the SWW content affects the development of rigid amorphous phases in PLA-SWW biocomposites.

TGA measures the mass of a sample as a function of temperature. Therefore, changes in the mass of the sample (gain or loss) are detected. TGA relies on a high degree of precision of three variables: weight, temperature, and heating rate. The changes in slope of the thermogram are associated with degradation or release of components of the sample. From the derivative weight-loss thermogram (DTGA), the temperature at which the weight loss is maximum can be determined. However, deconvolution process may be necessary if there are overlapping peaks in the DTGA [29].

Fig. 18.6A and B show the thermograms of TGA and DTGA of PLA-SWW with and without compatibilizer. In general, one degradation stage was observed in the thermograms (Fig. 18.6A). Nevertheless, the DTGA thermograms showed small shoulders associated with degradation of the components from the biocomposites. Neat PLA had a decomposition temperature (T_{dp}) of 346°C; however, incorporation of 20 wt% SWW shifted this temperature to 350 and 379°C in samples with codes C and F, respectively. Moreover, the thermal stability of the biocomposites filled with compatibilizer showed a significant improvement, which could be due to the compatibility and interfacial bonding between the components.

In DMA, the properties of the materials are measured when they are deformed under periodic stress. The sample is subjected to a sinusoidal mechanical stress. Parameters such as force amplitude, displacement (deformation amplitude), and phase shift are determined as a function of temperature or frequency. DMA enables the detection of thermal effects based on changes in the modulus or damping behavior. Most of the polymeric materials are viscoelastic and exhibit phase differences. Typical DMA curves described by amorphous thermoplastic polymer are shown in Fig. 18.7A. At low temperature, the polymer chains are immobile, unable to resonate, and therefore stiff. However, at high temperatures, the polymer chains become

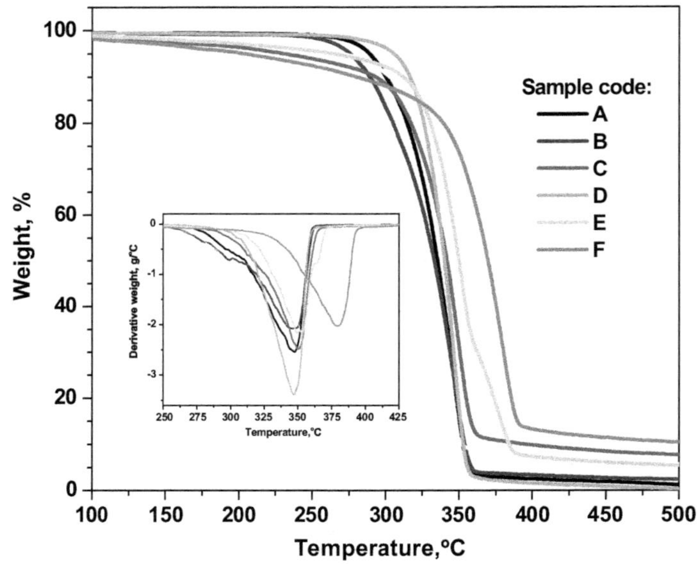

FIGURE 18.6

TGA thermograms of PLA-SWW biocomposites. Inset: DTGA thermograms of biocomposites. *PLA*, polylactic acid; *SWW*, seaweed waste; *TGA*, thermogravimetric analysis.

mobile and can resonate with the band. A polymer will be in the glassy state or energy elastic state at low temperature, but at high temperature, it will be in the rubber or entropy elastic state [30]. Fig. 18.7B shows that the height of the tan δ peak decreased with the agar content. It means that the agar particles hinders the chain mobility which produces a reduction of sharpness and height of this parameter [12].

18.9 MECHANICAL PROPERTIES

Polymers are described as viscoelastic materials (between viscous liquids and elastic solids). At low temperature, or high frequencies, polymers behave like a glass, but at high temperatures, they display a rubber-like behavior. Therefore, for the intermediate temperature of frequency range, the polymer is neither glassy nor rubbery. Under tension load, polymer deforms, and the stress increases with strain. A typical strain curve in tension is built (Fig. 18.8). The properties of the polymer can be related to the characteristic of their stress–strain curves. The mechanical parameters to be determined are the tensile strength, elongation at break, and tensile modulus. Some definitions of the transition from the elastic to the plastic region are shown in Fig. 18.8. The yield point or limit of proportionality is defined as the linear behavior of the stress–strain curve. The yield strength is defined as the stress resulting from the intersection of a parallel straight line of the stress–strain curve, with an

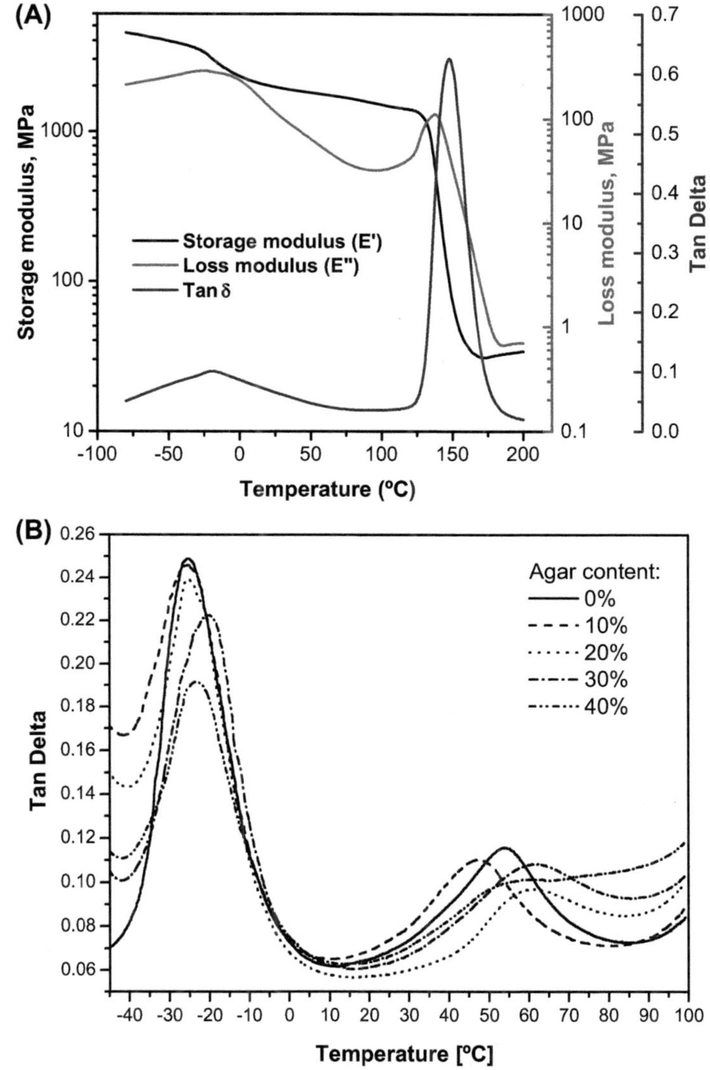

FIGURE 18.7

DMA curves of an amorphous polymer (A), and tan δ curves (B) of PBAT-agar biocomposites. *DMA*, dynamic mechanical analysis; *PBAT*, polybutylene adipate-co-terephthalate.

offset from the origin of this curve by a defined amount of strain ($\varepsilon = 0.001$ or 0.002). The elastic modulus or Young's modulus is calculated from the slope in the elastic region.

The tensile properties of the PLA-SWW biocomposites are shown in Fig. 18.9A and B. The tensile strength and elongation at break decreased with the addition of

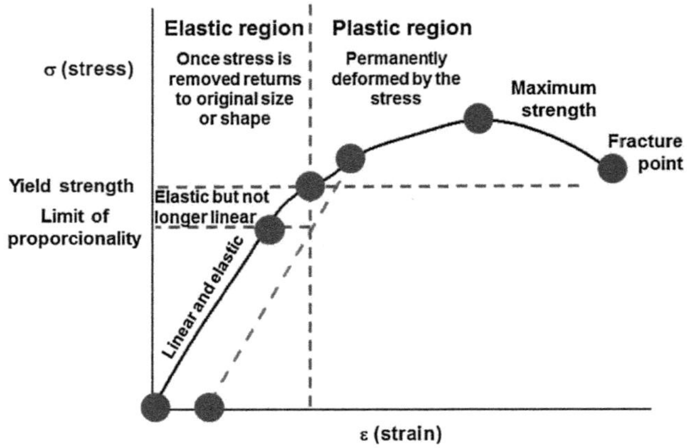

FIGURE 18.8

Typical stress–strain curve.

From www.cyberphysics.co.uk.

SWW. A decrease in tensile strength was observed at an SWW content of 5 wt% (Sample code B); however, at an SWW content 20 wt% (Sample code C), the value of this parameter is much lower than that of the PLA control. The compatibilization treatment performed on the biocomposites significantly increased ($P < .05$) the tensile strength, mainly in the formulation with Sample code C. This result indicates good adhesion between the SWW filler and PLA matrix. In contrast, the elongation at break of the untreated and treated biocomposites showed a drastic decrease reaching a minimum value (2.8%) when the SWW content was 20 wt% (Sample code C). The decrease in elongation at break is due to the increased stiffness of the biocomposites when the SWW content is increased, and this is a common characteristic of reinforced thermoplastic composites. The average tensile modulus increased with the addition of an SWW in formulation with code B (5 wt%); however, further addition of SWW filler decreased the modulus. At 20 wt% of SWW filler (Sample code C), the value of this property was 42% and 49% lower in comparison with neat PLA for biocomposites with and without compatibilization treatment, respectively. Several studies in which neat PLA and organic fillers were mixed without coupling agents were conducted to produce biocomposites [31,32], although these studies showed a marked reduction in the tensile strength and elongation at break and an increase in the tensile modulus as the filler concentration increased. From our results, the use of compatibilization (i.e., maleation) improves the mechanical properties of the resultant composites.

18.10 STRUCTURAL PROPERTIES

Infrared spectroscopy is a characterization technique used to elucidate the structure of matter at the molecular scale [33]. FTIR spectrometers give the infrared (IR)

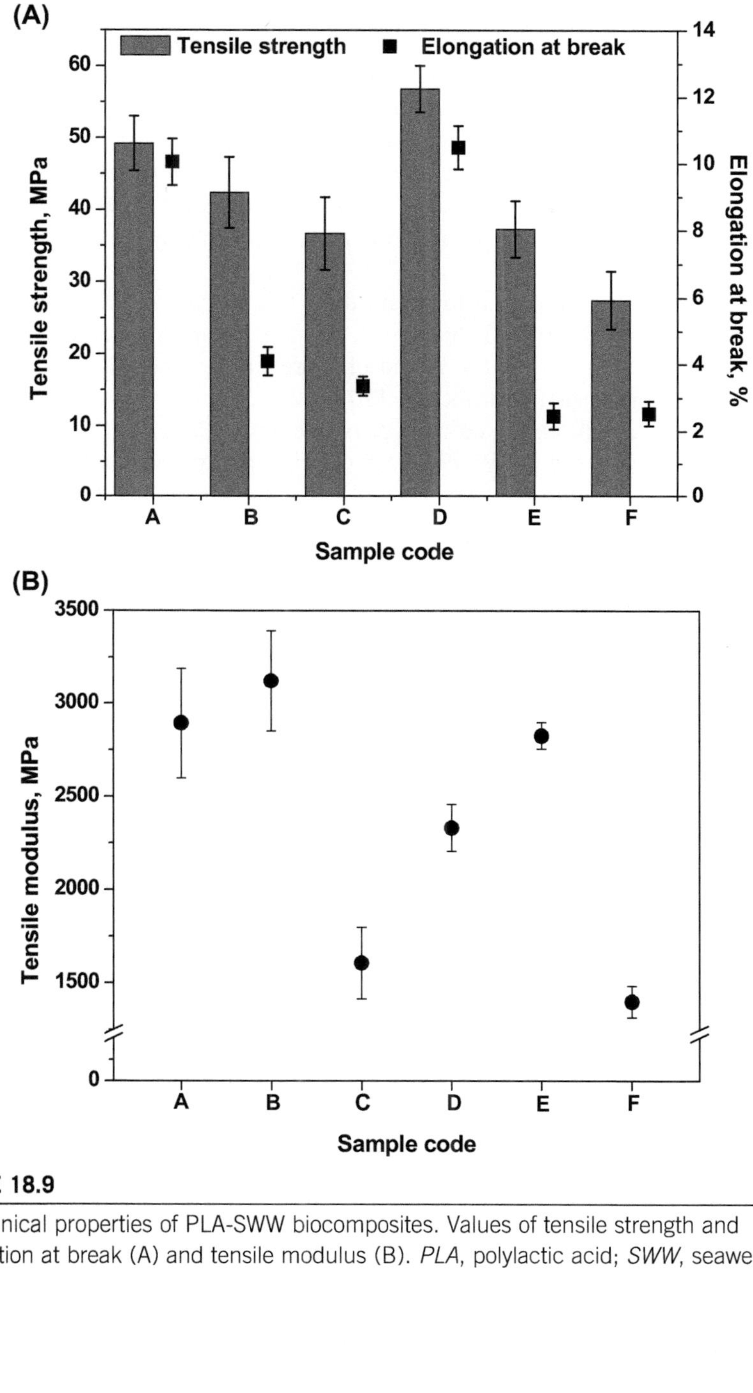

FIGURE 18.9

Mechanical properties of PLA-SWW biocomposites. Values of tensile strength and elongation at break (A) and tensile modulus (B). *PLA*, polylactic acid; *SWW*, seaweed waste.

spectrum by Fourier transformation of the signal from an interferometer. The FTIR technique can be used to analyze gases, liquids, and solids with minimal preparation [34]. The spectrometer can operate in transmission, reflection, or attenuated total reflection (ATR) mode. The ATR objective is a polished face of diamond, germanium, or zinc selenide (ZnSe) that is pressed into contact with the sample [35]. The infrared reflection is attenuated by absorption to a depth of few micrometers with the surface layer. In this case, achieving a good contact is necessary, although it is easy for most polymers. The main advantage of this technique is that no sample preparation is required.

The ATR technique was used to obtain the FTIR spectra for PLA-SWW biocomposites in the range of 2000–800 cm^{-1}, and the spectral results are shown in Fig. 18.10. The spectra of PLA-SWW and PLA are similar, and it appears that the SWW filler was encapsulated inside the PLA matrix. However, the FTIR spectra are useful to elucidate the existence and type of interaction that might occur between the molecules of PLA and SWW. Fig. 18.5 clearly reveals that the neat PLA and PLA-SWW biocomposite show six peaks (1745, 1445, 1350, 1265, 1183, and 1085 cm^{-1}). The band at 1755 cm^{-1} represents the C=O stretch and the O–C=O stretch at 1183 cm^{-1}, and both peaks are characteristic of ester bonds [36]. The bonds at 1445 and 1350 cm^{-1} are attributed to bending deformation of the CH_3 groups. The peaks situated at 1265 and 1085 cm^{-1} are related to the valence vibration of anhydrides (C–CO–O–CO–C), as well as to the 3,6-anhydro-galactose bridge of agar [37]. The band originating from the CO stretching vibration was situated at 1745 cm^{-1} for neat PLA and showed alternations as the SWW content

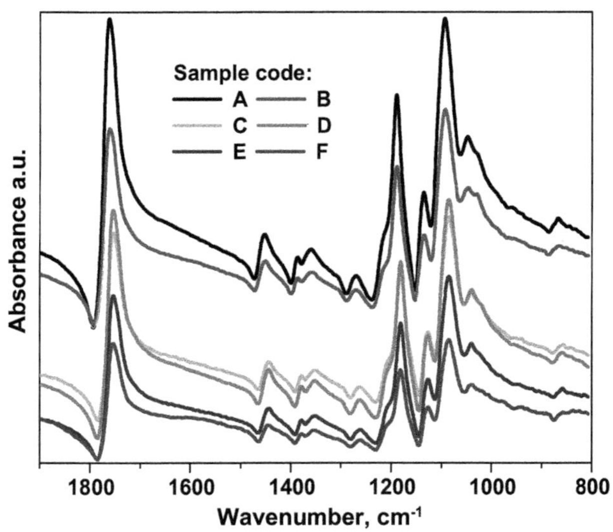

FIGURE 18.10

FTIR spectra of biocomposites based on SWW in biodegradable matrix of PLA. *FTIR*, Fourier transform infrared spectroscopy; *PLA*, polylactic acid; *SWW*, seaweed waste.

increased and when chemical modification of the biocomposite was performed. The amount of MAH present in the biocomposites (samples with codes D, E, and F) is small; the stretching of MAH is very weak and might overlap with the carbonyl group of PLA. Perhaps the extent of grafting is too low and the FTIR was not sensitive enough to identify the changes, a similar feature has been reported previously [36].

Several authors have discussed the theory of X-ray diffraction (XRD). However, Warren [38] has dealt with the application of X-ray methods to amorphous materials such as polymers. The general principles for the analysis of the X-ray scattering from amorphous materials were developed many years ago by Debye in 1915 and Zernicke and Prins in 1927 [39,40]. For partially crystalline polymers, the X-ray scattering curve shows sharp peaks due to scattering from the crystalline regions together with a broad diffuse scattering from the amorphous phase. For quantitative measurements, it is necessary to separate these two forms of scattering to determine crystalline/amorphous ratios and to obtain accurate intensity measurements for determination of the crystal structures. The interpretation of a two-phase model (crystalline/amorphous) is satisfactory for polycrystalline partially amorphous polymers. In fact, XRD is a primary technique for determining the degree of crystallinity in polymers, even though they are never 100% crystalline.

The effect of the SWW filler on the crystallization behavior of the PLA was studied using XRD as shown in Fig. 18.11. The diffraction pattern of PLA shows one broad peak with a maximum at $2\theta = 18.8$ degrees and a small peak at $2\theta = 29.45$ degrees. The pattern of the former peak remained unaffected upon the

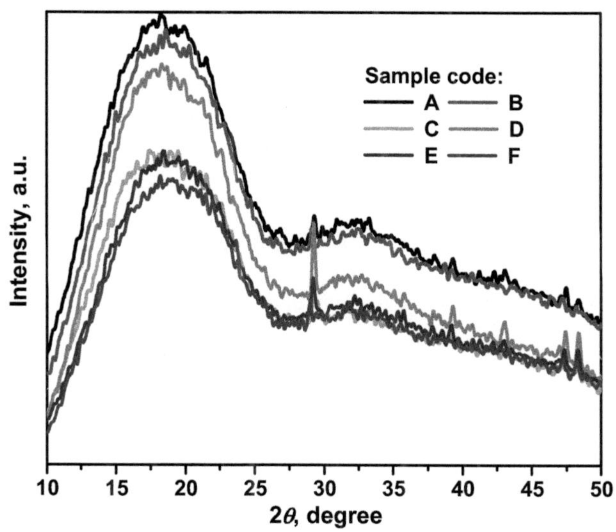

FIGURE 18.11

XRD curves of PLA-SWW biocomposites. *PLA*, polylactic acid; *SWW*, seaweed waste; *XRD*, X-ray diffraction.

addition of SWW filler, and it is fairly similar to the results reported in the literature [41,42] in which the diffraction peak is centered at 2θ angle of 16.4–25.3 degrees. The gradual reduction in the intensity among the samples was due to the difference in the relative crystallinity. In contrast, the peaks at 29.4, 47.4, and 48.4 degrees could be attributed to the diatomite earth (which contains calcite) used during the agar extraction process and the clay used to hold the sample during analysis.

18.11 CONCLUSION

Algal biomass represents a tremendous source as a renewable feedstock for composite materials applications. Moreover, the industrial production of biopolymers (phycocolloids) from seaweeds produces huge amounts of residues, and these large quantities of SWWs can be used as filler to produce biocomposites. The motivation for this research is the ecological concern and economic advantages. Besides the environmental concern, natural fillers used to reinforce polymer biocomposites can combine high performance, versatility, easy processing, and low-cost production. The introduction of seaweed fibers or powders into a polymeric matrix allows specific properties (degradability, flame retardancy, chemical resistance, etc.) to be achieved. Nevertheless, the performance of the seaweed fiber or filler is governed by particle size, surface chemistry/activity, degree of interaction between filler-polymer matrix, chemical composition, and shape of the filler.

In this chapter, the characterization techniques used to determine the physicochemical properties of biocomposites based on algae (seaweed) fillers and by-products produced from the industrial production of agar have been reported. The advantage of using SWW as filler has demonstrated that the PLA-SWW biocomposites have good thermal and mechanical properties. Furthermore, the use of SWW as filler in PLA provides an interesting alternative for the production of low-cost and ecologically friendly biocomposites.

REFERENCES

[1] Craige J. Cell wall. In: Cole KM, Sheath RG, editors. Biology of the red algae. Cambridge University Press; 1990 [Chapter 10].

[2] Lahaye M, Robic A. Structure and functional properties of ulvan, a polysaccharide from green seaweeds. Biomacromolecules 2007;8:1765–74.

[3] Hernández-Carmona G, Freile-Pelegrín Y, Hernández-Garibay E. Conventional and alternative technologies for the extraction of algal polysaccharides. Funct Ingred Algae Foods Nutraceut 2013:475–516.

[4] McHugh DJ. A guide to the seaweed industry. In: FAO fisheries technical paper, vol. 441; 2003. Rome.

[5] Murano E. Chemical structure and quality of agars from *Gracilaria*. J Appl Phycol 1995;7:245–54.

[6] Imeson AP. Carrageenan. In: Phillips GO, Williams PA, editors. Handbook of hydrocolloids. 2nd ed. Cambridge (UK): Cambridge Press; 2000.
[7] Van der Velde F, De Ruiter G. Carrageenan. In: Steinbüchel A, editor. Biopolymers. New York (USA): John Wiley & Sons-VCH; 2005 [Chapter 6].
[8] Haug A. Composition and properties of alginates. Throndheim: Norwegian Institute Seaweed Research; 1964.
[9] Haug A. The influence of borate and calcium on the gel formation of a sulfated polysaccharide from *Ulva lactuca*. Acta Chem Scand B 1976;30(6):562–6.
[10] Lahaye M, Ray B. Cell-wall polysaccharides from the marine green alga *Ulva "rigida"* (Ulvales, Chlorophyta) – NMR analysis of ulvan oligosaccharides. Carbohydr Res 1996;283:161–73.
[11] Crawford RJ. Plastics engineering. 3rd ed. Oxford (UK): Butterworth-Heinemann; 2002.
[12] Madera-Santana TJ, Misra M, Drzal LT, Robledo D, Freile Pelegrín Y. Preparation and characterization of biodegradable agar/poly(butylene adipate-co-terephthalate) composites. Polym Eng Sci 2009;49(6):1117–26.
[13] Odian G. Principles of polymerization. 4th ed. (New Jersey, USA): John Wiley & Sons; 2004.
[14] Bernhardt EC, Mckelvey JM. Polymer processing – new engineering specialty. Mod Plast 1958;35:154–5.
[15] Tadmor Z, Gogos CG. Principles of polymer processing. 2nd ed. (New Jersey, USA): John Wiley & Sons; 2006.
[16] Akay M. An introduction to polymer-matrix composites. 1st ed. Bookboon Com; 2015.
[17] Lee MW, Han SO, Seo YB. Red algae fibre/poly(butylene succinate) biocomposites: the effect of fibre content on their mechanical and thermal properties. Comp Sci Technol 2008;68(6):1266–72.
[18] Luan L, Wu W, Wagner MH, Mueller M. Seaweed as novel biofiller in polypropylene composites. J Appl Polym Sci 2010;118:997–1005.
[19] Mohanty AK, Misra M, Drzal LT. Sustainable bio-composites from renewable resources: opportunities and challenges in the green materials world. J Polym Environ 2002;10:19–26.
[20] Auras R, Lim LT, Selke SE, Tsuj H. In: Poly(lactic acid): synthesis, structure, properties, processing and applications. (New Jersey, USA): J Willey & Sons; 2010.
[21] Huda MS, Drzal LT, Misra M, Mohanty AK. Wood-fiber-reinforced poly(lactic acid) composites: evaluation of the physicomechanical and morphological properties. J Appl Polym Sci 2006;102(5):4856–69.
[22] Petersson L, Kvien I, Oksman K. Structure and thermal properties of poly(lactic acid)/cellulose whiskers nanocomposite materials. Comp Sci Technol 2007;67:2535–44.
[23] Wang Y, Tong B, Hou S, Li M, Shen C. Transcrystallization behavior at the poly(lactic acid)/sisal fibre biocomposite interface. Comp Part A Appl Sci Manuf 2011;42:66–74.
[24] Madera-Santana TJ, Freile-Pelegrín Y, Encinas JC, Ríos-Soberanis CR, Quintana-Owen P. Biocomposites based on poly(lactic acid) and seaweed wastes from agar extraction: evaluation of physicochemical properties. J Appl Polym Sci 2015; 132(31). http://dx.doi.org/10.1002/app.42320.
[25] Masudul Hassan M, Mueller M, Wagners MH. Exploratory study on seaweed as novel filler in polypropylene composite. J Appl Polym Sci 2008;109:1242–7.

[26] Jang YH, Han SO, Sim IN, Kim H-II. Pretreatment effects of seaweed on the thermal and mechanical properties of seaweed/polypropylene biocomposites. Comp Part A Appl Sci Manuf 2013;47:83–90.

[27] Schick C. Differential scanning calorimetry (DSC) of semicrystalline polymers. Anal Bioanal Chem 2009;395:1589–611.

[28] Zuza E, Ugartemendia JM, Lopez A, Meaurio E, Lejardi A, Sarusua JR. Glass transition behavior and dynamic fragility in polylactides containing mobile and rigid amorphous fractions. Polymer 2008;49(20):4427–32.

[29] Hatakeyama T, Quinn FX. Thermal methods of analysis: principles applications and problems. 1st ed. England: Blackie Academic and Professional; 1994.

[30] Menard KP. Dynamic mechanical analysis: a practical introduction. 1st ed. Boca Raton (FL): CRP Press LLC; 1999.

[31] Finkenstadt VL, Tisserat B. Poly(lactic acid) and osage orange wood fiber composites for agricultural mulch films. Ind Crops Prod 2010;3(2):316–20.

[32] Sutivisedsak N, Cheng NH, Dowd MK, Selling GW, Biswas A. Evaluation of cotton byproducts as fillers for poly(lactic acid) and low density polyethylene. Ind Crops Prod 2012;36:127–34.

[33] Bhargava R, Wang S, Koening JL. FTIR microscopy of polymeric systems. Adv Polym Sci 2003;163:137–91.

[34] Lee LH. Characterization of metal and polymer surfaces: polymer surfaces. New York (USA): Academic Press; 1977.

[35] Sawyer LC, Grubb DT, Meyers GF. Polymer microscopy. 3rd ed. Germany: Springer Verlag Berlin Heidelberg; 2008.

[36] Orozco VH, Brostow W, Chonkaew W, Lopez B. Preparation and characterization of poly(lactic acid)-g-maleic anhydride + starch blends. Macromol Symp 2009;277: 69–80.

[37] Freile-Pelegrín Y, Madera-Santana T, Robledo D, Veleva L, Quintana P, Azamar JA. Degradation of agar films in a humid tropical climate: thermal, mechanical, morphological and structural changes. Polym Degrad Stab 2007;92:244–52.

[38] Warren BE. X-ray diffraction. 1st ed. New York (USA): Dover Publications Inc.; 1969. Reprinted 1990.

[39] Debye P. Zerstreuung von Röntgenstrahlen. Ann Phys 1915;46:809–23.

[40] Zernicke F, Prins JA. Die beugung von Röntgenstrahlen in flüssigkeiten als effekt der molekühlanordnung. Z Phys A Hardron Nucl 1927;41:184.

[41] Kuwabara K, Gan Z, Nakamura T, Abe H, Doi Y. Crystalline/amorphous phase structure and molecular mobility of biodegradable poly(butylene adipate-co-butylene terephthalate) and related polyesters. Biomacromolecules 2002;3(2):390–6.

[42] Yeh JT, Tsou CH, Huang CY, Chen KN, Wu CS, Chai WL. Compatible and crystallization properties of poly(lactic acid)/poly(butylene adipate-co-terephthalate) blends. J Appl Polym Sci 2010;116(2):680–7.

CHAPTER

Processing Techniques of Algae-Based Materials 19

Shahzad Z. Iqbal[1], Fazal-ur-Rehman[1], Noeen Malik[2], Kanwal Rehman[3], Muhammad R. Asi[4]

Government College University Faisalabad, Faisalabad, Pakistan[1]; New York University, New York, NY, United States[2]; University of Agriculture, Faisalabad, Pakistan[3]; Nuclear Institute of Agriculture and Biology, Faisalabad, Pakistan[4]

19.1 PROCESSING OF BIODIESEL FROM ALGAE

19.1.1 INTRODUCTION

To meet the demands of climate change and food–feed security, algae-based fuels are proved to be the most renewable, sustainable, viable, effective, and environment-friendly. It also provides the only renewable energy resource that has the ability to take on the global demand for fuels for the long-term utilization. Currently, the whole world is dependent on energy and, thus, faces future energy challenges which will ultimately boost the prices of fossil fuel. It can also relay on the development of less accessible reserves, increasing the amount of greenhouse gas (GHG) emissions and release of other toxic and harmful pollutants. Therefore, to conduct a research on renewable energy is a key challenge of 21st century. The alternative resources are being used by different countries such as wind, hydro, solar, geothermal, and energy from biomass and waste [1]. Almost 25% of global energy requirements could be fulfilled using biomass [2]. Biodiesel can not only be obtained from biomass but can also play a vital role in expanding the global sources of energy. Furthermore, it is documented as a green and substitute renewable diesel fuel [1] due to the advantages such as nontoxicity, biodegradable, and lower GHG emission in diesel engines [1,3]. The sources of biodiesel production are crops such as soybeans, palm oil, sunflowers, and corn. However, about 70%–85% of the total biodiesel production costs are because of the consumption of these expensive crops. Considering the current circumstances, it would be impossible to meet the demands of biodiesel using first-generation sources, i.e., crops [4]. A report demonstrated that in 2010, the United States almost consumed 220 billion m^3 of diesel and to meet this entire demand using first-generation sources, i.e., soybeans, would require almost 367 million ha, considering the available area of 178 million ha in the United States [4]. Ultimately, the use of second-generation sources such as animal fats, *Jatropha*, and waste cooking oil are on the rise to produce biodiesel in the United States [5]. Recently, algae have appeared as a third-generation source of biodiesel because of these organisms' rates, noncompetitive nature in food markets, and ability to grow using nutrient waste streams [6].

Algae Based Polymers, Blends, and Composites. http://dx.doi.org/10.1016/B978-0-12-812360-7.00019-7

19.2 PROCESSING STEPS INVOLVED FOR THE PRODUCTION OF BIODIESEL FROM ALGAE WASTEWATER

Different processing steps are involved for conversion of algal biomass to biodiesel as shown in Fig. 19.1.

19.2.1 CULTIVATION

Algal biomass cultivation technology from wastewater exists since the 1950s with the practice of wastewater stabilization ponds or high-rate algal ponds for wastewater treatment [2]. William J. Oswald later developed an advanced integrated wastewater pond system [7].

19.2.2 HARVESTING

Harvesting is the second step involved in algae biodiesel production. The microalgae harvesting process poses challenges due to small size of algal cells (generally 1–20 mm) coupled with the need to handle large liquid volumes on account of the relatively low density of algal cells grown in open ponds [1,8]. The lack of efficient and cost-effective harvesting methods is the main reason why algal-based wastewater treatment is not used extensively by the wastewater industry [8].

19.2.3 COAGULATION/FLOCCULATION

This process aggregates the cells and increases their size before the flocculation. The flocks can be removed by sedimentation. The most common coagulants are aluminum sulfate and ferric iron chloride. Polymers and other natural coagulants (*Moringa*, chitosan, and cactus mucilage) were also tested. The pros of this technique involve the feasibility of well-established processing method and observed the recovery of more than 90% algae. However, the cons are that it is expensive

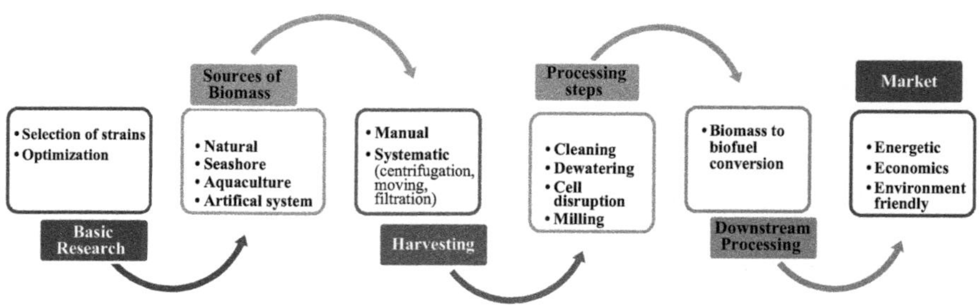

FIGURE 19.1

Supply chain considerations for algae to biofuels.

Modified from Bruton T, Lyons H, Lerat Y, Stanley M, Rasmussen MB. 2009. A review of the potential of marine algae as a source of biofuel in Ireland. Sustainable Energy Ireland, Dublin (Ireland).

and each coagulant has an optimum dosage, and among all coagulants some contribute to the GHG emission while others pose a threat to health [8–11].

19.2.4 FLOTATION

This method combines a chemical flocculent and algae with bubbling air to transport the flocks over the surface, where they are removed by conventional skimming methods or by dissolved air flotation (DAF). The advantages of this technique are the increase in algae flocculation from 31% to 55% because of ozone and the large-scale processing as DAF is a cheap method, handling more than 10,000 m^3/day [2,12,13].

19.2.5 SEDIMENTATION

Removal of suspended solids from liquid using gravity is known as sedimentation. The process has several advantages, e.g., most common harvesting method, easy to control, and very economical (low cost). Amenable to the large-scale biomass harvesting, recycled water rich in nutrients can be returned to the algae growth chamber due to the absence of chemicals. However, it has some drawbacks such as slowness of the process and deterioration of biomass at high temperature [8,14].

19.2.6 IMMOBILIZATION

Polymer encapsulation technique is used during cultivation of microalgae. The addition of these substances depends on the reaction conditions used prior to or during the cultivation. The most common immobilizing agent is alginate. Algal biomass can be recovered from an immobilized matrix. However, it is unclear whether immobilized algae could be used for oil production. The disadvantage of this process is that the efficiency of lipid extraction from the algal-polymer matrix is undetermined [8,15].

19.2.7 OIL EXTRACTION

If concentrated algal biomass is to be transformed into biodiesel, the oils must be extracted. The biomass is generally dried first. Solar drying is assumed to be the best method in tropical countries due to intensive sunlight [1].

19.2.8 CHEMICAL SOLVENT EXTRACTION

To extract microalgal lipids, a chemical method is used in which a solvent is diffused across microalgal cell walls. The most common solvents are *n*-hexane, methanol, ethanol, and mixed methanol–chloroform (2:1 v/v). The method gives an effective extraction of microalgal lipids [1,2].

19.2.9 HOMOGENIZATION

A process in which algae is strapped through small valves to disrupt the cell walls. The method can be used as pretreatment for further extraction [16].

19.2.10 SONIFICATION

To disrupting the cell walls of microalgae, a thermal method is used in which cavitation causes the formation and collapse of microscopic bubbles and thus, producing tremendous energy in the form of heat, pressure, and mechanical shear [17,18].

19.2.11 MICROWAVING

The cell walls of microalgae are shattered using the shock of high-frequency waves, and hence this process is known as microwaving [17–19].

19.2.12 AUTOCLAVING

It is a thermal process in a closed apparatus that produces temperature and pressure conditions sufficiently elevated for sterilization. The process is easy, can handle wet samples, and has low cost [17–19].

19.2.13 PYROLYSIS

It is performed in the absence of oxygen, using a thermal treatment at atmospheric pressure which converts dry biomass into charcoal, bio-oil, and a gaseous fraction at temperatures ranging 400–600°C.

19.2.14 HYDROTHERMAL CARBONIZATION

In this thermochemical process, slower temperatures are used which are maintained at 200°C and pressure below 2 MPa. The final product results in an insoluble char and an aqueous phase [20].

19.2.15 HYDROTHERMAL LIQUEFACTION

The process can be carried at intermediate temperature at 280–370°C and high pressure (10–25 MPa) for an aqueous feedstock slurry that contains 10%–20% algae by mass. The energy value of produced oil is very close to crude oil. In algae cultivation, the aqueous phase coproduct is recycled. The process has advantage that it is carried out in water and requires no drying after harvesting. Furthermore, it is not necessary to promote cellular lipid accumulation or to extract the lipids, and the produced oil has high heating value [21].

19.3 PROS AND CONS OF ECONOMIC AND ENERGETIC EVALUATIONS OF ALGAE BIOENERGY

The biomass production cost is the most important economic factor in biomass supply chain during the production of bioenergy from microalgae. The previous research reports have shown that the steps involved during the cultivations of microalgae biomass account for 75%–80% of the total costs of biofuel production from this feedstock [22]. To be more comparative, economical, and reliable with other biofuel feedstocks, the production cost of biofuel from microalgae will therefore have to be decreased significantly. To minimize the production cost, the use of wastewater as a nutrient supply and utilization of other high-value coproducts might help to expand the economical use of microalgae biomass for energy production. However, considering the current situation of biomass costs or its conversion conditions, the large-scale production of biofuel from microalgae appeared to be unfeasible [23]. An extensive economic examination of biodiesel production depicted that bioenergy production from microalgae could only be feasible if all of the major biochemical components in the algal cells are also treated and marketed [24].

Considering technically, the use of algae biomass for the production of bioenergy carriers shows similar techniques to those discussed above for the economic considerations, where the single production of one biofuel product is seen to be energetically disadvantageous with the use of this biomass. This gives the idea and basis for comparing the energy content of the biofuel product to the overall energy inputs required for the production of a unit of the biofuel. Ehimen [25] and Ehimen et al. [26] concluded that usage of current processing techniques for the production of biodiesel from microalgae biomass (*Chlorella* sp.) is not renewable because the input energy is over three times (from fossil fuels) more than required for its production from the other sources of fuel product. Henneberg et al. [27] documented unfavorable energetics for the production of fuels from macroalgae. The energy equilibrium of the production of algae fuels was discussed to be vastly improved with the coproduction or separation of other valuable products. Furthermore, the energy carriers from this biomass source will be utilized for the energy requirements of the harvesting, processing, conversion aspects of the algae biofuel production, and the use of organic-derived nutrients for the biomass cultivation [28].

19.4 BIOETHANOL PRODUCTION FROM ALGAE

In 1970s during the fuel crisis, a new concept of bioethanol production was initiated worldwide and the capacity of ethanol reached to 1 billion L in 1975, 39 billion L in 2006, and 100 billion L in mid-2015 [29,30] due to its extensive applications in many sectors, mainly in transport. The production of bioethanol is characterized based on various feedstocks and development of production technologies into first, second, and third generations [31–38]. The production and synthesis of each generation bioethanol from respective feedstocks are presented in Fig. 19.2. First-generation crops such as corn, wheat, barley, and sweet sorghum which produce ethanol are termed as grain alcohol, whereas ethanol produced from agro-residue such as wheat straw, rice straw, and grasses, and lignocellulose biomass is known as biomass ethanol or bioethanol [39].

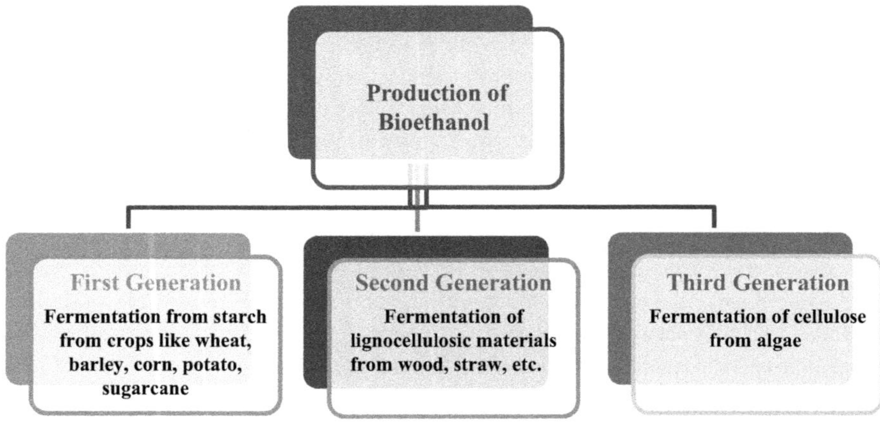

FIGURE 19.2

Production and synthesis of bioethanol from various feedstocks.

19.5 BIOETHANOL AN IDEAL SOURCE FOR THIRD GENERATION PRODUCED FROM ALGAE

The problems with first- and second-generation crops include not only its impact on food chain but the usage of more arable land for cultivation. This is the biggest challenge for the commercial production of bioethanol. For example, in Malaysia, due to increased demand and utilization of palm for food as well as for the production of biodiesel, urged the need to utilize more land for cultivating oil palm plantations which caused rapid deforestation and destruction of habitat of many animals and birds [40]. Therefore, without disrupting the environment, the researchers now focus on bringing up new potential alternative feedstock for commercial-scale production of biofuel. Algae, rich in lipids and carbohydrates, are considered as the only substitute to food crops for renewable fuel production. Microalgal cell walls are composed of mannans, xylans, cellulose, and sulfated glycans. These polysaccharides can be broken down enzymatically or chemically into simple units of sugars which can be converted into ethanol [29,30]. Morphologically, algae are divided into macro- and microalgae and are the fastest growing plant in the world. To differentiate between macro- and microalgae, macroalgae or "seaweeds," are larger, oceanic, multicellular plants which grow up to 60 m in length [41], while microalgae are microscopic, mostly existing as small cells of about 2–200 mm found in freshwater, seawater, and even in wastewater [42,43]. In photosynthetic process, they efficiently convert solar energy to biomass by utilizing sunlight, CO_2, and nutrients from water [44]. Algae can competently eliminate CO_2 and synthesize polysaccharides or oil which can be used for the production of biofuel [45]. Due to the absence of lignin, algal carbohydrates can be used for bioethanol fermentation after comparatively easy saccharification process [35]. Oil can be transesterified into biodiesel

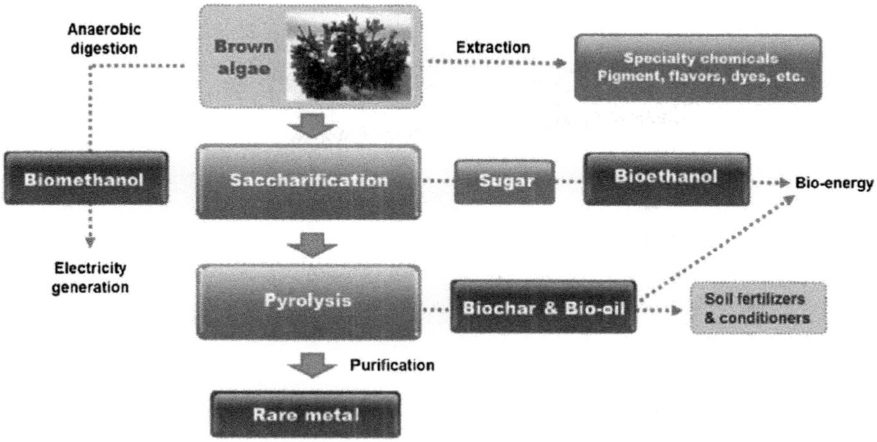

FIGURE 19.3

Process of production of biofuels and biochemicals from algae.

Courtesy Lee OK, Lee EY. Sustainable production of bioethanol from renewable brown-algae biomass. Biomass Bioener 2016; 92:70–75.

[46] and algal biomass can be used for heat and power generation using anaerobic digestion and pyrolysis [47–49] (Fig. 19.3).

19.6 BIOETHANOL AS RENEWABLE AND SUSTAINABLE BIOMASS

The production of biofuels from brown algae has several advantages over other feedstocks. First, brown algae have no competition with food production. Second, regardless to the conventional agriculture, which requires expensive nutrients, fertilizers, land and freshwater, the brown algae require no arable land and can be cultivated in the ocean using aquaculture. Due to the absence of lignin in brown algae, no extensive pretreatment is required to release sugars, but just milling or crushing can do the job. Also no moral issues are linked with the use of brown algae for the production of biofuel. If compared to conventional crops, brown algae can offer high biomass yields per acre [50,51]. Almost 55% (w/w) of the dry biomass contained the content of carbohydrates using brown algae [52]. Therefore, due to its abundance and high sugar levels that can be used for production of bioethanol and chemicals, brown algae are an ideal renewable and sustainable biomass. Worldwide, nearly 70 million dry tons of macroalgae are cultivated/harvested in offshore and nearshore coastal farms [33,37]. This production of algae is utilized mostly for food applications. Both manual and mechanical methods are being used for harvesting cultured brown algae. After harvesting, brown algae are normally processed into milling to reduce biomass sizes for efficient saccharification or alginate extraction [34,38] (Fig. 19.4). The saccharified broth can be used for bioethanol fermentation.

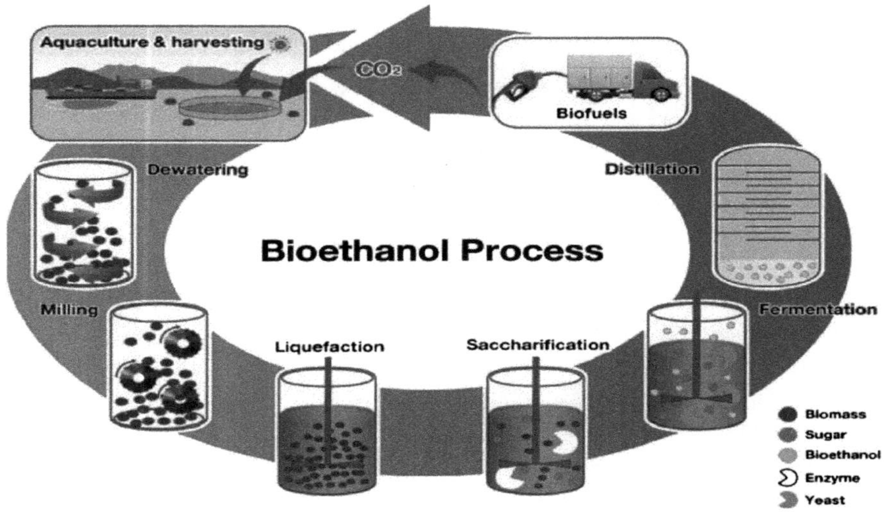

FIGURE 19.4

Bioethanol production from brown algae.

19.7 PROCESSING STEPS INVOLVED DURING PRODUCTION OF BIOETHANOL FROM ALGAE

The richest sugars in brown algae are mannitol, alginate, and laminarin. Glucose and mannitol derived from laminarin are normal sugars that are professionally used for production of bioethanol fermentation [53,54]. It is reported that bioethanol containing a yield of 0.4 g ethanol/g of sugars was produced with ethanogenic *Escherichia coli* KO_{11} from *Laminaria japonica* hydrolyzates mostly containing mannitol after chemo-enzymatic saccharification [55]. The simultaneous saccharification and fermentation (SSF) method in *Saccharina japonica* biomass produced ethanol of 7.7 g/L with a theoretical yield of 33.3% [56]. Alginate is classified as structural polysaccharide in the cell wall of brown algae [57,58] and is present in the insoluble calcium salt form [34,38]. The calcium alginate is converted to alginic acid in an acidic pre-extraction step of alginate-extraction process whereas in alkaline extraction alginic acid is converted to soluble sodium alginate [59]. Alginate can be converted into unsaturated uronate monosaccharides using alginate lyase [60] which is an enzyme that catalyzes the β-elimination breakage of the glycosidic bond of alginate. Various alginate lyases have been cloned and characterized [61–63]. Recently, alginate lyase–based saccharification was developed by employing exolytic alginate lyase as the biocatalyst with a 30% yield [64,65].

19.7.1 CULTIVATION OF MICROALGAE

The photosynthetic growth of microalgal biomass requires CO_2, light, organic nutrients, and water. The microalgal biomass can be achieved using a different cultivation process such as closed-culture systems, e.g., photobioreactors (PBRs) or open ponds that refer to the raceway pond and the hybrid system [49,66].

19.7.2 PHOTOBIOREACTOR

A bioreactor used for cultivating algae in a controlled environment is termed as an algae bioreactor or PBR. The PBR allows the growth of chosen species for long duration without varying their biological or physical characteristics unlike raceway ponds [49,67]. However, closed bioreactors are designed mostly as tubular reactors, bubbled column reactors, or plate reactors [66]. The reactor consists of vertically arranged or inclined rectangular boxes and is often divided into two parts to impact the agitation of the reactor fluid known as plate reactor. A reactor which consists of vertically arranged cylindrical columns made of transparent material is termed as bubble column PBR, in which air is passed at the bottom in this reactor. The most common diameter of bubbled reactor is 20–30 cm to distribute sunlight energy for algal growth [67]. According to Bajhaiya et al. [68], among different PBRs, the most satisfactory bioreactor for producing algal biomass is tubular PBR. A tubular PBR provides continuity to culture because fresh culture medium is fed at a constant rate and the same quantity of microalgal broth is withdrawn continuously. An array of straight transparent tubes that are typically made of plastic or glass is known as tubular PBR. This tubular array, or the solar collector, where the sunlight is captured, is generally 0.1 m or less in diameter. During night, the temperature can be maintained at a lower rate to avoid the loss of biomass due to respiration.

19.7.3 RACEWAY PONDS

A raceway pond is defined as an open pond in which usually algae are cultivated and water, algae, and nutrients circulate around a race track. Currently, open pond systems are most widely used for small as well as commercial and large-scale production of algal. Raceway ponds are used for mass culture of microalgae are used since the 1950s. The structure of pond consists of a closed loop recirculation channel with a water depth of 15–20 cm containing a paddle wheel. During daylight, the culture is fed at the front of paddle wheel and the broth can be harvested from the same area [66]. The advantage of using paddle wheel is that it can mix and circulate the algal biomass, and prevent the sedimentation.

19.7.4 NONSUSPENDED METHOD OF MICROALGAE CULTIVATION

Microalgae are cultivated both in an open pond system and PBR as suspensions in the culture medium. Using these processes, microalgae can flow freely inside the medium with the help of mixing to ensure even distribution of cells. However, this process only provides 1% of dry algal biomass due to high dilution of microalgal cells in the medium because remaining 99% of culture volume consists of water. To date, harvesting a large volume of microalgae is extremely expensive process. The importance of supplying water to maintain microalgae production is also necessary. To obtain 1 kg of biodiesel, about 3800 kg of water is required. Therefore, in suspended cultivation systems, a huge amount of water is needed for processing microalgal growth [69]. Liu et al. [70] has introduced a process to reduce the water consumption as well as simultaneously produced high amount of microalgal biomass using nonsuspended method. The experiment stated that if microalgae are placed on vertically arranged substratum with water supplied only to keep the surfaces wet, this process provides 70.9 g/m^2 of biomass per day for *Scenedesmus obliquus*. Currently, maximum of few grams of dry biomass per liter of media can be produced by suspended mode of cultivation system. For example, using suspended mode of cultivation, 2.8 g/L of biomass from *Nannochloropsis oculata* using PBR was achieved [71].

The nonsuspended way of cultivation has advantages such as (1) when microalgae accumulate significant quantities of biomass in smaller area, it is much easier to separate microalgae biomass from the medium; (2) in nonsuspended cultivation, the costs associated with water consumption are lower (economical). For example, in suspended cultivation, to manufacture one ton of microalgae required around 200 mt of water whereas in an attached cultivation method, only 17 t of water is needed for circulation with four tons consumed at surfaces to sustain appropriate wetness level; (3) as lights can be positioned in the spaces between surfaces, there would be no need for glass reactors. The light distribution is forecasted to be the highest among all existing methods because there would be no free-floating particles reflecting and absorbing light. It will only need air rich in CO_2 and humid conditions between the attachment surfaces [69].

19.7.5 HARVESTING OF MICROALGAE

Because of their diminutive size of about 3−30 mm in diameter, harvesting of microalgae is the principal step in the overall process for the production of biofuel. The basic challenge during large-scale production of microalgae is the development of a well-organized technique that not only separates the cells but preserves their viability and bioactivity before it can be used in the field [72]. The reports have documented that almost 20%−30% costs occurred during the harvesting step for production of biodiesel [73,74]. The method, either physical, chemical, or biological whichever is suitable during harvesting process, is used to remove the large amount of water for separating microalgae [49]. In previous studies, different harvesting process are documented for separating microalgae from culture medium such as gravity sedimentation, centrifugation, filtration, immobilization, flotation, electrolytic method, ultrasonic aggregation, chemical flocculation, magnetic nanoparticle separation, and bioflocculation. However, each method has advantages over other methods which ultimately affect the overall economics of the process. The scientists have reported that flocculation, among several harvesting methods, was found to be an effective and easy method for harvesting of microalgae. However, flocculant chemicals are nonbiodegradable and their intermediate secondary products are

[58] Wong TY, Preston LA, Schiller NL. Alginate lyase: review of major sources and enzyme characteristics, structure-function analysis, biological roles, and application. Annu Rev Microbiol 2000;54:289−340.

[59] Hernandez-Carmona G, McHugh DJ, Arvizu-Higuera DL, Rodriguez-Montesinos YE. Pilot plant scale extraction of alginate from *Macrocystis pyrifera*. Effect of pre-extraction treatments on yield and quality of alginate. J Appl Phycol 1999;10:507−13.

[60] Ochiai A, Yamasaki M, Mikami B, Hashimoto W, Murata K. Crystal structure of exotype alginate lyase Atu 3025 from *Agrobacterium tumefaciens*. J Biol Chem 2010;285:24519−28.

[61] Kim HS, Lee CG, Lee EY. Alginate lyase: structure, property, and application. Biotechnol Bioprocess Eng 2011;16:843−51.

[62] Lee SI, Choi SH, Lee EY, Kim HS. Molecular cloning, purification and characterization of a novel polyMG-specific alginate lyase responsible for alginate MG block degradation in *Stenotrophomas maltophilia* KJ-2. Appl Microbiol Biotechnol 2012;95: 1643−53.

[63] Rahman MM, Inoue A, Tanaka H, Ojima T. Isolation and characterization of two alginate lyase isozymes, AkAly28 and AkAly33, from the common sea hare *Aplysia kurodai*. Comp Biochem Physiol B 2010;157:317−25.

[64] Park HH, Kam N, Lee EY, Kim HS. Cloning and characterization of a novel oligoalginate lyase from a newly isolated marine bacterium *Sphingomonas* sp. MJ-3. Mar Biotechnol 2012;14:189−202.

[65] Ryu MR, Lee EY. Saccharification of alginate by using exolytic oligoalginate lyase from marine bacterium *Sphingomonas* sp. MJ-3. J Ind Eng Chem 2011;17:853−8.

[66] Khan SA, Rashmi, Hussain MZ, Prasad S, Banerjee UC. Prospects of biodiesel production from microalgae in India. Renew Sustain Energy Rev 2009;13:2361−72.

[67] Chisti Y. Biodiesel from microalgae: a review. Biotechnol Adv 2007;25:294−306.

[68] Bajhaiya AK, Mandotra SK, Suseela MR, Toppa K, Ranade S. Algal biodiesel: the next generation biofuel for India: review article. Asian J Exp Biol Sci 2010;1(4):728−39.

[69] Katarzyna L, Sai G, Singh OA. Non-enclosure methods for non-suspended microalgae cultivation: literature review and research needs. Renew Sustain Energy Rev 2015;42: 1418−27.

[70] Liu T, Wang J, Hu Q, Cheng P, Ji B, Liu J, et al. Attached cultivation technology of microalgae for efficient biomass feedstock production. Bioresour Technol 2013;127: 216−22.

[71] Surendhiran D, Vijay M, Sivaprakash B, Sirajunnisa A. Kinetic modeling of microalgal growth and lipid synthesis for biodiesel production. 3 Biotech 2014. http://dx.doi.org/10.1007/s13205-014-0264-3.

[72] McKendry P. Energy production from biomass (part 2): conversion technologies. Bioresour Technol 2002;83:47−54.

[73] Surendhiran D, Vijay M. Influence of bioflocculation parameters on harvesting *Chlorella* salina and its optimization using response surface methodology. J Environ Chem Eng 2013;1:1051−6.

[74] Salim S, Bosma R, Vermue MH, Wijffels RH. Harvesting of microalgae by bioflocculation. J Appl Phycol 2011;23:849−55.

[75] Teixeira CMLL, Kirsten FV, Teixeira PCN. Evaluation of *Moringa oleifera* seed flour as a flocculating agent for potential biodiesel producer microalgae. J Appl Phycol 2012;24: 557−63.

[76] Ganesh KC, Joo HS, Kumar R, Choi JW, Chang CS. Characterization of an extra cellular biopolymer flocculant from a haloalkalophilic *Bacillus* isolate. World J Microbiol Biotechnol 2004;20:837–43.

[77] Surendhiran D, Vijay M. Micro algal biodiesel – a comprehensive review on the potential and alternative biofuel. Res J Chem Sci 2012;2(11):71–82.

[78] Baliga R, Powers SE. Sustainable algae biodiesel production in cold climates. Int J Chem Eng 2010. http://dx.doi.org/10.1155/2010/102179.

[79] Li Y, Han D, Sommerfeld M, Hu Q. Photosynthetic carbon partitioning and lipid production in the oleaginous microalga *Pseudochlorococcum* sp. (Chlorophyceae) under nitrogen-limited conditions. Bioresour Technol 2011;102:123–9.

[80] Xue F, Gao B, Zhu Y, Zhang X, Feng W, Tan T. Pilot-scale production of microbial lipid using starch waste water as raw material. Bioresour Technol 2010;101:6092–5.

[81] Georgianna DR, Mayfield SP. Exploiting diversity and synthetic biology for the production of algal biofuels. Nature 2012;488:329–35.

[82] Kim HM, Wi SG, Jung S, Song Y, Bae HJ. Efficient approach for bioethanol production from red sea weed *Gelidium amansii*. Bioresour Technol 2015;175:128–34.

[83] Enquist-Newman M, Faust AME, Bravo DD, Santos CN, Raisner SRM, Hanel A, et al. Efficient ethanol production from brown macroalgae sugars by a synthetic yeast platform. Nature 2014;505:239–43.

[84] Meinita MDN, Marhaeni B, Winanto T, Jeong GT, Khan MNA, Hong YK. Comparison of agarophytes (*Gelidium*, *Gracilaria*, and *Gracilariopsis*) as potential resources for bioethanol production. J Appl Phycol 2013;25:1957–61.

[85] Kim H, Cho Y, Kim SK. Ethanol production from seaweed (*Undaria pinnatifida*) using yeast acclimated to specific sugars. Biotechnol Bioprocess Eng 2013;18:533–7.

[86] Trivedi N, Gupta V, Reddy CRK, Jha B. Enzymatic hydrolysis and production of bioethanol from common macrophytic green alga *Ulva fasciata* Delile. Bioresour Technol 2013;150:106–12.

[87] Ho SH, Huang SW, Chen CY, Hasunuma T, Kondo A, Chang JS. Bioethanol production using carbohydrate-rich microalgae biomass as feedstock. Bioresour Technol 2013;135:191–8.

[88] Song M, Pham HD, Seon J, Woo HC. Marine brown algae: a conundrum answer for sustainable biofuels production. Renew Sustain Energy Rev 2015;50:782–92.

CHAPTER

Future Prospects of Algae-Based Materials 20

Saima Rehman[1], Muhammad Ali[1], Mohammad Zuber[1], Khalid M. Zia[1], Rehana Iqbal[2]

Government College University Faisalabad, Faisalabad, Pakistan[1]; Bahauddin Zakariya University, Multan, Pakistan[2]

20.1 INTRODUCTION

Owing to their versatile features, synthetic polymers have immense importance for the improved quality of human life. According to recent reports, the global plastic production has risen to more than 300 million tons annually [1]. In spite of numerous applications in several spheres of human life, the use of petroleum-based polymeric materials is associated with certain problems. Depletion of fossil fuels, i.e., the raw material for the synthesis of synthetic polymers, has become a major issue related with the production and utilization of these polymers. Moreover, being nonbiodegradable the synthetic polymers accumulate in the environment, creating unmanageable pollution [2]. To combat these problems, there was rapid progress in the development of processes for the production of natural/bio-based polymers. The global production of natural polymers is predicted to enhance from 5.1 million tons in 2013 to 17 million tons in 2020 [3].

The increased number of researches on the development and applications of bio-based polymers depicts the accelerated interest of scientific community in this field [4]. Owing to certain characteristics, such as biodegradability, bio-renewability, and biocompatibility, worldwide demand of renewable bio-based raw materials has increased to replace the petroleum-based raw materials and polymers. The first-generation biopolymers were derived from the carbohydrate feedstocks such as potatoes and corn [5]. However, in the recent years, nonfood resources are under consideration for the production of monomeric raw materials and polymers. In addition to bio-based polymers, there are certain biomolecules, such as cellulose, starch, chitin, chitosan, polylactic acid, and alginates, which can be directly blended or can form composites with synthetic polymers. Such blends and composites have wide range of applications in the medical field, e.g., biosensor, gene therapy, drug delivery, tissue engineering, bone-grafting substitutes, surgical applications, and clinical diagnosis. The high-tech applications include water conditioning, wastewater treatment, barrier membrane, reinforcing material, and food packaging materials [6].

Algae Based Polymers, Blends, and Composites. http://dx.doi.org/10.1016/B978-0-12-812360-7.00020-3

Algae have attracted a great deal of attention among nonfood feedstocks for the production of biopolymers, blends, and composites owing to their high yields, short cultivation times, and diverse culture conditions. Algae occur in a wide variety of environments including brackish, marine, freshwaters, hot and cold springs, desert soils, and bare rocks. Diverse unicellular and multicellular algae groups exist in several shapes and forms. Phenolic compounds, polysaccharides, phycobiliproteins, polyhydroxyalkanoates, alkaloids, sterols, bioactive peptides, and lipids are the major constituents of algae [7−10].

20.2 ALGAE-BASED POLYMERS, BLENDS, AND COMPOSITES AND THEIR APPLICATIONS

As discussed previously, algal biomass and biomolecules play a significant role in the synthesis of biomaterials with unique properties and remarkable applications. Algae-based materials can exist in the form of algal polymers, blends, and composites of algal biomass or biomolecules with other polymeric materials.

Important algal polymers include polysaccharides and sulfated polysaccharides such as carrageenan, agars, alginates, and polyhydroxyalkaonates. In addition to the conventional use of algal polymers in food industry, because of their viscosifying and gelling abilities, these polymers have also found tremendous applications in numerous fields owing to their biological and pharmacological activities such as antioxidant, anticoagulation, antiviral, antioxidant, antitumor, antiinflammatory, antimicrobial or biocidic, immunological, and anticancer activities [10].

Algal biomass and biomolecules are successfully used as precursors of several biopolymers such as polyethylene, polyurethane, and polyesters. [11,12]. Incorporation of algal polymers and biomass into synthetic polymeric systems resulted in improved mechanical properties, biocompatibility, and biodegradability. In addition to the environmental concerns, algal fillers are used to reinforce polymer biocomposites to combine high performance, versatility, easy processing, and low-cost production. Algal polymers and their blends are regarded for medical, packaging, agriculture, dairy, and food pertinences such as icings, drug delivery, glazes, wound-coverage material, low fat sausages, and cell encapsulation/grafting based on their exceptional properties such as thickeners, stabilizers, emulsifiers, gelling, and film-making properties [5,13−16].

Moreover, alginate-based, hybrid, nanocomposites present another class of biomaterials with unique properties. Nanofillers, such as graphene, graphene derivatives, and titania, have modified the thermomechanical, electric, conductive, and biological properties of algal polymers, such as alginate, dramatically prompting the emerging applications such as adsorbents (separation of metal ions, dyes, antibiotics etc.), membranes for separation, encapsulating systems for biomolecules, electronic sensors/biosensors, and coatings [17,18].

20.3 FUTURE PROSPECTS OF ALGAE-BASED MATERIALS

Utilization of bio-based feedstocks for the production of bio-based materials is not a new concept. However, the cost-effective fossil fuels have dominated the market for last century. The limitations of fossil fuels and environmental concerns have accelerated the demands of bio-based materials in past few decades. It took more than 100 years to establish the fossil fuel–based chemical industry, while bio-based materials were not prioritized. Although, biomaterial industry is now catching up with fossil fuel–based chemicals, it may still require more than 20 years for the commercial development of chemicals based on the biomaterials [6,19].

Owing to the characteristics such as biocompatibility, biodegradability, and therapeutic activities, algae-based materials have attracted great attention from scientific community. Several polymeric systems based on algal biomass and biomolecules are synthesized and applied to several fields. However, the research is still mostly at the lab scale and requires a great deal of work for the production and application of these materials on a commercial scale. To prosper the algae-based bio materials, certain practices need to be prioritized.

On the basis of desired algal products, culture conditions should be optimized and other cultivation technologies should be developed. Moreover, yields of required products can be enhanced by genetic and metabolic engineering. The cost-effectiveness and cost-efficiency of algal monomers and polymers is another challenge which has to be answered in near future. An economically viable supply is also necessary for the commercialization of algae-based materials. Hence the efficient and easy downstream processing techniques and new manufacturing routes are required.

In some recent studies, genetically engineered microalgae were successfully utilized for the production of relevant polymers (PHB). However, there is a need for large-scale screening, process improvements for better yields, and synthesis of other important polymers [20].

Integrated and hybrid algal biorefineries are also under consideration for the production of algae-based biofuels and biomaterials, which may tremendously enhance the cost-efficiency of the process along with the least waste production.

The bioactivities of most of the algae-based materials were analyzed in vitro. Since only rats were used for the assessment of in vivo activities, there is a need to analyze the activities in human subjects to enhance the applications to different fields in human life. Moreover, there is also a need to elucidate the detailed chemical structures of algal polymers and to establish the structure–activity relationship. For the desired characteristics, the algal polymers can be modified by using suitable chemical and biocatalytic modifications.

There is a need for the development of polymers that have the properties and performances better than those based on fossil fuels. Algal biomass and biomolecules such as polysaccharides are utilized as fillers for the reinforcement of synthetic as well as other natural polymers for improved thermal, mechanical, and biological properties. Moreover, the use of nanoparticles such as graphene and its derivatives were shown to dramatically improve the thermomechanical, electric, and biological properties of algal polymers. Thus such reinforcement of algal polymers with a variety of nanofillers is an attractive route for the new functional algal materials with numerous applications.

20.4 CONCLUSION

In replacing the conventional polymers, bio-based polymeric materials are closer to the reality more than ever before. Algae-based biomaterials are promising candidates for the replacement of conventional, nonrenewable, fossil fuel–based polymeric materials. Owing to the advancements in biotechnology, algae-based polymers, blends, and composites have found many applications in numerous fields of human life, i.e., from food and medical to high-tech applications. However, in spite of these innovative advancements, there are some hurdles to wider commercialization and applications of these biomaterials. Among which most important challenges are performance and cost compared to their conventional counterparts.

REFERENCES

[1] Plastics Europe. Plastics–the facts 2014/2015. An analysis of European plastics production, demand and waste data. 2015. Available from: http://www.plasticseurope.org/documents/document/20150227150049-final plastics the facts 2014 2015260215.pdf.

[2] Iwata T. Biodegradable and bio-based polymers: future prospects of ecofriendly plastics. Angew Chem Int Ed 2015;54(11):3210–5.

[3] Mozejko-Ciesielska J, Kiewisz R. Bacterial polyhydroxyalkanoates: still fabulous? Microbiol Res 2016;92:271–82.

[4] Chen GQ, Martin KP. Plastics derived from biological sources: present and future: a technical and environmental review. Chem Rev 2012;112:2082–99.

[5] Lencina MMS, Andreucetti NA, Gómez CG, Villar MA. Recent studies on alginates based blends, composites, and nanocomposites. In: Thomas S, et al., editors. Advances in natural polymers, advanced structured materials; 2013. p. 193–254.

[6] Babu RP, O'Connor K, Seeram R. Current progress on bio-based polymers and their future trends. Prog Biomater 2013;2(8):1–16.

[7] Mutanda T, Ramesh D, Karthikeyan S, Kumari S, Anandraj A, Bux F. Bioprospecting for hyper-lipid producing microalgal strains for sustainable biofuel production. Bioresour Technol 2011;102:57–70.

[8] Munir N, Sharif N, Naz S, Manzoor F. Algae: a potent antioxidant source. Sky J Microbiol Res 2013;1:22–31.

[9] Klok AJ, Martens DE, Wijffels RH, Lamers PP. Simultaneous growth and neutral lipid accumulation in microalgae. Bioresour Technol 2013;134:233–43.

[10] Wijesekara I, Pangestuti R, Kim S-K. Biological activities and potential health benefits of sulfated polysaccharides derived from marine algae. Carbohydr Polym 2011;84:14–21.

[11] Chen GQ, Patel MK. Plastics derived from biological sources: present and future: a technical and environmental review. Chem Rev 2011;112(4):2082–99.

[12] Pawar MS, Kadam AS, Dawane BS, Yemul OS. Synthesis and characterization of rigid polyurethane foams from algae oil using biobased chain extenders. Polym Bull 2016; 73:727–41.

[13] Chiellini E, Cinelli P, Ilieva VI, Zimbardi F, Kanellopoulos N, Wilde B. Hybrid composites based on fibres of marine origin. Int J Mater Prod Technol 2009;36(1–4): 47–61.

[14] Kuilla T, Bhadra S, Yao D, Kim NH, Bose S, Lee JH. Recent advances in graphene based polymer composites. Prog Polym Sci 2010;35(11):1350–75.

[15] Azeez AA, Rhee KY, Park SJ, Hui D. Epoxy clay nanocomposites—processing, properties and applications: a review. Compos Part B 2013;45(1):308–20.

[16] Daemi H, Barikani M, Barmar M. Highly stretchable nanoalginate based polyurethane elastomers. Carbohydr Polym 2013;95(2):630–6.

[17] Cao K, Jiang Z, Zhao J, Zhao C, Gao C, Pan F, Wang B, Cao X, Yang J. Enhanced water permeation through sodium alginate membranes by incorporating graphene oxides. J Membr Sci 2014;469:272–83.

[18] Wan Y, Chen X, Xiong G, Guo R, Luo H. Synthesis and characterization of three-dimensional porous graphene oxide/sodium alginate scaffolds with enhanced mechanical properties. Mater Exp 2014;4(5):429–34.

[19] Michael C, Dirk C, Harald K, Jan R, Joachim V. Policy paper on bio-based economy in the EU: level playing field for bio-based chemistry and materials. 2011. Available from: www.bio-based.eu/policy/en.

[20] Hempel F, Bozarth AS, Lindenkamp N, Klingl A, Zauner S, Linne U, Steinbüchel A, Maier UG. Microalgae as bioreactors for bioplastic production. Microb Cell Fact 2011;10(81):1–6.

Index

'*Note*: Page numbers followed by "f" indicate figures and "t" indicate tables.'

B

D

E

R

S

T

U

V

W

X

Z

CPI Antony Rowe
Chippenham, UK
2017-07-31 21:24